# physica status solidi (a)

## applied research

**Volume 108 Number 1 July 1988**

**AKADEMIE-VERLAG BERLIN**

ISSN 0031-8965 phys. stat. sol. (a), Berlin **108** (1988) 1, 1–478, K1–K84, A1–A8

## International Classification System for Physics*)

60. Condensed matter: structure, mechanical and thermal properties

  61. Structure of liquids and solids; crystallography
  (see also 68.20. Solid surface structure, 71. Electron states)

  62. Mechanical and acoustical properties of condensed matter
  (see also 61.70. Defects in crystals, 68.30. Surfaces and interfaces)

  63. Lattice dynamics and crystal statistics
  (see also 65. Thermal properties, 66.70. Thermal conduction, 68.30. Dynamics of surface and interface vibrations, 78.30. Infrared and Raman spectra)

  64. Equations of state, phase equilibria, and phase transitions

  65. Thermal properties of condensed matter
  (see also 63. Lattice dynamics; for thermodynamic properties of quantum fluids, see 67.40; for thermal properties of solid helium, see 67.80)

  66. Transport properties of condensed matter (nonelectronic)

  67. Quantum fluids and solids; liquid and solid helium

  68. Surfaces and interfaces; thin films and whiskers
  (for impact phenomena, see 79; for crystal growth, see 61.50)

70. Condensed matter: electronic structure; electrical, magnetic, and optical properties

  71. Electronic states
  (see also 63. Lattice dynamics, 73. Electronic structure and electrical properties of surfaces, interfaces, and thin films)

  72. Electronic transport in condensed matter
  (for surfaces, interfaces, and thin films, see 73)

  73. Electronic and electrical properties of surfaces, interfaces, and thin films

  74. Superconductivity

  75. Magnetic properties and materials

  76. Magnetic resonances and relaxation in condensed matter; Mössbauer effect

  77. Dielectric properties and materials
  (for conductivity phenomena, see 72.20 and 72.80)

  78. Optical properties and condensed-matter spectroscopy and other interactions of matter with particles and radiation
  (for phonon spectra, see 63)

  79. Electron and ion emission by liquids and solids; impact phenomena

85. Devices**)

---

*) Excerpt; reproduced with permission of International Council for Scientific and Technical Information (ICSTI).

**) Outside the ICSTI Classification for Physics.

(The Substance Classification is given on cover three)

# Author Index

# physica status solidi (a)

## applied research



Volume 108 · Number 1 · Pages 1 to 478, K1 to K84, and A1 to A8

July 16, 1988

AKADEMIE-VERLAG · BERLIN

Subscriptions and orders for single copies should be directed
in the GDR:

to the Postzeitungsvertrieb or
to the Akademie-Verlag Berlin, Leipziger Str. 3–4, DDR-1086 Berlin;

in the other socialist countries:

to a book-shop for foreign language literature or
to the competent news-distributing agency;

in the FRG and BERLIN (WEST):

to a book-shop or
to the wholesale distributing agency Kunst und Wissen, Erich Bieber, OHG, Wilhelmstr. 4–6, D-7000 Stuttgart 1;

in the other Western European countries:

to Kunst und Wissen, Erich Bieber GmbH, General Wille-Str. 4, CH-8002 Zürich;

in USA and CANADA:

to VCH Publishers, Inc., 303 NW 12th Avenue, Deerfield Beach, FL 33442–1788, USA;

in other countries:

to the international book- and journal-selling trade,
to Buchexport, Volkseigener Außenhandelsbetrieb der Deutschen Demokratischen Republik, Postfach 160, DDR-7010 Leipzig, or
to the Akademie-Verlag Berlin, Leipziger Str. 3–4, DDR-1086 Berlin.

**Editorial Note:**

"physica status solidi (a)" undertakes that an original paper *accepted for publication* before the 23rd of any month will be published within 50 days of this date unless the author requests a postponement. In special cases there may be some delay between *receipt* and *acceptance* of a paper due to the review and, if necessary, revision of the paper.

Schriftleiter und verantwortlich für den Inhalt:
Professor Dr. E. Gutsche, Leipziger Straße 3–4, Postfach 1233, DDR-1086 Berlin.

Verlag:
Akademie-Verlag Berlin, Leipziger Str. 3–4, DDR-1086 Berlin; Fernruf: 2236221 und 2236229; Telex-Nr.: 114420; Bank: Staatsbank der DDR, Berlin, Kto.-Nr.: 6836-26-20712

Chefredakteur:
Dr. H.-J. Hänsch.

Redaktionskollegium:
Prof. Dr. E. Gutsche, Dr. H.-J. Hänsch, Dr. H. Lange, Dr. S. Oberländer.

Anschrift der Redaktion:
Leipziger Straße 3–4, Postfach 1233, DDR-1086 Berlin, Fernruf: 2236279.

Veröffentlicht unter der Lizenznummer 1620 des Presseamtes beim Vorsitzenden des Ministerrates der Deutschen Demokratischen Republik.

Gesamtherstellung:
VEB Druckerei „Thomas Müntzer", DDR-5820 Bad Langensalza.

Erscheinungsweise:
Die Zeitschrift „physica status solidi (a)" erscheint jeweils am 16. eines jeden Monats. Jährlich erscheinen 6 Bände zu je 2 Heften.

Bezugspreis:
Je Band 220,– DM zuzüglich Versandspesen; je Jahr 1320,– DM. Der gültige Jahresbezugspreis für die DDR ist der Postzeitungsliste zu entnehmen.

Bestellnummer dieses Bandes: 1085/108.



Printed in the German Democratic Republic.
AN (EDV) 20735
06500

phys. stat. sol. (a) **108**, No. 1 (1988)

# *Contents*

## Review Article



## Original Papers and Short Notes

### *Structure; crystallography*



### *Defects; nonelectronic transport*

***Lattice properties***

***Surfaces, interfaces, thin films; lower-dimensional systems***

***Localized electronic states***



***Electronic transport; superconductivity***

***Device-related phenomena***



**Pre-Printed Titles**



physica status solidi (a) is indexed in Current Contents/Physical, Chemical & Earth Sciences.

## Systematic List

**Contents of Volume 108 Continued on Page 481**

# *Review Article*

phys. stat. sol. (a) **108**, 11 (1988)

Subject classification: 73.40 and 85; S10

*Institut für Angewandte Physik der Technischen Hochschule Darmstadt*[1])

# Bistable Switching in Electroformed Metal–Insulator–Metal Devices

By

H. PAGNIA and N. SOTNIK

Dedicated to Prof. Dr. E. GUTSCHE on the occasion of his 60th birthday

## Contents



---

[1]) Schloßgartenstr. 7, D-6100 Darmstadt, FRG.

## 1. Introduction

Since the first discovery of switching effects in amorphous materials and the development of memory devices ("ovonics") by Ovshinsky [1] in 1962 there have been published numerous reports on switching phenomena in various mostly low-conducting or insulating (denoted "I") materials frequently sandwiched between two conductor or metal (denoted "M") electrodes. In the following we shall term such two-terminal devices "MIM diode", in a more general sense, even if they are constructed in a planar or any other configuration.

Switching provides at least two states of different conductivity, low (the OFF state) and high (the ON state), which may be mono- (volatile) or bi-stable (non-volatile), respectively. The first one, called "threshold switching", requires a holding current in order to remain in its ON state, while "memory switching" is denoted to two states, stable without applied voltage. In this review we shall restrict our attention mainly on memory switching.

An overview on amorphous alloy properties with emphasis on chalocogenide glasses, containing besides elemental semiconductors a large amount of chalcogen elements (S, Se, Te) was given e.g. by Adler et al. [2]. Besides some controversial discussions at large generally accetable models are established, explaining the OFF-to-ON memory switching as a phase transition from an amorphous (glassy) to a crystalline filamentary material made up e.g. of a metallic component.

Starting with Hickmott's investigations [3] on oxide insulators, after an electroforming process under vacuum, unusual electronic features were discovered, including switching phenomena. The long list of publications on this topic shall not be repeated here. They may be found in one of the following reviews. A review of early investigations on MIM-sandwich diodes applicable as cold cathodes was given by Eckertová [4]. Dearnaley et al. [5] described mainly amorphous oxides exhibiting besides switching effects, non-ohmic current–voltage ($I$–$U$) characteristics with a negative differential (NDR) branch (evidently correlated with switching), and different mechanisms of electron emission (EE) and of electroluminescence (EL). A paper on non-ohmic $I$–$U$ characteristics with NDR and further anomalous features, especially EL, of a planar configuration type MIM diode appeared in 1973 (Morgenthaler and Pagnia [6]). In a succeeding paper by Bischoff and Pagnia [7] the similarity between results from both sandwich and planar configurations was firstly stated. The comprehensive review by Biederman [8] should be mentioned covering the same field as treated in [5], but including also first results of planar diodes. Moreover an article by Oxley [9] reported on memory effects of (mainly) oxide films, based on the large body of literature compiled up to this date on both configurations.

Switching in organic insulators has first been reported by Szymansky et al. [10], later e.g. by Gundlach and Kadlec [11], who, for the first time, supposed that this case of switching should be independent of the insulator material. From experimental findings it was evident that in MIM diodes with an organic insulator different switching mechanisms must be taken into account, e.g. due to Pender and Fleming [12] a type of low voltage switching at applied voltages $U < 20$ V and high voltage switching in the range above. The latter effect will not considered here further on.

Switching in the chalcogenide glasses greatly stimulated research on other amorphous materials as well. In spite of the wide-ranging work on amorphous Si (a-Si), memory switching was not exhibited, leading to a general believe that this and other tetrahedrally bonded materials should not possess properties required for those phenomena, Gabriel and Adler [13]. However, in the early eighties it was found that heterogeneous junction layers of a-Si (having e.g. $p^+ni$ structure) could be formed to exhibit reliable, fast and polarity dependent memory switching, Owen et al. [14]. A phase transformation mechanism like in the ovonics was excluded, and although the nature of the ON state current paths is not yet known a formation associated with high fields and/or trapped charges was assumed to be likely, Le Comber et al. [15].

Bistable switching has been observed in ZnSe–Ge heterojunctions by Hovel [16] and in various GaAs and Si Schottky diodes by Moser [17], in which the switching events appear in regions with the Schottky contact and an ohmic contact in close proximity near the surface. In the latter case, the current paths were proved to exhibit a very small negative or near-zero temperature coefficient of resistivity, and they became superconducting (with a transition temperature for small-grain granular Ga), Moser and Rohrer [18]. Joule heating and material transport were claimed not to be involved in switching, lacking a reliable explanation like in most other cases.

It should be mentioned that already in 1969 Chapman [19] reported on fast bistable switching of two metal electrodes in close proximity ($\approx 50$ nm) cycled in an up to 10 to 100 V range in air. Metallic bridges between the electrodes were produced which could be ruptured by Joule heating and regrow from the electrode material in less than 20 ns. We shall show later that the question whether a device operates in air or not is one criterion to distinguish between different classes of switches.

The aim of this article is to give both an overview of representative switching systems and to discuss in more detail a special class of them, not reliably identified in the past. The wide-ranging work on the chalcogenide glasses and also on those systems where metal filaments are unequivocally recognized will, except for their general behavior, be omitted. Planar MIM diodes were expected to have some advantages over the (closed) sandwich configuration. The whole structure is open and directly accessible to the environmental gas atmosphere or the vacuum, which strongly affects the switching behavior. These structures are suitable also for electron microscope investigations. Of course, these advantages made it possible to enlarge the state of information of the fundamental mechanisms significantly. Therefore, conclusions in this review shall mainly be based on results from planar diodes, not neglecting those from sandwich ones if suitable.

## 2. General Behavior of Memory Switching MIM Diodes

### *2.1 The MIM diode family — configurations and materials*

The usual method to prepare a sandwich diode (Fig. 1a) is first to deposit a metal (or any other conductor) stripe on an insulating substrate (frequently of glassy consistence). This may be either oxidized or covered with an insulator (or any low conducting) material of thickness 5 nm to 1 μm (we shall not treat thinner samples,

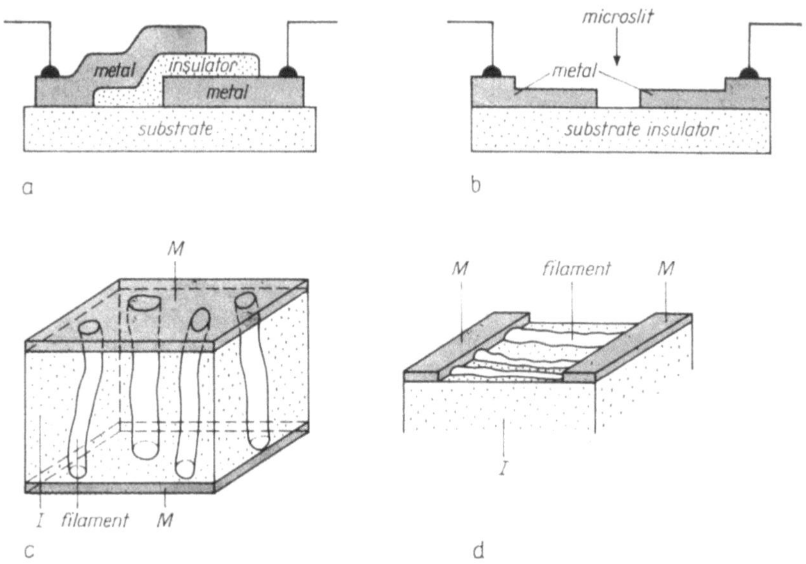


Fig. 1. Sketch of the two metal–insulator–metal (MIM) diode configurations: a) sandwich, b) planar, and c) and d) how filamentary paths may bridge the insulating gap

revealing single barrier quantum mechanical tunneling), and further by another metal or conductor. Point contacts for one or both of them are also common. In the open planar diode (Fig. 1 b) the substrate insulator may be identical with the gap insulator, while the upper half space is vacuum or any environmental gas atmosphere. The gap may, on the other hand, also be closed by another protecting insulator.

In Table 1 elemental materials utilized as ground or top electrodes of a sandwich configuration or as thin film materials in planar structures, respectively, are shown. Conducting compound materials like e.g. $In_2O_3/SnO_2$ (ITO), semiconducting compounds, metallic compounds and alloys, etc. have also been used. Because they behave like elemental materials we shall neither mention all of them nor give all special references (most of them are listed in the review articles [2, 4, 5, 8, 9, 20]), except when particular results are considered.

The number of insulators used is also large, and not all shall be mentioned here (see reviews above and e.g. [7, 11]). Only a representive (but random) selection is listed in Table 2 in order to get an impression. Air [19] and the various substrate glasses, quartz, and alkali halides may be mentioned, as well as multilayer compositions with insulators and semiconductors, too.

Table 1

Electrode and film materials (excerpt from Periodic Table)

| | I | II | | III | IV | | V | | VI | VII | VIII | | |
|---|---|---|---|---|---|---|---|---|---|---|---|---|---|
| 1 | | | | | | | | | | | | | |
| 2 | | Be | | | | C | | | | | | | |
| 3 | | Mg | | Al | | Si | | | | | | | |
| 4 | Cu | | Zn | | Ti | | | | Cr | Mn | Fe | Co | Ni |
| 5 | Ag | | | In | Zr | Sn | Nb | Sb | | | | | |
| 6 | Au | | | | | Pb | | Bi | W | | | Ir | Pt |

Table 2

Selected list of insulators utilized in the sandwich configuration or as substrate material, respectively

| | |
|---|---|
| chalcogenide glasses | $Te_{81}Ge_{14}Sb_2S_3$ (ON state: Te) [21], $Ge_{15}Te_{81}As_4$ (glass–crystalline transition) [22] |
| semiconductors | Si, CdS, GaP [17], ZnS [23], GaAs [17], Se [24] |
| oxides | $SiO_2$, $Al_2O_3$ [3], $Nb_2O_5$ [25], $Cu_2O$ [26], $TiO_2$, $CrO_2$ [3]; Ta, Cu, and Mg oxides [27], VO [28] |
| nitrides | AlN [29], BN [30] |
| fluorides | LiF [31, 8], $CaF_2$, $MgF_2$ [32] |
| polymerized films | styrene, acetylene, benzene, analine [12] |
| other organic films | collodion [32], Langmuir-Blodgett type [11] |

### 2.2 I–U characteristics and switching

*I–U* characteristics with negative differential resistance (NDR) are known from various semiconductor devices [33]. Even homogeneous semiconducting crystals exhibit due to current instabilities in principle two different NDR shapes. N-type (Fig. 2a) or voltage-controlled negative differential resistance (VCNR) is correlated with domains of different field strength. In S-type (Fig. 2b) or current-controlled negative differential resistance (CCNR) material filaments are created with different current density [34].

Considering mainly amorphous semiconductors, Adler [35] proposed that *I–U* characteristics with NDR may, including the two mentioned above, generally be separated into six types, altogether sketched in Fig. 2, depending on the control

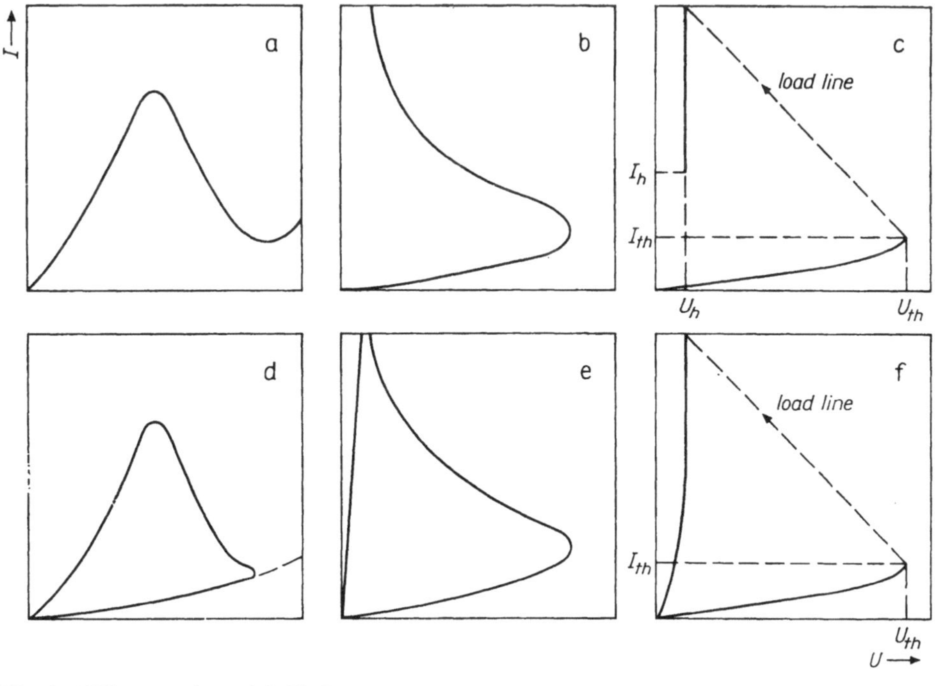


Fig. 2. Different types of *I–U* characteristics with negative differential resistance (after [35])

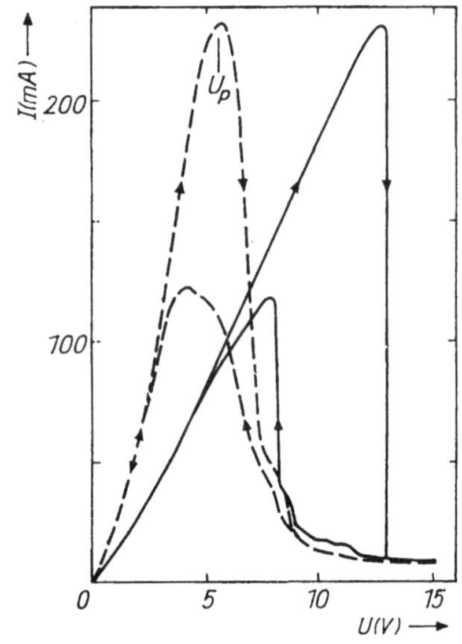


Fig. 3. $I$–$U$ characteristics of a planar MIM diode (Au on $Al_2O_3$): dashed curve without, full curve with series resistance (30 Ω) (after [36])

mechanism. Particularly, characteristics shown in Fig. 2c, f, are typical of the chalcogenide devices. Rising current lowers the voltage drop at the diodes. Including a series resistance switching becomes possible along a load line.

In devices exhibiting high field nonequilibrium effects of carrier density (or mobility) a holding current or holding voltage may be necessary to retain the high conducting ON state (Fig. 2c).

Besides these diodes with threshold switching there exist true memory devices with two stable states at zero voltage (Fig. 2d, e, f). This may be caused by electrically induced phase transitions, material transport, or other processes leading to stable states. Switching between these two states depends on circulating current or/and applied voltage and is described by the switching times. As in other cases insufficient knowledge on contact resistances is a serious problem in all those devices, for it may obscure shapes distinctly. Influence of a series resistance for a VCNR behaviour with memory (case 2d) is shown for one example in Fig. 3 [36].

Since the switching times (ON-to-OFF and vice versa) may be different by orders of magnitude, cycling a device reveals hysteresis effects between forward and backward traces. Switching in both directions may occur abrupt leading to sharp N-type shapes as has been exhibited by Borziak et al. [37] (see also Fig. 8 [38]).

### *2.3 Different models describing switching*

Memory states may have lifetimes of the order of years. A variety of models have been proposed concerning the stability of both states, also explaining switching mechanisms in both directions. A few typical examples and models are listed in Table 3 (out of the variety only one example is presented for each case). In all cases (known to the authors) the ON state was imagined to be related to spatially separated current paths of irregular shape, which, in the following, we nevertheless shall call filaments in accordance with other authors. These can be imagined to bridge the contact gaps as sketched in an idealized manner in Fig. 1c, d. The OFF state may be described either by the background insulator, or by a low-conducting state of the filaments, either materially or e.g. induced by any field effect. The latter is, of course, correlated with charging, and it is hard to imagine that a nonequilibrium charge distribution would exist for such long times in thin film devices of the usual quality; though in high quality thermal oxides of silicon such storage times are available at least at not too high temperatures. It is more likely that both states differ significantly in materials, their composition, or phase state. The simplest case is, without doubt, one in which a filament is intact or interrupted like an electrical fuse. An example is, as already mentioned, metal filament growth through an air gap between two metal tips [19], thus establishing an ON state, and rupturing by melting or explosion. Current paths in sandwiched insulators were sometimes explained with metal filaments [41], too.

Dearnaley [47] was the first to propose filamentary current paths of other material than metal from the electrodes or the original stoichiometric oxide. He suggested that at some nucleation sites at the gold–silicon oxide interface (in the particular system) chains of conducting material should start to grow under the influence of the intense electric field established successively at their free ends. He speculated about possible materials and structures, and came to the conclusion that it might be excess silicon showing twofold co-ordination bound to two oxygen neighbors. Chains of alternate Si and O should thus be formed, along which electronic conduction may be caused like in analoguous conjugated resonance structures with free electron $\pi$-orbitals. Even though this could not be verified, mainly because various other MIM systems containing no silicon and oxygen were proved to show the same features, the latest ideas for carbon filament material are similar.

In the review paper [5] Dearnaley et al. did not further on speculate upon filament materials but presented a global model of high acceptance, based on few terms, for it yielded good fits to the measurable $I$–$U$ characteristics and its frequency dependence. These were thermal rupturing by Joule heating and recoverage within a dead time as a re-growth process. The NDR branch originates from rupturing filaments with a distribution $P(\varrho)$ of different resistances $\varrho$. The total current is then given by the characteristics of all intact filaments at voltage $U$,

$$I(U) = \int_{\varrho_{\min}}^{\infty} U\varrho^{-1} P(\varrho) \, \mathrm{d}\varrho \,. \tag{2.1}$$

Assuming that rupture of the filaments occurs when a maximum filament temperature $T_{\max}$ is exceeded, the minimum resistance $\varrho_{\min}$ of intact filaments at voltage $U$ and the temperature rise in the uniform filament may be calculated. Dearnaley derived $P(\varrho)$ from $I(U)$ by noting

$$\frac{P(\varrho)}{\varrho} \sim \frac{\mathrm{d}(I/U)}{\mathrm{d}(U^2)} \,. \tag{2.2}$$

The calculated distributions $P(\varrho)$, deduced from experimental $I$–$U$ characteristics, were approximately triangular in shape.

Vice versa, it is possible to synthesize $I$–$U$ curves for a given distribution of resistances. Even though (at least) the assumption of ohmic behavior was too simple in the model, it is always possible to fit measured curves with properly elected distributions. If the individual resistances are voltage dependent, the treatment should be modified as was done by Sutherland et al. [23] leading to more complicated formulae.

Dearnaley et al. [5] also considered the dynamic characteristics and were able to formulate reasonable relations for the high frequency behavior based on the same assumptions as in the dc case. Introducing a dead time $\tau_d$ it was expected that no filaments once ruptured would be able to re-form for cycling frequencies above $1/\tau_d$, and, vice versa none of the filaments left intact would be able to rupture. The $I$–$U$ curves are thus dominated by intact filaments and the traces are smooth, showing no switching and NDR.

Completely different models (shortly characterized in Table 3) shall be mentioned, which are based on the insulator band structure and a particular broad impurity band [39, 40, 48]. Some of them may in principle also account for a relatively homogenous insulator material, others evidently require filamentary current paths. Though different in the details, long-term stability of the states is deduced to charge storage.

Table 3

Switching systems and models

| system | ON state | OFF state | ON–OFF | OFF–ON | ref. |
|---|---|---|---|---|---|
| Al–$SiO_x$–Au | broad impurity band ($\approx$3 eV with $N \approx 10^{18}/cm^3$), depletion layers at both electrodes | stored charge in the center of the insulator builds up space charge | trapping of electrons injected when $U$ exceeds $U_{max}$ (trap center shift relative to Fermi level) | removing of stored charge | [39] |
| MIM sandwich | filament of unknown material | ruptured filament and/or insulator leakage resistance | thermal rupture | growth mechanism, involving dielectric relaxation and ion motion | [5] |
| MIM sandwich | impurity band in amorphous material with fluctuating density causing constrictions in the mobility band widths (isoelectronic tunneling) | Fermi level shifted into the localized states of the constriction | 'hot' electron trapping outside the filament | de-trapping by field emission into the filament | [40] |
| PC–$Te_{81}Ge_{14}Sb_2S_3$–PC (PC: pyrolytic graphite) | crystalline Te filament | amorphous state of matrix | self-heating after threshold ON switching causes phase separation | melting and re-absorption into the alloy | [21] |
| W–$Ge_{15}Te_{81}As_4$–W | polycrystalline filament | amorphous material | thermal structure phase transition glass–melt–crystal (GMC) or glass–crystal dependent upon pulse form | thermal reverse process: CMG and quenching | [22] |
| Cu–SiO/$V_2O_5$–Cu | Cu filament | amorphous matrix | melting by Joule heating | metal diffusion from electrode | [41] |

| | | | | | |
|---|---|---|---|---|---|
| Au–air–Au | Au filament | air | rupture | growth in high electrical field | [19] |
| Al–$SiO_x$–Au | metal island chain activated tunneling: low barrier | same: high barrier | scattering of electrons into the insulator increasing metal–oxide barrier | removing of stored charge | [9, 45] |
| Cr–a-Si($p^+$ni)–Al | dispersed metal particles | amorphous silicon | charge trapping and de-trapping | high field enhanced in-diffusion | [15] |
| Rh–GaAs–Au/Te | Ga filament unique structure (superconducting) | metal cluster in dielectric matrix (activated transport) | (no Joule heating and/or material transport) | | [18] |
| Al–polystyrene–Al | carbon filament | ruptured filament, matrix | rupture | regrowth | [12] |
| Au on quartz glass | carbonaceous filament | ruptured filament, leakage current of substrate | rupture | two-stage process: regrowth and switching | [46] |

Yet no experimental proof for the existence of such systems with the considered anomalous properties, however, have been confirmed.

Another explanation on the same line was put forth by Emmer [49]. From studies of the oxygen influence on the $I$–$U$ characteristics (driving the diodes into the OFF state, Sections 3.1 and 4.4), he concluded that oxygen vacancies deliver the responsible energy states for ON conduction, performed via electron hopping.

A model which links the observations with the field of discontinuous metal films and their three-dimensional analogue, the cermets, was proposed by Thurstans [45]. In those systems, under certain geometrical conditions, carrier transport may be performed by activated tunneling due to the increasing electrostatic energy of adjacent islands (or clusters) by charging from the tunneling electrons. Some special features of such a mechanism, for example a zero-bias anomaly of the $I$–$U$ curve (noticeable rather in the first and second derivatives), was observed by the author in the ON state current. Stabilization of both states was again imagined to be caused by charging (or release) of the insulator, responsible in this model for changes in the tunneling barrier heights.

Similarly, LeComber et al. [15] considered that in their a-Si indiffused material from possibly isolated metal particles dispersed in the dielectric medium, like in the above-mentioned cermets (see e.g. Abeles et al. [50]). They believed that the initial stages of memory states were associated with high fields and/or trapped charges. But they also discussed other models, so that the mechanism working here is still uncertain.

Asymmetric ON state and, similarly, a rectifier characteristic in the OFF state have been observed in Schottky diodes on epitaxially grown n-type GaAs [17, 18]. Avalanche breakdown at the blocking contact was assumed to create the ON state filaments via melting both GaAs and contact metal which then penetrates into the semiconductor. Granular filaments should be formed exhibiting superconductivity in the ON state and a small negative or near-zero temperature coefficient of resistance, while in the OFF state carrier transport was found to be thermally activated. The question how both states are stabilized was not answered.

Switching filaments were electroformed also in various glow discharge grown polymer films sandwiches between different metals. Pender and Fleming [12] came to the conclusions that filaments in polystyrene were made up of carbon which grew by pyrolysis. Carbon filaments could rupture by Joule heating, and recover from the surrounding material. Such mechanism is, of course, full on the line with Dearnaley et al.'s model [5].

It is now well established that semicontinuous metal films, in which in a basic forming step a discontinuity, a microslit was introduced, can be electroformed only in a vacuum with organic residual gases to show the same memory switching and related features [7, 46, 51, 52] discussed in the case of sandwich MIM diodes with an oxide insulator (see e.g. Dearnaley et al.'s review [5]). From findings with these planar MIM diodes it was proposed that the filaments should consist mainly of carbon. These shall be in short named "carbon filaments", although other elements may be incorporated to some minor degree. Rupture and regrowth was assumed like in Dearnaley et al.'s model. But additionally it was discovered that regeneration (Section 4.1) (OFF–ON) is evidently a two-stage process (first: re-growth and second: on-switching). It was also found that off-switching can be induced by low energy electrons (Section 4.6), so that a pure thermal explanation is at least incomplete.

It is known that some MIM diodes are sensitive to moisture. Generally these influences shall not be treated here in detail, but a few remarks seem useful, for water influence was assumed in one case to be responsible for adsorption assisted regeneration [53]. Humidity dependent bistable switching was observed mainly with diodes containing silver (see e.g. Kimura et al. [54], Tanaka and Iwata [55, 56], and Gengenbach et al. [57]), but also, with less sensitivity, with gold [58]. The *I–U* curves and frequency dependences are so different from usual MIM switching, exhibiting their maximum (in the ON state) near or below 1 V, that the mechanism can easily be recognized, and it is also possible to omit those forming in dry environments. Those effects are, on the other hand, possibly useful for applications as humidity sensors. But that is out of the scope of this review.

Obviously, there may exist quite different filaments and switching mechanisms. But after this systematical survey about possible explanations, we shall come to the concrete type of "carbon filaments", which are the most numerous studied, but not uniquely recognized and treated.

## 3. MIM Diodes with Carbonaceous Current Paths

### *3.1 Definition of the particular class*

However, a variety of models were discussed in literature, often incompatible with one another, for numerous differently composed devices, experimentally observable features are qualitatively identical. Frequently, not all possible information is available later, and these punctual lacks cause difficulties when the results are compared. Nevertheless, we shall try to summarize phenomenologically significant features and behaviors in order to fence in the particular class of MIM diodes, which we will consider further on. That is of course an unconventional way, but carries the advantage, to easily decide whether or not an individual device is a member of this class. Evidently, only one model should then be applied accordingly.

Let us first look at a MIM diode with VCNR-like *I–U* characteristic and memory behavior (similar to Fig. 2d). The following more detailed features must then be fulfilled provided they contain carbon filaments, as demonstrated for one example in Fig. 4 (see also Fig. 37):

(i) In their high conducting ON state the *I–U* curves are non-ohmic at slow voltage sweep velocities (dc curves) and smooth at low voltages. After onset of primarily off switching events (traces are then "noisy"), and passing by this a current maximum, a negative differential resistance (NDR) branch results. Forward and return curves exhibit a hysteresis (single-arrowed curves in Fig. 4a, b, c).

(ii) Return curves traced with a high sweep velocity (starting from any point of the NDR branch; the lowest we call the low-conducting OFF state) are smooth showing no maximum. (The dashed curve sketches the principal shape, measured curves are presented in Fig. 10.) The next slow velocity forward curve remains on this trace for the present (and so do all curves under ac conditions), only up to a distinct threshold voltage ($U_{th} > 1.5$ V) from where the current rapidly increases towards the ON state (double-arrowed curve in Fig. 4a).

The onset of recoverage is phenomenologically introduced here as a threshold voltage. The question whether this behavior is due to a distinct physical process shall be discussed in more detail in Section 5.2. The border between slow and high sweep velocities under common high vacuum conditions (HV: $p \approx 10^{-4}$ Pa) at room temperature (RT) lies at about 1 to 10 V/s.

(iii) Exposure to air (or oxygen) causes switching from the ON to the OFF state under applied voltage beyond $U_{th}^*$ (triple-arrowed curve in Fig. 4b). Under those conditions no recoverage of the ON state is possible until HV conditions are re-established (double-arrowed curve in Fig. 4c).

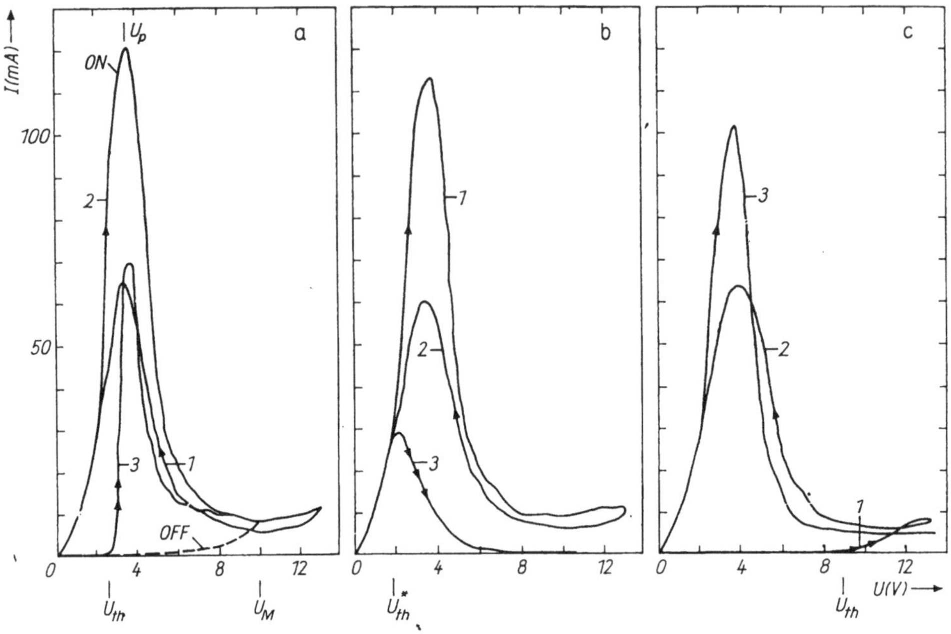


Fig. 4. $I$–$U$ features characterizing a MIM diode (here a planar one: Au on quartz glass) with carbonaceous current paths ("carbon filaments"): a) in HV, memory behavior: slow voltage rise to $U_M$ (curve 2), fast sweep to zero voltage (OFF state, dashed curve), regeneration at threshold voltage $U_{th}$ (curve 3); b) (1), (2) in HV, (3) after air inlet of atmospheric pressure; c) (1), (2), (3) after re-establishing HV (numbers at the curves note succession)

(iv) One further criterion, not so easily realizable, is the lack of on-switching in an UHV ambient (see Fig. 26). The reason is that no forming occurs in absence of organic molecules. On the other hand, a diode once formed in HV and then transferred to an UHV system switches off (traces are closely akin to those in air) and does not regenerate within moderate times. Since the partial pressure of the organic residuals is of relevance, proper conditions to cease effects related to the ON state may also be provided by freezing out the organic gases by a liquid nitrogen cold trap near (or around) the sample [51, 59].

Presence or absence of adsorbed organic molecules or fragments are evidently the fact which decides whether forming and OFF–ON switching is possible or not. That was the reason for introducing the term A-type forming [51] (Section 3.2) and claiming a carbonaceous filament material [60]. But it must be plainly stated that all of those proposals remain speculative as long as straightforward analysis of chemical composition and the structure properties are lacking. Nevertheless, there obviously exists a uniform class of MIM devices with filaments grown from material out of an external source, and they are recognizable when applying the above listed checks. That can only be a recipe for future studies; treating old results again should, hence, only be performed with some intuition.

### *3.2 Basic and adsorption-assisted electroforming*

Before MIM devices show the typical anomalous behavior described in Section 3.1 they must undergo an electroforming process. This may be divided into two steps. The first one, the basic or shortly B-forming creates free surfaces in or on the insulator

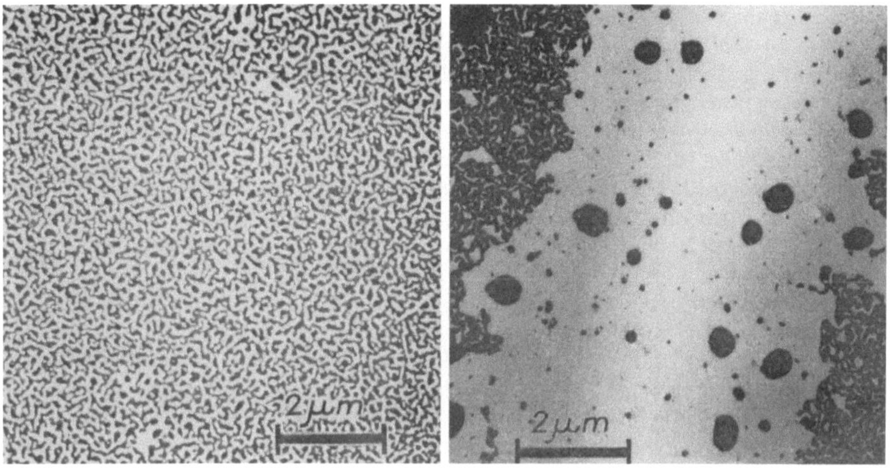


a b

Fig. 5. Electron micrographs of planar MIM diodes (Au on $SiO_2$) with a) the semicontinuous metal film and b) the "basic" for medmicroslit (see text); in both cases the microslit has about the same direction; in a) it can be recognized by the bright speckles only (after [62])

and may proceed even in laboratory air. The second one is an adsorption-assisted or A-forming and is usually done in high or rough vacuum containing hydrocarbons. Here filaments assumedly grow in a high electric field at voltages below 20 V. Although both forming processes are necessary they are not always recognized because they may be part of the device preparation.

Planar diodes are usually prepared by depositing a thin or semicontinuous metal film between two planar metal electrodes. B-forming, the production of the actual MIM element, is then performed by thermal melting and/or rupturing of the metal film creating a microslit with a free and possibly disturbed insulator surface up to some μm broad. These microslits may be very different in width (Fig. 5 [62]) without having influence on the properties of the device. Usually the microslit contains residuals of the metal meanders coagulated to islands.

The average power needed for B-forming of a diode on fused silica with 5 mm electrode breadth and an about 10 nm thick metal film was for the frequently used metlas Cu, Ag, Au between 3 and 6 W, for a low melting metal (In) an order of magnitude lower

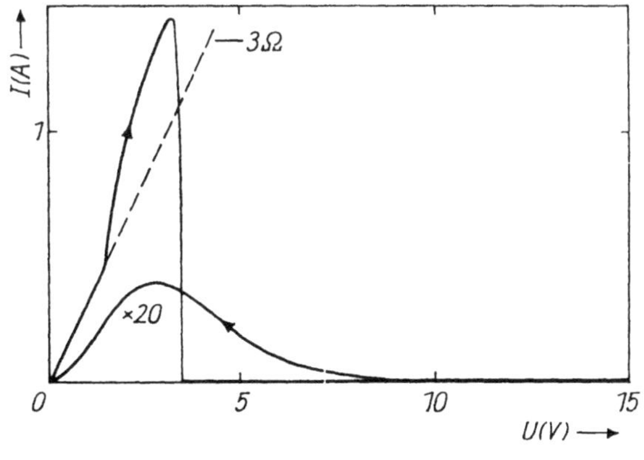


Fig. 6. $I$–$U$ curves of the forming steps of a planar MIM diode: B-type (forward trace) and A-type (return trace) (after [61]). After breakdown $I < 1$ mA

and an order higher for Cr (and C) [36]. One example of a B-forming curve is shown in Fig. 6 [61]. At the kink one can observe in an electron microscope the onset of coagulation making itself felt by thickening of the meanders (causing a reduced ohmic resistance), before pinching and rupture at flaws.

Preparation and forming of sandwich-type MIM diodes have been described often (see e.g. in the reviews [5, 20]). The need of B-forming depends on the quality of the insulator. It is known [5] that sandwich-type diodes with good quality dense insulators are hardly to form and with crystalline mica not at all [63]. On the other hand, porous insulators as achieved by usual deposition and anodic oxidation techniques and those with nonstoichiometric composition appear to be more easy to form. It is likely that pores or channels must be present, or if necessary built in by an electrical breakdown [64] creating at the same time holes in the top electrode (Fig. 12 [49]).

The critical voltage for B-forming or preforming was found to be dependent on electrode and insulator material and the polarity [5] like in other destructive breakdown of insulators. But one essential point is that ultimate (A-type) forming is in all cases independent of the environmental atmosphere, during which basic forming proceeds.

A MIM diode, once basically formed may change its electrical properties drastically during A-forming. This is done by operating the diode in a vacuum vessel containing hydrocarbons from rotary or oil diffusion pumps. Voltages used may be applied by pulses, pulse trains, or in any other form. But slowly cycling the voltage in the range 0 to 10 or 15 V seems to be most effective. It should be mentioned that forming voltages $U_F$ do neither depend uniquely on the insulator thickness in a sandwich diode, nor has any dependence on the microslit width been found. As was already discussed by Ray and Hogarth [20] the range of scattering of the values is broad, $2\text{ V} < U_F < 5\text{ V}$ or even up to 50 to 150 V [65] in some special configurations of Al–$SiO_x$–Au (or C) tip (anode) point contacts. The forming rate increases with voltage up to values, where the reverse process of ON–OFF switching prevails.

Since a growth process is dependent upon an external material supply, its rate is strongly governed by conditions of the source, mainly the residual gas composition. The forming process is no spontaneous event, for it takes time to grow the filaments. The time dependence of forming $I$–$t$ of a sandwich diode is shown for example in Fig. 7 (Hogarth and Taheri [66]).

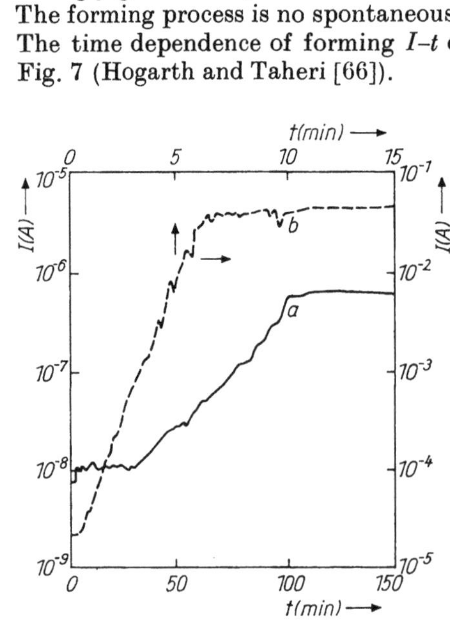


Fig. 7. Electroforming $I$–$t$ of sandwich MIM diodes at constant voltages, a) Al–$SiO_x/B_2O_3$–Al at 40 V, b) Ag–$SiO_x/B_2O_3$–Ag at 8 V (after [66])

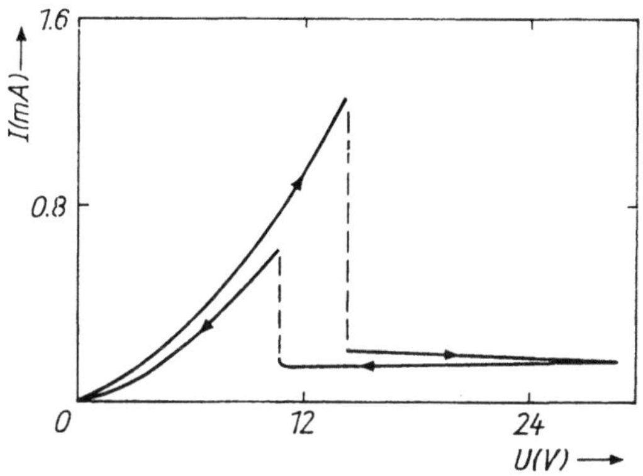


Fig. 8. N-type $I$–$U$ characteristic of a planar MIM diode (Au on quartz glass) with 4 μm electrode; redrawn from oscilloscope screen (after [38])

It seems to be possible that the smallest value ($U_F \approx 2$ V) is the characteristic one for the underlying mechanism, and that is of course a value about equal to the regeneration threshold voltage $U_{th}$ (Section 5.2).

### 3.3 *Current-voltage behavior*

It is generally accepted that the VCNR-like shape of the $I$–$U$ characteristics is caused by superimposed characteristics of individual filaments, having each a certain condition (which may change during repeated operation) at which they switch off. Single filaments should exhibit sharp N-type curves. Possibly experimental curves (Fig. 8) reveal properties of a bunch of minute fibers. This is seen in the return trace, where often stair-like changes instead of a direct jump [38] were found (Fig. 9). The dimensions of such fibers can be estimated roughly for individual smallest current steps in the order 20 nA. Assuming a conductivity of $10^2$ $\Omega^{-1}$ cm$^{-1}$, a value of nonfully graphitized material [67], and a length of 100 nm (usual electrode distance in sandwich structures) cross-sections of molecular size are possible.

The overall current of many filaments or branches connected parallel delivers the common dc characteristics (e.g. Fig. 4). The distribution of the filaments with their critical off-switching voltages are mainly responsible (as long as $R_{contact} \ll R_{filament}$) for the individual shape, and so for the location of the current maximum. Based on the Dearnaley model [5] one can estimate a distribution of the filament resistances from the shapes of the $I$–$U$ curves (Section 2.3). Even provided this would be a constant quantity (but $I$–$U$ curve shapes often change during repeated cycling) for one device, it is different from any other, even from the same preparation charge. Evidently, except accidently, current maxima of $I$–$U$ curves are neither symmetrical

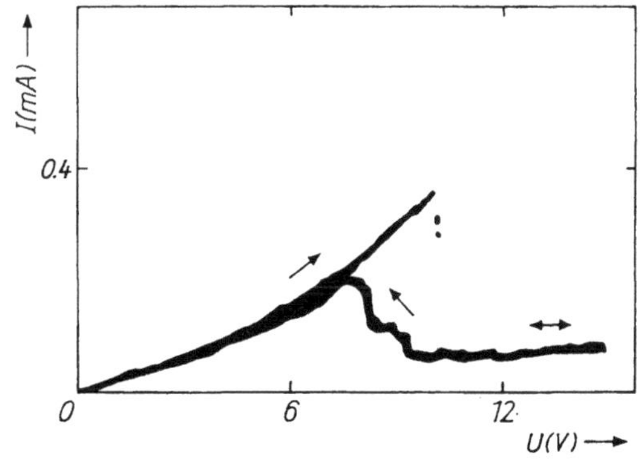


Fig. 9. Current–voltage characteristic of a planar MIM diode (Au on quartz glass) with small electrodes (4 μm) showing abrupt off-switching in the forward sweep and step-like current recoverage in the return sweep (after [38])

nor in the decaying branch exponential, as it was sometimes proposed [64]. Return curves have always a lower maximum (Fig. 4), and when fastly sweeping the voltage it is even absent. Ac voltage causes smooth curves, and they are symmetrical to zero voltage if swept also to negative values. The finding (e.g. [39, 68]) led to the assumption that a dc trace would on principle be an envelope for the set of ac curves, only dependent on the voltage amplitude. But that turned out to be true only for moderate frequencies. Thurstans [45] for example found dynamic curves exceeding the dc boundary (Fig. 10). This is probably a question of the pulse form and the ratio of introduced to dissipated power.

Under "fast sweeping conditions" (with any pulse form) a device remains in its momentary state, being either the OFF state when all the filaments are switched off or an intermediate state with some filaments switched off, the rest of them intact. It must be mentioned that the designations ON and OFF here account for the device as a whole. Even though one may also speak of the respective states of single filaments, we will omit such a notation in order to avoid confusion.

Some remarks concerning other relevant results may be useful. One of the discussed mechanisms of filament rupturing is a thermal process, being naturally controlled by the filament lattice temperature $T_l$. The delayed heating has the result that, e.g. during one voltage sweep of an $I$–$U$ measurement a filament is able to carry a higher current than a stationary value without causing rupture. This fact explains Thurstans' results of 300 Hz ac curves penetrating the quasistatic VCNR-like $I$–$U$ curve.

For a diode in the OFF state or an intermediate state, during the first slow velocity voltage sweep, at a characteristic threshold voltage $U_{th}$, first on-switching events occur, and the current starts to rise in the direction towards the ON state curve (Fig. 4). The measurable onset as well as the slopes of these edges are strongly dependent upon the sweep velocity, and also on the prehistory, e.g. the amplitude of the voltage applied last. The latter influences the number of off-switched filaments, but additionally also, as we shall see, the degree of filament destruction. One example (Thurstans [45]) is shown in Fig. 11, and empirically a relation $U_M = 5.6U_{th} - 4.7$ V was found in the range 4 V $< U_M <$ 15 V, and hence the lowest threshold voltage measured was about 1.6 V.

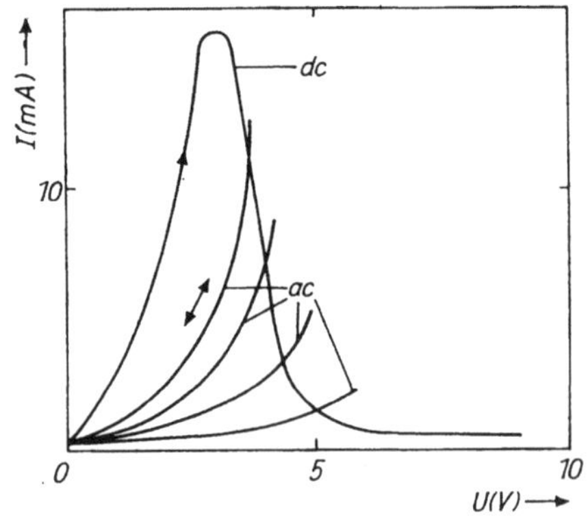


Fig. 10

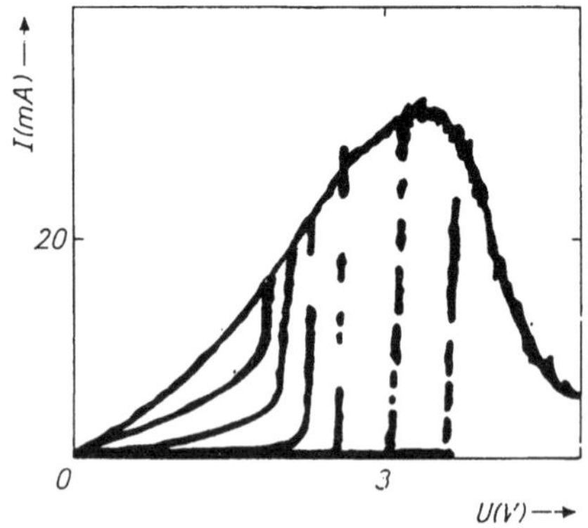


Fig. 11

Fig. 10. Quasistatically and dynamically measured $I$–$U$ characteristic of electroformed sandwich MIM diodes (Al–$SiO_x$–Au) (after [45])

Fig. 11. Shift of the regeneration threshold voltage with the foregoing turn-off voltage $U_M$ from left to right: 3.5, 4, 5, 6, 8, and 10 V (forward sweeps; after [45])

Recoverage of the ON state can also be observed at a fixed voltage $U_R > U_{th}$. In accordance with the existence of a high number of single filaments, current increases essentially step-like as a function of time towards its ON value as has been exhibited by Green et al. [69]. Similar, extended studies were performed later (Bach et al. [46], Pagnia et al. [52]) revealing more concrete information about the regeneration of the filaments (Section 4.1).

Many researchers have observed the appearance of several maxima in the $I$–$U$ curve (see e.g. [43]). A trace of a second maximum is, for example, to be seen in Fig. 36 [36]. As has been shown by the Kiev group [70], a second maximum at $U_{p2} \approx 2U_{p1}$ and possibly also a third and further maxima can be gained by distinct pulse forms and voltage cycling. In terms of Dearnaley et al.'s model [5] (off-switching by overheating) this can be explained by a complete redistribution of the filaments with their discrete (changed) critical resistances. Bischoff et al. [71] were able to show that at the maximum voltage of repeated sweeps (1 Hz) always an enrichment of filaments stable up to this particular voltage with a simultaneously decreasing peak value occurs. A real change of the filament stability can be assumed. The filament lattice temperature before off-switching is high (at maximum about 1500 K) so that an enhancement by pyrolytic effects is at least imaginable.

Concerning multiple maxima, it is not easy to imagine why single filament properties should change in equidistant instead of continuous steps. More likely is the following picture. Provided filaments are about equal in their dimensions and in the material, $n$ filaments in series naturally require $n$ times the voltage before one of them can switch off. This has some further implications since rather a network must be assumed than the before discussed model of filaments parallel bridging the effective electrode gap. A network model is naturally more realistic. Because in such systems non-switching filaments act as load resistance shifting the switch-off voltage and lowering the maximum current several maxima may occur.

### *3.4 Bistable states — the OFF and ON state conduction*

$I$–$U$ characteristics of the ON state are non-ohmic and smooth up to values approaching the maximum. That starts to be dramatically different after onset of off-switching events. In the voltage region where switching occurs, there exists no straightforward correlation between current and voltage. The main reason is that an individual off-switching event depends not only on the voltage but is also dependent upon events happening in the neighborhood; we shall see that some stimulation effects are working. Additionally the competing fact exists that filaments may, at equal voltage, also switch on. The result is a statistical fluctuation.

In the OFF state carrier transport is due to the common mechanisms known from insulators, being either the matrix, substrate, or any insulating phase of the filament material. Ionic conduction can be neglected for different reasons in most cases. One is the observation that down to temperatures of liquid helium (LHT) considerable conductivity remains. Hence, provided electrons being the relevant carriers, transport may be ohmic, space-charge-limited, tunneling, Poole-Frenkel-type, or contact controlled, e.g. by Schottky emission. Most of them are thermally activated processes, and the respective energies are found to be larger than some 10 to 100 meV [73]. In the case of semiconductors, of course, holes must be considered, too. Transport in insulators has been treated so often that we will not repeat it here. A discussion referring to MIM devices in the pre-formed state was given for example by Ray and Hogarth [20].

In the ON state of these devices, current density (at a fixed voltage) is not uniform everywhere in (or on) the insulator, but mainly confined in filamentary paths. The conductivity of the surrounding matrix and in the substrate insulator is low compared to that in these filaments, so that we shall neglect it. ON state conductivities are, compared to the OFF state, surprisingly, nearly temperature independent in a wide temperature range, possibly between 3 and 750 K. Hickmott [72] stated independence within 10% from RT to pumped LHT. $I$–$U$ curves up to 0.5 V were about identical at RT, 4.2 K, and 3 K, weakly decreased at 77 K within the range of experimental error between repeated measurements even at a fixed temperature. Simmons and Verderber [39] reported a weak temperature dependence between 94 K and RT following the relation $\sigma = \sigma_0(1 + \alpha T^2)$ with $\alpha \approx 10^{-6}\ \mathrm{K}^{-2}$. A weakly increasing conductivity between 77 and 750 K was also observed by Blessing [73] with a temperature coefficient of resistance (TCR) $\alpha = -10^{-3}\ \mathrm{K}^{-1}$. Values are, if expressed by an activation energy, roughly 1 to 5 meV [74].

A remark is needed concerning the apparent discrepancy between the proposed activated change of carrier density and the experimental finding of a nearly temperature independent conductivity. One should keep in mind that a changed temperature of a sample holder primarily tunes only the substrate temperature to about the same value while the filament ($T_l$) is coupled to it by thermal resistance and may be heated by the electron gas system at $T_e$ to at maximum an about equal temperature, $T_l < T_e$. In the case of bad heat dissipation the filament lattice temperature $T_l$ (during current flowing), may vary far less than that of an external reservoir. So it is possible that in an effective small temperature range (taking also in consideration the competeing tendency of mobility change), measurable effects are far lower than expected from the evaluated activation energies.

From voltage pulse experiments [75] (Section 6.4) it turned out that the filament material should be at RT a p-type semiconductor with a band gap of about 0.15 eV. Applying a steep voltage pulse rapid heating of the electron gas occurs within picoseconds. This is followed by a delayed heat dissipation via momentum relaxation of the phonon system of the filament lattice with the consequence of lowering $T_e$, while increasing $T_l$. The latter causes a thermally activated increase of the electron–hole pair density, driving the filament to an intrinsic state, simultaneously rising the filament current and the radiative interband recombination rate (Section 6.4).

The derived activation energies were found to lie in a range between 0 and 0.25 eV, most probably at 0.15 eV [133]. The latter value well accounts for the estimation that $E_g > 0.15$ eV should reveal a current instability leading to filament destruction, i.e. off-switching [75]. Such a range of energy gaps $E_g$ should be expected for amorphous carbon prepared by different means, like evaporation and annealing, thermal treatment or irradiation of organic material, or pyrolysis.

These results would in principle reliably rule out all other proposed transport mechanisms, e.g. metallic, tunneling, hopping, or impurity band conduction. Nevertheless, a few of them shall be discussed here, at least for two reasons. Firstly, we cannot exclude the presence of other filament materials than carbon in all cited investigations. Secondly, the carbonaceous material must not be a homogeneous semiconductor. Barriers may be included obscuring carrier transport in distinct voltage and temperature ranges.

The few cases in which a positive TCR, as an indication of metallic transport, was observed shall not be treated further. They are evidently related to metallic filaments [76], and the resistance was then ohmic.

Different correlations have been proposed describing the $I$–$U$ dependence in the "smooth" region. They are $I \sim \exp(K_1 U^{1/2})$ [51] in the Poole-Frenkel or Schottky case, $I \sim \sinh K_2 U$ [39] based on tunneling transport, and $I \sim U^n$ [77], ohmic with $n = 1$ at low voltage and $n > 1$ at higher voltages deduced to space-charge-limited currents (SCLC). Adjusting the constants or properly electing the exponent, all of them may fairly fit measured curves, delivering no reasonable information upon the actual mechanism. Some authors moreover claimed to fit their results better by two or more laws in different voltage ranges [102]. As in the considered voltage range no relevant information can obviously be gained, neither from dc measurements nor from temperature dependences, more sensitive investigations are necessary.

One example was performed by Thurstans [45, 74]. Below 1 V and in a low temperature range (4.3 to 30 K) he observed an ohmic behavior up to a few mV followed by a $\sigma \sim \exp(KU^{1/2})$ dependence analogous to Neugebauer and Webb's results [78] on island films. When a bias of more than 1 V was applied, $I$–$U$ curves obeyed a relation $I \sim \sinh KU$, and the temperature dependence of conductivity varied like $\sigma = \sigma_0(1 + \alpha T^2)$, where $\sigma_0$ and $\alpha$ are constants. Such a behavior can be expected from electron transport by activated tunneling (e.g. [79]). Thurstans came to the conclusion that the sandwich diodes must contain metal islands (in the filaments) through which current should proceed.

Estimated activation energies $\delta E$ (1 to 5 meV) delivered [45, 74] diameters of (assumed) spheres of 20 to 50 nm by a supposed interdistance of $s = 1$ nm, reasonable for tunneling. Without any doubt metal indiffusion occurs into the sandwich insulators [42], and diffusion on the substrate surface is of course possible. It should be mentioned that the so-calculated activation energy $\delta E$ is idendical with Abeles et al.'s value [50] developed for the case of cermets. But although that model seems to explain the results reasonably, there remain significant objections. In the case of planar diodes metal islands with the linear dimensions mentioned above should be easily recognizable in transmission electron microscopes (TEM). But all attempts failed [46, 73, 81].

If the islands are smaller, possibly below the resolution limit of the used TEM's, then the activation energies should be greater than estimated from the temperature dependence. Other arguments against a metallic nature of filament material, e.g. necessity of organic adsorbate for growth and regeneration, shall be mentioned here. On the other hand, activated tunneling may occur in other inhomogeneous materials, too, provided chargeable parts are separated closely by potential barriers.

One serious problem in quantitative analyses of the measurements in a minute voltage scale is the uncertainty of knowing the voltage drop in the filament itself, and particularly the dimensions of the section controlling current transport.

### *3.5 Electron microscopical search for the current paths*

Electroformed MIM sandwich diodes displaying VCNR-like $I$–$U$ characteiistics are frequently studied for structural changes possibly being correlated with their electrical behavior ([8] and references therein). Generally it is to state that in sandwich structures the top metal electrode exhibits after forming statistically distributed holes of usually some μm diameter. Fig. 12 shows one example [49]. These defects in the electrode film are assumedly open mouths of pipes, penetrating through the dielectric. It was frequently confirmed, e.g. by Biederman [8, 31], that electrons and photons are emitted out of them.

Since emissions are caused by the circulating current, carrier flow between the base and top electrodes should be confined in these small areas. Nevertheless, it was not clear whether there exist also other current paths with invisible impingement on the electrode. The structure of the assumed filaments, arranged perpendicular to the electrode plane, are hardly observable in the case of sandwich structures.

Therefore, attempts have been made to discover current paths in planar diodes. Taking in mind that B-forming usually leads to a structure, as shown in Fig. 5b, with metal islands randomly distributed in location and size (as the thermally coagulated

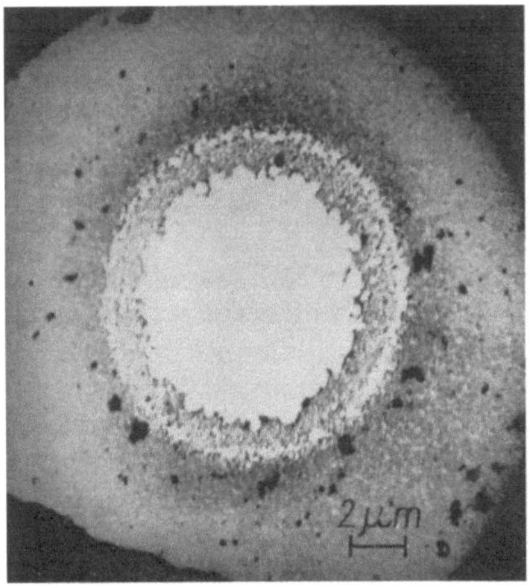


Fig. 12. Characteristic defect (identified as source of electron emission) in the top metal electrode of a electroformed sandwich MIM diode (Al–$Al_2O_3$–Au; after [49])

residuals of the prior meandered ultrathin metal films), a narrow relation to the field of metal island films seemed to be evident (see e.g. Morris and Coutts' review [79]).

The non-ohmic *I–U* behaviour similar to that of particular island films (activated tunneling [45]) supported these ideas. One could imagine current paths percolating between the electrodes, realized by island chains with proper interdistances allowing quantum mechanical electron tunneling. That would require interdistances of about 1 nm and, following Thurstans' estimation, an island size of at least 20 nm (due to measured activation energy, Section 3.4). But such island chains (or other current paths) have never been observed in TEM studies (see e.g. [81]). Distances between islands and their sizes are, except sometimes in small areas nonbridging the microslit gaps, far too large. One can argue that there could exist islands smaller than the TEM resolution (which would lead to wrong values of activation energies) or a "carpet" of metal atoms or any contamination [82]. Nevertheless, on account of the discreteness of the current paths required by the electrical features, conducting material must be inhomogeneously distributed. Therefore decoration methods were applied, but reveal-

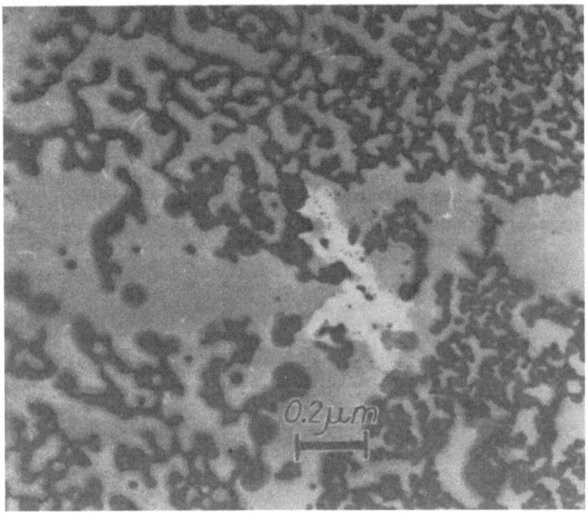


Fig. 13. Au-decorated region (previously identified as emission source) in a microslit of a planar MIM diode (Au on SiO) (after) [53]

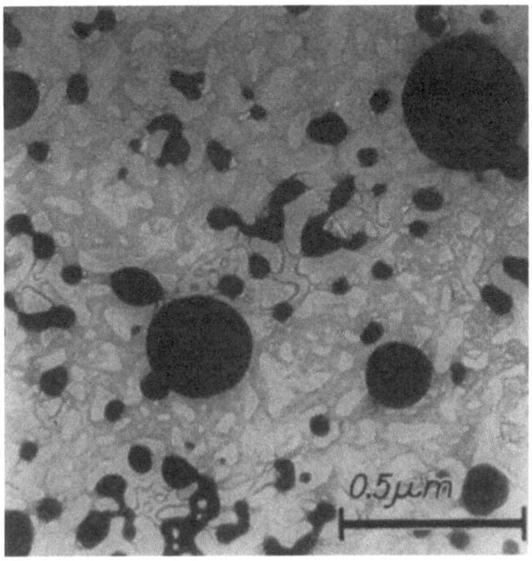


Fig. 14. Electron micrograph of the network structure in the microslit of an electroformed planar MIM diode (Au on $SiO_x$) after operation in $C_2HCl_3$ vapor of about $5 \times 10^3$ Pa; carbonaceous covers are argued to decorate filaments (after [83])

ed no reasonable success. One example of gold decoration adopted from Alekseenko and Fedorovich [53] is shown in Fig. 13. It should be mentioned that the bright region was not recognizable before the decoration (gold deposition), becoming visible by the small dark spots on the insulator surface. The authors observed a correlation between regions in which the effective microslit width is very small, due to the disturbed electrode edges and the island distribution, with emission spots (not being resolved on the same scale). No information upon the current path structures and involved material was however obtained.

During the course of investigations to produce more definite filament material in organic gas environments at low pressure ($p < 10^4$ Pa) a network structure was discovered (Fig. 14 [83]). Although conductivity was strongly increased under those conditions, it was not yet settled if the observed thick filaments really carry the current. It was argued that they possibly decorate the invisible filaments [46], in contrast to which they are no longer able to switch under the common voltage conditions. Nevertheless, a network structure of the filaments should be taken into consideration.

### *3.6 Localizing the voltage drop in planar MIM diodes*

The existence of active regions has been proved by experiments of Nepijko and Styopkin [84] with a scanning electron microscope (SEM) with an option additionally to probe potential distributions. They observed regions of μm dimensions in which scanned 0.33 V contours within an about 2 V drop are markedly different in the ON and OFF states (see Fig. 15). But microscopic resolution was insufficient for any reliable information on the particular structure.

Because of the low electron scattering factor of the filament material (C), and since we must assume the mass concentration in the possibly very small dimensions (on a molecular scale) to be very low compared to the substrates (e.g. $SiO_x$, $Al_2O_3$), it is not surprising that filaments are hardly to observe. Further decoration studies by Schlemper [62] have neither been successful. But new results, supporting the filament rupturing model, were obtained by the potential-modified defocusing-phase contrast [85]. By this method it is possible to localize the voltage drop by a sharp bright line in the defocused image (Fig. 16a, b is the focused image). Applying a turn-off voltage $U_M$ (= 15 V) nearly the whole voltage drops in a narrow area, while at a low applied voltage, with intact filaments, voltage drops smoothly, and does not noticeably affect the phase

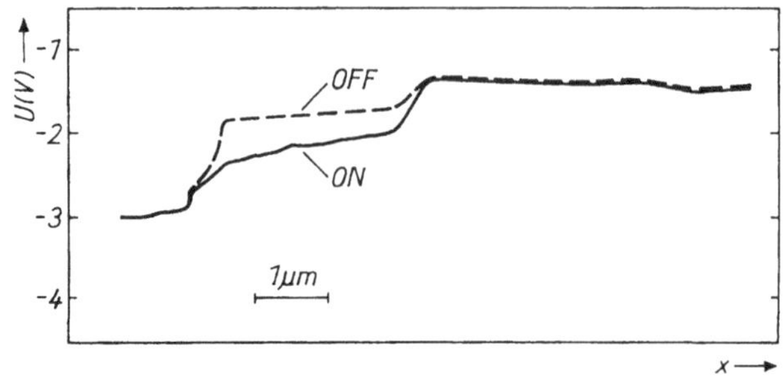


Fig. 15. Potential profiles in an active region of a microslit of a planar MIM diode (discontinuous Au film) deduced from equipotential line scan distributions measured in a scanning electron microscope ($x$: electron current direction) (after [84])

contrast. It should be mentioned that this rupturing line is usually thin compared with the microslit width, and does not touch the metal islands. Although still invisible, filaments are assumedly arranged perpendicularly to this line, hence are located between the islands on the free surface and there is no necessity of islands to be involved in current transport. That is confirmed by the fact that microslits containing no (visible) islands exhibit identical electroforming and switching behavior [85]. Since, by this method, the ruptured filament sections can be localized, there is a better chance for searching the filaments by more refined methods.

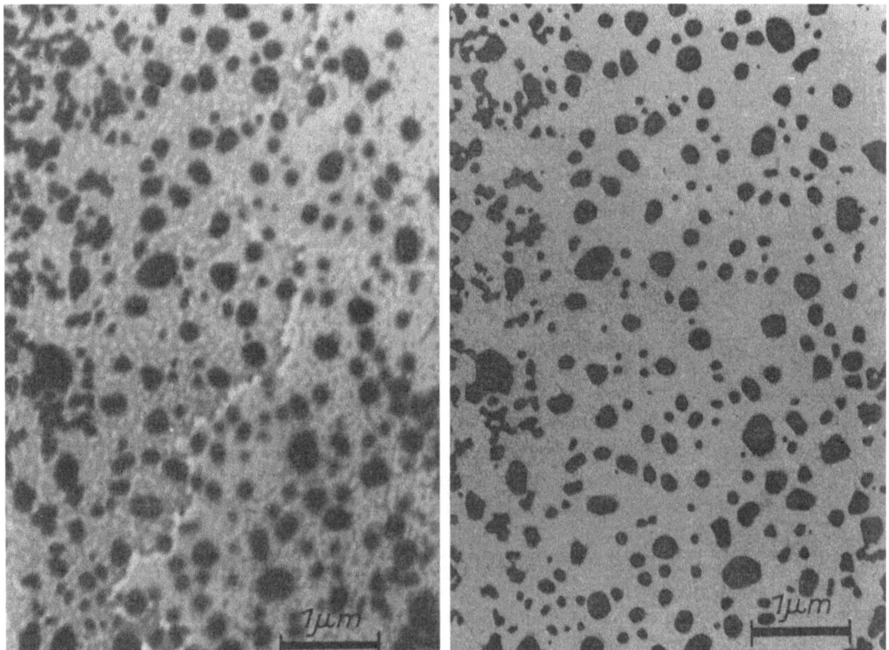


a b

Fig. 16. a) Defocused electron micrograph (TEM) of a part of the microslit of a planar MIM diode (Au on $SiO_x$) with applied voltage (15 V, OFF state), b) focused image (after [85])

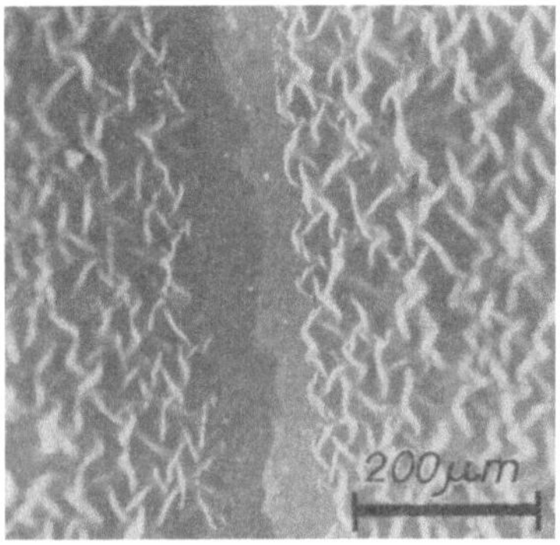


Fig. 17. SEM image of the microslit in a planar MIM diode (C on quartz glass) with applied voltage (5V); the dark–bright contrast line exhibits the voltage drop (after [36])

It should also be mentioned, that a voltage drop line (with naturally less resolution) can also be seen in a SEM (Fig. 17) and be used as a tool to find a badly recognizable microslit in those planar diodes.

The voltage drop may naturally also be measured with a scanning tunneling microscope (STM). Murald and Pohl [86] performed preliminary investigations with a modified STM allowing simultaneous measurement of the surface topography and the electric potential distribution. Further attempts to obtain more direct information upon the current path in planar MIM diodes are in progress. The development in this field promises reliable results by this method and simultaneously applied tunneling spectroscopy in the near future.

## 4. External Influences on Switching

### *4.1 Studies of the regeneration process — a two-stage mechanism*

The characteristic behavior of switching devices with memory are the stable states at or near zero voltage. When raising the voltage on-switching gets possible above a threshold voltage $U_{th}$. The increase of current versus time shall be called regeneration, and may be measured as follows [52]. Starting with a diode in the ON state, the voltage is raised (OAB in Fig. 18). The current follows a dc characteristic. At a memory voltage $U_M$ the voltage is turned off (BO in Fig. 18). Then a constant regeneration voltage $U_R$ is applied and the current increase recorded. A waiting time at zero voltage may be introduced before switching on $U_R$. Surrounding conditions like pressure, temperature, etc. remain unaltered. The usual procedure is to apply (repeated) pulse trains like those shown in Fig. 18b.

The averaged $I$–$t$ curve shapes do not follow a simple mathematical law. On the other hand, they may be characterized by a "half-recovery" time, or as we shall call it, regeneration time constant $\tau_R$. Hogarth and Kompany [87] performed similar studies with sandwich diodes. Based on a charging model, they used terms as loading (corresponding with regeneration) and a relaxation time constant. Nevertheless, from their $I$–$t$ curves (Fig. 33 in [20]) one can roughly estimate a dependence $\tau_R \sim \sim \exp(-aU_R)$ (with $5\ V^{-1} < a < 13\ V^{-1}$), which reliably fits into the results gained with planar diodes. A monotonous increase of $\tau_R$ with decreasing $U_R$ was always exhibited, tending to very large values compared with moderate measuring times near the threshold voltage $U_{th}$, and may hence be described by $\tau_R \sim \exp(U_R - U_{th})^{-1}$ (Fig. 19) [52].

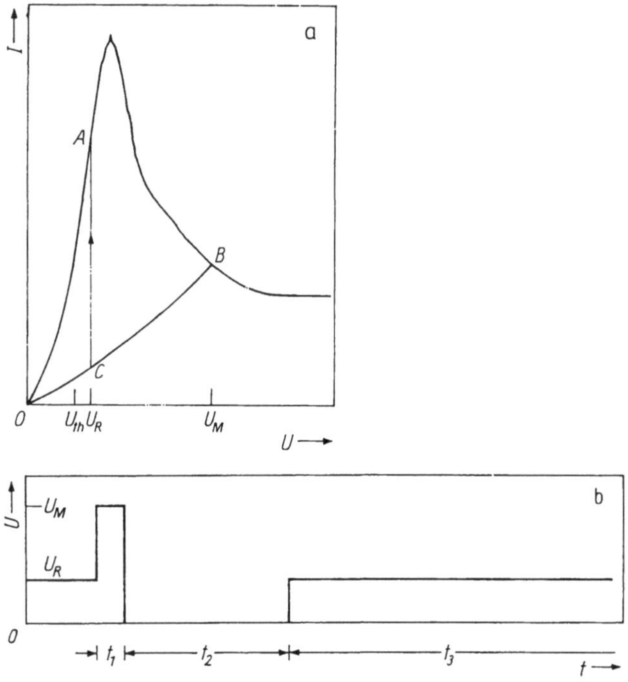


Fig. 18. a) $I$–$U$ curves illustrating the regeneration experiments and b) the applied voltage–time trains; $t_1$ pulse duration, $t_2$ waiting time ($U = 0$) (after [52])

The rise of the regeneration curve is much steeper applying $U_R$ after a waiting time ($t_2$) as shown for an example in Fig. 20 (curve b). Following Pagnia et al. [52] we can imagine that an off-switched filament is destroyed partly (Fig. 21). These sections ("holes") must be re-built during regeneration. This is evidently a two-stage process; one is attributed to filling of holes in the filament, or any re-growth process, the other due to a re-organizing effect, effectively switching the material to a conductive state. Both of the processes may be characterized by individual time constants $\tau_f$ and $\tau_s$, which are separable by distinct measuring and evaluation procedures. $\tau_s$ may be measured after a sufficient waiting time and depends similarly on $U_R$ as $\tau_R$ (Fig. 22 [52]).

Using a few evident assumptions the authors [52] could derive a formula describing the regeneration current with and without a waiting time. Here we shall treat only the two simplest cases.

(i) Assuming no waiting time and the on-switching being faster than refilling ($\tau_s \ll \tau_f$) which is the case for regeneration voltages near the current maximum (Fig. 18) we may write the probability for finding the filament conductive as

$$p(t) = \left\{1 - \exp\left(-\frac{t}{\tau_f}\right)\right\}^{\alpha}. \tag{4.1}$$

This is the product of equal probabilities for finding individual holes filled at time $t$.

Taking $I(t) \sim p(t)$ which means all filaments are connected parallel between the electrodes, one can fit experimental curves quite well for moderate switch-off voltages ($U_M < 10$ V at RT) (Fig. 23).

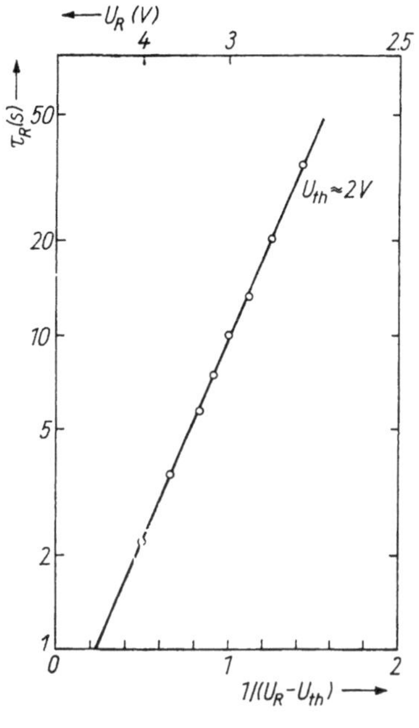


Fig. 19

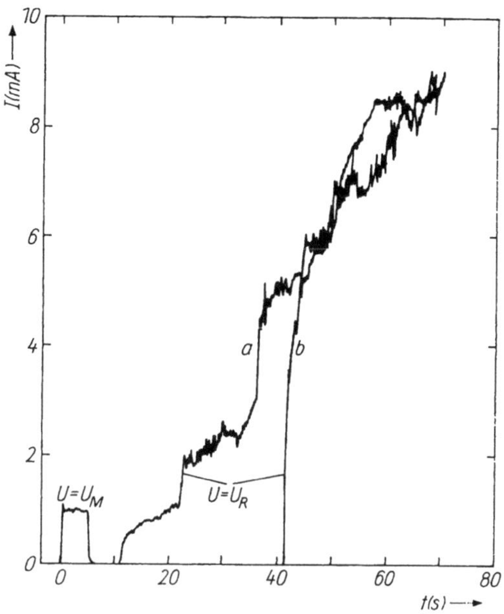


Fig. 20

Fig. 19. Regeneration time constant $\tau_R$ vs. reciprocal voltage difference ($U_R - U_{th}$) measured on a planar MIM diode (Au on quartz glass); $U_M = 6$ V, $T = 77$ K, $p = 10$ Pa (after [52])

Fig. 20. Regeneration behaviour of a sandwich MIM diode (Au–$SiO_x$–Au) (a) without, (b) with a waiting time $t_2$; $U_M = 10$ V, $U_R = 4$ V (after [46])

(ii) When introducing a sufficient waiting time ($t_2 \gg \tau_f$) the on-switching process may be investigated. Because of the lack of a physical model the authors assumed for the first a spontaneous switching process without filament–filament interaction. By fitting the experimental curves to an exponential function, $\tau_s$ the time constant for on-switching could be derived and showed a very strong dependence on regeneration voltage. Better fits to experimental curves could be achieved by allowing a filament to filament interaction. It could be shown [134] that switching filaments stimulate switchable filaments with a strong interaction coefficient.

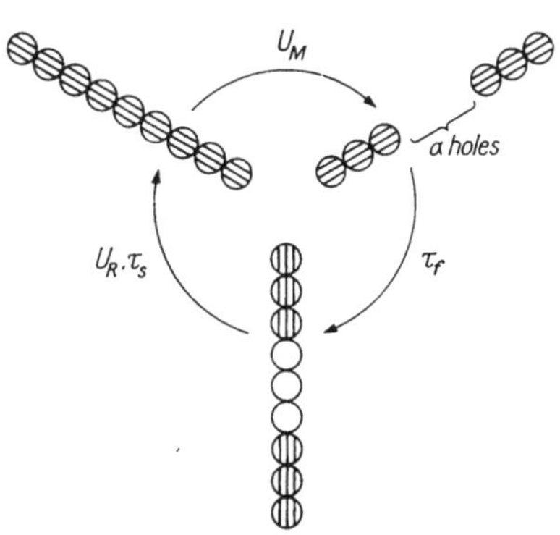


Fig. 21. Illustration of the filament–hole model par ameters (see text)

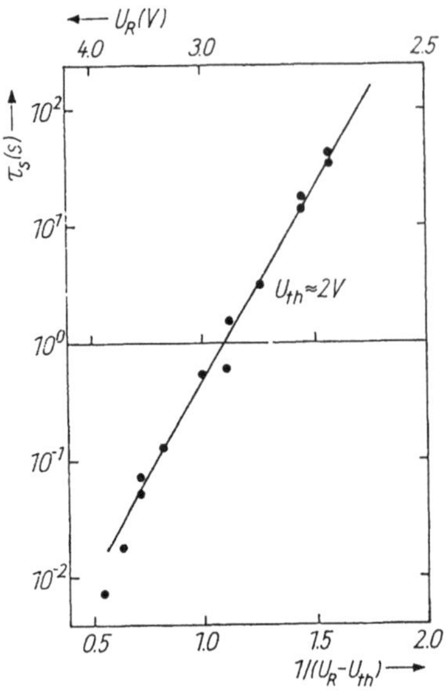


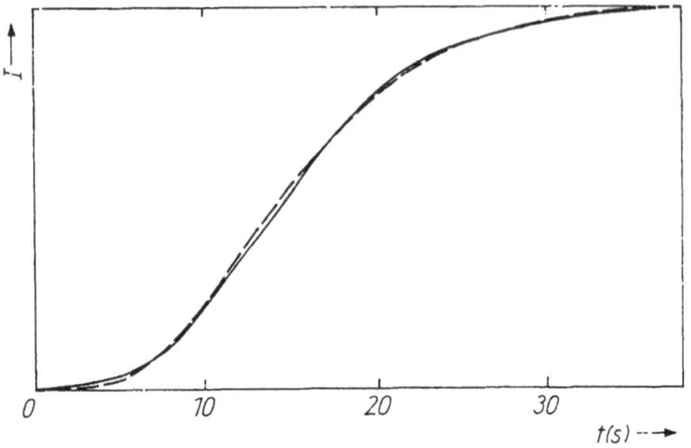


Fig. 22 Fig. 23

Fig. 22. Switching time constant $\tau_s$ vs. reciprocal voltage difference ($U_R - U_{th}$) measured like in Fig. 19

Fig. 23. Comparison of an experimental regeneration $I$–$t$ curve of a planar MIM diode (Au on quartz glass) (———) with a computer simulated one (FH model) (– – –); $\alpha = 8.4$, $\tau_F = 6.4$ s (after [52])

Besides the different time constants, the number of holes $\alpha$ is a measure for the filament destruction. It is a function of the pulse duration $t_1$ and the turn-off voltage $U_M$, making itself felt via $\tau_R \sim \exp(bU_M)$ ($b$ is constant at $T$ = const) (Fig. 24 [44]). A material loss mechanism must be assumed during such a process. Re-delivering the proper molecules naturally takes longer times for a higher degree of destruction, corresponding to a larger $\alpha$ in the model. Because of the exponential dependence of the regeneration times $\tau_R$ on the pulse height $U_M$ one may assume that already ruptured filaments are even more destroyed by raising the memory voltage $U_M$.

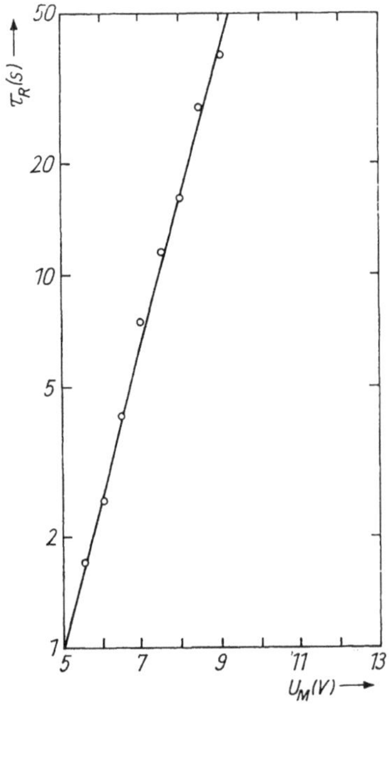


Fig. 24. Regeneration time constant $\tau_R$ as a function of the turn-off voltage $U_M$ from a planar MIM diode (Au on quartz glass); $U_R = 4$ V, $T = 77$ K, $p = 10$ Pa (after [44])

Although the derived model seems to work, that means at least the chosen parameters are physically meaningful. But what merits and deficiencies does it hold? The model describes well the steep increase of current during regeneration, but it does not reasonably describe the long-term regeneration showing a weak current increase instead of a saturation. This point, of course, requires a modification of the treatment and easily may be done by introducing passive molecules or a filament network consisting of two parallel filaments in series with a third one as a load resistance. On-switching of one filament lowers the voltage drop at the other and the regeneration is slowed down.

The regeneration process itself is affected by various further parameters. Environmental influences are mainly the partial gas pressure and compositions controlled by the pumping system. Not all of these parameters were studied unequivocally, and, hence, it was not yet possible to take them reliably into account. Nevertheless, the known examples shall be discussed in more detail.

### *4.2 Search for the adsorbed species*

The first results exhibiting the requirement for adsorbed molecules in order to allow A-type forming as well as regeneration were presented in 1980 by Alekseenko and Fedorovich [80, 89] and Blessing et al. [90]. Both groups showed ceasing of the VCNR-like characteristics in ultra-high vacuum (UHV). But in these preliminary studies they drew somewhat different conclusions concerning a proper gas responsible for regeneration. Although Blessing et al. [51] showed already in 1981 the importance of organic species with further gas inlet experiments, the problem is in so far unsolved as it is even now not clear whether there is only one proper molecule (or its fragment) responsible for filament growth or more, and what precisely are the growth conditions.

Here we must conclude that there exist different molecule species affecting the filament differently. Taking also into consideration all the other molecules present in a vacuum system (and also those introduced by aimed experiments), one should distinguish between three species:

(i) constructive (organic) molecules building ultimately the filament material;
(ii) destructive molecules (e.g. $O_2$ or others with high electro-negativity) destroying the ON state;
(iii) passive molecules (e.g. $N_2$ and the noble gases), which may be built in temporarily instead of (i).

An identification of the constructive molecules seems to be easy. Each usual HV environment performed by rotary pump and oil diffusion pump delivers the required conditions in a wide pressure range (10 to $10^{-5}$ Pa), and mass spectrometry is a common method analysing residual gases. The amount of inorganic and organic species and their fragments, however, causing a too high number of mass peaks, prevented direct identification until now.

Gas inlets into oil-free vacuum systems (unfit for A-type forming) have been performed by Pagnia et al. [91]. About 30 organic species, saturated as well as unsaturated hydrocarbons and alcohols exhibited no satisfactory success. After the usual (B-type) preforming, most of them caused a medium to strong increase in conductivity during voltage cycling. But none of them displayed the typical VCNR-like $I$–$U$ curves. The supply of organic molecules was possibly too high, hence the growing carbonaceous material was stable against voltage or current stresses which usually cause off-switching.

Nevertheless, alkyl di-halides and alkenes, all containing C–C double bonds were most effective, while singly bonded molecules were less useful to grow conductive materials. Alcohols were found to cease a formerly formed VCNR-like maximum as known from $O_2$ [49] and $Cl_2$ [92]. None of the pumping oils (at RT) brought directly into the vacuum chamber had any influence. This may be explained by the extremely low partial pressure of the pumping oils at RT, being orders of magnitude lower than the total pressure of the vacuum vessel used.

The findings of the Kiev group [70] should be mentioned who performed an UHV experiment (glass apparatus) with controlled heating of crystalline anthracene as the source of organic species, enabling them to produce the quantities of expected properties successfully. Their new result from an experiment with naphthalene is shown in Fig. 25 [93].

The present situation is of course not satisfactory, but a simple experiment shall be mentioned, clearly demonstrating the correlation between organic species (in the mass spectra) and diode behavior [91]. In the experiment cited a short pipe out of the UHV system was first flanged to the HV system (and so contaminated), and then to the UHV system again. The diode was mounted inside the pipe, which could be mildly heated. Under RT conditions the vacuum was good enough not to allow regeneration (curves 1 and 4 in Fig. 26), while thermal desorption of organic gases from the pipe wall and condensation on the colder diode properly allowed it (curves 2 and 3). This procedure could be repeated many times. A further proof for the responsibility of organic species will be discussed in the following.

### *4.3 Influence of partial pressures and residual gas composition on regeneration*

Qualitatively it is known that residual gases affect the $I$–$U$ characteristics and the regeneration process. An attempt to find quantitative correlations was recently performed by Borbonus et al. [94]. A vacuum system with a cylindrical cold trap having

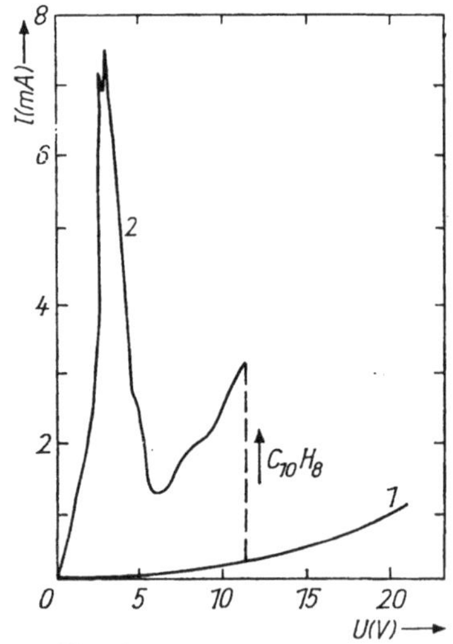


Fig. 25

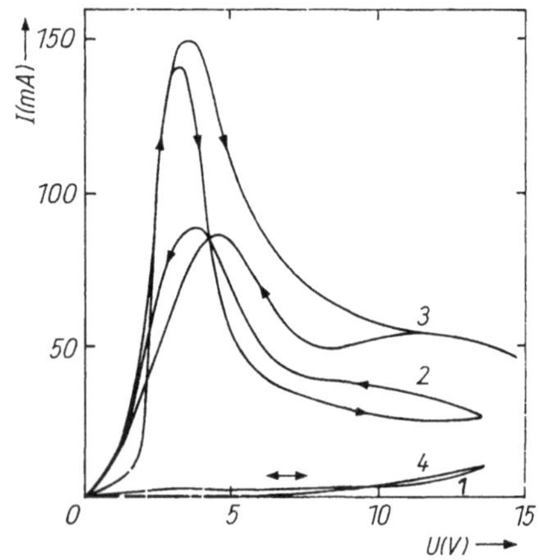


Fig. 26

Fig. 25. Development of the VCNR-like $I$–$U$ characteristic of a planar MIM diode (Au on dielectric substrate): (1) in UHV, (2) after exposure to $C_{10}H_8$ (and $E < 2 \times 10^5$ V/cm) (after [93])

Fig. 26. $I$–$U$ characteristic of a planar MIM diode (Au on quartz glass) in an oil-free vacuum vessel (1, 4) and after contaminating with organic residual gases (2, 3) (numbers at the curves note succession) [91] a surface area large compared with the diode surface, could be cooled or heated ($80\,K < T_c < 400\,K$) independent of the diode inside held at a different temperature $T_d$. In such a cryopump-like system two different sorts of gases must be distinguished, those being permanent like $N_2$, $O_2$, Ar, and $H_2$, whose partial pressure is mainly caused by an equilibrium between pumping and leakage rate, and the others which may potentially be frozen out like $H_2O$, $CO_2$, and all organic residuals. This delivered a tool to control the partial pressure $p_i$, particularly of the organic species, by variation of $T_c$.

The characteristic time for regeneration $\tau_R$ was found to change strongly and to be relatable to the organic gas pressure, corresponding with the cold trap temperature $T_c$, while the sample itself was held at RT.

The impingement rate $z = 1/\tau_f$ of molecules of area $A$ may be expressed in the kinematical gas theory as

$$z = \frac{p}{C(MT_c)^{1/2}}, \tag{4.2}$$

where $C$ is a constant. Integrating the Clausius-Clapeyron equation for an ideal gas and choosing the experimental conditions so that $\tau_f \approx \tau_R$ a correlation between $\tau_R$ and $T_c$ can be derived,

$$\log\left(\frac{\tau_R}{\sqrt{T_c}}\right) = \left(\frac{A^*}{T_c} - B\right) + \log\sqrt{\frac{M}{A}} + C^*, \tag{4.3}$$

where $B$, $C^*$ are constants, $A^*$ is the fluid–gas transition enthalpy. This equation could be verified as shown in Fig. 27 (triangles). The cold trap was cooled slowly with small temperature steps. By cooling faster and with wider steps and hence a smaller number of performed voltage cycles, a deviation from (4.3) could be observed at low temperatures (Fig. 27, full circles). Under those conditions the increase of $\tau_R$ was less steep.

Borbonus et al. [94] came to the conclusion that three temperature ranges of the cold trap $T_c$ must be considered:

(i) In a temperature range $T_c$ above RT and of some ten K below, the impingement rate controls the growth rate (filling, regeneration) and hence $\tau_R$.

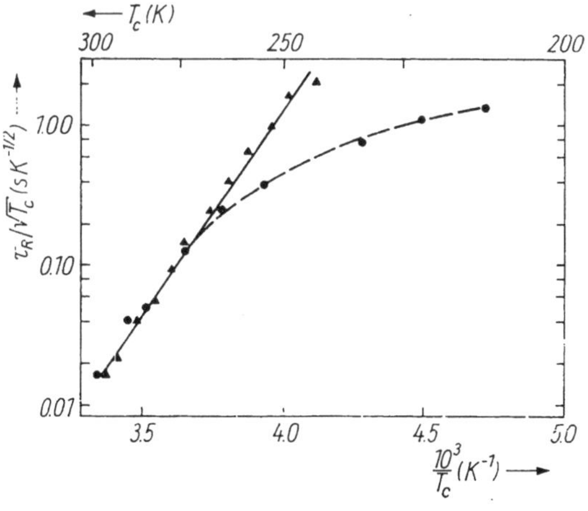


Fig. 27. Dependence of the regeneration time constant $\tau_R$ on the partial pressure controlled by the cold trap temperature $T_c$ during stepwise cooling; planar MIM diode of Au on quartz glass at RT (after [94])

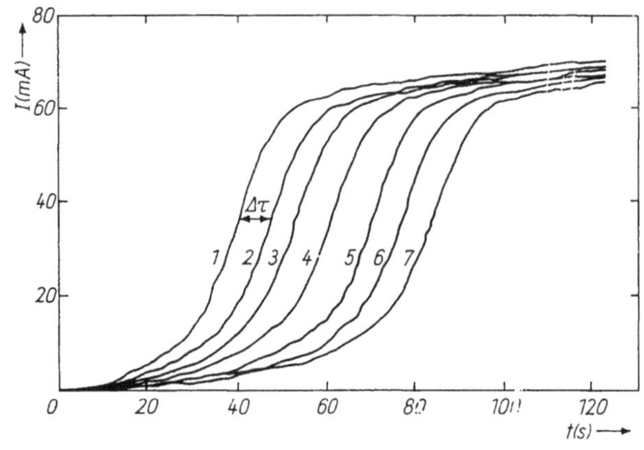


Fig. 28. Drifting regeneration *I–t* curves under reduced supply of organic molecules (see text); the numbers 1 to 7 note subsequent cycles (after [94])

(ii) Lowering the temperature $T_c$, without exhausting the supply from the gas space, has some storage effect. Impingement and diffusion deliver the required particles, and the increase of $\tau_R$ is slowed down.

(iii) At low temperatures $T_c$ the supply comes essentially from diffusing particles (stored between successive cycles) out of areas around the destroyed sections of the filaments.

Results of (iii) are drifting the *I–t* curves as shown for an example in Fig. 28.

A striking observation was that the regeneration process was independent of the diode temperature $T_d$ provided that the supply rate of organic molecules was sufficiently high. According to (i) it is probable that impinging particles are built in directly, and no temperature dependent process like diffusion is involved. This observation provides e.g. repeated regeneration even at $T_d = 77$ K and contradicts to earlier findings of a voltage–temperature memory effect (see e.g. [20, 49]).

### *4.4 Influence of gases with electronegative molecules*

Starting from the ON state with a diode operated in a HV chamber, the OFF state or intermediate states may be reached by operating the diode in a changed surrounding atmosphere. Examples are UHV or by any means clean (oil-free) vacuum conditions. Similar results may be achieved by raising the partial pressure of some gas drastically. This can be performed for example by an air inlet (Fig. 4b).

From various investigations we must conclude that oxygen is the responsible component for lowering the current. But it is not simply an adsorption effect. MIM diodes are stable in air if a voltage threshold $U_{th^*}$ is not exceeded before. The influence of oxygen on the *I–U* characteristic was firstly shown by Barriac et al. [64]. Emmer [49] published data from $Al–Al_2O_3–Au$ diodes stationarily operated under different $O_2$ pressures (Fig. 29). Characteristic was the current lowering as a function of the oxygen pressure. In the case of the investigated sandwich diode at voltages lower than $U_{th^*} = 1.8$ V the conductivity of formed samples was practically independent of the ambient gas pressure and composition. It was shown that after one full voltage cycle the sample was in its OFF state, and could not regenerate under maintained exposure of $O_2$. A re-establishment of HV conditions was necessary. This was reproduced by Tanaka et al. [29], who also explored a similar influence of chlorine $Cl_2$ [92]. An about linear relative current decrease with the voltage with a threshold behavior at fixed oxygen pressure has also been observed on planar diodes (Fig. 30) [88].

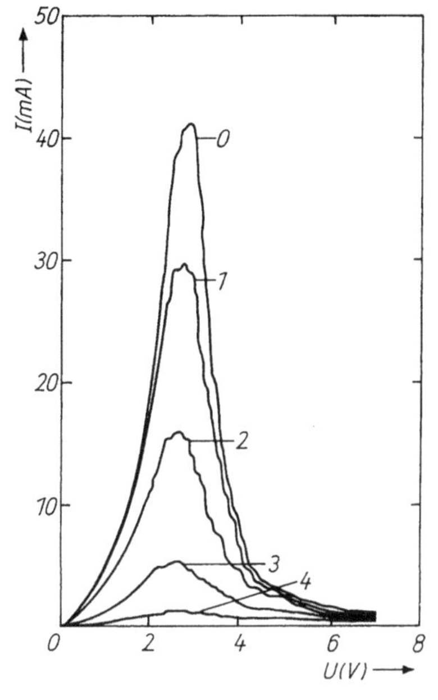


Fig. 29. $I$–$U$ curves of a sandwich MIM diode (Al–$Al_2O_3$–Au) in HV ($10^{-3}$ Pa, curve 0) and under different oxygen pressures between about 1 and $10^3$ Pa (curves 1 to 4) (after [49])

Alcohols were found to reduce the ON state conductivity in the same manner, too [91]. All these species have a high electronegativity (EN), which should be taken into account in order to explore the underlying microscopic processes. In the following we shall reduce the discussion on experiments with $O_2$ as a representative.

The particular influence of $O_2$ (and the above-mentioned other gases with high EN) was reproduced also by other investigators. Gengenbach et al. [95] claimed that this behavior should be one of the easily performable tests in order to recognize this special class of switching systems with carbonaceous filaments (Section 3.1).

Here we must point out a detail not taken into account reasonably in the literature. Most experiments were done by driving the diode in oxygen to voltages beyond the NDR branch. This allows the gas to act on filaments in the ON state as well as on destroyed filaments. Equal reactions in both cases cannot be expected. The difference is seen in the first voltage sweep after the oxygen has been removed. Samples with filaments exposed only in ON state or at low off-switching voltages regenerate already in the forward sweep at low voltages, while samples treated with higher voltages during oxygen exposition require higher voltages for regeneration. Here a repeated A-forming seems to be necessary. Keeping these interpretations in mind we will further on use the terms on- and off-switching in these cases, too.

The stationarity and reversibility of intermediate states between the full ON and OFF states caused by oxygen, are evidence that at least two competing effects must be involved. Borbonus et al. [94] proposed oxygen induced off-switching and normal regeneration. This is seen in experiments where the partial pressure of oxygen was

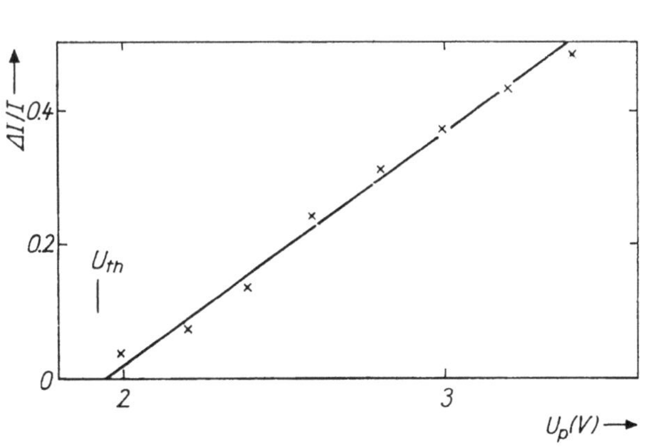


Fig. 30. Relative current decrease $\Delta I/I$ vs. the applied voltage at fixed oxygen pressure (1.5 Pa); planar MIM diode (Au on quartz glass) (after [88])

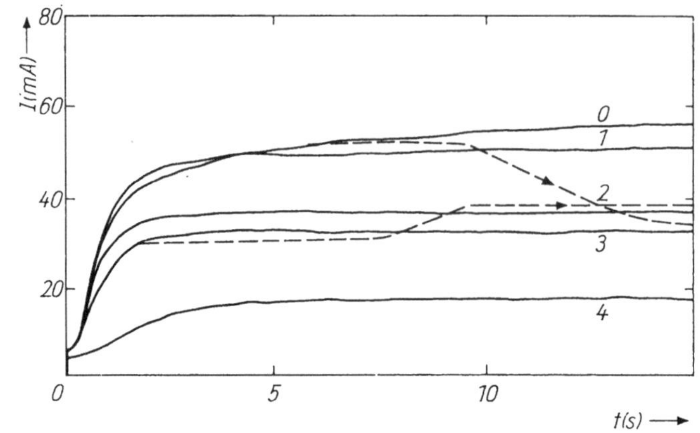


Fig. 31. Regeneration behavior of planar MIM diodes (Au on quartz glass) under vacuum ($10^{-3}$ Pa, curve 0) and various oxygen pressures between 10 and $10^2$ Pa (after [94]). Dashed curves: pressure was changed during recording

varied while the voltage at the diodes was held fixed as well as in experiments investigating regeneration. The stationary saturation values of the current of the $I$–$t$ trace could be reached from higher or lower states of conductivity (Fig. 31).

Effects of the destructive molecules like oxygen can be characterized, extending the model discussed in Section 4.1, through an enhanced off-switching probability $1/\tau_a$, being proportional to the partial pressure of oxygen in the whole composition. As has been shown [88], regeneration in an environment thus poisoned as well as the behavior of a former stable working ON state at fixed applied voltage after changing the gas composition can be described reasonably.

Drawing a conclusion from experiments one may assume oxygen to affect only filaments carrying current in the ON state. The problem of evaluating the number $n$ of stationary on-switched filaments dependent on changes of $O_2$ pressure may be solved by a usual reaction formulation for $n$ of $n_0$ filaments being conductive,

$$\frac{dn}{dt} = \frac{n_0 - n}{\tau_+} - \frac{n}{\tau_a}. \tag{4.4}$$

Taking into consideration an undisturbed OFF–ON switching time constant $\tau_+$, not necessarily equal to the on-switching time constant $\tau_s$, the influence of oxygen may be simulated not only for the quasistatic but also for the regeneration current. Comparisons with experimental results indicate short stay times (or low sticking coefficients). But more detailed experiments are necessary to determine the individual time constants.

At a fixed voltage $U > U_{th}$ switching events in both directions can be observed. It is an unsolved question whether oxygen is built in, on or in instead of normal filament substance thus destroying the effective conductivity, or if it really makes reconnection impossible.

Since at a fixed partial pressure a stationary state of the ratio of on- and off-switched filaments is established, a mechanism of the first kind is more likely, and we must assume oxygen desorption and by this a re-establishing of conductivity.

### *4.5 Influence of passive gas species on regeneration*

As already mentioned experiments have been performed to study a possible environmental influence on conductivity of formed MIM diodes. Emmer [49] for example

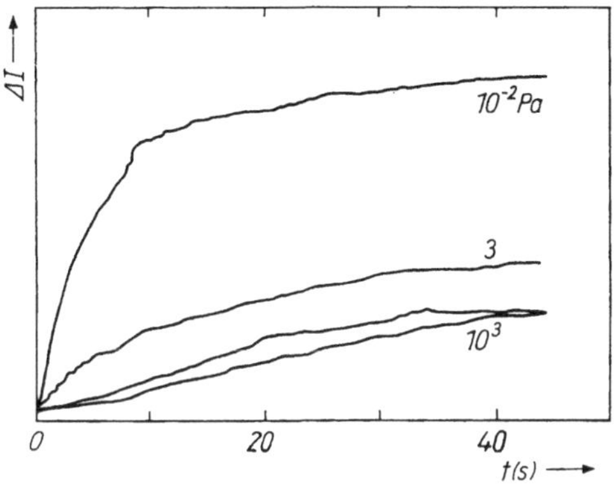


Fig. 32. Regeneration curves in nitrogen atmosphere of different pressure, planar MIM diodes (Au on quartz glass); RT, $U_M = 8$ V, $U_R = 3$ V (after [88])

found no effect of argon even at atmospheric pressure, and only a small decrease of $I$–$U$ maxima for $N_2$ and $H_2$. Further species also present in a residual gas of a vacuum system, like $H_2O$, CO, $CO_2$, seem to be passive. Our experiments neither revealed an appreciable influence of the above-mentioned gases. So we may conclude that noble gases and gases with molecules of low EN do not affect diodes at any voltage.

Nevertheless, passive gases may affect regeneration by temporarily filling holes in the filament by species hindering the re-growth of the filament material. Qualitatively one would expect a delayed $I$–$t$ behavior. This is observed as shown for an example of regeneration at different nitrogen pressures in Fig. 32.

The behavior of passive molecules may be characterized by a filling rate $1/\tau_{p_+}$ and a desorption rate $1/\tau_{p_-}$, which can be introduced into an extended filament–hole model (Section 4.1).

From computed dependences, [88] it turns out that the influence of passive molecules may be roughly divided into two groups. At a low vapor pressure and a long stay time regeneration curves show a reduced steep increase followed by a weak long-term rise of current. Experimental curves in HV show this behavior indicating that not all residual gas components with high sticking coefficients are active. Measured regeneration times $\tau_R$ are not affected by this group of gases.

In the opposite case (due to high vapor pressure and short stay time), sites occupied by passive molecules feign a lower vapor pressure of the constructive gases. This results in longer regeneration times. The regeneration curves are flattened with increasing pressure, so it may be doubtful whether complete regeneration can be achieved within moderate times.

### *4.6 Charged particle induced off-switching*

Even though it was possible to influence neither the ON nor the OFF state by irradiation of light from near uv to near ir [46] nor by operation in a transmission electron microscope ($W_e > 50$ keV) [73], it is now established that low energy electrons stimulate ON–OFF switching. This was shown for the first time by Nepiiko et al. [84, 96] in an energy range 1 to 39 keV (in a SEM) revealing the lowest energy to be most effective. Optimum bias voltage during exposure was $U < U_{th}$ and the degree of off-switching was dependent only on the electron dose. Berg et al. [97] observed an energy threshold at $W_{th} = 6.2$ eV, and a dip at about $W_d = 17.2$ eV (Fig. 33). Yield curves, or the decrease $\Delta I/I$, are similarly shaped to those of electron induced desorption [98]. The underlying basic energy transfer processes may be of similar character.

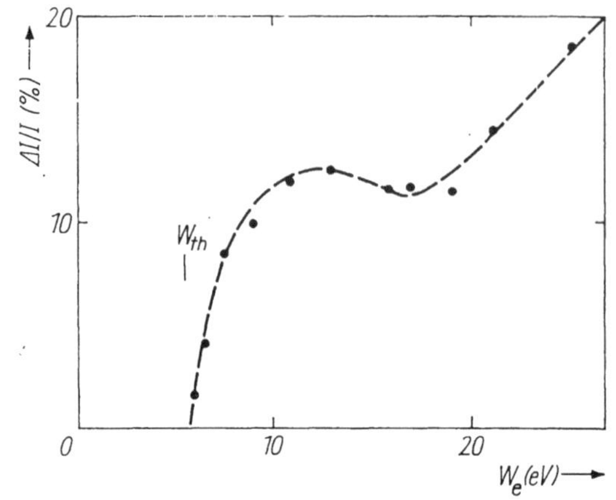


Fig. 33

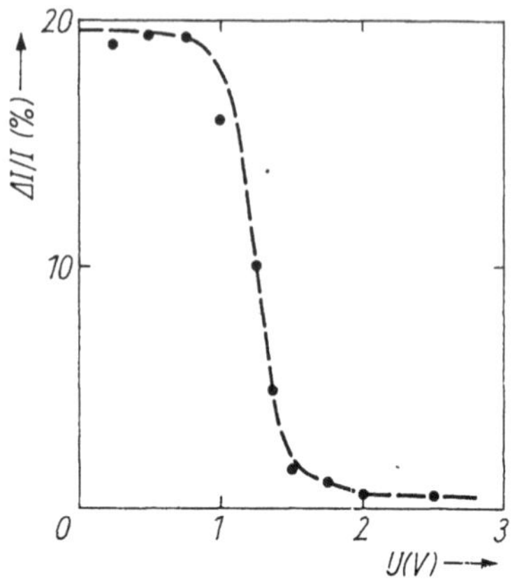


Fig. 34

Fig. 33. Relative current decrease (off-switching) of a planar MIM diode (Au on quartz glass) during electron bombardment; energy $W$; diode bias voltage $U = 0.5$ V $< U_{th}$ (after [97])

Fig. 34. Relative current decrease (like in Fig. 33) vs. the bias voltage $U$ (fixed at each measuring point) at fixed electron energy $W_e = 10$ eV (after [97])

Looking at the voltage dependence of the current decrease (Fig. 34) the steep slope of the efficiency at about the threshold voltage $U_{th}$ is remarkable. Above $U_{th}$, where regeneration occurs under HV conditions, stationary intermediate states due to both competing mechanisms, electron induced off-switching and normal regeneration, may be gained by properly choosing the incident electron current density.

Off-switching can also be stimulated by ions, as was shown by Berg et al. [99]. An impact desorption (sputtering) of atoms building the filament material may be the reason.

### *4.7 Can metal re-covering cause regeneration?*

So far regeneration was discussed as an effect where re-filling of interrupted filament sections by organic species would occur followed by on-switching.

The latter process could be imagined as a phase transition of grown nonconductive organic material, some ordering mechanism like an alignment of molecules, or a mere mechanical mechanism of material connecting the filament tip with an electrode.

To get more insight into the regeneration process post-deposition experiments were done. During post-deposition of a small amount of Au (not of Al) onto the Au electrode Thurstans [45] observed a stepwise recovery of current from the OFF state to about half the ON state current (applied voltage $U = 1.4$ V $< U_{th}$) at about 30% of the investigated MIM sandwich structures. Recorded regeneration curves were very similar to the usual ones (Section 4.1). This was claimed to be a proof of the metallic nature of the filaments.

As was shown in preliminary studies of Gengenbach et al. [95], planar open MIM diodes behave similarly. Post-deposition of small amounts of metal (Au, Ag, In, Bi, Pb, Cr [36]) after foregoing common electroforming and operation enhance the VCNR-like $I$–$U$ characteristics significantly. But repeated voltage cycling was required, before a new reproducible characteristic at a higher level than the original one was gained.

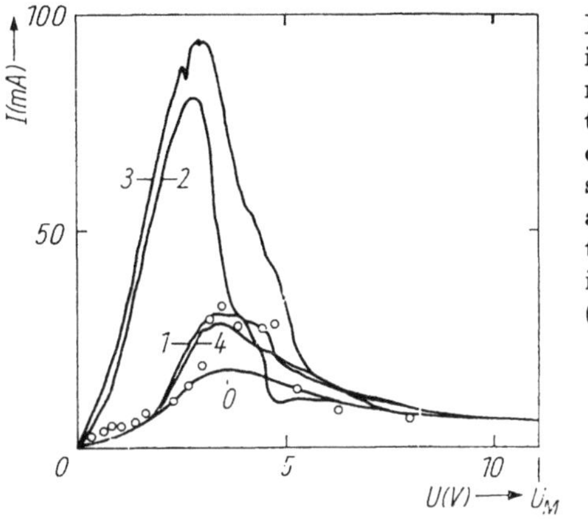


Fig. 35. $I$–$U$ characteristics of a planar MIM diode illustrating metal (Au) post-deposition. Curve 0: return sweeps, (1) forward sweeps before metal treatment, (2) after off-switching at $U_M$ and metal evaporation (diode in the OFF state), return sweep on trace 0, (3) after metal evaporation of approximately equal amount on the diode in the ON state, (4) control forward sweep. Circles indicate roughly the difference of curves 3 and 2 (after [118])

Some new investigations on planar diodes [118] shall be discussed here on an example illustrated in Fig. 35. The $I$–$U$ characteristic showed forward and return curves like 0 and 1 under repeated voltage cycling. Prior to exposure to the metal evaporation source the diode was switched to its OFF state. Choosing the metal amount properly (less than necessary for durable metallic short-circuiting) condensed at zero applied voltage, the first run was non-ohmic ON state-like having even a higher maximum than the original $I$–$U$ trace (curve 2). An NDR branch sets on at a lower voltage and falls more steeply than curve 1. After one cycle the $I$–$U$ traces approach the prior shape on traces like 0 and 1 (which can again be repeated many times).

After a slow reverse sweep to zero voltage (the diode then is in its ON state) a nominally equal amount of metal was deposited again. The next run is shown as curve 3, the following return trace equals curve 0, and the forward one curve 4. That means within experimental error a complete disappearance of the metal effect. It was generally shown that subsequently repeated voltage cycling and metal deposition could not alter the underlying basic behavior significantly.

By what mechanism metal is lost during operation is still an open question. Possibly it is due to Joule heating and in principle a B-type electroforming is performed. This was confirmed by the fact that with fast single voltage sweeps, and hence reduced input power, $I$–$U$ characteristics of metal-covered diodes were ohmic (without the breakdown branch).

It is evident from the difference of curves 3 and 2 that metal deposition mainly affects the diode simply through a bypassing conductance. The shape (circles) is very similar to curve 1, even though scattering is strong. Causes for that are the uncertainty of deposition of exactly equal amounts of metal, as well as the usually bad reproducibility of repeated cycling (e.g. curves 1 and 4), even under nominally identical conditions.

Obviously, post-deposited small amounts of metal cannot stimulate true regeneration of the original filaments. It can only temporarily increase the resulting conductance by an unstable metal cover connected parallel to the underlying ON and/or OFF state material.

With higher amounts of post-deposited metal, B-type and A-type electroforming occur simultaneously during voltage cycling. Such effects may explain Gengenbach et al.'s findings of an increased maximum [95].

One remaining problem should be mentioned. Common regeneration seems to be suppressed in the presence of metal traces as seen e.g. in curve 2 where the minimum is lower than the expected regeneration current. Possibly, even this is merely caused by shunting the ruptured filament sections, thus preventing the required voltage drop $U > U_{th}$ locally. Electrochemical effects may be also the reason.

## 5. Memory Effects and Material Influence

### 5.1 Memory states

The literature usually distinguishes between various memory effects ([20], and citations therein). The known conditions to transfer a MIM diode (of either configuration) from the ON state to the OFF state are:

(i) applying a memory voltage $U_M$,
(ii) operating the diode beyond a threshold voltage $U_{th*}$ (Section 4.4) in an environment of highly electronegative molecules,
(iii) operating the diode in UHV or other vacuum free of organic residuals (Section 4.3),
(iv) irradiating the diode by charged particles of proper energy at $U < U_{th}$ (Section 4.6).

In all known cases the joint procedure to recover the ON state is to re-establish proper environmental conditions (if changed) and apply a voltage $U_R > U_{th}$. Nevertheless, regeneration curves may be strongly different because they are all governed by the time constant $\tau_R$. This time constant not only depends on the actual regeneration conditions, but also on the degree of destruction. At the moment, there is no evidence that the different methods (i) to (iv) should yield qualitatively distinguishable results. It should be pointed out again that in terms of a charging model regeneration was explained as relaxation (see Section 4.1).

From the actual point of knowledge there is no need for particular explanations of different memory states. Emmer [49] used different $O_2$ pressures to switch his diodes to intermediate or full OFF states, but he needed re-established HV conditions (or a decreased $O_2$ component) and a voltage beyond the threshold voltage ($U_{th} = 2.5$ V) for regeneration.

A lowered temperature of the diode itself has no influence on the regeneration process, as was discussed in Section 4.3. With an unaltered $\tau_R$, switching behaves like at RT. Temperature–voltage memory effects apparently occur only in cases where the temperature of the cold trap affects the gas composition.

There is naturally no need to introduce an UHV or an irradiation memory as well. Presumably there exists only one memory effect, and it is characterized by a threshold voltage $U_{th}$. Below $U_{th}$, both states are stable (an exception is (iv)), thus revealing the relevant behavior of a memory device.

### 5.2 The regeneration threshold voltage

Without taking particular care, the measurable value of the (formerly called memory) threshold voltage $U_{th}$ depends strongly on the conditions by which an OFF state was established, as well as on the actual measuring conditions. Measurements of $U_{th}$ were already performed by Simmons et al. [39, 68, 100] and later by Thurstans [45], who came to different conclusions. Evidently $U_{th}$ is not a constant; its particular lowest value should be defined as the real relevant quantity. But even this definition is rather uncertain since a threshold behavior may be pretended by a steeply (e.g. exponentially) increasing variable (here as a function of applied voltage). In spite of this situation it remains a useful quantity for characterizing this type of MIM diodes (Section 3.1).

On the other hand, considering the discussion on regeneration (Section 4.1), $U_{th}$ is identical with the lowest onset voltage at which regeneration from the OFF to the ON state is possible, i.e. $U_{th} = U_{R,min}$. This corresponds with the first on-switching event measured as an upward current step during a forward voltage sweep only when distinct conditions are obeyed. The voltage sweep velocity must be chosen very slow and a waiting time must be kept between turning off at $U_M$ and starting the sweep. The reason is according to the filament hole model (Section 4.1) that filling must be accomplished before on-switching is possible. The filling time depends evidently upon the degree of filament destruction, i.e. on $U_M$, as well as on the impingement rate of "constructive" molecules (Section 4.2). Values as low as $U_{th} = 1.5$ V have been observed under the above-mentioned conditions [61].

It seems therefore useful to distinguish between $U_{th}$ as the lowest voltage stimulating on-switching events and characterizing a microscopical physical mechanism and, on the other hand, the onset of measurable regeneration dependent upon the diode prehistory, the actual regeneration conditions, and on the measuring procedure. Values of this quantity are frequently about $U_{th} = 2$ V (e.g. in HV at RT with a voltage sweep velocity of 0.1 V/s and $U_M < 10$ V [57]) for both planar and sandwich diodes.

Various values of $U_{th}$ were thus observed due to the (mostly unknown) particular regeneration conditions ranging between 1.5 and 3.5 V for diodes of different configurations and material compositions. Until now there is no evidence of the existence of any correlation between $U_{th}$ and the materials used [36].

As expected, but in contrast to Bernède's finding [101], all planar diodes in Gengenbach et al.'s investigations [95] showed, during a forward sweep, switching to the OFF state in air at atmospheric pressure with an onset at $U_{th*}$ (Section 4.4). The lowest value of $U_{th*}$ found was 1.6 eV for a planar diode with silver electrodes (for an Al–$Al_2O_3$–Au sandwich diode 1.8 V [49]). Possibly, the lowest $U_{th*}$ values are identical with the lowest $U_{th}$ achieved, which would indicate that oxygen assumedly attacks the filament under similar conditions as on-switching events are possible.

Provided there exists a threshold effect, e.g. due to electron excitation to a higher energy level or penetration through a barrier, such an electron transfer can be a prestep to on-switching (e.g. building up a bond) as well as for chemical bonding of (high electronegative) molecules.

Again, no plain correlation of the measured values with material parameters was found; it rather seems a question of choosing the right measuring (and contacting) procedure to get one unique threshold voltage for all the diodes ultimately. This seems, on the other hand, not to be unreasonable taking into account the uniform filament material.

### *5.3 Material parameters and long-term stability*

Many attempts were made to explore correlations between typical parameters of the current–voltage features and parameters of insulator or metal, without any reasonable success. The behavior of MIM diodes (with $Al_2O_3$ as insulator) under air exposure was investigated for example by Bernède [101] as one aim to distinguish between various electrode metals. He concluded the existence of different groups of utilized metals having an about equal or lower electronegativity $\chi$ than the filament material carbon ($\chi_C = 2.6$).

In contrast, Gengenbach et al. [57] found that all investigated planar diodes with various metals (and carbon) exhibited a sensitivity to air as was discussed in Section 4.4. The materials are listed in Table 4. Since Gengenbach et al. [36, 57] used planar MIM diodes, B-forming was required. It turned out that producing the discontinuity was expectedly dependent upon the respective melting points $T_m$.

Table 4

Data of the forming processes of planar MIM diodes [36, 57, 95, 102]

| element | $T_m$ (K) | B-forming reliability | A-forming $U_p$ (V) | | $I_p$ (mA) | $I_p/I_m$ | remarks on $I$–$U$ characteristics |
|---|---|---|---|---|---|---|---|
| In | 429 | a | 5 to 8 | 2.5*) | 20 to 30 | 1 to 1.5 | degrade |
| Bi | 544 | b | 8 | 3.5**) | 1 to 30 | 1 to 2 | well |
| Pb | 600 | b | 6.5 | 3**) | 10 to 20 | 2 to 2.5 | well |
| Sb | 904 | b | 6 | — | 10 to 20 | 2 to 2.5 | broaden |
| Al | 933 | b | 4 to 5 | 3.5*) | 3 to 10 | 1.5 to 10 | unstable |
| Ag | 1234 | c | 3.7 | 3.5*) | 1 to 100 | 7 | well |
| Au | 1336 | c | 3 to 3.5 | 1.7 to 3.5 | 200 | 10 | well |
| Cu | 1356 | c | 3.8 | 3.5 | 50 to 120 | 7 | well |
| Cr | 2163 | d | 6 | — | 0.1 to 4 | 1 to 5 | N-type, fluctuating |
| C | ≈4000 | d | 9 to 14 | — | 6 to 120 | 3 to 5 | frequently N-type |

a: Critical power dose. b: Easily formable. c: Optimum film thickness required. d: Not easily formable; other notations see text.
The $U_p$-values of column 5 are examples of A-formed sandwich diodes with the element as top electrode adopted from*) [101], **) [3], and [39] (without mark).

Electron micrographs often exhibited metal islands within the microslit (but see also figures in Section 3.2), strongly disturbed microslit edges for example in case of In and Al, and the well-known meandering structures with Au and Cu (see e.g. [81]). The low melting In showed material transport under operation (after A-forming) making itself felt by a droplet accumulation at both electrode edges. A-forming being once performed, the VCNR-like $I$–$U$ characteristic degraded after a few sweeps. Cr and C electrodes were, on the other hand, hardly to attack by Joule heating and they exhibited no appreciable material transport, but in the case of Cr cracking occurred with the consequence of increasing the contact film resistance. These qualitative features are listed in Table 4.

Taking in mind that after B-forming the mere role of the metals is assumed to act as contact to the carbon filaments, it is evident that chemical and structural properties of the metals strongly influence the contact quality. Only few materials are useful for revealing sharp $I$–$U$ characteristics at a low peak voltage $U_p$ and a large peak to minimum OFF current ratio $I_p/I_m$ [95]. Further treatments performed are presented in Table 5.

Some of the diodes exhibited also a partly reversible air influence of another kind than discussed before. The current below $U_{th*}$ was already lowered after air exposure (Fig. 36; $\Delta I/I$ at fixed 0.5 V). Aging, which usually expresses itself by a shift of the peak voltage $U_p$ to higher values, lowering of $I_p$, and broadening of the $I$–$U$ shape, was tested by an average of ten cycles in air and subsequent regeneration in HV.

Long-term aging means storage of some months to years in air of atmospheric pressure, and checking of the $I$–$U$ characteristic in HV, particularly to what degree regeneration is possible.

From this finding it is obvious that carbon would yield the optimum material concerning any application. That is easily explained by the fact that filament and contact material at least approach identity. On the other hand, preparation is complicated and the high contact resistance reveals distorted $I$–$U$ characteristics. The graphs of Fig. 37 chosen here are from a planar carbon device, which was prepared in order to check whether metal is needed at all for this type of switching devices [57]. Although

Table 5
Air influence on the $I$–$U$ characteristics and aging after [36] (column 3 after [101])

| element | ON to OFF in air | | $\Delta I/I$ (%) at 0.5 V | aging | long-term aging | regeneration ability |
|---|---|---|---|---|---|---|
| In | yes | no | 40–70 | weak | strong | incomplete |
| Bi | yes | no | 15–50 | medium | strong | incomplete |
| Pb | yes | | 60 | medium to strong | extreme | no |
| Sb | yes | | 5–20 | medium | medium | yes |
| Al | yes | yes | 20 | weak | medium | yes |
| Ag | yes | no | 0–15 | strong | medium | incomplete*) |
| Au | yes | yes | 0–15 | weak | very weak | yes |
| Cu | yes | no | 0–30 | medium to strong | strong | weak |
| Cr | yes | | 0 | weak | weak | incomplete |
| C | yes | | 0 | no | no | yes |

*) Dependent on storage in humid or dry environment.

dilated in the voltage scale due to a high contact-film resistance no principle difference obviously exists, when compared with the criteria discussed in Section 3.1. In spite of a few less perfect features, gold is evidently the most promising material.

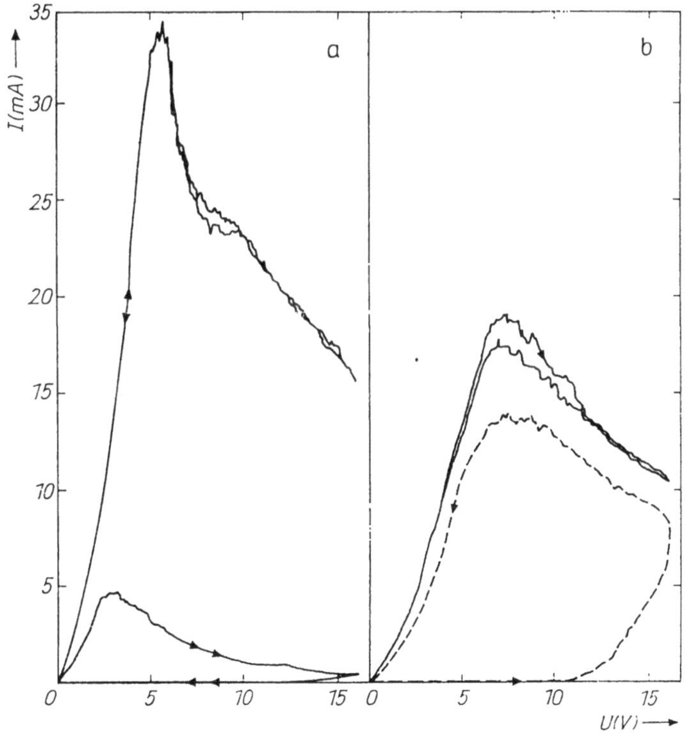


Fig. 36. a) Air influence on the $I$–$U$ characteristic (double arrowed curve) of a planar MIM diode (Bi on quartz glass); b) after long-term operation under restored vacuum conditions (dashed curve: onset of regeneration) (after [36])

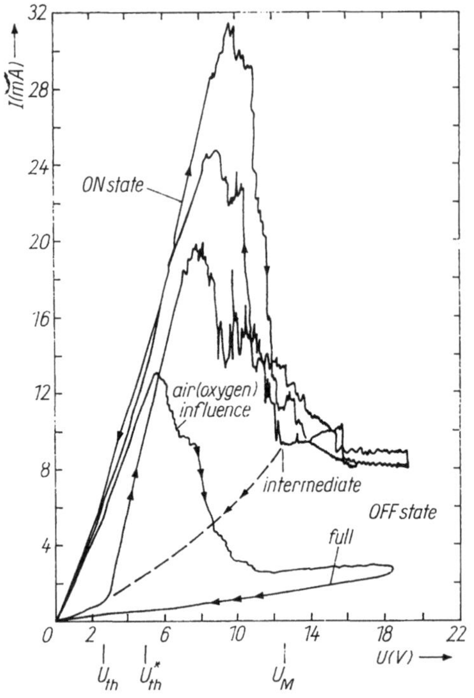


Fig. 37. Typical features of a planar diode made completely of carbon (on quartz glass substrate): B-type formed Bradley film plus graphitic contact paste; the voltage axis is dilated due to a high contact resistance (after [57])

## 6. Emission Phenomena

### 6.1 Electron emission

Electron emission (EE) from MIM sandwich diodes is long known. Already in 1963 Hickmott [103] published emission current–voltage $I_e$–$U$ characteristics with typical shapes. Such curves have an onset at about 2 V, a maximum at the NDR branch, and after having passed a minimum, ascent steeply in the high voltage branch. This general behavior was reproduced by various authors with different material combinations (see the reviews cited, especially [4, 5]). One example of Barriac et al. [104] is shown in Fig. 38. Similar behavior was found with planar devices by Bischoff et al. [105]. Emission characteristics of these diodes were more smeared out compared to those from sandwich structures. Earlier publications of Borzyak et al. [106] and Dittmer [107] on EE were concerned only with the high voltage branch (they had no NDR in the $I$–$U$), which was identified by Dittmer by means of Fowler-Nordheim plot to be caused by field emission.

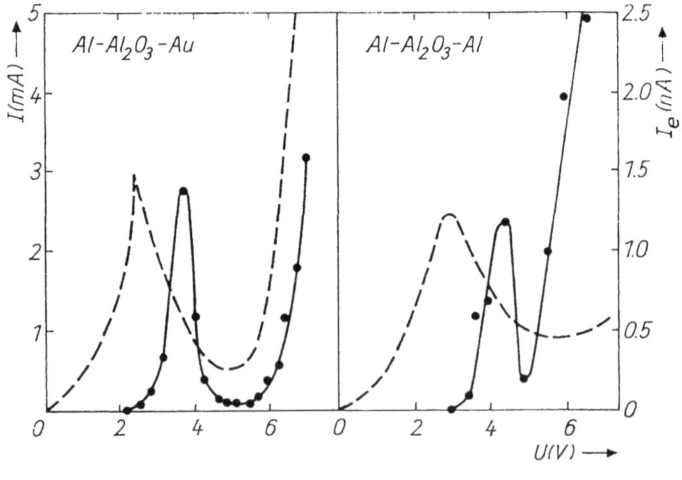


Fig. 38. Circulating current $I$ (— — —) and electron emission current $I_e$ (———) vs. applied voltage of sandwich MIM diodes (after [104])

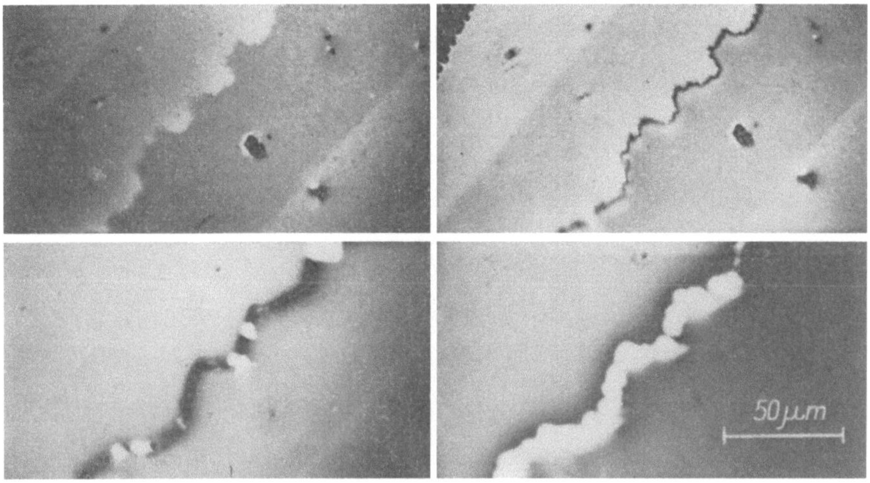


Fig. 39. Electron optical image of electron emission spots from a planar MIM diode by use of a photo emission electron microscope; applied voltage a) $U = 0$, b) 3 V, c) 6.5 V, d) 14 V (Au on quartz glass) (after [109])

Sources of EE are small areas (spots). In planar diodes they are located within the microslit like a chain parallel to the electrode edges (Fig. 39). They are statistically distributed over the slit length or, in the case of sandwich diodes, over the whole electrode area, nonuniform in density and varying from sample to sample in both cases. In the sandwich-type diodes channels act as sources of electrons. The microscopic mechanisms and filament configurations within them are not exactly known.

By means of retarding field measurements Hickmott [108] has shown that two distinct groups of electrons are emitted, one, high energy electrons, with energies approaching the maximum applied voltage and one of low energy electrons appearing additively beyond applied voltages of about 5 V. Similar dependences with a characteristic inflection, bordering these two groups, were measured with planar diodes [105]. Hickmott concluded that at applied voltages greater than about 1.8 V to exceed the work function barrier (of the gold electrode) at least part of the high energy electrons must have gained energy within the insulator by some mechanism.

More improved measurements by Blessing and Pagnia [109] on planar MIM diodes in an emission electron microscope (EEM), with an electrostatic energy analyser confirmed these findings. In their equipment it was possible to study the features of single emission spots. One $I_e$–$U$ characteristic is presented in Fig. 40a which separately shows the two voltage regions. The energetic position relative to the energy positions of uv photo emitted electrons from both electrodes (sketched in Fig. 40b) clearly demonstrates the two different mechanisms. At high applied diode voltage field-emitted electrons at the lower energy position are detected in the energy spectrum. This was confirmed by a Fowler-Nordheim plot. The electrons detected near the cathode photoelectron peak (at low applied voltage) must have surmounted the barrier by means of some energy gain or excitation, assumedly a thermal process of a hot electron gas. In addition one fact should be pointed out, the single sources can only emit electrons of either of both types alternatively.

The complicated structure of the source with its high internal field strength, having strong components perpendicular to the detection direction in the planar diodes, the

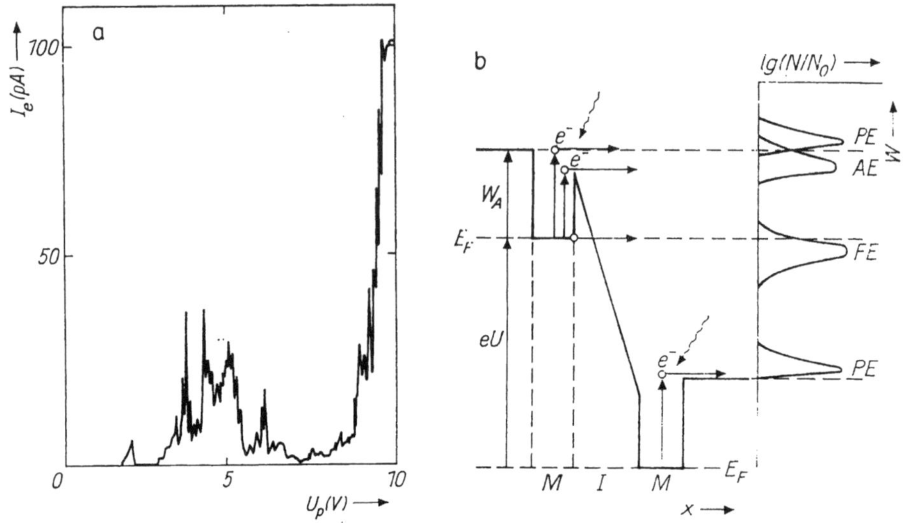


Fig. 40. a) Electron emission characteristic $I_e$–$U$ from a single spot of a planar MIM diode (Au on quartz glass) and b) an energy band model sketching the different emission processes: photo emission PE for calibrating, field emission FE, and excitation over a barrier AE (after [109])

statistical fluctuations of the emission process, and, not at last, the limited resolution of the analysers used allowed no detailed interpretation of the experimental results.

One possible consequence of the electron emission should be mentioned. As has been discussed before, the main part of off-switching events responsible for the NDR branch of the $I$–$U$ characteristics (Section 3.3), is due to the electrical power input and hence are of thermal nature. Because free electrons are present we may assume that also electron-induced switching takes place in this voltage range. This would enhance the off-switching rate, and stable filaments could be switched off, too. Possibly, electrons are the coupling medium responsible for the frequently observed NDR branches with an avalanche-like off-switching of neighboring filaments. One example of collective off-switching in a planar diode which contains only a few filaments in close proximity (due to the small electrode breadth of 4 μm), is shown in Fig. 9 while regeneration nevertheless proceeds in discrete steps.

### *6.2 Photon emission spectra*

In 1964 Hickmott detected electroluminescence (EL) from various diodes [48]. The analyses of the spectra were better suited to gain information on the basic mechanism. Spectral measurements, however, were made only roughly using optical filters, and the spectral range was small. He reported a maximum intensity measured at RT at photon energies of about 1.3 eV vanishing above 1.6 eV. Also visible radiation was detected. Visible EL was reported in the publication on planar diodes by Borzyak et al. [106], too, and on sandwich diodes by other authors (see [5]). Well resolved spectra from sandwich structures (Pong et al. [111]) and planar diodes (Bischoff and Pagnia [7]) were published in 1975. They were obtained to some amount in different spectral ranges, and showed differences in details, but displayed at least the same general tendency.

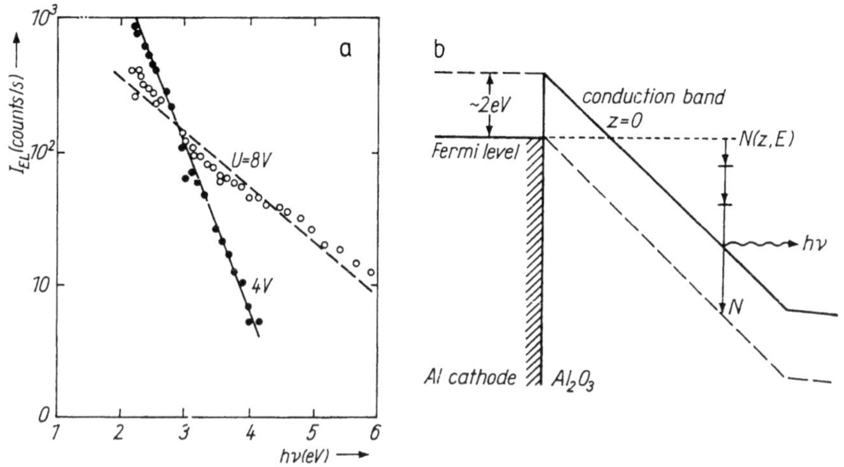


Fig. 41. a) Photon emission spectra at two applied voltages from a sandwich MIM diode (Al–$Al_2O_3$–Au) (with correction of the absorption loss in the Au electrode) and b) the emission mechanism explained in an energy band diagram (after [111])

EL spectra exhibit no pronounced structure in the visible and near ir. On the average, they are monotonic curves (Fig. 41a) well fitted by

$$I_{EL} \sim \exp\left(-\frac{h\nu}{a}\right). \tag{6.1}$$

Interpretations of the spectra were different. Pong et al. utilized an insulator band model (Fig. 41b) with recombination of the injected electrons. Evidently, spectra should be cut off at maximum energy $eU$ ($e$ elementary charge, $U$ applied voltage), which contradicts the fairly well confirmed findings [7, 112]. Photons with even $h\nu > eU$ appear in the spectra (Fig. 42a).

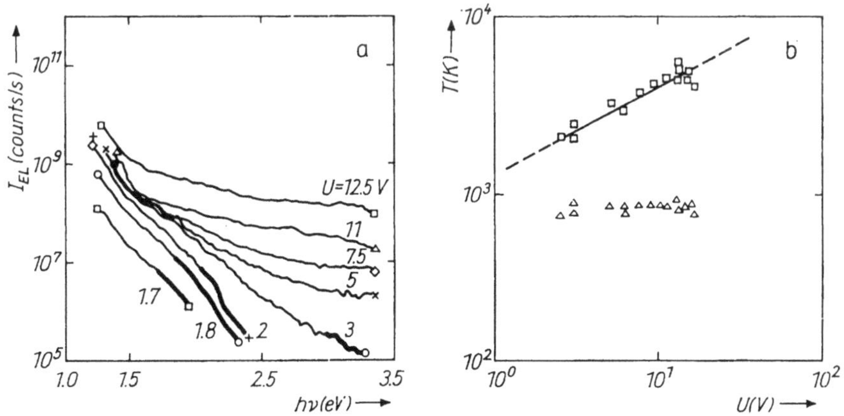


Fig. 42. a) Photon emission rate spectra of a planar MIM diode (Au on quartz glass) at different bias voltages, strengthened branches characterize $h\nu > eU$ ranges; b) the evaluated temperatures of the electron gas $T_e$ (squares) and the substrate $T_s$ (triangles) (after [112])

To search for any processes permitting a broad energy distribution of the photons, most likely to work is a thermalization process, yielding light emission at high electron gas temperatures $T_e$. Equation (6.1) is an approximation of the Planck formula provided $h\nu \gg kT$ and neglecting the $\nu^3$ term. The estimated electron temperatures $\left(kT_e = a \text{ in } (6.1)\right)$ were between 1200 and 6000 K for applied voltages increasing from about 4 to 10 V (Au on quartz) [7]. Consequently it was proposed to be a hot electron effect. The gold lattice temperature would rarely exceed the melting point without irreversibly destroying the diodes. New investigations [112] in a somewhat wider energy range confirmed the older findings. One example of these spectra is shown in Fig. 42, which exhibits two different branches. As has been shown by Bischoff and Pagnia [7], they can be fitted by Wien's formula

$$I_{EL} \sim \nu^3 \exp\left(-\frac{h\nu}{kT_e}\right). \tag{6.2}$$

Not neglecting the $\nu^3$ term, the steep slopes in the visible red to near ir range deliver temperatures from RT to about 1000 K. There is some evidence that this part of radiation essentially came from the heated substrates ($T_s$). The flat branches were deduced to correlate with the hot electron temperature range always holding $T_e > T_s$. Though a carbon filament could withstand partly such temperatures without destruction, hot electron effects must be supposed. Recent investigations [75] (Section 6.4) revealed the electron gas really to be heated practically independent of the filament lattice temperature $T_l$.

Quite similar spectra under dc conditions have been observed by Delaunay and Despujols [113] with Al–SiO–Au sandwich diodes (Fig. 43). They derived electron temperatures $T_e = 3000$ to 6400 K, dependent on voltage as

$$T_e = T_0 + aU^{1/2} \tag{6.3}$$

with $T_0 = 300$ K and $a = 1300\ \text{KV}^{-1/2}$, and explained their radiation originating from bremsstrahlung (which contradicts $h\nu > eU$). As has been shown by Bischoff et al. [112] the $U^{1/2}$ dependence is a consequence of the Stefan-Boltzmann law ($I_{EL} \sim T^4$), under stationary dc conditions the filaments loose the input power $P \sim U^2$ (assuming approximately ohmic behavior) mainly by radiation (Fig. 42b). Under ac conditions, the dependence of temperature on voltage is rather like $T_e = \alpha eU/k$; $\alpha = 0.026$ [7]. In this case, the lattice temperature depends upon the input power at a given voltage amplitude. Only the filaments heated to temperatures $T_l$ being close

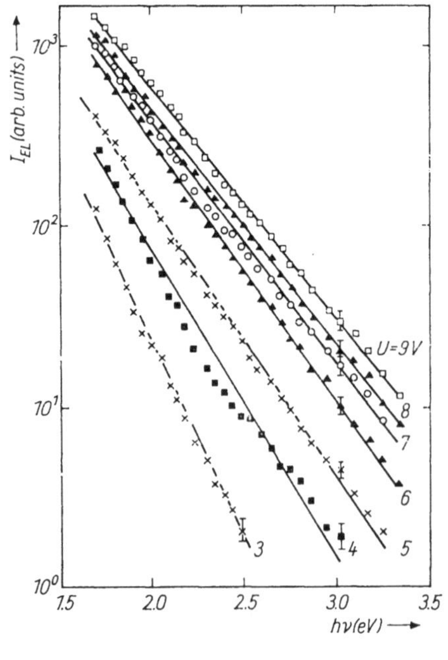


Fig. 43. Photon emission spectra of sandwich MIM diodes (Al–SiO–Au) at different bias voltages (after [113])

to the off-switching conditions take part in detectable luminescence due to the simultaneously high electron gas temperature $T_e$. Computer simulations approximately confirm this empiric relation, at least at low voltages [112].

In 1977 Biederman and Plasek [114] published spectra from sandwich and planar diodes displaying similar curves as in Fig. 42 with the two characteristic ranges, but also showing some deviations and fluctuations. Generally, this seems to confirm the other results.

We may conclude that photon spectra are predominantly caused by black-body-like radiation. One source are the minute filaments in which the hot electron gas is confined, the other the much larger area of the substrate, covered mainly by the metal electrodes in case of the planar diodes, or the top electrode of the sandwich structures.

### *6.3 Correlation between switching events of the circulating current and the emission processes*

From the results mentioned it is evident that there should exist a coincidence between the circulating current and the emission processes. Borziak et al. [115] investigated samples exhibiting no VCNR-like shape. They applied voltages corresponding to the field emission range, and were the first to state fluctuations of EE and EL occurring simultaneously. They also stated coincidence between sources of both emissions, and suggested a common cause of these phenomena.

Bischoff et al. [71] observed two different forms of fluctuations in current and EL as shown in Fig. 44. Solitary δ-shaped pulses were found, lying in a few cases (within 1 ms) in the flank of plateau-shaped ones. An upward current step corresponds with

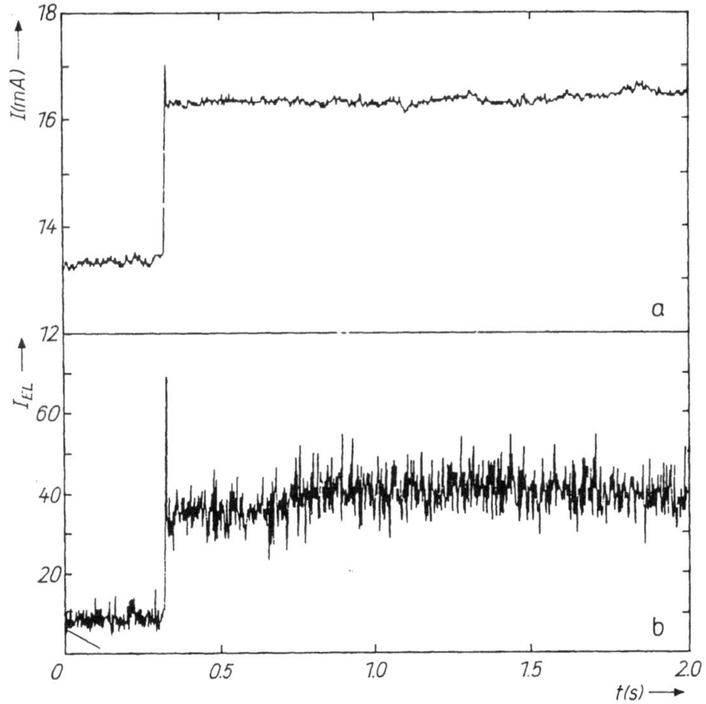


Fig. 44. Coincidence of a) circulating current $I$ and b) photon emission $I_{EL}$ (photon count rate in kHz) from a planar MIM diode (Au on quartz glass, $U = 4.5$ V) (after [71])

on-switching of a filament, so we can conclude that on the average most emitted photons originate from intact filaments. The average efficiencies (photons per electron in the current) were of the order of $\eta_{EE} = 10^{-6}$ in a single ON pulse, compared with $5 \times 10^{-8}$ in the background due to all the other on-switched filaments.

Reasonable correlation $I$–$I_e$ and $I_e$–$I_{EL}$ was also observed by Pagnia et al. [116]. The efficiency of emitted electrons per travelling electron was measured to be uniformly $\eta_{EE} = 10^{-8}$, and a correspondence of both emissions to the ON state current was confirmed.

In Fig. 45 circulating and emission currents obtained with a planar MIM diode with small electrodes (and hence few filaments [117]) are shown. The emission current rise sets on before the switch-off event occurs. This behavior can well be explained with the hot electron conception (Section 6.2). During the increasing branch of a voltage cycle the electron gas is heated ($T_e$). At the rupturing point $T_e$ is at its maximum, and so will, of course, be the emission rate.

An indication that there could also exist transient emission mechanisms can possibly be seen in the fact that EE maxima were observed normally at applied voltage of the steepest fall of the NDR branch [104] (Fig. 38), where the most off-switching events occur. An assumption that each off-switching event sets free a number of electrons (possibly one or two per ruptured C–C bond) is at least conceivable. This would be some kind of exoemission, frequently studied with mechanically ruptured

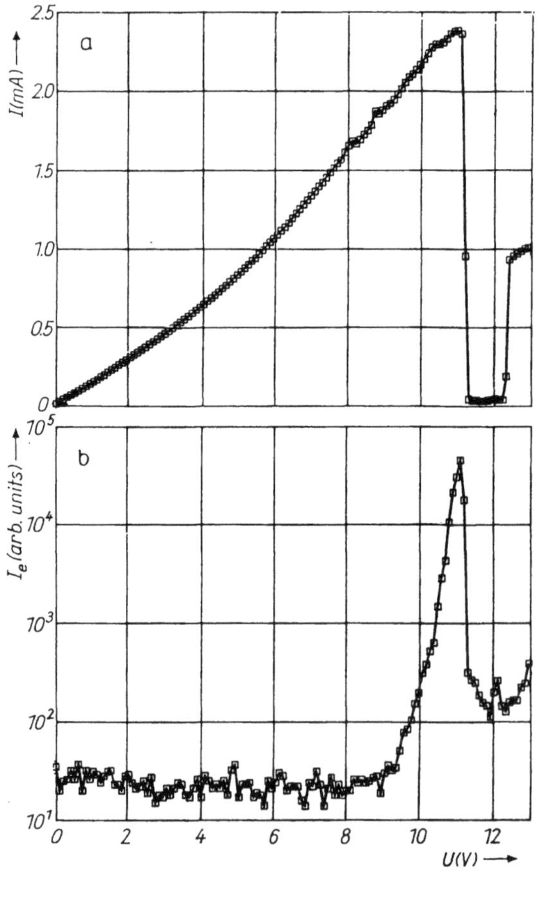


Fig. 45. Correlation between a) circulating current and b) the electron emission during a forward sweep; planar MIM diode (Au on quartz glass, electrode breadth 4 μm) (after [117])

polymer foils. But the number of potentially free electrons estimated with realistic geometric dimensions of the ruptured surface is too low by orders of magnitude compared with measured emission current eruptions.

### *6.4 Voltage pulse experiments*

Some striking features extracted from photon spectra (Section 6.2), led to the assumption of a hot electron gas. This was deduced mainly from the black-body radiation with high color temperatures compared with the melting points of diode material and from the evidence of photons with energies $h\nu > eU$, the maximum energy gain possible by field acceleration [7]. With his new thoroughly performed and analysed pulse experiments Bischoff [75] was able to confirm these explanations and to get more data of the filament substance so far unknown.

Only a few results shall be discussed here. In the upper part of Fig. 46 the time dependence of the voltage pulse and sample current are shown, below the simultaneously emitted photons at different wavelengths. This was done in order to evaluate the black-body temperatures, provided validity of Wien's law (Section 6.2). A time dependent factor $R(t)$ was to be introduced into the pre-exponential factor. $R(t)$ turned out to be either a changing radiating area or volume, respectively, or more likely, the carrier recombination rate of a hot electron gas. As expected for a system with a given heat capacity, the intensity rise is strongly delayed in comparison with the current increase. But the behavior of the radiator temperature is completely different. Within the time resolution ($< 1$ ns), it starts with a temperature peak (Fig. 47a) decaying within 100 ns to a quasi-equilibrium state. Such a behavior is in agreement with the assumption of an electron gas heated by an electrical field, and cooled by optical phonons. Electron heating only occurs in nearly empty energy bands of a nonmetal.

Taking into consideration electron scattering by acoustic phonons, responsible for momentum relaxation and energy relaxation by nonpolar optical phonons a relation can be derived,

$$kT_e = kT_l\left\{1 + \frac{(\beta_n eU)^2}{(kT_l)^n}\right\}, \tag{6.4}$$

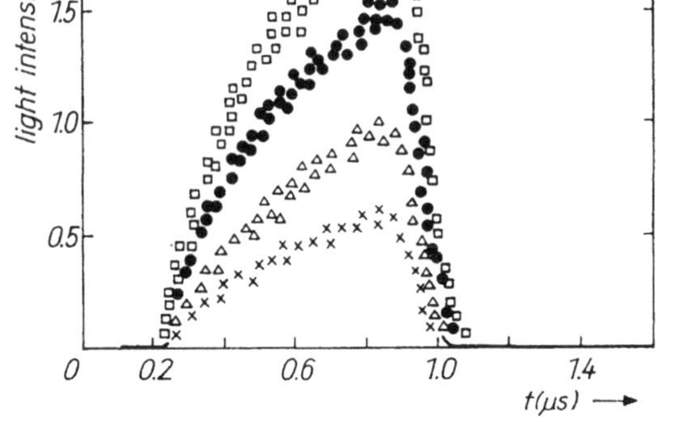


Fig. 46. a) Pulse shape vs. time (dashed: voltage, full: circulating current); b) light intensity emitted by a planar MIM diode (Au on quartz glass) at four wavelengths: □ 835, ● 700, △ 600, × 535 nm; $U = 2$ V, $I = 25$ mA (after [75])

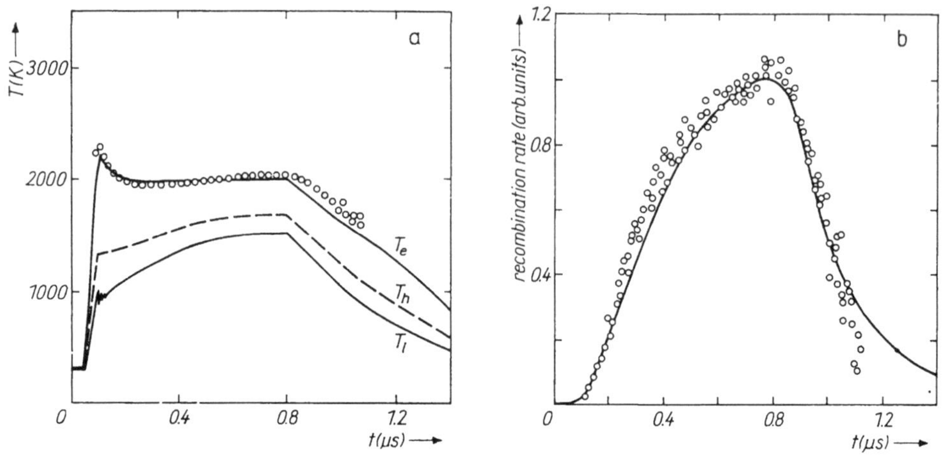


Fig. 47. Comparison of computer simulations and measured values of a) temperatures (notation: e electrons, h holes, l filament lattice) and b) carrier density (proportional to the recombination rate) during a voltage pulse (sample as in Fig. 46; after [75])

with $\beta_n$ a constant depending on the scattering mechanism, and $n = 2$ for temperatures below and $n = 3$ beyond the Debye temperature. One example is drawn in Fig. 48 showing the general tendency of such relations whose concrete shape is determined by the exponent $n$. Already in 1960 Dykman and Tomchuk [119] treated a similar problem with about the same result. At the beginning of the voltage pulse the electron temperature $T_e$ rapidly follows the broken curve up to high values. Since heat is transferred to the host lattice, its temperature increases, while $T_e$ is now decreasing. Well fitted and measured electron temperatures are shown for one example in Fig. 47, which also depicts the evaluated lattice temperatures $T_l$. They are not to be confused with the radiation of the substrate detected in experiments with periodic or quasi-static voltage sweeps.

Since the pre-exponential factor $R(t)$ was assumed to represent the carrier recombination rate (and by this the proportional carrier density), its temperature dependence should give further information upon the filament material. Computer

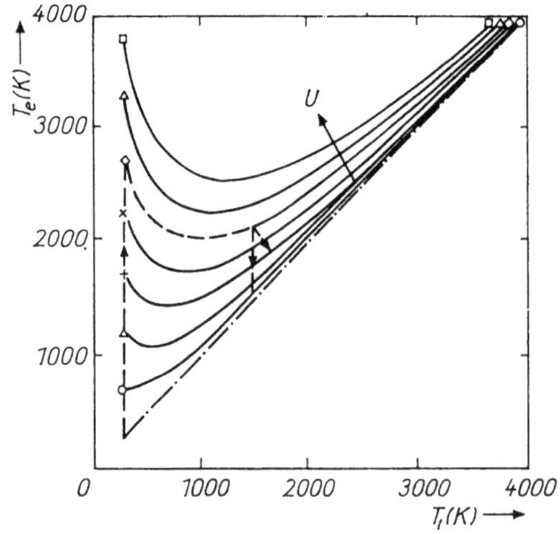


Fig. 48. Calculated electron temperature $T_e$ as a function of lattice temperature $T_l$ from (6.4), parameter: applied voltage $U$ (after [75])

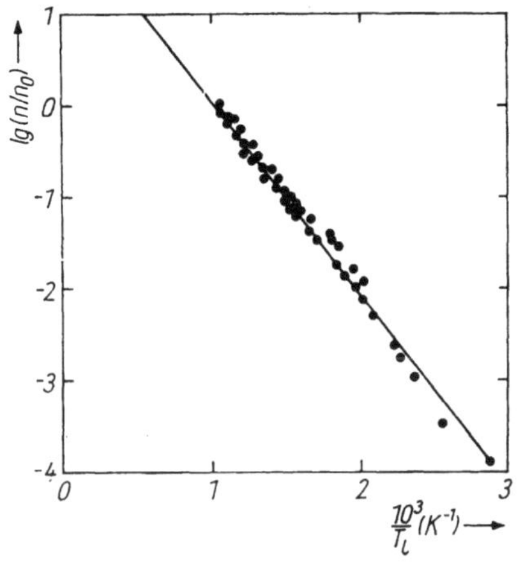


Fig. 49. Carrier density vs. reciprocal filament lattice temperature (after [75])

simulations of a p-type extrinsic semiconductor ($\mu_n/\mu_p > 1$) with a lattice temperature above the Debye temperature could be fitted best to the experimental points (Fig. 47b).

These experiments clearly demonstrate that the filament substance must be a semiconductor, in which hot electrons (and holes with $T_h < T_e$ provided $\mu_n > \mu_p$) can be yielded. As long as this occurs during a short voltage pulse the temperature of the electron gas is always higher than that of the host lattice.

Naturally, with increasing pulse length both the temperatures converge to constant, but different, values dependent upon the introduced power and the loss mechanisms. One of these is the heat transport to the substrate, whose temperature $T_s$ is also increased to some amount (Section 6.2) so that, neglecting the holes, three relevant temperatures must be considered in the sequence $T_e > T_l > T_s$. No appreciable change of $T_s$ is expected in such a pulse experiment, while temperatures up to 1000 K were observed under stationary conditions [112].

Since the carrier generation is an activated process the respective energies were derived from the Arhenius plot (Fig. 49). Usual values were about 0.15 eV, which may be either a gap energy or the energy of a doping level. The existence of electron emission out of the filaments (Section 6.1) proves that they are the high mobility carriers, but only a basic p-doping then reveals the right magnitude of the evaluated dependence of carrier density on temperature.

## 7. Conclusions

### *7.1 What is known about the carbonaceous filament switches?*

It is still an unsolved problem, whether filaments possess a distinguishable ON and OFF state, or if the latter merely represents a ruptured current path. We shall therefore only discuss the ON state. The state-of-art data are presented in Table 6. Though there is a lack of essential points like physical and chemical material analysis and the filaments have yet not been seen, circumstantial evidence strikingly shows that discrete current paths exist and they are predominantly composed of carbon. But it is unknown whether they really are of filamentary shape. A plane arrangement is conceivable as well.

The high conductivity of the material is an indication that delocalized electrons must be available due to $sp^2$ or $sp^1$ hybridization of carbon atoms. Plane configurations like in graphite or even linear structures of polyacethylene $(CH)_x$ or a polyyne may be possible.

Systems of amorphous carbon, based on Hückel theory arguments, were recently treated by Robertson and O'Reilly [120]. They considered different sizes of clusters of either compact spatial or linear fused sixfold ring configurations of $sp^2$ hybridized

Table 6
Material data of the filaments

| | |
|---|---|
| the ON state, stable below $U_{th} = 1.5$ to 2 V | |
| material: mainly carbon (out of contaminated vacuum apparatus) | |
| conductivity: $\sigma > 10^2$ (possibly $>10^4$) $\Omega^{-1}$ cm$^{-1}$ | |
| majority carrier density $10^{18}$ to $10^{19}$/cm$^3$ (RT) | |
| $n = p \leqq 10^{21}$/cm$^3$ (1500 K) | |
| possibly p-type with $\mu_n/\mu_p \geqq 1$ | |
| band gap: $E_g \approx 0$ eV (from $T$), $0 < E_g < 0.25$ eV (from pulse experiments) | |
| photoconduction: not observed (uv, visible, near ir) | |
| temperature coefficient of resistivity: $\alpha = -10^{-3}$/K (2 to 1100 K) | |
| thermal stability: up to 1100 K (possibly higher, limited by contact metal or substrate) | |
| gas sensibility (ON-to-OFF switching): by electronegative molecules above $U_{th} \approx 1.5$ to 2 V | |
| : no forming and regeneration in UHV | |
| maximum temperatures observed during operation (from optical spectra) | electron gas: $T_e < 6000$ K |
| | filament lattice: $T_l < 1500$ K |
| | substrate and electrode film: $T_s < 1000$ K (quartz and Au film) |
| irradiation effects (ON-to-OFF switching) | electrons: $W > 6.2$ eV below $U_{th}$ |
| | ions: $W > 50$ eV (possibly lower) below $U_{th}$ |

carbon. Due to their results it is possible that carbon filaments are composed of clusters of the mentioned linear type. In this case $E_g$ is low, especially e.g. 20 rings per cluster delivering $E_g \approx 0.15$ eV. Cluster interconnections were supposed to be made up of tetrahedral sp$^3$ bonds (at least in case of the compact clusters). The above theoretical results confirm the empirically long known facts of a decrease of carrier activation energy as well as optical gap due to the preparation conditions. Gap energies between some electron volts (e.g. for a a-C:H film 1.8 eV [121]), typically 0.5 eV for an evaporated carbon film [122], an about zero gap after heat treatment [67] and approaching the "negative" gap of the semimetal graphite (in the layer plane) [123] have been prepared.

The question arises whether filament growth (and regeneration) can be a pyrolytic process. Filament temperatures in the ON state may be high enough for such a process, but in the low conductivity OFF state, the electrical power density seems to be far too low. The observed forming threshold voltage also contradicts such an explanation. High field growth processes with tunneling as one step to establish the proper bond configuration seem to be more likely.

On the other hand, some effects may be of pyrolytic nature. One is the aging due to long-term operation making itself felt by flattening and shifting the current peak and a remarkable current increase at higher voltages. This may be caused by pyrolytic growth of stable (at the given voltage amplitude nonswitching) carbonaceous material on or beside the hot filaments.

Similarly, pyrolytic growth can also be the explanation for the creation of shunting material during the application of many repeated fast sweeps changing the $I$–$U$ characteristics significantly. In this case, the created material is stable up to the applied voltage amplitude, too, and can be destroyed, increasing the amplitude by a small amount. This is an evidence for its nature to be almost identical with common filament material.

Some further comments are stated in Table 6. Concerning the temperatures, originating from pulse experiments, some remarks should be made. The high temperature of the electron gas is of the common order of magnitude as found in other solid state microlight sources, e.g. the microplasmas [124]. Maximum filament lattice temperatures, which in dc measurements should even be higher, exceed the melting point of the contact and the ultrathin metal film (e.g. $T_m(\mathrm{Au}) \approx 1336$ K). A carbonized material is of course stable enough. The lattice temperature of filaments is comparable with annealing temperatures of carbonized organic material of equal gap energy [125]. Possibly there really exists a correlation, which would mean that stabilizing the filament material during the A-forming process could be of thermal nature at least partly.

The thermal stability of the whole switching device is naturally limited by the component with the lowest melting point, i.e. in most cases by the electrode material. Higher stability should be achieved e.g. with carbon contacts and using substrates more stable than quartz glass.

The threshold energy of incident electrons for OFF switching (Section 4.6) reasonably agrees with values known from desorption experiments [98] as well as with electron energy losses in nonperfect carbon [126]. In this case energy may be released in a way which changes the system durably. This may e.g. be a change of the bond configuration, an atomic rearrangement, desorption of outer atoms, or even rupturing of current carrying paths.

Many details are now known about the systems, but further data are needed to complete a full characterization.

### *7.2 Application aspects and an outlook*

The small dimensions of the filament (at least smaller than 1 $\mu m^2$, possibly in the nm range), and electron and photon emissions at low applied voltages, offer a variety of applications in various fields. None of the early trials extensively discussed in the reviews (Section 1) were, however, successful mainly because of the long-term instability and a lack of reproducibilty within reasonable tolerances [5, 9].

These problems may be overcome in the near future with an improved knowledge about preparation conditions and the contact features. One useful application could be a memory element with a high package density, at least below the limit known from Si devices due to the average space charge layer width ($2.5 \times 10^7$ bits/cm$^2$). But the involved time constants are relatively long because an external source of organic molecules is required for regeneration. Provided links of a destroyed filament remain, e.g. in a pore, instead of desorbing or diffusing away or enhancing the supply of the yet unknown organic molecules, the regeneration rate can be shortened. Attempts were made to prevent any particle exchange with the environment by additional coating, but the results were unsatisfactory [130]. An uncontrolled reduced material loss was likely to be in the case of both types of diodes. An absolutely dense coating brought on before A-forming should prevent it at all. A few data are given in Table 7 for illustration. The shortest pulse lengths for switching the diodes from the ON to the OFF state were found to be about ns [5] (which is in discrepancy with the ac measurements). The regeneration times ($\tau_R$) are longer in the open devices, but in the sandwich diodes they were found to be of the same order of magnitude as the switching times ($\tau_s$) of open diodes (Section 4.1). The number of repeated switching events seems to be limited, but values of the order $10^3$ are satisfactory considering read-mostly (or read-only) memory devices, suggesting themselves from the point of view of the discussed time constants.

On the other hand, such a memory element could be attractive due to its high thermal stability, and since there exist emission phenomena related to both the different states, the reading mechanism could for some special purpose be decoupled from the electric addressing procedure.

Table 7

Switching behavior of MIM diodes, all measurements at RT in high vacuum (if not otherwise stated); all predicted to exhibit filamentary current paths (in their ON states); switching times $\tau_s$ (Section 4.1) in parantheses

| insulator/ substrate | configu- ration*) | switching type**) | shortest pulse length ON-OFF (s) | min. or average recovery (OFF–ON) time (s) | remarks and information | ref. |
|---|---|---|---|---|---|---|
| quartz | p | m | | 0.2 ($10^{-3}$) | $U_M = 6$ V | [52] |
| | | | $<10^{-5}$ | | $U_M = 8$ V | [59] |
| | | | | 0.2 | $10^3$ events, $U_M = 12$ V | [71] |
| SiO | s | m | | $10^{-3}$ to $2 \times 10^{-2}$ | | [39] |
| Langmuir-Blodgett | s | m | | 20 | 77 K forming and switching below 225 K | [11] |
| poly-styrene | s | m | $10^{-2}$ to $10^{-5}$ | $10^{-5}$ or "slow" | "low voltage switching", $10^3$ to $10^6$ events frequently eratic | [12] |
| chalcogenide glass | s | m<br>th | $5 \times 10^{-6}$ | $10^{-3}$ to $10^{-4}$<br>$10^{-6}$ | amorphous–crystalline transformation | [127] |
| $VO_2$ | p | th | | $10^{-6}$ | $10^8$ events, semicond.–metal transition 68 °C | [128] |
| $TiO_2$ | s | th | | $3 \times 10^{-7}$ | $8 \times 10^{11}$ events | [129] |

*) p planar, s sandwich.
**) m memory, th threshold.

According to the general discussion in the Introduction data of ovonics and a few threshold switching systems are added in Table 7. The latter are much faster to operate, and they exhibit long-term stability, but they do not remain in their ON state at zero applied voltage.

Since there also exist intermediate states between full ON or OFF, respectively, graded (or analogue) information may be stored corresponding with different voltage amplitudes. At small voltages the resistance of a diode may be lowered (regeneration) and increased at high voltages (off-switching). The diode is able to adapt itself to the circumstances, and is usable as a learning device in a future neurocomputer. Also this may be an interesting feature.

Whether or not small emission sources or panels built up of MIM diodes (one early attempt is known [131]) can gain interest is still an open question.

Furthermore the filament dimensions do not seem to be limited concerning minimization compared with those of the necessary contact stripes. Nanometer lithography [132], e.g., or even direct electron beam organometallic deposition, which were developed in recent years, let envisage very small switching systems as some pre-step to molecular electronics.

Moreover it should be mentioned that carbonaceous switches are cheap to produce although great effort will be necessary to develop procedures required for commercial applications.

***Acknowledgements***

The authors acknowledge the very stimulating atmosphere in the solid state electronics group engaged in this subject for so many years. We are grateful to both present and former collaborators for their work and helpful discussions. We also wish to thank Mrs. E. Grün and Mr. K.-H. Vetter for technical assistance, Mrs. H. Schäfer for typing the manuscript, and Mrs. B. Hackel for drawing the figures.

## References

[1] S. R. Ovshinsky, U.S. Patent 3, 052, 830, 1962; J. non-crystall. Solids **2**, 99 (1970).
[2] D. Adler, H. K. Henisch, and N. Mott, Rev. mod. Phys. **50**, 209 (1978).
[3] T. W. Hickmott, J. appl. Phys. **33**, 2669 (1962).
[4] L. Eckertová, phys. stat. sol. **18**, 3 (1966).
[5] G. Dearnaley, A. M. Stoneham, and D. V. Morgan, Rep. Progr. Phys. **33**, 1129 (1970).
[6] H. Morgenthaler and H. Pagnia, Phys. Letters A **42**, 529 (1973).
[7] M. Bischoff and H. Pagnia, Thin Solid Films **29**, 303 (1975).
[8] H. Biederman, Vacuum **26**, 513 (1976).
[9] D. P. Oxley, in: Oxide and Oxide Thin Films, Ed. A Vigh, Dekker, New York 1981.
[10] A. Szymansky, D. C. Larson, and M. M. Labes, Appl. Phys. Letters **14**, 88 (1969).
[11] K. H. Gundlach and J. Kadlec, phys. stat. sol. (a) **10**, 371 (1972).
[12] L. F. Pender and R. J. Fleming, J. appl. Phys. **46**, 3426 (1975).
[13] M. C. Gabriel and D. Adler, J. non-crystall. Solids **48**, 297 (1982).
[14] A. E. Owen, P. G. LeComber, W. E. Spear, and J. Hajto, J. non-crystall. Solids **59/60**, 1273 (1983).
[15] P. G. LeComber, A. E. Owen, W. E. Spear, J. Hajto, A. J. Snell, W. K. Choi, M. J. Rose, and S. Reynolds, J. non-crystall. Solids **77/78**, 1373 (1985).
[16] H. J. Hovel, Appl. Phys. Letters **14**, 141 (1970).
[17] A. Moser, Appl. Phys. Letters **20**, 244 (1972).
[18] A. Moser and H. Rohrer, Solid State Commun. **17**, 939 (1975).
[19] R. Chapman, Electronics Letters **5**, 246 (1969).
[20] A. K. Ray and C. A. Hogarth, Internat. J. Electronics **57**, 1 (1984).
[21] H. K. Henisch, Thin Solid Films **83**, 217 (1981).
[22] P. T. Orechkin, S. P. Vikhrov, A. S. Glebor, A. S. Borshchevskii, A. L. Denisor, and S. I. Malchenko, Fiz. Tekh. Poluprovod. **10**, 511 (1976) (Soviet Phys. – Semicond. **10**, 305 (1976)).
[23] R. R. Sutherland, K. O. Legg, and R. A. Collins, Thin Solid Films **6**, R39 (1970).
[24] G. Jones and R. A. Collins, phys. stat. sol. **53**, 339 (1979).
[25] T. W. Hickmott, J. Vacuum Sci. Technol. **6**, 828 (1970).
[26] E. L. Cook, J. appl. Phys. **41**, 551 (1970).
[27] G. S. Kreynina, Soviet Phys. – Radioengng. and Electronics **7**, 1949 (1962).
[28] C. N. Berglund, IEEE Trans. Electron. Devices **16**, 432 (1969).
[29] K. Tanaka, Y. Uemura, and M. Iwata, Thin Solid Films **50**, L25 (1978).
[30] T. Kimura, K. Yamamoto, T. Shimizu, and S. Yago, Thin Solid Films **70**, 351 (1980).
[31] H. Biederman, Ph. D. Thesis, Charles University, Prague 1974.
[32] T. Lewoski, S. Sendechi, and B. Sujak, Acta phys. Polon. **28**, 343 (1965).
[33] S. M. Sze, Physics of Semiconductor Devices, Wiley & Sons, New York 1969.

[34] F. STÖCKMANN, Festkörperprobleme (Advances in Solid State Physics), **9**, 138 (1969).
[35] D. ADLER, Amorphous Semiconductors, Butterworths, London 1972.
[36] T. GENGENBACH, Diplomarbeit, Technical University Darmstadt, 1986.
[37] B. BORZIAK, YU. KULYUPIN, and S. NEPIJKO, Thin Solid Films **36**, 235 (1976).
[38] R. BLESSING, K. H. GÜRTLER, and H. PAGNIA, Phys. Letters A **84**, 341 (1981).
[39] J. G. SIMMONS and R. R. VERDERBER, Proc. Roy. Soc. **A301**, 77 (1967).
[40] J. E. RALPH and J. M. WOODCOCK, J. non-crystall. Solids **7**, 236 (1972).
[41] F. A. S. AL-RAMADHAN and C. A. HOGARTH, J. Mater. Sci. **19**, 1939 (1984).
[42] R. RAHMAN, Thin Solid Films **94**, 199 (1982).
[43] R. D. GOULD and C. A. HOGARTH, Internat. J. Electronics **37**, 167 (1974).
[44] R. BLESSING, J. SCHNELLBÄCHER, H. PAGNIA, and N. SOTNIK, Ninth Internat. Vacuum Congr. and Fifth Internat. Conf. Solid Surfaces, Madrid (Spain) 1983 (p. 145, abstract).
[45] R. E. THURSTANS, Ph. D. Thesis, Leicester Polytechnic, 1976.
[46] TH. BACH, R. BLESSING, H. PAGNIA, and N. SOTNIK, Thin Solid Films **103**, 283 (1983).
[47] G. DEARNALEY, Phys. Letters A **25**, 760 (1967).
[48] T. W. HICKMOTT, J. appl. Phys. **35**, 2679 (1964).
[49] I. EMMER, Thin Solid Films **20**, 43 (1974).
[50] B. ABELES, PING SHENG, M. D. COUTTS, and Y. ASIS, Adv. Phys. **24**, 407 (1975).
[51] R. BLESSING, H. PAGNIA, and R. SCHMITT, Thin Solid Films **78**, 397 (1981).
[52] H. PAGNIA, J. SCHNELLBÄCHER, and N. SOTNIK, phys. stat. sol. (a) **87**, 709 (1985).
[53] B. V. ALEKSEENKO and R. D. FEDOROVICH, Thin Solid Films **92**, 253 (1982).
[54] T. KIMURA, K. YAMAMOTO, T. SHIMIZU, and S. YAGO, Thin Solid Films **70**, 351 (1980).
[55] K. TANAKA and M. IWATA, Thin Solid Films **86**, 279 (1981).
[56] K. TANAKA, Electronics and Commun. Japan **67**, C–68 (1984).
[57] T. GENGENBACH, H. PAGNIA, and N. SOTNIK, phys. stat. sol. (a) **100**, K199 (1987).
[58] R. MÜLLER and H. PAGNIA, Mater. Letters **2**, 283 (1984).
[59] M. BORBONUS, Diplomarbeit, Technical University Darmstadt, 1986.
[60] R. BLESSING, H. PAGNIA, and N. SOTNIK, Thin Solid Films **85**, 119 (1981).
[61] F. BERG, Diplomarbeit, Technical University Darmstadt, 1986.
[62] K. SCHLEMPER, Diplomarbeit, Technical University Darmstadt, 1985.
[63] L. ECKERTOVÁ and J. JISKRA, Czech. J. Phys. **B18**, 131 (1968).
[64] C. BARRIAC, F. GIRAUD-MERAUD, P. PINARD, and F. DAVOINA, C.R. Acad. Sci. (France) **262**, 900 (1966).
[65] L. S. MILLER and R. KERSHAW, Thin Solid Films **29**, 165 (1975).
[66] C. A. HOGARTH and E. H. Z. TAHERI, Internat. J. Electronics **37**, 145 (1974).
[67] H. PAGNIA, phys. stat. sol. **1**, 499 (1961).
[68] J. G. SIMMONS and R. R. VERDERBER, Radio Electronic Engng. **34**, 81 (1967).
[69] P. D. GREEN, E. L. BUSH, and I. R. RAWLINGS, Proc. Symp. Thin Film Dielectric Materials, Montreal 1969 (p. 167).
[70] R. D. FEDOROVICH, private communication (1985).
[71] M. BISCHOFF, H. PAGNIA, and K. WEYRICH, Thin Solid Films **139**, 25 (1986).
[72] T. W. HICKMOTT, Thin Solid Films **9**, 431 (1972).
[73] R. BLESSING, Thesis, Technical University Darmstadt, 1981 (D17).
[74] R. E. THURSTANS, Thin Solid Films **57**, 153 (1979).
[75] M. BISCHOFF, Thesis, Technical University Darmstadt, 1988 (D17).
[76] S. A. Y. AL-ISMAIL and C. A. HOGARTH, phys. stat. sol. (a) **76**, 559 (1983).
[77] T. W. HICKMOTT, Thin Solid Films **3**, 85 (1969).
[78] C. A. NEUGEBAUER and M. B. WEBB, J. appl. Phys. **33**, 74 (1962).
[79] J. E. MORRIS and T. J. COUTTS, Thin Solid Films **47**, 3 (1977).
[80] B. V. ALEKSEENKO and R. D. FEDOROVICH, Thin Solid Films **92**, 253 (1982).
[81] R. BLESSING and H. PAGNIA, Thin Solid Films **52**, 333 (1978).
[82] P. A. TICK and R. P. FEHLNER, J. appl. Phys. **43**, 362 (1972).
[83] H.-E. KNAUS, Wissenschaftliche Hausarbeit, Technical University Darmstadt, 1982.
[84] S. A. NEPIJKO and V. I. STYOPKIN, Poverkh. Fiz. Khim. Mekh. **1**, 51 (1983).
[85] H. PAGNIA, K. SCHLEMPER, and N. SOTNIK, Mater. Letters **5**, 260 (1987).
[86] P. MURALD and D. W. POHL, Appl. Phys. Letters **48**, 514 (1986).

[87] C. A. Hogarth and A. Kompany, Internat. J. Electronics **53**, 301 (1982).
[88] H. Pagnia and N. Sotnik, to be published.
[89] B. V. Alekseenko and R. D. Fedorovich, 3rd All-Union Symp. Properties of Small Particles and Discontinuous Metal Films, Lvov (USSR) 1980 (in Russian) (p. 97).
[90] R. Blessing, H. Pagnia, and R. Schmitt, see [89] (p. 104).
[91] H. Pagnia, N. Sotnik, and H. Stauss, phys. stat. sol. (a) **90**, 771 (1985).
[92] K. Tanaka and G. Matsumoto, Thin Solid Films **67**, L41 (1980).
[93] B. V. Alekseenko, R. D. Fedorovich, and P. M. Tomchuk, Mater. Sci. (Wrocław) **13**, 161 (1987).
[94] M. Borbonus, H. Pagnia, and N. Sotnik, Thin Solid Films **151**, 333 (1987).
[95] T. Gengenbach, H. Pagnia, and N. Sotnik, Mater. Letters **5**, 215 (1987).
[96] S. A. Nepijko, H. Pagnia, and V. I. Styopkin, phys. stat sol. (a) **90**, K113 (1985).
[97] F. Berg, H. Pagnia, and N. Sotnik, phys. stat. sol. (a) **95**, K91 (1986).
[98] D. Menzel, Ultramicroscopy **114**, 175 (1984).
[99] F. Berg, H. Pagnia, and N. Sotnik, phys. stat. sol. (a) **94**, K89 (1986).
[100] J. G. Simmons, Endeavour **27**, 138 (1968).
[101] J. C. Bernède, phys. stat. sol. (a) **80**, 85 (1983).
[102] M. S. Raven, Proc. Eighth Int. Vacuum Congress, Canners 1980 (p. 619).
[103] T. W. Hickmott, J. appl. Phys. **34**, 1569 (1963).
[104] C. Barriac, P. Pinard, and F. Davoine, phys. stat. sol. **34**, 621 (1969).
[105] M. Bischoff, R. Holzer, and H. Pagnia, Phys. Letters A **62**, 512 (1977).
[106] P. G. Borziak, O. G. Sarbei, and F. D. Fedorovich, phys. stat. sol. **8**, 55 (1965).
[107] G. Dittmer, Thin Solid Films **9**, 317 (1972).
[108] T. W. Hickmott, J. appl. Phys. **36**, 1885 (1965).
[109] R. Blessing and H. Pagnia, phys. stat. sol. (b) **110**, 537 (1982).
[110] T. W. Hickmott, Thin Solid Films **3**, 85 (1969).
[111] W. Pong, C. Inouye, F. Matsunaga, and M. Moriwaki, J. appl. Phys. **46**, 2310 (1975).
[112] M. Bischoff, M. Olt, and H. Pagnia, Thin Solid Films (1988), to be published.
[113] G. Delauney and J. Despujols, Thin Solid Films **51**, 227 (1978).
[114] H. Biederman and J. Plasek, Thin Solid Films **47**, L1 (1977).
[115] P. Borziak, Yu. Kulyupin, and P. Tomchuk, Thin Solid Films **30**, 47 (1975).
[116] H. Pagnia, J. Schnellbächer, and N. Sotnik, phys. stat. sol. (a) **88**, K209 (1985).
[117] J. Sanchez-Moreno, Diplomarbeit, Technical University Darmstadt, 1988.
[118] H. Pagnia and N. Sotnik, to be published.
[119] I. M. Dykman and P. M. Tomchuk, Soviet Phys. – Solid State **2**, 1988 (1961).
[120] J. Robertson and E. P. O'Reilly, Phys. Rev. B **33**, 2946 (1987).
[121] F. W. Smith, J. appl. Phys. **55**, 764 (1984).
[122] J. Hauser, Solid State Commun. **17**, 1577 (1975).
[123] D. R. Lovett, Semimetals and Narrow-Bandgap Semiconductors, Pion Ltd., London 1977, and references therein.
[124] A. G. Chynoweth, in: Semiconductors and Semimetals, Vol. 4, Ed. R. K. Willardson and A. C. Beer, Academic Press, New York 1968 (p. 263).
[125] E. E. Loebner, Phys. Rev. **102**, 46 (1956).
[126] L. B. Leder and J. A. Suddeth, J. appl. Phys. **31**, 1422 (1960).
[127] H. Frizsche, in: Electronic and Structural Properties of Amorphous Semiconductors, Ed. P. G. LeComber and J. Mort, Academic Press, London 1973.
[128] G. B. Adam and J. C. Duchene, IEEE Trans. Electron. Devices **19**, 820 (1972).
[129] G. Taylor and B. Lakevic, Solid State Electronics **19**, 669 (1976).
[130] H. Stauss, private communication (1985).
[131] R. W. Lomax and J. G. Simmons, Radio Electronic Engng. **35**, 265 (1968).
[132] H. G. Craighead, J. appl. Phys. **55**, 4420 (1984).
[133] M. Bischoff, Mater. Sci. (Wrocław) **13**, 49 (1987).
[134] N. Sotnik, Mater. Sci. (Wrocław) **13**, 125 (1987).


*(Received May 3, 1988)*

# *Original Papers*

phys. stat. sol. (a) **108**, 67 (1988)

Subject classification: 61.10; 61.70; S5.12

*Institute of Metal Physics, Academy of Sciences of the Ukrainian SSR, Kiev*[1])

## The Thickness Dependence of the X-Ray Diffuse Scattering Intensity for Crystals with Microdefects at Laue-Case Diffraction

By

V. V. Kochelab, V. B. Molodkin, S. I. Olikhovskii, and M. E. Osinovskii

The theory of the dynamical X-ray diffuse scattering (DS) in single crystals with homogeneously distributed microdefects (point defect clusters, inclusions, new phase nuclei, dislocation loops, etc.) is developed. The thickness dependence of the DS intensity distribution in reciprocal lattice space is investigated by constructing both, iso-intensity contours and plane projections of three-dimensional images. Possibilities to obtain a detailed information concerning various-kind microdefects characteristics by using distribution patterns of the DS intensity are discussed.

Развита теория динамического диффузного рассеяния рентгеновских лучей в монокристаллах, содержащих однородно распределенные микродефекты (кластеры точечных дефектов, включения, зародыши новой фазы, дислокационные петли и т.п.). Путем построения изодиффузных линий и плоских проекций объемных изображений исследована толщинная зависимость распределения интенсивности диффузного фона в пространстве обратной решетки. Обсуждаются возможности извлечения детальной информации о характеристиках микродефектов различного типа по картинам распределения интенсивности диффузного рассеяния.

### 1. Introduction

The investigation of the X-ray diffuse scattering (DS) in crystals containing microdefects allows one to obtain detailed information concerning their characteristics, i.e., concentration, size, orientation, etc. [1]. The information reliability is increased essentially due to the recently developed method of triple-crystal diffractometry [2 to 6]. It allows to measure precisely the DS intensity distribution in the near vicinity of reciprocal lattice points. However, applications of the kinematical theory of DS [1] are limited to cases of thin crystals or polycrystalline materials having grain size $t \ll \Lambda$ ($\Lambda$ is the extinction length).

When investigating the characteristics of microdefects in single crystals with thickness $t \geqq \Lambda$, the coherent DS effects are to be taken into account. These distort the kinematical patterns of the intensity distribution but, in general, allow to obtain the same information about defects as the kinematical analysis. The general formulae for both the DS amplitude and the DS intensity in various perturbation-theory approximations are obtained in earlier papers on the dynamical theory of DS in crystals with static defects [7, 8]. Also typical features of the dynamical DS have been described. Subsequent developments in the theory [9 to 11], including triple-crystal diffractometry [12], have allowed to determine interesting quantitative regularities for coherent rescattering of the diffuse background by using, as a model, defects with spherically symmetrical displacement fields of Coulomb type ($\boldsymbol{u}(\boldsymbol{r}) =$

[1]) Prospekt Vernadskogo 36, SU-252680 Kiev 142, USSR.

$= C\boldsymbol{r}/r^3$, $C$ is the defect power). In the Bragg-case diffraction this rescattering results in an extinction effect, whereas in the Laue-case diffraction it gives an anomalously transmitted diffuse background. The experimental confirmation of above-mentioned effects [13, 14] has proved the validity of the main conclusions of the theory and stimulated subsequent theoretical investigations of DS in real single crystals with microdefects [15, 16].

In this paper (see also [17]) we develop the dynamical theory of the X-ray DS taking into account both the anisotropy of single crystals and the asymmetry of the displacement fields of the microdefects (point defect clusters, new phase nuclei, low-size dislocation loops, etc.).

## 2. Diffuse Scattering Intensity

A general expression for the dynamical DS intensity of the diffraction of X-rays and electrons in single crystals with randomly distributed defects was obtained in [8]. In the two-beam case of Laue diffraction with an incident linearly polarized X-ray wave $E_0 \exp(-i\boldsymbol{K}\boldsymbol{r})$, the DS intensity in the vicinity of a reciprocal lattice point H, which corresponds to the reciprocal lattice vector $\boldsymbol{h}$, is given by the following expression [10]:

$$I_{\boldsymbol{h}}^{\mathrm{DS}}(\boldsymbol{K}', \boldsymbol{K}) = \langle | \sum_{g=0,h} \sum_{\delta=1,2} D_g^\delta F_{hg}^\delta(\boldsymbol{K}', \boldsymbol{K})|^2 \rangle \,, \tag{1}$$

where $\boldsymbol{K}' = \boldsymbol{K} + \boldsymbol{h} + \boldsymbol{k}$ is the wave vector of a diffusely scattered wave in vacuum ($|\boldsymbol{K}'|) = K$), the vector $\boldsymbol{k}$ is the deviation of the vector $\boldsymbol{K}'$ from the reciprocal lattice point H, $\langle ... \rangle$ means averaging over the defect distribution. The amplitudes of strong Bragg waves in the crystal

$$D_0^\delta = (-1)^\delta E_0 \frac{-2\gamma_0 \Delta_{\delta'} + \chi_0}{2\gamma_0(\Delta_1 - \Delta_2)} \,, \tag{2}$$

$$D_{\boldsymbol{h}}^\delta = -D_0^\delta \frac{-2\gamma_0 \Delta_\delta + \chi_0}{P_s \chi'_{-h}} \qquad (\delta' \neq \delta = 1, 2)$$

are functions of the excitation errors $\varepsilon_0^\delta$ which are determined by the deviation $w$ of the incident wave vector $\boldsymbol{K}$ from that satisfying exactly the Bragg condition $(\boldsymbol{K} + \boldsymbol{h}/2) \cdot \boldsymbol{h} = \boldsymbol{0}$ and for the symmetrical Laue case are equal

$$\varepsilon_0^\delta \equiv \gamma_0 \Delta_\delta = \frac{\chi_{0r}}{2} + \pi\gamma_0 \frac{w - (-1)^\delta \sqrt{w^2 + P_s^2}}{K\Lambda} - \frac{i\,[\mu_0 + (-1)^\delta \mu_h P_s^2/\sqrt{w^2 + P_s^2}]}{2K} \,, \tag{3}$$

where $w = (|\boldsymbol{K} + \boldsymbol{h}| - K)/K \sqrt{\chi'_{hr}\chi'_{-hr}}$, $\mu_0 = -K\chi_{0i}$, $\mu_h = K\sqrt{\chi'_{hr}\chi'_{-hr}}\,\mathrm{Re}\,(\chi'_{hi}/\chi'_{hr})$. Here the index $\delta = 1, 2$ denotes the dispersion surface sheets, $\chi_0$, $\chi_{\pm h}$ are Fourier components of the perfect crystal polarizability, $\chi'_{\pm h} = \chi_{\pm h} \exp(-L_h)$ are the Fourier components of the distorted crystal polarizability, ($\exp(-L_h)$) is the static Debye-Waller factor [1]), the indices r and i correspond to the Fourier components of real and imaginary parts of the polarizability, respectively, $K = 2\pi/\lambda$ with $\lambda$ the X-ray wavelength, $\gamma_0 = \cos\theta$ with $\theta$ the Bragg angle, $\Lambda = \gamma_0\lambda/\sqrt{\chi'_{hr}\chi'_{-hr}}$ is the extinction length for the distorted crystal for $\sigma$-polarized radiation, $P_s$ is the polarization factor ($P_1 = 1$, $\sigma$-polarization; $P_2 = |\cos 2\theta|$, $\pi$-polarization).

The functions $F_{hg}^\delta(\boldsymbol{K}', \boldsymbol{K})$ are partial scattering amplitudes of the strong Bragg waves $D_g^\delta$ with the crystal wave vectors $\boldsymbol{K}_g^\delta = \boldsymbol{K} + \boldsymbol{g} + K\Delta_\delta\boldsymbol{n}$ ($\boldsymbol{n}$ is the unit inner normal to the input crystal surface) into the diffusely scattered wave in vacuum with the wave vector $\boldsymbol{K}'$ [9, 10]. They depend on the Fourier components of the fluctuation of crystal polarizability $\chi_{\boldsymbol{g}+\boldsymbol{q}}$, where $\boldsymbol{q}$ is the wave vector of the fluctuation wave of static displacements from crystal defects [1]. For microdefects having Coulomb-type displacement fields we can assume that $|\chi_{\boldsymbol{q}}| \ll |\chi_{\boldsymbol{q}\pm\boldsymbol{h}}|$ at $q \ll h$, and according to [10] we have

$$F^{\delta}_{\boldsymbol{hg}} = V[8\pi\gamma_0(\Delta_1' - \Delta_2')]^{-1} \sum_{\tau=1,2} (-1)^{\tau} M^{\delta\tau}_{\boldsymbol{hg}} \exp(-iK\Delta_\tau' t)\,, \tag{4}$$

$$M^{\delta\tau}_{\boldsymbol{h}0} \approx K^2(2\gamma_0\Delta_\tau' - \chi_0)\, P_s\chi_{\boldsymbol{h}+\boldsymbol{q}_{\delta\tau}}\,, \qquad M^{\delta\tau}_{\boldsymbol{hh}} \approx K^2 P_s^2 \chi_h' \chi_{-\boldsymbol{h}+\boldsymbol{q}_{\delta\tau}}\,.$$

Here $t$ and $V$ are the thickness and the volume of a plane-parallel single crystal plate, respectively; $\tilde{\varepsilon}_0^{\tau} \equiv \gamma_0\Delta_\tau'$ are excitation errors for diffusely scattered waves in the crystal as in (3), which are functions of the deviation $w' = (|\boldsymbol{K}' - \boldsymbol{h}| - K)/K\sqrt{\chi_{h\tau}'\chi_{-h\tau}'}$ of the diffusely scattered wave in vacuum from the exact Bragg direction.

The wave vectors of static displacement fluctuation waves are related with the deviation $\boldsymbol{k}$ from the reciprocal lattice point as follows: $\boldsymbol{q}_{\delta\tau} = \boldsymbol{k}_{||} + q_{z\delta\tau}\boldsymbol{n}$, where $\boldsymbol{k}_{||} = k_x\boldsymbol{e}_x + k_y\boldsymbol{e}_y$, $q_{z\delta\tau} = K(\Delta_\tau' - \Delta_\delta - \boldsymbol{k}_{||}\boldsymbol{K}/K^2)$, $\boldsymbol{n} \perp (\boldsymbol{e}_x, \boldsymbol{e}_y)$. Here basic vectors $\boldsymbol{e}_x$, $\boldsymbol{e}_y$ are parallel to the input crystal surface and $\boldsymbol{e}_x$ lies in the scattering plane $(\boldsymbol{K}, \boldsymbol{h})$. At fixed orientation of the wave vector $\boldsymbol{K}$ of the incident wave, four vectors $\boldsymbol{q}_{\delta\tau}$ in the crystal correspond to one wave vector $\boldsymbol{K}'$ of the diffusely scattered wave in vacuum. This fact is one of the causes distorting the DS intensity distribution patterns as compared with those according to the kinematical theory.

As (4) shows the calculation of the dynamical DS intensity (1) is reduced to the determination of products of fluctuation Fourier components of polarizability averaged over a defect distribution: $\langle \chi_{\boldsymbol{g}+\boldsymbol{q}_{\delta\tau}}\chi^*_{\boldsymbol{g}'+\boldsymbol{q}_{\varkappa\sigma}} \rangle$, where $\boldsymbol{g}, \boldsymbol{g}' = \pm\boldsymbol{h}$; $\delta, \tau, \varkappa, \sigma = 1, 2$. The fluctuation Fourier component of polarizability for the distorted crystal is [1] $\chi_{\boldsymbol{h}+\boldsymbol{q}} = \sum_s \gamma f^s_{\boldsymbol{h}+\boldsymbol{q}} \exp[i(\boldsymbol{h} + \boldsymbol{q})(\boldsymbol{R}_s^0 + \boldsymbol{u}_s)]$, where $f^s_{\boldsymbol{h}+\boldsymbol{q}}$ is the structure factor of the $s$-th unit cell, $\boldsymbol{R}_s^0$ is the radius vector of the first atom of the $s$-th unit cell in a perfect crystal, $\boldsymbol{u}_s$ is the static displacement of this cell caused by defects; $\gamma = -r_e\lambda^2/\pi v$, $r_e$ is the classical electron radius, $v$ is the unit-cell volume. At low concentration ($c \ll 1$) of defects the displcacement $\boldsymbol{u}_s$ may be described as a superposition $\boldsymbol{u}_s - \boldsymbol{u}_s^0 = \sum_t (c_t - c)\,\boldsymbol{u}_s^{(t)}$, where the sum is taken over all $N$ unit cells, $\boldsymbol{u}_s^0 = \langle \boldsymbol{u}_s \rangle$ is the average displacement of the $s$-th cell, $\boldsymbol{u}_s^{(t)}$ is the displacement of the $s$-th cell caused by the defect lying in the position $t$; $c = \langle c_t \rangle$, $c_t = 1$ or 0, if the position $t$ is occupied by a defect or not. If local distortions caused by defects are small ($\max |\boldsymbol{h} \cdot (\boldsymbol{u}_s - \boldsymbol{u}_s^0)| \ll 1$), then, neglecting the difference between the structure factors of different unit cells, we may put approximately [1] $\chi_{\boldsymbol{h}+\boldsymbol{q}} \approx iN\chi_{\boldsymbol{h}}\boldsymbol{h} \cdot \boldsymbol{u}(\boldsymbol{q}) \sum_t (c_t - c) \exp(i\boldsymbol{q}\boldsymbol{R}_t)$, where $\chi_h = \gamma f_h$, $\boldsymbol{R}_s = \boldsymbol{R}_s^0 + \boldsymbol{u}_s^0$, and the Fourier component of the displacement field of a single defect is introduced by $\boldsymbol{u}(\boldsymbol{q}) = N^{-1} \sum_s \boldsymbol{u}(\boldsymbol{R}_s - \boldsymbol{R}_t) \exp[i\boldsymbol{q}(\boldsymbol{R}_s - \boldsymbol{R}_t)] \approx (Nv)^{-1} \int_V \mathrm{d}\boldsymbol{r}\, \boldsymbol{u}(\boldsymbol{r}) \exp(i\boldsymbol{q}\boldsymbol{r})$. Since the wave vectors $\boldsymbol{q}_{\delta\tau}$ and $\boldsymbol{q}_{\varkappa\sigma}$ differ only in their $z$-components and, further, deviations $k \ll h$ will be considered, we obtain after averaging over a random defect distribution

$$\langle \chi_{\boldsymbol{g}+\boldsymbol{q}_{\delta\tau}}\chi^*_{\boldsymbol{g}'+\boldsymbol{q}_{\varkappa\sigma}} \rangle \approx cN\chi_{\boldsymbol{g}}\chi^*_{\boldsymbol{g}'}\,[\boldsymbol{g} \cdot \boldsymbol{u}(\boldsymbol{q}_{\delta\tau})]\,[\boldsymbol{g}' \cdot \boldsymbol{u}(\boldsymbol{q}_{\varkappa\sigma})]^*\, p(q_{z\delta\tau} - q^*_{z\varkappa\sigma}) \tag{5}$$

with $p(q) = [\exp(iqt) - 1]/iqt$ and $\boldsymbol{g}, \boldsymbol{g}' = \pm\boldsymbol{h}$.

When the local distortions caused by the defects are large ($\max |\boldsymbol{h} \cdot (\boldsymbol{u}_s - \boldsymbol{u}_s^0)| \geqq 1$), then expansion of the exponent into powers of small displacements is no longer possible. Nevertheless, for weakly distorted crystals ($L_h \ll 1$), and at small $\boldsymbol{q}$ values, i.e. $q \ll (hC)^{-1/2} \ll R^{-1}$ for clusters and $q \ll R^{-1}(hb)^{-1/2}$ for dislocation loops ($R$ is the cluster or loop radius, $C$ the cluster power, $b$ the Burgers vector modulus), in the sum over $s$ those terms give a main contribution for which $|\boldsymbol{h} \cdot (\boldsymbol{u}_s - \boldsymbol{u}_s^0)| \ll 1$ (see [1]). Therefore, by imposing restrictions (5) is valid even at large local distortions, and approximating expressions for the displacement fields of defects may be used.

We shall determine the "diffuse" reflectivity of distorted crystal as $R_{\mathrm{DS}}(\boldsymbol{k}) = I_{\boldsymbol{h}}^{\mathrm{DS}}(\boldsymbol{K}', \boldsymbol{K})/(|E_0|^2 S\gamma_0)$, where $S = V/t$ is the input surface area of the crystal. Substituting (2), (4) into (1) and using (5) we obtain

$$R_{\mathrm{DS}}(\boldsymbol{k}) \approx P_0(KP_s|\chi_h|)^2 (|\Delta_1 - \Delta_2| \, |\Delta_1' - \Delta_2'|)^{-2} \times \times \sum_{\delta\tau\varkappa\sigma} (-1)^{\delta+\tau+\varkappa+\sigma} D_{\delta\tau\varkappa\sigma} \Pi_{\delta\tau\varkappa\sigma} [\boldsymbol{h}_0 \cdot \boldsymbol{u}(\boldsymbol{q}_{\delta\tau})] [\boldsymbol{h}_0 \cdot \boldsymbol{u}(\boldsymbol{q}_{\varkappa\sigma})]^* , \tag{6}$$

$$\Pi_{\delta\tau\varkappa\sigma} = \{\exp[-iKt(\Delta_\delta - \Delta_\varkappa^*)] - \exp[-iKt(\Delta_\tau' - \Delta_\sigma'^*)]\} \times \times [iKt(\Delta_\tau' - \Delta_\sigma'^* - \Delta_\delta + \Delta_\varkappa^*)]^{-1} ,$$

$$D_{\delta\tau\varkappa\sigma} = (\Delta_\tau' - \Delta_\delta)(\Delta_\sigma' - \Delta_\varkappa)^* (2\gamma_0\Delta_{\delta'} - \chi_0)(2\gamma_0\Delta_{\varkappa'} - \chi_0)^* .$$

Here $P_0 = ctvh^2/16\gamma_0^3\lambda^2$, $\boldsymbol{h}_0 = \boldsymbol{h}/h$, $\delta' \neq \delta$, $\varkappa' \neq \varkappa$, and the crystal was assumed to be centrosymmetric ($\chi_{-h}^* = \chi_h$).

The "diffuse" reflectivity (6) of the distorted crystal and, correspondingly, the dynamical DS intensity are principally different from those calculated in the kinematical approximation. The main difference is that for the DS in crystals with microdefects the anomalous transmission effect takes place, which for the first time was predicted in [10] for spherically symmetric Coulomb-type defects and was confirmed experimentally in [14]. This effect is, firstly, due to the anomalous transmission (Borrmann effect) of the strong Bragg wave being a source of diffusely scattered waves and, secondly, due to the anomalous transmission of diffusely scattered waves undergoing multiple coherent rescattering between both the transmitted and diffracted beam directions.

As seen from (6), the anomalous transmission of the diffuse background is described by the factor $\Pi_{\delta\tau\varkappa\sigma}$ containing interference absorption coefficients of strong Bragg waves and diffusely scattered waves. Depending on the index combination $\delta, \tau, \varkappa, \sigma$, this factor is equal, in order of magnitude, to $\Lambda/t \exp(-\mu_0 t/\gamma_0)$ at $\varkappa \neq \delta$, $\sigma \neq \tau$, or $\Lambda/t \exp[-(\mu_0 \pm \mu_h) t/\gamma_0]$ at $\varkappa \neq \delta$, $\sigma = \tau$ or $\varkappa = \delta$, $\sigma \neq \tau$, or $t_a/t \exp[-(\mu_0 \pm \mu_h) t/\gamma_0]$ at $\varkappa = \delta$, $\sigma = \tau$, where $t_a = 1/\mu_0$ is the absorption length, $\mu_0$ is the photoelectric absorption coefficient. Since usually the inequalities $\mu_0 - \mu_h \ll \mu_0$ and $t_a \gg \Lambda$ are valid, the expression for $\Pi_{\delta\tau\varkappa\sigma}$ may be rewritten for $\mu_0 l > 1$ ($l = t/\gamma_0$) as

$$\Pi_{\delta\tau\varkappa\sigma} \approx \delta_{\varkappa\delta}\delta_{\sigma\tau} \frac{[\exp(2Kt \operatorname{Im} \Delta_\delta) - \exp(2Kt \operatorname{Im} \Delta_\tau')]}{2Kt \operatorname{Im}(\Delta_\delta - \Delta_\tau')} , \tag{7}$$

where $\delta_{\varkappa\delta}$ and $\delta_{\sigma\tau}$ are Kronecker symbols. Therefore, for single crystals of both intermediate and large thicknesses ($\mu_0 l > 1$) (6) may be simplified by using (7) as follows:

$$R_{\mathrm{DS}}(k) \approx P_0(K_s P_h |\chi_h|)^2 (K\Lambda/2\pi)^2 (w^2 + P_s^2)^{-1} (w'^2 + P_s^2)^{-1} \times \times \sum_{\delta,\tau} |\Delta_\tau' - \Delta_\delta|^2 \, |2\gamma_0\Delta_{\delta'} - \chi_0|^2 \, \Pi_{\delta\tau\delta\tau} \, |\boldsymbol{h}_0 \cdot \boldsymbol{u}(\boldsymbol{q}_{\delta\tau})|^2 . \tag{8}$$

One of four terms in (8) ($\delta = 2, \tau = 2$) corresponds to anomalously absorbing strong Bragg waves and diffusely scattered waves, and for thick crystals ($\mu_0 l > 5$) may be neglected. The remaining terms correspond to anomalously transmitted Bloch waves, i.e., to the strong Bragg wave and the diffusely scattered one ($\delta = 1, \tau = 1$), to the strong Bragg wave only ($\delta = 1, \tau = 2$), and to the diffusely scattered wave only ($\delta = 2, \tau = 1$). Hence, contributions to $R_{\mathrm{DS}}(\boldsymbol{k})$ of each of the four fluctuation waves of statical displacements having wave vectors $\boldsymbol{q}_{\delta\tau}$ are different at fixed $\boldsymbol{k}$ (consequently, at fixed $w$ and $w'$) and functions of the crystal thickness $t$. The investigation of intensity distribution patterns of the dynamical DS at various thicknesses $t$ will be carried out further in more details for two microdefect models.

## 3. Clusters in an Elastically Anisotropic Crystal

Let us consider a model of clusters of cubic symmetry (e.g., inclusions, new phase nuclei, etc.) with an effective radius $R$ in an elastically anisotropic, cubic matrix. In this case the Fourier component of the cluster displacement field is given by [1]

$$\boldsymbol{u}(\boldsymbol{q}) = \frac{\boldsymbol{A}_{\boldsymbol{q}} s(\boldsymbol{q})}{v}, \qquad \boldsymbol{A}_{\boldsymbol{q}} = \frac{L \boldsymbol{\alpha}_{\boldsymbol{q}}}{q}, \qquad L = \frac{1}{3v} \frac{\mathrm{d}v}{\mathrm{d}c}, \tag{9}$$

where

$$\alpha_{\boldsymbol{q}x} = (c_{11} + 2c_{12})\,(1 + \xi q_{0y}^2)\,(1 + \xi q_{0z}^2)\,\frac{q_{0x}}{D(\boldsymbol{q}_0)},$$

$$D(\boldsymbol{q}_0) = c_{11} + \xi(c_{11} + c_{12})\,[(q_{0x}q_{0y})^2 + (q_{0x}q_{0z})^2 + (q_{0y}q_{0z})^2] + \\ + \xi^2(c_{11} + 2c_{12} + c_{44})\,(q_{0x}q_{0y}q_{0z})^2 \,.$$

$s(\boldsymbol{q})$ is the Fourier component of the cluster form factor $s(\boldsymbol{r})$, $\boldsymbol{q}_0 = \boldsymbol{q}/q$, $\xi = (c_{11} - c_{12} - 2c_{44})/c_{44}$, $L\delta_{ij}$ is the self-deformation tensor, $c_{11}$, $c_{12}$, $c_{44}$ are elastic stiffness constants, expressions for $\alpha_{\boldsymbol{q}y}$ and $\alpha_{\boldsymbol{q}z}$ are obtained from that for $\alpha_{\boldsymbol{q}x}$ by the cyclic permutation of indices. At low $\boldsymbol{q}$ values ($q \ll 1/R$) $s(\boldsymbol{q})/v \approx s(0)/v = n_0$, where $n_0$ is the number of unit cells in the matrix being substituted by the cluster volume.

In the symmetrical case of Laue diffraction ($\boldsymbol{h} \perp \boldsymbol{n}$) variables $k_x$, $k_z$ describing the deviation of the wave vector $\boldsymbol{K}' = \boldsymbol{K} + \boldsymbol{h} + \boldsymbol{k}$ in the scattering plane ($\boldsymbol{K}$, $\boldsymbol{h}$) from the reciprocal lattice point H are simply connected with deviations of both the incident wave and the diffusely scattered one in vacuum from their exact Bragg directions,

$$k_x = \frac{\pi(w' - w)}{\Lambda \operatorname{tg} \theta}, \qquad k_z = \frac{\pi(w' + w)}{\Lambda}. \tag{10}$$

By using (6) (or (8) at $\mu_0 l > 1$) and (10) we may calculate the dynamical DS intensity distribution in the reciprocal lattice space for single crystals with randomly distributed clusters taking into account both the matrix anisotropy and photoelectric absorption of X-rays. To study the "diffuse" reflectivity distribution $R_{\mathrm{DS}}(k_x, k_y = 0, k_z)$ in the scattering plane we use two methods of graphical representation for functions of two variables. One of them is plotting of iso-intensity contours. $k_x^{\mathrm{m}}$ and $k_z^{\mathrm{m}}$ in the relevant figures represent limits for $k_x$ (from $-k_x^{\mathrm{m}}$ to $+k_x^{\mathrm{m}}$) and $k_z$ (from $-k_z^{\mathrm{m}}$ to $+k_z^{\mathrm{m}}$) expressed in terms of the reciprocal extinction length $1/\Lambda$. $d$ is equal to the maximum of the function divided by the highest level as well as equal to the ratio of each two neighbouring levels. Function maxima are denoted in the figures as $\varrho_{\mathrm{m}} = R_{\mathrm{DS}}^{\max}(\boldsymbol{k})/(cC^2 \times 10^5)$ where $C = n_0 L$ for the cluster model.

The second method is a planar projection of three-dimensional figures being described by functions of two variables. As compared with iso-intensity contours, it allows to represent a distribution pattern of the DS intensity in the reciprocal lattice plane more clearly and in more detail. An absolute function maximum is denoted in the relevant figures by $\varrho_{\mathrm{m}}$.

The DS iso-intensity contours (isodiffuse contours) plotted in Fig. 1 for the case of Ge single crystal with randomly distributed cubic symmetrical clusters allow to observe the effect of the crystal thickness on the intensity distribution character.

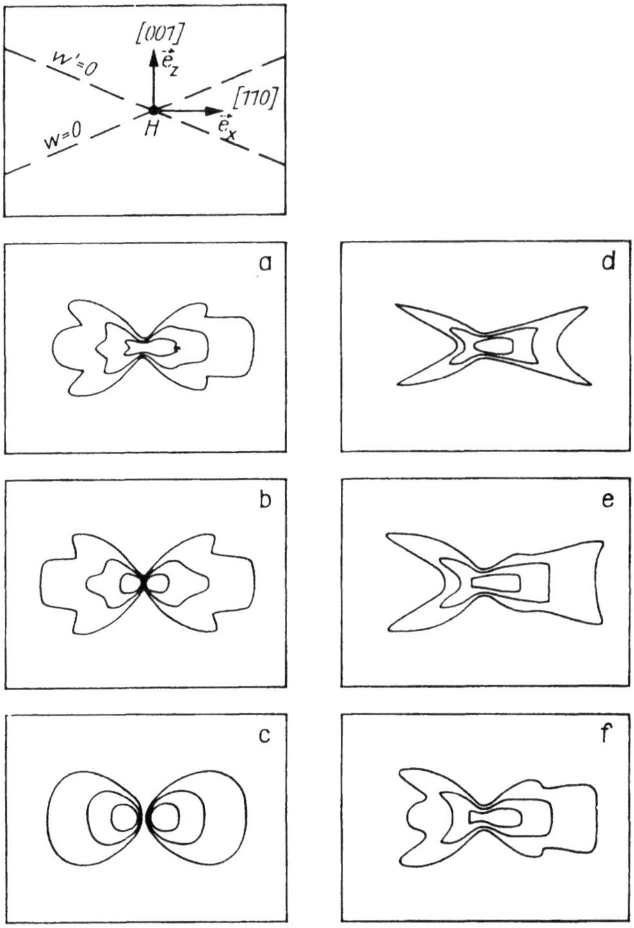


Fig. 1. Iso-intensity contours ($CuK_\alpha$) for the "diffuse" reflectivity of Ge (220) crystals of various thickness containing small-radius clusters ($k_x^m = 85\Lambda^{-1}$, $k_z^m = 60\Lambda^{-1}$, $\Lambda = 7.0\,\mu m$; $\boldsymbol{n} = [001]$, $\theta = 22.6°$). a) $\mu_0 l = 0.67$, $\varrho_m = 100$, $d = 2.8$; b) 0.13, 30, 2.8; c) 0.03, 8, 2.8; d) 5.36, 325, 4.5; e) 2.68, 215, 3.8; f) 1.34, 140, 3

It was assumed that the effective cluster radius is small as compared with the extinction length ($R \ll \Lambda$) and the crystal is weakly distorted ($L_h \ll 1$).

As seen in Fig. 1, for thin crystals with thickness $t = \Lambda/10$ ($\mu_0 l = 0.03$) the iso-intensity contour shape follows that given by the kinematical theory [1] (elastic constants for Ge are taken from [18], and the Fourier components of the polarizability from [19]). Really, for thin crystals, $t \ll \Lambda$, in the range $|w|$, $|w'| \gg 1$, (6) is transformed into the kinematical expression

$$R_{DS}^{kin}(\boldsymbol{k}) = 4\gamma_0^2 P_0 (K P_s |\chi_h|)^2 [\boldsymbol{h}_0 \cdot \boldsymbol{u}(\boldsymbol{k})]^2 . \tag{11}$$

In this case the difference between iso-intensity contours plotted by using both the dynamical expression (6) and the kinematical one (11) may be observed only in the range $|w|$, $|w'| \leqq 1$ ($k \leqq 2\pi/\Lambda$). This fact is seen from Fig. 2a, which represents the planar projection of the three-dimensional image of the DS intensity distribution at $t = \Lambda/10$.

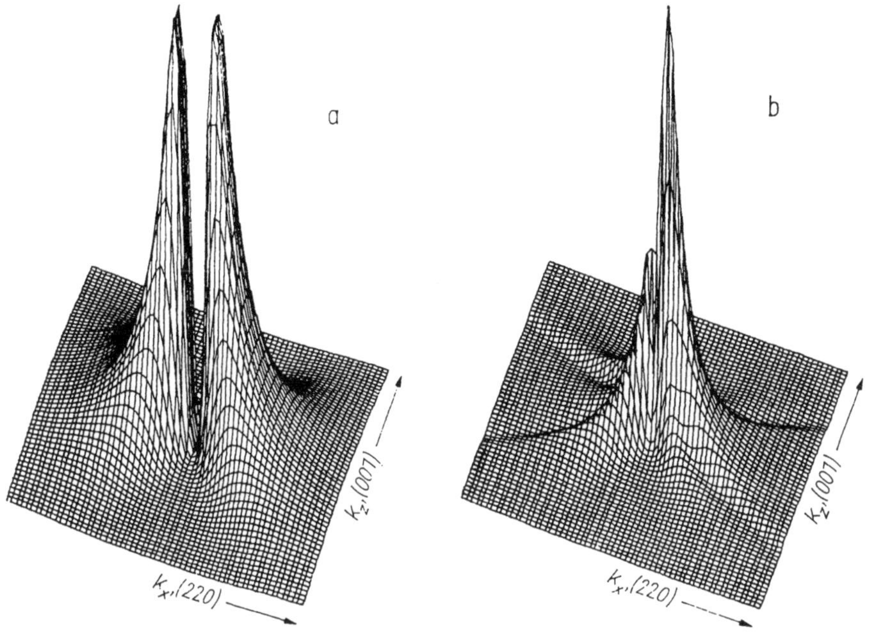


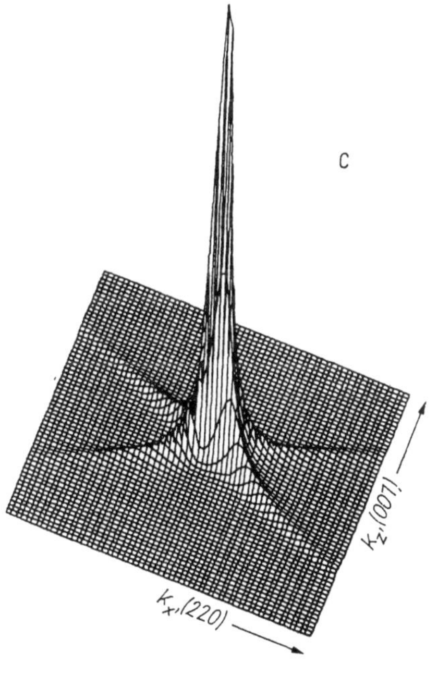


Fig. 2. Planar projections of three-dimensional images for the "diffuse" reflectivity of crystals of various thickness containing small-radius clusters (Ge (220), $CuK_\alpha$, $\boldsymbol{n} = [001]$; $k_x^m = 100\Lambda^{-1}$; $k_z^m = 80\Lambda^{-1}$, $\Lambda = 7.0\,\mu m$). a) $\mu_0 l = 0.027$, $\varrho_m = 8$; b) 1.34, 140; e) 5.36, 325

For single crystals with thicknesses between the extinction and absorption lengths ($\Lambda \leqq t \leqq t_a$), the shape of isodiffuse contours differs essentially from that being described by the kinematical theory. This can be seen in Fig. 1, where also isodiffuse contours for crystals with thicknesses $t = 0.5\Lambda$ ($\mu_0 l = 0.13$), $t = 2.5\Lambda$ ($\mu_0 l = 0.67$ and $t = 5\Lambda$ ($\mu_0 l = 1.34$) are plotted.

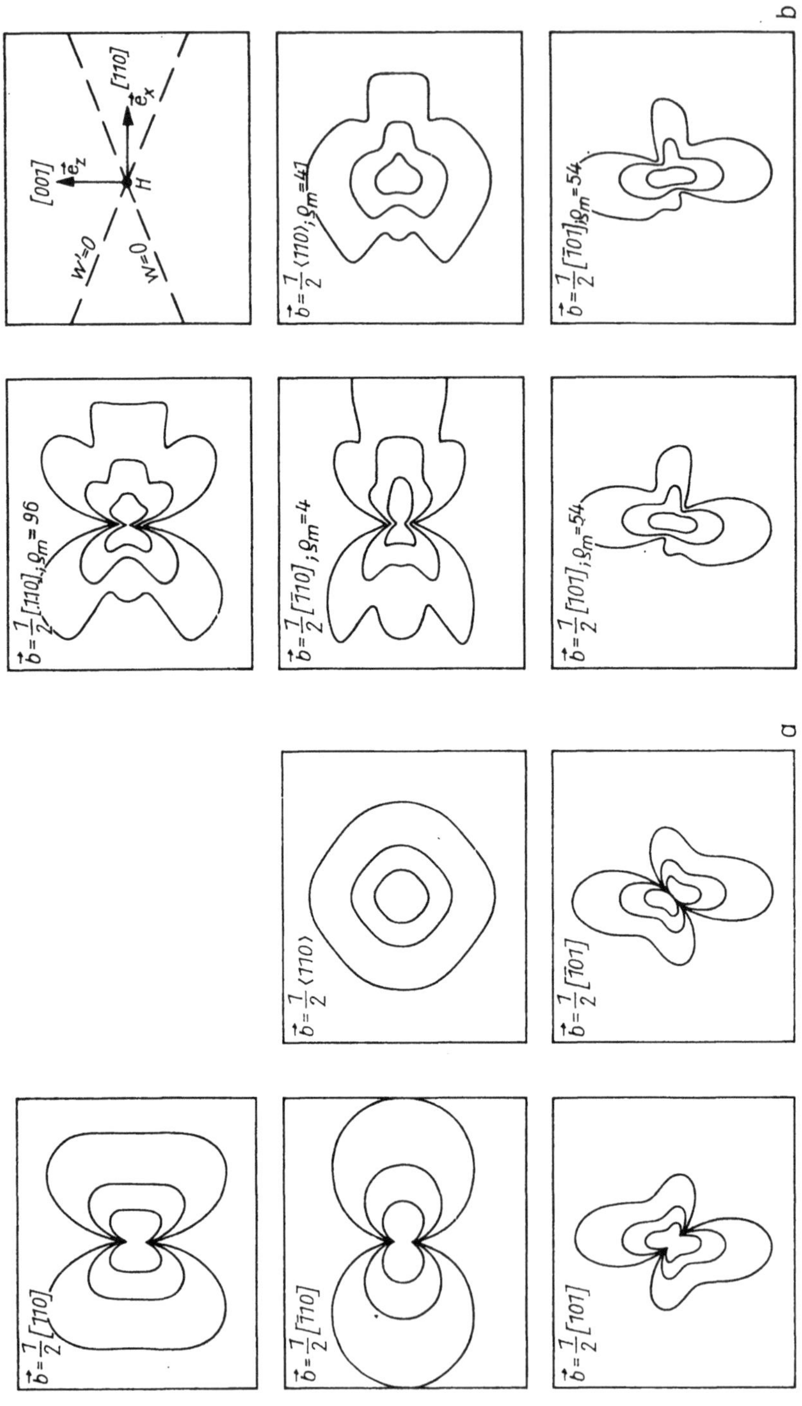
b
[001]
$\vec{e}_z$
$\vec{e}_x$
$[\bar{1}10]$
H
W'=0
W=0
$\vec{b}=\frac{1}{2}[110];\ \varrho_m=96$
$\vec{b}=\frac{1}{2}\langle\bar{1}10\rangle;\ \varrho_m=41$
$\vec{b}=\frac{1}{2}[\bar{1}10];\ \varrho_m=4$
$\vec{b}=\frac{1}{2}[\bar{1}0\bar{1}];\ \varrho_m=54$
$\vec{b}=\frac{1}{2}[10\bar{1}];\ \varrho_m=54$
a
$\vec{b}=\frac{1}{2}[110]$
$\vec{b}=\frac{1}{2}\langle\bar{1}10\rangle$
$\vec{b}=\frac{1}{2}[\bar{1}10]$
$\vec{b}=\frac{1}{2}[\bar{1}0\bar{1}]$
$\vec{b}=\frac{1}{2}[10\bar{1}]$

Fig. 3

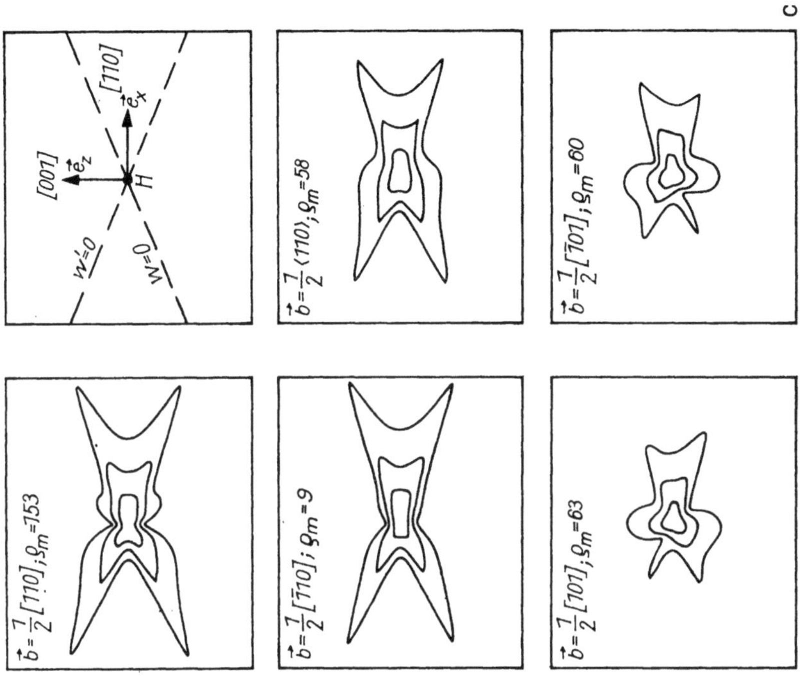


Fig. 3. Iso-intensity contours (Cu$K_\alpha$) for the "diffuse" reflectivity of Ge (220) crystals containing small-radius prismatic dislocation loops at various orientations of the Burgers vector $\boldsymbol{b}$ ($k_x^m = 70 \Lambda^{-1}$; $k_z^m = 60 \Lambda^{-1}$, $\Lambda = 7.0$ μm, $\boldsymbol{n} = [001]$): a) kinematical theory ($d = 3.5$); b) dynamical theory, $t = 5\Lambda$ ($\mu_0 t = 1.34$, $d = 3.5$, $\theta = 22.6°$); c) dynamical theory, $t = 20\Lambda$ ($\mu_0 t = 5.36$, $d = 5$, $\theta = 22.6°$)

In this thickness range behaviour of the dynamical DS is affected by two factors. As mentioned above, one of them is that four fluctuation waves of static displacements give contributions to the DS intensity in each point of the reciprocal lattice space as distinct from the kinematical theory where only one fluctuation wave gives a contribution.

The second factor is associated with the known asymmetry of the anomalously transmitted and anomalously absorbed strong Bragg wave fields in the direction of the transmitted beam with respect to the sign of their deviation $w$ from the exact Bragg direction. Namely, at $w < 0$ the anomalously absorbed wave field is effective, i.e. one has a dominant effect on the resulting amplitude of the transmitted wave, whereas at $w > 0$ the anomalously transmitted wave field is effective [19]. For wave fields formed by diffusely scattered waves in the diffracted beam direction the same asymmetry takes place with respect to $w'$. It is seen easily from (6) for $R_{DS}(\boldsymbol{k})$, which includes factors $D_{\delta\tau\varkappa\sigma}$ with odd-power terms in $w$ and $w'$. Together with the factors $\Pi_{\delta\tau\varkappa\sigma}$ they cause the asymmetry pointed out above. The asymmetry appears in iso-intensity contours (Fig. 1) as a specific deformation along a family of straight lines $k_z = k_x \operatorname{tg} \theta + 2\pi w/\Lambda$ at $|w| \leqq 1$ and $k_z = -k_x \operatorname{tg} \theta + 2\pi w'/\Lambda$ at $|w'| \leqq 1$. (See insert in Fig. 1. The dotted straight lines at $w = 0$ and $w' = 0$ making the Bragg angle $\theta$ with the diffraction vector $\boldsymbol{h}$ correspond to exact Bragg conditions for both the incident wave and diffusely scattered one, respectively. These lines will be called further "anomalous transmission lines".) The iso-intensity contour deformation means that they are tightened to the reciprocal lattice point H along the anomalous transmission lines with negative values of $w'$ and $w$ (from the left to H) and are expanded from the H point along the same lines to positive values of $w'$ and $w$ (from H to the right). In other words, for crystal thicknesses $\Lambda \leqq t \leqq t_a$ in the DS intensity distribution at smaller Bragg angles there are anomalous absorption areas, whereas at larger Bragg angles there are anomalous transmission areas. These areas are more expressed for thicknesses $t \approx t_a$ (Fig. 2b).

In the case of thick crystals (Fig. 1, $t = 10\Lambda$ or $\mu_0 l = 2.68$, and $t = 20\Lambda$ or $\mu_0 l = 5.36$) isodiffuse contours are strongly expanded along the anomalous transmission lines. Their shape depends essentially no longer on defect type and crystal anisotropy, being only determined by the anomalous transmission of the diffuse background. As seen from Fig. 2c, the DS intensity distribution forms then a narrow central peak

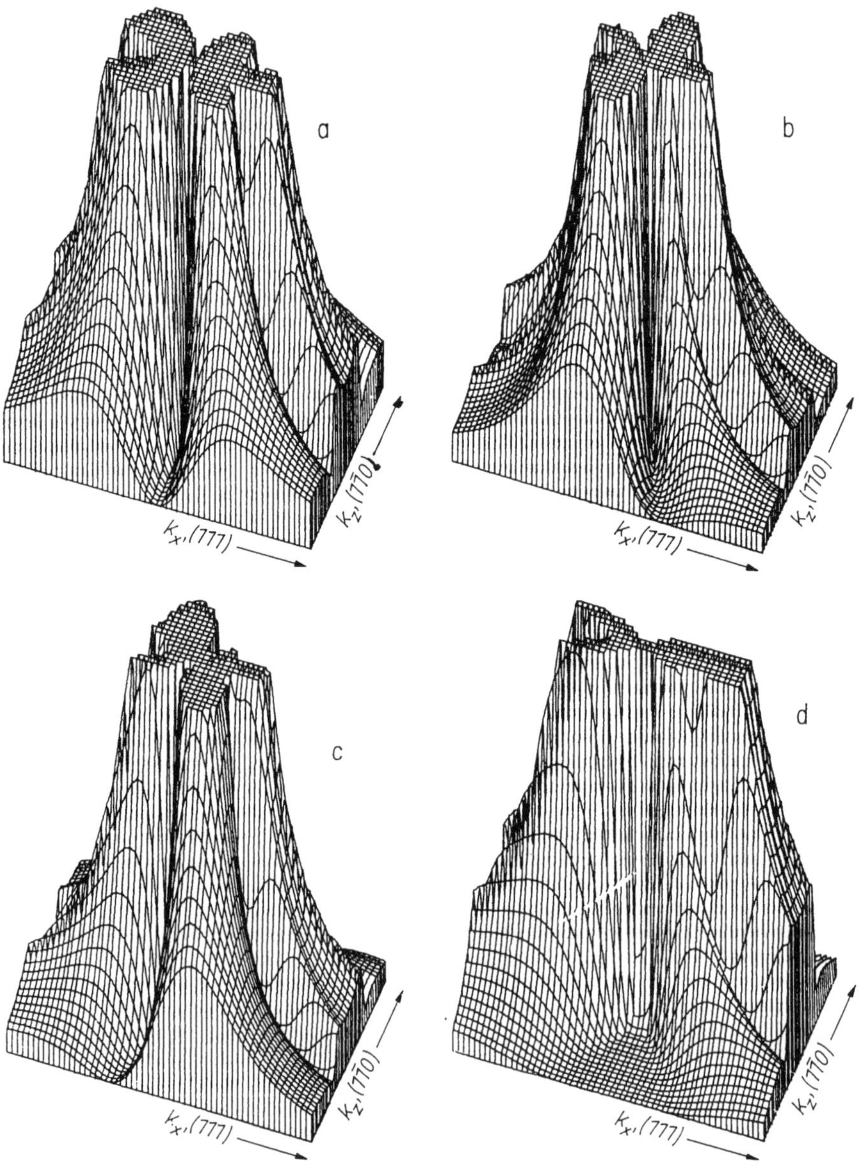


Fig. 4

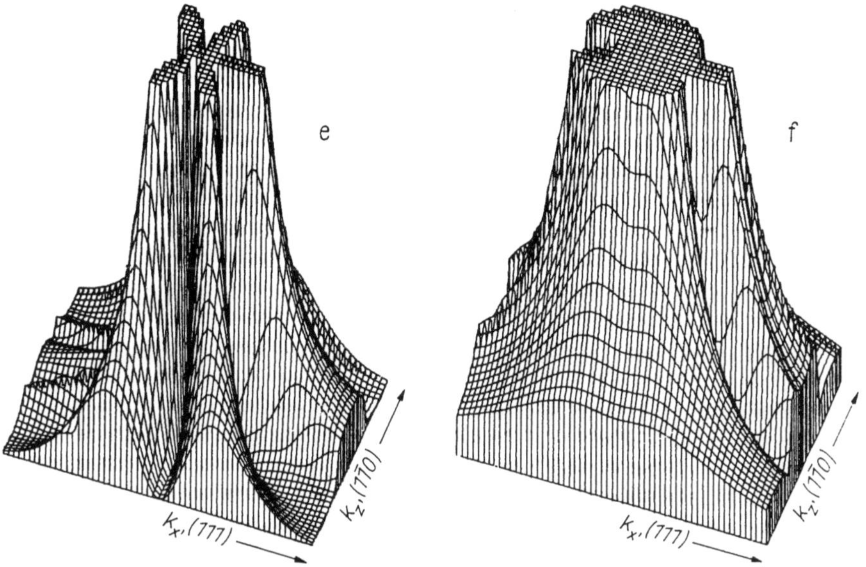


Fig. 4. Planar projections of three-dimensional images for the "diffuse" reflectivity of crystals containing small-radius prismatic dislocation loops at various orientations of the Burgers vector $\boldsymbol{b}$, the crystal thickness is $t = 10\Lambda$ ($\mu_0 l = 3.14$), the section level is $\varrho_c = \varrho_m/40$ (Ge (111), CuK$_\alpha$, $\boldsymbol{n} = [1\bar{1}0]$; $k_x^m = 120\Lambda^{-1}$, $k_z^m = 80\Lambda^{-1}$, $\Lambda = 8.7$ µm). a) $\boldsymbol{b} = (1/2)\,[110]$, $\varrho_m = 11$; b) (1/2) [101], 20; c) (1/2) [011], 20; d) (1/2) [0$\bar{1}$1], 0.5; e) (1/2) [$\bar{1}$10], 2.4; f) (1/2) ⟨110⟩, 6.8

with decreasing "mountain ranges" placed along the anomalous transmission lines. It should be noted here that in the thickness range $\Lambda \leqq t \leqq t_a$ for the defect type considered the intrinsic shape of isodiffuse contours is preserved, being analogous in some details to that in the kinematical theory.

## 4. Dislocation Loops in an Elastically Isotropic Crystal

In a general case of arbitrary mutual orientation of the Burgers vector $\boldsymbol{b}$ and the normal $\boldsymbol{n}_L$ to the loop plane the expression for the displacement field of the dislocation loop is very complicated even in the approximation of an elastically isotropic medium [20, 21]. Therefore, to investigate the characteristic features of the dynamical DS in crystals with randomly distributed small-radius dislocation loops we consider the simple case of prismatic dislocation loops ($\boldsymbol{b} \parallel \boldsymbol{n}_L$).

The Fourier component of the displacement field of the prismatic dislocation loop in the approximation of an elastically isotropic medium is given by [22]

$$\boldsymbol{u}(\boldsymbol{q}) = \frac{\pi b R^2}{v} \frac{1}{q} \left[ 2\boldsymbol{b}^0(\boldsymbol{b}^0 \cdot \boldsymbol{q}^0) + \frac{\nu}{1-\nu} \boldsymbol{q}^0 - \frac{1}{1-\nu} (\boldsymbol{b}^0 \cdot \boldsymbol{q}^0)^2 \boldsymbol{q}^0 \right], \tag{12}$$

where $\boldsymbol{q}^0 = \boldsymbol{q}/q$, $\boldsymbol{b}^0 = \boldsymbol{b}/b$, $\nu$ is the Poisson ratio, $R$ the loop radius. Expression (12) is valid when the condition $q \ll 1/R$ is fulfilled. In the present treatment, when the relation $q \ll 1/(R\sqrt{hb})$ is required (see above), this condition is certainly satisfied.

In Fig. 3 calculated results are shown for the "diffuse" reflectivity of a Ge single crystal with randomly distributed prismatic dislocation loops at fixed orientation of the Burgers vector $\boldsymbol{b}$. The shape of the kinematical isodiffuse contours (Fig. 3a) is uniquely determined by the square of the Fourier component of the loop displacement field (11). The shape of the dynamical isodiffuse contours is more complicated, as shown above. In the relevant Fig. 3b, c the absolute maxima of the "diffuse" reflectivity are given in relative units $\varrho_m = R_{DS}^{max}(\boldsymbol{k})/(cC^2 \times 10^5)$, where $c$ is the loop concentration and $C = \pi b R^2/v$. As seen, the shape of the dynamical isodiffuse contours is also very sensitive to the mutual orientation of the diffraction vector $\boldsymbol{h}$ and the Burgers vector $\boldsymbol{b}$ (see also Fig. 4). The magnitude of the DS intensity is maximum at $\boldsymbol{h} \parallel \boldsymbol{b}$ and minimum at $\boldsymbol{h} \perp \boldsymbol{b}$ (see $\varrho_m$ in Fig. 3b, c and Fig. 4). The sensitivity of the DS intensity magnitude to the mutual orientation of $\boldsymbol{b}$ and $\boldsymbol{h}$ corresponds to various degrees of reflection-plane distortion in the asymmetrical displacement field of dislocation loops.

The thickness dependence of the DS intensity distribution and of the isodiffuse contour shape is analogous to that described above for the cluster model. At the same time for crystals having intermediate thickness ($1 \leqq \mu_0 l \leqq 5$) or even a large one ($\mu_0 l > 5$) the isodiffuse contours retain in some cases the shape corresponding to the given loop orientation. They occur, when the anomalous transmission lines do not coincide with the maximum intensity directions of the kinematical DS, i.e. with those directions in the reciprocal lattice plane along which the Fourier components of the defect displacement field are maximum. In this case, however, the DS intensity is small because of the photoelectric absorption described by $\mu_0$ (there is no anomalous transmission along directions of the diffuse background maxima). On the contrary, when anomalous transmission lines coincide with the directions of kinematical DS maxima, the "diffuse" reflectivity is only slightly decreased due to the smallness of the interference absorption coefficient $\mu_i = \mu_0 - \mu_h \ll \mu_0$.

## 5. Summary

The DS intensity distribution for single crystals of thickness $t \gtrsim \Lambda$ with randomly distributed microdefects was considered in the near vicinity of a reciprocal lattice point, which corresponds to the Huang DS in the kinematical theory. To represent graphically the DS intensity distribution besides iso-intensity contours also planar projections of three-dimensional images are used providing more complete information about the distribution obtained.

The analysis carried out for the dynamical DS intensity as well as for the "diffuse" reflectivity of a single crystal with microdefects shows that during dynamical diffraction the kinematical picture of the DS intensity distribution is distorted because of the coherent rescattering of diffusely scattered beams. Nevertheless, the results obtained indicate that also in this case detailed information on the microdefect characteristics may be obtained by using DS intensity distribution patterns. In experimental investigations, by successfully using the anomalous transmission effect for the diffuse background and by appropriate choice of radiation wavelength and reflections investigated, one can avoid considerably the difficulties caused by the photoelectric absorption of diffusely scattered beams at Laue-case diffraction in real single crystals.

## References

[1] M. A. KRIVOGLAZ, Theory of X-ray and Thermal Neutron Scattering by Real Crystals, Plenum Press, New York 1969.

[2] B. C. LARSON and W. SCHMATZ, Phys. Rev. B **10**, 2307 (1974).

[3] A. IIDA and K. KOHRA, phys. stat. sol. (a) **51**, 533 (1979).

[4] A. M. AFANASEV, M. V. KOVALCHUK, E. F. LOBANOVICH, R. M. IMAMOV, P. A. ALEXANDROV, and M. K. MELKONYAN, Kristallografiya **26**, 28 (1981).

[5] E. K. Kovev, V. V. Ratnikov, and L. M. Sorokin, Fiz. tverd. Tela **23**, 1626 (1981).
[6] P. Zaumseil and U. Winter, phys. stat. sol. (a) **70**, 497 (1982).
[7] E. A. Tikhonova, Fiz. tverd. Tela **9**, 516 (1967).
[8] V. B. Molodkin, Metallofiz. **3**, No. 4, 27 (1981).
[9] V. B. Molodkin, S. I. Olikhovskii, and M. E. Osinovskii, Metallofiz. **5**, No. 1, 3 (1983).
[10] V. B. Molodkin, S. I. Olikhovskii, and M. E. Osinovskii, Metallofiz. **5**, No. 5, 3 (1983).
[11] V. B. Molodkin, S. I. Olikhovskii, and M. E. Osinovskii, Metallofiz. **5**, No. 6, 112 (1983).
[12] V. B. Molodkin, S. I. Olikhovskii, M. E. Osinovskii, V. V. Kochelab, A. Yu. Kazimirov, M. V. Kovalchuk, and F. H. Chukhovskii, Metallofiz. **6**, No. 3, 7 (1984).
[13] V. V. Ratnikov, E. K. Kovev, and L. M. Sorokin, Fiz. tverd. Tela **26**, 2155 (1984).
[14] R. N. Kyutt and V. V. Ratnikov, Metallofiz. **7**, No. 1, 36 (1985).
[15] V. M. Kaganer and V. L. Indenbom, Metallofiz. **8**, No. 1, 25 (1986).
[16] B. F. Zhuravlev, V. B. Molodkin, S. I. Olikhovskii, and M. E. Osinovskii, Metallofiz. **7**, No. 5, 24 (1985).
[17] B. F. Zhuravlev, V. V. Kochelab, V. B. Molodkin, S. I. Olikhovskii, and M. E. Osinovskii, Preprint 17.85, Institute Met. Phys., Academy of Sciences of the Ukrainian SSR, Kiev 1985 (in Russian).
[18] C. Teodosiu, Elastic Models of Crystal Defects, Springer-Verlag, Berlin/Heidelberg/New York 1982.
[19] Z. G. Pinsker, Dynamical Scattering of X-rays in Crystals, Springer-Verlag, Berlin/Heidelberg/New York 1978.
[20] M. A. Krivoglaz and K. P. Ryaboshapka, Fiz. Metallov i Metallovedenie **16**, 641 (1963).
[21] R. I. Barabash and M. A. Krivoglaz, Metallofiz. **4**, No. 3, 3 (1982).
[22] P. H. Dederichs, Phys. Rev. B **4**, 1041 (1971).

*(Received December 17, 1987)*

phys. stat. sol. (a) **108**, 81 (1988)

Subject classification: 61.50; 66.30; 72.80; S5.11

*Department of Physics, Nanjing University*[1])

# Effect of the Structure of Laser Crystallized Silicon-on-Insulator on Impurity Diffusion[2])

By

QING GU and XI-MAO BAO

The laser crystallization mechanism of a-Si:H and the diffusion behaviour of implanted impurities in the crystallized film are studied. A new region liquid-phase laser crystallization (LP-LCR), is found at low scan velocity. To investigate the diffusion mechanism, the regions LP-LCR, OD and FCR-2 are implanted with $As^+$ in a dose of $5 \times 10^{15}$ ions/cm$^2$ at 80 keV followed by rapid thermal annealing (RTA). The spreading-resistance measurement and secondary ion mass spectroscopy (SIMS) show that the diffusions in grains and along grain boundaries occur in all of the three regions. There is another diffusion way, diffusion along defects, in OD. It is the fastest among the three diffusion ways.

Der Laserkristallisationsmechanismus von a-Si:H und die Diffusionseigenschaften der implantierten Störstellen in der kristallisierten Schicht werden untersucht. Eine neue Kristallisationszone, Flüssigphasen-Kristallisation (LP-LCR), wird bei niedriger Laserabtastgeschwindigkeit gefunden. Zur Untersuchung des Diffusionmechanismus der Störstellen werden alle drei Zonen: LP-LCR, OD und FCR-2 mit 80 keV-$As^+$ bei Dosen von $5 \times 10^{15}$ Ionen/cm$^2$ implantiert und anschließend durch RTA (rapid thermal annealing) ausgeheilt. Messungen des Ausbreitungswiderstandes und SIMS-Analysen zeigen, daß in allen drei Zonen Diffusion in den Körnern und entlang der Korngrenzen auftreten kann. Außerdem tritt in OD noch Diffusion entlang Defekten auf, die am schnellsten unter den drei Diffusionen ist.

## 1. Introduction

The wide-ranging developments in the area of poly-Si SOI (silicon on insulator) structures have offered the prospect of important advances in semiconductor devices. Some current research is focused on the crystallization of hydrogenated amorphous silicon SOI. The remaining hydrogen content in SOI after laser crystallization is very helpful for passivation of the grain boundaries and defects, and consequently for an improvement on the electrical properties of SOI [1, 2]. In some experiments only three crystallization zones can be found when the scan velocity is less than 30 cm/s [3]. They are the optical damage zone (OD) and the furnace-like crystallization zones (FCR-1 and FCR-2). In this work a new crystallization zone, liquid-phase laser crystallization (LP-LCR), has been found. The crystallization mechanisms of the four regions are investigated. Arsenic implantation is carried out on laser crystallized SOI. The behaviour of the dopants during rapid thermal annealing is studied by using spreading-resistance measurement and SIMS. Some explanations of the results are given combined with the structure analysis.

---

[1]) Nanjing, People's Republic of China.

[2]) This work was funded by grants from Ion Beam Laboratory of Shanghai Institute of Metallurgy, Academia Sinica.

## 2. Results and Discussion

a-Si : H films of 0.6 μm thick were prepared at 250 °C on fused quartz by glow discharge. The films were crystallized by a cw $Ar^+$ laser. With controlled laser power $P$, scan velocity $V$, energy distribution in laser spot, and overlap, LP-LCR, OD, FCR-2, and FCR-1 can be obtained individually.

### *2.1 Structure of laser crystallized SOI*

Fig. 1 shows the scan trace of the laser. Four crystallized regions are visible. LP-LCR which lies in the middle of the trace has a chevron structure. OD, lying adjacently to LP-LCR, has many projecting points on the surface. FCR-2 and FCR-1 have smooth surfaces and different colours. FCR-2 is the most narrow one among the four regions.

When the local temperature reaches the melting point of crystalline silicon, the regular liquid-phase crystallization takes place. Fig. 2a and b are the planar and cross-sectional TEM micrographs of LP-LCR, respectively. They show that the film of LP-LCR consists of large crystalline grains greater than 40 μm in size and has a well ordered crystalline structure. The fine interface indicates good adhesion of silicon on quartz. In the films a few grain boundaries exist and penetrate straightly the whole thickness of the crystallized layer.

The zone where the local temperature exceeds the melting point of a-Si but is still below regular melting point of crystalline silicon is in a metastable supercooled state, hence fast crystallization will inevitably occur. In this case the grain contains a lot of defects and the grain size cannot be large, about several μm, as shown in Fig. 3. OD has random grain orientation and a rough surface. The defects are large and penetrate the film as the grain boundaries.

The planar and cross-sectional TEM micrographs of FCR-2 are given in Fig. 4a and b. They show that FCR-2 is a well crystallized poly-Si region with small grain size (about 100 nm), smooth surface, and few defects. The film consists of many grains, therefore it has zigzag grain boundaries. FCR-2 is a solid-phase crystallized region.

Electron diffraction indicates that FCR-1 is a mixture of small-grain polycrystalline Si and a-Si. According to the structural features, it must be some kind of microcrystal.

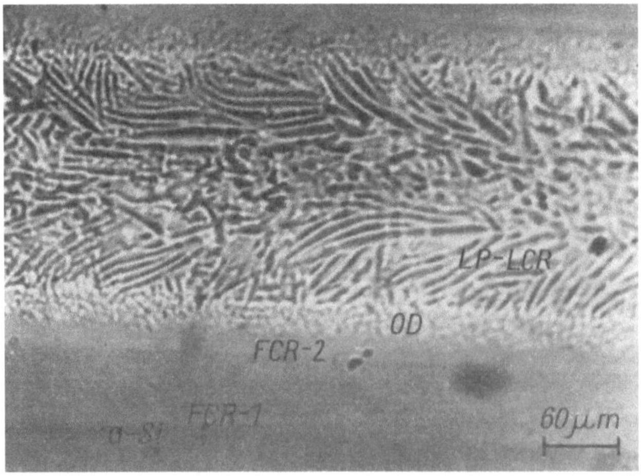


Fig. 1. Transmission optical micrograph of the laser scan line on a-Si : H SOI

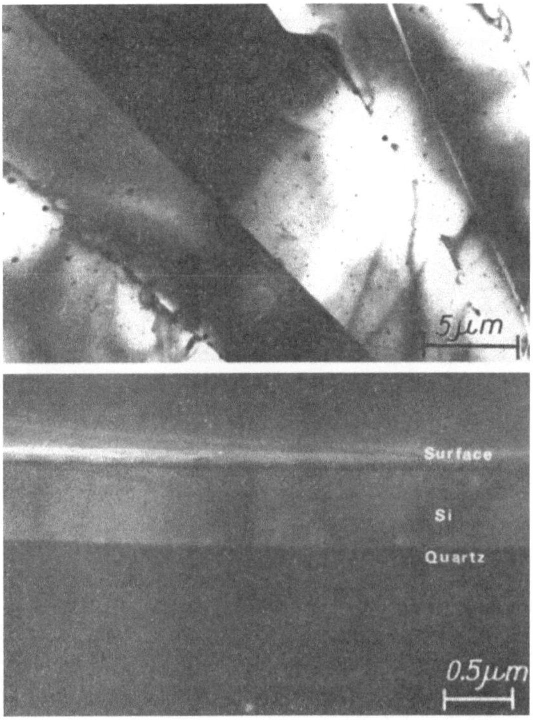


Fig. 2. Transmission electron micrographs of LP-LCR; a) planar TEM; b) cross-sectional TEM

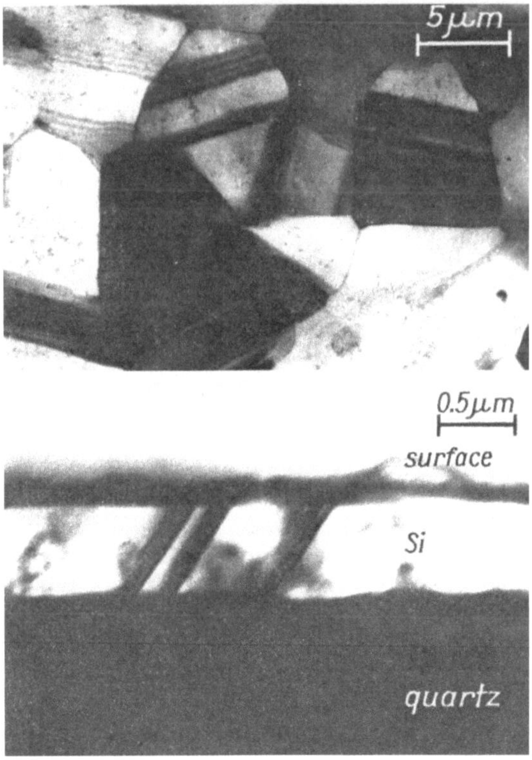


Fig. 3. Transmission electron micrographs of OD; a) planar TEM; b) cross-sectional TEM

a

b

Fig. 4. Transmission electron micrographs of FCR-2; a) planar TEM; b) cross-sectional TEM

### *2.2 The diffusion behaviour of arsenic implanted in SOI*

LP-LCR, OD, and FCR-2 were implanted with $As^+$ in a dose of $5 \times 10^{15}$ ions/cm$^2$ at 80 keV. Rapid thermal annealing (RTA) was then carried out at temperature 1150 °C in nitrogen ambient. The diffusion behaviour of implanted impurities was investigated by spreading-resistance analysis and SIMS.

The conductivity profiles obtained by spreading-resistance measurement for different annealing times are shown in Fig. 5. It can be seen that a shoulder appears in LP-LCR after 5 s RTA. As annealing time increases, it moves towards the inside of the film and its amplitude grows. When the annealing time lasts for 15 s, the shoulder reaches the interface and a notable enhancement of the amplitude occurs. A peak appears as annealing time increases continuously. Two shoulders are seen in OD after 5 s RTA, both of them move inside the film as annealing time increases. The first shoulder lying near the surface is similar to that in LP-LCR, namely, its amplitude grows with increasing annealing time. In contrast, the second one does not change obviously in amplitude, but gradually moves towards the interface with annealing time and disappears after 35 s. No shoulder appears in FCR-2. The conductivity of FCR-2 increases slowly in the upper layer of the film and drops sharply at a certain depth. Although a marked displacement in the profile can be seen as annealing time increases, it is slower than those in the previously described regions. For example, 35 s RTA is not long enough for arsenic to reach the interface. It is reasonable that the conductivity of FCR-2 is low owing to the presence of more grain boundaries.

To observe the extent of dopant redistribution resulting from RTA, SIMS measurements were made. As shown in Fig. 6, an obvious displacement of the impurity profile

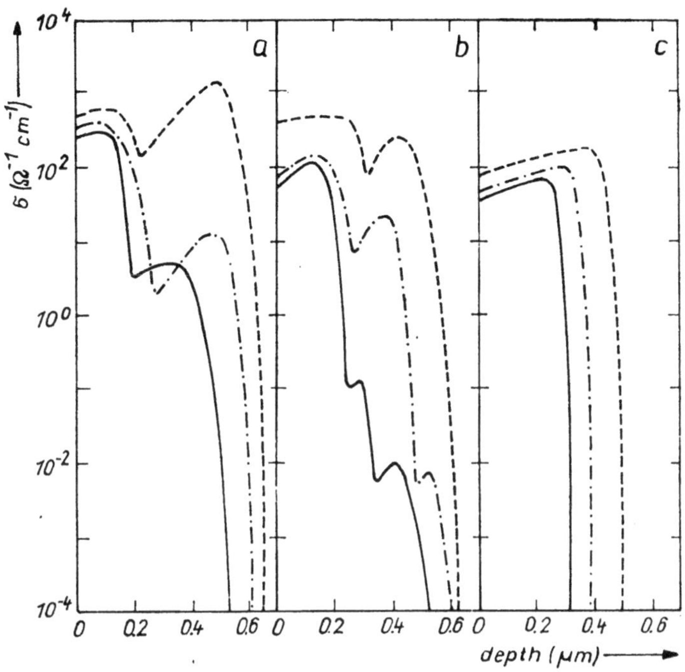


Fig. 5. Development of the conductivity ($\sigma$) profiles of three crystallization zones: a) LP-LCR, b) OD, c) FCR-2 with rapid thermal annealing time at 1150 °C. Conductivities were measured by spreading-resistance analysis, ——— 5, —·—· 15, – – – 35 s

is observed in all three crystallization regions for 15 s annealing. The displacement rapidly increases with increasing annealing time, as illustrated by the curves for LP-LCR. It should be pointed out that an accumulation of impurities near the interface for 15 s annealing has been measured in both LP-LCR and OD, but not in FCR-2. As seen from the curves of LP-LCR, the longer the annealing time is, the more impurities accumulate. These results excellently correspond with the measurements by the spreading-resistance analysis.

The redistribution of dopants in SOI is structure related. Grain boundaries play an important role in this case. On the one hand, As diffuses into grains, like in single-crystal silicon. On the other hand, they diffuse rapidly along grain boundaries. Arsenic coming into grain boundaries simultaneously diffuses into grains. When the diffusion front reaches the interface, the dopants begin to accumulate, then they diffuse laterally, and go back taking the arsenic accumulated at the interface as source. As reported by some researchers, the diffusivity in the grain boundary is two to three orders of magnitude larger than that in single-crystal silicon [4, 5]. It is believed that the shoulder in the conductivity of LP-LCR is caused by fast diffusion along grain boundaries [6]. As shown in Fig. 5, the shoulder extends to the interface and then rapidly increases with increasing annealing time. These observations can be attributed to the accumulation of impurities at the interface and diffusing from here into the film. In OD, the first shoulder is also due to the fast diffusion along grain boundaries. As for the second one, it may reasonably be suggested from the diffusion along defects. However, the conductivity due to defects is low because its number is much less than that of the grain boundary. Therefore, the second shoulder will be merged by the first one when impurities accumulate near the interface.

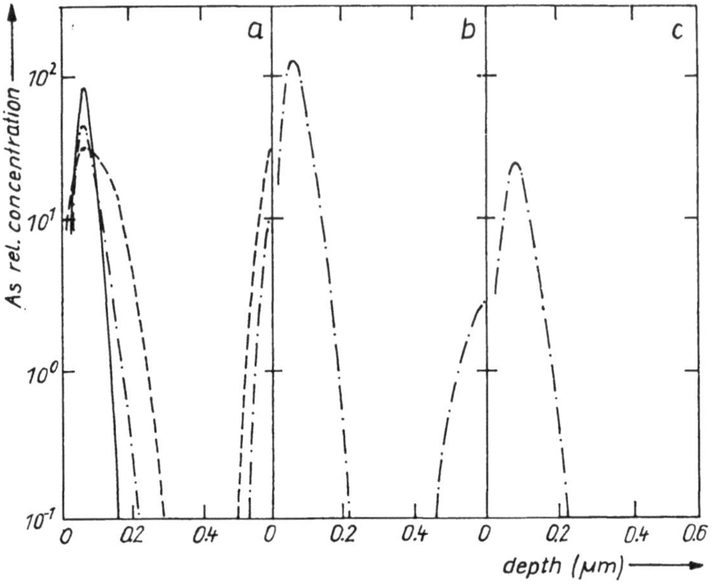


Fig. 6. Arsenic profiles of three crystallization zones: a) LP-LCR, b) OD, c) FCR-2 with 15 s rapid thermal annealing time at 1150 °C. The profiles were measured by SIMS. The arsenic profiles of unannealed and 35 s RTA of LP-LCR were also shown. ——— Unannealed, ·—·—· 15 s, — — — 35 s

There are many reports on the diffusion along the grain boundary and along the defect. However, either of the two diffusion ways was observed separately, so that the direct comparison between the two ways may be difficult. Our experiment is a direct way to investigate these diffusions simultaneously. In the given case we conclude that diffusion along a defect is faster than along a grain boundary.

FCR-2 is a layer piled up of small grains and essentially free of large defects. The dopants are easily distributed in grains for short annealing times. Hence no shoulder is formed in the conductivity profile. Diffusion along grain boundaries just increases the diffusivity.

## 3. Conclusions

Four crystallization regions LP-LCR, OD, FCR-2, and FCR-1 for a-Si:H have been observed. LP-LCR is a liquid-phase crystallized region with large grain size (about 40 µm) and few defects. OD is a supercooling liquid-phase crystallized region with lots of defects. Its grain size is about several µm. FCR-2 and FCR-1 are solid-phase crystallized regions with small grain size. There are two diffusion mechanisms in LP-LCR, diffusion in grains and diffusion along grain boundaries. The latter is much faster than the former. The dopants can accumulate near the interface and diffuse back into the film. In OD, three diffusion mechanisms are observed, the diffusion in grains, along grain boundaries, and along defects. In our experiment, diffusion along defects is the fastest among the three mechanisms. This is the first time to compare these three diffusions directly. In FCR-2, arsenic diffusion along grain boundaries just increases the effective diffusivity.

## References

[1] C. H. SEAGER and D. S. GINLEY, Appl. Phys. Letters **34**, 337 (1979).
[2] J. L. BENTON, C. J. DOHERTY, S. D. FERRIS, D. L. FLAMM, L. C. KIMERLING, and H. J. LEAMY, Appl. Phys. Letters **36**, 670 (1980).
[3] G. AUVERT, D. BENSAHEL, A. PERIO, F. MORIN, G. A. ROZGONYI, and V. T. NGUYEN, Laser and Electron-Beam Interactions with Solids, Ed. B. R. APPLETON and G. K. CELLER, North-Holland Publ. Co., 1982 (p. 535).
[4] B. SWANINATHAN, K. C. SARASWAT, R. W. DUTTON, and T. I. KAMINS, Appl. Phys. Letters **40**, 795 (1982).
[5] H. BAUMGART, H. J. LEAMY, L. E. TRIMBLE, C. J. DOHERTY, and G. K. CELLER, Grain Boundaries in Semiconductors, Ed. H. J. LEAMY, G. E. PIKE, and C. H. SEAGER, North-Holland Publ. Co., 1982 (p. 311).
[6] LIN CHANGLU and TSOU SHIHCHANG, Nuclear Instrum. Methods Phys. Res. **B21**, 627 (1987).

*(Received December 28, 1987)*

phys. stat. sol. (a) 108, 89 (1988)

Subject classification: 61.10; 64.70; S8

*Institute for High Pressure Physics, Troitsk*[1])

# X-Ray Studies of Thallium Chalcogenides TlS and TlSe up to 37 GPa

By

G. B. DEMISHEV, S. S. KABALKINA, and T. N. KOLOBYANINA

The crystal structures of thallium sulfide and thallium selenide in the pressure range of 0 to 37 GPa are studied. Three first order phase transitions are found in TlS: TlS I (TlSe-type) → TlS II ($\alpha$-$NaFeO_2$-type) → TlS III (distorted $\alpha$-$NaFe_2O$-type) → TlS IV (CsCl-type) at $p = 5$, 10, and 25 GPa, respectively. Phase TlS II is metastable under normal conditions and reveals semiconductor properties. A first order transition to the CsCl-type is found in TlSe at 21 GPa. The equations of state for TlS and TlSe are determined.

Изучена кристаллическая структура сульфида и селенида таллия в диапазоне давлений 0 до 36 GPa. В TlS обнаружены три фазовых перехода первого рода: TlS I (тип TlSe) → TlS II (тип $\alpha$-$NaFeO_2$) → TlS III (искаженный тип $\alpha$-$NaFeO_2$) → → TlS IV (тип CsCl) при $p = 5$, 10 и 25 GPa соответственно. Фаза TlS II метастабильна в нормальных условиях и обладает полупроводниковыми свойствами. В TlSe при 21 GPa обнаружен переход первого рода в тип CsCl. Определены уравнения состояния TlS и TlSe.

## 1. Introduction

Recently much attention has been paid to studies of optical and other physical properties of $A^{III}B^{VI}$ semiconductors with TlSe-type structure [1]. Vibrational properties of crystals of this type, infrared reflection spectra (IR), Raman spectra (RS) [2 to 4], and the effect of hydrostatic pressure on RS of TlS up to 1 GPa [5] have been studied. The thermal expansion coefficients of TlS and TlSe in the range of 78 to 350 K [6] as well as the heat capacity of TlS [7] have been determined. The phase diagrams for TlS and TlSe are not available.

The aim of the present paper is to study the stability of the TlSe structure type under high pressure (HP), the possibility of a semiconductor-to-metal transition in TlS and TlSe, as well as to determine the equation of state.

The tetragonal structure of the TlSe type (space group $D_{4h}^{18}$-I 4/mcm) is built up of chains of $TlSe_2$, parallel to [001], where the $Tl^{3+}$ ions are surrounded by four $Se^{2-}$ ions, $Tl^+$ ions are in the centres of tetragonal antiprisms at a distance of $c/2$ from each other. The TlSe formula may be presented as $Tl^+ (Tl^{3+}Se_2^{2-})^-$. The interatomic distances are in a good agreement with an ionic-covalent bonding character [8].

## 2. Experimental

In the present work the structures of TlS and TlSe were studied by means of X-ray diffraction under pressure up to 37 GPa. Polycrystalline samples were used with the cell parameters: TlS $a = 0.777$ nm, $c = 0.679$ nm; TlSe $a = 0.802$ nm, $c = 0.70$ nm.

[1]) SU-142092 Troitsk, Moscow Region, USSR.

The experiments were carried out in a diamond anvil cell (DAC) [9]. The pressure was measured by the fluorescence shift of the ruby $R_1$-line. A mixture of ethanol–methanol in the ratio of 1:4 was added to the TlS (TlSe) sample to get a hydrostatic pressure up to 10 GPa and to diminish the X-ray absorption of the sample. Several experiments were carried out in an HP "toroid" chamber [10] at $p = 10$ GPa and $T = 300$ °C, the tempering time was 1 to 3 h. The "quenched" samples were analyzed by X-ray method under normal conditions.

## 3. Results

### 3.1 Thallium sulfide

If pressure up to 36 GPa is applied, TlS reveals three structural transitions occurring in the sequence: TlSe-type (TlS I) → α-$NaFeO_2$-type (TlS II) → distorted α-$NaFeO_2$-type (TlS III) → CsCl-type (TlS IV); if the pressure is lowered gradually the transition sequence is inverted. The diffraction pattern of TlS under normal conditions (after unloading the cell) corresponds to a phase mixture of I + II. The same picture was observed for samples obtained by quenching (on compressed samples of TlS at $p \approx \approx 10$ GPa and $T = 300$ °C) in the "toroid" cell. The content of phase II in the mixture I + II in the quenched samples depends on the tempering time of TlS under high pressure and temperature: after 3 h tempering the samples contain more than 80% and after 1 h about 40% of that phase. After pressure of 30 GPa in DAC at room temperature was applied the TlS samples contain no more than 40% of phase II. Under normal conditions TlS II is preserved during two weeks, gradually converting into TlS I. Thus, the I → II transition is of diffusion type. The diffraction pattern of TlS II is indexed using α-$NaFeO_2$ as a model structure; the cell parameters under normal conditions are: $a = 0.3945$ nm, $c = 2.1788$ nm, $Z = 6$; space group $D_{3d}^5$-R$\bar{3}$m

Table 1

X-ray data for TlS II (α-$NaFeO_2$ type, space group $D_{3d}^5$-R$\bar{3}$m, $a = 0.3945$ nm, $c = = 2.1788$ nm, $c/a = 5.52$, $Z = 6$)

| h k l | $d_{obs}$ ($10^{-1}$ nm) | $d_{cal}$ ($10^{-1}$ nm) | $I_{obs}$*) |
|---|---|---|---|
| 0 0 6 | 3.638 | 3.631 | m |
| 0 1 2 | 3.254 | 3.258 | s |
| 1 0 4 | 2.891 | 2.894 | vs |
| 0 1 8 | 2.133 | 2.130 | s |
| 1 1 0 | 1.971 | 1.972 | m |
| 1 0 10 | 1.842 | 1.837 | w |
| 1 1 6 | 1.735 | 1.733 | w |
| 2 0 2 | 1.687 | 1.688 | w |
| 0 2 4 | 1.635 | 1.630 | w |
| 2 0 8 | 1.455 | 1.447 | w |
| 0 2 10 | 1.344 | 1.344 | vw |
| 1 1 12 | 1.335 | 1.336 | vw |
| 1 0 16<br>2 1 4 | 1.259 | 1.265<br>1.257 | vw |
| 1 2 8 | 1.175 | 1.167 | vw |
| 2 1 10 | 1.110 | 1.111 | vw |

*) vw very weak, w weak, m middle, s strong, vs very strong.

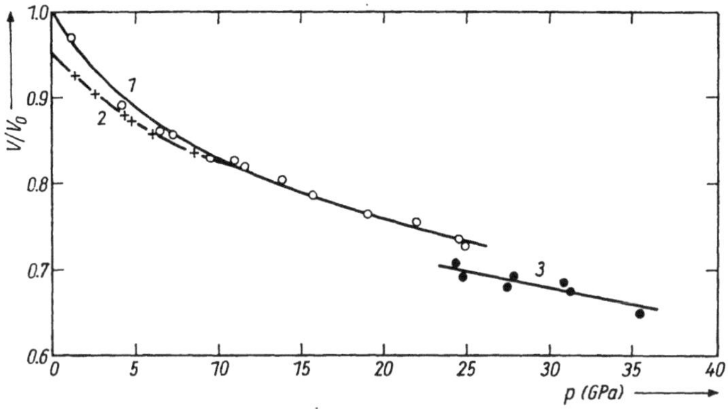


Fig. 1. The pressure dependence of the relative volume $V/V_0(p)$ for TlS: (1) TlS I, (2) TlS II, (3) TlS IV

(Table 1); the volume ratio $V_{II}/V_I = 0.955$. The intensity distribution on the TlS II X ray patterns agrees with the presumed structure type, which is an ordered variant of NaCl-type structure [11].

The experimental dependences $V/V_0(p)$ for phases I and II were fitted to the Mournagan-Birch equation (Fig. 1, solid lines); the corresponding bulk moduli and their pressure derivatives are given in Table 2. The volume reduction $\Delta V/V$ at the I → II transition (4.5% under normal conditions) is gradually decreased to zero; at $p = 10$ GPa the $V/V_0(p)$ dependences corresponding to phases I and II intersect. X-ray data for TlS obtained in DAC unambiguously prove the existence of a reversible transition II → III, the pressure of the direct transition II → III is 10 GPa, that of the III → II transition is 5.5 GPa. A comparison of the X-ray patterns of phases II and III implies the structure of distorted $\alpha$-$NaFeO_2$ type for TlS III. TlS IV appears near 25 GPa and exists mixed with phases I + III in the pressure range of 25 to 30 GPa. The volume jump at the transition I → IV is $\Delta V/V = 4\%$ (30 GPa); at 35.5 GPa the X-ray patterns correspond to the pure phase IV with structure of CsCl type: $a = 0.3202 \pm 0.0005$ nm; the total volume change is $V/V_0 = 0.64$ (Table 3). A decrease in pressure to 24 GPa leads to a weakening of the diffraction pattern of

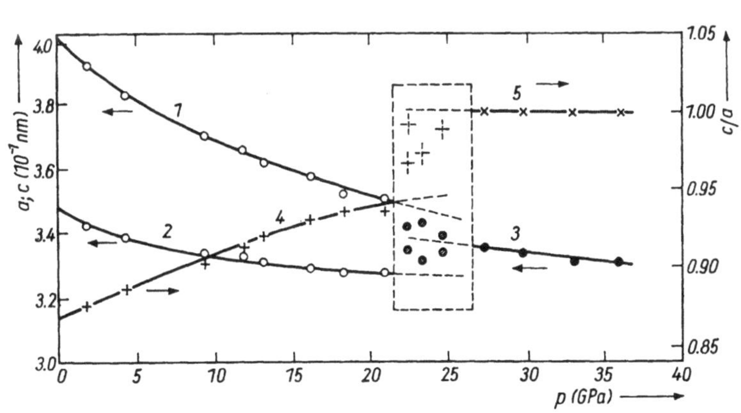


Fig. 2. The pressure dependence of cell parameters and $c/a$ ratio for tetragonal TlSe I and cubic TlSe II phases: (1) $a_t/2$, (2) $c_t/2$, (3) $a_{cub}$, (4) $c_t/a_t$, (5) $c_{cub}/a_{cub}$. The two-phase area is shown by the dashed lines

Table 2
Bulk moduli $B_0$ and its pressure derivatives $B_0'$ for TlS and TlSe

| compound | structure type | $p$ (GPa) | $B_0$ (GPa) | $B_0'$ |
|---|---|---|---|---|
| TlS | TlSe | 0 | 29 | 6.7 |
| TlS | $\alpha$-$NaFeO_2$ | 0 | 47.5 | 4.5 |
| TlSe | TlSe | 0 | 26.8 | 4.7 |

Table 3
X-ray data for CsCl-type modifications: TlS IV ($p = 35.5$ GPa, $a = 0.3202 \pm 0.0004$ nm) and TlSe II ($p = 35.7$ GPa, $a = 0.3310 \pm 0.0004$ nm)

| $h\ k\ l$ | TlS IV | | | TlSe II | | |
|---|---|---|---|---|---|---|
| | $I_{obs}$ | $d_{obs}$ ($10^{-1}$ nm) | $d_{cal}$ ($10^{-1}$ nm) | $I_{obs}$ | $d_{obs}$ ($10^{-1}$ nm) | $d_{cal}$ ($10^{-1}$ nm) |
| 1 0 0 | m | 3.201 | 3.197 | m | 3.316 | 3.310 |
| 1 1 0 | s | 2.269 | 2.264 | vs | 2.338 | 2.340 |
| 1 1 1 | w | 1.844 | 1.845 | vw | 1.907 | 1.911 |
| 2 0 0 | w | 1.590 | 1.598 | w | 1.657 | 1.655 |
| 2 1 0 | w | 1.432 | 1.429 | | | |

phase IV and the appearance of weak reflections corresponding to a mixture of I + III; at $p = 22 \pm 1$ GPa TlS IV disappears; the composition of the mixture I + III depends on the unloading procedure: a fast unloading favours the appearance and the growth of phase I.

### 3.2 *Thallium selenide*

The X-ray study revealed a certain regularity in the change of the diffraction pattern upon compression of thallium selenide to 35.7 GPa. The doublet lines (202) and (220) gradually come closer with increasing pressure: at 20.9 GPa there exists a sharp

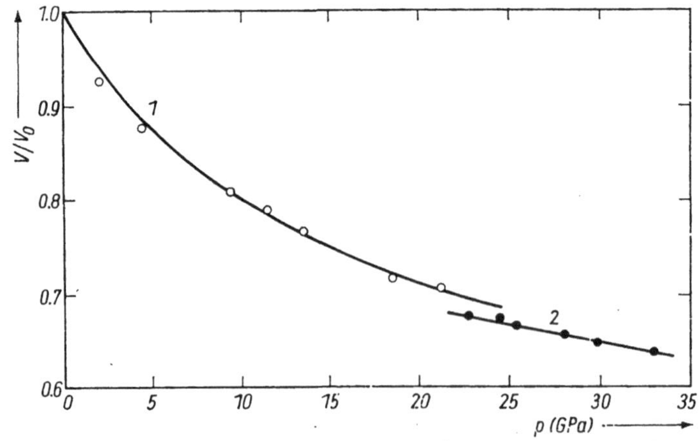


Fig. 3. The pressure dependence of the relative volume $V/V_0$ for TlSe: (1) TlSe I, (2) TlSe II

Table 4

Cell parameters for TlSe type and α-$NaFeO_3$ type for TlS, TlSe, $KInS_2$, and $TlInS_2$ (normal conditions)

| compound | $a$ ($10^{-1}$ nm) | $c$ ($10^{-1}$ nm) | $c/a$ | ref. |
|---|---|---|---|---|
| TlSe-type | | | | |
| TlS I | 7.77 | 6.79 | 0.87 | [8] |
| TlSe I | 8.02 | 7.00 | 0.87 | [8] |
| $KInS_2$ II | 7.769 | 6.672 | 0.86 | [12] |
| α-$NaFeO_2$-type | | | | |
| TlS II | 3.945 | 21.788 | 5.52 | this work |
| $KInS_2$ III | 3.875 | 21.794 | 5.62 | [12] |
| $TlInS_2$ II | 3.83 | 22.23 | 5.80 | [13] |

doublet, at 22 to 23 GPa there is a broad line, and at 27, 33, and 36 GPa a singlet is observed. Changes in cell parameters $a/2$, $c/2$, and $c/a$ are shown in Fig. 2, in Fig. 3 the pressure dependence of $V/V_0(p)$ in thallium selenide is seen (up to 20 GPa the $V/V_0$ dependence is approximated by the Mournagan-Birch equation, the bulk modulus $B_0$ and its pressure derivative $B_0'$ are given in Table 2). In Table 4 the diffraction data of TlSe corresponding to the CsCl type ($p = 35.7$ GPa) are given (a very weak superstructure line near (110) is seen on the X-ray patterns). Thus, the compression of thallium selenide results in the structural transition TlSe-type → CsCl-type (TlSeI → TlSeII) near 27 GPa followed by a transformation of the tetragonal cell to the cubic one ($a_{T/2} = c_{T/2} = a_{cub}$) and by a shift of the selenium atoms from the position with $x = 0.18$, $y = 0.68$, $z = 0$ to that with $x = 0.25$, $y = 0.75$, $z = 0$ (considered in the TlSe-type structure).

## 4. Discussion

Using the pressure dependences of Raman frequencies in TlS [5] and isothermal compressibility of TlS from X-ray experiments (Table 2), one can estimate the values

$$\Gamma_j = \frac{1}{\beta \nu_j} \left( \frac{\partial \nu_j}{\partial p} \right)_T ,$$

where $\Gamma_j$ is the Grüneisen parameter for the $j$-th phonon with frequency $\nu_j$, $\beta = 1/B_0$ is the isothermal compressibility. A set of mode-Grüneisen parameters gives information about the character of the binding between atoms in crystals: in covalent $A^{IV}$-type crystals $\Gamma_j \approx 1$, in ionic-covalent $A^{III}B^{V}$ and $A^{II}B^{VI}$ crystals $\Gamma_j \approx 2$, in alkali halides of $A^{I}B^{VII}$ type $\Gamma_j \approx 3$ [11]. According to an estimation made for thallium sulfide $\Gamma_j$ values ranging from 0.3 to 1.3 correspond to the interchain $(Tl^{3+}S_2^{2-})^-$ atomic vibrations, while $\Gamma_j \approx 3.2$ corresponds to an interaction between the positive $Tl^+$ ions and negatively charged chains. Thus, $\Gamma_j$ values conform with the covalent character of bonding between the atoms in the chains as well as the ionic bonding between these chains and $Tl^+$ ions; the ionic covalent bond character must be retained in the stability area of the TlSe-type phase.

Upon phase transitions TlSe-type → α-$NaFeO_2$-type (TlS) and TlSe-type → CsCl-type (TlS, TlSe) the covalent chains are destroyed. The structure of α-$NaFeO_2$ is an ordered variant of the NaCl structural type. The transition TlSe-type → α-$NaFeO_2$-type under HP was found in a quenched HP phase of the ternary compound $KInS_2$

[12]. The $\alpha$-$NaFeO_2$-type phase was obtained from the monoclinic phase of $TlInS_2$ by the same procedure [12]. Cell parameters for the modifications of TlSe type and those of $\alpha$-$NaFeO_2$ type for $KInS_2$, $TlInS_2$ and TlS are given in Table 4. The interatomic distances K–S, Tl–S, and In–S in $KInS_2$ III and $TlInS_2$ II ($\alpha$-$NaFeO_2$-type) correspond to the ionic bonding character in these crystals [12, 13]. Thus, the bonding character in $KInS_2$ changes at the transition TlSe type → $\alpha$-$NaFeO_2$ type. Since one-valent and three-valent thallium ions in TlS II (as well as in TlSe type) occupy different crystallographic positions, the charge transfer from $Tl^+$ to $Tl^{3+}$ is structurally forbidden, hence the TlS II phase has semiconductor properties. According to our estimation made by means of optical absorption measurements, the width of the forbidden zone in TlS II is $E_g = (1.0 \pm 0.1)$ eV under normal conditions.

The phase TlS I exists as a metastable state in the pressure range from 5 to 30 GPa. An analogous phenomenon was observed earlier for InTe [14] and was explained as a result of a very small change in the volume at the I → II transition which cannot provide shear strains needed to give rise to a plastic deformation favouring the completion of a phase transition. It might be assumed that an essential area of the metastable phase of TlS I is the result of a small volume change occurring at the I → II and I → III transitions.

The structural transition TlSe type → CsCl type in thallium selenide is a first-order transition with a volume jump $\Delta V/V = 2$ to 2.5% (Fig. 3). In the pressure range of 21 to 27 GPa $c/a$ increases from 0.935 to 1.0, the distances between Se and $Tl^{3+}$ atoms and those between Se and $Tl^+$ atoms become equal to 0.29 nm at 27.2 GPa and, hence, the bond character is changed. The range of 21 to 27 GPa corresponds to the two-phase area I + II, where broad diffraction lines appear in the X-ray patterns as a result of overlap of similar diffraction patterns. Calculated intermediate values of $c/a$ amount from 0.935 to 1.0 (Fig. 2, the two-phase area is shown by the dashed line). A similar structural transition TlSe-type → CsCl-type is observed in TlS in the metastable area of phase I ($p = 24$ GPa), it is followed by the volume jump $\Delta V/V = 4\%$. The structural transformation is analogous to that in TlSe (Fig. 4).

As follows from the X-ray studies of TlS and TlSe under HP the structure of TlSe type is stable in a limited pressure range. The shortening of the distance between the neighbouring ions $Tl^+$ to $r = c/2 \leq 0.330$ to 0.328 nm leads to an increase in the electrostatic repulsion which results in the instability of TlS I and TlSe I phases at $p > 5$ GPa and $p > 21$ GPa, respectively. The densest modifications of thallium chalcogenides under consideration are those of the CsCl type. One might assume metal-

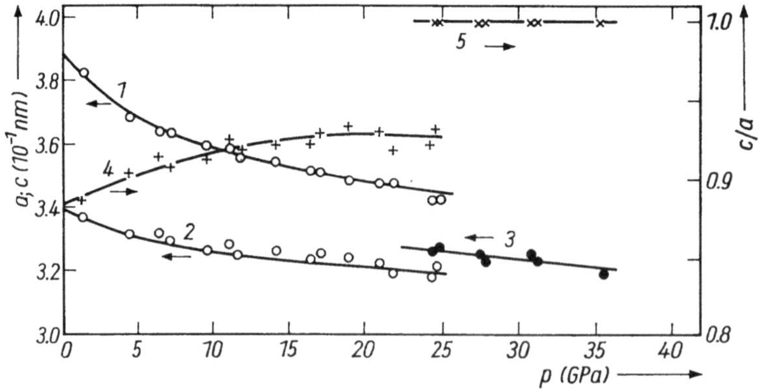


Fig. 4. The pressure dependence of cell parameters and $c/a$ ratio for tetragonal TlS I and cubic TlSe IV phases: (1) $a_t/2$, (2) $c_t/2$, (3) $a_{cub}$, (4) $c_t/a_t$, (5) $c_{cub}/a_{cub}$

lic properties for TlS IV and TlSe II since the charge transfer from $Tl^+$ to $Tl^{3+}$ is structurally allowed in CsCl type. To verify this hypothesis additional studies of electric resistivity and of other physical properties of TlSe in the pressure range up to 30 GPa are needed.

### *Acknowledgements*

The authors thank L. G. Khovostantsev and O. B. Ciok for their help in high pressure synthesis and A. N. Stepanov for the assistance in the optical measurements.

### References

[1] L. I. Man, R. M. Imamov, and S. A. Semiletov, Kristallografiya **21**, 632 (1976).
[2] K. R. Allakhverdiev, E. A. Vinogradov, N. N. Melnik, M. A. Nizametdinova, E. Yu. Salaev, and R. M. Sardarly, phys. stat. sol. (b) **87**, K115 (1978).
[3] K. R. Allakhverdiev, M. A. Nizametdinova, E. A. Vinogradov, E. Yu. Salaev, R. M. Sardarly, N. Yu. Safarov, and G. N. Zhizhin, Solid State Commun. **36**, 527 (1980).
[4] K. R. Allakhverdiev, E. Yu. Salaev, N. Yu. Safarov, L. K. Vodopyanov, and L. Golubev, phys. stat. sol. (b) **92**, K91 (1979).
[5] K. R. Allakhverdiev, N. Yu. Safarov, M. A. Nizametdinova, E. A. Vinogradov, N. N. Melnik, A. F. Goncharov, and S. I. Subbotin, Solid State Commun. **42**, 485 (1982).
[6] M. M. Kurbanov, E. M. Godzhaev, and E. G. Mamedov, Izv. Akad. Nauk SSSR, Ser. neorg. Mater. **22**, 2059 (1986).
[7] M. A. Aldzhanov and D. A. Guseinov, Fiz. tverd. Tela **26**, 1521 (1984).
[8] O. Muller, G. Eulenberger, and H. Hahn, Z. anorg. allg. Chem. **398**, 207 (1973).
[9] T. N. Kolobyanina, G. B. Demishev, T. I. Dyuzheva, V. G. Losev, and S. S. Kabalkina, Prib. i Tekh. Eksper. No. 2, 194 (1984).
[10] L. G. Khvostantsev, L. F. Vereshchagin, and A. P. Novikov, High Temp.-High Pressures 9. 637 (1977).
[11] S. S. Mitra, O. Brafman, W. B. Daniels, and R. K. Grawford, Phys. Rev. **186**, 942 (1969).
[12] K. J. Range and G. Mahlberg, Z. Naturf. **30b**, 81 (1975).
[13] K. J. Range, G. Engert, W. Muller, and A. Weiss, Z. Naturf. **29b**, 181 (1974).
[14] G. B. Demishev, S. S. Kabalkina, and T. N. Kolobyanina, Fiz. tverd. Tela **29**, 3463 (1987).

*(Received January 25, 1988)*

phys. stat. sol. (a) 108, 97 (1988)

Subject classification: 61.50; 72.15; S4

*Laboratory of Metallurgy, Delft University of Technology*[1])

# Change of Electrical Resistance during Crystallization of Icosahedral $Al_{86}Mn_{14}$

By

M. HEYTING, B. J. THIJSSE, and A. VAN DEN BEUKEL

The isothermal change of the electrical resistance of icosahedral $Al_{86}Mn_{14}$ is measured in the temperature range 450 to 650 K. After a small increase of about 1% the resistance drops 20%, and after passing through a minimum it steeply rises again by about 20%. From X-ray diffraction data it is concluded that crystallization of the icosahedral phase takes place during the steep rise. A Johnson-Mehl-Avrami analysis of the data yields $E = (190 \pm 10)$ kJ/mol for the activation energy of crystallization and $n = 1.9 \pm 0.1$ for the Avrami parameter.

Die isotherme Änderung des elektrischen Widerstandes von ikosaedrischem $Al_{86}Mn_{14}$ wird im Temperaturbereich 450 bis 650 K gemessen. Nach einem geringen Anstieg von 1% fällt der Widerstand um 20% ab, und nachdem er durch ein Minimum gegangen ist, steigt er wieder schnell um etwa 20% an. Aus Röntgenbeugungsdaten wird geschlossen, daß die Kristallisation der ikosaedrischen Phase während des schnellen Anstiegs stattfindet. Eine Johnson-Mehl-Avrami-Analyse der Daten liefert $E = (190 \pm 10)$ kJ/Mol für die Aktivierungsenergie der Kristallisation und $n = 1{,}9 \pm 0{,}1$ für den Avramiparameter.

## 1. Introduction

The icosahedral phase first discovered by Shechtman et al. [1] in $Al_{86}Mn_{14}$ is a metastable phase. Upon heating it transforms to the stable crystalline state (orthorhombic $Al_6Mn$). The transformation gives rise to an evolution of heat, which shows up as an exothermic peak in a differential scanning calorimetry experiment as has been shown by several authors (e.g. [2 to 4]). Depending on heating rate, the reported peak temperatures for $Al_{86}Mn_{14}$ are in the range 600 to 700 K. The kinetics of the crystallization process was investigated by Chen et al. [4] by submitting their calorimetric data to an Avrami analysis.

Data of changes in physical properties other than internal energy upon crystallization are very scarce. Pavuna et al. [5] showed that in a continuous heating experiment the electrical resistivity of $Al_{86}Mn_{14}$ exhibits a minimum near 600 K. This was attributed to crystallization. In a recent summary of a conference lecture Kelton and Holzer [6] have reported isothermal resistivity changes during crystallization of the same material. Finally, Chen and Chen [7] observed a considerable increase of Young's modulus during crystallization of icosahedral $Al_{86}Mn_{14}$. This was preceded by a rather gradual and small increase, ascribed by the authors to structural relaxation in the icosahedral phase.

In this paper a series of isothermal resistance measurements are presented on as-quenched icosahedral $Al_{86}Mn_{14}$ in the temperature range 450 to 650 K. The results reveal the existence of several processes. In order to identify one of these processes as crystallization, additional X-ray diffraction experiments were carried out. This

[1]) Rotterdamseweg 137, NL-2628 AL Delft, The Netherlands.

enabled us to analyze the effect of crystallization on the resistance and to obtain the main kinetic parameters of the process. Furthermore the effect of structural relaxation on the electrical resistance is indicated.

## 2. Experimental

The material was prepared by melt spinning (at 40 m/s) of Al (14% Mn). X-ray diffraction analysis using a Guinier camera showed that the as-quenched material consisted of two phases: f.c.c. Al with a few per cent Mn solute and the icosahedral phase. The $d$-values for the icosahedral reflections were taken from the paper by Bancel et al. [8].

The electrical resistance was measured using a computer-controlled self-balancing 293 Hz Thompson bridge. For the annealing treatments the specimen was quickly brought into a furnace with 0.2 K stability. The time required for the specimen to adjust to the furnace temperature was about 5 s. The resistance of the specimen was continuously recorded during the anneal with an estimated accuracy of 10 ppm.

Because of the rather small estimated changes of the specimen dimensions during the processes studied, and because of the relatively large resistance changes observed, the relative changes in resistance can be identified with relative changes in resistivity.

## 3. Results and Discussion

A representative result is given in Fig. 1. The isothermal change of the resistance at 600 K is shown as a function of logarithm of time. The complete set of resistance isothermals is presented in Fig. 2a (low temperature range, 475 to 550 K) and b (high temperature range, 575 to 650 K).

The observed resistance isothermals can be divided into four distinct parts (Fig. 1):

1. AB, a relatively small resistance increase of the order of 1%. As this part is hardly visible on the scale of Fig. 2a, it is represented on an enlarged scale in Fig. 3a. The maximum B shifts to lower times and decreases with increasing temperature.
2. BC, a large resistance decrease of about 20%, slightly decreasing with increasing temperature.
3. CD, a sharp increase of the resistance, again on the order of 20%.
4. DE, an increase of the resistance slower than CD.

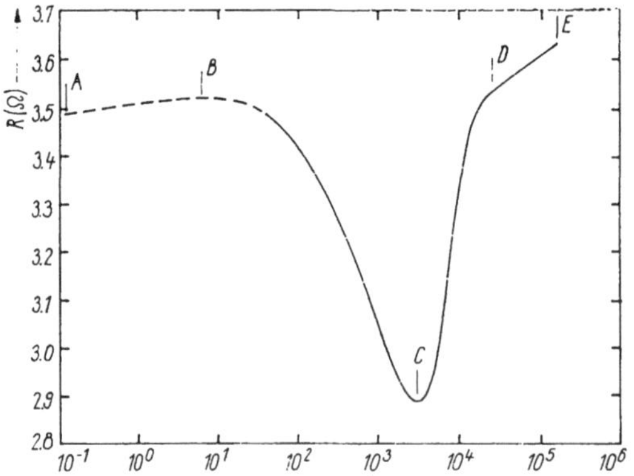


Fig. 1. Resistance versus time at 600 K

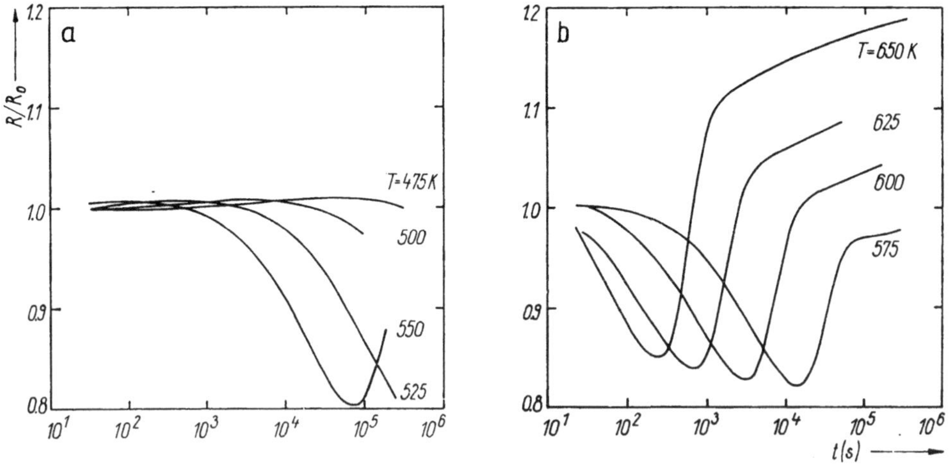


Fig. 2. Relative resistance changes versus time at the temperatures indicated

Diffraction patterns of the samples taken in various stages of the annealing yield the following results:

1. At the beginning (point A, Fig. 1) the lines corresponding to the icosahedral phase are clearly visible.
2. In the minimum (point C) these lines are about 10 to 20% weaker than in A.
3. At point D they have completely disappeared.

It is concluded that the transformation from the icosahedral to the crystalline state takes place during the steeply rising part CD of the resistance isothermals. As the minimum C is caused by two processes, represented by the parts BC and CD in the resistance isothermals, the onset of crystallization will not exactly coincide with C, but will take place somewhat earlier. This is in agreement with the X-ray results, which show that in the minimum a small fraction of the transformation has already taken place.

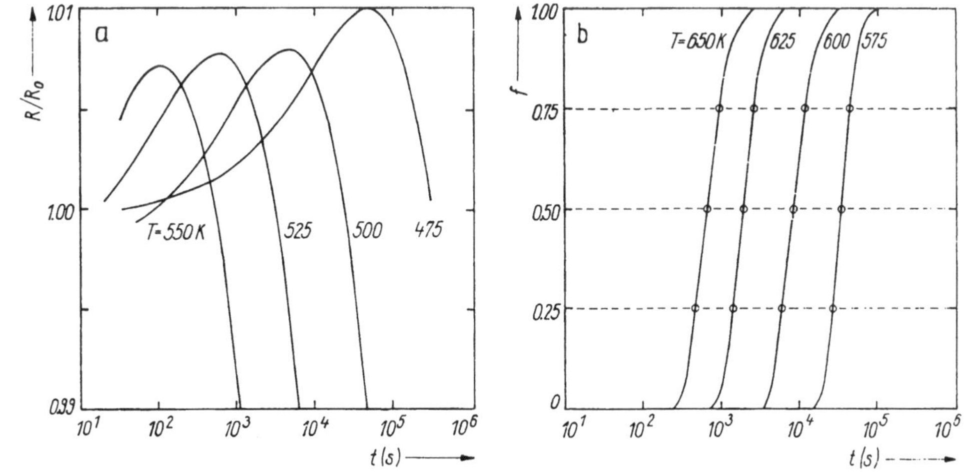


Fig. 3. a) Initial parts of the isothermals of Fig. 2a on an enlarged scale. b) Part CD of the isothermals of Fig. 2b on a reduced scale

In order to analyze the crystallization process further, the parts CD of the isothermals between 575 and 650 K (Fig. 2b) are represented in Fig. 3b on a reduced scale. The resistance change CD is taken as 100%. It is assumed that the resistance changes represented in this way describe the fractional transformation of the quasicrystalline phase. Then the data of Fig. 3b can be subjected to the well-known Johnson-Mehl-Avrami analysis, describing a transformation by a nucleation and growth process. The transformed fraction is given by

$$f = 1 - \exp(-kt^n), \tag{1}$$

where

$$k = k_0 \exp(-nE/RT). \tag{2}$$

$E$ is the activation energy of the process, $n$ the so-called Avrami parameter. From (1) it follows that a plot of $\ln \ln 1/(1-f)$ versus $\ln t$ should yield a straight line with slope $n$ and intercept $\ln k$. This has been done for the data of Fig. 3b, and the results are shown in Fig. 4. The data points clearly lie on curved lines. In this analysis, however, the fact that in C a small fraction $f_0$ (about 10 to 20%) had transformed already was not taken into account. If this is done, one should take $f' = f_0 + f(1 - f_0)$ instead of $f$. For $f_0 = 10\%$ or $20\%$ the results are also shown in Fig. 4. In the latter case good straight lines are obtained with a fairly constant slope of $1.9 \pm 0.1$. According to (2) a plot of the intercepts $\ln k$ in Fig. 4 versus $1/T$ will yield a straight line with slope $nE$. This is shown in Fig. 5. The result is $E = (190 \pm 20)$ kJ/mol. It follows from (1) and (2) that the activation energy can be obtained more directly if $n$ is constant, by making intersections at constant $f$ in Fig. 3b and plotting the $(t, T)$ values of the intersection points as $\ln t$ versus $1/T$. When this is done (not shown here) it yields an activation energy of $(165 \pm 20)$ kJ/mol.

The results obtained so far can be compared with scarce literature data. From DSC measurements of the transformation Chen et al. [4] report $E = 220$ kJ/mol, which is higher than the value reported here. These authors find for the Avrami parameter $n = 2.1 \pm 0.1$. On the other hand, Kelton and Holzer [6] obtain $E = (165 \pm 20)$ kJ per mol from DSC results, using the Kissinger plot. These authors, however, obtain from similar data as presented here the values $E = (230 \pm 20)$ kJ/mol and $n = 2.3 \pm 0.2$. It is concluded that the available data agree fairly well in the value of $n$ (about 2.0), but that a rather large scatter exists in the reported activation energies.

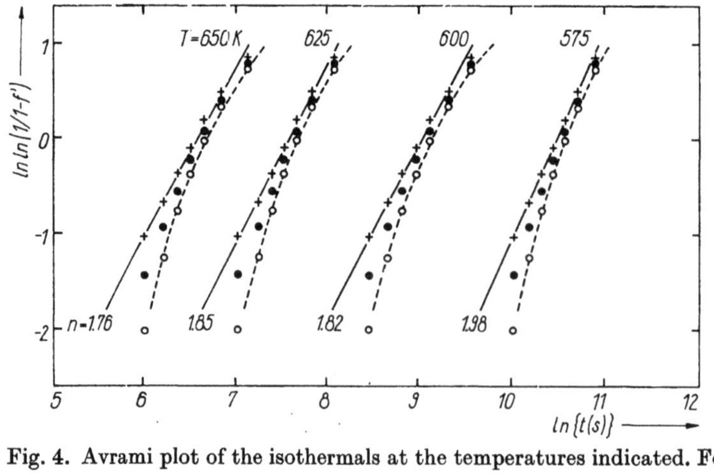


Fig. 4. Avrami plot of the isothermals at the temperatures indicated. For an explanation see text

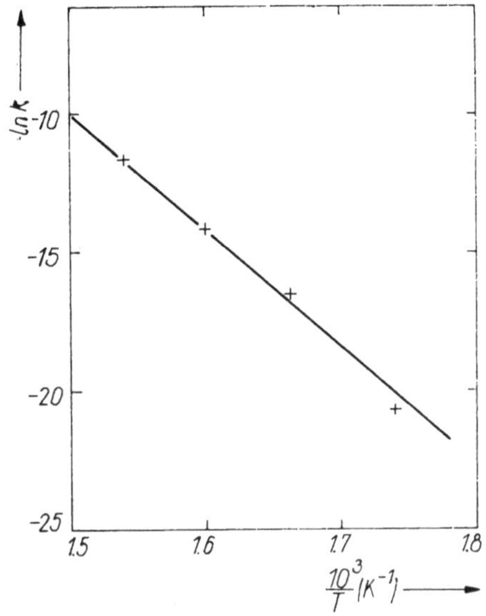


Fig. 5. Intercepts ln $k$ from Fig. 4 as a function of $1/T$

The nature of the other three processes observed in the resistance isothermals AB, BC, and DE cannot be derived from the present data alone. The small rising part AB (Fig. 3a) strong ly reminds of structural relaxation as observed in metallic glasses (see e.g. [9]) and it might be due to structural relaxation in the icosahedral phase. It is very similar to the change in Young's modulus preceding crystallization as reported by Chen and Chen [7] in the same material and ascribed to structural relaxation. It is highy improbable that this is also true for the large resistance decrease in part BC, because of the magnitude of the effect. As suggested by Kelton and Holzer [6], the part BC is more likely due to a precipitation process in the AlMn crystalline matrix. Finally, the part DE might be due to a phase transformation in the crystallized quasicrystal. All these phenomena, however, must be investigated further before they can be identified with certainty.

***Acknowledgements***

The authors are much indebted to R. Werkman of the University of Groningen for the preparation of the samples and to N. M. van der Pers and J. F. van Lent for assistance with the X-ray diffraction work. H. van der Zee designed and built the hardware and software for the resistance measurements.

## References

[1] D. Shechtman, I. Blech, D. Gratias, and J. W. Cahn, Phys. Rev. Letters **53**, 1951 (1984).
[2] K. F. Kelton and T. W. Wu, Appl. Phys. Letters **46**, 1059 (1985).
[3] J. Sekhar, K. M. Rao, T. Rajasekharan, and V. Seshu Bai, Mater. Res. Bull. **20**, 1109 (1985).
[4] H. S. Chen, C. H. Chen, A. Inoue, and J. T. Krause, Phys. Rev. B **32**, 1940 (1985).
[5] D. Pavuna, C. Berger, F. Cyrot-Lackman, P. Germi, and A. Pasturel, Solid State Commun. **59**, 11 (1986).
[6] K. F. Kelton and J. C. Holzer, 6th Internat. Conf. Rapidly Quenched Materials (Summary), Montreal 1987 (p. 64).
[7] H. S. Chen and C. H. Chen, Phys. Rev. B **33**, 668 (1986).
[8] P. A. Bancel, P. A. Heiney, P. W. Stephens, A. I. Goldman, and P. M. Horn, Phys. Rev. Letters **54**, 2422 (1985).
[9] A. van den Beukel, in: Rapidly Quenched Materials, Ed. P. W. Lee and S. Carbonara, ASM, 1986 (p. 193).

*(Received September 14, 1987)*

phys. stat. sol. (a) 108, 103 (1988)

Subject classification: 61.55; 64.70; S4

*Sektion Mathematik der Pädagogischen Hochschule Halle*[1]) *(a)*
*and Polytechnical Institute, Hanoi*[2]) *(b)*

# On the Existence Field of Guinier-Preston Zones in the Al–Zn System

By

CHU THIEN TRUONG (b), G. WENDROCK (a), and H. LÖFFLER (a)

The existence field of Guinier-Preston (GP) zones, viz. the upper limit temperatures of their formation $T_{hn}^{GPZ}$ as well as stability $T_{diss}^{GPZ}$, are investigated by SAXS experiments in five Al-(10 to 18) at% Zn alloys. The main results are the following: (i) GP zones are doubtlessly the first product of decomposition homogeneously nucleated during continuous cooling from the range of homogeneity. (ii) The temperature stability of the GP zones is increasing with their sizes. (iii) $T_{diss}^{GPZ}$ is larger than $T_{hn}^{GPZ}$. (iv) The two curves, viz. $T_{diss}^{GPZ}$ versus $x_{Zn}$ and the zinc content of the matrix $x_{Zn}^{m}$ in metastable equilibrium with GP zones versus $T$, are in good agreement.

Das Existenzfeld der Guinier-Preston-(GP-)Zonen, nämlich die obere Grenztemperatur ihrer Bildung $T_{hn}^{GPZ}$ als auch ihrer Stabilität $T_{diss}^{GPZ}$ wurde mittels RKWS-Untersuchungen in fünf AlZn (10 to 18)–At%-Legierungen ermittelt. Die Hauptergebnisse sind folgende: 1.) Das erste während der Abkühlung aus dem Homogenitätsgebiet gebildete Produkt der homogenen Entmischung sind zweifelsfrei GP-Zonen. 2.) Die Temperaturstabilität der GP-Zonen nimmt mit ihrer Größe zu. 3.) Es gilt $T_{diss}^{GPZ} > T_{hn}^{GPZ}$. 4.) Die Kurven, $T_{diss}^{GPZ}$ über $x_{Zn}$ und der Zinkgehalt der Matrix $x_{Zn}^{m}$ im metastabilen Gleichgewicht mit den GP-Zonen über $T$ aufgetragen, stimmen gut überein.

## 1. Introduction

Although the Al–Zn system is mainly of interest from the viewpoint of basic research, it belongs to the best investigated ones. For recent reviews see, e.g., [1 to 3]. Nevertheless, several problems remained unsolved, such as the limits of the existence field of the Guinier-Preston (GP) zones which are, by definition, fully coherent to the ambient matrix.

In the recent past, the authors of this paper contributed to solve this open problem by performing small-angle X-ray scattering (SAXS) investigations, on determining the positions of the upper limit temperatures (i) of rapid homogeneous nucleation (rhn) of precipitates $T_{rhn}$ [4 to 6] by slow cooling of alloys with zinc content between 10 and 54 at% from the range of homogeneity, (ii) of homogeneous nucleation of precipitates $T_{hn}$ [7 to 10] by extrapolating $T_{rhn}$ to zero cooling rate or by isothermal ageing of the alloys (zinc content between 4.5 and 18 at%) in the vicinity of the supposed position of $T_{hn}$, and (iii) of the dissolution of GP zones $T_{diss}^{GPZ}$ [11] grown in an Al-4.5 at% Zn alloy at room temperature (RT) to sizes of $R_K = (2.1 \pm 0.2)$ nm, where $R_K = (5/3)^{1/2} r_G$ and $r_G$ is Guinier's radius.

---

[1]) Kröllwitzer Str. 44, DDR-4050 Halle, GDR.

[2]) K10a, F202, T/T, DHBK Hanoi, Vietnam.

Additionally, it was proved [6] that in the Al–15 at% Zn alloy the first products of homogeneous decomposition at $T \leqq T_{rhn}$ are GP zones. Furthermore, the zinc content of both the GP zones $x_{Zn}^{GPZ}$ and the ambient matrix $x_{Zn}^{m}$ was determined in an Al–15 at% Zn alloy for temperatures $T_a$ ranging between 80 and 180 °C [9, 12]. It is expected that the $x_{Zn}^{m}$–$T_a$ curve is identical with the position of $T_{diss}^{GPZ}$ within the limits of experimental error.

Since the information about both the position of $T_{diss}^{GPZ}$ and the nature of the first product of decomposition below $T_{hn}$ is fairly scarce, it is the aim of this paper to contribute to filling these gaps.

## 2. Experimental

Five Al–Zn alloys with $x_{Zn} = 10$, 12, 14, 16, and 18 at%, produced from compounds of 4 N purity, were investigated. The decomposition states of the alloys were followed by SAXS investigations. In most cases the SAXS-intensity ($I$–$\varepsilon$) curves were measured at angles $\varepsilon = 2\theta$ ranging between 0.1 and 4.2° in steps of 0.04° by means of a camera of KRATKY type using $CuK_\alpha$ radiation. More details are given in [8]. Details about the heat treatment applied are given in the respective sections.

## 3. Results and Discussion

### 3.1 *Nature of the first product of decomposition below $T_{hn}$*

As was shown in [6], in the Al–15 at% Zn alloy the precipitates, nucleated and grown during the slow cooling with $v_c = 10$ K/min from the range of homogeneity ($T_q = 490$°) to $T_a = 220$ °C (that is just below the temperature of the onset of rapid homogeneous nucleation, viz. $T_{rhn} = (225 \pm 3)$ °C) and kept at 220 °C for 10 min, attained a mean size of $\bar{R}_K = (9.0 \pm 0.5)$ nm. Since the transition of GP zones to more stable phases takes place at a critical radius of $R_c \approx 5$ nm at RT, but only at $R_c \approx 8$ nm at $T_a = 120$ °C [13], the precipitates nucleated in the Al–15 at% Zn alloy after passing $T_{rhn}$ are doubtlessly GP zones; thus, $T_{rhn} = T_{rhn}^{GPZ}$.

The question remains, whether this statement also holds for $T_a$ situated between $T_{rhn}$ and the upper limit temperature $T_{hn}$ of the homogeneous nucleation of precipitates. In order to contribute to solving this problem, specimens of four Al–Zn alloys with zinc contents between $x_{Zn} = 12$ and 18 at% were cooled with $v_c = 10$ K/min from 490 °C to $T_a$ situated close to, but below, $T_{hn}$, and stored at $T_a$ for 10 min. Then, the SAXS intensity curve was measured directly at $T_a$, in order to avoid additional heat treatments which could falsify the measured mean size of the decomposition products. The time to measure the whole scattering curve lasted about 10 min. The immediately repeated measurement showed that the increase of the size of the GP zones during the two measurements was not larger than about 10%. The obtained results are summarized in Table 1. As the measured values of $\bar{R}_K$ do not exceed 12.5 nm, one can be sure

Table 1

Mean radius $\bar{R}_K(T_a)$ of the precipitates grown during the ageing at the respective $T_a$ for 10 min (SAXS intensity curves measured "in-situ" at $T_a$). $T_a$ is situated between $T_{rhn}$ [1 to 6] and $T_{hn}$ [9, 10]

| $x_{Zn}^{alloy}$ (at%) | 12 | 14 | 16 | 18 |
|---|---|---|---|---|
| $\bar{R}_K(T_a)$ (nm) | 9.8 ± 0.5 | 11.6 ± 0.5 | 10.7 ± 0.5 | 12.3 ± 0.5 |
| $T_a$ (°C) | 200 | 220 | 230 | 250 |
| $T_{rhn}$ (°C) | 196 ± 3 | 218 ± 3 | 229 ± 3 | 248 ± 3 |
| $T_{hn}$ (°C) | 203 ± 3 | 225 ± 3 | 236 ± 3 | 254 ± 3 |

that, also after passing $T_{hn}$, the first product of decomposition are GP zones. Note that at 250 C°, in comparison to RT, the zinc content of the GP zones is diminished from $x_{Zn}^{GPZ}(RT) = (72.0 \pm 0.5)$ at% to $x_{Zn}^{GPZ}(250\ °C) = (60 \pm 1)$ at% and the respective zinc content of the matrix in equilibrium with the zones increases from $(1.8 \pm 0.2)$ at% to $(14 \pm 1)$ at% (see the results presented in Fig. 4). That means, the difference of $x_{Zn}^{GPZ}$ and $x_{Zn}^{m}$ is considerably lower at 250 °C than at RT and thus, the coherency strains, accommodated at the periphery of the zones, too. Additionally, the ductility of the aluminium-rich matrix is increasing with rising temperature and thus, as the consequence of both effects, the driving forces for the break-down of the full coherency of the GP zones are considerably lowered with increasing $T_a$.

The statement that GP zones are growing after passing $T_{hn}$ was proved by another experiment. As recently shown by Hübner et al. [14], in the Guinier plot (lg $I$ versus $\varepsilon^2$) occasionally two fairly well-separated straight lines appear (see Fig. 1 in [14]), indicating that the X-rays (slit screen) are scattered by an assembly of precipitates offering two mean lengths perpendicularly directed to the slit. This effect can be due to two limiting cases. (i) Two sets of spherical GP zones are present, having, on the average, the two measured $R_K$. This explanation, however, can be excluded, because high-resolution transmission electron microscopic (HTEM) investigations, performed with these alloys over a wide range of $T_a$, delivered in each case a uniform size distribution of the GP zones [15]. But, in full agreement with X-ray results of other authors, by means of HTEM a transition from the spherical to a disk-like shape of the GP zones was stated [15], if a certain critical radius (which is increasing with $T_a$) is exceeded. As a consequence, Hübner et al. [14] assumed another limiting case, i.e. (ii) the two measured $R_K$ are associated with the major ($a$) and minor half axes ($c$) of the diks-like (ellipsoidal) GP zones. The values of $c/a$ in dependence on the mean size of the ellipsoidal GP zones, averaged by a sphere with radius $R_m = (a^2c)^{1/3} = (R_{K1}^2 R_{K2})^{1/3}$ (the values of $R_{K1}$ and $R_{K2}$ are evaluated from the slopes of the two straight lines occurring in the Guinier plot), fit well the results obtained by other authors using methods different from that applied in [14]. Thus, we also used the method applied in [14] for our purposes.

Samples of the Al–12 at% Zn and Al–16 at% Zn alloys were homogenized at 490 °C for 1 h, cooled with a rate of $v_c = 50$ K/min to $T_a = 190\ °C < T_{hn}$ and $T_a = 220\ °C < T_{hn}$, respectively, and kept at $T_a$ for 17 min (time necessary to measure the SAXS intensity curve). Then, the specimens were cooled from the respective $T_a$ with a rate of $v_c = 10$ K/min to 120 °C and, once more, the SAXS intensity curve was measured "in-situ" at 120 °C. It turned out that in the Guinier plot of each of the obtained curves, two straight lines were present, from the slope of which the values of $c$ and $a$ could be determined in the usual manner. The respective results are shown in Fig. 1. The solid and dotted lines drawn in the $c/a$ versus $R_m$ plot interpolate the values measured by various authors (see the references given in Table 1 of [14]) for GP zones grown at RT and 120 °C, respectively. The $c/a$-values indicated by triangles and spheres belong to this work. The most interesting results are the following. (i) As expected, the deviation of the GP zones from the spherical shape is further decreasing with the rise of $T_a$ above 120 °C, due to the diminution of the coherency strains. (ii) The mean size $R_m$ of the GP zones, pre-grown at evaluated $T_a$, is decreasing during cooling to and ageing at 120 °C and is accompanied by a drastic decrease of the $c/a$-value (increasing deviation from the spherical shape). The decrease of $R_m$ is explained by two effects, viz. the decrease of the lattice constants and the increase of the zinc content of the zones (the zinc atoms are smaller than the aluminium atoms) with decreasing $T_a$. The decrease of the $c/a$-value is due to the increasing rigidity of the aluminium matrix with decreasing $T_a$. With rising deviation from the spherical shape, the coherency stresses are lowered, if

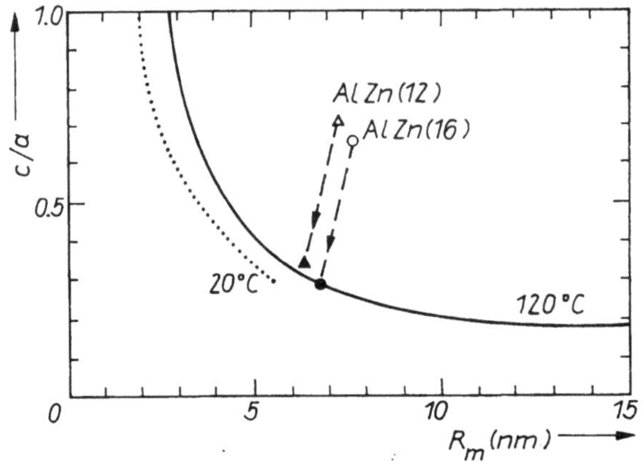


Fig. 1. Deviation of the shape of GP zones from spherical ($c/a = 1$) at various temperatures $T_a$. The solid and dotted lines interpolate the values obtained by various authors (for references see Table 1 in [14]) for GP zones grown at RT and 120 °C, respectively. This work: Al–12 at% Zn alloy, △ measured at 190 °C and ▲ after subsequent cooling to 120 °C; Al–16 at% Zn alloy, ○ measured at 220 °C and ● after subsequent cooling to 120 °C, respectively

considering GP zones with constant value of $R_m$. (iii) After cooling the specimens from elevated $T_a$ (190 and 220 °C, respectively) to 120 °C, the measured values of $c/a$ fit within the range of experimental error the $c/a$-$R_m$ curve obtained for specimens directly aged at 120 °C. This behaviour can only be understood, if the specimens pre-grown at elevated $T_a$ are indeed GP zones, because otherwise another $c/a$-value has to be expected after cooling the specimens to 120 °C than that belonging to GP zones grown at 120 °C. Note that a transition of not fully coherent precipitates into GP zones (fully coherent by definition) is totally impossible from the thermodynamic viewpoint.

### *3.2 Reversion of GP zones grown at RT to various sizes*

It is the aim of this section to obtain information on the reversion behaviour of GP zones grown at RT to sizes of $R_K = 2$ and 4 nm, respectively, particularly on the mechanism of their reversion, as well as the position of the upper limiting temperature $T_{diss}^{GPZ}$ for their stability in dependence on alloy composition and zone size. GP zones grown at RT to sizes of $R_K = 2$ nm are doubtlessly spherical in shape, those with $R_K = 4$ nm, however, ellipsoidal (see Fig. 1), but they are still GP zones, because their transition into more stable phases occurs at a critical radius of $R_c \approx 5$ nm.

We investigated two alloys with $x_{Zn} = 10$ and 12 at%, respectively. The specimens were homogenized at 490 °C for 1 h, quenched in water of RT, and then stored at RT until the wanted $R_K$-value was reached. Two different but mutually supporting methods of investigation were applied. (i) The specimens were kept at their respective reversion temperatures $T_{rev}$ during the time $t_{rev}$. Then, the specimens were quenched in water at RT and after that the full SAXS intensity curve was measured immediately at RT. From the measured intensities $I_{meas}$ the contributions of the background scattering intensities $I_{bg}$ were subtracted ($I_{corr} = I_{meas} - I_{bg}$). For details see [6], where it is also outlined that the contributions of the scattering effects, due to the very small clusters of zinc atoms formed during the quench to RT and the time needed to measure the SAXS intensity curve, can be distinguished well from the decomposition state attained after the end of the reversion treatment. This experience was additionally checked by method (ii): The changes of the SAXS intensities during $t_{rev}$ were followed at $T_{rev}$ "in-itu" by measuring the intensities at a fixed angle $\varepsilon_{fix}$ in distances of 20 s. The selected position of $\varepsilon_{fix}$ is indicated in Fig. 2a.

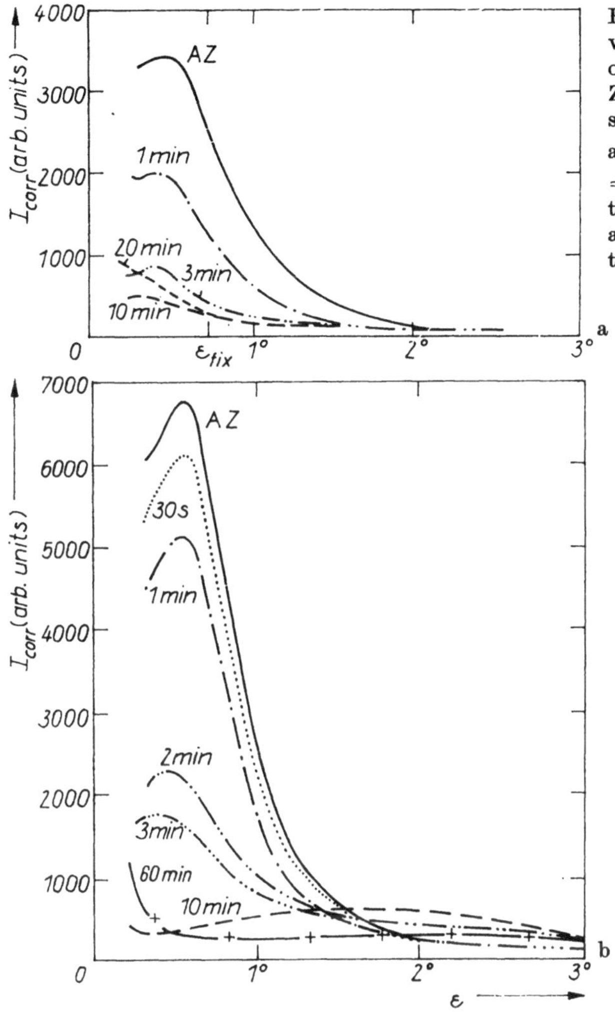


Fig. 2. SAXS intensity curves ($I_{corr}$ vs. $\varepsilon$) obtained during the reversion of GP zones, grown in an Al–10 at% Zn alloy during ageing at RT to sizes of $R_K = (4.0 \pm 0.2)$ nm, at a) $T_{rev} = 195\ °C < T_{diss}^{GPZ}$; b) $T_{rev} = 200\ °C > T_{diss}^{GPZ}$. AZ means the intensity course after room temperature ageing and before reversion, respectively

The general features of the curves obtained by the two applied methods can be taken from Fig. 2 and 3. Fig. 2 a and b show the SAXS intensity curves obtained by reverting the GP zones grown in the Al–10 at% Zn alloy at RT to sizes of $R_K = (4.0 \pm 0.2)$ nm, during the reversion at $T_{rev} = 195$ °C (Fig. 2a) and 200 °C) (Fig. 2b), respectively. The main results are the following. (i) Considering Fig. 2a, the SAXS intensities do not drop to zero. They reach their minimum after $t_{rev} = 10$ min and then they are increasing again at small scattering angles, indicating the coarsening of the GP zones having survived the reversion treatment at $T_{rev} = 195$ °C. Thus, 195 °C $< T_{diss}^{GPZ}$ ($R_K = 4$ nm). Note that no indication of the formation of small clusters of zinc atoms is visible in the scattering curves. (ii) If reverting the same specimens at 200 °C, the GP zones grown at RT to sizes of about $R_K = 4$ nm survive the reversion treatment only up to $t_{rev} \approx 10$ min. At higher $t_{rev}$, however, they become completely dissolved, i.e., 200 °C $\geqq T_{diss}^{GPZ}$ ($R_K = 4$ nm) and, at the expense of the zones dissolved at $T_{rev} = 200$ °C, during the quench to RT and the subsequent measurement of the SAXS-intensity curve, new small GP zones (or clusters of zinc atoms) are formed, as

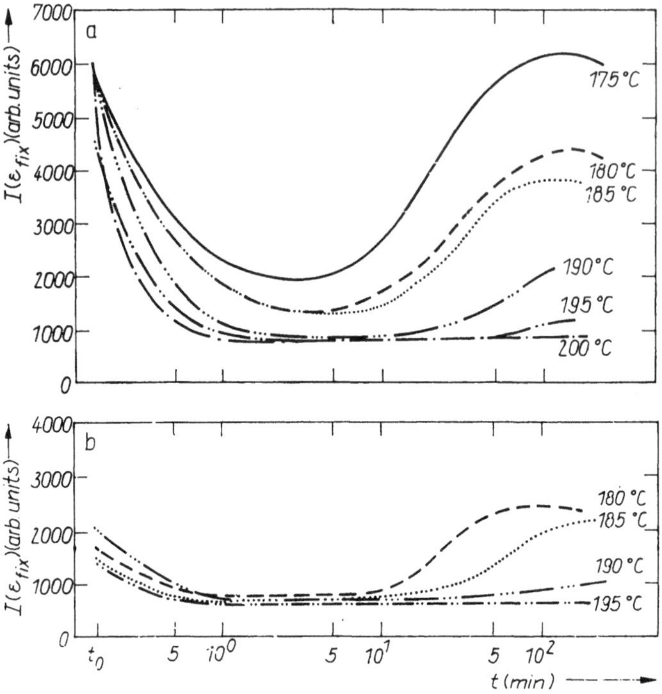


Fig. 3. Course of the SAXS intensities measured at a fixed angle $\varepsilon_{fix}$ at various $T_{rev}$ ("in-situ" measurements) vs. $t_{rev}$ for GP zones grown in an Al–10 at% Zn alloy at RT to various sizes $R_K$. $t_0 = 10$ s. a) $R_K = (4.0 \pm 0.2)$ nm; b) $R_K = (2.0 \pm 0.2)$ nm

reflected in the $I_{corr}$–$\varepsilon$ curve by the intensity maximum forming at about 1.7° (see the curve belonging to $t_{rev} = 10$ min). Note that, as proved by TEM investigations, the increase of the scattered intensities near the primary beam between $t_{rev} = 10$ and 60 min is due to the heterogeneous nucleation of the β(Zn)-equilibrium phase at grain boundaries and at dislocation loops. (iii) As proved by the results presented in Fig. 2a and b, independent of $T_{rev} \lesssim T_{diss}^{GPZ}$, in the first stage of the reversion treatment, i.e. for $t_{rev} \lesssim 1$ min, the decrease of the zinc content of the GP zones is the dominating process. This statement can be proved in the following manner. The position of the maximum of the scattered intensities $\varepsilon_{max}^{I}$ is at least a rough measure for the mean distance of the GP zones (see, e.g., [12]). According to Fig. 2 a and b, $\varepsilon_{max}^{I}$ is not changed up to $t_{rev} = 1$ min. In addition, within this time span, the mean radius $R_K$ of the GP zones remains constant within the range of experimental error. Thus, the only reasonable explanation for the diminution of the SAXS intensities up to $t_{rev} = 1$ min is the decrease of the difference of the zinc content between the GP zones and the ambient matrix. (iv) Generalizing the results shown in Fig. 2 a and b, $T_{diss}^{GPZ} = (196 \pm 3)$ °C in the Al–10 at% Zn alloy for GP zones grown at RT to sizes of $R_K = 4$ nm. This conclusion is supported by the results presented in Fig. 3a, where the course of the SAXS intensities $I(\varepsilon_{fix})$, measured "in-situ" at various $T_{rev}$ at a constant angle $\varepsilon_{fix}$, are plotted versus $t_{rev}$. At the beginning of the reversion treatment $I(\varepsilon_{fix})$ is decreasing due to the diminution of the zinc content of the GP zones. The increase of $I(\varepsilon_{fix})$ after passing the minimum occurring for $T_{rev} < T_{diss}^{GPZ}$, is explained by the coarsening of the GP zones having survived the reversion treatment. Applying $T_{rev} =$

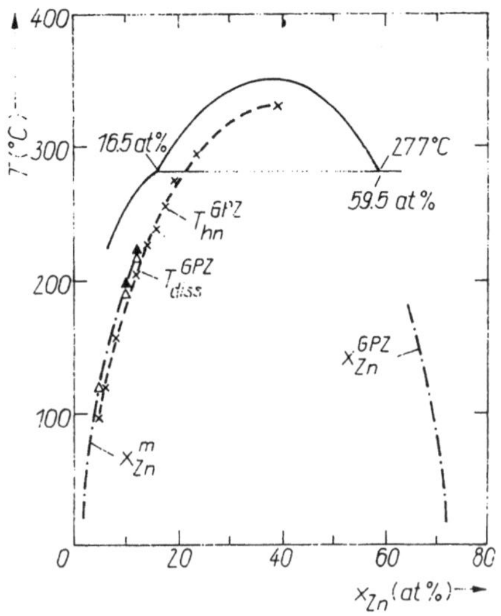


Fig. 4. Positions of $T_{diss}^{GPZ}$, $T_{hn}^{GPZ}$, $x_{Zn}^{m}$, and $x_{Zn}^{GPZ}$ (for details see the text)

$= 200$ °C $> T_{diss}^{GPZ}$, no increase of $I(\varepsilon_{fix})$ was observed. (v) From the results presented in Fig. 3b it can be concluded that for GP zones grown in the same alloy at RT to smaller sizes, namely to $R_K = (2.0 \pm 0.2)$ nm, $T_{diss}^{GPZ}(R_K = 2\text{ nm}) = (190 \pm 3)$ °C. This result is supported by the other method applied in this paper.

For the alloy with $x_{Zn} = 12$ at% the results obtained by the two methods applied give $T_{diss}^{GPZ}(R_K = 2\text{ nm}) = (216 \pm 3)$ °C and $T_{diss}^{GPZ}(R_K = 4\text{ nm}) = (222 \pm 3)$ °C, respectively.

### *3.3 Summarizing Discussion*

In Fig. 4, the curves for $T_{diss}^{GPZ}$, $T_{hn}^{GPZ}$, $x_{Zn}^{m}$, and $x_{Zn}^{GPZ}$ are presented. In order to avoid confusion, for the zinc content of the GP zones $x_{Zn}^{GPZ}$ and the matrix $x_{Zn}^{m}$ in metastable equilibrium with the GP zones, the values measured by the various authors are omitted and only the interpolating curves are drawn. For details see [9, 12]. The following features are of particular interest. (i) As should be expected, $T_{diss}^{GPZ} > T_{hn}^{GPZ}$ in the whole concentration range where both temperatures are determined. (ii) Within the limits of experimental error, the positions of $x_{Zn}^{m}(T)$ and $T_{diss}^{GPZ}(x_{Zn})$ are the same. If this is true, we conclude, taking into account that $T_{diss}^{GPZ}$ is increasing with rising zone size (see the open and the full triangles), that $x_{Zn}^{m}$ should decrease with increasing zone size.

## References


[1] J. L. Murray, Bull. Alloy Phase Diagrams **4**, 55 (1983).
[2] M. Simerska, P. Bartuška, and V. Syneček, Acta cryst. **A34**, 304 (1978).
[3] H. Löffler, V. Syneček, M. Simerská, G. Wendrock, P. Bartuška, and R. Kroggel, phys. stat. sol. (a) **65**, 197 (1981).
[4] G. Wendrock, phys. stat. sol. (a) **78**, 497 (1983).
[5] G. Wendrock, Cryst. Res. Technol. **20**, 91 (1985).
[6] G. Wendrock, H. Löffler, O. Kabisch, and Chu Thien Truong, phys. stat. sol. (a) **91**, 453 (1985).

[7] G. WENDROCK, H. LÖFFLER, I. RAPHELT, and CHU THIEN TRUONG, Crystal Res. Technol. **22**, 457 (1987).
[8] G. WENDROCK, H. LÖFFLER, E. NÜSSLER, and S. REIMANN, Crystal. Res. Technol. **21**, 387 (1986).
[9] CHU THIEN TRUONG, Thesis, Pädagogische Hochschule Halle, GDR, 1988.
[10] CHU THIEN TRUONG, G. WENDROCK, and H. LÖFFLER, Crystal Res. Technol. **21**, K180 (1986).
[11] H. LÖFFLER, G. WENDROCK, and K. PAWELZYK, phys. stat. sol. (a) **99**, 73 (1987).
[12] CHU THIEN TRUONG, V. SYNEČEK, G. WENDROCK, and H. LÖFFLER, Makromol. Chem., Makromol. Symp. **15**, 241 (1988).
[13] R. RAMLAU and H. LÖFFLER, Crystal Res. Technol. **19**, 1279 (1984).
[14] G. HÜBNER, H. LÖFFLER, and G. WENDROCK, Crystal Res. Technol. **21**, K152 (1986).
[15] R. RAMLAU and H. LÖFFLER, phys. stat. sol. (a) **79**, 141 (1983).

*(Received February 19, 1988)*

phys. stat. sol. (a) **108**, 111 (1988)

Subject classification: 61.14 and 64.80; S10.15

*Laboratoire de Chimie Minérale Structurale, associé au C.N.R.S., U.A., n° 320, Faculté des Sciences, Limoges*[1]) *(a)*
*Department of Physics (b), and Department of Materials Science (c), University College, Cardiff*[2])

# Diffraction Patterns from Ordered Intergrowth Phases in the $Bi_2O_3$–$TeO_2$ System

By

D. MERCURIO (a), B. FRIT (a), G. HARBURN (b), B. H. PARRY (b, c), R. P. WILLIAMS (b), and R. J. D. TILLEY (c)

Idealised structural models are proposed for a set of complex intergrowth phases which form in the $Bi_2O_3$–$TeO_2$ system between the composition limits $Bi_2TeO_5$ and $Bi_2Te_2O_7$. Expressions are obtained for the Fourier transforms of these models and masks representing them are used to produce optical Fraunhofer diffraction patterns. Electron diffraction patterns are in agreement with these results and this has allowed the microstructures of the phases to be inferred without recourse to electron-optical imaging.

Idealisierte Strukturmodelle werden für einen Satz von komplexen Zwischenwachstumsphasen vorgeschlagen, die sich im $Bi_2O_3$–$TeO_2$-System zwischen den Zusammensetzungsgrenzen $Bi_2TeO_5$ und $Bi_2Te_2O_7$ bilden. Ausdrücke werden für die Fouriertransformierten dieser Modelle erhalten und Masken, die sie repräsentieren, werden benutzt, um optische Fraunhoferbeugungsbilder zu erzeugen. Die Elektronenbeugungsbilder befinden sich mit diesen Ergebnissen in Übereinstimmung und dies erlaubt, die Mikrostrukturen der Phasen zu bestimmen, ohne Rückgriff auf elektronenoptische Abbildung.

## 1. Introduction

El Farissi et al. [1] have suggested that the solid solution $Bi_{1-x}Te_xO_{(3+x)/2}$ $(0.33 \leq x \leq 0.50)$ is composed of a series of microphases with a superstructure consisting of ordered intergrowths of the $Bi_2TeO_5$ and $Bi_2Te_2O_7$ structures. The structure of $Bi_2TeO_5$ has been determined and can be considered to be derived from a fluorite-type subcell, of side $a_F$, and to have a unit cell given by $a = 1.1602$ nm, $b = 1.6461$ nm, $c = 0.5523$ nm, i.e. $a \approx 2a_F$, $b \approx 3a_F$, $c \approx a_F$ [2]. In this compound, the metal atoms are in the positions normally occupied in the fluorite structure, but ordered columns of "vacancies" occur on the anion sublattice to reduce the metal-to-oxygen ratio below that obtained in the fluorite structure. Ordering of the defects produces the supercell with a lattice parameter of $3a_F$ in the $\boldsymbol{b}$-direction. The structure of the phase $Bi_2Te_2O_7$, with lattice parameters $a = 2.2974$ nm, $b = 2.2065$ nm, $c = 0.5526$ nm, i.e. $a \approx 4a_F$, $b \approx a_F$, has not yet been solved in detail. However it is known that it is similar to that of $Bi_2TeO_5$ in that all the metal atoms are found in the positions normally occupied with respect to the parent fluorite structure and that ordered arrays of anion "vacancies" account for the stoichiometric imbalance. A notable difference is that the unit cell has a lattice parameter of $4a_F$ in the $\boldsymbol{b}$-direction rather than $3a_F$ for $Bi_2TeO_5$.

[1]) 123, Av. A. Thomas, F-87060 Limoges, Cédex, France.

[2]) Cardiff, CF2 1TA, Great Britain.

The diffraction patterns of the phases are dominated by that of the fluorite substructure, but contain numerous weak superlattice reflections, and are of considerable complexity [1]. The determination of the structures of these intergrowth phases has proved to be difficult using X-ray diffraction techniques because the major contribution to the diffracted intensities comes from the perfect array of heavy metal atoms, which does not change greatly across the whole of the composition range. In addition, the phases and all have long unit cells. The imaging of the intergrowths by means of high-resolution electron microscopy is similarly difficult because different arrangements of weakly scattering oxygen "vacancies" in the perfect array of heavy metal atoms need to be distinguishable in the images.

Despite these difficulties electron diffraction patterns from very small regions of crystal are readily obtained. In order to determine the extent to which the structures could be predicted from electron diffraction patterns we have investigated, by the use of Fourier methods, the form that such patterns would take. It is the purpose of this paper to obtain expressions for the Fourier transforms of the proposed structures and to prepare optical transforms [3] for them, that is, optical Fraunhofer diffraction patterns from masks which represent the structures. These are then compared with experimental electron diffraction patterns. It is shown that despite the simplicity of the model chosen the experimental diffraction patterns are well reproduced and the structural intergrowth proposed [1] is verified.

## 2. Theory

Although the $a$ parameters of both $Bi_2TeO_5$ and $Bi_2Te_2O_7$ are multiples of the fluorite parent, no complex ordering patterns have been found in this direction. In the $\boldsymbol{b}$-direction, however, a variety of long unit cell structures occurs. This direction is therefore the one of particular interest. The $\boldsymbol{c}$-direction is invariable and the same for both phases.

In Fig. 1 idealised models of the intergrowth phases $Bi_2TeO_5$ and $Bi_2Te_2O_7$ are illustrated. In the projection shown, which is the only one we will consider in this paper, we can ignore the heights of the atoms. If there were no oxygen deficiency, the structure, say $f(x, y)$, would be, in two dimensions, square, with parameter $a'$ equal to $a_F/2$. Let the structure of the missing atoms be $g(x, y)$, then the defective crystal actually has the structure $f(x, y) - g(x, y)$. Since we have no great interest in the perfect structure $f(x, y)$ we may ignore it for our purposes and concentrate on the defects, represented by $g(x, y)$. Furthermore, as we are interested only in the evolution of superlattice reflections along the $\boldsymbol{b}$ axis, we can consider the defect to be essentially one-dimensional and represent it by a one-dimensional function, say, $g(y)$, and ignore the variation in the $x$-direction. Accordingly, we define the nature of the defect, $g(y)$, and find its Fourier transform, (FT). The Fraunhofer diffraction pattern (FDP) of $g(y)$ is then the square of the modulus of its FT.

The FT theory needed for the present purpose is given by Bracewell [4]. The aspects we need are briefly indicated here. The one-dimensional space in which the profile of the FDP is of interest is denoted by $\nu$. The FT, $G(\nu)$ of $g(y)$ may be defined by

$$G(\nu) = F\{g(y)\} = \int_{-\infty}^{\infty} g(y) \exp(-2\pi i \nu y)\, dy . \quad (1)$$

The shift theorem states that

$$F\{g(y - y_0)\} = \exp(-2\pi i \nu y_0)\, G(\nu) , \quad (2)$$

where $y_0$ is a constant. The convolution theorem states that if two functions, say $g_1(y)$ and $g_2(y)$ are convolved in $y$-space, then the FT of the convolution is the product of

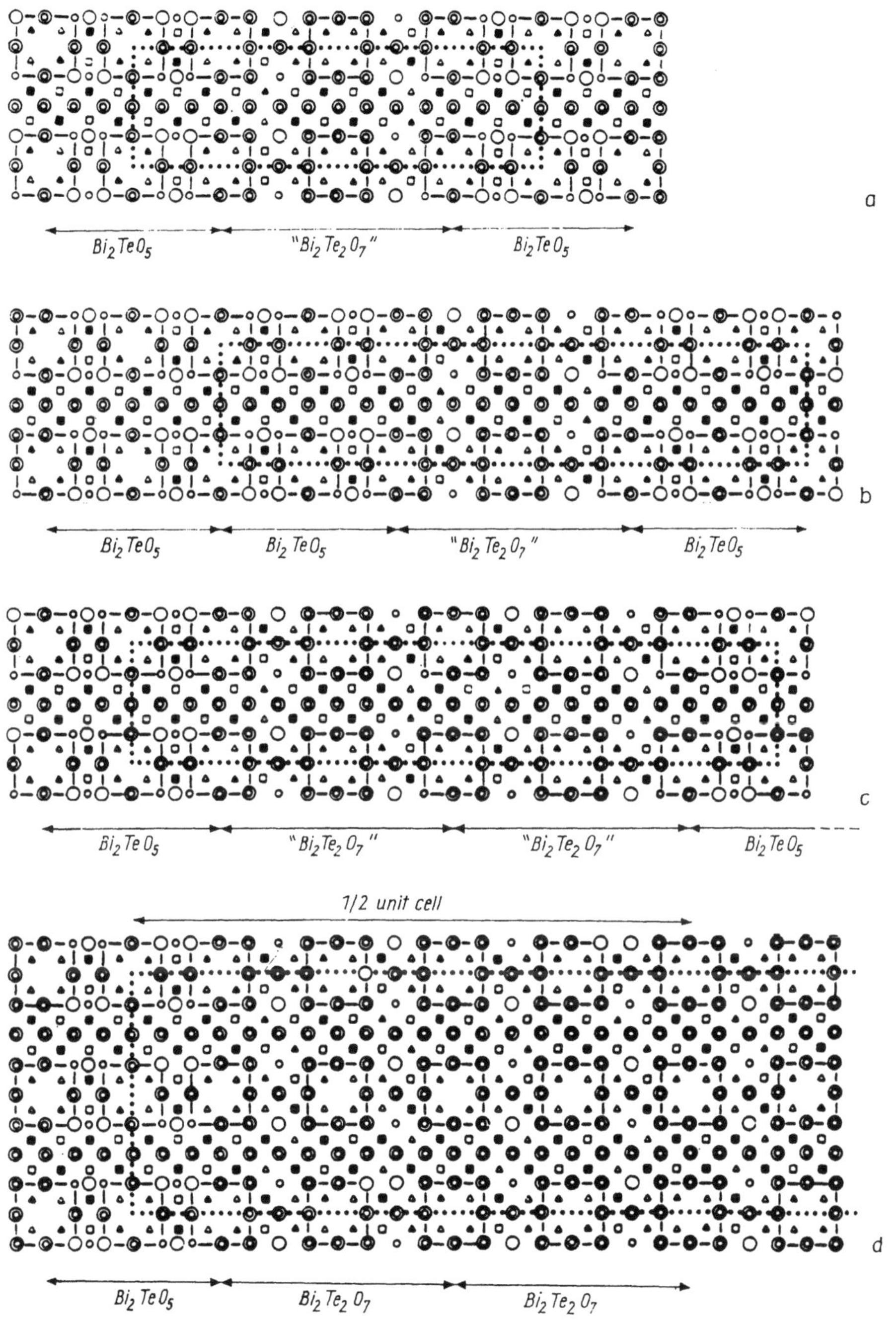


Fig. 1. Projection of the idealised intergrowths down (001). a) $Bi_2TeO_5$ + $Bi_2Te_2O_7$ ($3^24^2$); b) 2 $Bi_2TeO_5$ + $Bi_2Te_2O_5$ ($3^44^2$); c) $Bi_2TeO_5$ + 2 $Bi_2Te_2O_7$ ($3^24^4$); d) $Bi_2TeO_5$ + 4 $Bi_2Te_2O_7$ ($3^24^8$)

the individual FT's. That is to say,

$$F[g_1(y) * g_2(y)] = G_1(y)\, G_2(y)\ . \tag{3}$$

The FT of the "comb" function is another "comb" function. Specifically,

$$F\{\sum_{k=-\infty}^{\infty} \delta(y - k\beta)\} = \beta^* \sum_{k=-\infty}^{\infty} \delta(\nu - k\beta^*)\ ,$$

where $\beta$ is the distance between neighbouring "teeth" and $\beta^* = 1/\beta$. Lastly, the intensity $I(\nu)$ of the FDP is

$$I(\nu) = G(\nu)\, G^*(\nu)\ ,$$

where the asterisk here denotes the complex conjugate.

### *2.1 Simple model*

We come now to a representation of the defects of an intergrowth structure. We may represent each defect by an appropriately-positioned delta function. In order to generate the structure of the $Bi_2Te_2O_7$ we must introduce a defect in every fourth unit cell. These defects are then separated by $4a'$. The true cell is then given by $b = 2 \times 4a'$. Similarly, to generate one cell of $Bi_2TeO_5$ one defect must be placed in every third unit cell, at a separation on $3a'$. The true cell is then $b = 2 \times 3a'$. The simplest model of an intergrowth phase that we can then construct will consist of defects in every fourth unit cell in a sequence numbering, say, $M$ defects. Following this there are defects in every third unit cell with a total of, say, $N$ defects. The whole pattern is then repeated indefinitely. The periodicity of an intergrowth is therefore $(4M + 3N)\, a'$. A typical pattern of delta functions is depicted in Fig. 2.

The mathematical form of $g(y)$ for this model is

$$g(y) \equiv [\sum_{m=1}^{M} \delta(y - 4ma') + \sum_{n=1}^{N} \delta(y - (4M + 3n)\, a')] *$$

$$* \sum_{k=-\infty}^{\infty} \delta(y - k(4M + 3N)\, a')\ . \tag{4}$$

With $F\{\delta(y - a')\} = \exp(-2\pi i a'\nu)$, which we may deduce immediately from (1) using the basic property of the delta function in an integral, and using (3) and (4), we

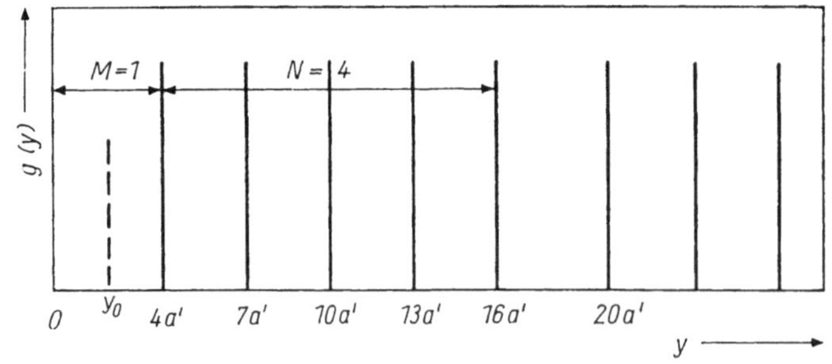


Fig. 2. The pattern of delta functions depicting the defects in the simple model. Two sequences alternate to form a periodic array of $M$ cells of parameter $4a'$ and $N$ cells of parameter $3a'$. In this example $M = 1$ and $N = 4$

may obtain $G'(\nu)$ as

$$G(\nu) = \frac{1}{(4M+3N)\,a'}\,[\sum_{m=1}^{M} \exp(-8\pi i m a'\nu) + \sum_{n=1}^{N} \exp(-2\pi i\{4M+3n\}\,a'\nu)] \times$$

$$\times \sum_{k=-\infty}^{\infty} \delta\left[\nu - \frac{k}{(4M+3N)\,a'}\right] \tag{5}$$

$$= \frac{1}{(4M+3N)\,a'}\left[\sum_{m=1}^{M} \exp\left(-\frac{8\pi i m k}{4M+3N}\right) + \sum_{n=1}^{N} \exp\left(-\frac{2\pi i(4M+3n)\,k}{4M+3N}\right)\right] \times$$

$$\times \sum_{k=-\infty}^{\infty} \delta\left[\nu - \frac{k}{(4M+3N)\,a_F}\right]. \tag{6}$$

Equation (6) follows from (5) because the comb is zero at all values of $\nu$ other than $k/(4M+3N)\,a'$.

Now, (6) is not, in general, either real or even. It would facilitate computation if it were so. However, by redefining the origin of $y$-space to be at $y_0$ half way along the $M$ sequence, as shown in Fig. 2, we see that $g(y)$ is real and even with the consequence that $G(\nu)$ will also be real and even. (This may be deduced from (1).) So we consider the defect sequence to be $g'(y) = g(y - y_0)$ and we deduce from the shift theorem (2) that

$$G'(\nu) = \frac{1}{(4M+3N)\,a'} \exp\left[\frac{+4\pi i M k}{(4M+3N)}\right] \times$$

$$\times \left[\sum_{m=1}^{M} \exp\left(-\frac{8\pi i m k}{4M+3N}\right) + \sum_{n=1}^{N} \exp\left(-\frac{2\pi i(4M+3n)k}{4M+3N}\right)\right] \times$$

$$\times \sum_{k=-\infty}^{\infty} \delta\left[\nu - \frac{k}{(4M+3N)\,a'}\right]. \tag{7}$$

We note that, although the function $g(y)$ has been shifted with a consequent change in its FT, there is no change whatsoever in the FDP intensity pattern, since

$$I(\nu) = G(\nu)\,G^*(\nu) = G'(\nu)\,G'^*(\nu)\,.$$

### *2.2 General model*

Although the simple theory just presented has been found to suffice for many situations, ambiguity will arise for even quite small values of $M$ and $N$. In Table 1 we list the possibile sequences for $(4M + 3N) = L$ up to $L = 20$. It is seen for example, that a value of $(4M + 3N) = 17$ can be due to two alternative arrangemenst of "vacancies" or delta functions, viz. (33344), i.e. $(3^3 4^2)$ or (34334), i.e. $(343^2 4)$. Similarly, if $(4M + 3N) = 20$, we can have the arrangements $(3^4 4^2)$, $(343^3 4)$, $(4^5)$, and $(3^2 43^2 4)$. This latter arrangement is identical to that for $(4M + 3N) = 10$ and so can be ignored, but the other two still exist as possible arrangements.

In order to distinguish between these possibilities it is necessary to expand the restricted theory given above to the more general situation. The principles and the model, however, remain the same. We show in Fig. 3a pattern of delta functions which represents an example of the general case. In this situation, the blocks of $M$

Table 1
Values of $(4M + 3N)$ and their possible sequences

| $(4M + 3N) = L$ | sequences |
|---|---|
| 3 | (3) |
| 4 | (4) |
| 5 | — |
| 6 | $(3^2)$ $\|Bi_2TeO_5\|$ |
| 7 | (34) |
| 8 | $(4^2)$ $\|Bi_2Te_2O_7\|$ |
| 9 | $(3^3)$ |
| 10 | $(3^24)$ |
| 11 | $(34^2)$ |
| 12 | $(3^4)$, $(4^3)$ |
| 13 | $(3^34)$ |
| 14 | $(3^24^2)$ |
| 15 | $(3^5)$, $(34^3)$ |
| 16 | $(4^4)$, $(3^44)$ |
| 17 | $(3^34^2)$, $(343^24)$ |
| 18 | $(3^6)$, $(3^24^3)$, $(434^23)$ |
| 19 | $(3^54)$, $(34^4)$ |
| 20 | $(4^5)$, $(3^44^2)$, $(343^34)$ |

and $N$ are split into the series

$$M = M_1 + M_2 + M_3 + \dots + M_p\,,$$
$$N = N_1 + N_2 + N_3 + \dots + N_p\,.$$

For a particular value of $4M + 3N$, $p$ will always be equal to the minimum value of either $M$ or $N$, e.g. $4M + 3N = 20$, then $M = 2N = 4$, so $p$ will be 2. The defects in this general case are represented by the function $g(y)$ where

$$\begin{aligned} g(y) = [ & \sum_{n_1=1}^{N_1} \delta(y - n_1 3a') + \sum_{m_1=1}^{M_1} \delta(y - a'\{3N_1 + 4m_1\}) + \\ & + \sum_{n_2=1}^{N_2} \delta(y - a'\{3N_1 + 4M_1 + 3n_2\}) + \\ & + \sum_{m_2=1}^{M_2} \delta(y - a'\{3N_1 + 4M_1 + 3N_2 + 4m_2\}) + \dots] * \\ & * \sum_{n=-\infty}^{\infty} \delta(y - n\{4M + 3N\}\, a')\,. \end{aligned} \tag{8}$$

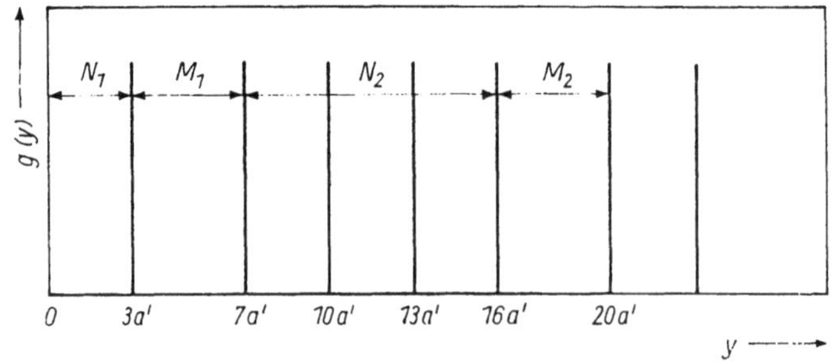


**Fig. 3.** The pattern of delta functions depicting the defects in a more general arrangement than in Fig. 2. In this example $M_1 = 1$; $N_1 = 1$, $M_2 = 1$, $N_2 = 3$

If we put $(3N_r + 4M_r) = P_r$ and define $P_0 = 0$, then the summations of $N_1$, $N_2$, up to $N_p$ will be equal to

$$\sum_{n_1=1}^{N_1} \delta[y - n_1 3a'] + \sum_{n_2=1}^{N_2} \delta[y - P_1 a' - 3n_2 a'] +$$

$$+ \sum_{n_3=1}^{N_3} \delta[y - P_1 a' - P_2 a' - 3n_3 a'] +$$

$$+ \sum_{n_4=1}^{N_4} \delta[y - P_1 a' - Pa_2' - Pa_3' - 3n_4 a'] +$$

$$+ \dots + \sum_{n_p=1}^{N_p} \delta[y - P_1 a' - P_2 a' \dots - P_{p-1} a' - 3n_p a'] .$$

This can be simplified to give the $i$-th term as

$$\sum_{n_1=1}^{N_1} \delta[y - 3n_1 a'] + \sum_{n_2=1}^{N_2} \delta[y - P_1 a' - 3n_2 a'] +$$

$$+ \sum_{n_3=1}^{N_3} \delta[y - a'(P_1 + P_2) - 3n_3 a'] + \sum_{n_4=1}^{N_4} \delta[y - a'(P_1 + P_2 + P_3) - 3n_4 a'] +$$

$$+ \dots + \sum_{n_i=1}^{N_i} \delta[y - a'(P_1 + P_2 + \dots + P_{i-1}) - 3n_i a'] + \dots$$

$$\equiv \sum_{i=1}^{p} \sum_{n_i=1}^{N_i} \delta[y - 3n_i a' - a' \sum_{r=0}^{i-1} P_r] .$$

Similarly for the $M$ terms,

$$\sum_{m_1=1}^{M_1} \delta[y - a'(3N_1 + 4m_1)] + \sum_{m_2=1}^{M_2} \delta[y - a'(3N_1 + 4M_1 + 3N_2 + 4m_2)] +$$

$$+ \sum_{m_3=1}^{M_3} \delta[y - (3N_1 + 4M_1 + 3N_2 + 4M_2 + 3N_3 + 4m_3)] +$$

$$+ \dots + \sum_{m_i=1}^{M_i} \delta[y - a'(3N_1 + 4M_1 + 3N_2 + 4M_2 + \dots + 3N_i + 4m_i)] + \dots =$$

$$= \sum_{i=1}^{p} \sum_{m_i=1}^{M_i} \delta[y - a'(P_1 + P_2 + \dots + P_{i-1} + 3N_i + 4m_i)] =$$

$$= \sum_{i=1}^{p} \sum_{m_i=1}^{Mi} \delta[y - 3a'N_i - 4a'm_i - a' \sum_{r=0}^{i-1} P_r] .$$

So (8) now can be simplified to give

$$\{ \sum_{i=1}^{p} \sum_{n_i=1}^{N_i} \delta[y - 3n_i a' - a' \sum_{r=0}^{i-1} P_r] + \sum_{i=1}^{p} \sum_{m_i=1}^{M_i} \delta[y - 3a'N_i - 4a'm_i - a' \sum_{r=0}^{i-1} P_r]\} *$$

$$* \sum_{k=-\infty}^{\infty} \delta[y - k(4M + 3N)\, a'] . \qquad (9)$$

The Fourier transform of (9) yields

$$G(\nu) = \frac{1}{(4M + 3N)\, a'} \left[ \sum_{i=1}^{p} \sum_{n_i=1}^{N_i} \exp\left(-2\pi i\nu\{3n_i a' + a' \sum_{r=0}^{i-1} P_r\}\right) + \right.$$

$$\left. + \sum_{i=1}^{p} \sum_{m_i=1}^{M_i} \exp\left(-2\pi i\nu\{3a'N_i + 4a'm_i + a' \sum_{r=0}^{i-1} P_r\}\right)\right] \sum_{n=-\infty}^{\infty} \delta\left(\nu - \frac{n}{(4M+3N)\, a'}\right),$$

where $\nu = k/(4M + 3N)\, a'$, with $k$ taking integral values from 0 to $\pm\, \infty$.

Previously we made $G(\nu)$ real and even by the use of the shift theorem. In this case the shift method cannot be applied, and the form of $G(\nu)$ will be complex. This does not have any effect on the overall intensity pattern, since

$$I(\nu) = G(\nu)\, G^*(\nu)\, .$$

## 3. Experimental

Each sample was prepared by a solid state reaction between $\alpha$-$Bi_2O_3$ and $TeO_2$ oxides which were high-purity commerical products, each of which was first heated separately under an oxygen flow for 24 h at 550 °C. The two oxides, carefully mixed and ground in an agate mortar, were heated under a pure nitrogen atmosphere in a gold crucible at 650 °C for at least 2 d, and slowly cooled.

Electron microscope samples were prepared by grinding small quantities of material in an agate mortar under n-butanol and allowing a drop of the resultant suspension to dry on a porous carbon support film. The samples were examined by transmission electron microscopy using a JEOL 100B microscope fitted with a top entry goniometer stage capable of $\pm$ 30° tilt.

Masks of the model structures were made by punching holes in a sheet of opaque material, each hole representing a vacancy, that is a delta function of Fig. 1. The diffraction patterns were obtained using a high-quality optical diffractometer [3].

## 4. Results

In Fig. 4 we present four typical results, and include optical diffraction patterns and intensities calculated from (7). We have chosen the arrangements $M = N = 1$ (i.e. 43) and $M = N = 2$ (i.e. $4^2 3^2$) to correspond to intergrowths at the half unit cell and whole unit cell level for the phase 1 $Bi_2Te_2O_7$:1 $Bi_2TeO_5$. We have also chosen a sequence which can be considered to represent a $Bi_2TeO_5$-rich composition, viz. $M = 1$, $N = 3$, ($43^3$), corresponding to the phase 1 $Bi_2Te_2O_7$:3 $Bi_2TeO_5$ and a sequence rich in $Bi_2Te_2O_7$, viz. $M = 3$, $N = 1$ ($4^3 3$) corresponding to 3 $Bi_2Te_2O_7$:1 $Bi_2 \cdot TeO_5$. All these figures show a similar overall intensity pattern. A group of strong reflections occurs bordered by weaker reflections which are not arranged symmetrically about each stronger reflection. The separation of the reflections is proportional to $1/(4M + 3N)\, a'$.

In the diagrams shown, because we take $a' = a_F/2$, the first intense reflection corresponds to the $020_F$ reflection of the fluorite subcell. The simple model chosen, however, does not take into account the fact that the symmetry of the fluorite cell leads to systematic absences, which results in the extinction of the $010_F$ reflection. This point must be noted when relating experimental diffraction patterns with the patterns just produced. For phases rich in $Bi_2TeO_5$ the intense superlattice reflections lie near to 1/3 and 2/3 of the separation between the 000 and 020 reflections. As the quantity of $Bi_2Te_2O_7$ increases, these strong superlattice reflections move towards 1/4 and 3/4 of this separation.

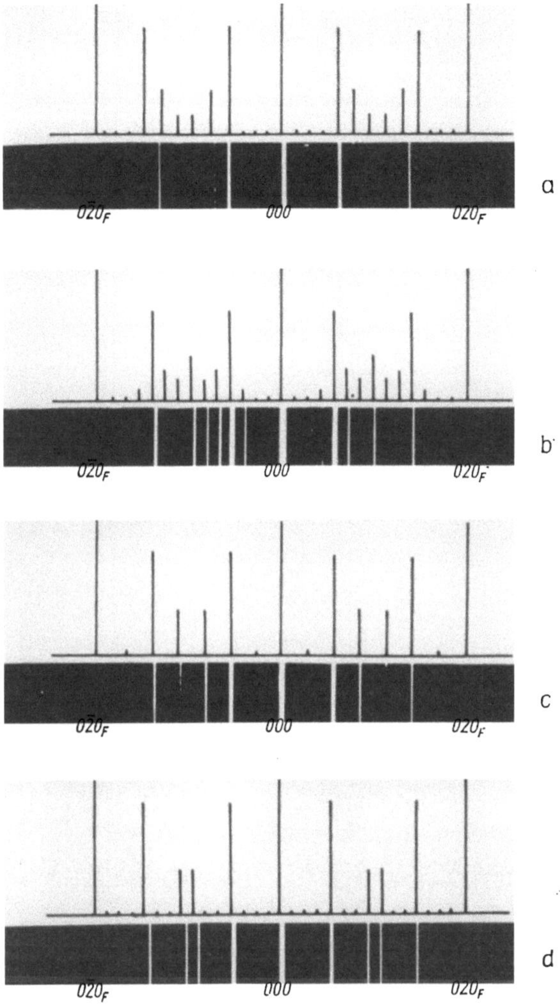


Fig. 4. Optical diffraction patterns and calculated intensities for several ordered intergrowths of $Bi_2TeO_5$ ($3^2$) and $Bi_2Te_2O_7$ ($4^2$). a) ($43^3$), b) ($4^23^2$), c) (43), d) ($4^33$)

The results of the general theory are very similar to those found in the simple theory. Once again the space between 000 and $020_F$ contains two intense superlattice reflections, one at a position between (1/4 and 1/3) $d^*_{020}$ and the other between (2/3 and 3/4) $d^*_{020}$. In Fig. 5 we show the results for values of $(4M + 3N)$ equal to 17, 18, and 20.

In Fig. 6 we show a typical electron diffraction pattern. The square array of intense fluorite type reflections is clearly visible, as are the superlattice reflections. It is immediately apparent that the overall intensity pattern of the superlattice spots is in good agreement with the theoretical models presented in Fig. 4 and 5. The spacing between the superlattice spots gives the total $4M + 3N$ by the simple relationship $d^*_{020}/d^*_{SL} = 4M + 3N$. Thus in Fig. 6a, $4M + 3N = 14$. It is possible to factorise this into the sequence ($3^24^2$). The calculation of the intensities of the superlattice (SL) spots for this structure is shown in Fig. 4b. Comparison of the intensity profile and the diffraction pattern reveals that our interpretation is unequivocal.

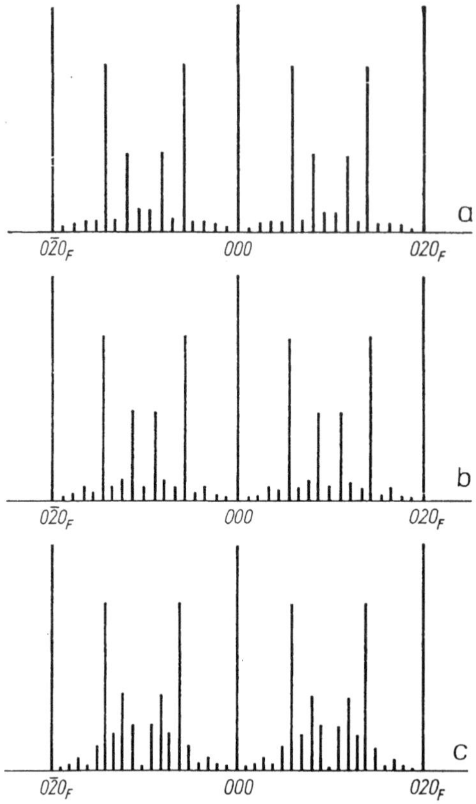


Fig. 5. Calculated intensities for the ordered intergrowths a) $343^24$, $(4M + 3N) = 17$; b) $3434^2$, $(4M + 3N) = 18$; and c) $343^34$, $(4M + 3N) = 20$

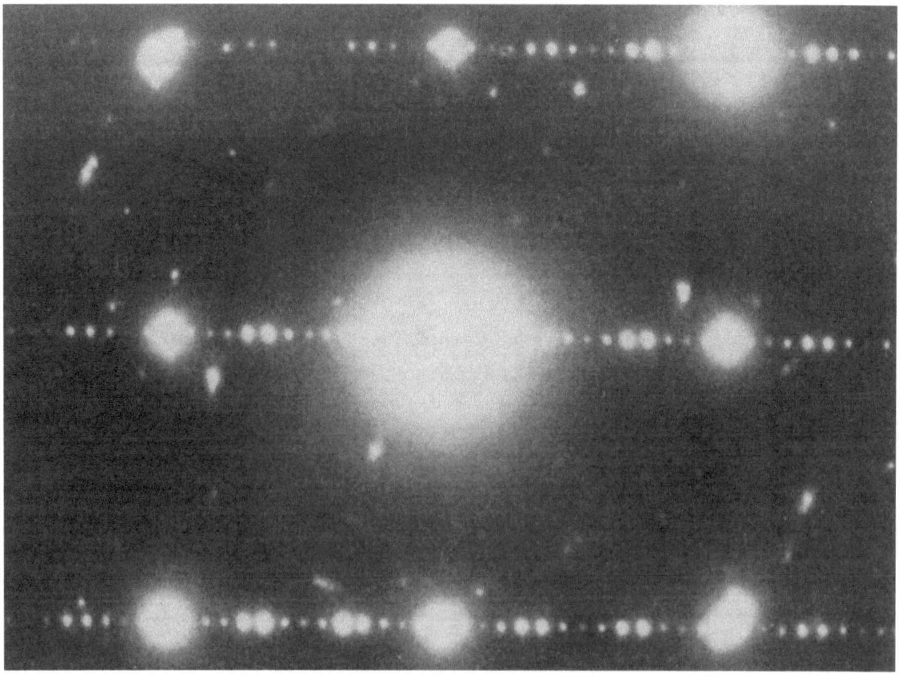

Fig. 6. Experimental electron diffraction pattern for which $(4M + 3N) = 14$. The sequence is $3^24^2$ which corresponds to 1 $Bi_2TeO_5$: 1 $Bi_2Te_2O_7$

In Fig. 7a we show an electron diffraction pattern in which the superlattice spacing is given by a value of $(4M + 3N) = 18$. The two possible stacking sequences are shown in Table 1 as ($4^3 3^2$) and ($3434^2$). It is possible to calculate the intensity variations for the first of these arrangements with the simple theory and the second with the general theory. Whilst the intensity fit is not perfect, it is still clear that the cor-

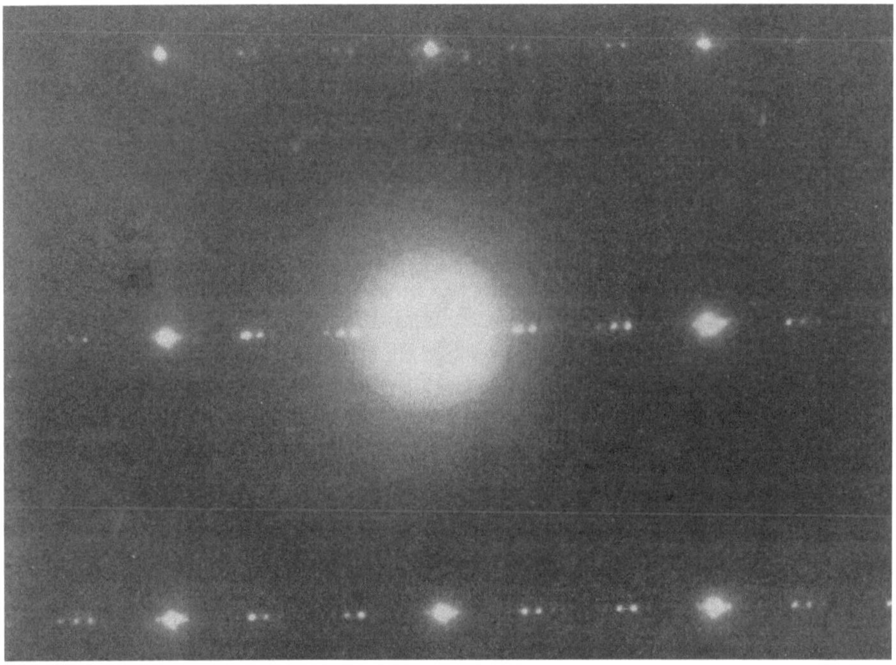

a

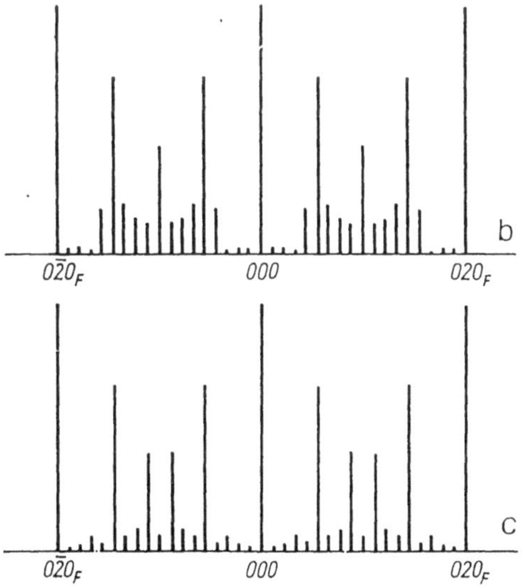


Fig. 7. a) Experimental electron diffraction pattern and calculated intensity plots for the arrangements b) $3^2 4^3$ and c) $3434^2$ which correspond to $(4M + 3N) = 18$. It is seen that the arrangement b) is correct

rect sequence is ($4^33^2$), shown in Fig. 7b, provided that the spurious peaks close to $010_F$ are ignored.

## 5. Discussion

The results of the theoretical and optical analysis, despite the simplicity of the one-dimensional model chosen, are in excellent agreement with the experimentally obtained electron diffraction patterns. The principle feature of these is the two strong superlattice reflections which lie between the (000) and (020) spots of the fluorite sublattice. The calculations show that the positions of these spots varies from 1/3 and 2/3 of $d_{(020)}$ for $Bi_2TeO_5$ to 1/4 and 3/4 of $d_{(020)}$ for $Bi_2Te_2O_7$. This variation is made in discrete steps and depends upon $(4M + 3N)$, but in practice the variation can be considered to be continuous because of the large number of combinations of $M$ and $N$ which are possible. Exactly the same variation is found in the experimentally determined patterns.

The positions of the weaker superlattice spots is also well accounted for in our model, and a measurement of the superlattice spacings gives the total value of $(4M + 3N)$. The combination of data in the positions of the most intense superlattice spot spacing often leads to an unambiguous assignment of $M$ and $N$.

In conclusion, therefore, it is possible to say that the intergrowth model proposed by El Farissi et al. [1] is completely supported by this study. However, some electron diffraction patterns show further complexities, not treated in the simple model described here. These need a more sophisticated treatment that will be published in the future.

***Acknowledgements***

One of the authors (B. H. P.) acknowledges with gratitude the tenure of an SERC research grant. R. J. D. Tilley is grateful to the University of Limoges for support.

## References

[1] M. El Farissi, D. Mercurio, and B. Frit, Mater. Chem. Phys. **16**, 133 (1987).
[2] D. Mercurio, M. El Farissi, B. Frit, and P. Goursat, Mater. Chem. Phys. **9**, 467 (1983).
[3] G. Harburn, C. A. Taylor, and T. R. Welberry, An Atlas of Optical Transforms, G. Bell & Sons, London 1975.
[4] R. N. Bracewell, Fourier Transform and Its Applications, McGraw-Hill Publ. Co., 1978.

*(Received March 8, 1988)*

phys. stat. sol. (a) 108, 123 (1988)

Subject classification: 61.60; S7.16

*Sektion Physik*[1]) *(a) und Sektion Chemie*[2]) *(b) der Karl-Marx-Universität Leipzig*

# Internal Distortion in GaInAsP Quaternary Alloys Studied by Ion Channeling

By

R. FLAGMEYER (a), V. GOTTSCHALCH (b), and H. FREY[3]) (a)

Dedicated to Prof. Dr. E. GUTSCHE on the occasion of his 60th birthday

First ion channeling analyses of quaternary GaInAsP films on InP (100) substrates are reported. The influence of the distorted crystal lattice on the channeling effect is demonstrated by Rutherford backscattering of $He^+$ ions channeled along the [100] direction and by angular scans of proton induced $GaK_\alpha$ and $AsK_\beta$ radiation.

Erste Channeling-Analysen von quaternären GaInAsP-Schichten auf InP(100)-Substrat werden vorgestellt. Anhand der Rutherfordrückstreuung in [100]-Richtung kanalisierter $He^+$-Ionen und der Winkelverteilung protoneninduzierter $GaK_\alpha$- und $AsK_\beta$-Strahlung wird der Einfluß des verzerrten Kristallgitters auf den Channeling-Effekt nachgewiesen.

## 1. Introduction

The local structure of ternary and quaternary III–V semiconductors has begun to attract much attention since it is indispensable for an understanding of their characteristic properties. As first shown by EXAFS [1, 2] the atoms are systematically dislocated from their regular lattice positions. This is because the bond lengths in pseudobinary alloys tend to remain close to those in the constituent binary compounds, deviating from the average VCA values.

In an (In, Ga)As alloy the difference in In–As and Ga–As bond lengths results in a mean displacement of the As atoms, e.g. vertically from ⟨100⟩ atomic strings, of about 10 pm. Such distances, however, are of the order of the thermal vibration amplitudes as well as the potential screening length to which ion channeling is known to be very sensitive. As expected RBS/channeling measurements showed increasing dechanneling of the $He^+$ ions which was much stronger than in binary single crystals [3]. Even In dopants in GaAs crystals result in a considerable increase of the channeling backscattering yield [4].

In this paper a proof of the atomic displacements in quaternary alloys will be given using the RBS and PIXE channeling technique for the first time.

The results are compared with those of Haga et al. [3] for ternary layers, and the experimentally [2] and theoretically [5] determined bond lengths in quaternary alloys. The $In_{1-x}Ga_xAs_yP_{1-y}$ layers were grown by a step cooling technique on InP(100) substrates with smooth and oxide free surfaces [6].

[1]) Linnéstr. 5, DDR-7010 Leipzig, GDR.

[2]) Liebigstr. 18, DDR-7010 Leipzig, GDR.

[3]) Present address: VEB Mikroelektronik „Karl Marx", Technikum Mikroelektronik, Rudolfstr. 47, DDR-5010 Erfurt, GDR.

## 2. RBS/Channeling Analysis

The Rutherford backscattering investigations [7] were performed on the van de Graaff accelerator of the Physics Department of the Karl-Marx-University Leipzig using 1.2 to 1.6 MeV $He^+$ ions. Generally, the surface normal [100] direction was used for channeling measurements with an angular divergence of the incident ion beam of $\Delta\psi \approx 0.02°$. The characteristic lattice distortion as well as the crystalline quality of the epilayer are related to the normalized channeling yield $\chi(t)$, which is defined as the ratio of the [100] aligned yield to the random yield. In one case (sample Q2) an RBS angular scan across the [100] axis was taken using 1.72 MeV $H^+$ ions to monitor the PIXE angular scans.

The depth-dependent dechanneling fractions $\chi(t)$ for three quaternary samples are plotted in Fig. 1. Also shown are the normalized channeling minimum yields of the binary crystals GaAs and InP. (The $\chi(t)$ curves of GaP and InAs are known to be just as low.)

Whereas the structure of sample Q3 probably includes some disorder [7] (cf. Table 1) samples Q1 and Q2 are nearly perfect lattice matched to the substrates with dislocation

Table 1

As-grown thickness $d$, perpendicular mismatch $(\Delta a/a)_\perp$, and $\Delta\chi(t)$ values at depths $t$ of 0.1 and 0.6 μm

| sample No. | $d$ (μm) | $10^4\ \Delta a/a_\perp$ | $\Delta\chi(0.1)$ (%) | $\Delta\chi(0.6)$ (%) |
|---|---|---|---|---|
| Q1 | 1.1*) | −2.5 to 1.5 | 2.2 | 8.5 |
| Q2 | 0.28*) | 1 to 2 | 1.9 | — |
| Q3 | 2.8 | 3 to 8 | 3.4 | 11.5 |
| T [8] | 5.5 | −2 | 2.2 | 6.5 |
| T [3] | 1.62 | 2 | 2.3 | 5.5 |

*) RBS results.
The experimental error of $\Delta\chi$ is ±0.2%.

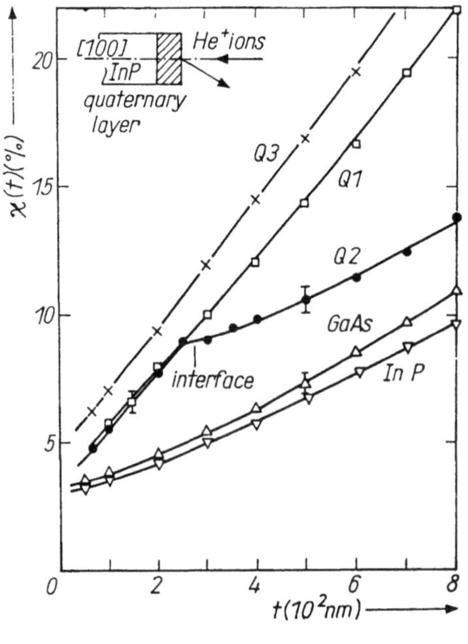


Fig. 1. Depth dependence of the normalized [100] channeling minimum yield $\chi(t)$ of the backscattered $He^+$ ions

densities below the detection limit of the channeling technique. Thus the difference of $\chi(t)$ between the quaternary alloys and the binary crystals is attributed to the scattering of the channeled $He^+$ ions by the displaced lattice atoms. The decrease of the dechanneling rate when the ions had reached the InP substrate (sample Q2) gives evidence for an additional dechanneling mechanism in the quaternary layer. — It is noteworthy that the $\chi(t)$ curves of Q1 and Q2 lie only somewhat higher than those of $Ga_{0.47}In_{0.53}As$/InP(100) samples [3, 8]. For comparison Table 1 contains the parameters of some ternary samples, where $\Delta\chi(t)$ is given by the difference of the minimum yields of the quaternary (ternary) and LEC-grown binary crystals, $\Delta\chi(t) = \chi(t)^{Q,T} - \chi(t)^{B}$.

Let us consider the $\chi(t)$ values at the depth of $t = 50$ nm (named $\chi_{min}$ in the following) of Q1, Q2, and the binary compounds. According to the semi-empirical calculation [9], $\chi_{min}$ is given by

$$\chi_{min} = Nd\pi(3u^2 + 0.2a_{TF}^2) , \quad (1)$$

where $N$ is the atomic density, $d_{\langle hkl \rangle}$ the distance between the atoms in the $\langle hkl \rangle$ string, $u^2$ commonly the two-dimensional amplitude of the lattice thermal vibrations $u_{th}^2$, $a_{TF} = 0.8853a_0/(Z_1^{2/3} + Z_2^{2/3})^{1/2}$ the Thomas-Fermi screening length with the Bohr radius $a_0 = 53$ pm and the atomic numbers of the ion and the lattice atom, $Z_1$ and $Z_2$, respectively. Equation (1) is applied to $\chi_{min}$ of monoatomic or binary crystals with straight atomic strings. In the present case $u^2$ can be replaced by

$$\bar{u}^2 = u_{th}^2 + \bar{u}_{st}^2 , \quad (2)$$

where $\bar{u}_{st}$ is the mean value of all atomic displacements normal to the string. For a rough estimation, we assume that the area density $Nd$ and the thermal vibration amplitude of the quaternary crystals are equal to those of the binary one. In this way, from (1) and (2) it follows that

$$\bar{u}_{st}^2 = \frac{u_{th}^2}{0.07a_{TF}^2} \left( \frac{\chi_{min}^{Q}}{\chi_{min}^{B}} - 1 \right) . \quad (3)$$

Using $u_{th} = 13$ pm [10] and the experimental values of $\chi_{min}^{Q} = (4.6 \pm 0.2)$ and $\chi_{min}^{B} = (3.3 \pm 0.2)\%$ we get $\bar{u}_{st} = (8.4 \pm 1.5)$ pm. This average value is comparable to the experimental results for $Ga_{0.47}In_{0.53}As$ of $\bar{u}_{st} = 7.8$ [8] and 8.3 pm (using the $\chi_{min}$ (50 nm) values in [3]), and it is in the order of the bond-length variations measured by EXAFS [2].

## 3. PIXE and RBS Angular Yield Profiles

Ion channeling combined with particle-induced X-ray emission (PIXE) offers the possibility to distinguish between the displacements of different crystal components under the following conditions: (i) The greater part of the characteristic X-rays is produced in the first 100 to 200 nm. — This is necessary since the model of separate interaction of the channeled ions with the (Ga, In) and (As, P) [100] atomic rows is only valid as a surface approximation. (ii) The emission depths of the X-rays are equal.

Taking into account that In and P are present both in the layer and in the substrate, and their K or L X-ray production cross-sections and adsorption lengths are very different we restricted the experiment to the comparison between the $GaK_\alpha$ (9.25 keV) and $AsK_\beta$ (11.80 keV) X-ray yields. Although there is a very small difference in the absorption lengths of Ga and AsK radiation, the reduction in the X-ray yield due to absorption is much less than that associated with the energy-dependent cross-section [11]. This results in effectively the same mean emission depths for the Ga and AsK X-rays, which differ only by 0.2% for the sample configuration used here.

The PIXE measurements were carried out with 1.72 MeV $H^+$ ions. Setting an angle of 45° to the sample surface we detected the X-rays in a Si(Li) detector with a resolution of 190 eV. Using the area under the peaks of $GaK_\alpha$ and $AsK_\beta$ two [100] angular-yield profiles were recorded simultaneously with an angular resolution of 0.02°. Fig. 2 shows the Ga and As channeling dips normalized to the random yield level.

As is well known the angular half-width $\psi_{1/2}$ of a channeling dip is a function of $u_{th}$ and $a_{TF}$,

$$\psi_{1/2} = \alpha \left(\frac{u_{th}}{a_{TF}}\right) \psi_c \,, \tag{4}$$

where $\psi_c$ is Lindhard's critical angle for channeling [12]. Since the critical distance of closest approach to the [100] rows is in the order of the screening length, $r_c \approx a_{TF}$, that is about ten times larger than the adiabatic distance for X-ray production ($r_{ad} \approx \approx 1$ pm for Ga and AsK) the X-rays are induced predominantly by the dechanneled fraction of the $H^+$ beam [13]. In other words, the interaction of the ions with the screened Coulomb potentials of the rows determines the K X-ray yield in almost the same way as the RBS yield.

The ratio of the critical angles for the $In_{0.73}Ga_{0.27}$ ($\bar{Z}_2 = 44.1$) and the $As_{0.63}P_{0.37}$ rows ($\bar{Z}_2 = 26.3$) is $\psi_c(\text{In, Ga}) : \psi_c(\text{As, P}) = 1.29$. Using the experimental values (Fig. 2) of the PIXE angular scans $\psi_{1/2}(\text{Ga}) = 0.44° \pm 0.02°$ and $\psi_{1/2}(\text{As}) = 0.39° \pm \pm 0.02°$ we get from (4)

$$\alpha_{Ga} : \alpha_{As} = 0.87 \pm 0.08 \,.$$

This means the Ga dip is too *narrow* for the As dip. If we introduce a static displacement as we did for $\chi_{min}$ in (2) and regard the different $a_{TF}$-values of the strings we get, from the universal function $\alpha(u/a_{TF})$ [14],

$$u_{Ga} \approx 1.15 u_{As} \,.$$

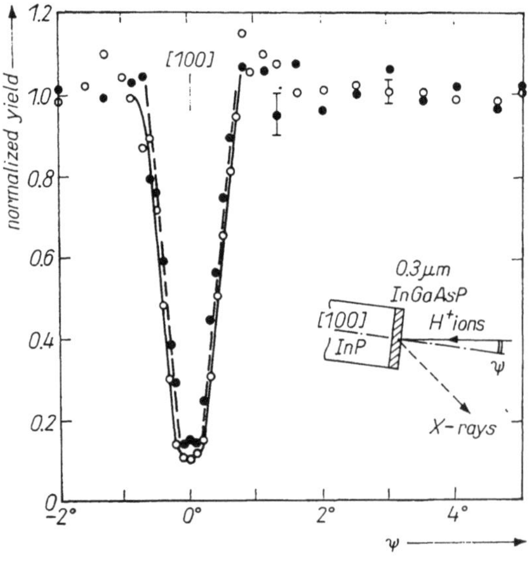


Fig. 2. Normalized angular yield profiles along the [100] axis for the proton-induced $GaK_\alpha$ (○) and $AsK_\beta$(●) X-rays

The RBS angular-scan simultaneously recorded (not shown in Fig. 2) yielded a similar $\chi_{min}$-value of 10% and is identical with the $AsK_\beta$ distribution in the half-width, $\psi_{1/2}(RBS) = 0.39° \pm 0.02°$. The yield within the backscattering energy window corresponding to the film thickness of 280 nm contains about 70% backscattering events from the In atoms (and 10% from the Ga atoms). It follows that

$$\alpha_{(In)RBS} : \alpha_{As} \approx 0.8$$

and, finally,

$$u_{(In)RBS} \approx 1.5 u_{As} .$$

There is no information available about the P displacement.

## 4. Discussion

Calculated [5, 15] and EXAFS experimental [1, 2] bond lengths of ternary and quaternary compounds lattice matched to InP are listed for comparison in Table 1. The ternary alloy is known to be composed of about 50% $(In_2Ga_2)As$ tetrahedron cells and about 50% $(In_1Ga_3)As$ or $(In_3Ga_1)As$ tetrahedra [16]. The internal distortion is effected by displacement of only the As central atoms in ⟨100⟩ or ⟨111⟩ directions, which results in a mean displacement from the ⟨100⟩ rows of $\bar{u}_{st}(As) \approx 10$ pm. These clear conditions allow computer simulations for a study of the effect of As displacement on the ion channeling [8]. Because of the lack of a model for the quaternary crystal structure which consists of a variety of tetrahedron cells of different compositions, there is no simple connection of the measured $\bar{u}_{st}$-value and the $u$-relations for Ga, As, and In atoms with the bond lengths. Additionally, the fact that in compound semiconductors each kind of lattice atoms can vibrate with its own amplitude makes the quantitative determination more difficult. Experimental $u_{th}$ values are not available.

It is somewhat surprising that in the quaternary alloy the displacement of the In atoms is larger than that of the Ga and As atoms. But assuming a mean square thermal vibrational amplitude of the In atoms nearly twice as large as that of the Ga or As atoms which has been established for the ternary (In, Ga)As compounds [17], and using (2) the relation of the atomic displacements decreases to

$$u_{st}(In) \approx u_{st}(Ga) \gtrapprox u_{st}(As) .$$

Considering the nearly identical differences of the bond lengths from the VCA value (Table 2) of 14.1 and 13.4 pm for the $(In_nGa_{1-n})P$ and $(In_nGa_{1-n})As$ tetrahedron,

Table 2

Experimental and calculated bond lengths in quaternary and ternary alloys in comparison with the VCA lattice constant. For the samples in question the values were taken from the contour chartes (Fig. 2) in [5]

| composition $Ga_xIn_{1-x}As_yP_{1-y}$ | | bond length (pm) Ga–P | Ga–As | In–As | In–P |
|---|---|---|---|---|---|
| $x = 0.26$, $y = 0.58$ | exp. [2] | 240 | 247 | 259 | — |
| $x = 0.27$, $y = 0.63$ | calc. [5] | 240.3 | 247.2 | 260.6 | 254.4 |
| VCA | | | 254.1 | | |
| diff. from VCA | | −13.8 | −6.9 | +6.5 | +0.3 |
| $x = 0.47$, $y = 1$ | exp. [1, 2] | | 247 | 260.4 | |
| $x = 0.47$, $y = 1$ | calc. [5, 15] | | 247.2 | 260.3 | |
| diff. from VCA | | | −7.0 | +6.3 | |

respectively, in the quaternary alloy and of 13.3 pm for the $(In_nGa_{1-n})As$ tetrahedron in the ternary one the similar $\chi_{min}$- and $\bar{u}_{st}$-values can be explained. On the other hand, the $u_{st}$-relation can be accounted for by tetrahedron cells of the types $Ga(As_mP_{1-m})$ and $In(As_mP_{1-m})$ with the *cations* as central atoms. From this view point, considerable displacements of the In and Ga atoms can be expected [18]. Summarizing the discussion we suggest a mean displacement of the Ga, As, In, and, probably, also the P atoms from the ⟨100⟩ atomic strings of 8.5 pm which is somewhat lower than the As displacement in ternary compounds. Whereas in the latter the (In, Ga) sublattice was found to be an undistorted virtual crystal lattice, in our case, obviously, both sublattices are distorted from a f.c.c. lattice.

***Acknowledgements***

The authors wish to thank Dr. U. Pietsch for the XRD results and stimulating discussions, U. Lehmann for technical assistence, Dr. J. Vogt and Dr. K. Gärtner for useful comments, and Prof. Dr. G. Otto for his support.

## References

[1] J. C. MIKKELSEN and J. B. BOYCE, Phys. Rev. B **28**, 7130 (1983).
[2] H. OYANAGI, Y. TAKEDA, T. MATSUSHITA, T. ISHIGURO, and A. SASAKI, J. Physique **47**, C8–423 (1986).
[3] T. HAGA, T. KIMURA, Y. ABE, T. FUKUI, and H. SAITO, Appl. Phys. Letters **47**, 1162 (1985).
[4] M. SATOH, K. KURIJAMA, and C. KIM, in: Semi-Insulating III–V Materials, Ohmsha, Ltd., 1986 (p. 35).
[5] M. ICHIMURA and A. SASAKI, Japan. J. appl. Phys. **26**, 246 (1987).
[6] E. KURTH, A. REIF, V. GOTTSCHALCH, J. FINSTER, and E. BUTTER, Crystal Res. Technol. **23**, 117 (1988).
[7] R. FLAGMEYER and V. GOTTSCHALCH, to be published.
[8] R. FLAGMEYER, K. GÄRTNER, and U. PIETSCH, to be published.
[9] J. H. BARRETT, Phys. Rev. B **3**, 1527 (1971).
[10] LANDOLDT-BØRNSTEIN, New Series, Vol. 22a, Ed. O. MADELUNG, Springer-Verlag, 1987.
[11] R. S. BHATTACHARYA, J. appl. Phys. **54**, 5039 (1981).
[12] J. LINDHARD, Kong. Danske Vid. Selsk., mat.-fys. Medd. **34**, No. 14 (1965).
[13] R. KALISH and G. BAHIR, Nuclear Instrum. and Methods **218**, 415 (1983).
[14] J. U. ANDERSEN and E. LAEGSGAARD, Radiat. Eff. **12**, 3 (1972).
V. GEIST and R. FLAGMEYER, Thesis, Karl-Marx-Universität Leipzig, 1976.
[15] T. FUKUI, Japan. J. appl. Phys. **23**, L208 (1984).
[16] A. SASAKI and M. ICHIMURA, Superlattices and Microstructures **3**, 127 (1987).
[17] U. PIETSCH, phys. stat. sol. (b) **107**, 185 (1981).
[18] U. PIETSCH, phys. stat. sol. (b) **133**, 483 (1986).

*(Received February 22, 1988)*

phys. stat. sol. (a) **108**, 129 (1988)

Subject classification: 61.10; 68.20; S5.11

*Sektion Physik der Ludwig-Maximilians-Universität München*[1])

# Grazing Incidence X-Ray Diffraction from Si with an Implantation Induced Amorphous Surface Layer[2])

By

G. Wallner, E. Burkel, H. Metzger, and J. Peisl

Si single crystals are implanted with As ions of 80 keV to a dose of $1 \times 10^{15}$ cm$^{-2}$. Under conditions of total external reflection Bragg diffraction of X-rays is measured. Intensity distributions $I_H$ as a function of angles of grazing exit $\alpha_H$ are determined with high resolution at different angles of grazing incidence $\alpha$. The experimental results are compared to calculations on the basis of dynamical diffraction theory. It is found that implantation produces an amorphous layer at the surface, the thickness $t_{am}$ of which is determined from a least-squares fit of several experimental profiles $I_H$ to theoretical curves as $t_{am} = (144.5 \pm 5)$ nm. Complete agreement is achieved only if a variation in $t_{am}$ of 10% across the sample is allowed for. At incidence angles smaller than the critical angle amorphous diffuse scattering is observed.

Silizium-Einkristalle werden mit Arsen-Ionen einer Energie von 80 keV zu einer Dosis von $1 \times 10^{15}$ cm$^{-2}$ implantiert. Unter Totalreflexionsbedingungen wird die Bragg-Streuung von Röntgenstrahlen gemessen. Bei verschiedenen Winkeln des streifenden Einfalls werden Intensitätsverteilungen $I_H$ als Funktion des Winkels streifenden Ausfalls $\alpha_H$ mit hoher Auflösung bestimmt. Die experimentellen Ergebnisse werden mit Rechnungen auf der Basis der dynamischen Streutheorie verglichen. Es zeigt sich, daß die Implantation eine amorphe Schicht an der Oberfläche der Probe erzeugt. Deren Dicke wird aus einer Anpassung von mehreren experimentellen $I_H$ Profilen an die theoretischen Kurven zu $t_{am} = (144{,}5 \pm 5)$ nm bestimmt. Vollständige Übereinstimmung wird erst erzielt, wenn eine Variation in $t_{am}$ von 10% über die Probe berücksichtigt wird. Bei Einfallswinkeln kleiner als der kritische Winkel wird amorph diffuse Streuung beobachtet.

## 1. Introduction

In recent years new methods for the investigation of thin subsurface layers in solids by X-ray diffraction have been developed. Among them grazing incidence diffraction [1 to 5] is one of the most powerful: X-rays incident onto a crystal under grazing angle $\alpha$ are diffracted by lattice planes perpendicular to the surface. The diffracted intensity is observed at grazing angles of exit $\alpha_H$. By varying the angles $\alpha$ and $\alpha_H$ at the same time in a controlled way one disposes of the unique possibility to scan the depth dependence of structural parameters with high resolution. In a recent publication we have shown that dynamical diffraction theory can in detail describe precision measurements for perfect Si single crystals [6]. An essential point in this study was the observation of diffracted Bragg intensity as a function of $\alpha_H$. In this case the use of very intense X-ray sources is mandatory because of the stringent requirements upon

[1]) Geschwister-Scholl-Platz 1, D-8000 München 22, FRG.

[2]) Dedicated to Prof. Dr. U. Bonse on the occasion of his 60th birthday.

collimation of incident and exit beams. In this communication we report the application of this method to the study of a disturbed layer produced by ion implantation. We again focus on the dependence of diffracted intensity on $\alpha_H$ — thus extending previous work [7] — and on a detailed comparison of experiment with predictions of dynamical diffraction theory.

## 2. Experimental

The experimental conditions were the same as those described in more detail in [6]. The measurements were conducted at Hasylab (DESY) at the three-axis diffractometer [8]. A wavelength of 0.132 nm was used. The diffraction geometry is schematically shown in Fig. 1. The divergences $\delta\alpha = 0.3$ mrad and $\delta\theta = 2.9$ mrad of the incident beam were determined by spatial collimation and the natural characteristics of synchrotron radiation. Diffraction from (220) planes of Si single crystals with [100] surface normal was observed. The divergences $\delta\alpha_H = 0.1$ mrad and $\delta\theta_H = 2.9$ mrad were determined by spatial collimation. The intensity distributions $I_H$ (intensity as a function of $\alpha_H$) were recorded in a one-dimensionally position sensitive detector.

The surface of two high-purity Si single crystals was prepared by usual methods to industrial standard. The (220) lattice planes formed an angle of $\psi = 4.9$ mrad with the surface normal. All measurements were carried out at negative tilt angles $\psi$, i.e. the (220) lattice plane normal pointed out of the sample surface. One sample was implanted at ambient temperature with 80 keV As ions to a total dose of $1 \times 10^{15}$ cm$^{-2}$. Both samples were measured in the identical experimental arrangement. The reference measurement on the unimplanted sample served to set the absolute intensity scale: from the comparison of calculated Bragg intensities with the measured ones a scaling factor was derived. By this factor the experimental intensities of the implanted sample were scaled, and thus could be compared with calculated intensities on an absolute scale.

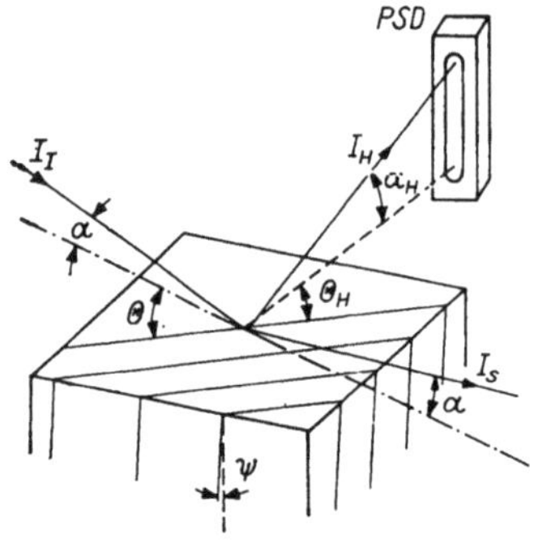


Fig. 1

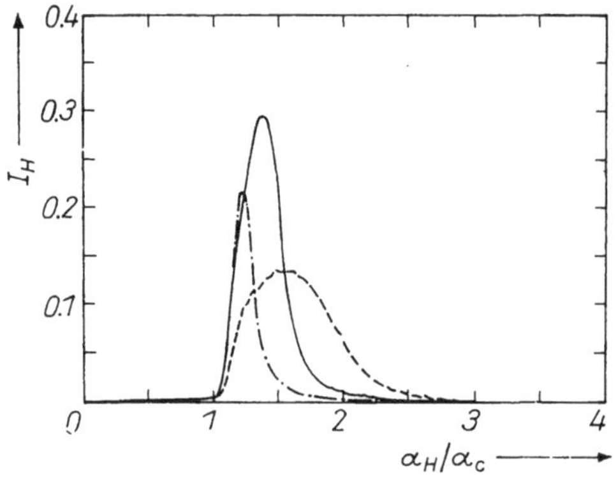


Fig. 2

Fig. 1. Diffraction geometry. The incident beam with intensity $I_I$ forms the glancing angle $\alpha$ with the sample surface, and the angle $\theta$ with the (220) planes. These are tilted by $\psi$ relative to the [100] surface normal. The specular beam with intensity $I_s$ is reflected under $\alpha$. The diffracted and reflected beam with intensity $I_H$ forms an angle $\theta_H$ with the lattice planes, and is observed as a function of $\alpha_H$ in the position sensitive detector PSD

Fig. 2. Experimental intensity distributions $I_H$ (normalized) as a function of exit angle $\alpha_H$ for different angles of incidence $\alpha$, for a Si single crystal implanted with $1 \times 10^{15}$ As ions of 80 keV. $\alpha/\alpha_c = 1.06$ (dashed), 1.24 (solid), 1.57 (dash-dotted)

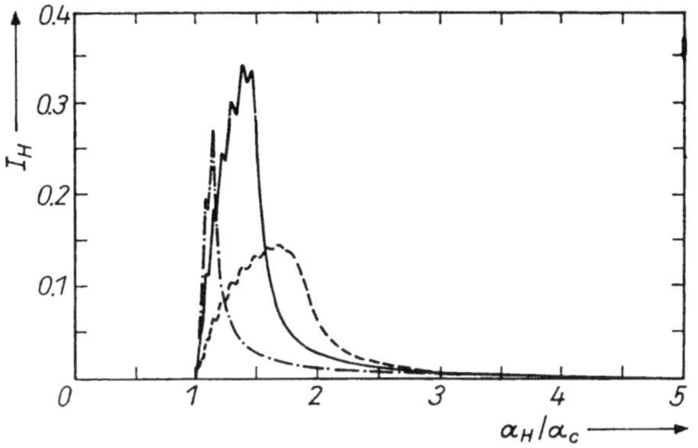


Fig. 3. Theoretical intensity distributions $I_H$ (normalized) for different angles of incidence $\alpha$. Calculations performed with a thickness of the implantation induced amorphous layer $t_{am} = 144.5$ nm. $\alpha/\alpha_c$ as in Fig. 2

## 3. Experimental Results and Comparison with Theory

Distributions $I_H$ of diffracted intensity were recorded for 35 different angles of incidence $\alpha$. Typical examples are shown in Fig. 2. The most apparent feature is the absence of Bragg-diffracted intensity for $\alpha$ and $\alpha_H$ smaller than the critical angle $\alpha_c$ (= 3.33 mrad for these experiments). It can be explained by the presence of an amorphous surface layer of thickness $t_{am}$ considerably greater than the minimum penetration depth $t_{min} = 6.12$ nm.

The experimental results were compared with theoretical calculations of $I_H$, as given by (17) and (14) of [6]. Ten experimental intensity profiles with different $\alpha$ between $0.9\alpha_c$ and $2.5\alpha_c$ were selected. They were simultaneously fitted to the respective theoretical profiles. The only free parameter in the least-squares fit procedure was $t_{am}$, the absolute intensity scale was fixed by the reference measurement. The result of the best fit is $t_{am} = (144.5 \pm 5)$ nm.

Theoretical intensity profiles for this value of $t_{am}$ are shown in Fig. 3. Their overall form and normalized absolute intensity agree very well with experiment. The oscillations superimposed on the theoretical distributions are not detected by experiment. They arise from interference between diffracted, and diffracted and reflected waves. Here the interfaces between amorphous layer and vacuum, and between amorphous layer and crystalline material, at which reflection occurs, were assumed ideally sharp, and the thickness of the amorphous layer of uniform thickness. However, an amorphous layer of uniform thickness is not expected to be produced by implantation. Accordingly, calculations were performed assuming a 10% variation of $t_{am}$ laterally across the sample. The resulting theoretical distributions (Fig. 4) do not show oscillations, and agree satisfactorily with experiment.

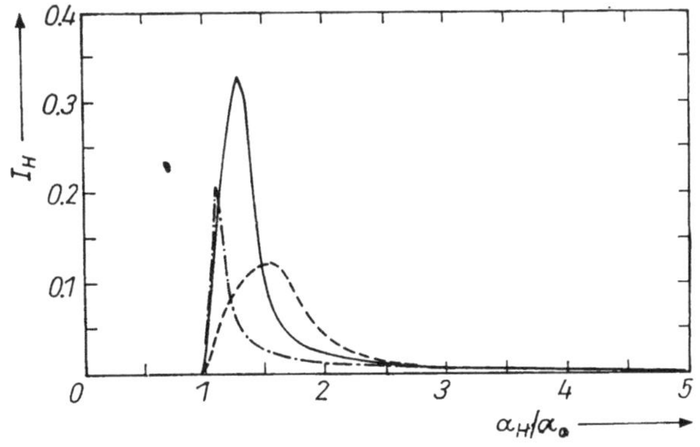


Fig. 4. Same as Fig. 3, with a 10% variation in $t_{am}$ across the sample face

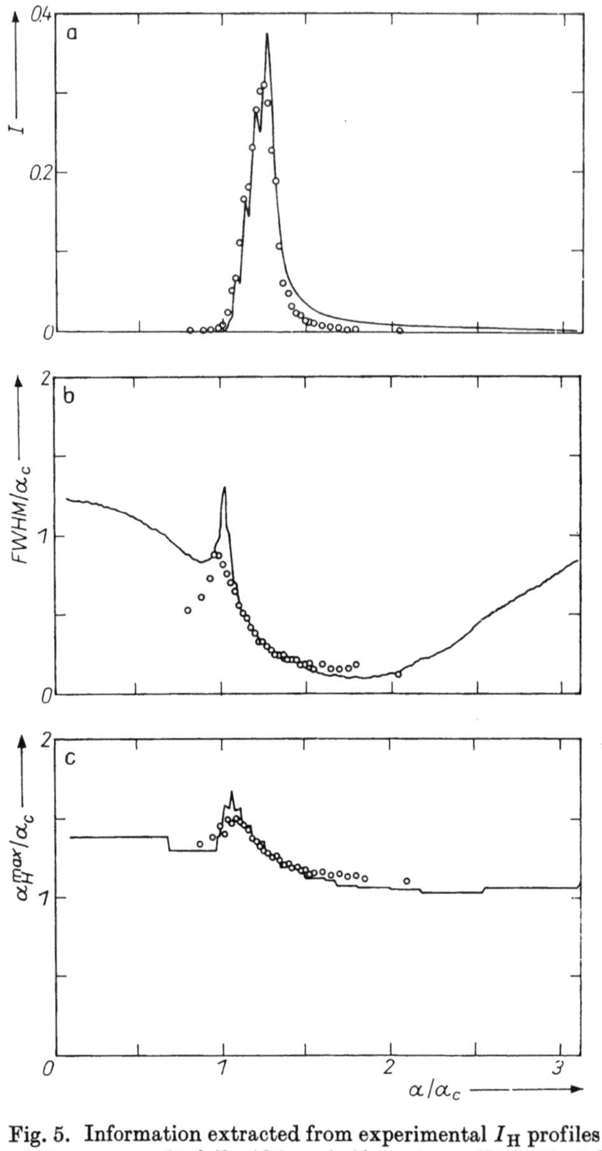


Fig. 5. Information extracted from experimental $I_H$ profiles for different $\alpha$. a) intensity (normalized) for $\alpha = \alpha_H$, b) full width at half maximum FWHM of $I_H$ profiles, c) exit angle of maximum intensity $\alpha_H^{max}$, all as function of $\alpha$

In order to make possible a more detailed and illustrative comparison of experiment with theory characteristic quantities were extracted from the experimental curves and compared to theory: intensity for $\alpha = \alpha_H$ (Fig. 5a), full width at half maximum FWHM (Fig. 5b), exit angle of maximum intensity $\alpha_H^{max}$ (Fig. 5c), all as a function of $\alpha$. As discussed in [6], the behaviour of these quantities clearly reflects the importance of dynamical diffraction effects.

Again the overall agreement between experiment and theory is satisfactory. Sharp structures in the theoretical curves are not observed experimentally. That once more

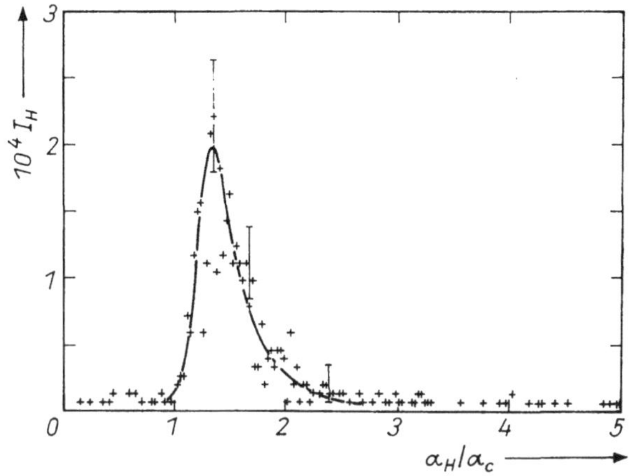


Fig. 6. Distribution $I_H$ (normalized) of diffuse scattering for $\alpha = 0.81\alpha_c$. Crosses represent the data points, the solid line is an eye guide only

is mainly due to the smearing of $t_{am}$, and in part also to resolution effects. The disagreement for $\alpha < \alpha_c$ in the $\alpha_H^{max}$ and FWHM curves (Fig. 5 b and c) is artificial. The measured intensity for $\alpha < \alpha_c$ is dominated by diffuse scattering (Fig. 6) and not representative for Bragg diffracted intensity. Dynamical diffraction effects are manifest in the distributions shown in Fig. 5: kinematical theory, for example, predicts no variation of $\alpha_H^{max}$ and the FWHM with $\alpha$.

The intensity of the diffuse scattered intensity for $\alpha = 0.81\alpha_c$ is smaller than Bragg diffracted intensity by a factor of about $10^{-4}$. At these intensity levels even measurements at bending magnet synchrotron sources are difficult to conduct, and Wiggler sources will have to be used. Nevertheless, the study of diffuse scattering under grazing incidence conditions will open new possibilities to obtain information on the structural and dynamical properties of surfaces and near surface regions. A first diffuse scattering study of implantation damage in Si was carried out in our group and will be published elsewhere.

***Acknowledgements***

We thank the Hasylab at DESY, Hamburg, for kind hospitality and support. We are indebted to J. Als-Nielsen, M. Nielsen, R. Feidenhansl, F. Grey for help with the experimental set-up, to H. Wagner, G. Gompper, U. Adler, N. Bernhard, H. Dosch for discussions and experimental help, and to H. Kakoschke, Siemens-Munich, who supplied the implanted sample. This work has been funded by the German Federal Minister for Research and Technology (BMFT) under contract number 03PE1LMU/2.

## References

[1] W. C. Marra, P. Eisenberger, and A. Y. Cho, J. appl. Phys. **50**, 6927 (1979).
[2] A. M. Afanasev and M. K. Melkonyan, Acta cryst. **A39**, 207 (1983).
[3] P. A. Aleksandrov, A. M. Afanasev, M. K. Melkonyan, and S. A. Stepanov, phys. stat. sol. (a) **81**, 47 (1984).
[4] P. A. Aleksandrov, A. M. Afanasev, A. L. Golovin, R. M. Imamov, D. V. Novikov, and S. A. Stepanov, J. appl. Cryst. **18**, 27 (1985).
[5] H. Dosch, Phys. Rev. B **35**, 2137 (1987).
[6] N. Bernhard, E. Burkel, G. Gompper, T. H. Metzger, J. Peisl, H. Wagner, and G. Wallner, Z. Phys. B **69**, 303 (1987).
[7] A. L. Golovin, R. M. Imamov, and E. A. Kondrashkina, phys. stat. sol. (a) **88**, 505 (1985).
[8] J. Als-Nielsen and P. S. Pershan, Nuclear Instrum. and Methods **208**, 545 (1983).

*(Received March 21, 1988)*

phys. stat. sol. (a) 108, 135 (1988)

Subject classification: 64.70 and 66.30; S8

*Department of Physics of Solids, Bashkir State University, Ufa*[1])

# Phase Transformations and Ionic Transport in the $Cu_{2-\delta}Te$ Superionic Conductor

By

R. A. Yakshibaev, N. N. Mukhamadeeva, and R. F. Almukhametov

Results are presented for the structural features of $Cu_{2-\delta}Te$ and the ionic transport as a function of temperature and nonstoichiometricity degree in the interval $T = 20$ to 450 °C and $\delta = 0$ to 0.08. The phase transformations are shown to be first-order phase transitions accompanied by changing the activation energy for ionic conduction. The ionic conductivity at 400 °C equals $0.8\ \Omega^{-1}\ cm^{-1}$ and decreases with deviation from stoichiometry. The chemical diffusion coefficient decreases from $8 \times 10^{-4}\ cm^2\ s^{-1}$ to $4.4 \times 10^{-4}\ cm^2\ s^{-1}$ when $\delta$ changes from 0.015 to 0.075 and $T = 400$ °C. The phase transition is also found in the anomalous dependence of the ionic thermo-e.m.f. on temperature. The ionic transport heat, as calculated from the ionic thermo-e.m.f., equals 0.23 at $\delta =$ $= 0.015$.

Представлены результаты исследований структурных особенностей $Cu_{2-\delta}Te$ и явлений ионного переноса как функция температуры и степени нестехиометричности в интервале $T = 20$ до 450 °C и $\delta = 0$ до 0,08. Показано, что фазовые переходы являются переходами 1-го рода и сопровождаются изменением энергии активации ионной проводимости. Ионная проводимость равна $0{,}8\ \Omega^{-1}\ cm^{-1}$ при 400 °C и уменьшается с отклонением от стехиометрии. Коэффициент химической диффузии уменьшается с $8 \times 10^{-4}\ cm^2\ s^{-1}$ до $4{,}4 \times 10^{-4}\ cm^2\ s^{-1}$ при изменении $\delta$ с 0,015 до 0,075 и $T = 400$ °C. Фазовый переход обнаруживается также в аномальной зависимости ионной термо-э.д.с. от температуры. Теплота переноса ионов, расчитанная исходя из ионной термо-э.д.с., равна 0,23 eV при $\delta = 0{,}015$.

## 1. Introduction

Copper telluride is a compound of variable composition with copper deficit [1]. The homogeneity region of $Cu_{2-\delta}Te$ reaches $\delta = 0.08$ at 400 °C [2]. With increasing temperature $Cu_{2-\delta}Te$ undergoes a number of phase transformations at 305, 360, and 555 °C [3]. Being a p-type degenerated semiconductor in its electrical properties copper telluride exhibits an anomalously high cation conductivity. Estimated from the build-up and decay curves of the concentration polarization, the ionic conductivity is equal to $30\ \Omega^{-1}\ cm^{-1}$ [4]. The present paper shows the results of investigations of the structural transformations in $Cu_{2-\delta}Te$ and the ionic transport features in the range 325 to 450 °C as a function of the degree of non-stoichiometry.

## 2. Experimental Technique

Using the ampoule-synthesis method, the specimens for the investigations were prepared from elements of exceptional purity by coagulating them in a vacuum at 600 °C with subsequent homogenization at 400 °C. The specimens were single-phase

[1]) Ul. Frunze 32, SU-450074 Ufa, USSR.

as was controlled by X-ray analysis. Powder diffractograms were taken by means of a diffractometer having a high-temperature attachment with $CuK_\alpha$ radiation using a β-filter. The recording of the diffraction line profile to estimate the lattice parameters was made at steps of 0.025°. The measuring temperature was kept constant with a precision to ±0.5 K.

The ionic conductivity, the chemical diffusion coefficient, and the ionic thermo-e.m.f. of the single-phase specimens were studied with the help of a cell of type (i)

$$\begin{array}{l} Pt_1 \mid Cu \mid CuBr \mid Cu_{2-\delta}Te \mid CuBr \mid Cu \mid Pt_4 \, . \\ Pt_2 \longrightarrow\uparrow \quad \uparrow\longrightarrow Pt_3 \end{array}$$

The use of CuBr having a unipolar copper ion conductivity in the temperature range 300 to 450 °C [5] allows only the ionic current to pass through the $Cu_{2-\delta}Te$ specimen studied. Stationary values of the potential difference $V_i^i$ between two potential probes Cu/CuBr estimated from the difference of the chemical potentials of copper ions allow the specific ionic conductivity to be calculated [6]. The time dependence $V_i^i(t)$ relates to the build-up and decay processes of the concentration polarization. Along the specimen length $L$ the dropping of the concentration polarization after switching-off the ionic current, is expressed by the time dependence $V_i^i(t)$ [7]

$$V_i^i(t) = V_i^i(0) \exp\left(-\frac{\pi^2 \tilde{D} t}{L^2}\right), \tag{1}$$

where $V_i^i(0)$ is the potential difference at the moment of switching-off the current, $\tilde{D}$ is the chemical diffusion coefficient.

The degree of deviation from stoichiometry ($\delta$) was controlled by the e.m.f. value $E$ of the part of the cell (ii)

$$Pt_2 \mid Cu_{2-\delta}Te \mid CuBr \mid Cu \mid Pt_4 \, ,$$

using the previously obtained curves of the Coulometric titration $E = f(\delta)$, $\delta$ being changed by passing a definite quantity of charge through the cell (ii) during the experiment. Since CuBr is a purely ionic conductor [5], the quantity of charge ($It$) through the cell (ii) was used as a measure of the composition change. According to the Faraday law,

$$\Delta\delta = \frac{Lt}{n_x F}, \tag{2}$$

where $n_x$ is the number of non-metallic gram-atoms in the specimen and $F$ the Faraday number.

To measure the ionic thermo-e.m.f. along the specimen in cell (i) a temperature gradient of 20 to 25 K/cm was established.

## 3. Results and Discussion

The X-ray spectra of the powder preparations are well indexed in the hexagonal lattice with the unit cell parameters $a = 1.2605$ nm and $c = 2.1750$ nm. Fig. 1 illustrates the temperature dependences of unit cell volume and lattice parameters.

Jumping changes of both the parameter $a$ and the unit cell volume occur at 300 and 350 °C. Therefore, the phase transitions at these temperatures are first-order phase transformations.

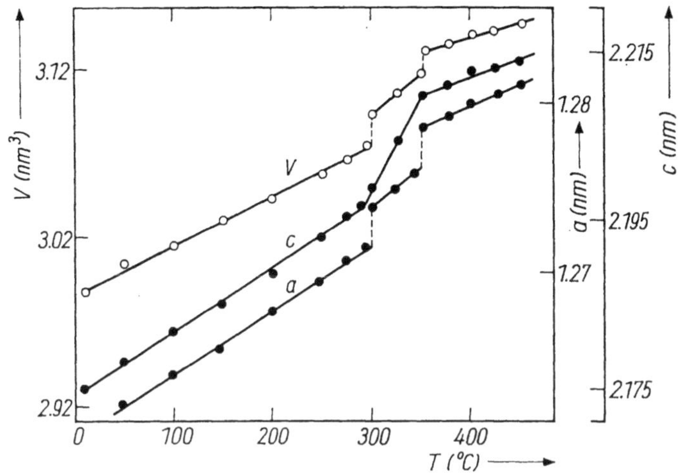


Fig. 1. Dependence of the lattice parameters $a$, $c$, and volume $V$ on temperature for $Cu_2Te$

The dependences of the e.m.f. of the electrochemical cell (ii) on the degree of non-stoichiometry at various temperatures are shown in Fig. 2. The plateau of the dependence $E = f(\delta)$ corresponds to the two-phase region. It is clear from that figure that the extension of the homogeneity region of copper telluride increases with increasing temperature.

Fig. 3 and 5 show the dependences of the ionic conductivity and the chemical diffusion coefficient as a function of the degree of deviation from stoichiometry at various temperatures. Temperature dependences of the ionic conductivity and the thermo-e.m.f. are shown in Fig. 4 and 6.

### *3.1 Ionic conductivity*

The ionic conductivity of $Cu_{2-\delta}Te$ and its functional dependence on the degree of non-stoichiometry are specified by two features. First, the high values of $\sigma_i$ being a consequence of structural features of the system involved. Copper telluride belongs

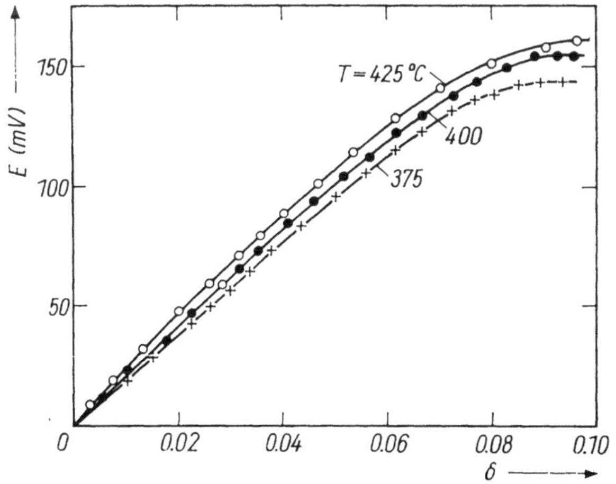


Fig. 2. Dependence of the cell-e.m.f. on the degree of non-stoichiometry at various temperatures

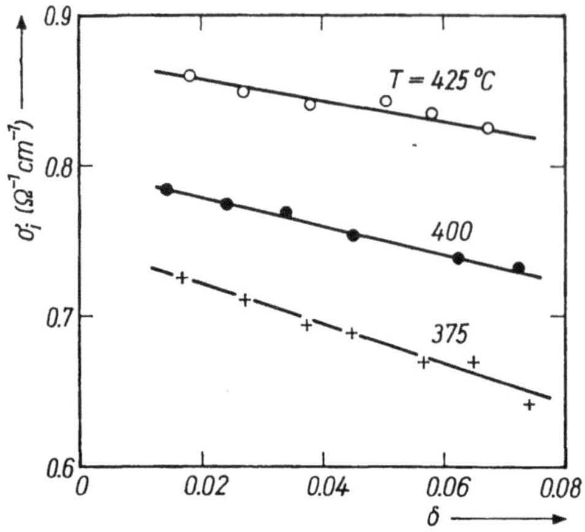


Fig. 3

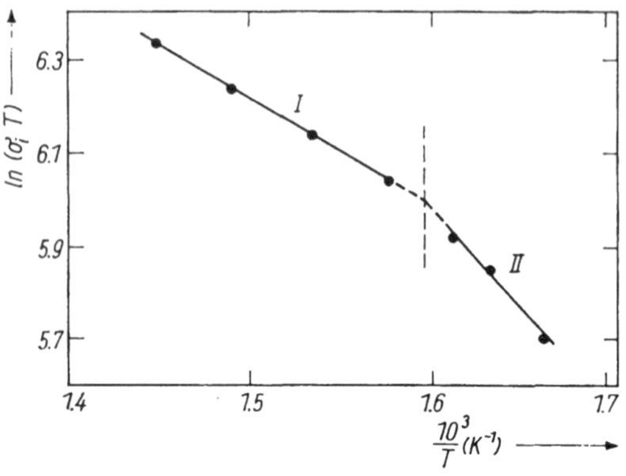


Fig. 4

Fig. 3. Dependence of the ionic conductivity on the degree of non-stoichiometry for $Cu_{2-\delta}Te$ at various temperatures

Fig. 4. Dependence of the ionic conductivity (in $\Omega^{-1}$ $cm^{-1}$) on reciprocal temperature for $Cu_{1.96}Te$

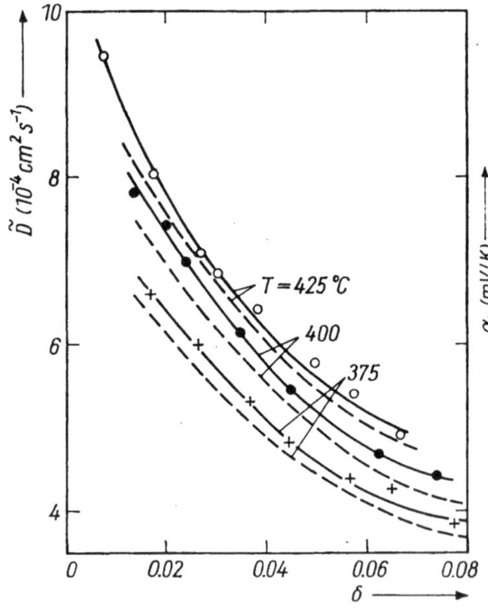


Fig. 5

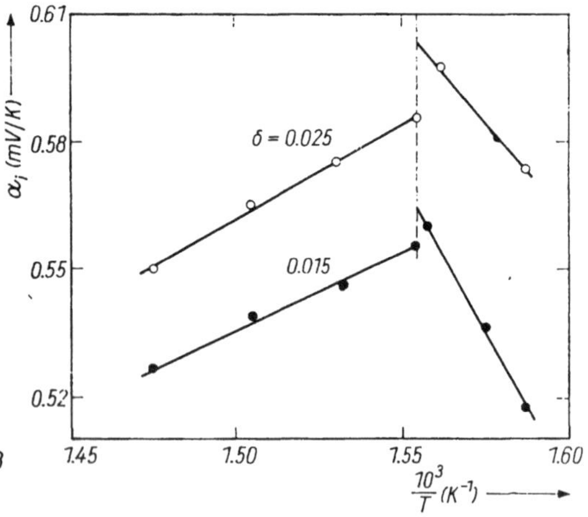


Fig. 6

Fig. 5. Dependence of the chemical diffusion coefficient on the degree of non-stoichiometry for $Cu_{2-\delta}Te$ at different temperatures. Symbols experimental, dashed curves calculation

Fig. 6. Dependence of ionic thermo-e.m.f. on temperature for $Cu_{2-\delta}Te$ at various $\delta$

to the class of structurally disordered superionic conductors whose high conductivity is due to the fact that part of copper ions have weak bonding with the rigid frame. Second, the ionic conductivity decreases with deviation from the stoichiometric composition (Fig. 3), which may be explained only by a decrease of the concentration of mobile ionic carriers. The concentration of mobile ions is determined by structural features of the phase and is independent of temperature. Therefore, the temperature changes of the ionic conductivity is fully conditioned by changes of the copper ion mobility.

The ionic conductivity of superionic conductors may be expressed as [8]

$$\sigma_i = \frac{q^2 v_0 f a^2}{kT} \varkappa N_i \exp\left(-\frac{E_m}{kT}\right), \tag{3}$$

where $a$ is the jump length, $f$ the correlation factor, $v_0$ the frequency of the ion oscillations in the vicinity of the equilibrium state, $q$ the ion charge, $N_i$ the number of ions per unit volume, $\varkappa$ the part of mobile ions, and $E_m$ the activation energy for migration.

Fig. 4 shows the dependence $\ln(\sigma_i T) = f(1/T)$ for the specimen of composition $Cu_{1.96}Te$. At the phase transition point at 350 °C a jumping change of the activation energy occurs. Above 350 °C the activation energy is equal to $E_a^{(i)} = (0.20 \pm 0.02)$ eV and below it $E_a^{(ii)} = (0.40 \pm 0.03)$ eV. It should be mentioned that copper telluride shows superionic properties both above and below the phase transition temperature, this transition being a superionic–superionic phase transition. Within the accuracy of experiment, the value of $\sigma_i$ shows no jump near the phase transition point. For superionic conductors the following empirically established relation between the ionic conductivity value and the degree of disorder of the mobile ion sublattice $n/n_i$ is true [8]:

$$\frac{\sigma_i}{N_{Cu^+}} = A \exp\left(B \frac{n}{n_{Cu^+}}\right), \tag{4}$$

where $N_{Cu^+}$ is the copper ion density per unit volume, $n$ the number of possible positions of mobile ions in the unit cell, $n_{Cu^+}$ the number of mobile ions in the elementary unit cell, $A$ and $B$ are constants. In comparison with other superionic conductors of the group of copper chalcogenides $Cu_2S$ and $Cu_2Se$ [9, 10], for $Cu_{2-\delta}Te$ (4) gives a comparatively low value of the ionic conductivity as a consequence of the smallness $n/n_{Cu^+}$. The hexagonal modification of $Cu_2Te$, specifically, yields the value 2 for $n/n_{Cu^+}$, whereas for copper sulfides and selenides with cubic symmetry the value is 6.

### *3.2 Chemical diffusion*

When the ionic component of conductivity is measured, the use of electrodes blocking the electron current results in the development of concentration polarization of a gradient of the non-stoichiometry degree along the specimen. The rate of the concentration polarization build-up and decay is determined by the value of the chemical diffusion coefficient. In the general case, the chemical diffusion describes diffusion processes under the action of the chemical potential gradients of the particles. Copper telluride is a compound of variable composition with mainly electron conductivity and a high cation conductivity $\sigma_e \gg \sigma_{Cu^+}$, $\sigma_{Te^{2-}} = 0$. For similar systems the relation between the coefficients of chemical diffusion and self-diffusion $D_i$ may be written as [11]

$$\tilde{D} = \frac{N_i D_i}{kT} \frac{d\mu_{Cu}}{dN_i}, \tag{5}$$

where $\mu_{Cu}$ is the metal atom chemical potential in the phase concerned, $N_i$ the ion concentration. The change of the metal atom chemical potential may be expressed by the e.m.f. of cell (ii) [2],

$$d\mu_{Cu} = e\,dE\,, \tag{6}$$

where $e$ is the electron charge.

Expressing the copper ion concentration in $Cu_{2-\delta}Te$ as

$$N_i = \frac{N_A}{V_m}(2-\delta)\,, \tag{7}$$

where $N_A$ is the Avogadro number and $V_m$ the molar volume, (5) yields for the chemical diffusion coefficient

$$\tilde{D} = \frac{(2-\delta)\,D_i e}{kT}\,\frac{dE}{d\delta}\,. \tag{8}$$

The dashed lines in Fig. 5 shows dependences $\tilde{D}(\delta)$ estimated by (8) at various temperatures using experimental data on $D_i$ and $E(\delta)$. The self-diffusion coefficient was calculated by the Nernst-Einstein relation from the ionic conductivity. There is a fairly satisfactory agreement between the experimental and calculated dependences of the chemical diffusion coefficients on the degree of non-stoichiometry. As compared to $D_i$, higher values of $\tilde{D}$ are due to the fact that copper ions are accelerated by the internal electric fields developing due to the difference between the mobilities of ions and electrons.

### *3.3 Ionic thermo-e.m.f.*

The ionic thermo-e.m.f. is determined as the stationary electromotive force $\Delta\varphi_i$ measured between two ionic probes Cu/CuBr, if there is a temperature gradient along the specimen. According to [8], the coefficient of ionic thermo-e.m.f. is determined by

$$\alpha_i = \frac{\Delta\varphi_i}{\Delta T} = \frac{1}{e}\left(\frac{\partial\eta_i}{\partial T} + S^0_{Cu}\right), \tag{9}$$

where $S^0_{Cu}$ is the copper atom entropy in metallic copper, $\eta_i$ the copper ion electrochemical potential in $Cu_{2-\delta}Te$. The first term in (9) represents the homogeneous component of the ionic thermo-e.m.f. The second term is conditioned by the temperature dependence of the contact potential difference of electrode and phase and is a heterogeneous component of the ionic thermo-e.m.f. For the stationary value of $\alpha_i$ the phenomenological theory of the ionic thermo-e.m.f. of the superionic conductors yields

$$\alpha_i = -\frac{1}{e}\left(S_{Cu^+} - S^0_{Cu} + \frac{Q_{Cu^+}}{T}\right), \tag{10}$$

where $S_{Cu^+}$ is the copper ion entropy in $Cu_{2-\delta}Te$, $Q_{Cu^+}$ the copper ion transport heat in $Cu_{2-\delta}Te$.

The assumption of the temperature independence of the entropy difference is true for the majority of superionic conductors. Consequently, it is possible to determine the ion transport heat from the slope of the dependence $\alpha_i = f(1/T)$. The transport heat values thus determined are in satisfactory agreement with the values of the activation energy for ionic conduction of the silver superionic conductor group [8]. On the basis of such assumptions it was possible to determine the ion transport heat values for silver chalcogenides and solid solutions on their basis [12, 13] exhibiting mixed ion–electron conductivity, which are also in satisfactory agreement with the values of the activation energy for ionic conductivity. The dependence $\alpha_i(1/T)$ of $Cu_{2-\delta}Te$ for two values of $\delta$ is shown in Fig. 6.

At high temperatures a straight line in the coordinates $\alpha_i(1/T)$ is observed. At the phase transition point the ionic thermo-e.m.f. undergoes an anomaly. The anomalous dependence of $\alpha_i$ on temperature below the phase transition point seems to indicate that the specimen is not single-phase owing to a great temperature gradient along the specimen.

The calculation of the transport heat from the slope of the straight line $\alpha_i(1/T)$ for the high-temperature region yields 0.4 to 0.5 eV, this essentially exceeding the activation energy for ionic conduction. This is explained by the dependence of the entropy difference on temperature for copper telluride. Therefore, the research of the electrochemical cell (ii) at various temperatures (Fig. 2) has resulted in the determination of the value $(S_{Cu^+} - S^0_{Cu})$. Further, the transport heat was estimated from the known value $\alpha_i$ using (10). The copper ion transport heat thus estimated is $Q_{Cu^+} = (0.23 \mp 0.03)$ eV for $\delta = 0.015$ and $Q_{Cu^+} = (0.25 \mp 0.03)$ eV for $\delta = 0.025$. The values of transport heats in the high-temperature superionic phase correlate well with the values of the activation energy for ionic conduction.

## 4. Summary

The experimentally obtained results for copper telluride and their analysis show that $Cu_{2-\delta}Te$ undergoes a number of first-order phase transformations in the temperature range 300 to 450 °C. These transformations, however, do not result in a change of the lattice type or the degree of the cation sublattice disorder. In the range 300 to 450 °C copper telluride preserves the superionic properties and at the phase transition point the values of the activation energy for ionic conduction change spasmodically. The ionic conductivity in copper telluride is conditioned by the fact that part of cations have weak bonding with the rigid frame. The concentration of these cations is actually independent of temperature. Therefore, the activation energy for ionic conduction involves only the activation energy for mobility and the ion transport heat estimated from the ionic thermo-e.m.f. approximates the activation energy for ionic conduction. The deviation from stoichiometric composition results in a decrease of the concentration of mobile ions and, consequently, of the ionic conductivity at constant temperature. The process of establishing equilibrium in the specimen with the gradient of the degree of non-stoichiometry is determined from the chemical diffusion coefficient. A satisfactory agreement has been obtained for the experimental and phenomenologically estimated dependences of the chemical diffusion coefficients on the degree of non-stoichiometry.

## References

[1] I. Kh. Abrikosov (Ed.), Poluprovodnikovye khalkogenidy i splavy na ikh osnove, Izd. Nauka, Moskva 1975 (p. 220).
[2] G. Lorenz and C. J. Wagner, J. chem. Phys. **26**, 1607 (1957).
[3] I. Bougnot, F. Guostavino, H. Luquet, and D. Sodini, Mater. Res. Bull. **5**, 763 (1970).
[4] R. Routie, M. Mannoni, and J. Mahene, Rev. Chim. Mineral. **9**, 357 (1972).
[5] J. B. Wagner and C. Wagner, J. chem. Phys. **26**, 1597 (1957).
[6] I. Jokota, J. Phys. Soc. Japan **16**, 2213 (1961).
[7] K. Weiss, Berichte Bunsengesellsch. Phys. Chem. **73**, 338 (1969).
[8] K. Shahi, phys. stat. sol. (a) **41**, 11 (1977).
[9] R. A. Yakshibaev, V. N. Konev, and M. Kh. Balapanov, Fiz. tverd. Tela, **26**, 3641 (1984).
[10] R. A. Yakshibaev, M. Kh. Balapanov, and V. N. Konev, Fiz. tverd. Tela **28**, 1566 (1986).
[11] V. N. Chebotin, Uspekhi khimii **55**, 914 (1986).
[12] M. Koch, H. Rickert, and G. Schlechtriemen, Solid State Ionics, **9/10**, 1197 (1983).
[13] R. A. Yakshibaev, M. Kh. Balapanov, and V. N. Konev, Fiz. tverd. Tela, **29**, 937 (1987).

*(Received December 18, 1987)*

phys. stat. sol. (a) 108, 143 (1988)

Subject classification: 61.80; 71.55; S5.11

*A. N. Sevchenko Scientific Research Institute for Applied Physics Problems, V. I. Lenin Belorussian State University, Minsk*[1])

# Influence of Irradiation Temperature on Radiation Defect Accumulation in Silicon

By

V. I. KUZNETSOV, P. F. LUGAKOV, and V. V. LUKYANITSA

The processes of radiation defect formation in n-type silicon are studied at various irradiation temperatures ($T_{irr}$ = 20 to 400 °C) during 640 MeV proton or 10 MeV electron bombardment. The results are obtained from the analysis of the temperature dependences of charge carrier concentration in crystals irradiated at various temperatures and investigating samples (which are not irradiated but subjected to similar thermal treatment). It is established that production efficiencies for main radiation defects (A-, E-centres, divacancies, complexes of carbon–oxygen-divacancy) are complicated functions of $T_{irr}$, however the form of these functions appear to be different for the two types of irradiation used. The results obtained are interpreted taking into account the fact that during cascade defect formation (in the case of proton irradiation) the effect of $T_{irr}$ is due to the probabilities of secondary processes (dissociation and formation of complexes) changing with temperature, while in the case of diffusive mechanism (electron irradiation) the irradiation temperature $T_{irr}$ has also an influence on the direct annihilation of Frenkel pairs generated by irradiation.

Изучены процессы образования радиационных дефектов в кремнии n-типа в зависимости от температуры ($T_{обл}$ = 20 до 400 °C) кристаллов при облучении протонами 640 MeV или электронами 10 MeV. Результаты получены из анализа температурных зависимостей концентрации носителей заряда в облученных при различных температурах и контрольных (прошедших аналогичную термообработку без облучения) кристаллах. Установлено, что эффективность введения основных радиационных дефектов (A-, E-центры, дивакансии, комплексы углерод–кислород–дивакансия) сложным образом зависят от $T_{обл}$, однако характер этих зависимостей существенно различается для использованных видов облучения. Полученные результаты интерпретированы с учетом того, что при каскадном механизме дефектообразования (облучение протонами) влияние $T_{обл}$ обусловлено изменением с температурой вероятностей вторичных процессов (распад и формирование комплексов), а при диффузионном (электронное облучение) — наряду с этим и влиянием $T_{овл}$ на вероятность прямой аннигиляции генерируемых облучением пар Френкеля.

## 1. Introduction

The irradiation temperature ($T_{irr}$) is known to have a substantial influence on the processes of radiation defect formation in silicon at different kinds of bombarding particles [1 to 8]. This is so because the irradiation temperature defines the charge state of components participating in radiation defect (RD) formation as well as formation and dissociation cross-sections for various complexes [3]. It was suggested in [3] that the threshold energy $E_d$ of Frenkel pair formation is also dependent on the irradiation temperature, though there is a diametrically opposite point of view [5].

[1]) SU-220064, Minsk, USSR.

The available information on what role $T_{irr}$ plays in RD accumulation is ambiguous, and in some cases contradictory [3, 4, 6]. While interpreting such results various model insights were used. These insights are based on the suggestion that either primary or secondary processes play the dominant role in the RD accumulation [1, 3]. In the present paper the influence of high-temperature irradiation on efficiency of stable RD accumulation is studied for the cases of cascade and diffusive mechanisms of defect formation [9].

## 2. Experimental Procedure

The n-type Si crystals (with initial resistivity $\varrho = 70$ to $150\ \Omega$ cm) obtained by the Czochralski method (crucible crystals) and by the floating-zone melting (zone crystals) were used. The samples were irradiated with 640 MeV protons or 10 MeV electrons. The integrated proton flux was $\Phi = 1.7 \times 10^{12}$ cm$^{-2}$ while flux intensity was $I = 2.0 \times 10^{9}$ cm$^{-2}$ s$^{-1}$, for electrons $\Phi = 4.0 \times 10^{13}$ cm$^{-2}$ and $I = 5.0 \times 10^{11}$ cm$^{-2}$s$^{-1}$. The irradiations were carried out in the temperature interval from 20 to 400 °C. The control crystals were subjected to similar thermal treatment without irradiation. The results were obtained from the Hall coefficient measurements. The energy levels, RD concentrations, and production rates were defined from measured Hall coefficients.

## 3. Experimental Results

Fig. 1 shows $T_{irr}$ dependences of production rates $\eta_{0.40}$ obtained by summation over all RD with levels $\geqq E_c - 0.40$ eV (curves 1, 2) as well as of production rates $\eta_{0.17}$ for RD with level at $E_c - 0.17$ (curve 3). These rates were normalized to their values at $T_{irr} = 20$ °C. Curves 1 were obtained in zone crystals and curves 2, 3 were obtained in crucible crystals irradiated with protons (Fig. 1a) and electrons (Fig. 1b). We use the following notations $\eta_{0.17} = N_{0.17}/\Phi$ and $\eta_{0.40} = (n_0 - n_f)/\Phi$, where $N_{0.17}$ is the concentration of RD with level $E_c - 0.17$ eV; $n_0$ and $n_f$ are the charge carrier concentrations measured at $T = 250$ K before and after irradiation, respectively. It is seen that the RD production rates are complicated nonmonotonic functions of $T_{irr}$ and

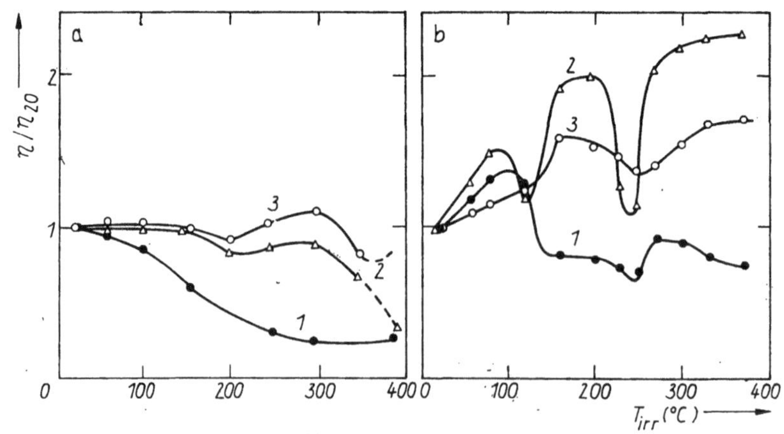


Fig. 1. Radiation defect production rate vs. irradiation temperature. a) 640 MeV proton irradiation, b) 10 MeV electron irradiation. (1) $\eta_{0.40}$ in zone silicon, (2) $\eta_{0.40}$ in crucible silicon, (3) $\eta_{0.17}$ in crucible silicon

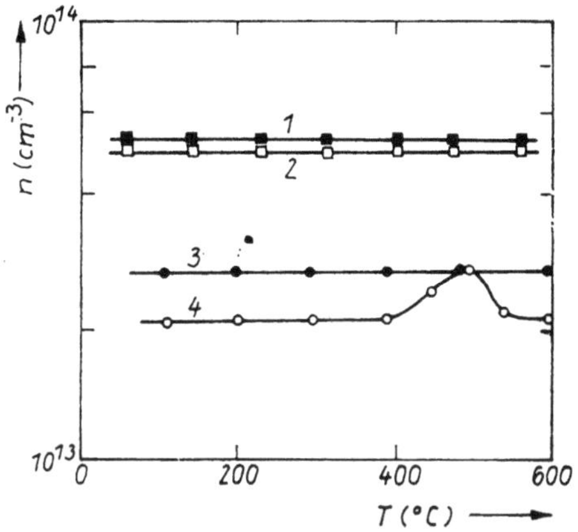


Fig. 2. Charge carrier concentrations in control crystals vs. temperature of thermal treatment for (1), (2) 80, and (3), (4) 850 s; (1), (3) flat-zone Si, (2), (4) crucible Si

are dependent on the type of irradiation (mechanism of radiation defect formation), as well as on the method by which the crystals were obtained (impurity composition). Fig. 2 shows the dependence of charge carrier concentration on annealing temperature in the control crystals. From these results it follows that in the temperature interval under investigation the charge carrier concentration increases only in crucible silicon after thermal treatment at $T > 400$ °C with duration corresponding to that of proton irradiation and being approximately equal to 15 min. This increasing of charge carrier concentration is due to thermodonor formation [3].

## 4. Discussion

The RD with levels $\geqq E_c - 0.40$ eV produced during irradiation of n-type silicon are predominantly vacancy complexes. Their production rate is dependent on the impurity composition of crystals as well as on irradiation type. In zone crystals the phosphorus–vacancy complexes (E-centres) stable up to 150 °C [5] are among those RD which have high production rates. In the case of crucible silicon the E-centres are produced with a much lower rate. In this material the carbon–oxygen–divacancy complexes ($COV_2$) annealed at $T \approx 450$ °C are among the major RD. Besides the above-mentioned RD the divacancies ($V_2$) stable up to 280 °C [5] are produced in both zone and crucible crystals. The RD with levels $E_c - 0.17$ eV in crucible silicon are the vacancy–oxygen complexes (A-centres) annealed at $T = 350$ °C [5]. All above-mentioned RD are produced during irradiation as secondary defects, i.e. they are the results of vacancy migration and capture by impurities, impurity associations, or by each other. It should be noted that in the case of proton irradiation the divacancies unlike other RD are more effectively produced as primary RD. Their annealing in such a case begins at lower temperatures ($T > 150$ °C) [10].

During annealing of these RD their reconstruction can occur along with additional formation of other more stable RD. During annealing of E-centres the complexes of phosphorus–two vacancies ($PV_2$) (which are stable up to $T \approx 450$ °C) with level $> E_c - 0.45$ eV [11] are produced, while during annealing of divacancies and other complex RD the complexes of several vacancies with acceptor levels in the lower part of the forbidden zone (which are annealed at $T \geqq 600$ °C) are produced [12].

Observed distinctions in the $T_{irr}$ dependences of RD production rates during proton and electron irradiation can be connected with various mechanisms of defect formation which are characteristic of these two types of bombardment particles. When silicon is irradiated with electrons of relatively low energies ($<10$ MeV) the diffusive mechanism of defect formation is realized. In this mechanism the formation of stable complexes is controlled by diffusion of primary RD, which are generated by irradiation almost uniformly over all crystal volume. In the cascade mechanism (proton irradiation) the RD produced are predominantly localized within defect clusters [13, 14] formed in the place of the displacement cascade, which appears as a result of multiple interatomic collisions during stopping of primary knock-on atoms. Divacancies are included in the central region (nucleus) of clusters, while secondary RD are included in the periphery region of clusters. The cluster formation process consists of several stages [13, 14]. At the first stage the cross annihilation of vacancies and intinsic interstitials occurs within the cascade region, and nonannihilated interstitials migrate out of the cascade region. The migrated atoms are captured by impurities and form the stable RD. At the second stage vacancies migrate out of the cascade region and then interact with impurities localized near the cascade region, and as the result the RD are produced. Therefore in the cascade mechanism of defect formation free vacancies and intrinsic interstitials appear to be spatially separated at the first stage yet. From this fact it is followed that in the cascade mechanism an influence of irradiation temperature on RD formation efficiency can be connected with a temperature variation of the probability of interaction between components reacting during complex formation.

During proton irradiation the formation probability for divacancies as primary RD should not be dependent on temperature. As a consequence of this fact the value of $\eta_{0.40}$ in crucible silicon appears to be constant in the temperature interval from 20 to 150 °C. However, in the cascade mechanism of defect formation due to high RD concentration within clusters the divacancies appear to be unstable at $T_{irr} > 150$ °C [10, 16]. The value of $\eta_{0.40}$ is observed to be decreased just at such temperatures (Fig. 1a). Annealing of these divacancies causes the $COV_2$-complex and A-centre formation probabilities to be increased when the irradiation temperature rises from 200 to 300 °C (Fig. 1a, curves 2 and 3). From our point of view a decrease of $\eta_{0.40}$ in the temperature interval from 300 to 400 °C is due to the formation of thermal defects. Indeed, as one can see from the dependences presented in Fig. 2 (curves 3 and 4), the charge carrier concentration increases in the control crucible crystals, due to the thermodonor formation as known from [3]. Accumulation of these defects during the thermal treatment with simultaneous high-energy particle irradiation can also cause the temperature interval of their formation to be shifted toward lower temperatures [17]. This process is apparently responsible for $\eta_{0.40}$ increasing at $T_{irr}$ in the interval from 300 to 400 °C.

Decreasing of $\eta_{0.40}$ in zone silicon when $T_{irr}$ rises from 20 to 150 °C is due to the instability of E-centres, whereas in the temperature interval from 150 to 300 °C the decrease is due to a change in the stability of divacancies as is the case in crucible crystals. The slight increase of $\eta_{0.40}$ at $T_{irr} = 300$ to 400 °C is caused by the formation of complexes of several vacancies with acceptor levels localized in the lower part of the forbidden zone [12].

An analysis of the results obtained allows to consider that in the cascade mechanism of defect formation due to the spatial separation of Frenkel pair components a situation is realized that follows from the model in which secondary processes predominate in RD formation [3]. If this model is valid in the diffusive mechanism of defect formation, then the $T_{irr}$ dependences of RD production efficiencies should be identical for proton and electron irradiation. However, the experimental results (Fig. 1 a and b) show that the $T_{irr}$ dependences of RD production rates are substantially distinct for electron and proton irradiations. Moreover, in the case of electron irradiation the dependences $\eta(T_{irr})$ have some characteristic features which cannot be explained in the framework of above model. This is concerned with RD production efficiency in the temperature intervals from 20 to 100 °C and from 200 to 300 °C, in which the production rates for various RD of vacancy type are observed to be analogously changed. Indeed, according to the model of [3], if the concentration of some RD of vacancy type decreases with $T_{irr}$, then the concentration of other such RD should be increased due to drains competing with vacancy capture. Since the present experiment does not confirm the above conclusion (Fig. 1b), then the results obtained in the case of electron irradiation can be described, in our opinion, in the framework of the model which suggests that metastable Frenkel pairs participate in the diffusive mechanism of defect formation [1]. The fact that the observed $\eta(T_{irr})$ dependences have similar forms for various RD can be explained by the influence of $T_{irr}$ on the probabilities of annihilation and separation of Frenkel pair components. Obviously, at the temperature interval from 20 to 100 °C the components of Frenkel pairs have such charge states, that at higher temperatures the rate of their annihilation appears to be much lower than that rate at lower $T_{irr}$. Therefore, the probability of free vacancy and interstitial formation increases with $T_{irr}$. These vacancies and interstitials are able to participate in the formation of complexes which are stable under the present experimental conditions. This is the cause of the observed increase of the $\eta(T_{irr})$ dependences for various RD.

An analogous situation occurs at temperatures $T_{irr}$ between 200 and 300 °C. However, in this case either a double change in the charge state of a Frenkel pair component or a sequential change in the charge state of both components takes place. This causes the probability of primary RD annihilation to be firstly increased and then decreased. This process defines the minimum of production rates for all types of secondary defects at the temperatures under consideration. The production efficiencies for various secondary RD achieve their minimum values when vacancies and interstitials are in opposite charge states and Coulomb attraction facilitates the annihilation of these defects. It should be noted that similar nonmonotonic dependences of secondary complex formation rates have been also observed in the case of increasing intensity of 10 MeV electron irradiation [18]. The influence of irradiation conditions (particle intensity) on annihilation and separation of Frenkel pair components generated by irradiation have been also observed in the experiments of [18]. These results confirm that phenomena which proceed during the variation of irradiation temperature (at constant intensity) and of irradiation intensity (at constant temperature) appear to be similar.

The $\eta_{0.40}(T_{irr})$ dependences obtained in the temperature intervals from 100 to 200 °C and from 300 and 400 °C in crystals irradiated with electrons can be explained, as in the case of proton irradiation, by transformations of less stable RD into RD which are stable up to higher temperatures. Decreasing production rate for RD with levels $E_c - 0.40$ in zone silicon as well as increasing corresponding rate in crucible silicon at $T_{irr} = 100$ to 200 °C are connected with the thermal instability of E-centres. Therefore, A-centre and $COV_2$-complex accumulation efficiency increases. At $T_{irr} \approx$ $\approx 300$ °C divacancies are unstable and can participate in the formation of either $COV_2$ complexes or other defects containing phosphorus ($PV_2$ complexes), which are stable up to higher temperatures ($\approx 450$ °C). This fact explains the dependence $\eta_{0.40}(T_{irr})$ observed at $T_{irr} > 300$ °C.

In the present paper the interpretation of experimental results on the influence of irradiation temperature on RD accumulation processes is given without accounting for the temperature dependence of $E_d$. This means that for the irradiations used, the mechanism of impact atomic displacement in which $E_d$ should be independent of irradiation conditions [5], is realized. This is confirmed by the fact that when in the crucible silicon the temperature of proton irradiation is increased from 20 to 150 °C the values of $\eta_{0.40}$ and $\eta_{0.17}$ appear to be independent of $T_{irr}$, whereas if $E_d$ decreases with $T_{irr}$, the production rate for all RD should increase.

## 5. Conclusion

The results obtained show that radiation defect accumulation processes in both diffusive and cascade mechanisms of defect formation appear to be substantially dependent on irradiation temperature. The temperature dependences of RD production rates are of various character in the two cases. In the cascade mechanism the Frenkel pairs are spatially separated into their components due to knock-on atoms generated in the cascade gaining high energies. Therefore, the $T_{irr}$ dependences of RD production rates are conditioned by the temperature dependence of the secondary process (complex formation) probability. The distinction observed in functions $\eta(T_{irr})$ for diffusive and cascade mechanisms of defect formation allows to suggest that in the diffusive mechanism the annihilation of genetically connected Frenkel pairs plays a main role in the disappearance of primary defects. The efficiency of annihilation is dependent on the charge state of Frenkel pair components. Therefore, the $T_{irr}$ dependences of production rates of stable RD in the diffusive mechanism are defined by not only the probability of secondary processes, but also by the probability of primary ones (annihilation of genetically connected Frenkel pairs).

## References

[1] G. K. WERTHEIM, Phys. Rev. **115**, 568 (1959).
[2] J. W. CORBETT, J. P. KARINS, and T. V. TAN, Nuclear Instrum. and Methods **182/183**, 457 (1981).
[3] L. S. SMIRNOV (Ed.), Physics Processes in Irradiated Semiconductors, Izd. Nauka, Novosibirsk 1977 (in Russian).
[4] L. S. MILEVSKII and T. A. PAGAVA, Fiz. Tekh. Poluprov. **9**, 1201 (1975).
[5] V. V. EMTSEV and T. V. MASHOVETS, Impurities and Point Defects in Semiconductors, Izd. Radio i Svyaz, Moscow 1981 (in Russian).
[6] F. P. KORSHUNOV and I. G. MARCHENKO, Fiz. Tekh. Poluprov. **17**, 540 (1983).
[7] G. GÖTZ, K.-D. KLINGE, F. SCHWABE, and V. SOLOVEV, Exper. Tech. Phys. **25**, 71 (1977).
[8] R. G. ELLIMAN, J. S. WILLIAMS, W. L. BROWN, A. LEIBERICH, D. M. MAHER, and R. V. KNOELL, Nuclear Instrum. and Methods **B19/20**, 735 (1987).
[9] D. N. SEIDMAN, R. G. AVERBACK, P. R. OKAMOTO, and A. C. BAILY, Phys. Rev. Letters **58**, 900 (1987).
[10] V. I. KUZNETSOV, P. F. LUGAKOV, and V. V. SHUSHA, Radiat. Eff. Letters **86**, 199 (1986).
[11] P. F. LUGAKOV and V. V. LUKYANITSA, phys. stat. sol. (a) **84**, 457 (1984).
[12] V. I. KUZNETSOV and P. F. LUGAKOV, phys. stat. sol. (b) **112**, 457 (1982).
[13] N. A. UKHIN, Fiz. Tekh. Poluprov. **6**, 931 (1972).
[14] V. L. VINETSKII and G. A. KHOLODAR, Radiation Physics of Semiconductors, Izd. Naukova Dumka, Kiev 1979.
[15] V. I. KUZNETSOV and P. F. LUGAKOV, phys. stat. sol. (a) **70**, 387 (1982).
[16] L. J. CHENG and J. LORI, Phys. Rev. **171**, 586 (1968).
[17] I. M. STARCHIK, Fiz. Tekh. Poluprov. **3**, 153 (1969).
[18] P. F. LUGAKOV and V. V. LUKYANITSA, phys. stat. sol. (a) **83**, 521 (1984).

*(Received July 13, 1987; in revised form March 14, 1988)*

phys. stat. sol. (a) **108**, 149 (1988)

Subject classification: 66.30; 72.20; S8

*Department of Chemistry, Friedrich-Schiller-University, Jena*[1])

# Model of Photostimulated Diffusion of Ag in Amorphous Chalcogenides

By

H. KAHNT, F. SCHIRRMEISTER, and A. FELTZ

A model for photoenhanced Ag diffusion and formation of step-like Ag profiles in chalcogenide glasses is presented. The model is deduced from well known properties of amorphous chalcogenides such like low electronic mobility, insensitivity against doping, and strong photoconductivity.

Es wird ein Modell für die photostimulierte Ag-Diffusion und die Ausbildung von stufenförmigen Ag-Profilen in Chalkogenidgläsern vorgeschlagen. Das Modell leitet sich von solchen bekannten Eigenschaften der amorphen Chalkogenide ab wie der geringen elektronischen Beweglichkeit, der Unempfindlichkeit gegenüber Dotierungen und der ausgeprägten Photoleitung.

## 1. Introduction

Photostimulated Ag diffusion has been reported for a wide range of compositions in amorphous As and Ge chalcogenides [1]. This process is useful for the application as high-resolution photoresist in silicon microfabrication technology [2]. In spite of the practical interest in this phenomenon the process of photodoping is not completely clear. Ishikawa [3] argues that charged intrinsic defects of the valence alternation pair type (VAPs) are responsible for the photo-enhanced diffusion. However, the mechanism proposed in [3] cannot explain the step-like Ag profiles that are usually observed. Moreover the concentration of VAPs is $< 10^{20}$ cm$^{-3}$, whereas the Ag concentration in the Ag-rich side of the diffusion profile is much higher. An interpretation of photodiffusion in terms of a solid state reaction was first given by Malinowski and Buroff in [4]. Janai suggested a solid state reaction, too [5]. In the presence of a sufficiently high amount of chalcogen the formation of $Ag_2X$ (X chalcogen) is thermodynamically favoured [6]. If assuming that the absorption of phonons leads to a local increase of temperature [7] the photodiffusion process could be considered like thermal diffusion. However the formation of step-like Ag profiles remains unclear. Lakshmikumar [8] discusses the formation of a heterojunction between the Ag-rich and the Ag-poor regions of the sample. He attributes the stimulating effect of light on the Ag diffusion to some photovoltaic effect at this junction. In a recent paper Kluge [9] gave a new interpretation of the photodoping effect. He showed that the intercallation of Ag into the chalcogenide network is a solid state reaction that is accompanied with the formation of additional negatively charged defect states. Moreover, he pointed out that internal fields would hinder the intercallation and that charge neutrality can only be ensured by movable charge carriers.

In the present paper the influence of photoexcited carriers on the Ag diffusion is considered in detail and a model is discussed that takes into consideration the specific

---

[1]) Am Steiger 3, Haus 3, DDR-6900 Jena, GDR.

transport properties of amorphous chalcogenide materials. Starting with an analysis of the processes that control Ag diffusion in chalcogenide glasses the influence of illumination is demonstrated.

## 2. Ag Diffusion in Chalcogenide Glasses

Let us first consider the diffusion process without illumination. As mentioned above there is a thermodynamically favoured tendency of $Ag_2X$ formation. This solid state reaction is expected to take place at the interface of metallic Ag and the chalcogenide glass. For further progressing of this reaction the Ag atoms have to cross this layer. However, the diffusion of Ag through this region is not the limiting process for the velocity of Ag propagation into the glass because of the high mobility of Ag in Ag chalcogenides. That is why the $Ag_2X$-rich interlayer may be considered as Ag source for the Ag diffusion into the chalcogenide glass. The limitation arises at the diffusion front in the region of small Ag concentration as is shown below. Here some specific properties of chalcogenide glasses play an important role that will be discussed now.

First this is the effect of doping on the conductivity. From electromigration studies [10] of Ag tracer atoms it is well established that silver is ionized in the chalcogenide in accordance with the reaction

$$Ag = Ag^+ + e^- . \tag{1}$$

The silver atoms therefore act as donors. However, there will be only little effect on the electronic conductivity as far as the Ag concentration is smaller than the concentration of the charged intrinsic defects (VAPs). Calculations of the position of the Fermi energy $E_F$ in amorphous chalcogenides show that there is an effective pinning of $E_F$ near midgap caused by the VAPs. This position is shifted closer to one of the band edges only when the concentration of donors (or acceptors) becomes comparable to that of the VAPs [11]. This stability against doping was one of the first established characteristics of chalcogenide glasses [12]. In the region of low Ag concentration the electrons resulting from the reaction (1) are effectively trapped by the VAPs without changing the electronic conductivity in the system.

Alternatively, according to Kluge [9] additional negatively charged defects may be created by the reaction

$$C_2^0 + e^- + M^+ \rightarrow C_1^- M^+ \tag{2}$$

following the notation of [9] (C chalcogen, M metal). However, when the concentration of these defects is smaller than that of the intrinsic VAPs there will be no change in the position of the Fermi energy.

The low mobility of electronic charge carriers in amorphous semiconductors is well documented by drift mobility measurements [13]. By reasons of electroneutrality the diffusion of $Ag^+$ is coupled with the motion of electrons. $Ag^+$ ions move faster into the sample than electrons. An electric field is built up stopping the $Ag^+$ motion [14] similar to the coupling of electrons and holes in the Dember effect. Consequently, the Ag diffusion must be considered as a bipolar diffusion process of the coupled system of $Ag^+$ ions and electrons. The velocity of the diffusion front is controlled by the low mobility of the electrons.

When the Ag concentration exceeds the density of VAPs the electronic conductivity increases steeply by several orders of magnitude [11]. Then the space charge created by $Ag^+$ is quickly compensated by mobile electronic charge carriers. In this region the $Ag^+$ motion is not hindered by the trapped space charge of electrons. The high $Ag^+$ mobility in the Ag-rich region on one hand and the small $Ag^+$ mobility in the low concentration region at the diffusion front on the other hand is the origin of step-like diffusion profiles. Calculations of Sebastian and Süptitz [15] demonstrated that such profiles are consistent with the assumption of a concentration dependent diffusion coefficient.

### 3. Photostimulated Diffusion

Next let us consider the influence of illumination on the diffusion process. It is well known that chalcogenide glasses show photoinduced phenomena such as photoinduced luminescence and EPR or photostimulated structural changes which are extensively studied [16]. The fact that chalcogenide glasses exhibit strong photoconductivity seems to be nearly forgotten today. However, many investigations in this field are known from the early years of amorphous semiconductor research showing that photoconductivity in chalcogenide glasses usually exceeds dark conductivity by several orders of magnitude [12]. Excess carriers that are generated by illumination produce a breakdown of the electric field that hinders the motion of $Ag^+$ ions. In other words, photoconductivity supports a large number of mobile electronic charge carriers allowing for the quick compensation of the inner space charge that was established because of the different mobility of $Ag^+$ ions and electrons. That is why the velocity of Ag motion into the chalcogenide glass is strongly increased by illumination.

### 4. Model Calculations

The validity of the ideas mentioned above is demonstrated now on a simple model. The following considerations are restricted to the region of the diffusion front where the Ag concentration is small. The ionic and electronic particle flows resulting from concentration gradients and the inner field are given by

$$j_n = -D_n \frac{\partial n}{\partial x} + \frac{n D_n e}{kT} \frac{\partial \varphi}{\partial x}, \tag{3}$$

$$j_A = -D_A \frac{\partial A}{\partial x} - A \frac{D_A e}{kT} \frac{\partial \varphi}{\partial x}. \tag{4}$$

Here $n$ is the concentration of electrons, $A$ the concentration of $Ag^+$ ions, $D_n$ and $D_A$ the diffusion coefficient for electrons and ions, respectively, $\varphi$ the electrical potential, and $e$, $k$, $T$ have their usual meaning. Assuming stationary conditions, i.e. both $Ag^+$ and $n$ diffuse together coupled by the electric field with equal velocity, we can determine $\partial\varphi/\partial x$ through

$$j_n = j_A. \tag{5}$$

Then

$$\partial\varphi/\partial x = \frac{D_n\, \partial n/\partial x - D_A\, \partial A/\partial x}{n D_n e/kT + A D_A e/kT}. \tag{6}$$

Inserting (6) in (4) we obtain

$$j_A = -D_{bi} \frac{\partial A}{\partial x}. \tag{7}$$

The bipolar diffusion coefficient $D_{bi}$ is given by

$$D_{bi} = D_A \left[1 - A \frac{D_A - D_n\, \partial n/\partial A}{n D_n + A D_A}\right]. \tag{8}$$

As far as the ions are strongly coupled to the trapped electronic space charge, $\partial n/\partial A$ will be nearly unity. From (8) we find that in the limit $D_n \to 0$ the bipolar diffusion coefficient $D_{bi} \to 0$. Obviously in the case of a low diffusion coefficient for electrons $D_n < D_A$ the motion of the diffusion front is mainly controlled by electrons. To demonstrate the influence of a large number of mobile photogenerated excess carriers on $D_{bi}$ we refer to the transport mechanism of electrons (or holes) in amorphous semiconductors. The electronic conductivity can be understood either as the movement of a small number of "free" (in bands or hopping states near the band edges) electrons $n_f$ with relatively high mobility $\mu_f$ or as the motion of all electrons $n$, including free $n_f$ and trapped $n_t$ ones with a small mobility $\mu_n$. This distinction is similar to that of conductivity and drift mobility,

$$\sigma_n = e n_f \mu_f = e n \mu_n\,. \tag{9}$$

Using Einstein's equation

$$\mu = \frac{De}{kT} \tag{10}$$

we find

$$n D_n = n_f D_f\,, \tag{11}$$

and with $n = n_t + n_f$ the bipolar diffusion coefficient $D_{bi}$ is rewritten

$$D_{bi} = D_A \left[1 - A \frac{D_A - D_f/(1 + n_t/n_f)}{n_f D_f + A D_A}\right]. \tag{12}$$

Under illumination $n_f$ is increased by several orders of magnitude. Thus $D_n$ is increased, too, because $D_f$ remains unchanged. These excess carriers result either from the generation of electrons from the valence band or from defect states (decrease of $n_t$) into the conduction band. Both processes lead to a decreasing difference

$$D_A - \frac{D_f}{1 + n_t/n_f} \tag{13}$$

and an increasing sum

$$n_f D_f + A D_A\,. \tag{14}$$

The bipolar diffusion coefficient (11) then increases reaching $D_A$ when (13) becomes zero. With further increase of $n_f$ the polarity of the inner field changes and Ag diffusion is even stimulated.

In the model considerations hole conduction was neglected for simplicity. However, a contribution of holes is easily enclosed into the basic system of currents (3) and (4). As far as the total electronic conductivity $\sigma_n + \sigma_p$ is small in comparison to the ionic conductivity, there will be no change in the general conclusions of the model. Photoexcited holes act in the same way as electrons yielding a lowering of the coupling potential $\varphi$.

***Acknowledgement***

The authors wish to thank Prof. M. Frumar for stimulating discussions on the problems of photoenhanced diffusion.

## References

[1] M. Yamaguchi, I. Shimizu, and E. Inoe, J. non-crystall. Solids **47**, 341 (1982).
[2] A. Yoshikawa, H. Nagai, and Y. Mizushima, Japan. J. appl. Phys., Suppl. **16**-1, 67 (1977).
[3] R. Ishikawa, Solid State Commun. **30**, 99 (1979).

[4] J. MALINOWSKI and A. BUROFF, Contemp. Phys. **19**, 99 (1978).
[5] M. JANAI, Phys. Rev. Letters **47**, 726 (1981).
[6] T. WAGNER, M. FRUMAR, and L. BENES, J. non-crystall Solids **90**, 517 (1987).
[7] J. C. PHILLIPS, J. non-crystall. Solids **64**, 81 (1984).
[8] S. T. LAKSHMIKUMAR, J. non-crystall. Solids **88**, 196 (1986).
[9] G. KLUGE, phys. stat. sol. (a) **101**, 105 (1987).
[10] A. BUROFF, E. NEBAUER, and P. SÜPTITZ, phys. stat. sol. (a) **48**, K109 (1978).
[11] D. ADLER and E. JOFFA, Phys. Rev. Letters **36**, 1197 (1976).
[12] N. F. MOTT and E. A. DAVIS, Electronic Processes in Noncrystalline Materials, Clarendon Press, Oxford 1971 (pp. 229, 324).
[13] W. E. SPEAR, Adv. Phys. **23**, 523 (1974).
[14] M. FRUMAR, A. P. FIRTH, and A. E. OWEN, Proc. Internat. Conf. Amorphous Liquid Semiconductors, Gabrowo 1984, Vol. 1, Bulgarian Acad. Sci., Sofia 1984 (p. 216).
[15] N. SEBASTIAN and P. SÜPTITZ, Proc. Internat. Symp. Solid State Chemistry, Karlovy Vary, Ed. M. FRUMAR and L. KOUDELKA, Dum Techniky CSVTS, Usti nad Labem 1986 (p. 279).
[16] For instance see Proc. 8th Internat. Conf. Non-Crystall. Semiconductors, Balatonszeplak 1986, in: J. non-crystall. Solids **90** (1986).

*(Received September 28, 1987; in revised form February 17, 1988)*

phys. stat. sol. (a) **108**, 155 (1988)

Subject classification: 66.30 and 72.10; S10

*Departamento de Química, Instituto Superior Técnico, Lisboa*[1]) *(a)*
*and Escola Superior de Tecnologia de Tomar*[2]) *(b)*

# Transport Equations in Common Solid State Electrochemical Situations

By

C. A. C. SEQUEIRA (a) and J. M. B. FERNANDES DINIZ (b)

The transport equations describing conduction processes which occur in oxide solid electrolytes are applied to some common electrochemical situations. The determination of the transport number for ions in either the current supply or open circuit case is discussed. Mathematics are presented that will allow one to calculate the electron and hole conductivities directly from polarization measurements.

Les équations de transport pour la conduction électrique dans les oxydes électrolytes solides sont appliquées à des situations ordinaires d'électrochimie. La détermination des nombres de transports ioniques dans les cas des contacts conducteur électronique/électrolyte solid réversibles et bloquantes est examinée. La résolution mathématique permettant la détermination des conductivités par des électrons et des trous d'électrons par suite de la polarisation des électrodes est presentée.

## 1. Introduction

In a previous study [1], the quasi-chemical description of the defect structure in an oxide solid electrolyte was considered. It was shown how variations in the environment give rise to variations in the concentrations of defects which participate in the conduction process of the oxide. Then, the relevant transport equations that describe the conduction processes which occur in oxide solid electrolytes were developed.

To use the electrical transport equations, the nature of the interfaces between the oxide conductor and the applied electrodes must be considered since the transfer of holes, electrons, or ions at interfaces is not always easy. In general, there are two types of interfaces, ohmic, i.e., those that have the same potential drop independent of current flow, and blocking or rectifying, i.e., those that allow significant current of some type to pass in only one direction.

Not only must the physical nature of the interface be considered, but the chemical environment which the electrodes define for the oxide is also important. The conductivity that was defined for an arbitrary species in the description of the transport equation was a function of both the concentration of the species and its mobility [1]. We shall assume here that as long as the defect concentrations are low and there is no significant interaction between defects, the mobility of any particular species will be independent of its concentration. This may be expressed as

$$\sigma \equiv \nu_k n_k \, |z_k| \, q \sim K' n_k \tag{1}$$

or

$$\nu_k \neq \nu(n_k) \, . \tag{2}$$

---

[1]) 1096 Lisboa Codex, Portugal.

[2]) 2300 Tomar, Portugal.

We have seen that the oxygen partial pressure has a significant effect on the concentrations of the various defects; if the above assumption is correct, then the conductivity of a particular species under various boundary conditions is defined by changing oxygen pressure and should behave in a manner similar to the dependence of concentration under pressure.

Let us now apply the electrical transport equations to some common electrochemical cases, namely that where current is supplied to reversible electrodes at the interfaces of an oxide conductor and that of the ionic conductor with ohmic contacts but no current supply. Moreover, let us see what information may be obtained from this application. This is the purpose of this paper.

## 2. Ohmic Contacts with a Current Supply

The first situation to be examined is that of ohmic contacts, reversible electrodes, and a defined oxygen pressure. Experimental geometries that meet these requirements are given in Fig. 1. Fig. 1a shows a case in which both electrodes are comprised of a coexist ence mixture of a metal B and its oxide BO. Case b in Fig. 1 is that where inert contacts are made with the oxide and the general environmental oxygen pressure is controlled. In either case, the oxide of interest will be $MO_2$ that has been doped with a significant amount of another oxide, AO.

For the above condition the oxide is everywhere in equilibrium with a defined oxygen pressure; as such, no concentration gradients exist in the oxide. We have seen that the electrochemical potential for the participating species 1 [1] is given by

$$\tilde{\mu} = \mu_1 + |z_1|\, q\psi \tag{3}$$

and

$$\mu_1 = \mu_1^0 + kT \ln n_1\,, \tag{4}$$

where $k$ is the Boltzmann constant, $T$ the absolute temperature, and $n_1$ the concentration of species 1.

The gradient of the electrochemical potential is

$$\text{grad}\, \tilde{\mu}_1 = \frac{kT}{n_1} \text{grad}\, n_1 + |z_1|\, q\, \text{grad}\, \psi\,, \tag{5}$$

but the condition of equilibrium requires that $\text{grad}\, n_1 = 0$. Therefore,

$$\text{grad}\, \tilde{\mu}_1 = |z_1|\, q\, \text{grad}\, \psi = -|z_1|\, qE\,, \tag{6}$$

where $E$ is the electric field given by

$$E = -\text{grad}\, \psi\,. \tag{7}$$

As such, the electronic current density may be written as

$$J_e = -\frac{\sigma_e}{q} \text{grad}\, \tilde{\mu}_n = \sigma_e E\,, \tag{8}$$

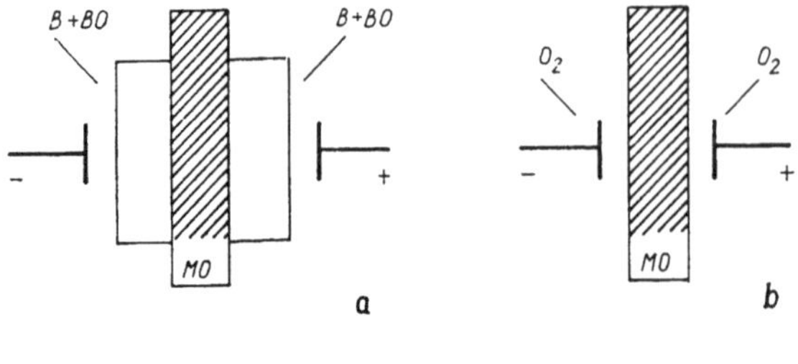


Fig. 1. Ionic conductor with ohmic contacts

which is just ohmic behavior. Likewise, for the ionic contribution to the conductivity (charge transported by two ionic species),

$$J_{\mathrm{I}} = -\frac{\sigma_1 + \sigma_2}{|z_1|\, q} \operatorname{grad} \tilde{\mu}_1 = (\sigma_1 + \sigma_2)\, E\,. \tag{9}$$

The above expressions indicate that as long as the electrodes do not become current limiting due to material transport problems, the current through the oxide is linearly proportional to the applied voltage.

The determination of the relative magnitudes of the partial ionic or electronic current densities could be accomplished by measuring the uptake or output of oxygen at the cathode or anode, and comparing this, through the Faraday constant, to the total current flowing in the external circuit. The measurement could be easily done by monitoring the weight change of the metal–metal oxide electrodes, the cathode losing weight, and the anode gaining weight. The relationship between the magnitudes of the various current densities may be expressed in terms of the transference number. Let us define the transference number of a species as that fraction of the total current carried by that species. As such, the transference number for species of type $k$ is expressed by

$$t_k = \frac{J_k}{\sum\limits_k J_k} = \frac{\sigma_k}{\sum\limits_k \sigma_k}\,. \tag{10}$$

Transfer numbers are subject to the constraint $\sum\limits_k t_k = 1$.

### 3. Ohmic Contacts with no Current Supply

Let us now consider the geometry defined by Fig. 2.

In this case, the oxide specimen is between two fixed chemical potentials of oxygen, defined by the coexistence of the two different metal–metal oxide mixtures, and there is no current supplied by an external source. As such, the total current through the specimen is zero so that

$$J_{\mathrm{tot}} = 0 = \frac{\sigma_{\mathrm{T}}}{q} \operatorname{grad} \tilde{\mu}_{\mathrm{p}} - \frac{\sigma_i}{|z_2|\, q} \operatorname{grad} \tilde{\mu}_{\mathrm{X}} \tag{11}$$

or

$$\operatorname{grad} \tilde{\mu}_{\mathrm{p}} = \frac{\sigma_i}{\sigma_{\mathrm{T}}} \frac{1}{|z_2|} \operatorname{grad} \tilde{\mu}_{\mathrm{X}} \tag{12}$$

in terms of the anion and hole electrochemical potential gradients.

Using the intrinsic electronic equilibrium [1],

$$\operatorname{grad} \tilde{\mu}_{\mathrm{n}} = -\operatorname{grad} \tilde{\mu}_{\mathrm{p}}, \tag{13}$$

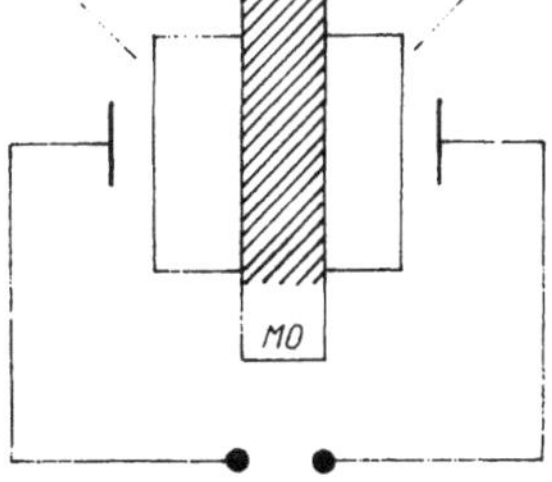


Fig. 2. Ionic conductor with ohmic contacts but no current supply

to express (12) in terms of all negatively charged species, we have

$$\text{grad}\, \tilde{\mu}_n = -\frac{1}{|z_2|} t_{ion}\, \text{grad}\, \tilde{\mu}_X \,, \tag{14}$$

having remembered that $\sigma_i/\sigma_T = t_{ion}$.

Assuming that no injection or extraction processes are occurring at the boundaries, we can write for grad $\tilde{\mu}_n$

$$\text{grad}\, \tilde{\mu}_n = q\, \text{grad}\, \psi \,. \tag{15}$$

Substituting this in (14), converting to planar geometries so that there is only a one-dimensional variation to $\tilde{\mu}_X$, and integrating between the boundaries, we have

$$\int_{I}^{II} d\psi = -\frac{1}{|z_2|\, q} \int_{I}^{II} t_{ion}\, d\tilde{\mu}_X \,. \tag{16}$$

However, the left-hand integral

$$\int_{I}^{II} d\psi = \psi^{II} - \psi^{I} = \varepsilon \tag{17}$$

is just the difference in electrical potential between the two boundaries. As such, (16) is given by

$$\varepsilon = -\frac{1}{|z_2|\, q} \int_{I}^{II} t_{ion}\, d\tilde{\mu}_X \,. \tag{18}$$

In terms of an oxide conductor with $|z_2| = 2$ and $\tilde{\mu}_X = \tilde{\mu}_0 = \frac{1}{2}\mu_{O_2}$, we arrive at the well-known equation of Wagner [2] for electrolytes of mixed conduction,

$$\varepsilon = -\frac{1}{4q} \int_{\mu_{O_2}^{I}}^{\mu_{O_2}^{II}} t_{ion}\, d\mu_{O_2} \,. \tag{19}$$

If the potential difference is small so that $t_{ion}$ is approximately constant and can be replaced with an average value, we can integrate the right-hand side to give

$$\varepsilon = -\frac{\bar{t}_{ion}}{4q} (\mu_{O_2}^{II} - \mu_{O_2}^{I}) \,. \tag{20}$$

Using for the chemical potential of oxygen

$$\mu_{O_2} = \mu_{O_2}^{0} + kT \ln P_{O_2} \tag{21}$$

we write for the e.m.f. generated by a chemical potential difference

$$\varepsilon = -\frac{\bar{t}_{ion}}{4q} kT \ln \frac{P_{O_2}^{II}}{P_{O_2}^{I}} . \tag{22}$$

Equation (22) now affords another possibility for determining the relative magnitudes of the electronic and ionic conductivities. If one were to use two metal–metal oxide mixtures with known thermodynamic potentials, an expected $\varepsilon$, based upon the assumption that $t_{ion}$ was unity, could be calculated. By then comparing this with the actually measured $\varepsilon$, the transference number for ionic conduction could be calculated from

$$t_{\text{ion}} = \frac{\varepsilon_{\text{meas}}}{\varepsilon_{\text{theor}}}. \tag{23}$$

However, we again encounter the problem that when $t_{\text{ion}} \gg 0.99$, errors in the thermodynamic potential determinations can negate the ability to accurately determine transference numbers of low level electronic conduction.

## 4. Blocking or Polarizing Electrodes

The direct determination of electronic conductivities can be accomplished by use of a geometry employing a blocking or polarizing electrode. Such a geometry is shown in Fig. 3.

In the blocking electrode case, a potential less than the decomposition potential of the specimen is applied between an inert, polarizing electrode and one that behaves reversibly at a fixed chemical potential of oxygen. Initially, some ionic current flow exists as the interface between the polarizing electrode and the electrolyte is depleted of oxygen and as a shift occurs in the ionic distribution within the electrolyte. However, as the ionic charge redistribution occurs, a reverse potential builds up to eventually prevent any further ionic current; at this time, the steady-state current flowing in the external circuit will be due to electron or hole conduction through the electrolyte.

The decomposition voltage for an oxide electrolyte is that voltage corresponding to the energy for the reaction

$$O_O \rightarrow V_O^{\cdot\cdot} + 2e' + \tfrac{1}{2} O_2 . \tag{24}$$

For the case of calcia-stabilized zirconia, for example, the decomposition voltage is approximately 2.16 V.

We shall now develop an expression for the voltage–current relationship during steady-state, polarized conduction for an oxide electrolyte. The method used shall be similar to that first put forward by Patterson et al. [3].

The partial current density for a species is given by

$$j_i = -\frac{\sigma_i}{|z_1|\, q} \operatorname{grad} \tilde{\mu}_1 . \tag{25}$$

At steady state, in the polarized condition, the current of ions is zero, i.e., $j_{\text{ion}} = 0$ so that grad $\tilde{\mu}_{O''} = 0$. As such, the steady-state current is given by the sum of the partial current densities of the holes and electrons

$$J_{ss} = j_p - j_n = \frac{\sigma_n}{q} \operatorname{grad} \tilde{\mu}_n - \frac{\sigma_p}{q} \operatorname{grad} \tilde{\mu}_p . \tag{26}$$

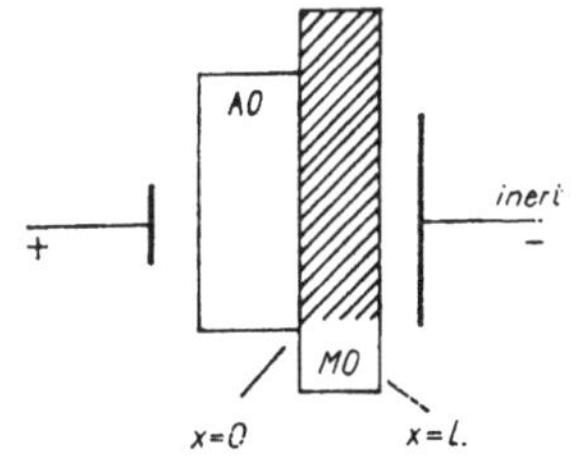


Fig. 3. Blocking or polarizing electrode

From the equilibrium of the oxide electrolyte with its environment we can write

$$\tfrac{1}{2}\,O_2 + 2e' \rightarrow O'' \tag{27}$$

giving

$$\tfrac{1}{2}\mu_{O_2} + 2\tilde{\mu}_n = \tilde{\mu}_{O''} \tag{28}$$

and

$$\tfrac{1}{2}\,\mathrm{grad}\,\mu_{O_2} + 2\,\mathrm{grad}\,\tilde{\mu}_n = \mathrm{grad}\,\tilde{\mu}_{O''}\,. \tag{29}$$

However, due to the polarized condition, we have seen

$$-\tfrac{1}{4}\,\mathrm{grad}\,\mu_{O_2} = \mathrm{grad}\,\tilde{\mu}_n = -\,\mathrm{grad}\,\tilde{\mu}_p\,. \tag{30}$$

The last term in (30) arises from the intrinsic equilibrium which states

$$\mathrm{grad}\,\tilde{\mu}_n + \mathrm{grad}\,\tilde{\mu}_p = 0\,. \tag{31}$$

Putting (30) into the expression for the steady-state current (26), we have

$$J_{ss} = -\frac{\sigma_n}{4q}\,\mathrm{grad}\,\mu_{O_2} - \frac{\sigma_p}{4q}\,\mathrm{grad}\,\mu_{O_2}\,. \tag{32}$$

If we now consider planar geometries with only an $x$ component to the gradient, then

$$J_{ss} = -\frac{\sigma_n}{4q}\frac{d\mu_{O_2}}{dx} - \frac{\sigma_p}{4q}\frac{d\mu_{O_2}}{dx}\,, \tag{33}$$

$$J_{ss} = -\frac{(\sigma_n + \sigma_p)}{4q}\frac{d\mu_{O_2}}{dx}\,. \tag{34}$$

Since, at steady state, the current density is independent of position, the above is integrated to give

$$J_{ss}(L) = -\frac{1}{4q}\int\limits_{\mu^0_{O_2}}^{\mu^L_{O_2}} (\sigma_n + \sigma_p)\,d\mu_{O_2}\,, \tag{35}$$

where $L$ is length of the specimen and the limits of integration are the chemical potentials of oxygen at the two boundaries. Equation (30) may be written in terms of a one-dimensional variation,

$$d\tilde{\mu}_n = -d\tilde{\mu}_p = -\tfrac{1}{4}\,d\mu_{O_2}\,, \tag{36}$$

and if one considers the case of no external field and low concentrations of defects, then the above may be expressed as follows:

$$\begin{aligned} d\tilde{\mu}_n &= kT\,d\ln n = -\tfrac{1}{4}\,d\mu_{O_2}\,,\\ d\tilde{\mu}_p &= kT\,d\ln p = \tfrac{1}{4}\,d\mu_{O_2}\,. \end{aligned} \tag{37}$$

These may be integrated between limits corresponding to the concentrations and oxygen potentials at the boundaries; then, if the concentration of the upper limit is expressed as a function of the oxygen potential, we have

$$\begin{aligned} C_n &= C^0_n \exp\left[-\frac{1}{4kT}\,(\mu^L_{O_2} - \mu^0_{O_2})\right],\\ C_p &= C^0_p \exp\left[\frac{1}{4kT}\,(\mu^L_{O_2} - \mu^0_{O_2})\right]. \end{aligned} \tag{38}$$

If we assume the mobility of the carriers to be independent of concentration, the partial conductivities of the two species are proportional to their concentrations so that the sum of the conductivities is expressed by

$$\sigma_n + \sigma_p = \sigma_n^0 \exp\left[-\frac{\mu_{O_2}^L - \mu_{O_2}^0}{4kT}\right] + \sigma_p^0 \exp\left[\frac{\mu_{O_2}^L - \mu_{O_2}^0}{4kT}\right]. \tag{39}$$

This may now be put in the expression for the steady-state current (35), and the integral evaluated to yield

$$J_{ss}(L) = \frac{kT}{q}\,\sigma_n^0\left[\exp\left(-\frac{\mu_{O_2}^L - \mu_{O_2}^0}{4kT}\right) - 1\right] + \sigma_p^0\,\frac{kT}{q}\left[1 - \exp\left(\frac{\mu_{O_2}^L - \mu_{O_2}^0}{4kT}\right)\right]. \tag{40}$$

From the previous example of no ionic current flowing through the electrolyte, we have (20) which describes the electromotive force generated by a partial pressure gradient. Assuming that $t_{ion} = 1.0$ and imposing an EMF such that $J_{ion} = 0$, i.e., making the reversible electrode positive, the above also describes an imposed oxygen potential generated by an applied voltage. For this situation, the oxygen potential at the polarized interface is lower than that at the reversible electrode. For our case, this oxygen potential difference may be written as

$$\mu_{O_2}^L - \mu_{O_2}^0 = -4\varepsilon q\,. \tag{41}$$

Putting the above in (40) and converting to practical units for the constants, gives the desired result,

$$J_{ss} = \frac{RT}{LF}\,\sigma_n^0\left[\exp\left(\frac{\varepsilon F}{RT}\right) - 1\right] + \frac{RT}{LF}\,\sigma_p^0\left[1 - \exp\left(-\frac{\varepsilon F}{RT}\right)\right], \tag{42}$$

which is identical to the result of Patterson et al. [3].

Equation (42) may be simplified by combining some of the constants,

$$U = \frac{\varepsilon F}{RT}, \qquad K = \frac{RT}{LF}, \tag{43}$$

so that the steady-state polarization current may be written as

$$J_{ss} = K\sigma_n^0\,[(\exp U - 1)] + K\sigma_p^0\,[1 - \exp(-U)]\,, \tag{44}$$

where $\sigma_n^0$ and $\sigma_p^0$ are the partial electron and hole conductivities that are defined for the oxide when it is equilibrated with an oxygen chemical potential defined by the reversible electrode. A polarizing electrode technique allows us to measure the partial electrical conductivities directly, and for the case of $t_{ion} \gg 0.99$ avoids the necessity of taking the difference between two large numbers to arrive at a value for the low level electronic contribution to the total conductivity.

## 5. Summary

In the present paper we have seen that under situations where current is supplied to reversible electrodes at the interfaces of an oxide conductor, the current is ohmic in nature. If a chemical potential gradient is applied across the oxide conductor and no external conduction circuit is employed, an e.m.f. will be generated by the electrolyte that is proportional to the chemical potential difference across the conductor. Brief discussions were given describing methods of determining the transference number for ions in either the current supply or open circuit case. Finally, mathematics were presented that allow one to calculate the electron and hole conductivities directly from polarization measurements.

A combination of these various applications and techniques will allow the conduction processes of an oxide conductor to be characterized over a wide range of oxygen partial pressures in both electronic and ionic conduction regimes. This, of course, will help in understanding the transport properties of various combinations of oxides, which can be universally used at low and high temperatures, as smartly emphasized by Goto, in his recent materials science monograph [4].

***Acknowledgements***

This work was supported by Junta Nacional de Investigação Científica e Tecnológica (Portugal), contract No. 87289, and Anglo-Portuguese Joint Research Programme POR/991/3(87/88)53.

## References

[1] C. A. C. Sequeira and J. M. B. Fernandes Diniz, phys. stat. sol. (a) **105**, 123 (1988).
[2] C. Wagner, Z. phys. Chem. B **21**, 25 (1935).
[3] J. W. Patterson, E. C. Bogren, and R. A. Rapp, J. Electrochem. Soc. **114**, 752 (1967).
[4] K. S. Goto, Solid State Electrochemistry and Its Applications to Sensors and Electronic Devices, Elsevier Publ. Co., New York 1988.

*(Received February 29, 1988)*

phys. stat. sol. (a) **108**, 163 (1988)

Subject classification: 61.70; 64.75; S1.2; S1.3

*Department of Metallurgy, Faculty of Engineering, Nagoya University*[1]) *(a)*
*and Japan Atomic Energy Research Institute, Ibaragi*[2]) *(b)*

# Effect of Alloying on the Void Swelling in Cu-Base and Ag-Base Alloys

By

M. HASHIMOTO (a), H. SAKA (a), T. IMURA (a), S. JITSUKAWA (b), and
K. SUZUKI (b)

Pure Ag, pure Cu, and a variety of Cu-base and Ag-base solid solution alloy single crystal are irradiated with 1 MeV electrons in the high-voltage electron microscopes in a temperature range from 100 to 300 °C, and void swelling behaviour of these metals and alloys is studied. It is shown that in Ag–Al, Cu–Ge, and Cu–Si alloys void swelling is much smaller than in pure Ag, pure Cu, Ag–In, Ag–Sn, and Cu–Al alloys. In the former group a segregation of solute atoms at the sinks of point defects is induced by irradiation, while in the latter group either such a segregation is not induced, or a segregation of solvent atoms is induced by irradiation. Possible mechanisms for the suppression of void swelling in those alloys where a segregation of solute atoms is induced by irradiation are discussed.

Im Hochspannungselektronenmikroskop und im Temperaturbereich von 100 bis 300 °C werden reine Ag, reine Cu und eine Vielzahl von Cu- und Ag-Legierungsmischeinkristalle mit 1 MeV-Elektronen bestrahlt und das Hohlraumschwellverhalten dieser Metalle und Legierungen untersucht. Es wird gezeigt, daß in Ag–Al-, Cu–Ge- und Cu–Si-Legierungen Hohlraumanschwellung viel geringer als in reinem Ag, reinem Cu, Ag–In-, Ag–Sn- und Cu–Al-Legierungen ist. In der ersten Gruppe wird eine Segregation der Lösungsatome an den Senken von Punktdefekten durch die Bestrahlung induziert, während in der letzteren Gruppe entweder eine solche Segregation nicht induziert wird oder eine Segregation der Solventatome durch Bestrahlung induziert wird. Mögliche Mechanismen für eine Unterdrückung der Hohlraumschwellung in solchen Legierungen, wo eine Segregation der Lösungsatome durch Bestrahlung induziert wird, werden diskutiert.

## 1. Introduction

Void swelling is one of the most serious problems arising in connection with structural materials used in nuclear engineering. Many experimental as well as theoretical studies have been carried out in order to identify those factors which affect the behaviour of void swelling of metals and alloys. Effects of alloying on the void swelling has been studied most extensively not only from the practical point of view but also from theoretical ones [1, 2].

However, systematic studies on alloying effects on void swelling using well-characterized alloy systems appear still lacking. In the present study, effects of alloying on the void swelling have been studied comprehensively using in-situ electron irradiation experiments in high-voltage electron microscopes on Cu-base and Ag-base solid solution alloys with a wide range of compositions. These materials were selected because the material parameters for the characterization are available.

---

[1]) Furo-cho, Chikusa-ku, Nagoya 464, Japan.
[2]) Tokai-mura, Ibaragi 316-11, Japan.

## 2. Experimental Procedures

Single crystals of a series of Cu-base and Ag-base alloys were grown by the Bridgman technique, and discs with surface orientations close to {111} of diameter 3 and thickness 0.2 to 0.3 mm were obtained by a multi-wire abraser saw. Foil specimens for transmission electron microscopy were prepared by a standard electropolishing technique using a Tenupol automatic thinner, and irradiated with 1 MeV electrons during observation in a temperature range between 100 and 300 °C either in a Hitachi HU-1000D microscope of Nagoya University or JOEL JEM-1000 microscope of JAERI, both operating at 1 MeV. In order to avoid possible effects caused by the crystal surfaces, regions thicker than 0.5 μm for Cu alloys and 0.3 μm for Ag alloys were observed. The specimen thickness was estimated from the number of the equi-thickness contours. Compositions of the alloys used for void swelling are given in Table 1.

Table 1

Specimens used for study of void swelling

| pure Cu | pure Ag |
|---|---|
| Cu–2.0 at% Al, Cu–7.0 at% Al, Cu–13.43 at% Al | Ag–5.0 at% Al, Ag–15.0 at% Al |
| Cu–2.6 at% Si, Cu–10.7 at% Si | Ag–2.0 at% Sn |
| Cu–5.0 at% Ge, Cu–10.8 at% Ge | Ag–10.0 at% In |

## 3. Results

Fig. 1 shows microstructures of pure Cu and Cu-base alloys irradiated by 1 MeV electrons to a dose of $7 \times 10^{23}$ electrons/cm$^2$ at 250 °C. Fig. 2 shows microstructures of pure Ag and Ag-base alloys irradiated by 1 MeV electrons to a dose of $8.5 \times \times 10^{23}$ electrons/cm$^2$ at 150 °C. In Fig. 3 a and b the void swelling of Cu-base and Ag-base alloys is plotted against the irradiation temperature, respectively. From these results it is clear that void swelling of Cu–Si, Cu–Ge, and Ag–Al alloys is much smaller than that of the other alloys studied.

In Table 2 void swelling in Cu-base and Ag-base alloys is summarized together with those factors which may affect void swelling of these alloys. The values of the stacking-fault energy (SFE) are cited from weak-beam studies of dissociated dislocations [3 to 8]. From Table 2 no correlation between the suppression of void swelling and SFE is observed. The same is true for the valence and the relative atomic size of solute atoms [9]. On the other hand, void swelling in those alloys where a segregation of solute atoms at sinks such as dislocations takes place during irradiation is much smaller than that in those where the segregation of solute atoms does not take place or in those where a segregation of solvent atoms (hence a depletion of solute atoms) takes place.

## 4. Discussion

In general the formation of voids requires a supersaturation of vacancies and/or a preferential sink for self-interstitials. Since the present study shows definitely that there is a correlation between the occurrence of a radiation-induced segregation of solute atoms and the suppression of void swelling, it may be worth reviewing the mechanisms of the radiation-induced segregation [8].

Fig. 4 shows weak-beam electron micrographs of dissociated dislocations in Ag–10.0 at% In, Ag–10.0 at% Al, Cu–5.1 at% Si, and Cu–5.3 at% Ge alloys irradiated

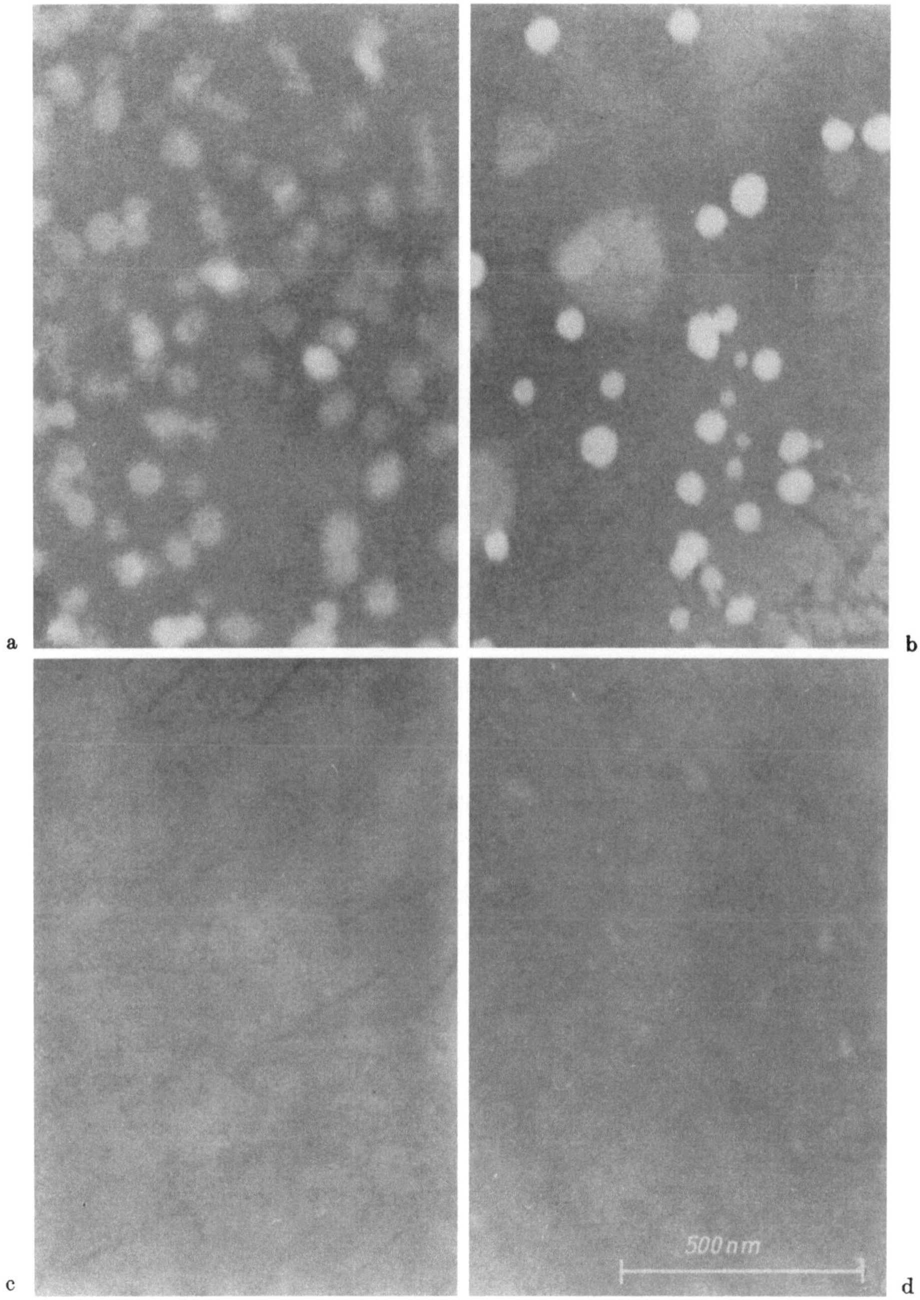


Fig. 1. Typical microstructures of pure Cu and Cu-base alloys irradiated with 1 MeV electrons to a total dose of $7 \times 10^{23}$ electrons/cm$^2$ at 250 °C. a) Pure Cu, b) Cu–13.43 at% Al, c) Cu–5.0 at% Ge, and d) Cu–10.7 at% Si

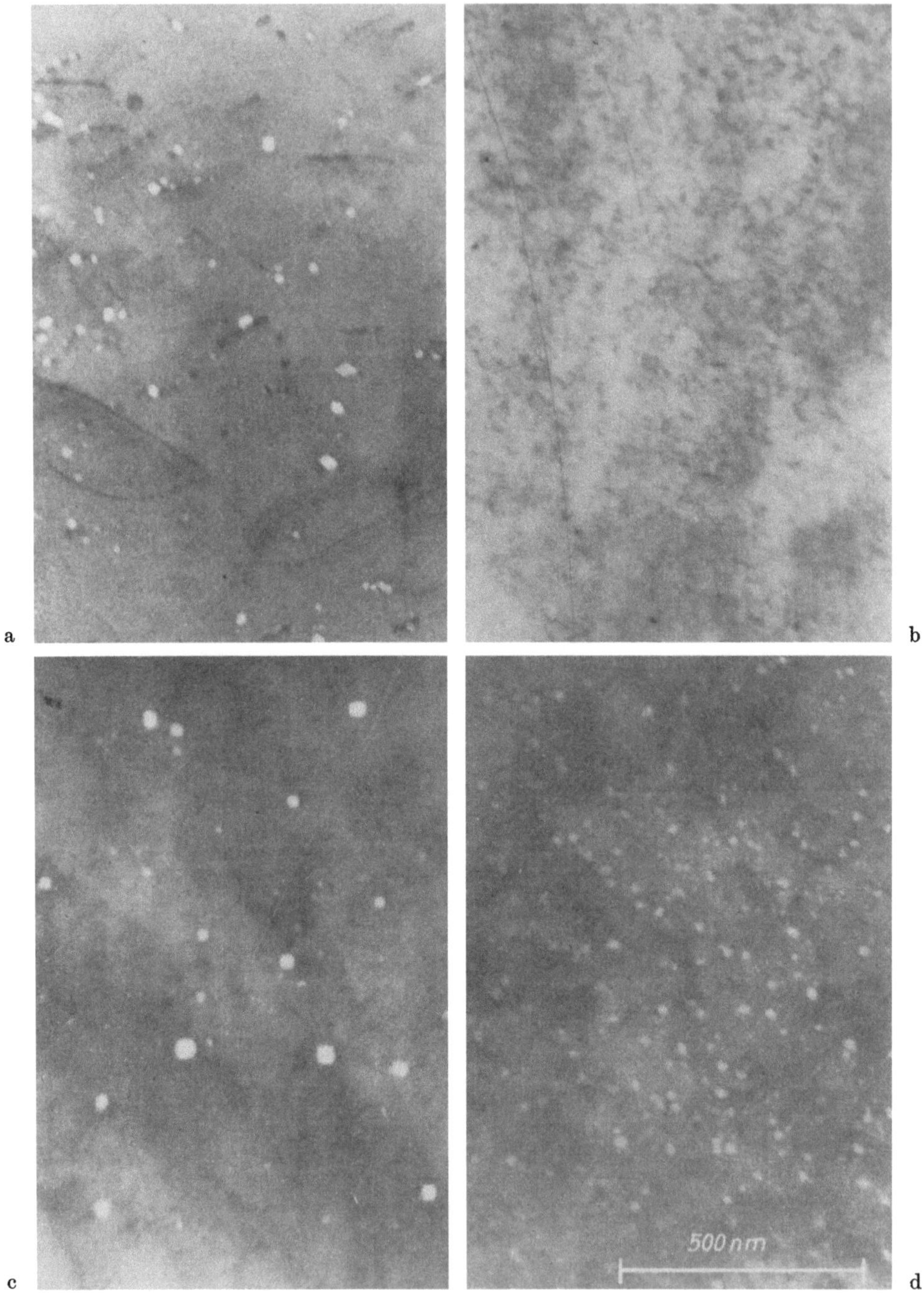


Fig. 2. Typical microstructures of pure Ag and Ag-base alloys irradiated with 1 MeV electrons to a total dose of $8.5 \times 10^{23}$ electrons/cm$^2$ at 150 °C. a) Pure Ag, b) Ag–5.0 at% Al, c) Ag–10.0 at% In, and d) Ag–2.0 at% Sn

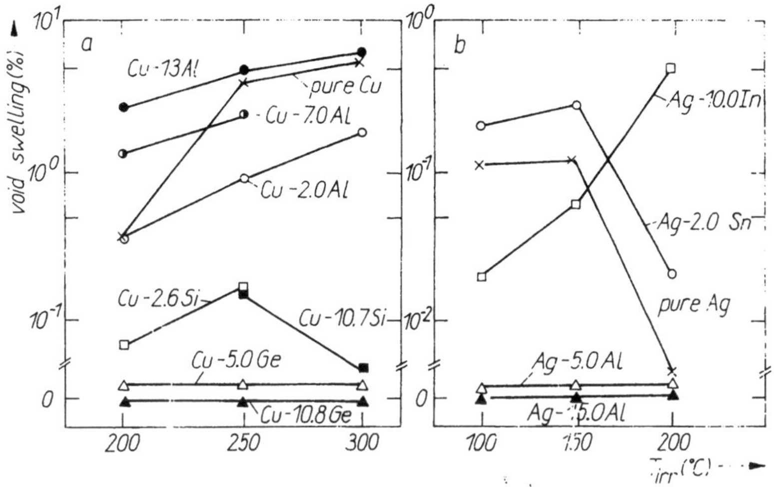


Fig. 3. Void swelling of a) Cu-base and b) Ag-base alloys as a function of the irradiation temperature

with 1 MeV electrons. Here, in Cu–Ge, Cu–Si, and Ag–Al alloys the width of dissociated dislocations increased on irradiation and the reverse was true in Ag–In [8]. This indicates that a segregation of solute atoms to the dislocations is induced by irradiation in Cu–Ge, Cu–Si, and Ag–Al, while a segregation of solvent atoms (hence a depletion of solute atoms) is induced in Ag–In. Among these alloys only Ag–Al has undersized solute; Ag–In, Cu–Si, and Cu–Ge alloys have oversized solutes [9]. This suggests that the mechanism for the segregation for Cu–Ge and Cu–Si is likely to be different from that in Ag–Al and Ag–In alloys. The segregation behaviour in the latter group can be explained by the usual size-effect, where the formation of complex between inter-

Table 2
Relation between void swelling and various material parameters

| alloy | SFE ($mJm^{-2}$) | valence of solute | relative atomic size | radiation-induced segregation of solute | void swelling (%) | normalized void swelling |
|---|---|---|---|---|---|---|
| pure Cu | 41 | | | | 3.8 | 1 |
| Cu–2.0 Al | 38 | 3 | over | no | 0.92 | 0.24 |
| Cu–7.0 Al | 18 | 3 | over | no | 2.4 | 0.63 |
| Cu–13.3 Al | 7 | 3 | over | no | 4.7 | 1.24 |
| Cu–2.4 Si | 29 | 4 | over | enrichment | 0.17 | 0.045 |
| Cu–10.7 Si | 5 | 4 | over | enrichment | 0.15 | 0.039 |
| Cu–5.0 Ge | 17 | 4 | over | enrichment | 0 | 0 |
| Cu–10.8 Ge | 5 | 4 | over | enrichment | 0 | 0 |
| pure Ag | 20 | | | | 0.12 | 1 |
| Ag–5.0 Al | 12 | 3 | under | enrichment | 0 | 0 |
| Ag–15.0 Al | 4 | 3 | under | enrichment | 0 | 0 |
| Ag–10.0 In | 4 | 3 | over | depletion | 0.06 | 0.50 |
| Ag–2.0 Sn | 13 | 4 | over | no | 0.27 | 2.25 |

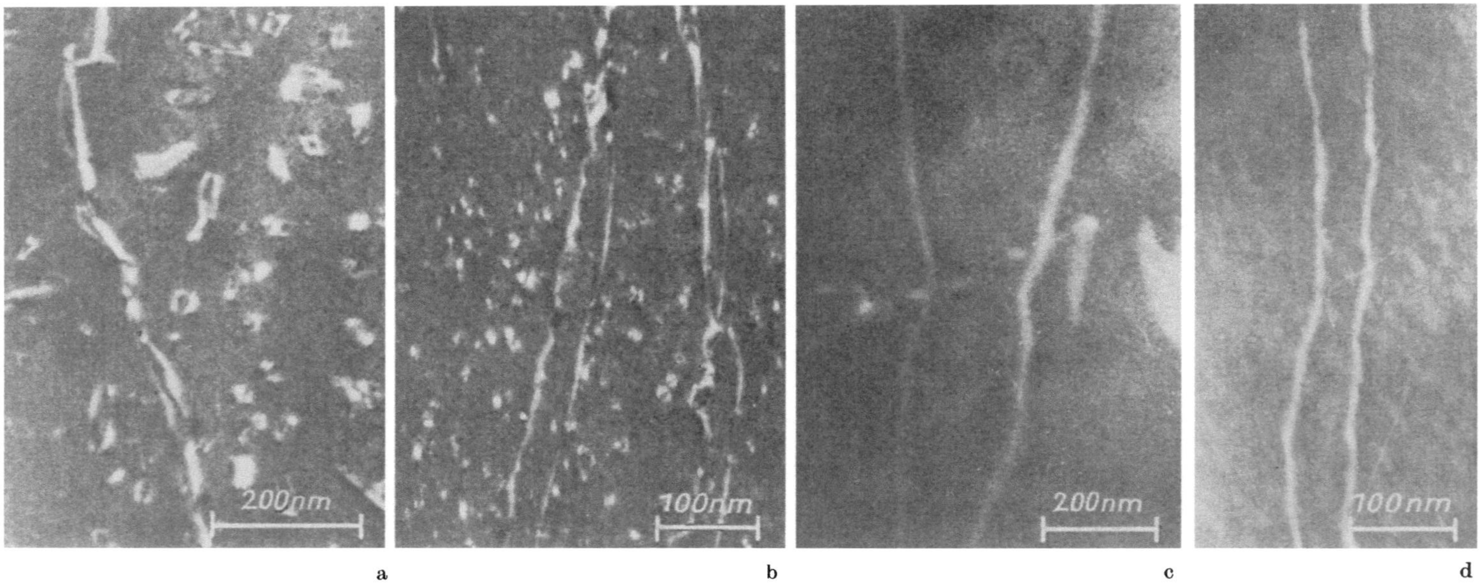


Fig. 4. Weak-beam electron micrographs of dissociated dislocations irradiated with 1 MeV electrons in Cu-base and Ag-base alloys: a) Ag–10.0 at% In irradiated at 150 °C, b) Ag–10 at% Al irradiated at 150 °C, c) Cu–5.1 at% Si irradiated at 300 °C, and d) Cu–5.3 at% Ge irradiated at 300 °C. The width of dissociated dislocation in Ag–In has decreased as a result of the irradiation, while that in the others has increased

stitials and atoms with smaller size (Al for Ag–Al and Ag for Ag–In) is assumed, while for Cu–Ge and Cu–Si the formation of a solute–vacancy complex must be invoked [8].

This may be confirmed from a detailed examination of the configurations of irradiated dislocations in these alloys shown in Fig. 4. In Ag–In and Ag–Al alloys many small dislocation loops were formed in the matrix, and the irradiated dislocations were heavily jogged, the density of jogs being much higher in Ag–In than in Ag–Al. On the other hand, in Cu–Ge and Cu–Si alloys a formation of such dislocation loops was not observed in the matrix and the dislocations are not jogged. The formation of jogs on initially jog-free dislocations indicates that those dislocations have climbed by absorbing preferentially either interstitials or vacancies [10]. In other words, there is a significant difference in flux between interstitials and vacancies.

In Ag–Al and Ag–In, interstitials, which are expected to migrate much faster than vacancies, flow to the dislocations, accompanying solute or solvent atoms, depending on the relative atomic size, and this results in a segregation of solute or solvent atoms near the dislocations. Since in Ag–In there is a strong binding between solvent atoms and interstitials and not between interstitials and solute atoms, the mobility of interstitials is unlikely to be affected by the presence of solute atoms. On the other hand, in Ag–Al there is a strong binding between interstitials and solute atoms, so that the motion of interstitials is most likely retarded by the solute atoms. Thus the vacancy–interstitial annihilation rate is increased, and this leads to a reduced steady-state concentration of vacancies. It is to be pointed out here that under these circumstances the mobility of interstitials is still expected to be much higher than that of vacancies in Ag–Al, and this leads to a net flux of interstitials to the dislocations. However, this net flux of interstitials in Ag–Al (where there is a strong binding between interstitials and solute atoms) must be smaller than that in Ag–In (where there is no strong binding between interstitials and solute atoms). This may explain why the irradiated dislocation in Ag–In is jogged more heavily than in Ag–Al, as is shown in Fig. 4 a and b.

On the other hand, in Cu–Ge and Cu–Si alloys there is a strong binding between solute atoms and vacancies and this reduces the mobility of vacancies. This reduced mobility of vacancies will lead to an increaase in the steady-state vacancy concentration, giving rise to an increased vacancy–interstitial annihilation rate and to a reduced flux of interstitials to dislocations. In fact, weak-beam micrographs of dissociated dislocations, to which solute atoms must segregate, show no evidence of climb, indicating that the fluxes of interstitials and vacancies to the dislocations are approximately equal.

***Acknowledgements***

The authors wish to express their heart-felt thanks to Dr. A. Hishinuma for discussion and for reading the manuscript, and to Mr. K. Fukai for his assistance with experiment. Part of the present study was supported by the Grant-in-Aid for Fundamental Scientific Research from the Ministry of Education, Science, and Culture of Japan.

## References

[1] Y. Adda, in: Radiation-Induced Voids in Metals, Ed. J. W. Corbett and L. C. Lanniello, National Technical Information Service, US Department of Commerce, Springfield (Va.) 1972 (p. 31).

[2] A. Hishinuma, A. Katano, and K. Shiraishi, J. nuclear Mater. (Amsterdam) **103/104**, 1063 (1981).

[3] D. J. H. Cockayne, M. L. Jenkins, and I. L. F. Ray, Phil. Mag. **24**, 1383 (1971).

[4] C. B. CARTER and I. L. F. RAY, Phil. Mag. **35**, 189 (1977).
[5] H. SAKA, Y. SUEKI, and T. IMURA, Phil. Mag. **A37**, 273 (1978).
[6] H. SAKA, T. KONDO, and T. IMURA, Phil. Mag. **A47**, 859 (1983).
[7] H. SAKA, T. IWATA, and T. IMURA, Proc. 7th Europ. Congr. Electron Microscopy 1980, Ed. P. BREDEROO and J. VAN LANDUYT, Electron Microscopy Foundation, Leiden 1980 (p. 326).
[8] H. SAKA, Phil. Mag. **A48**, 239 (1983).
[9] H. W. KING, J. Mater. Sci. **1**, 79 (1966).
[10] D. CHERNS, P. HIRSCH, and H. SAKA, Proc. Roy. Soc. **A371**, 213 (1980).

*(Received February 29, 1988)*

phys. stat. sol. (a) 108, 171 (1988)

Subject classification: 61.70; 61.55; 62.20; S4

*Sektion Mathematik/Physik der Pädagogischen Hochschule Halle*[1]) *(a) and Institut für Nichteisenmetalle Freiberg des Mansfeldkombinats*[2]) *(b)*

# The Effect of Additions of Mn and Cr and the Applied Solidification Technique on the Structure of an Al–2.2 at% Zn–1.4 at% Mg Alloy

By

P. CZURRATIS (a), E. KORNGIEBEL (a), G. JACOBI (b), and H. LÖFFLER (a)

Dedicated to Prof. Dr. E. GUTSCHE on the occasion of his 60th birthday

The structure and hardness of two Al–2.2 at% Zn–1.4 at% Mg alloys with additions of Mn and Cr are investigated in dependence of the applied production technique, viz. classical processing method or rapid solidification (RSP). The RSP samples reveal the following main features: (i) Metastable supersaturation of Mn and Cr atoms, (ii) finely distributed precipitates of intermetallic phases ($Al_6Mn$, $Al_7Cr$) in the grain interior and in the boundaries, (iii) the low as well as high angle boundaries of the very fine grains are pinned by precipitates, and (iv) the formation of small precipitates during the ageing at elevated temperatures. All these effects contribute to the fact that the RSP samples have in many respects significantly better properties than the classically produced ones.

Die Struktur und Härte von zwei AlZn(2,2 At%)Mg(1,4 At%)-Legierungen mit Zugaben von Mn und Cr werden in Abhängigkeit vom Herstellungsverfahren, nämlich klassisches Gießen und Schnellerstarrung (RSP), untersucht. Die RSP-Proben zeigen folgende Hauptmerkmale: (i) Metastabile Übersättigung der Mn- und Cr-Atome, (ii) fein verteilte Ausscheidungen intermetallischer Phasen ($Al_6Mn$, $Al_7Cr$) im Korninnern und in den Korngrenzen, (iii) die Klein- als auch die Großwinkelkorngrenzen der sehr kleinen Körner werden von Ausscheidungen festgehalten und (iv) die Bildung kleiner Ausscheidungen während der Alterung bei erhöhten Temperaturen. Alle diese Effekte tragen dazu bei, daß die RSP-Proben in vielen Beziehungen bessere Eigenschaften besitzen als die über das klassische Gießverfahren hergestellten.

## 1. Introduction

The technically widespread Al–Zn–Mg alloys always contain additional elements. Few of them are "natural joiners" of Al (e.g. Si and Fe), but some are purposely added, e.g. Mn or Cr, in order to improve their response against stress corrosion cracking. The impurity atoms may influence the decomposition processes proceeding in Al–Zn–Mg alloys in the following manner: (i) They may precipitate and sometimes consume alloying elements (e.g. Si by forming $Mg_2Si$, diminishing in this manner the strength of the alloy); (ii) they can stabilize the Guinier-Preston (GP) zones as, e.g., reported for additions of Ag [1] and Cu [2]; (iii) they may assist the heterogeneous nucleation of metastable phases more stable than GP zones, e.g. $\eta'(MgZn_2)$-phase, if the alloy is directly aged at temperatures above the upper limit temperature of the formation of GP zones, as known for additions of Si [3]; and (iv) they can capture vacancies, retarding in this manner the formation of GP zones, as known, e.g., for additions of Si, V, and Ti [4].

[1]) Kröllwitzer Str. 44, DDR-4050 Halle (Saale), GDR.

[2]) Lessingstr. 41, DDR-9200 Freiberg, GDR.

Recently, the effect of additions of a few elements, namely Si, Cr, Mn, and Fe, single and pairwise, on the mechanical properties of an Al–1.9 at% Zn–1.3 at% Mg mother alloy was investigated by Korngiebel and coworkers [3, 5 to 8]. Especially the effect of additions of Mn and/or Cr was of interest, because these elements considerably raised the hardness and in addition, according to the literature, they may improve the response of the Al–Zn–Mg alloys against stress corrosion cracking. However, it turned out that, if adding 0.43 at% Mn to the Al–Zn–Mg alloy, coarse precipitates of the $Al_6Mn$ phase were present in the samples within the grain boundaries as well as inside the grains, after applying the usual conventional two-step ageing heat treatment, viz. homogenization at 490 °C, quench in water of room temperature (RT), 7d ageing at RT, and then 10 h ageing at 160 °C. These coarse precipitates have a detrimental effect on the crack properties (low ductility) of the alloys.

These coarse precipitates are nucleated and grown during homogenization at 490 °C, since, actually, the maximum solubility of Mn in the Al matrix amounts to 0.76 at% at the eutectic temperature of 658.5 °C, but less than 0.15 at% at 490 °C. One can expect that the precipitates of the intermetallic $Al_6Mn$ phase do not coarsen during the ageing at 160 °C (or at least only to an insignificant extent), since the mobility of Mn in the Al matrix is too low at 160 °C (see e.g. the discussion in [9]). The amount of Mn added to the Al–Zn–Mg alloy was deliberately chosen so high, because (i) itcan be expected that the amount of impurity additions will drastically increase in the future as more scrap is used (recycling) for the production of Al alloys, and (ii) from the technological point of view, one is interested to add fairly high amounts of Mn, in order to bind elements (e.g. Fe and Si) with harmful effects on the properties of the Al–Zn–Mg alloys in phases containing Mn (e.g. $(FeMn)Al_6$, $Mn_3Si_2Al_{51}$).

As known for about three decades, in many cases, the capacity of the Al matrix to solve alloyed elements can be drastically increased (metastable supersaturation), if the melt is rapidly solidified (rapid solidification processing (RSP), see e.g. the reviews given in [10, 11]). From investigations performed with Al–Cu alloys with additions of Mn produced by RSP, it is known [12] that Mn additions significantly raise the warm strength properties of the Al–Cu mother alloy.

Thus, it is the aim of this paper to investigate the effect of RSP on the structure and hardness of two alloys of type Al–2.2 at% Zn–1.4 at% Mg containing 0.4 at% Mn or 0.4 at% Mn and 0.2 at% Cr. For comparison, additionally samples produced by the conventional processing method (CPM), namely casting in bars, homogenization in the solid state, quenching in water of RT, and ageing, are examined.

## 2. Experimental

The compositions of the investigated alloys (all concentrations are given in at%) are Al–Zn(2.1)–Mg(1.3)–Mn(0.43) and Al–Zn(2.3)–Mg(1.6)–Mn(0.43)–Cr(0.23), respectively. They were produced from components of 4N purity conventional casting of rods (diameter 50 mm) in a chilled copper mold at the Institut für Nichteisenmetalle Freiberg. As already mentioned at the end of Section 1, the specimens were prepared by two quite different procedures, viz.: (i) CPM, i.e. after a homogenization treatment of the cast bars of 8 h at 490 °C, slices of the rods were warm-rolled in several steps to sheets of a thickness of about 0.2 mm. Parts of these sheets were once more homogenized at a temperature $T_q$ ranging between 400 and 490 °C for various times $t_q$ (1 to 19 h), quenched in water of RT (1 h $T_q/H_2O$ (RT) for short). Afterwards the specimens were aged in the wanted manner at $T_a$ ($n$ h at $T_a$) and then quenched in water at RT. All investigations were performed at RT. (ii) RSP, i.e. about 3 g of the material of the cast rods were remelted and pushed through on orifice with a diameter of 1 mm by technically pure Ar (pressure about 1 MPa) on a well polished rotating copper disk. The melt-spun ribbons had a sthickness of about $D = 0.05$ mm. To evaluate the solidification rate $\dot{T}$, the relation $\dot{T} = B/D^2$ given in [13] is used. The parameter $B$ characterizes the heat transfer between the melted alloy and the cooled copper wheel or disk. In this case $B \approx 10^4$ mm² K/s and one obtains, for $D = 0.05$ mm, $\dot{T} \approx 10^5$ K/s. Details of the used apparatus for RSP are given elsewhere [14].

The structures of the specimens were investigated by means of a transmission electron microscope (TEM) of type TESLA BS 540 operating at 120 kV. The samples were prepared by means of the window technique using an electrolyte consisting of 400 ml $CH_3OH$, 125 ml $C_2H_5OH$, 300 ml $C_3H_8OH$, and 25 ml $HClO_4$ (60%). The Vickers' microhardness was measured by means of a microhardness attachment to the Neophot II due to Hanemann using a load of 0.1 N at RT. The thickness of the specimens used for the hardness measurements was about 0.2 mm for the CPM and 0.1 mm for the hardness measurements was about 0.2 mm for the CPM and 0.1 mm for the RSP samples. In each case, the depth of the indents did not exceed 30% of the sample thickness.

## 3. Results and Discussion

### *3.1 Al–Zn–Mg alloy with addition of 0.43 at% Mn*

In this section the structure and hardness of the Al–Zn(2.1)–Mg(1.3)–Mn(0.43) alloy are investigated in dependence on $T_q$, $t_q$, and the applied production method, namely CPM or RSP.

#### *3.1.1 Structure of the CPM samples after the quench in water of RT*

If considering the structure of the CPM samples, one has to take into account that at $T_q = 490$ °C (this $T_q$ is mostly applied in this work) the solubility of Mn in the Al matrix amounts only to about 0.2 at% and at 400 °C even only to 0.1 at%. $T_q$ was varied between 400 and 490 °C in steps of 30 K. Inside the grains, in each case, coarse precipitates having the shape of rhombs and hexagons, are visible (see Fig. 1 a and b). They belong to the $Al_6Mn$ phase, which has an orthorhombic lattice with $a = 0.650$, $b = 0.755$, and $c = 0.886$ nm (error $\pm 1$ pm) [15]. It is known from the literature that the precipitates of the $Al_6Mn$ phase have the shapes of rhombs and hexagons [16]. Dark field (DF) images performed with an extra spot of the $Al_6Mn$ phase prove that all these coarse precipitates belong to this phase.

In Table 1 the density and the size range (mean diameter) of the precipitates of the $Al_6Mn$ phase, found in the interior of the grains, are listed in dependence on $T_q$. As it has to be expected, the volume fraction of these precipitates is diminished with increasing $T_q$.

Table 1

Density and size range of the precipitates of the intermetallic $Al_6Mn$ phase found in the CPM samples of the Al–2.1 at% Zn–1.3 at% Mg–0.43 at% Mn alloy immediately after the quench from various $T_q$ in water of RT

| $T_q$ (°C) | density ($10^{12}$ cm$^{-3}$) | size range of the precipitates (nm) |
|---|---|---|
| 400 | 6.0 | 150 to 1000 |
| 430 | 4.5 | 150 to 800 |
| 460 | 4.0 | 150 to 550 |
| 490 | 4.0 | 160 to 550 |

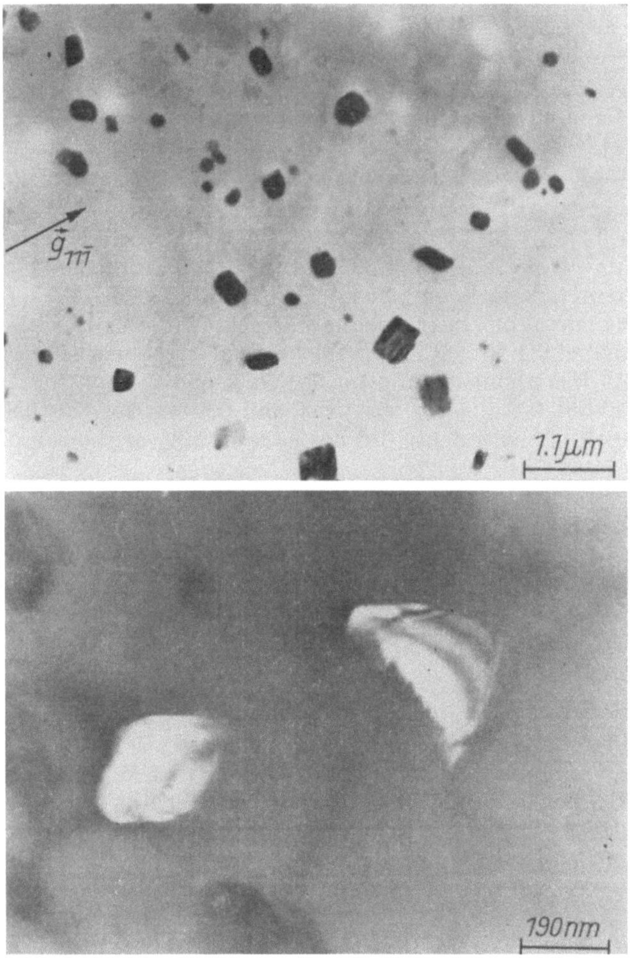


Fig. 1. Quasi-homogeneous distribution of precipitates of the $Al_6Mn$ phase in the CPM sample of the Al–2.1 at% Mn–1.3 at% Mg–0.43 at% Mn alloy, homogenized for 1 h at 490 °C and then quenched in water of RT. a) Bright field image (**B** = [112]), b) dark field image

The structure changes proceeding during the homogenization performed at $T_q =$ $= 490$ °C are investigated, i.e. $t_q$ was varied between 1 and 19 h. The main results are the following: (i) No significant change of the density of the precipitates of the $Al_6Mn$ phase could be found up to $t_q = 19$ h, due to the rather low mobility of the Mn atoms in the Al matrix. (ii) Striking, however, is the formation of "nests" of precipitates, beginning with $t_q = 5$ h (in Fig. 2a and b examples for $t_q = 19$ h are presented). (iii) After $t_q \approx 19$ h, in a few cases, big plate-like precipitates of the $Al_6Mn$ phase are grown (see Fig. 2c), with dimensions up to 0.05 mm.

We explain the formation of these "nests" of precipitates and of these big plates in the frame of the following conception. Often the close vicinity of the dislocation cores is enriched with solute atoms (Suzuki and Cottrell clouds, respectively). In addition, the diffusion of solute atoms along dislocations (pipe diffusion) is much easier than in the undistrubed matrix. Thus, at sufficiently high temperatures, precipitates can easily

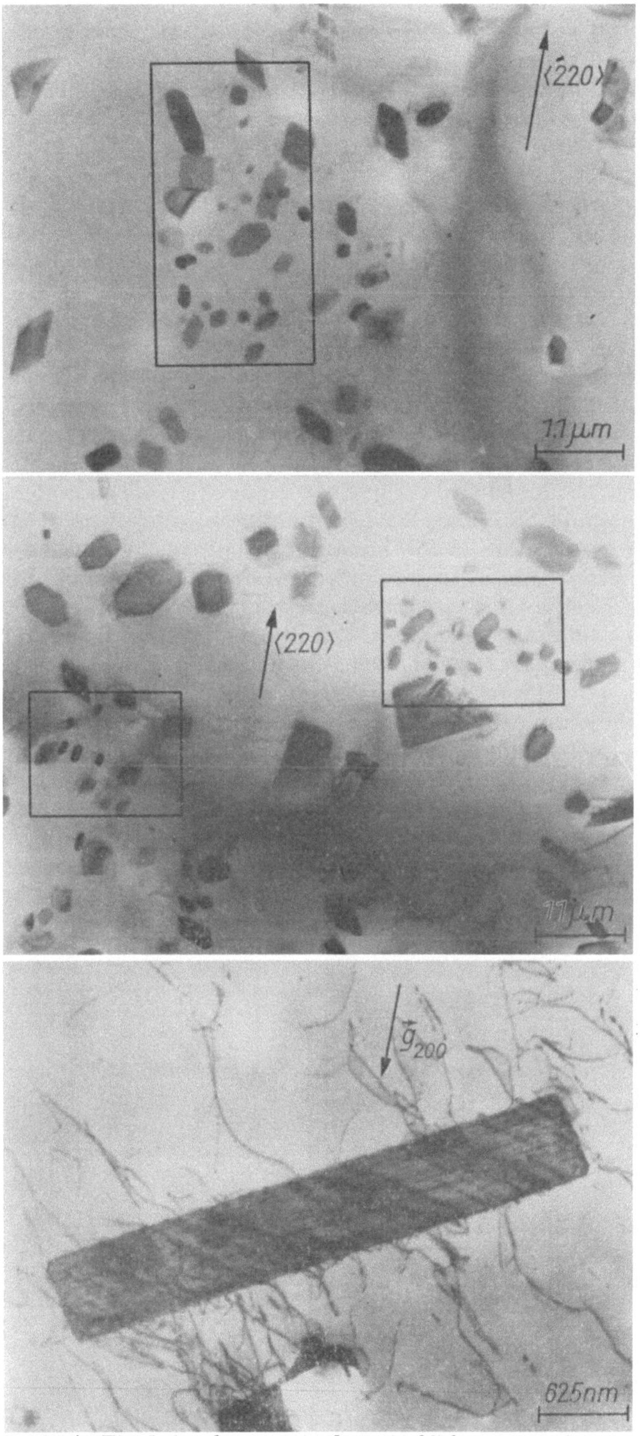

a
b
c

Fig. 2. As Fig. 1, but homogenized at 490 °C for $t_q$ = 19 h. a) $\boldsymbol{B}$ = [001] and b) [001] ("nests" of precipitates), c) $\boldsymbol{B}$ = [013] (big plate of the $Al_6Mn$ phase) form at dislocations. Due to their volume misfit to the ambient matrix, these precipitates are associated with strains, often resulting in the formation of extrinsic misfit dislocations in the adjoing parts of the matrix. As pointed out by Lorimer [17], the stress field present around the primarily formed precipitates may initiate the nucleation of further precipitates in its close vicinity (autocatalytic nucleation). This process might be supported by the misfit dislocations, acting as sites for the heterogeneous nucleation, formed around the precipitates. In Fig. 2b, precipitates, connected by dislocations, are visible. Furthermore, these big plates of the $Al_6Mn$ phase are most probably formed by agglomeration of precipitates present in such nests, favoured by the connecting dislocations via pipe diffusion.

If inspecting the precipitation structure in the grain boundaries, also coarse precipitates of the $Al_6Mn$ phase were found (Fig. 3) with diameters up to about 500 nm. These coarse precipitates were present in each type of grain boundaries, that is small and high angle boundaries with arbitrary as well as coincidence lattice orientation. If applying fairly low $T_q$ (about 400 °C), in addition, small precipitates of the $\eta(Zn_2Mg)$-equilibrium phase were found (see the arrows in Fig. 3), as proved by the respective diffraction pattern. Note that in the interior of the grains no indication of the presence of precipitates of the $\eta$-phase could be found. Also in the mother alloy free of Mn, precipitates of this phase were detected, if the homogenization was performed at 400 °C [18]. According to the ternary phase diagram of the Al–Zn–Mg system published by Mondolfo [19], however, for the alloy investigated in this paper, the lower limit of the range of homogeneity is situated at about 300 °C. But nevertheless, in the grain boundaries precipitates of the $\eta$-phase are formed during the homogenization at 400 °C. In our opinion, the main reason for this precipitation effect is the tendency of the Mg and Zn atoms to segregate at the grain boundaries. This conclusion is supported by the results presented in Fig. 4a and b, demonstrating that even at $T_q = 490$ °C (if $t_q \geqq 5$ h) distinct indications of segregation and nucleation processes are visible. In the random, high angle boundary, the misorientation is $\langle 111 \rangle$ 32°, the segregation effect being particularly distinctly pronounced at the steps and ledges of the grain boundary plane (see Fig. 4a). After fairly long $t_q$ ($\approx 20$ h) nucleation of precipitates

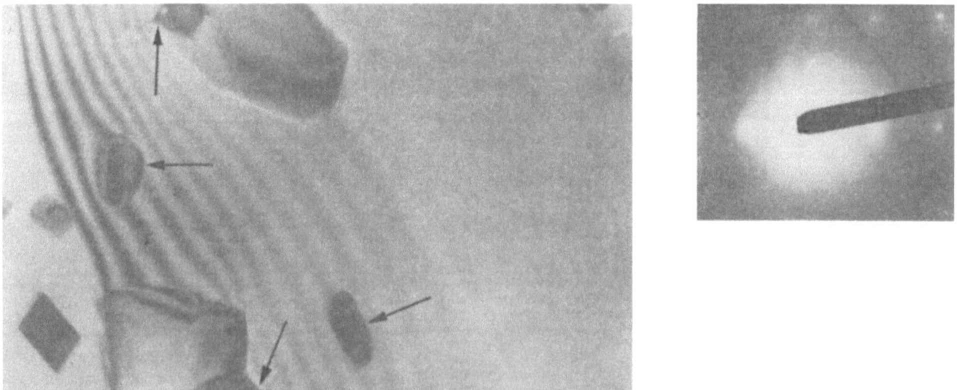


Fig. 3. Big precipitate of the $Al_6Mn$ phase placed in a wide angle boundary of the Al–Zn–Mg–Mn alloy (CPM sample) (1 h at 400 °C)

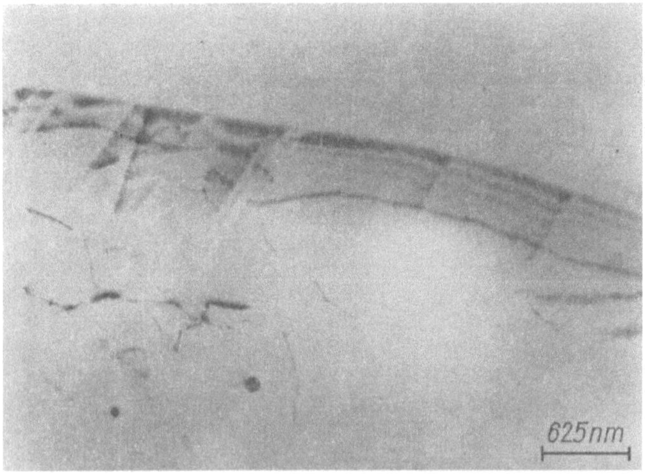


a

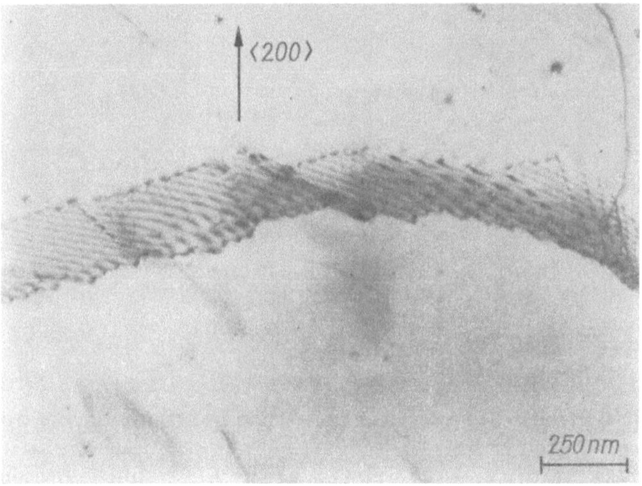


b

Fig. 4. Segregation of alloyed elements during the homogenization of the Al–Zn–Mg–Mn alloy (CPM sample) at 490 °C a) in a high angle grain boundary (misorientation: 32° with respect to the ⟨111⟩ axis; $t_q = 5$ h); b) in dislocation nodes of a subgrain boundary network ($t_q = 19$ h)

of the $Al_6Mn$ phase occurs at dislocation intersections in the dislocation network of low angle boundaries (see Fig. 4b).

If considering the structure of the grain boundaries of the Al–Zn–Mg–Mn alloy, the high density of defects in the grain boundaries, viz. steps, intrinsic dislocations, and broad ledges (Fig. 5 a and b) is striking. The density of these defects is considerably higher in comparison to the alloy free of Mn. We explain this difference in the frame of the following conception: (i) In order to diminish the quenched-in lattice strains, the coarse precipitates of the $Al_6Mn$ phase, present in the interior of the grains, tend to punch out dislocations. After being rearranged in the matrix, they are piled up at the grain boundaries and it can be expected that a considerable part of them becomes absorbed by the grain boundaries. (ii) Due to the strains generated during the quench, part of the grain boundaries may glide. These moving boundaries are trapped by the coarse precipitates and at these places their orientationrelationships to the adjoining

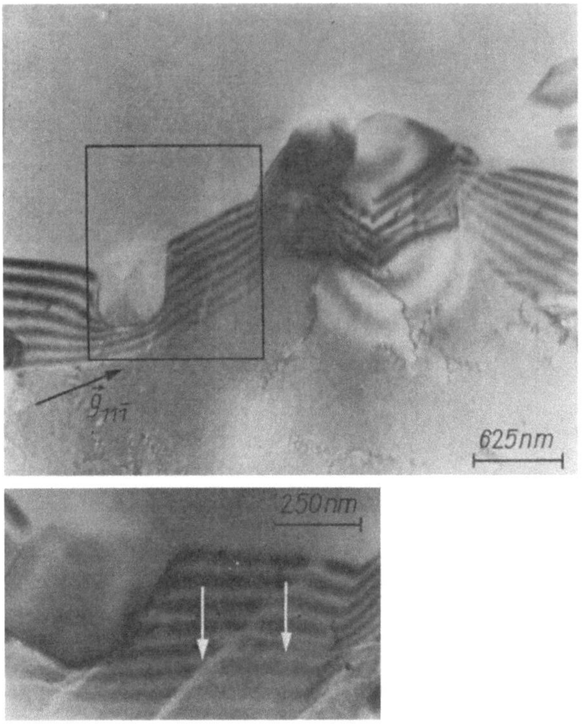


Fig. 5. Defect structure in parts of a high angle boundary of the CPM sample of the Al–Zn–Mg–Mn alloy a) around a coarse precipitate of the $Al_6Mn$ phase, b) in a bent region of the grain boundary (the area shown in b) is marked in a))

grains is changed. In these regions the density of defects is particularly high as demonstrated by the results presented in Fig. 6.

### *3.1.2 Structure of the RSP samples*

The typical structure of the RSP samples investigated immediately after solidification can be taken from the results presented in Fig. 7 a and b. The following features are of particular interest: (i) The great majority of precipitates of the $Al_6Mn$ phase present

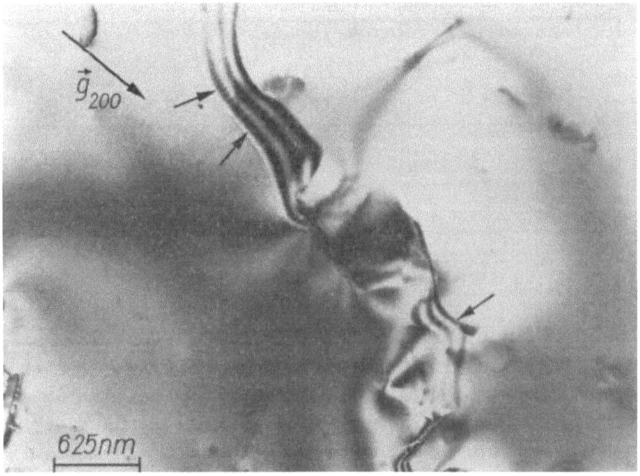


Fig. 6. Change of the periodicity of parallely arranged edge dislocations in a low angle tilt boundary within bent regions (see the arrows). The coarse precipitates of the $Al_6Mn$ phase pin the grain boundaries during quench-induced grain sliding ($\boldsymbol{B} = [001]$)

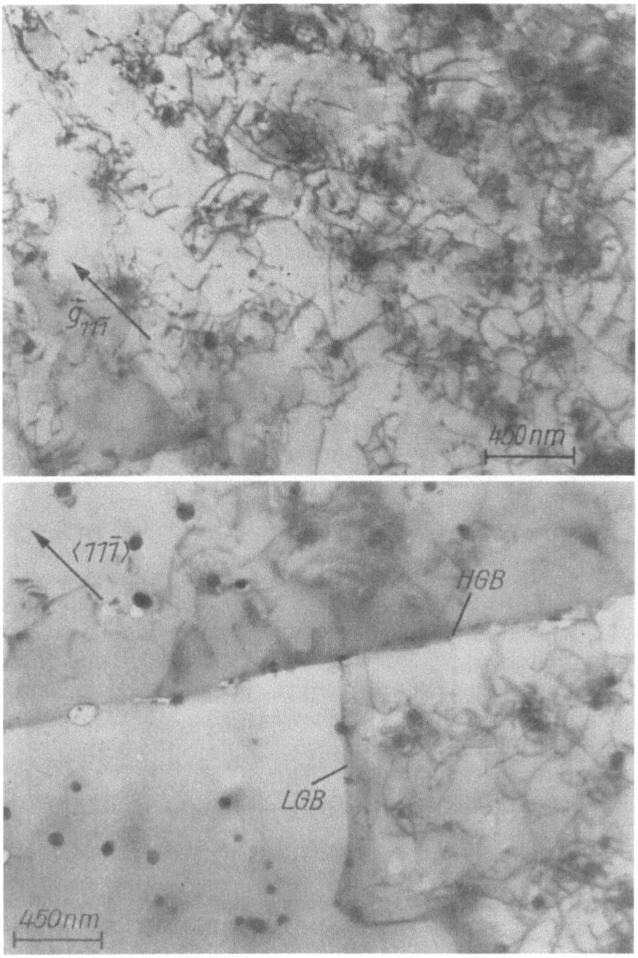


Fig. 7. Structure of the RSP sample of the Al–Zn–Mg–Mn alloy (as solidified state) a) in the interior of the grains ($\boldsymbol{B} = [011]$), b) in high angle (HGB) as well as low angle boundaries (LGB)

in the interior of the grains as well as in the grain boundaries have sizes ranging between 20 and 100 nm in diameter, i.e. no coarse precipitates are present in the RSP samples as it had been typical for the CPM samples. (ii) The mean density of these precipitates is $(3 \pm 1) \times 10^{13}/\text{cm}^3$. Together with the mean size of the precipitates, their composition and lattice spacing, one can evaluate that only about $\frac{1}{10}$ of the alloyed Mn atoms are inserted in the precipitates of the $Al_6Mn$ phase, i.e. the main part of the Mn atoms remains finely distributed in the matrix. (iii) Most of the precipitates show misfit dislocations at their interfaces (see Fig. 7a). Supposing the presence of sufficiently high stresses, from these incoherent parts of the particle–matrix interface dislocations can be punched out (source of dislocations), which lower the strains in the samples. By means of glide and climb processes of these punched out dislocations, subgrain as well as small angle grain boundaries may form, which are pinned by the precipitates of the $Al_6Mn$ phase (see Fig. 7b).

### *3.2 Structure of the RSP samples of the Al–Zn–Mg–Mn–Cr alloy*

The structure of this alloy in the as-solidified state resembles that of the other alloy investigated in this paper. In the alloy with additions of Cr, the mean size of the precipitates is the same, only their density is a bit higher, viz. $(5 \pm 2) \times 10^{13}/cm^3$ in the interior of the grains. This can be well understood, because beneath precipitates of the $Al_6Mn$ phase such of the $Al_7Cr$ phase are formed. On inspecting the micrographs, however, evidence for the action of classical dislocation multiplication mechanisms is found. In Fig. 8 examples for the punching of discrete dislocation loops from incoherent parts of the particle–matrix interface (see the small arrows) as well as of dislocations from grain boundaries (big arrows) are shown. In the alloy free of Cr, the latter effect was not observed so distinctly and often. We explain it in the following manner. In the alloy with Cr additions, the precipitation density in the grain boundaries is higher than in the alloy free of Cr, that means, in the alloy with Cr the strains are fairly high in the grain boundary regions; they are lowered by punching out dislocations from suitable parts of the boundaries, e.g. ledges and intrinsic dislocations. These punched out dislocations may form subgrain goundaries and even small angle grain boundaries, resulting in improved strength and good formability of the samples.

### *3.3 Hardness of the two alloys (RSP and CPM)*

For both alloys the microhardness HV was determined during isochronal annealing from RT to 450 °C is samples produced by RSP and CPM. Before starting the isochronal annealing, the CPM samples were homogenized at 490 °C for 1 h and quenched in water of RT. For each ageing temperature $T_a$ a new specimen was used. The samples were kept at $T_a$ for 1 h, quenched in water of RT and then the hardness was measured at RT. The respective results are presented in Fig. 9. They can be easily interpreted: (i) As a general feature, the hardness of the RSP samples is higher in comparison to the CPM ones. The reason is that in the RSP samples much more Mn (and Cr) remains solved in the matrix in comparison to the CPM specimens (solution hardening effect). (ii) In each case, the hardness is decreasing up to about $T_a = 250$ °C. This effect is due

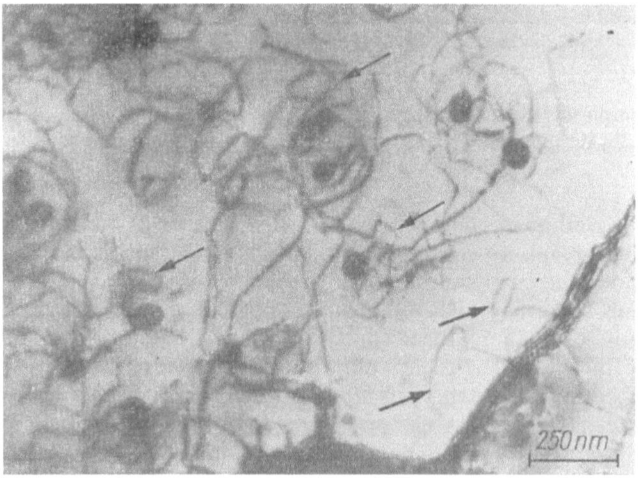


Fig. 8. Examples for the action of "classical" dislocation multiplication mechanisms in RSP samples of the Al–2.3 at% Zn–1.6 at% Mg–0.43 at% Mn–0.23 at% Cr alloy, viz. punching of (i) dislocation loops from incoherent parts of the particle–matrix interface (small arrows) and (ii) dislocations in strained parts of grain boundaries (big arrows) ($\boldsymbol{B} = [013]$)

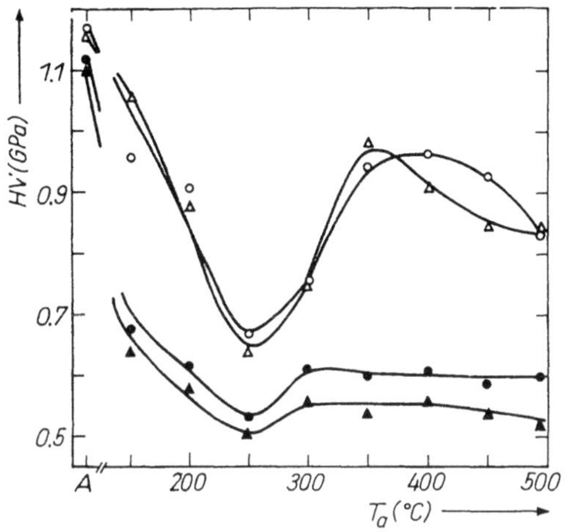


Fig. 9. Course of the microhardness HV (measured at RT) during the isochronal heating (kept for 1 h at the respective $T_a$) of the Al–Zn–Mg alloy containing Mn (triangles) and Mn and Cr (circles), respectively. The full symbols belong to the CPM samples and the open symbols to the RSP samples. The symbol A characterize the hardness of the as-solidified state (RSP samples) and of the state after homogenization at 490 °C for 1 h, subsequent quench in water of RT (CPM samples) and storing at RT for 90 d

to the proceeding dissolution of GP zones (containing Zn, Mg, and Al atoms) and precipitates of the metastable, partially coherent hexagonal $\eta'(Zn_2Mg)$-phase (diminution of the precipitation hardening). (iii) Of particular interest is the increase of the hardness observed in the RSP samples in the temperature range between about 250 and 375 °C. As proved by the results presented in Fig. 10, this effect is due to the nucleation and growth of small precipitates of the $Al_6Mn$ and( $Al_7Cr$) phase from the Mn (and Cr) atoms, present in excess in the Al matrix of the RSP samples, which become movable at elevated temperatures. Note that in the CPM samples only faint indications of such an increase of the hardness are visible, because the matrix contains only few Mn (and Cr) atoms solved in excess in the matrix. (iv) The diminution of the hardness, observed in the RSP samples at $T_a \geqq 375$ °C, is due to the coarsening of the precipitates of the $Al_6Mn$ (and $Al_7Cr$) phase as well as to the increasing solubility of Mn (and Cr) atoms in the matrix with rising temperature.

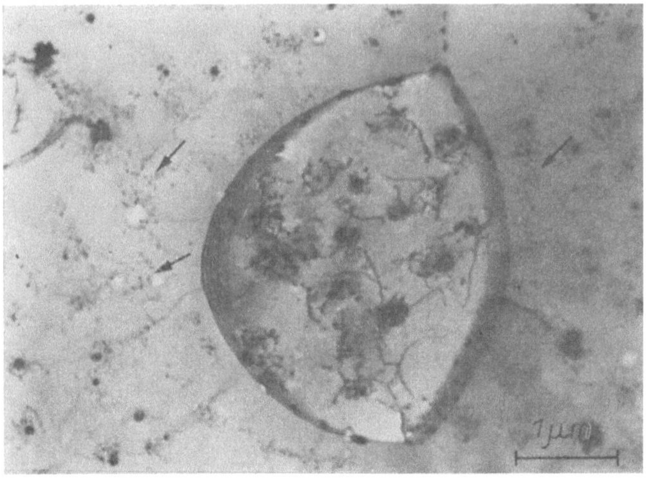


Fig. 10. Precipitation structure in the RSP sample of the Al–Zn–Mg–Mn alloy after ageing at 400 °C for 15 min. Of particular importance are the small precipitates (arrows) in the interior of the grains

***Acknowledgements***

The authors would like to thank very heartily Dr. Fiedler and Dr. Löser from the Zentralinstitut für Festkörperphysik und Werkstofforschung in Dresden of the Academy of Sciences of the GDR as well as Prof. W. Heyroth from the Pädagogische Hochschule Halle for supplying us with melt-spun ribbons (RSP samples), Mrs. M. Knorek for her careful hardness measurements and Mrs. I. Leonhardt (both from PHH) for preparing the TEM samples and making the photographs.

## References

[1] F. J. Kievitz and A. J. Zuithoff, Aluminium **46**, 760 (1970).
[2] R. W. Gould and E. A. Starke, Adv. X-ray Analysis **9**, 59 (1966).
[3] J. Dutkiewicz, J. Salawa, H. Löffler, and E. Korngiebel, phys. stat. sol. (a) **93**, 429 (1986).
[4] E. Szentirmay and G. Groma, Aluminium **57**, 534 (1981).
[5] E. Korngiebel and H. Grätz, Neue Hütte **26**, 299 (1981).
[6] E. Korngiebel, Neue Hütte **29**, 426 (1984).
[7] E. Korngiebel, Thesis (B), Pädagogische Hochschule Halle (GDR), 1985.
[8] H. Löffler, E. Korngiebel, J. Dutkiewicz, and J. Salawa, Cryst. Res. Technol. **21**, 619 (1986).
[9] H. Löffler, D. Schulze, G. Wendrock, and R. Kroggel, Cryst. Res. Technol. **22**, 615 (1987).
[10] T. R. Anantharamen and C. Suryanarayana, J. Mater. Sci. **6**, 1111 (1971).
[11] P. Furrer and H. Warlimont, Z. Metallk. **62**, 12 and 100 (1971).
[12] Lecong Dzuong, H. Löffler, Nguyen Quyet Thang, and Hoang Minh Tri, Cryst. Res. Technol. **22**, 1031 (1987).
[13] H. Jones, J. Mater. Sci. **9**, 1043 (1984).
[14] W. Heyroth, Forschungsber. Pädagogische Hochschule Halle (GDR) 1987.
[15] T. Onoshi and Y. Naketani, J. Japan. Inst. Light Metals **25**, 253 (1975).
[16] L. F. Mondolfo, in: Aluminium Alloys: Structure and Properties, Butterworths, London/Boston 1976 (p. 324).
[17] G. W. Lorimer, Proc. Residual Spring Conf. Phase Transformations, The Institute of Metals, London 1979 (p. 69).
[18] P. Czurratis, Thesis, Pädagogische Hochschule Halle (GDR), 1987.
[19] L. F. Mondolfo, Internat. Met. Rev. **153**, 95 (1971).

*(Received February 23, 1988)*

phys. stat. sol. (a) **108**, 183 (1988)

Subject classification: 66.30 and 77.50; S9.12

*Laboratoire d'Infrarouge, associé au C.N.R.S. n° 809, Université de Nancy I*[1])

# Dielectric Constant and ITC Measurements of $BaF_2:Dy^{3+}$

By

G. JEANDEL, A. WARIN, and G. MORLOT

The ionic thermocurrents (ITC) are measured for $BaF_2:Dy^{3+}$. It is known that the rare earth ($TR^{3+}$) associated with the interstitial ion ($F_i^-$) of charge compensation forms a dipole. The ITC current and the dielectric constant measurements give interesting information about the dipole orientation ($C_{3v}$) and their activation energy is obtained ($E = 0.57$ eV). It is possible to characterze the ionic conductivity induced by the $F_i^-$ ion migration ($E' = 0.973$ eV).

Nous avons mesuré les courants thermoioniques pour $BaF_2:Dy^{3+}$. Il est bien connu que la terre rare ($TR^{3+}$) forme un dipôle avec l'ion interstitiel ($F_i^-$) de compensation de charge. Le courant ITC et les mesures de constantes diélectriques permettent d'obtenir des renseignements intéressants sur l'orientation des dipôles ($C_{3v}$) et sur leur énergie d'activation ($E = 0{,}57$ eV). Nous avons pu aussi caractériser la conductivité ionique induite par la migration des ions $F_i^-$: ($E' = 0{,}973$ eV).

## 1. Introduction

Many electrical studies have already been carried out concerning fluorite structures. In particular, we know that when rare earth ions $TR^{3+}$ are introduced, they substitute for divalent cations, and the charge compensation is shown under our conditions of synthesis to be a fluorine ion which settles in interstitial position $F_i^-$.

In a previous study, we measured residual absorption due to the impurities ($Y^{3+}$, $Co^{2+}$) in $BaF_2$ [1, 2] at a very low temperature (1.4 K). We interpreted the IR transmission spectra by the Green's function formalism.

$F_i^-$ ions, associated with rare earths, form dipoles which can orient themselves under the influence of an electric field. Two types of dipoles have been identified in fluoride structures according to the rare earth position in comparison to that of the interstitial ion. (NN: nearest neighbour and NNN: next-nearest neighbour.)

For $BaF_2:TR^{3+}$, Laredo et al. [3] showed these two dipoles entail on ITC peak (ionic thermo current) at very different temperatures. For $SrF_2:Ce^{3+}$, Dorenbos et al. [4] recently complemented the ITC results with impedance measurements [5]. We present results on $BaF_2:Dy^{3+}$ (0.2 mol%) using both methods in the following.

## 2. Theory

When a sample is under the influence of an electric field, two phenomena may appear: dipole rotation and space charge displacements.

In the case of insulating electrodes, if the external voltage is applied in the presence of free charges, their displacement on the sample surface will progressively cancel the

[1]) B. P. 239, F-54506 Vandoeuvre Les Nancy Cédex, France.

local field [6]. This means that, if very mobile dipoles are present, they initially orient themselves under the influence of the applied field, and become progessively disorientated when the local field decreases.

While conserving the external voltage, we cooled the sample down to a sufficiently low temperature for the relaxation time of dipoles and free charges to be comparable to the deviation of the experiment at this temperature.

Replacing the external supply by an electrometer, we set up a new equilibrium: the space charges are fixed on the sample surface and create an electric field $E_s$ for which [7]

$$E_s = \frac{Q_0}{d_s(C_I + C_s)}, \tag{1}$$

where $Q_0$ is the quantity of fixed superficial charge, $d_s$ and $C_s$ are the thickness and the capacity of the monocrystal, $C_I$ is the fixed electrode capacity.

During linear heating ($T = T_0 + bt$), if the dipoles orient themselves more rapidly than the rate of displacement of "free" charge [6], we measure a current density $J(T)$ due to dipole orientation,

$$J(T) = -\frac{\mathrm{d}P}{\mathrm{d}t},$$

where $P$ is the polarisation.

The $J(T)$ function possesses a maximum at a temperature depending on the activation energy of the dipole family.

Under the same conditions, the space charge relaxation appears only at higher temperature according to the law

$$J(T) = (1-f)\,f\,\frac{P_0\sigma(T)}{\varepsilon_0\varepsilon_s}\exp\left(-\int_0^T \frac{\sigma(T')}{b\varepsilon_0\varepsilon_s}\,f\,\mathrm{d}T'\right) \tag{2}$$

with

$$f = \frac{2d_I\varepsilon_s}{d_s\varepsilon_I + 2d_I\varepsilon_s}; \tag{3}$$

$\varepsilon_I$ and $\varepsilon_s$ are the insulator and crystal dielectric constants, $d_I$ and $d_s$ are their thicknesses, $\sigma(T)$ is the conductivity due to the fixed charges.

These two phenomena, i.e. dipole rotation and free charge migration, may also be observed by impedance measurements. To introduce the quantities studied, we recall

$$\sigma\boldsymbol{E} + \frac{\partial\boldsymbol{D}}{\partial t} = j\omega\varepsilon_0\varepsilon^*\boldsymbol{E} \tag{4}$$

with

$$\varepsilon^* = \left[\varepsilon(\infty) + \frac{\varepsilon(0)-\varepsilon(\infty)}{1+(\omega\tau)^2}\right] - j\left[\frac{[\varepsilon(0)-\varepsilon(\infty)]\,\omega\tau}{1+(\omega\tau)^2} + \frac{\sigma}{\omega\varepsilon_0}\right],$$

where $\omega$ is the applied field pulsation, $\tau$ the dipole relaxation time: $\tau = \tau_0\,\mathrm{e}^{E/kT}$, $E$ being the activation energy, $\varepsilon(0) - \varepsilon(\infty)$ the increase of the dielectric constant in relaxation to the frequency of a dipole family.

The measurement of the capacitive and resistive section of admittance permits the calculation of $\varepsilon'$ and $\sigma' = \omega\varepsilon_0\varepsilon''$. These quantities lead to the determination of the activation energy.

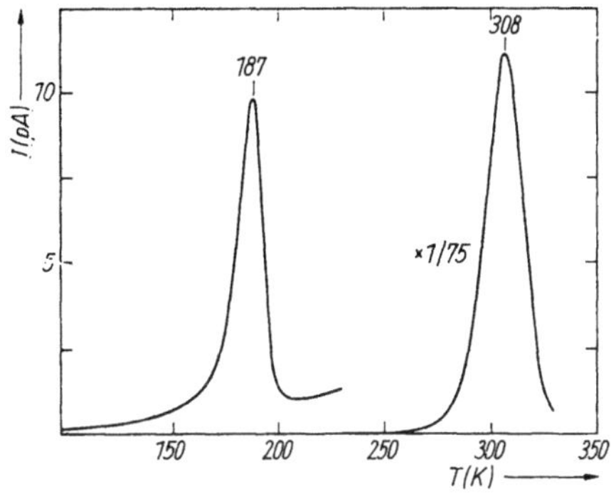


Fig. 1. Curve ITC of $BaF_2:Dy^{3+}$ (0.2 mol%)

## 3. Measuring Techniques and Experimental Results

We synthetize the fluorite structures using the Bridgman method. $BaF_2$ powder is placed in a graphite crucible, then melted in an induction oven under a controlled atmosphere.

Impurities are introduced as fluoride powder mixed before melting.

### *3.1 ITC*

For ITC measurements, we use cylinders 12 mm in diameter and about 1 mm thick. The insulating electrode used is made of sapphire ($d_1 = 36$ μm). The field applied at $T = 320$ K is equal to 1400 V and the rate of heating is 0.05 K/s.

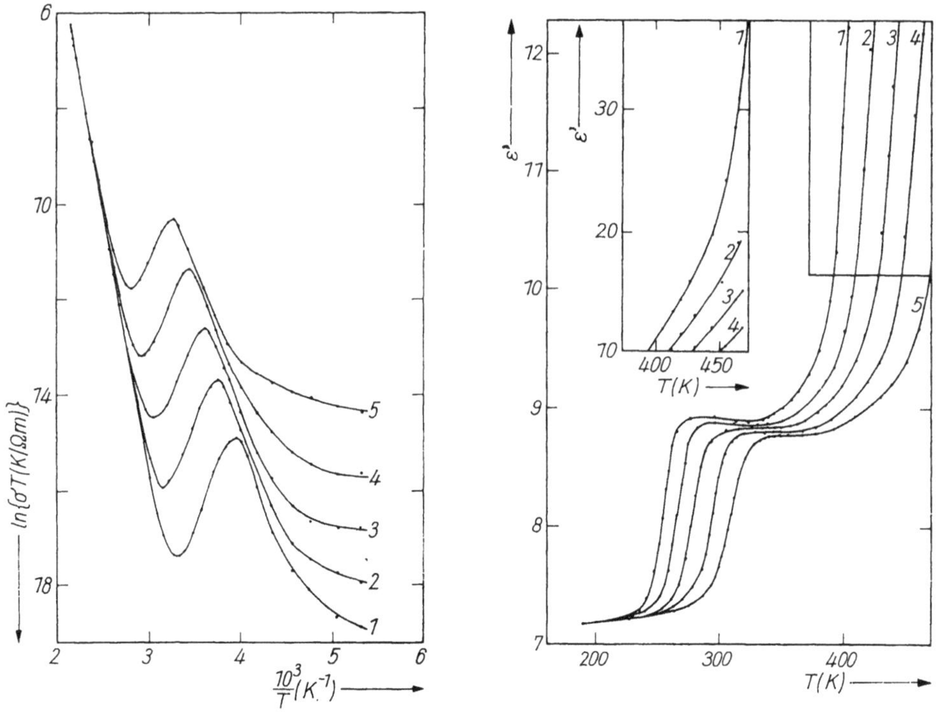


Fig. 2. a) The evolution of ln ($\sigma' T$) as a function of ($1/T$) for $BaF_2:Dy^{3+}$ (0.2 mol%) and b) variation of $\varepsilon'$ vs. $T$ for $BaF_2:Dy^{3+}$ (0.2 mol%) at the frequencies: (1) $\nu = 30$ Hz; (2) 100 Hz; (3) 300 Hz; (4) 1 kHz; (5) 3 kHz

The ITC graph (Fig. 1) shows unsymmetrical peaks located at 187 and 308 K. The high temperature peak is the more intense: for the purposes of representation, its amplitude has been divided by 75.

### *3.2 Dielectric constants*

The measurements of dielectric constants are carried out using a General Radio 1616 bridge for five frequencies: 30 Hz, 100 Hz, 300 Hz, 1 kHz, and 3 kHz and for temperatures from 77 to 470 K.

Fig. 2 describes the evolution of ln $(\sigma' T)$ as a function of $(1/T)$ and of $\varepsilon'$ versus $T$.

The two curves are complementary. An increase of $\varepsilon'$ (Fig. 2b) corresponds to the low temperature peak appearing on Fig. 2a.

For the highest temperatures, we observe a strictly linear evolution of ln $(\sigma' T)$ versus $(1/T)$, accompanied with an increase of $\varepsilon'$.

## 4. Discussion

Fig. 2 shows two different phenomena:

(i) A dipolar relaxation process is observed at low temperature. It appears as a maximum in the ln $(\sigma' T)$ curve and as an increase of $\varepsilon'$.

Resonance is observed at different temperatures as a function of the measured frequency [6].

Plotting the $\ln \tau$ versus $(1/T)$ curve enables the activation energy $E$ and the pre-exponential factor $\tau_0$ to be determined,

$$E = 0.573 \pm 0.015 \text{ eV},$$

$$\tau_0 = 10^{-13.7 \pm 0.5} \text{ s}.$$

It must be noted that $\varepsilon'$, the variation (Fig. 2b) accompanying the peak of ln $(\sigma', T)$, is not the same for all frequencies because it does not take place at the same temperature. Moreover, we can observe a slight decrease of $\varepsilon'$ after the rapid increase due to dipolar relaxation. It may be explained by the presence of interfacial polarisation.

(ii) At higher temperatures, we observe a linear evolution of ln $(\sigma' T)$ and a rapid increase of $\varepsilon'$. This is typical of an ionic conductivity, following the Arrhenius law $\sigma = (\sigma_0/T) \exp(-E'/kT)$. At these temperatures, the dipolar relaxation becomes negligible. Curve (2a) allows us to determine both the activation energy and $\sigma_0$,

$$E' = 0.973 \pm 0.010 \text{ eV},$$

$$\sigma_0 = (6.5 \pm 0.5) \times 10^7 \text{ K}/\Omega\text{m}.$$

Previous studies [5] showed that space charges are interstitial fluorine ions. To obtain these charge migrations, it is necessary to provide a sufficient energy to break the $F_i^-$–$R^{3+}$ bond ($E_d$) and to move the interstitial ($E_a$),

$$E = E_a + \frac{E_d}{2}.$$

Bollmann [8] gives the activation energy value $E_a$ for $BaF_2$ of 0.79 eV. We can then calculate the dissociation energy of the dipole $E_d = 0.28$ eV.

The ITC curve (Fig. 1) reveals the presence of two phenomena already observed using admittance measurements.

The low temperature peak may be attributed to inverted dipoles. Laredo et al. [3] has already brought to light a peak at approximately this temperature with other rare earths: he attributed it to the NNN symmetry.

The high temperature peak could be due to space charge ($F_i^-$) migration, initially at the surface when the external voltage was applied. The number of these charges is very small compared with the quantity of rare earth introduced. This observation is consistent with the results obtained with admittance measurements and conforms with Dorenbos' conclusions [5].

## 5. Conclusion

We have shown that ITC and impedance measurements are complementary because they highlight the same phenomena, rotation of a single dipole family, and migration of space charges ($F_i^-$) associated with a rare earth.

The activation energies calculated from the admittance measurements are consistent with the results obtained from ITC for $BaF_2:Dy^{3+}$ (0.2%). The value of the activation energy of the doped crystal charges $F_i^-$ compared with that of the pure crystal leads us to suppose that, initially, the $F_i^-$ ions are associated with a rare earth.

## References

[1] G. Jeandel, A. Warin, and G. Morlot, phys. stat. sol. (b) **118**, 167 (1983).
[2] G. Jeandel, A. Warin, and G. Morlot, Infrared Phys. **24**, 527 (1984).
[3] E. Laredo, D. R. Figueroa, and M. Puma, J. Physique **41**, C6-451 (1980).
[4] P. Dorenbos, S. Vrind, J. Dolping, and H. W. den Hartog, Phys. Rev. B **35**, 5766 (1987).
[5] P. Dorenbos, H. den Hartog, R. Kruizinga, and S. Vrind, Phys. Rev. B **35**, 5774 (1987).
[6] P. Müller, phys. stat. sol. (a) **67**, 11 (1981).
[7] P. Dorenbos and H. W. den Hartog, Phys. Rev. B **31**, 393 (1985).
[8] W. Bollmann, Crystal Res. Technol. **16**, 1039 (1981).

*(Received March 4, 1988)*

phys. stat. sol. (a) **108**, 189 (1988)

Subject classification: 64.60; 77.40; S11.1

*Department of Thermal Physics, Academy of Sciences of the Uzbekian SSR, Tashkent*[1])

# Mechanism of a Superion Phase Transition in α-$LiIO_3$

By

A. E. ALIEV, A. SH. AKRAMOV, L. N. FERSHTAT, and P. K. KHABIBULLAEV

The mechanism of a superion phase transition is studied by electro-physical and acoustical methods in lithium iodate of hexagonal modification. The role of structural factors in the mechanism of a phase transition is revealed and the contribution of a correlated motion of the disordered ions to the dielectric parameters is established.

Mittels elektrophysikalischer und akustischer Methoden wird der Mechanismus eines Superion-Phasenübergangs in Lithiumjodat hexagonaler Modifikation untersucht. Die Rolle des Strukturfaktors im Mechanismus des Phasenübergangs wird aufgezeigt und der Beitrag einer korrelierten Bewegung der fehlgeordneten Ionen zu den dielektrischen Parametern festgestellt.

## 1. Introduction

The major distinctive feature of superion conductors is an abnormally high ion conductivity. Development of ion conductivity by means of a phase transition is rather a typical phenomenon for superion conductors. Nevertheless the types of transitions and the ways of their display are quite different. The majority of superion crystals have a certain critical temperature at which the electric conductivity is abruptly changed. Such transformations are caused by a radical change in the structure — a high-temperature phase as a rule has a higher lattice symmetry and a more dense ion packing as for instance in AgI, $RbAg_4I_5$, $Li_2SO_4$, and others. The study of the behaviour of a disordered sublattice in such compounds is very complicated as such transformations are accompanied by significant variations in thermodynamical functions and they are typical phase transitions of first order.

In our opinion from the point of view of the study of a superion phase transition mechanism and structure factor analysis the most promising compounds are the ones belonging to another class like fluorite $CaF_2$ whose disordering does not cause structural changes. This category of compounds comprises the crystals of lithium iodate with a hexagonal modification (α-$LiIO_3$) that have the simplest lattice $P6_3$ [1] and are suitable model objects for the study of processes resulting in the appearance of a state with a high conductivity or of a superion state.

## 2. Experimental

To study the mechanism of a superion phase transition we investigated the electro-physical and acoustic properties of lithium iodate. Electroconductivity $\sigma_{ij}$, dielectric permeability $\varepsilon_{ij}$ (adiabatic dielectric permeability at constant press), and tangent of dielectric losses tg $\delta$ of α-$LiIO_3$ single crystals ($pH = 2$) were measured in a specially made thermostatic cell ($\pm 0.1$ K) in the atmosphere of an inert gas ($N_2$). α-$LiIO_3$ samples grown without a deliberate doping were cut from one strictly oriented single

---

[1]) Tashkent, USSR.

crystal in the form of plates with sizes of $7 \times 7 \times 0.5$ mm$^3$ (for $\varepsilon_{33}$) and in the form of bars with sizes of $7 \times 7 \times 10$ mm$^3$ (for $\sigma_{33}$). Prior to the measurement of $\varepsilon_{33}$ and tg $\delta$ the plates were polished and isolated from molybdenum electrodes by mica layers with a thickness of 0.01 mm.

Electroconductivity $\sigma_{33}$ and dielectric parameters $\varepsilon_{33}$, tg $\delta$ were measured at frequencies of 20 Hz to 50 MHz in the temperature range from 100 to 470 K above which lithium iodate of the $\alpha$-modification undergoes a destructive phase transition that causes its breakdown.

To study the acoustic properties of lithium iodate we measured the velocity and absorption of bulk ultrasonic waves in $\alpha$-$LiIO_3$ single crystals by the method of acoustic interferometer and freely decaying acoustic oscillations. Investigations were performed in the frequency range of 60 to 250 kHz at temperatures from 100 to 400 K.

The high sensitivity of the method of the amplitude interference of continuous acoustic waves for the first time permitted to observe very fine effects caused by the superion phase transition at $T \approx 243$ K.

## 3. Results

To reveal the mechanism of conductivity we studied electroconductivity on a direct current and the response of an electrochemical cell $C/LiIO_3/C$ with ideally polarized (blocked) electrodes to the pulsed electric signal. Along the $C^6$ axis there was a high ion conductivity and the contribution of the electronic component measured according to the given technique does not exceed the experimental error ($\approx 1\%$). The crystal electrolysis and the growth of lithium dendrites near one of the electrodes (—) in a constant electric field of $E_3 \approx 300$ V/cm (400 K) permits to conclude that the current carriers in $\alpha$-$LiIO_3$ are $Li^+$ cations. The conductivity along directions perpendicular to the $C^6$ axis ($\sigma_{11}$, $\sigma_{22}$) was not observed up to $10^{12}$ $\Omega^{-1}$ cm$^{-1}$.

The high mobility of lithium ions in the disordered phase leads to the fact that already at room temperature the conductivity of $\alpha$-$LiIO_3$ along the $C^6$ axis attains a value of $5.6 \times 10^{-4}$ $(\Omega$ m$)^{-1}$ with the activation energy of $E_\sigma = 0.26$ eV.

It should be noted that in the given group of superion crystals including the considered materials there is no jump-like increase of conductivity as it occurs for classical superionics (AgI, $AgRb_4I_5$, $Ag_2HgI_4$). The process of disordering in such a crystal according to the data of electrophysical [2] and acoustic [3, 4] studies starts at $T \approx 170$ to 180 K. At $T_c = 243$ K in the temperature dependence of conductivity with ac $\sigma_{33}(T)$, dielectric permeability $\varepsilon_{33}(T)$, and absorption of acoustic waves there are pronounced anomalies whose analysis together with the results of [5] permitted to classify the process as a superion phase transition (SPT). The behaviour of some parameters studied in the range of the SPT, i.e. the absence of the temperature hysteresis, jump-like changes of activation energy of conductivity $E_\sigma$ at continuous $\sigma_{33}(T)$, and the jump of $\varepsilon_{33}(T)$ permit to describe the given transition as a second-order PT.

In the range of the phase transition and the temperature dependence of the conductivity (Fig. 1) there is an appreciable drop in the activation energy from 0.29 to 0.14 eV. A similar behaviour of electroconductivity is characteristic of the whole class of superion crystals that do not undergo structural changes in a rigid sublattice and for some rare earth trifluorides ($LaF_3$, $CeF_3$, $NdF_3$, $PhF_3$) [6]. Such a considerable drop in $E_\sigma$ favours a substantial increase of conductivity at further temperature increase.

The high degree of disorder in $\alpha$-$LiIO_3$ caused by the structural peculiarities of the lattice results in abnormally high values of dielectric conductivity at low frequencies.

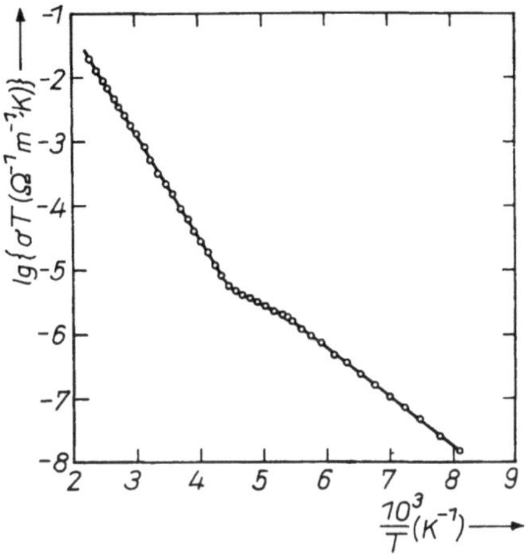


Fig. 1. Temperature dependence of electroconductivity measured along the hexagonal piezoactive axis of $\alpha$-LiIO$_3$ crystal ($pH = 2, f = 100$ Hz)

An idealized structure of the hexagonal modification of lithium iodate is given in Fig. 2. Oxygen atoms (II) form a very dense hexagonal packing, iodine atoms (III) occupy octahedral voids. Due to the strong interaction between the adjacent groups of IO$_3$ a strong three-dimensional net is formed where atoms make small thermal oscillations [1]. Lithium atoms (I) are also situated in the centre of oxygen octahedra forming a chain parallel to the sixth-order axis $C^6$. The small ion radius $r_{ion} = 0.068$ nm $[r_i(I^-) = 0.22$ nm, $r_i(O^{2-}) = 0.136$ nm$]$ [7], the high polarization of the rigid core formed by IO$_3$ groups, and the presence of open channels along the $z$-axis (shown by the dotted line) already at low temperatures of 170 to 180 K result in a considerable mobility of Li$^+$ ions. The motion in $x$- and $y$-directions was blocked by large I$^-$ ions and up to fusing temperatures no electroconductivity was observed along these axes.

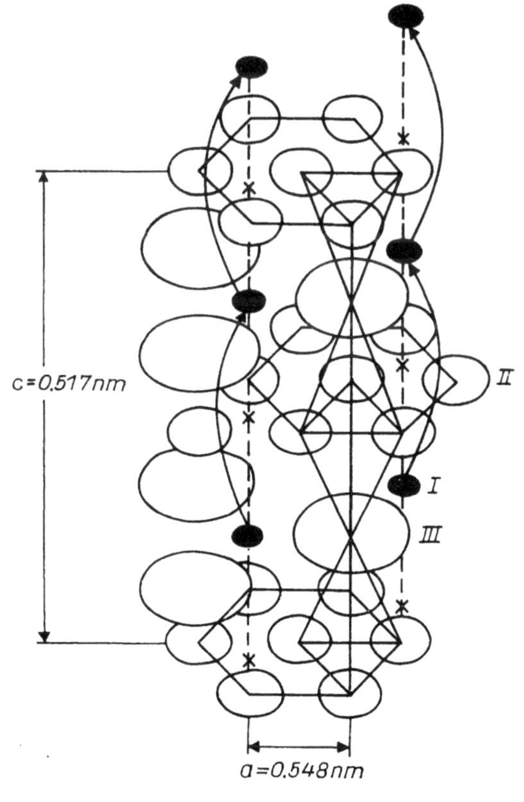


Fig. 2. Structure of $\alpha$-LiIO$_3$. I — lithium, II oxygen, III iodine

We assume that $Li^+$ ion disordering can result in a considerable growth of dielectric permeability on low frequencies. Actually, considering the simplest model of thermal ion polarization where a weakly bound ion can occupy two equally possible positions A and B being at a distance $\delta$ and separated by a potential barrier one can get the relation between the dielectric permeability $\varepsilon_{33}$, concentration of weakly bound ions $n_0$, and temperature $T$.

According to this model, when applying an external field $E$, there is a deformation of the curve of the total potential energy and thus the redistribution of the probability of particle location in positions A and B. It results in a dielectric polarization and each excessive transferred ion creates a dipole moment equal to $e\delta$ and the electric moment of the bulk unit will be equal to [8]

$$p = \frac{n_0 e^2 \delta^2}{12kT} E ,$$

where $e$ is the charge, $k$ the Boltsmann constant, $E$ the applied voltage.

Using the well-known relations of electrostatics and introducing the following parameter

$$\tau = \frac{\exp U/kT}{2\nu}$$

which indicates the relaxation time of weakly bound ions where $U$ is the potential barrier height and $\nu$ the frequency of ion oscillations in a fixed state, one gets the contribution to the dielectric permeability conditioned by the mechanism of thermal transfer of weakly bound ions in the direction of the external field,

$$\Delta\varepsilon = 1 + \frac{n_0 e^2 \delta^2}{12\varepsilon_0 kT} \left(1 - \exp\left(-\frac{t}{\tau}\right)\right) ,$$

where $t$ is the polarization time, $\varepsilon_0$ the dielectric permeability of vacuum. At $T < T_c$ thermal transfer seems to occur by single independent jumps. Considering the fact that at a frequency of 1 kHz (polarization time is $t = 5 \times 10^{-4}$ s) in the region of the phase transition $\omega\tau \lll 1$, the multiplier $(1 - \exp(-t/\tau)$ can be neglected ($\tau = 4.55 \times 10^{-6}$ s [2]) and it can be assumed that all lithium ions participate in the polarization process, i.e. $n_0 = 1.49 \times 10^{28}\ m^{-1}$. The contribution to $\varepsilon_{33}$ appears to be introduced only by the transfers along the chain of $Li^+$ ions situated at a distance of $\delta = a/2 = 0.258$ nm. Thus

$$\Delta\varepsilon = 72.6 .$$

Taking into account the polarization connected with the elastic shift of charges ($\varepsilon_s = 10$) one determines the total dielectric permeability of $\alpha$-$LiIO_3$ at the frequency of 1 kHz,

$$\varepsilon = \Delta\varepsilon + \varepsilon_e = 82.6 .$$

The experimental value of the dielectric permeability at $T < T_c$ is $\varepsilon = 57$. A good agreement between the calculated and the experimental values of $\varepsilon_{33}$ can be apparently attained, when considering the role of crystal acidity ($pH$) in the formation of lithium vacancies. A simple extrapolation of the results of the dependence of $\sigma_{33}$ on $pH$ [1] during the transition from acid ($pH = 2$) to neutral crystals ($pH = 7$) results in the value of $\varepsilon_{33} = 61$. Thus, a high dielectric permeability of $\alpha$-$LiIO_3$ at low frequencies is conditioned by the mechanism of thermal ion polarization by weakly bound lithium ions.

The high degree of disorder in lithium iodate along with the high dielectric permeability and electroconductivity can lead to the collective interaction of the whole set of disordered ions and vacancies left by them, both between them and "across" each other. In this case the variation in the thermodynamical potential of the crystal conditioned by the defect formation is given as [9]

$$\Delta G = n\varrho - \frac{1}{2}\frac{\lambda n^2}{N_0} - TS_{\text{conf}}\,,$$

where $n$ is the number of disordered ions, $\varrho$, $\lambda$ formation and interaction energy of a pair of defects, $N_0$ the number of lithium cations, $T$ the temperature, $S_{\text{conf}}$ the configurational entropy.

In the general case a smaller energy $\varrho$ is required for ion activation from the lattice site under the condition that nearby "the state has been changed" as a result of similar transitions. The value of this difference — the interaction energy — can be apparently estimated directly from the results of conductivity measurements (Fig. 1),

$$\lambda = E_\sigma^{\text{I}} - E_\sigma^{\text{II}} = 0.29 - 0.14 = 0.15\ \text{eV}\,,$$

where $E_\sigma^{\text{I}}$ and $E_\sigma^{\text{II}}$ are activation energies of conductivity under the conditions of single independent transfers and in the presence of interaction, respectively. In this case the ion activation energy (defect formation energy) $\varrho$ cannot be defined on the basis of these measurements as the conductivity activation energy $E_\sigma$ defined in the experiment consists of the energy of defect formation and the energy of motion activation $E_{\text{m}}$ caused by the presence of steric factors,

$$\sigma \approx n_{\text{d}}\nu_{\text{d}} \approx \left[\exp\left(-\frac{\varrho}{kT}\right)\right]\left[\exp\left(-\frac{E_{\text{m}}}{kT}\right)\right],$$

where $n_{\text{d}}$ is the number of defects, $\nu_{\text{d}}$ the jump rate. The value $\varrho$ can be determined from measurements of dielectric permeability in the microwave range or from the data on Raman scattering as the local motion of mobile ions inside their unit cells make the main contribution to the value at these frequencies. Usually the value $\varrho$ of crystals of the considered class is significantly less than $E_{\text{m}}$ and for lithium iodate, basing on our Raman measurements, it equals to $\approx 0.03$ eV. It permits to define the barrier height for $Li^+$ ion motion,

$$E_{\text{m}} = E_\sigma - \varrho = 0.29 - 0.03 = (0.26 \pm 0.01)\ \text{eV}\,.$$

The high value of motion activation energy $E_{\text{m}}$ prevents an avalanche-like growth of conductivity as is observed in AgI. The conductivity growth in $\alpha$-$LiIO_3$ occurs in a wide temperature range after the phase transition. Due to the fact that as a result of a strong interaction, ions having come to adjacent vacant sites, as if extract the other ones, form a united conglomerate. It is a peculiar ion liquid — a superion state.

The considerable growth of $Li^+$ ion mobility connected with collective low-frequency excitations caused by long-range interactions between diffusing ions results in anomalies of dielectric properties of $\alpha$-$LiIO_3$ in the region of a phase transition. The temperature dependence of the dielectric permeability exhibits a jump in the region $T_{\text{c}}$ (Fig. 3). Knowing the value of the jump $\varepsilon_{33}$ it is possible to estimate the degree of motion correlation in the superion phase assuming that at $T < T_{\text{c}}$ the motion of disordered ions is independent due to their small concentration.

Let us introduce the correlation factor $K$ as the ratio of the medium length of a disordered $Li^+$ ion path after $\bar{\delta}_2$ transition to the medium length of the path up to the

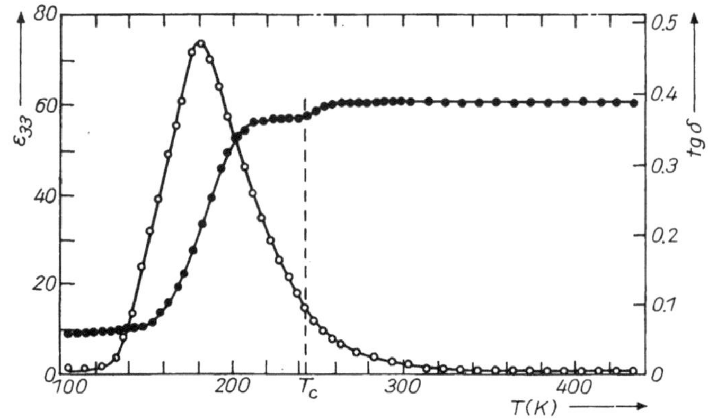


Fig. 3. Temperature dependence of $\varepsilon_{33}$ (●) and tg $\delta$ (○) along the $C^6$ axis of $\alpha$-$LiIO_3$ crystal ($f =$ $= 1$ kHz)

transition $\bar{\delta}_1 = 0.258$ nm,

$$K = \frac{\bar{\delta}_2}{\bar{\delta}_1} \,.$$

As in a superion state due to motion correlations the jumps are performed by an ion tandem simultaneously, the medium efficient path length is increased. It results in the growth of the dipole moment and the relevant growth of $\Delta\varepsilon_{33}$. Hence,

$$K = \frac{\bar{\delta}_2}{\bar{\delta}_1} = \sqrt{\frac{\Delta\varepsilon_{33,\, T > T_c}}{\Delta\varepsilon_{33,\, T < T_c}}} = 1.12 \,.$$

As now not one ion but $K$ ones participate in a single transfer process, the narrowing of the relaxation time spectrum should be also expected.

Fig. 4 shows the temperature dependence of the tangent of dielectric loss angle for some frequencies. It is clearly seen that in the $T_c$ region there is a narrowing of the line whose width characterizes the spectrum of the relaxation time in general.

The ratio of the line width in the $T < T_c$ region to the width at $T > T_c$ measured at the level of 0.5 is

$$\frac{\Delta T_1}{\Delta T_2} = 1.2$$

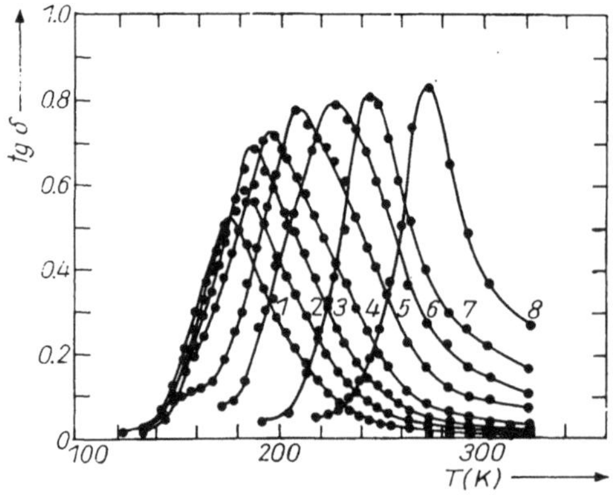


Fig. 4. Temperature dependence of the tangent of lithium iodate loss angle at different frequencies: (1) 0.5, (2) 1, (3) 2, (4) 4, (5) 10, (6) 20, (7) 40, (8) 100 kHz

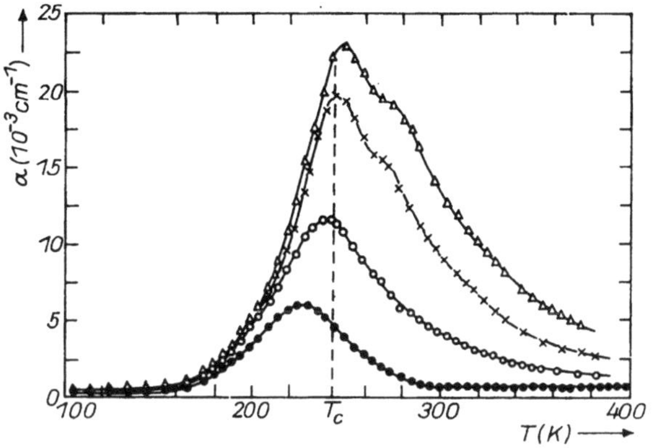


Fig. 5. Temperature dependence of absorption of longitudinal acoustic waves spreading along the piezoactive $C^6$ axis of lithium iodate crystal (● 63, ○ 116, × 165, △ 223 kHz)

which is in satisfactory agreement with the introduced correlation factor. The most pronounced correlations are revealed in one-dimensional systems including α-$LiIO_3$. In this case according to geometrical reasons it is not possible to shift the given ion to the distance $a$ without transferring energy to $n$ particles being in the same range. However, the presence of accidental barriers in channels conditioned by impurities and steric factors result in a considerable drop of cooperative motion.

While the α-$LiIO_3$ belongs to the piezoelectric class of materials and has a marked conductivity, the spreading acoustic waves should result in a redistribution of moving charge carriers and thus to the change of their absorption and phase rate. The study of the intrinsic acoustic resonance of lithium iodate showed that the phase transition caused by the process of Li sublattice disordering is accompanied by a peak in the temperature dependence of the absorption ($\alpha$) of longitudinal acoustic waves (Fig. 5). The analysis of relaxation times calculated according to the data of $\varepsilon$, tg $\delta$, and $\alpha$ measurements demonstrated that the acoustic wave absorption in SPT region is mainly promoted by short-range processes occurring inside a unit cell or within the nearest energetic barriers while the times of "conductivity relaxation" are one order higher.

Thus, the abnormally high values of dielectric permeability in α-$LiIO_3$ crystals are caused by the process of Li sublattice disordering by means of thermal ion polarization. Crystal structure peculiarities result in the one-dimensional motion of disordered Li ions along the channels. This should lead to a considerable correlation in the motions of Li ions. However, the existence of steric factors displayed in a high value of the activation energy of the motion causes a considerable slurring of the phase transition. Correlation effects having long-range character, but due to the above-mentioned reasons revealed within the nearest energetic barriers, result in the appearance of a new phase, i.e. a superion state.

## References

[1] K. I. Avdienko, S. V. Bogdanov, S. M. Arkhipov, B. I. Kidyarov, V. V. Lebedev, Y. E. Nevskii, V. I. Trunov, D. V. Sheloput, and R. M. Shklovskii, Lithium Iodate – Crystal Growing, Properties and Application, Izd. Nauka, Novosibirsk 1980.

[2] A. E. Aliev and A. Sh. Akramov, Izv. Akad. Nauk UzSSR, Ser. fiz. mat. Nauk No. 1, 76 (1987).

[3] A. E. ALIEV, A. SH. AKRAMOV, and L. N. FERSHTAT, Investigation of Acustation Mechanism of Absorption in Superion Crystals, Proc. 25th Conf. Acoustics, Ultrasound – 86, Vol. 1, Bratislava 1986 (p. 8).
[4] A. A. ABRAMOVITCH, V. A. SHUTILOV, T. D. LEVITSKAYA, B. I. KIDYAROV, and P. L. MITNITSKYI, Fiz. tverd. Tela **14**, 2585 (1972).
[5] J. DEL CERRO, S. RAMOS, and J. ISIDORO, phys. stat. sol. (a) **72**, K23 (1982).
[6] P. K. KHABIBULLAEV, L. N. FERSHTAT, and A. E. ALIEV, Phase transitions and domain formation in superion crystals, Dokl. Akad. Nauk SSSR, **281** 320 (1984).
[7] B. K. VAINSTEIN, V. M. FRIDKIN, and V. L. INDENBOM, Crystal Structure, Vol. 2, Izd. Nauka, Moskva 1979.
[8] I. S. ZHELUDEV, Crystal Dielectric Physics, Izd. Nauka, Moskva 1968 (p. 460).
[9] YU. YA. GUREVITCH and YU. I. KHARKATS, Uspekhi fiz. Nauk., **136**, 693 (1982).

*(Received September 28, 1987)*

phys. stat. sol. (a) **108**, 197 (1988)

Subject classification: 64.70; 76.80; S8.15

*Institute of Physics (a) and Institute of Semiconductor Physics (b), Academy of Sciences of the Lithuanian SSR, Vilnius*[1])

# Effect of a Deviation from Stoichiometry on the Mössbauer Parameters of $Sn_{0.8}Ge_{0.2}Te$ during the Phase Transition

By

D. BALTRŪNAS (a), B. VENGALIS (b), S. MOTIEJŪNAS (a), and N. SHIKTOROV (b)

The system of solid solutions $(Sn_{0.8}Ge_{0.2})_{1-y}Te_y$ with the values of $y = 0.499$; 0.505; 0.510 is investigated by means of the Mössbauer effect on the $^{119}Sn$ nucleus within the temperature range 160 to 300 K. The effective Debye temperature and mean-square displacements of tin atoms from equilibrium are calculated. In the region of phase transition (about 200 K) fractures on the temperature dependencies of the isomer shift and Debye temperature are observed, while no anomalies of the Mössbauer effect probability are noticed. No quadrupole interaction is observed with the symmetry decrease in the lattice. A correlation between the isomer shift and interatomic distance is revealed. The ionicity of the chemical bond in the system $Sn_xGe_{1-x}Te$ is $\approx 54\%$, when $x = 0.9$, and about 40%, when $x = 0.1$.

С помощью эффекта Мессбауэра на ядрах $^{119}Sn$ в интервале температур 160 до 300 K исследована система твердых растворов $(Sn_{0.8}Ge_{0.2})_{1-y}Te_y$ при значениях $y = 0{,}499$; 0,505; 0,510. Рассчитаны эффективные дебаевские температуры и среднеквадратичные смещения атомов олова от положения равновесия. В области фазового перехода (около 200 K) наблюдались изломы на температурных зависимостях изомерного сдвига и дебаевской температуры и не наблюдалось аномалий для вероятности эффекта Мессбауэра. С понижением симметрии решетки проявления квадрупольного взаимодействия не обнаружено. Замечена корреляция между изомерным сдвигом и межатомным расстоянием. Получено, что ионность химической связи в системе $Sn_xGe_{1-x}Te$ при $x = 0{,}9$ составляет около 54%, а при $x = 0{,}1$ — около 40%.

## 1. Introduction

Together with the temperature decrease in germanium and tin tellurides a displacement-type phase transition (PT) has been observed which is accompanied by a decrease of the crystal symmetry from cubic (NaCl structural type) to rhombohedral. The low-temperature phase is characterized by the rhombohedral distortion angle $\alpha$ and the relative shift of metal and chalcogene sublattices along the diagonal of the cube. Significant deviation from stoichiometry towards tellurium is characteristic of both materials and their solid solutions. As a result of crystalline field screening by free holes, supplied by doubly charged vacancies in metal, inhibition of ferroelectric features takes place together with the deviation from stoichiometry [1]. On the other hand, the extremely high concentration of free holes together with the multivalley valence band of the compounds may stimulate quite a different PT mechanism [2]. Its essence lies

---

[1]) K. Poželos 54, SU-232600 Vilnius, USSR.

in the spontaneous decrease of the crystal symmetry at the expense of the energetically profitable redistribution of free carriers among different groups of valleys.

Significant information on the PT mechanism can be obtained by means of the Mössbauer effect. The temperature dependence of the Mössbauer effect probability reflects the peculiarities of the crystal vibration spectra. Symmetry decrease of the crystal lattice can give rise in the region of the Mössbauer nucleus to an electric field gradient which can induce quadrupole interaction, splitting or broadening the Mössbauer line. The value of the isomer shift of the line partly depends on the crystal structure, too. The PT in the solid solutions $Sn_xGe_{1-x}Te$ has been studied by means of the Mössbauer effect many times [3 to 6], the data of different works, however, disagree. Thus, according to [3] the Mössbauer effect probability $f'$ decreases in the PT region for solid solutions when $x = 0.9$ and $x = 0.7$, while according to [4] $f'$ increases (when $x = 0.9$), the author of [5] has noticed no anomalies for $x = 0.95$ and 0.9. Besides, the authors of [3] have observed in the low-temperature phase a smooth broadening of the line with decreasing temperature.

Unlike in [3 to 5], the investigation in [6] was carried out at room temperature within the whole concentration range. With decreasing $x$ the area under the spectrum grew, the isomer shift in the rhombohedral phase reduced monotonously, while the line width did not change.

The supposition arose that many features of the crystals of the solid solutions $Sn_xGe_{1-x}Te$ in the PT region are determined by the deviation from stoichiometry, i.e. by their doping degree. Therefore, for the experimental studies of PT the system $(Sn_{0.8}Ge_{0.2})_{1-y}Te_y$ has been chosen with $y = 0.499; 0.505; 0.510$. The aim of the present paper is to determine whether the Mössbauer spectra parameters change in the region of the phase transition, and how the deviation from the stoichiometric composition of the samples towards increasing Te concentration influences these parameters.

## 2. Experimental

The solid solutions $(Sn_{0.8}Ge_{0.2})_{1-y}Te_y$ with $y = 0.499; 0.505; 0.510$ were prepared by means of direct ampoule synthesis from elementary components of high purity. The synthesis was carried out in a tilting resistive furnace providing a continuous mixing of the melt.

The temperature of the furnace was increased up to 1400 K and kept at that point for 3 h. Later it was lowered to 1100 K and kept another 3 h, then the furnace was slowly cooled. The substance obtained was annealed for 25 h in argon atmosphere at a temperature of 870 K, then the temperature was lowered to 770 K, and was kept for another 75 h.

The structure of the samples was checked by an X-ray diffractometer, and the crystal lattice parameter was measured at room temperature. All the crystals were of NaCl type structure. The concentration of free charge carriers was determined by the Hall effect, the following values of the hole concentration were found: $p = 5.4 \times 10^{20}$, $1.2 \times 10^{21}$, $3.0 \times 10^{21}$ cm$^{-3}$ for the samples with $y = 0.499$, 0.505, and 0.510, respectively.

For the Mössbauer measurements the crystals were crushed and deposited on a aluminium foil in alcohol solution of the BF-2. cement. The thickness of the absorbers was 10 mg/cm$^2$ in terms of natural tin. The compound $Ca^{119}SnO_3$ at room temperature served as $\gamma$-source. Between the source and the absorber a 60 $\mu$m thick palladium foil filter was placed.

The magnitude of the effect $\alpha$, line-width $\Gamma$ and shift of the line centre $\delta$ were calculated by the least-squares method. The area $S$ below the curve of the spectrum was calculated according to the formula

$$S = \frac{\pi \alpha \Gamma}{2}. \tag{1}$$

In the case of a single Lorentz line the area of the absorption line is expressed as in [7],

$$S = \varkappa \pi f t_a \exp\left(-\frac{t_a}{2}\right)\left[I_0\left(\frac{t_a}{2}\right) + I_1\left(\frac{t_a}{2}\right)\right], \tag{2}$$

where $\varkappa$ is the relative contribution of the resonance $\gamma$-quanta in the total counting rate, $f$ the probability of recoilless emission of $\gamma$-quanta, $t_a = \sigma_0 f' n_a$ the effective thickness of the absorber, $\sigma_0$ the maximum cross section of resonance absorption, $f'$ recoilless emission probability to be determined, $n_a$ the number of atoms of the isotope under study for the area unit of the absorber, $I_0$ and $I_1$ the Bessel functions of zero and first order, respectively.

The constant $A = \pi \varkappa f$ was determined by means of the absorber with the known probability $f'$. A β-Sn foil of thickness 1.46 mg/cm² of $^{119}Sn$ was used as such an absorber. The probability for β-Sn at room temperature $f' = 0.046$ [8].

Using the probability values the Debye-Waller factor has been calculated $2W(f' = \exp(-2W))$, which depends on the thermal vibrations of atoms. In the case of isotropic crystals in the harmonic approximation the mean square of thermal displacements of atoms $u^2$ can be calculated

$$f' = \exp(-2W) = \exp\left(-\frac{1}{3} k^2 \langle u^2 \rangle\right), \tag{3}$$

where $k$ is the wave vector of $\gamma$-quanta.

In the Debye approximation the index $2W$ is given by the Debye temperature [9],

$$2W = \frac{6R}{k_B \theta}\left[\frac{1}{4} + \int_0^{\theta/T} \frac{x\,dx}{e^x - 1}\right], \tag{4}$$

where $R$ is the recoil energy, $k_B$ the Boltzmann constant, $T$ the temperature of the absorber.

## 3. Results and Discussion

Table 1

Temperature dependence of mean-square displacements of tin atoms (error: ±0.0004)

| $y = 0.499$ | | $y = 0.505$ | | $y = 0.511$ | |
|---|---|---|---|---|---|
| $T$ (K) | $\langle u^2 \rangle^{1/2}$ (nm) | $T$ (K) | $\langle u^2 \rangle^{1/2}$ (nm) | $T$ (K) | $\langle u^2 \rangle^{1/2}$ (nm) |
| 170 | 0.0180 | 165 | 0.0172 | 168 | 0.0175 |
| 176 | 0.0186 | 186 | 0.0185 | 177 | 0.0177 |
| 186 | 0.0192 | 201 | 0.0199 | 186 | 0.0184 |
| 206 | 0.0204 | 206 | 0.0202 | 206 | 0.0197 |
| 216 | 0.0213 | 215 | 0.0208 | 211 | 0.0201 |
| 226 | 0.0219 | 221 | 0.0213 | 220 | 0.0205 |
| 231 | 0.0224 | 226 | 0.0214 | 231 | 0.0214 |
| 238 | 0.0228 | 241 | 0.0225 | 241 | 0.0220 |
| 246 | 0.0233 | 248 | 0.0230 | 251 | 0.0225 |
| 253 | 0.0236 | 253 | 0.0233 | 256 | 0.0227 |
| 294 | 0.0260 | 295 | 0.0250 | 297 | 0.0247 |

The Mössbauer spectra of all the samples consisted of single lines, the parameters of which changes with temperature. The experimental data are given in Fig. 1 to 5 and in Table 1.

### *3.1 Isomer and temperature line shifts*

According to [10] the PT for the investigated compounds takes place in the temperar ture range near 200 K. As the experimental data show, the Mössbauer isomer shift $\delta$ (Fig. 1) suffers the most evident change during the PT. The curves presented in Fig. 1 may be conventionally broken up into three parts. Part "a" corresponds to the high-temperature region in which the solid solutions have NaCl-type cubic structure. The second part "b" — the temperature region in which the PT takes place, and, finally, part "c" — rhombohedral, low-temperature region in which the rhombohedral angle narrows with the temperature decrease. From Fig. 1 it is evident that the temperature variations of all the three curves are different. Curve 3 undergoes the weakest changes. On it the PT is reflected only by a change in the slope of the line, which takes place approximately at $T = 220$ K. A more marked change is suffered by curve 1. According to the figure, the PT at this value of $y$ takes place in the temperature range 225 to 245 K. The greatest changes are seen on curve 2, they take place in the temperature range 200 to 225 K.

The temperature shift of the line (slope in Fig. 1) is mainly caused by the second-order Doppler effect (SODE). It is connected with thermal vibrations of atoms, and in Debye approximation it is determined by the temperature of the sample $T$, and the effective Debye temperature $\theta$ [11]. As the Debye temperature in the cubic phase is determined to be 130 K, for this $\theta$ value the ratio $\Delta\delta_{\text{SODE}}/\Delta T$ has been estimated making use of the approximate formula [12]

$$\delta_{\text{SODE}} = -\frac{9}{2}\frac{k_{\text{B}}T}{mc}\sum_{n=1}^{30}\frac{1}{n^4}\left[6 - \exp\left(-n\frac{\theta}{T}\right)\left\{\left(n\frac{\theta}{T}\right)^3 + 3\left(n\frac{\theta}{T}\right)^2 + 6n\frac{\theta}{T} + 6\right\}\right],$$

where $m$ is the atomic mass of the isotope under investigation, $c$ the velocity of light, $k_{\text{B}}$ the Boltzmann constant. The value obtained is $\Delta\delta_{\text{SODE}}/\Delta T = -3.42 \times 10^{-4}$ mm per s K.

Quite close to this value are the experimental values $\Delta\delta/\Delta T$ for the samples with $y = 0.499$ and $y = 0.505$ in the cubic phase, $-\Delta\delta/\Delta T = -2.9 \times 10^{-4}$ mm/s K and $-2.5 \times 10^{-4}$ mm/s K, respectively. For the sample with $y = 0.51$ in the cubic phase $\Delta\delta/\Delta T = -5.5 \times 10^{-4}$ mm/s K, which exceeds clearly the SODE by the absolute value. The stronger temperature dependence of the isomer shift may, evidently, be due to the manifestation of the dynamic isomer shift caused by electron–phonon interaction [11].

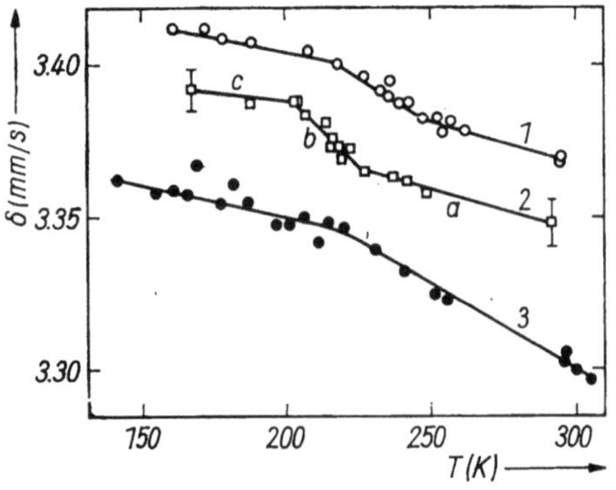


Fig. 1. Dependence of the isomer shift in the system $(Sn_{0.8}Ge_{0.2})_{1-y}Te_y$ on temperature. (1) $y = 0.499$, (2) 0.505, and (3) 0.510

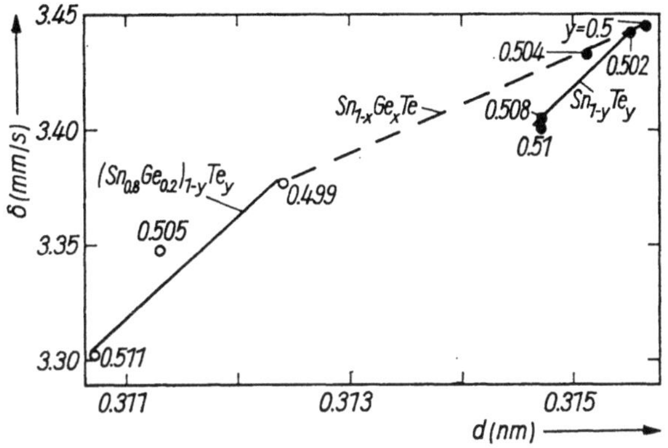


Fig. 2 Correlation of the isomer shift with the interatomic distance

In the low-temperature phase the isomer shift dependence becomes weaker for all three samples. The slope of the lines in this temperature region is caused by several reasons: firstly, the above mentioned second-order Doppler effect, because of which, evidently, the same slope as before the phase transition has to be fixed; secondly, the isomer shift is influenced by the change of the rhombohedral angle. As is shown in [13], the isomer shift decreases with the rhombohedral angle. Furthermore, in the rhombohedral phase the lengths of the chemical bonds change [14] with the temperature decrease, which may also influence the value $\delta$.

A correlation observed between the isomer shift $\delta$ and interatomic distance $d$ at room temperature is presented in Fig. 2. The values for the system $Sn_{1-y}Te_y$ are taken from [15]. From Fig. 2 it is seen that for the solid solutions $(Sn_{0.8}Ge_{0.2})_{1-y}Te_y$ as well as for $Sn_{1-y}Te_y$ the decrease of the interatomic distance causes the decrease of the isomer shift. If Sn atoms are replaced by Ge atoms, however, the dependence of $\delta$ on $d$ is weaker (dashed line) than if the Te concentration increases. A similar dependence between the isomer shift and interatomic distance was also observed in [16].

### 3.2 *Quadrupole interaction*

The width of the Mössbauer line $\Gamma$ in room temperature for all the samples was the same and equal to $0.81 \pm 0.02$ mm/s lowering the temperature caused a smooth broadening of the line-width, and at $T = 170$ K it reached the value $T = 1.00 \pm 0.02$ mm per s. A similar line broadening is caused by the increase of the effective thickness of the absorber, which may be estimated by the formula [17]

$$\Gamma = \Gamma_n(2 + 0.27t_a) ,$$

where $\Gamma_n$ is the natural Mössbauer line-width.

Hence, the conclusion can be drawn that, because of the symmetry decrease of the lattice, an observable electric field gradient does not appear, and quadrupole interaction does not take place (or is insignificant).

### 3.3 *Mössbauer effect probability*

As was mentioned earlier, literature data on the temperature dependence of the absolute value of the Mössbauer probability, especially in the PT region, are contradictory. Besides, it is well known from literature that PT is ferroelectric, while the frequency of the soft mode of the transverse optical vibrations falls when approaching the PT. Hence, one could expect the value of the Mössbauer probability to change in the PT region. Fig. 3 shows that no changes of the temperature dependence $f'$ in the PT region take place. This can be caused by the following reasons: 1. the temperature of

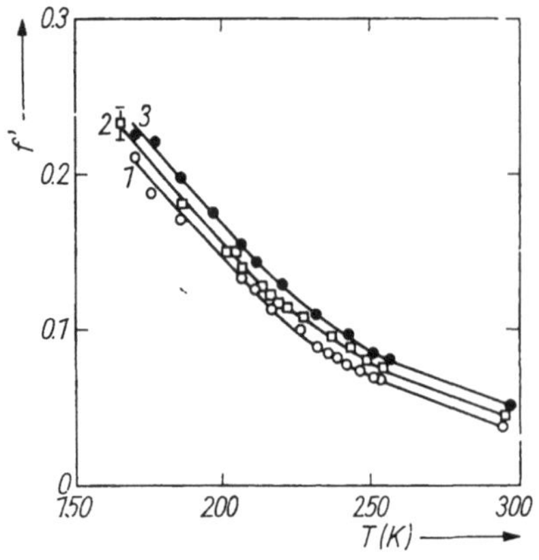


Fig. 3. Dependence of the Mössbauer probability on temperature in the system $(Sn_{0.8}Ge_{0.2})_{1-y}Te_y$. Parameters of the curves as in Fig. 1

the phase transition is rather high ($T > \theta$), and the main contribution to the thermal displacements of the Mössbauer atom is made by the high-frequency vibrations which have no anomalies in the vicinity of the PT; 2. the dependence $f'(T)$ gets smooth because of the strongly smeared PT (using the data from Fig. 1 the temperature width of the PT region is about 20 K).

### *3.4 Mean-square displacements of atoms*

Proceeding from the assumption that the solid solutions under study are isotropic (this, probably, holds true in cubic crystals), and the thermal vibrations of Sn atoms are harmonic, the amplitudes of the mean-square displacements of tin atoms from equilibrium have been calculated, $\langle u^2 \rangle^{1/2}$. The results of the calculations, carried out for the whole temperature range under study, are presented in the table, the results obtained for the rhombohedral phase, however, are less reliable. The table shows that the mean-square displacements of tin atoms decrease with increasing $y$ value.

This correlates well with the fact that the homogeneity region in the system Sn–Te is shifted towards the Te-rich side. Hence, the chemical bond in this region is stronger, while the mean-square displacements are smaller, which agrees with the conclusions of [15, 18].

### *3.5 Effective Debye temperature*

It is known that the Debye model, though applied strictly for the description of the vibrational spectrum of cubic crystals with one atom in the unit cell, is sometimes applicable for more complex systems. In that case the dependence of $2W$ on $T$ is

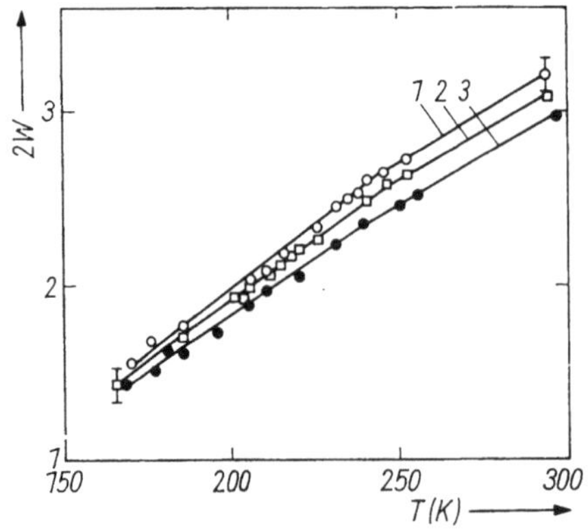


Fig. 4. Temperature dependence of the exponent 2W in the Debye-Waller factor. Parameter of the curves as in Fig. 1

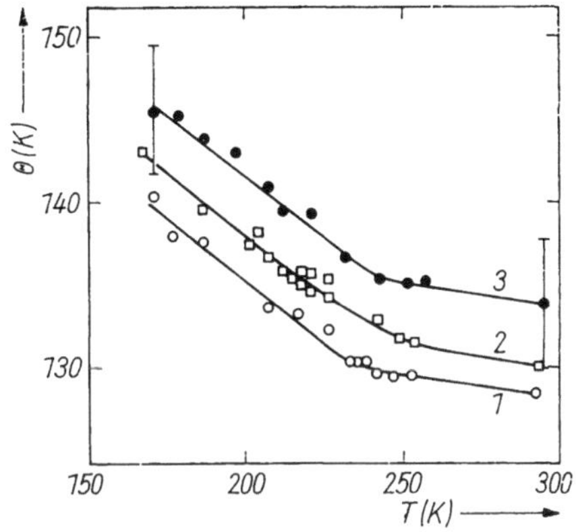


Fig. 5. Dependence of the effective Debye temperature on the temperature of the absorber. Parameters of the curves as in Fig. 1

linear, and the effective Debye temperature $\theta$ does not depend on $T$ [9]. From the data presented in Fig. 4 and 5, it follows that the Debye model is applicable for the description of thermal vibrations of Sn atoms in the cubic phase, which is not valid for the rhombohedral phase. At temperatures lower than those of the PT region, the effective Debye temperature begins rising with the decrease of $T$. Hence, it follows that with the growth of rhombohedral lattice distortions, the strength of the chemical bond of Sn atoms increases monotonously (as in the system $Sn_{1-x}Ge_nTe$ with the increase of $x$ [6]).

### 3.6 Chemical bond

One more parameter of the chemical bond, its ionicity, needs consideration. It is known that the Mössbauer isomer shift and its change reflect the change of the chemical bond ionicity. It could be evaluated for the system $Sn_{1-x}Ge_xTe$. According to [15] the Mössbauer shift at room temperature for the compound SnTe is 3.44 mm/s, and for the value of $x = 0.9$, $\delta = 3.05$ mm/s [6]. According to [19]

$$\delta = -0.45 + 3.01 n_s - 0.20 n_s^2 - 0.17 n_s n_p ,$$

where $\delta$ is the isomer shift relative to $CaSnO_3$, $n_s$ the number of 5s electrons, $n_p$ the number of 5p electrons.

From [3] it follows that for the totally ionic stannous compound $\delta = 4.77$ mm/s (s = 2, p = 0), and for the totally covalent one $\delta = 1.85$ mm/s (s = 1, p = 3).

The bond in a mixed ionic-covalent crystal being represented as $\delta = x\delta_{ion} + (1 - x)\,\delta_{cov}$, where $x$ is the ionic part, one can get that in the cubic phase (in which the isomer shift practically does not depend on the composition) the chemical bond ionicity makes up 54%, in good agreement with the conclusions drawn in [20]. For the solid solution $Sn_{0.1}Ge_{0.9}Te$ the bond ionicity will be about 40%. It is evident that in the rhombohedral phase the bond ionicity decreases rather strongly.

## References


[1] S. Sugai, K. Murase, S. Katayama, S. Takaoka, S. Nishi, and H. Kawamura, Solid State Commun. **24**, 407 (1977).

[2] V. A. Kochelap, V. N. Sokolov, and B. Yu. Vengalis, Phase Transitions in Semiconductors with the Deformed Electron–Phonon Interaction, Izd. Nauka, Kiev 1984 (in Russian).

[3] A. Rigamonti and G. Petrini, phys. stat. sol. **41**, 591 (1970).

[4] V. V. Chekin, O. P. Balkashin, and G. D. Sultanov, Ukr. fiz. Zh. **14**, 105 (1969).

[5] R. C. Knauer, Phys. Letters **A33**, 313 (1907).

[6] S. V. MOTIEJŪNAS and D. J. BALTRŪNAS, Abstracts of the 11th Conf. Ferroelectric Physics Vol. 2, Institute of Physics, Academy of Sciences of the Ukrainian SSR, Chernovtsy 1986 (p. 279).
[7] G. A. BYKOV and F. Z. HIEN, Zh. eksper. teor. Fiz. **43**, 909 (1962).
[8] Z. M. STADNIK, Proc. Internat. Conf. Mössbauer Spectroscopy, Vol. 1, Central Institute of Physics, Bucharest 1977 (p. 59).
[9] M. A. KRIVOGLAZ, Mössbauer Effect and its Application in Solid State Physics, Izd. Nauka, Moscow 1965 (in Russian).
[10] N. BIERLY, L. MULDAWER, and O. BECKMAN, Acta metall. **11**, 447 (1963).
[11] K. SHRIVASTARA, Hyperfine Interactions, **24/26**, 817 (1985).
[12] W. R. DUNHAM, C. T. WU, R. M. POLICHAR, R. H. SANDS, and L. J. HARDVING, Nuclear Instrum. and Methods **145**, 537 (1977).
[13] S. MOTIEJŪNAS and D. BALTRŪNAS, Abstracts Internat. Conf. Application of the Mössbauer Effect, September 16 to 20, 1985, I.K.S. K.U. Leuven (p. 3.58).
[14] N. H. ABRIKOSOV and L. E. SHELIMOVA, Semiconductor Materials on the Basis of the Compounds $A^{IV}B^{VI}$, Izd. Nauka, Moscow 1975 (in Russian).
[15] D. BALTRŪNAS, S. MOTIEJŪNAS, and E. I. ROGACHEVA, phys. stat. sol. (a) **97**, K131 (1986).
[16] A. DRAGUNAS, K. MAKARIUNAS, and M. BALČIUNIENE, phys. stat. sol. (a) **77**, 463 (1983).
[17] S. M. IRKAEV, R. N. KUZMIN, and A. A. OPALENKO, Nuclear Gamma-Resonance, Izd. Moscow University, Moscow 1970 (in Russian).
[18] V. A. VARNEK, L. N. MAZALOV, YU. P. SIDOROV, YU. O. KANTER, and V. I. MASHCHANOV, Izv. Akad. Nauk SSSR, Ser. neorg. Mater. **16**, 931 (1980).
[19] P. A. FLIN, in: Mössbauer Isomer Shifts, Ed. G. K. SHENOY and F. E. WAGNER, Chap. 9a, North-Holland Publishing Company Amsterdam, New York, Oxford 1978.
[20] P. ENDERS, phys. stat. sol. (b) **139**, K129 (1987).

*(Received November 24, 1987)*

phys. stat. sol. (a) 108, 205 (1988)

Subject classification: 62.20

*M. I. Kalinin Leningrad Polytechnical Institute*[1])

# Atomic Mechanisms of Fracture Nucleation and Fracture Development in Two-Dimensional Crystals in Thermodynamic Equilibrium

## I. One-Phase Systems

By

E. YA. BAIGUZIN, A. I. MELKER, and A. I. MIKHAILIN

The method of molecular dynamics is used to study the mechanism governing crack growth in two-dimensional crystals at low and high temperatures. It is shown that at low temperatures and small strains the fracture is due to fluctuations breaking interatomic bonds in the crack tip. At large strains the fracture becomes a three-stage process which is connected with dislocation emission, cleavage of dislocation cores, and nucleation of new cracks.

Die Methode der Molekulardynamik wird benutzt, um den das Rißwachstum in zweidimensionalen Kristallen bei niedrigen und hohen Temperature bestimmenden Mechanismus zu untersuchen. Es wird gezeigt, daß bei niedrigen Temperaturen und kleinen Spannungen der Bruch durch Fluktuationen, die interatomare Bindungen an der Rißspitze aufbrechen, hervorgerufen wird. Bei großen Spannungen geht der Bruch in einen Drei-Stufenprozeß über, der mit Versetzungsemission, Spalten von Versetzungskernen und Bildung neuer Rißkeime verbunden ist.

## 1. Introduction

Computer simulation of molecular dynamics revealed the following distinction in the structure of rupture fluctuations in systems consisting of the two coupled anharmonic chains with strong [1] and weak [2] interactions, both systems being in thermodynamic equilibrium. In a system with weak interaction, a rupture fluctuation appears at first only in one chain, while in a system with strong interaction, such fluctuation covers both chains simultaneously. As an important consequence, the time to failure of these systems depends on the rupture fluctuation structure [3]. These facts put two questions. Firstly, how the rupture fluctuation structure is connected with the number of interacting parallel chains under stress and secondly, could a rupture fluctuation arise in ideal stressed crystallites which contain a large number of strongly interacting close-packed rows of atoms. To answer these questions, the fracture of two-dimensional crystals was investigated at a given temperature and external force in two series of computer simulation experiments.

In the first case, the crystals were made from ten stressed anharmonic chains, each of them consisted of 40 atoms, the interaction between neighbouring chains being weak. Such chain crystals can be interpreted as a structure model of highly oriented polymers. It was found [4] that in the chain crystals in thermodynamic equilibrium the rupture fluctuations formed only in one chain. (The structure of these one-dimensional fluctuations and its connection to the time to failure were studied in detail elsewhere [5, 6].) The broken chain began to relax and slide, its atoms were pulling

[1]) Leningrad, USSR.

the atoms of neighbouring chains via the weak bonds. As a consequence, the tensile stress acting on the nearest strong bonds of neighbouring chains was increasing and one of these chains ruptured, too. It should be noted that breaking of the second strong bond is a two-stage process: at first an anomalously stretched bond is forming [7], its tensile strain exceeding the critical value, which corresponds to the tensile strength of an interatomic strong bond, and then, after several periods of the atomic vibration this bond breaks up under the influence of a momentum fluctuation.

During breaking of a strong bond part of the elastic tensile energy transforms into the kinetic energy of the broken chain, and, if this energy is large enough, the chain is chopped into pieces. Afterwards breaking of strong bonds and sliding of broken chains interchange so that the fracture is characterized by crack growth normal to the applied force, and the specimen splits along the lines parallel to a tensile axis. It is worth noting that another fracture centre arises in different parts of a chain crystal independent of the first one. This mechanism can explain splitting of polymers along the tensile axis of fibres [8].

A structural defect in this type of fracture is an anomalously stretched bond (or a transcritical dilaton [9]) which may be considered as a special form of solitary tensile waves [10, 11]. A momentum fluctuation which occurs during the breakage can be described as an explosion-type solution of the Ginzburg-Landau equation in which a correlated momentum of rupture plays the role of an order parameter [12]. Chopping of anharmonic chains was investigated earlier [13] and it was found that this phenomenon can explain thermal effects observed during polymer fracture. At present, we have preliminary results concerning the atomic mechanisms of sliding in chain crystals. Therefore, in the main, the elementary events which compose the breakage of a chain crystal are known and the problem is to construct an equation covering the whole fracture process.

In the second case we studied fracture of perfect two-dimensional crystals with a triangular lattice under constant load. The crystals contained three to ten close-packed rows of atoms, each row having 40 atoms [14]. It was found that fluctuational rupture of an interatomic bond on a free boundary caused a slip along the slip line passing through the broken bond. Approximately during one period of atomic vibrations prior to breakage, there appeared a region of correlated motion of atoms, the region consisting of three extreme close-packed rows, and the correlation was due to the transverse component of the atom momentum. The main contribution to the nucleation of fracture was due to a fluctuation in the potential energy. If the transverse size of the crystallite was several times larger than the breadth of the correlated momentum region in the direction normal to the applied load, the slip mechanism changed. After rupture of an interatomic bond on a free boundary, a dislocation was formed in the region of the broken bond. This dislocation moved to another free boundary and transformed into a jog on the boundary. The stress in the arising neck approached the breaking strength, which led to a simultaneous shift along the same slip line, and the crystallite separated into two non-interacting parts.

To ascertain the role of rupture fluctuations in the state of a break-up slip, after the nucleation of the first dislocation on a free boundary, the atoms of the opposite boundary were pinned [15]. The first dislocation with a trapezoidal core moved into the crystallite and stopped at a distance of the order of two to three interatomic distances from the pinned row of atoms. After that, the second interatomic bond on the free boundary ruptured and another dislocation was formed. The second dislocation moved along the slip line crossing the slip line of the first one. Then the interatomic bonds in the distorted region between the cores of these dislocations were breaking, and vacancies and sessile dislocations with pentagonal cores were formed [16]. Later on, the fracture was due to breaking of the interatomic bonds in the cores of sessile dislocations which led to nucleation of voids and cracks. The fracture was terminated by breakage of the perfect parts of the crystallite between these new defects. The mechanism considered explains the ductile rupture during wedging [17]. The role of rupture fluctuations increases if plastic deformation is constraint [18].

The two-dimensional structure of rupture fluctuations and nucleation of dislocations lead to more complex atomic mechanisms of fracture in perfect two-dimensional crystals in comparison with chain ones. Here, it is difficult to decompose fracture into elementary events. Besides, these events are not completely understood. At the same time, it is well known that the fracture of real solids is due to nucleation and growth of cracks [17]. Due to this reason, we picked out two fracture problems: crack growth in perfect two-dimensional crystals and crack nucleation and growth in two-phase two-dimensional crystals which contain low and high strength inclusions. In accordance with this division, we shall consider in the first part of the present investigation a molecular dynamical simulation of crack growth in perfect two-dimensional crystals at low and high temperatures. The second part is devoted to crack nucleation near inclusions. In both cases the results of computer simulation are compared with experimental ones.

## 2. Model and Method of Calculation

The crystallite investigated has the form of a rectangle and contained 392 atoms, which were arranged along 27 close-packed parallel rows of a triangular lattice. The atomic interactions in the crystallite were described by a Morse potential, the parameters of the potential corresponding to copper [19]. We chose a cut-off radius $r_m = 1.37r_0$ where $r_0$ is the equilibrium interatomic distance so that only the interactions of nearest neighbours were considered. We applied to the system the initial constant tensile strain $\varepsilon$ normal to the 27 close-packed rows of atoms by increasing the initial spacings between these rows by $3/2\, r_0\varepsilon$ with a simultaneous $\nu\varepsilon r_0$ decrease in the interatomic distances of the rows. Here $\nu$ is the Poisson ratio which is equal to 1/3 for a two-dimensional triangular lattice [20]. Then we let the system relax to its equilibrium static configuration using the molecular dynamics method with an artificial energy dissipation. The method consists in equating atomic velocities to zero whenever the kinetic energy of a system reaches its maximum. The equations of motion were integrated numerically by the Nordsieck method to the fifth order of accuracy on an ES 1052 computer.

We used boundary conditions which may be named "needle-type" ones. Here the distance between two extreme close-packed rows is fixed but the atoms of these rows can move along their rows. In this case the boundary atoms are strung to two parallel needles which hold the initial tensile strain of a crystallite. Such boundary conditions directly permit to take into consideration the dependence of elastic constants on strain and temperature.

A crack was introduced into a relaxed crystallite parallel to the "needles" by breaking nine interatomic bonds between two close-packed rows beginning with a free boundary. Then the crystallite was thermalized at a temperature $T$ determined by the Maxwell distribution of the longitudinal and transverse velocities of atoms. During the thermalization which took several periods of atomic vibrations we excluded the atomic interaction between the crack beaches.

### 3. Crack Growth

The atomic configurations near the tip of a crack which propagates in quasi-static and dynamic ways were investigated earlier [21 to 23]. It is found that increasing tensile strain leads to transition from a brittle to ductile growth. The latter is characterized by the emission of dislocations from a crack tip. Except for [22] these results are obtained at zero initial temperature, therefore, they cannot be used fully to explain the real fracture experiments. We consider below the results of our investigations concerning the atomic mechanisms of crack propagation in two-dimensional crystals in a wide temperature range.

#### *3.1 Low temperatures*

A crack growth at 100 K was observed if the initial tensile strain exceeded 6%. (At lower strain the crack collapses.) In a typical computer simulation at $\varepsilon = 6.5\%$, a crack grows along a straight line in a brittle way through the whole crystallite by successive breaking interatomic bonds at its tip. If the strain is increased to 7%, the propagation mechanism changes (Fig. 1). After several ruptures of interatomic bonds at the crack tip along the straight line (Fig. 1a), the bond which does not lie on this line can break (breaking bonds are crossed out in figures) and there appears a pentagonal defect (Fig. 1b) corresponding to the core of a sessile dislocation [16, 23, 24] (this dislocation will be denoted as B-dislocation). Then the bond in the core of this dislocation breaks and a new crack forms (Fig. 1c) which grows in the same way as the initial one (Fig. 1d). As a consequence, the fracture surface has a jog.

Strain increase up to 8% results in the emission of a dislocation (Fig. 2a) with a trapezoidal core (C-dislocation [14, 24]). After breaking the bonds in the core, this dislocation transforms into a new crack (Fig. 2b) which begins to grow along two directions: to a free boundary of the crystallite (Fig. 2c) and to the side of the initial crack (Fig. 2d). This growth creates jogs on the fracture surface. If the strain is increased to 9%, the crack tip emits almost simultaneously two dislocations along the

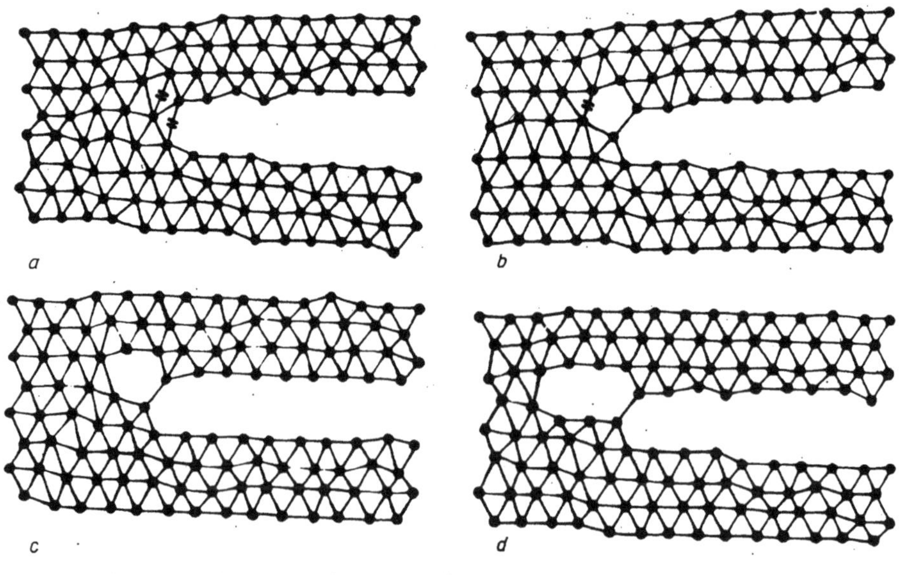

Fig. 1. Jog formation during brittle growth of a crack

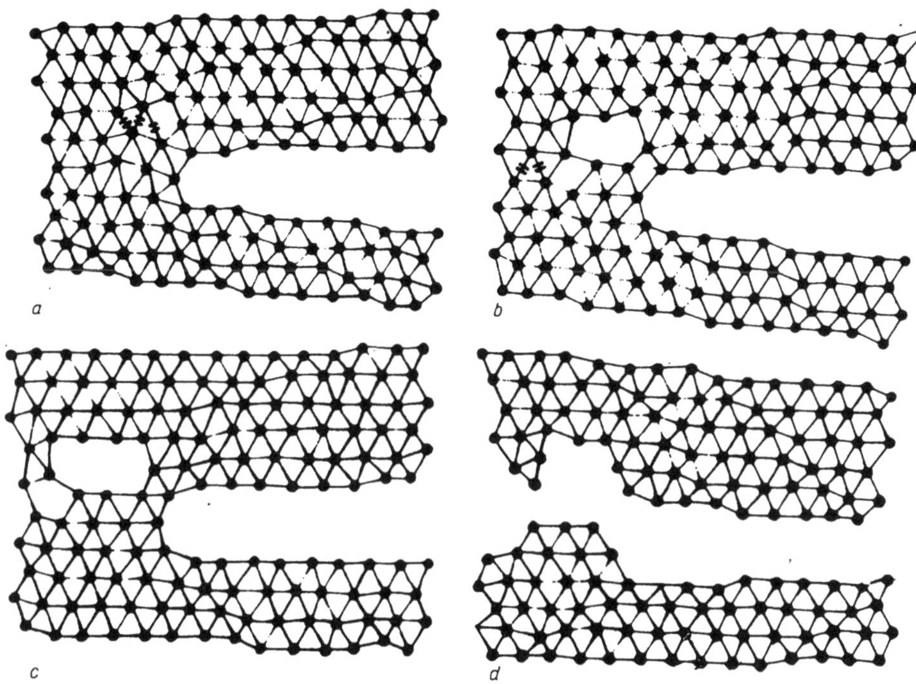


Fig. 2. Ductile growth of a crack

slip lines crossing at the tip (Fig. 3a). The crystal lattice in the vicinity of the dislocations relaxes by breaking some interatomic bonds (Fig. 3b). As a result, there forms a new crack in the upper part of the crystallite and pentagonal defects in the lower one (Fig. 3c). Further structural transformations lead to collapsing of the new crack into a pentagonal defect and the fracture localizes in the lower part of the crystallite (Fig. 3d) where the pentagonal defects transform into another crack (Fig. 3e) which begins to grow towards a free boundary (Fig. 3f).

Crack propagation at 200 K, as in the previous case, takes place if a strain exceeds 6%. Fig. 4 shows the development of plastic deformation and fracture near a crack tip for one of the realizations of computer simulation at 200 K and $\varepsilon = 7\%$. The picture is similar to that shown in Fig. 1. Parallel with this picture, we observed another fracture mechanism at the same temperature and strain (Fig. 5). Here the crack at first produces pentagonal defects and then grows by breaking interatomic bonds in their vicinity.

At large strains ($\varepsilon = 10\%$, Fig. 6) the fracture picture is in general similar to that discussed earlier and shown in Fig. 3. However, there is a new element: the nucleus of the C-dislocation (Fig. 6b, upper part of the crystallite) can transform into a vacancy near a free surface (Fig. 6c) that leads to intrusion (Fig. 6d). Further increasing initial strain up to 11% results in emitting a dislocation from a crack tip at an earlier stage of crack growth (Fig. 7). The emitted dislocation transforms into a new crack (Fig. 7a) which in its turn generates the second (Fig. 7b) and third (Fig. 7c) dislocations. Then these dislocations convert themselves into pentagonal defects (Fig. 7c). Subsequently, the fracture picture is analogous to the previous case, i.e. in the vicinity of a free boundary a pentagonal defect transforms into a crack or vacancy (Fig. 7d).

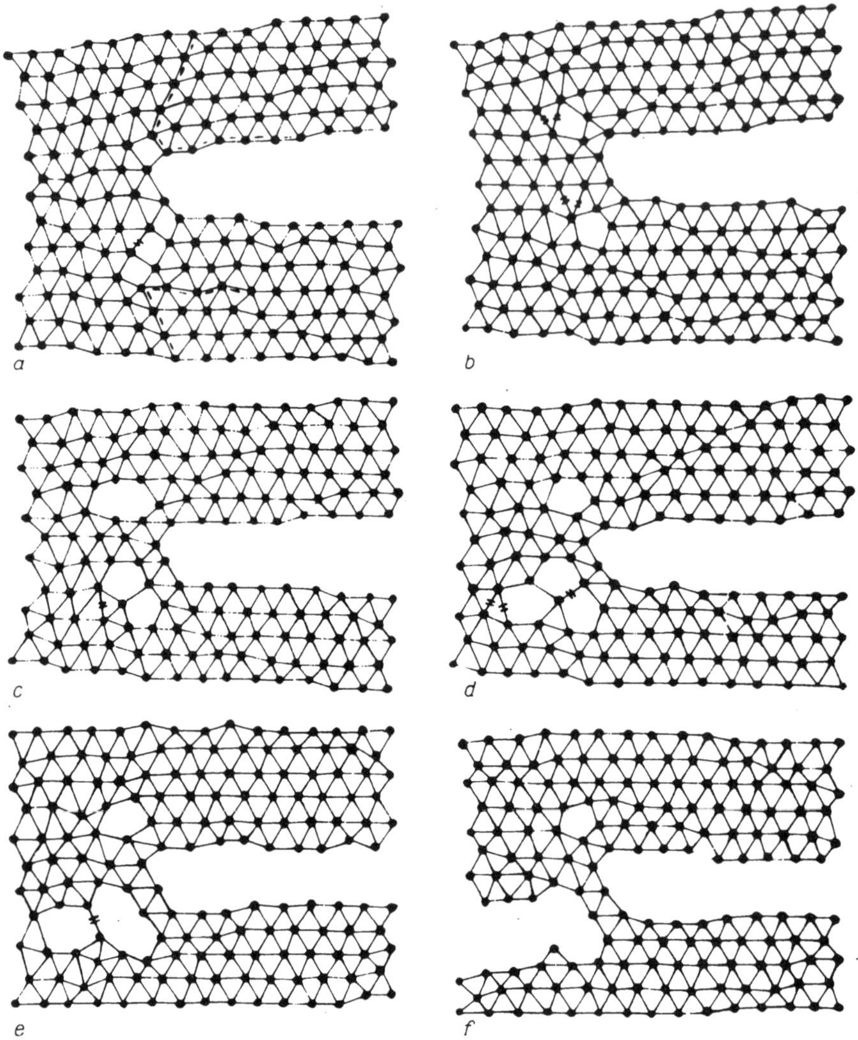

Fig. 3. Formation of the pre-fracture zone in the vicinity of a crack

### *3.2 Discussion*

It follows from the above results that in two-dimensional crystals two fracture mechanisms are realized at low temperatures:

1. successive breaking of interatomic bonds at a crack tip at small strains, i.e. brittle fracture in pure form;

2. many-stage process at large strains which include emitting of dislocations from a crack tip, cleavage of the cores of the emitted cracks, coalescence of new and old cracks. In the second case, the numerous defects in the vicinity of the initial crack break up the crystal and form a pre-fracture zone [25]. This mechanism can be also explain the density decrease in deformed materials [26].

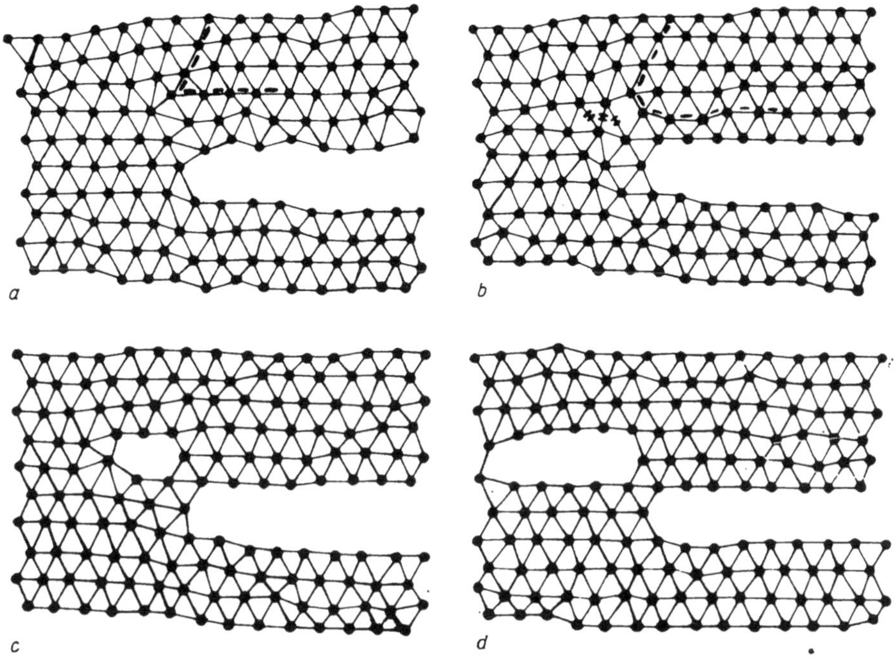


Fig. 4. Cleavage of a dislocation core and appearance of a new crack

Let us compare our results with those obtained at zero initial temperature [13, 21, 23]. It can be seen that in both cases there is a general tendency, i.e. the fracture mechanism depends on strain. However, temperature decreases the critical strain

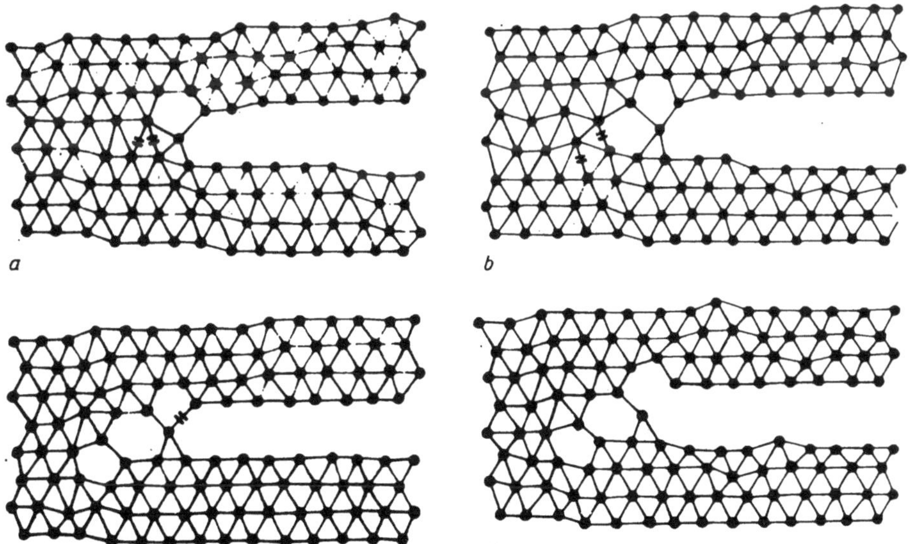


Fig. 5. Pre-fracture zone near a crack tip

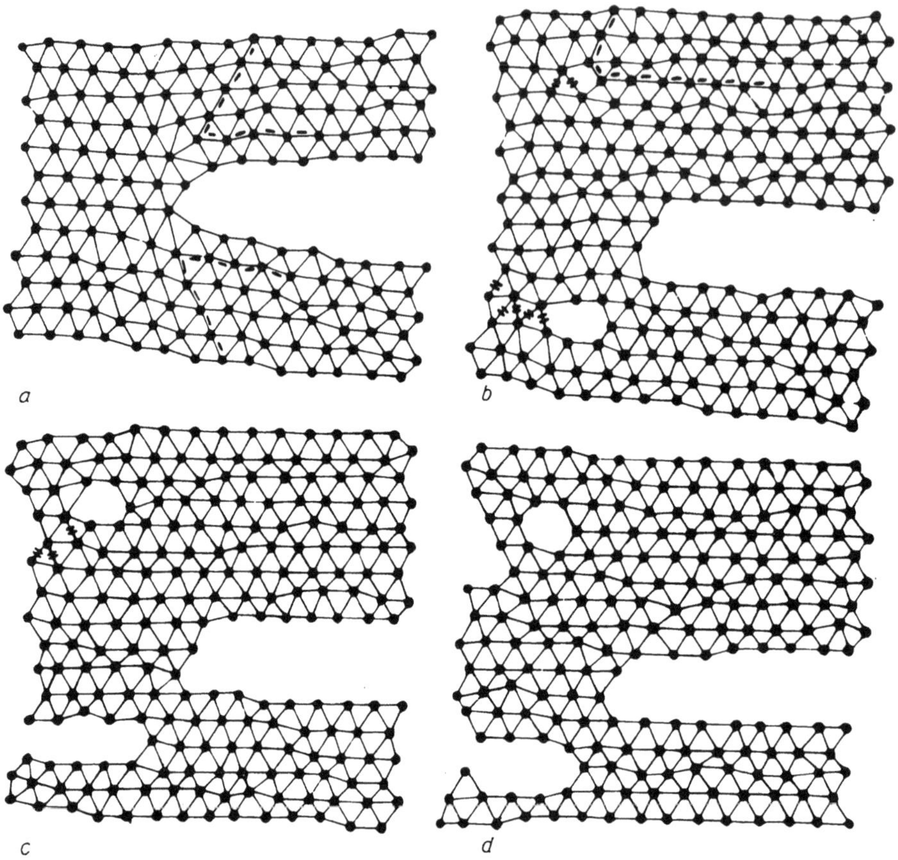


Fig. 6. Vacancy formation and intrusion due to plastic zone evolution near a crack tip

separating the brittle fracture mechanism from the ductile one, and the generation of dislocations begins at lower stress. Besides, the emitted dislocations pass smaller distances that can be induced by decreasing the mean free path of the dislocations.

According to [27, 28], ductile fracture is connected with the generation of dislocations at a crack tip. Our results show that it is not sufficient to prevent brittle fracture. The emitted dislocation must come out in the vicinity of the crack tip without producing new cracks. This is in agreement with experimental results [29, 30] which indicate that in ductile materials there are dislocation-free zones near the crack tips.

### *3.3 High temperatures*

Fig. 8 shows the dynamics of crack propagation at 350 K and $\varepsilon = 6.5\%$. The picture presented is typical for small strains below which the crack collapses. At first, the interatomic bond crossed out in Fig. 8a ruptures and the crack tries to emit an A-dislocation (Fig. 8b). (It was shown [24] that the A-dislocation is an unrelaxed configuration of a C-dislocation.) However, this dislocation has no time to run away and after two periods of atomic vibrations the bonds in the region between crack and dislocation break. As a result, the crack grows along the slip line of the dislocation

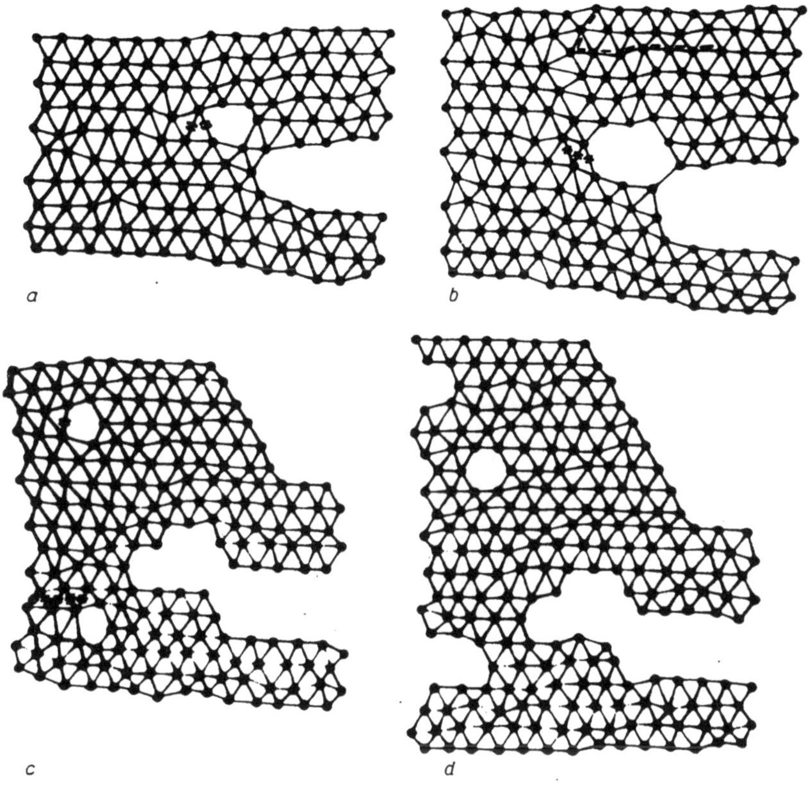


Fig. 7. Crack growth caused by breaking interatomic bonds and dislocation emission

generating a new A-dislocation (Fig. 8c) which is absorbed by its own crack (Fig. 8d). Simultaneously, the crack emits another A-dislocation and this dislocation manages to run away from the crack. The dislocation moves a distance of the order of three interatomic spacings and converts into a sessible B-dislocation (Fig. 8d). After that the crack does not grow, but the sessile B-dislocation splits into a C-dislocation and another B-dislocation, the latter having the Burgers vector normal to the tensile axis of the crystallite (Fig. 8e). Then the core of the C-dislocation transforms into a new crack (Fig. 8f).

The picture considered is conserved if the tensile strain is increased to 8% (Fig. 9). Further strain increasing to 9% causes the crack to emit and absorb its own defects (Fig. 10). If one observes the initial crack growth in large time intervals, one may come to the conclusion that the crack grows, as in the case of low temperatures, by successive breaking the interatomic bonds at its tip (Fig. 10a, b). However, in reality the crack emits dislocations (Fig. 10c, e) which cannot run away from it. As a result, the crack propagates along different slip lines and the fracture surface has numerous jogs (Fig. 10d, f). Increasing the temperature up to 500 K does not change the basic features of this mechanism (Fig. 11 and 12). However, a pre-fracture zone can be formed (Fig. 11d) in the front of a crack. Besides, dislocations (Fig. 11b, c, d) and cracks (Fig. 12c, d) are generated on a free boundary.

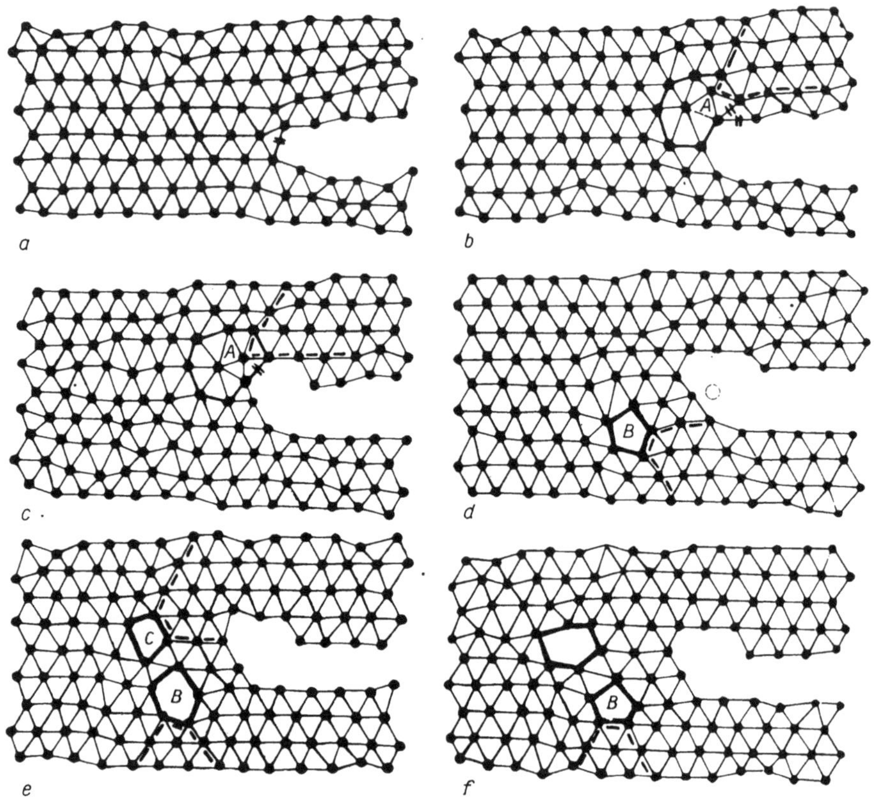


Fig. 8. Ductile crack growth at high temperatures

### *3.4 Discussion*

Let us consider the reasons according to which the new cracks observed are generated by breaking interatomic bonds in the cores of emitted dislocations. It is known [14, 16, 24] that in two-dimensional crystals there are two types of moving dislocations, i.e. A- and C-dislocations. The A-dislocation has a pentagonal core, whose centre contains one atom (e.g., Fig. 11a). Usually this dislocation arises under dynamic load when a crystal has no time to relax. The C-dislocation has a trapezoidal core (e.g., Fig. 11b) and is generated in crystals being in thermodynamic equilibrium. It is a relaxed configuration of the A-dislocation [14]. If to rupture the interatomic bond which is the largest side of the trapezoid, the C-dislocation becomes a pentagonal defect (or sessile B-dislocation, e.g., Fig. 11c). Further breaking of any bond in the core of the B-dislocation produces a crack nucleus [16, 23].

The question arises why the C-dislocation so easily converts into a crack. The reasons lie in its core structure. The core of the C-dislocation is a trapezoid, three sides of which are almost equal to each other and to the equilibrium interatomic distance. The fourth (largest) side is an anomalously stretched bond [7], its tensile strain exceeding the critical value corresponding to the tensile strength of an interatomic bond. According to [9], the energy of a rupture fluctuation during breaking an anomalously stretched bond is smaller than for an ordinary bond. For this reason, the C-dislocation easily transforms into a pentagonal defect. The quantitative analysis

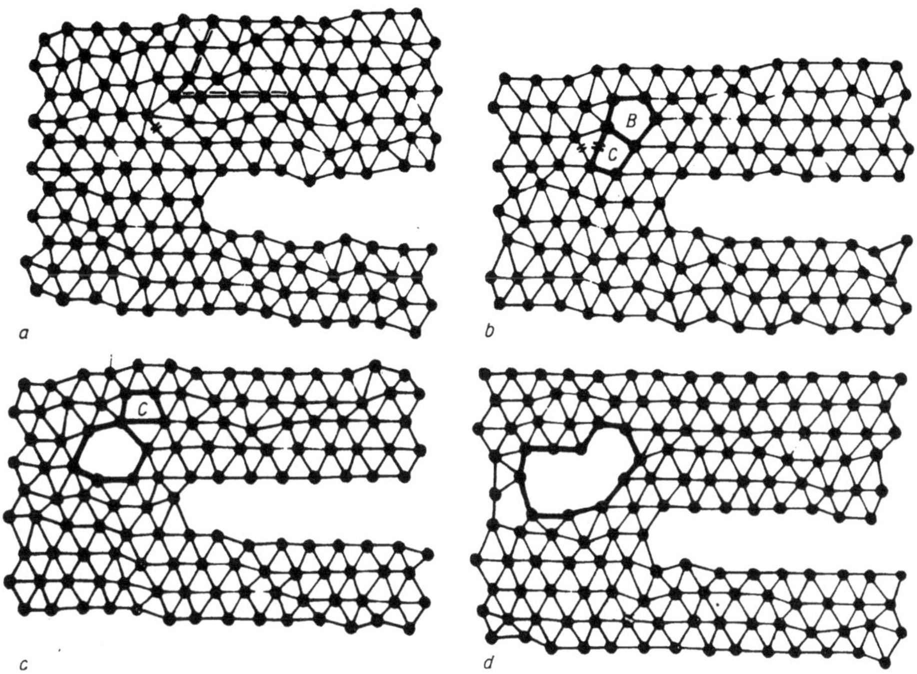


Fig. 9. Pre-fracture zone at high temperatures

of dislocations in two-dimensional crystals has shown [31] that the C-dislocation is a special form of solitary tensile waves. Therefore, the anomalously stretched bonds can be considered as elementary destructive excitations in two-dimensional solids. It is interesting to compare our results with the experiments in the electron microscope where fracture of thin foils is investigated in situ [32 to 36]). According to [35] during the ductile fracture of molybdenum, the dislocations generated by a crack determine the propagation trajectory of this crack. This is in accordance with our investigations. Besides, it is found that the emission of dislocations diminishes when the dislocations fill the plastic zone in the vicinity of a crack tip and the crack stops. Then a new crack appears in this region and this crack coalesces with the old one as a consequence of the fracture of the region between the cracks along the slip planes of dislocations. This is also observed in our molecular dynamics experiments. This mechanism takes place in the case of other materials (single crystals and polycrystals of iron, vanadium, gold, tungsten, and copper [32 to 34]). Therefore, in situ investigations lead to the same conclusions and are in good agreement with molecular dynamics simulation.

## 4. Conclusion

We have shown that in two-dimensional crystals at low temperatures and small strains a crack grows by means of breaking interatomic bonds at its tip. This mechanism characterizes brittle growth in a pure form. If the strain increases, fracture becomes ductile independent of temperature. The process includes generation of dislocations at the crack tip and their transformation into new cracks which coalesce

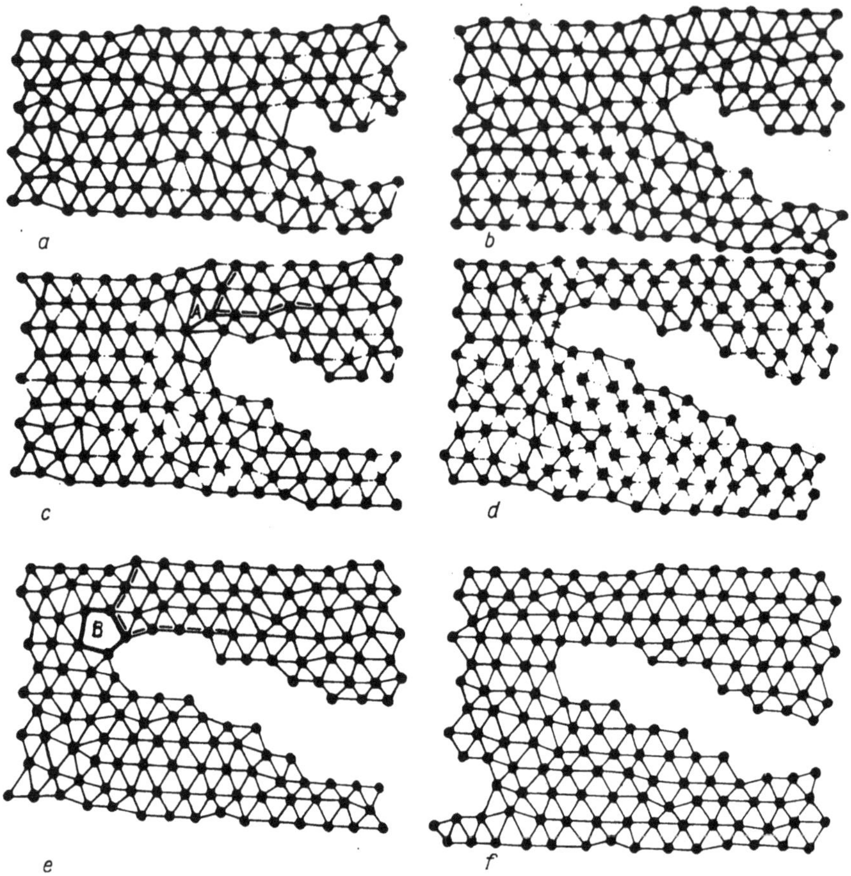


Fig. 10. Ductile crack growth due to the absorption of emitted dislocations

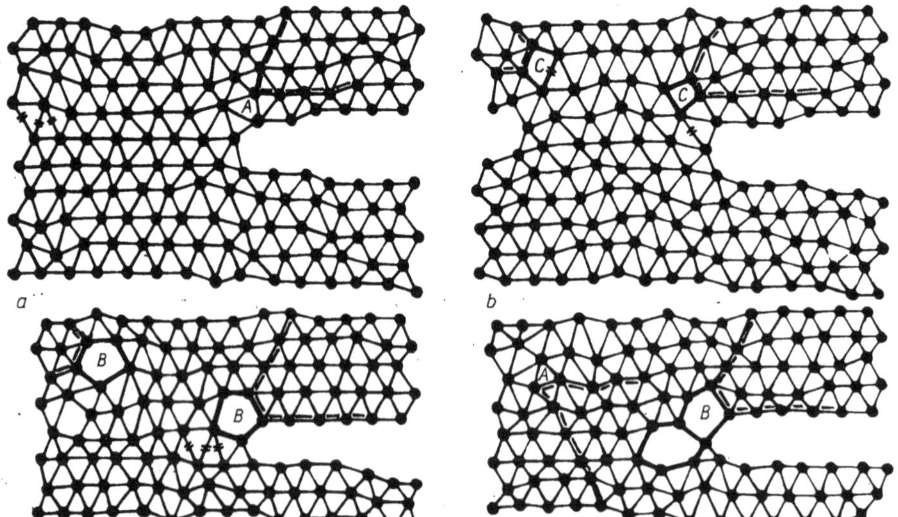


Fig. 11. Plastic zone in the vicinity of a crack tip

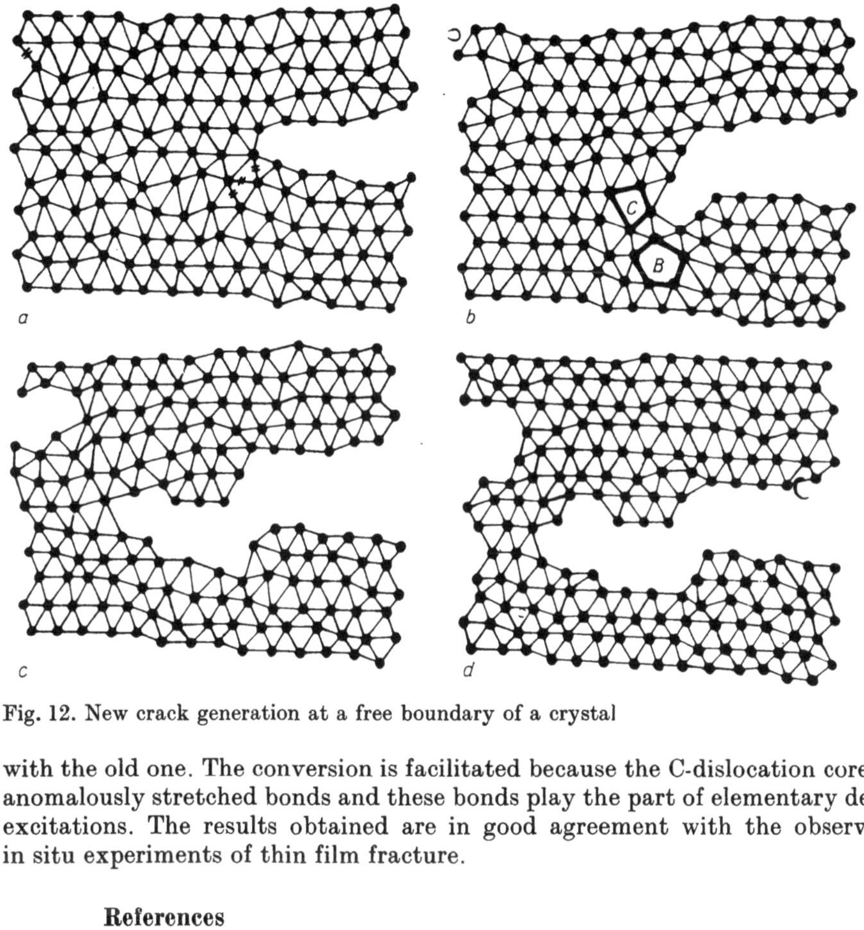


Fig. 12. New crack generation at a free boundary of a crystal

with the old one. The conversion is facilitated because the C-dislocation cores contain anomalously stretched bonds and these bonds play the part of elementary destructive excitations. The results obtained are in good agreement with the observations of in situ experiments of thin film fracture.

## References

[1] A. I. Melker and A. V. Ivanov, phys. stat. sol. (a) **84**, 417 (1984).
[2] A. V. Ivanov and A. I. Melker, phys. stat. sol. (a) **89**, K51 (1985).
[3] A. I. Melker and A. V. Ivanov, Fiz. tverd. Tela **28**, 3079 (1986).
[4] A. I. Melker and A. V. Ivanov, Fiz. tverd. Tela **29**, 1556 (1987).
[5] A. I. Melker and A. I. Mikhailin, Fiz. tverd. Tela **23**, 1746 (1981).
[6] A. I. Melker and A. I. Mikhailin, Fiz. tverd. Tela **26**, 1236 (1984).
[7] A. N. Orlov, Yu. M. Plishkin, and I. M. Shepeleva, Fiz. Metallov i Metallovedenie **4**, 540 (1957).
[8] H. H. Kausch, Polymer Fracture, Springer-Verlag, Berlin/Heidelberg/New York 1978 (p. 123).
[9] A. I. Mikhailin and A. I. Melker, Khim. Fiz. **4**, 15 (1985).
[10] A. I. Melker, Fiz. tverd. Tela **24**, 4086 (1982).
[11] A. I. Melker and A. V. Ivanov, Fiz. tverd. Tela **28**, 3396 (1986).
[12] A. I. Melker and I. A. Ovidko, Fiz. tverd. Tela **27**, 594 (1985).
[13] A. I. Melker, A. I. Mikhailin, and T. E. Kuznetzova, Mekhanika kompozit. Mater. No. 4, 720 (1979).
[14] A. V. Ivanov and A. I. Melker, Fiz. tverd. Tela **28**, 3637 (1986).
[15] A. I. Melker and A. V. Ivanov, Izv. Akad. Nauk SSSR, Ser. Met. No. 6, 128 (1987).
[16] S. V. Govorov, A. V. Ivanov, A. I. Melker, and A. I. Mikhailin, Izv. Akad. Nauk SSSR, Ser. Met. No. 6, 139 (1985).

[17] H. LIEBOWITZ (Ed.), Fracture, Vol. 1, Academic Press, New York/London 1968 (p. 21).
[18] A. I. MELKER and A. V. IVANOV, Mekhanika komposit. Mater. No. 6, (1987), in the press.
[19] L. A. GIRIFALKO and V. G. WEIZER, Phys. Rev. **114**, 687 (1959).
[20] P. D. ESBJORN and E. J. JENSEN, J. Phys. Chem. Solids **37**, 1081 (1976).
[21] G. J. DIENES and A. PASKIN, Proc. NATO Adv. Study Inst. Atomistics of Fracture, Corsica (France) 1981 Ed. R. M. LATANISON and J. R. PICKENS, Plenum Press, New York (p. 1)
[22] E. D. SHCHUKIN and V. S. YUSHCHENKO, J. Mater. Sci. **116**, 313 (1981).
[23] S. V. GOVOROV, A. V. IVANOV, A. I. MELKER, and A. I. MIKHAILIN, Problemy Prochnosti No. 7, 26 (1985).
[24] A. I. MELKER, A. I. MIKHAILIN, and E. YA. BAIGUZIN, Fiz. Metallov i Metallovedenie **64**, 1066 (1987).
[25] D. N. KARPINSKII and A. N. ORLOV, Evolutsiya Defektnoi Struktury Kristallov, FTI Akad. Nauk SSSR, Leningrad 1984 (p. 34).
[26] V. I. BETEKHTIN and V. I. VLADIMIROV, Problemy Prochnosti i Plastichnosti, Izd. Nauka, Leningrad 1979 (p. 142).
[27] A. KELLY, W. R. TYSON, and A. N. COTTRELL, Phil. Mag. **15**, 567 (1967).
[28] J. R. RICE and R. THOMPSON, Phil. Mag. **29**, 73 (1974).
[29] S.-J. CHANG and S. M. OHR, J. appl. Phys. **52**, 7174 (1981).
[30] S. M. OHR and S.-J. CHANG, J. appl. Phys. **53**, 5645 (1982).
[31] A. I. MELKER and S. V. GOVOROV, unpublished.
[32] D. W. PASHLEY, Proc. Roy. Soc. **A225**, 218 (1960).
[33] G. F. SARZHAN, V. I. TREFILOV, and S. A. FIRSTOV, Fiz. Metallov i Metallovedenie **38**, 76 (1974).
[34] G. F. WILSDORF, Kristall und Technik **14**, 1265 (1979).
[35] I. I. NOVIKOV, V. A. ERMISHKIN, V. S. GREBNEVA, V. G. ZHARKOV, and E. N. SAMOILOV, Fiz. tverd. Tela **25**, 2096 (1983).
[36] I. I. NOVIKOV, V. A. ERMISHKIN, V. G. ZHARKOV, E. N. SAMOILOV, and V. S. GREBNEVA, see [26] (p. 56).

*(Received December 5, 1987)*

phys. stat. sol. (a) **108**, 219 (1988)

Subject classification: 62.40; S4

*Faculty of Engineering, Ibaraki University*[1]) *(a),*
*Japan Atomic Energy Research Institute*[2]) *(b), and*
*Hitachi Research Laboratory, Hitachi Ltd.*[1]) *(c), Ibaraki*

# Internal Friction in Single-Crystal Al Alloys after Electron Irradiation at Low Temperature

By

M. Kobiyama (a), S. Takamura (b), and K. Nakata (c)

Internal friction in single crystal Al–Si, Al–Zn, Al–Ge, Al–Sn, and Al–Pb alloys are measured after electron irradiation at 78 K. The internal friction spectra in these alloys are very similar. Since the atomic size of Sn and Ge atoms is oversized as compared to the host atom, the internal friction peaks in these alloys are considered to be due to rearrangement of the type b first-neighbour configuration. The two peaks at 10 to 12 K and at 55 to 60 K are attributed to the "frozen-free split" phenomenon. The former has tetragonal symmetry and the latter the trigonal symmetry.

Die innere Reibung in Al–Si-, Al–Zn-, Al–Ge-, Al–Sn- und Al–Pb-Legierungseinkristallen wird nach Elektronenbestrahlung bei 78 K gemessen. Die Spektren der inneren Reibung sind sich sehr ähnlich. Da die atomaren Abmessungen von Sn- und Ge-Atomen im Vergleich zum Wirtsatom größer sind, wird angenommen, daß das Maximum der inneren Reibung durch eine Umverteilung der b-Konfiguration der ersten Nachbarn verursacht wird. Die beiden Maxima bei 10 bis 12 K und bei 55 bis 60 K werden dem Phänomen der „eingefrorenen Aufspaltung" zugeordnet. Das erstere hat tetragonale und das letztere trigonale Symmetrie.

## 1. Introduction

Internal friction and ultrasonic attenuation measurements supply knowledge on the symmetry, migration of defects, and the strain field around defects. The internal friction and ultrasonic attenuation measurements related to interstitial–solute complexes have been made for dilute alloys after electron and neutron irradiations [1 to 6]. Ultrasonic attenuation measurements in single crystal Al–Ag and Al–Si alloys showed that the peak due to the complex with ⟨100⟩ symmetry appears at about 20 K and the peak due to the complex with ⟨111⟩ symmetry at about 110 K [2]. In polycrystalline Al–Pb, Al–Si, Al–Zn, Al–Ag, Al–Sn, and Al–In alloys after neutron irradiation, the internal friction spectra are very similar. In these alloys, the main peaks in internal friction are situated around 10 to 12 and 55 to 60 K [3]. These peaks disappear after annealing at around 130 K and correspond to the peaks observed in the ultrasonic attenuation measurements. They seem to have very similar configuration for the complex in these Al alloys.

The aim of the present work is to study the symmetry of interstitial–solute complexes from measurements of internal friction using single crystals with the ⟨111⟩ and ⟨100⟩ directions of Al–Pb, Al–Sn, Al–Ge, Al–Zn, and Al–Si alloys after 3 MeV electron irradiation.

---

[1]) Hitachi, Ibaraki, Japan.

[2]) Tokai-mura, Ibaraki, Japan.

## 2. Experimental Procedures

Dilute Al alloys were made by melting together 99.999% Al and high purity metals of 99.999% Sn, Pb, Ge, and Si and 99.9999% Zn. The compositions (in at%) of the alloys were as follows: Al–Pb: 0.02 Pb, Al–Sn: 0.07 Sn, Al–Ge: 0.06 Ge, Al–Zn: 0.07 Zn, and Al–Si: 0.08 Si. The specimens with the ⟨111⟩ and ⟨100⟩ directions were spark cut from single crystals of these dilute Al alloys. The directions of the specimens were determined by means of Laue X-ray photographs. Each specimen had a sizes of about $0.1 \times 2 \times 20$ mm$^3$ with a thicker step where the specimen was gripped in the specimen holder. After they had been etched electrically, they were annealed at about 720 K for 2 h. The details of the experimental procedures were described in previous papers [3, 4, 7]. The specimens were set on a floating plate in liquid nitrogen. 3 MeV electron irradiations were performed to about $5 \times 10^{21}$ electrons/m$^2$ at 78 K. The irradiation fluence could not be determined accurately because the specimens were in liquid nitrogen and the distance from the specimen and the surface of liquid nitrogen was different in each specimen. After irradiation, the specimen was set on a specimen holder and on the measuring system. The measuring system was cooled down to 4.2 K and was evacuated. The flexural vibration of the specimens was measured during continuous warming up to a given temperature. The resonant frequency of flexural vibration in the specimens was 1000 to 2500 Hz.

## 3. Experimental Results and Discussion

Internal friction was measured as a function of temperature. The internal friction rises continuously to very large values with increasing temperature. The background seems to be attributed to the existence of dislocations. The internal friction peaks due to the radiation-induced defects were extracted by subtracting the broad peaks with Gaussian distribution from the measured internal friction curve. The relaxation peaks obtained by such a treatment are shown in Fig. 1 for single crystal with the ⟨100⟩ and ⟨111⟩ directions of Al–Si alloy. The internal friction peaks are caused by the stress-induced relaxation of defects.

The 40 K peak seems to be caused by a slight deformation which was induced during the time of installing the specimen into the specimen holder after irradiation. The occurrence of the peak may be due to the enhancement of double kink formation on a dislocation by the interaction with complexes [8].

Fig. 2a to d show the annealing behaviour of internal friction in Al–Zn, Al–Ge, Al–Sn, and Al–Pb alloys after irradiation. A very small peak is observed in the specimens with the ⟨111⟩ direction of Al–Si, Al–Zn, and Al–Ge alloys at 10 K. However, the peak temperature does not correspond to the peak temperature at 10 to 12 K in the specimen with ⟨100⟩ direction because the peak temperature does not agree with the latter peak. Internal friction spectra in these alloys are very similar although the relative height of two peaks at 10 to 12 and 55 to 60 K is different. Therefore, the defect configuration in these alloys seems to be the same in spite of the size difference of the solute. The two peaks at 10 to 12 and 55 to 60 K have the same annealing behaviour; the peaks are annealed at the same temperature of 120 to 130 K and are considered to be attributed to the simple interstitial–solute complex since the production rate of single interstitials is considered to be larger as compared with that of di-interstitials and tri-interstitials after 3 MeV electron irradiation. Therefore, the two peaks are due to the "frozen-free split" phenomenon of single interstitial–solute complex where the defect reorients through two different jump processes [9]. The peak at 10 to 12 K has tetragonal symmetry with ⟨100⟩ symmetry axis and the peak at 55 to 60 K has trigonal symmetry with ⟨111⟩ symmetry axis.

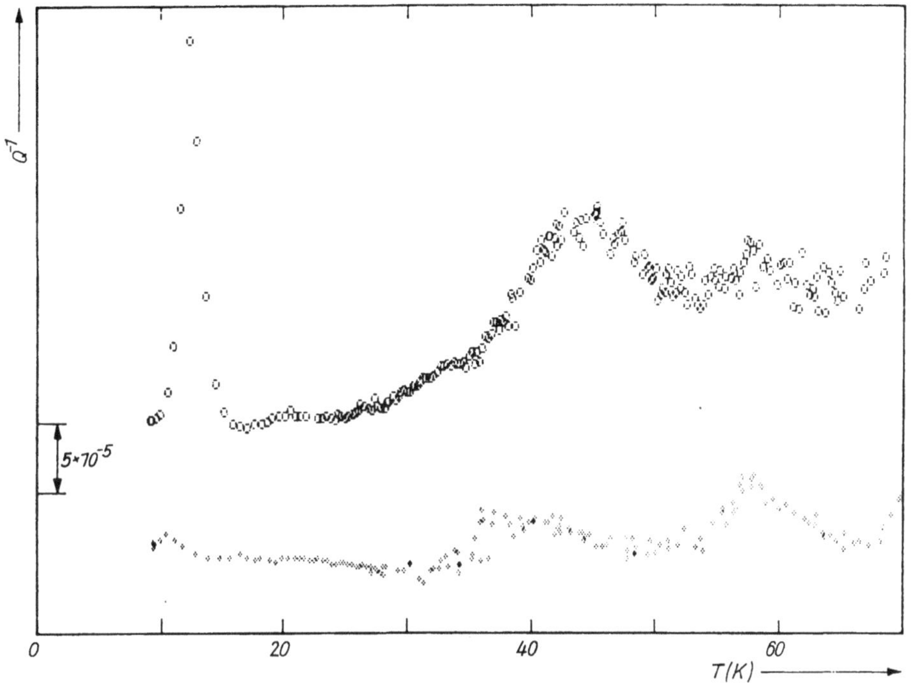


Fig. 1. Temperature dependence of internal friction in Al–Si alloy after 3 MeV electron irradiation at 78 K. The lower (+) and upper (o) curves show the spectrum for a single crystal with ⟨111⟩ and ⟨100⟩ directions, respectively. Curves are shifted vertically

Dederichs et al. [10] have estimated the configuration of interstitial–solute complexes for an oversized solute and the most stable complex is the type b first-neighbour one in which the centre of the dumbbell interstitial is situated at the first-neighbour sites of a solute. In the molecular dynamic calculation of the complex, the mixed dumbbell is unstable and the type b first-neighbour configuration is also the most stable one [11]. Robrock [5] proposed the model that the internal friction peaks occurred by jumps of the dumbbell to the equivalent sites around an oversized solute. Johnson and Granato [2] explained the feature of the two peaks in Al–Ag and Al–Si alloys by a tilted caged mixed dumbbell. This model is suitable to explain the internal friction structure of complexes with an undersized solute or a solute of nearly same atomic size. However, it seems to be difficult to explain the behaviour for the complex with oversized solute since the complex with oversized solute is the type b first-neighbour one [11].

The interstitial–solute complex in Al–Ge and Al–Sn alloys is considered to form the type b first-neighbour complex, since the volume size factors of Ge and Sn atoms are larger than those of the host atom. The internal friction spectra in Al–Si, Al–Zn, and Al–Pb alloys are very similar to those in Al–Ge and Al–Sn alloys, so that the configuration and the behaviour of the complex in these alloys seem to be same.

The linear size factor obtained by X-ray techniques of solute in Al alloys is −22.66, +7.45, +4.19, −1.95, and −5.56% for Pb, Sn, Ge, Zn, and Si atoms [12], respectively, where + sign means the oversized solute and − sign the undersized one. On the other hand, the values calculated from Seitz radii of the elements for each solute are +23.00, +17.70, +10.94, −2.78, and +5.50% for Pb, Sn, Ge, Zn, and Si atoms, respectively

[12]. In dilute Al alloys, it seems that the atomic size of solutes in interstitial–solute complexes could be better obtained from the Seitz radii of the elements, on the basis of the present internal friction experiments.

It is considered that the peak at 10 to 12 K is due to the movement of a type b first-neighbour complex to the equivalent positions around a solute and the peak

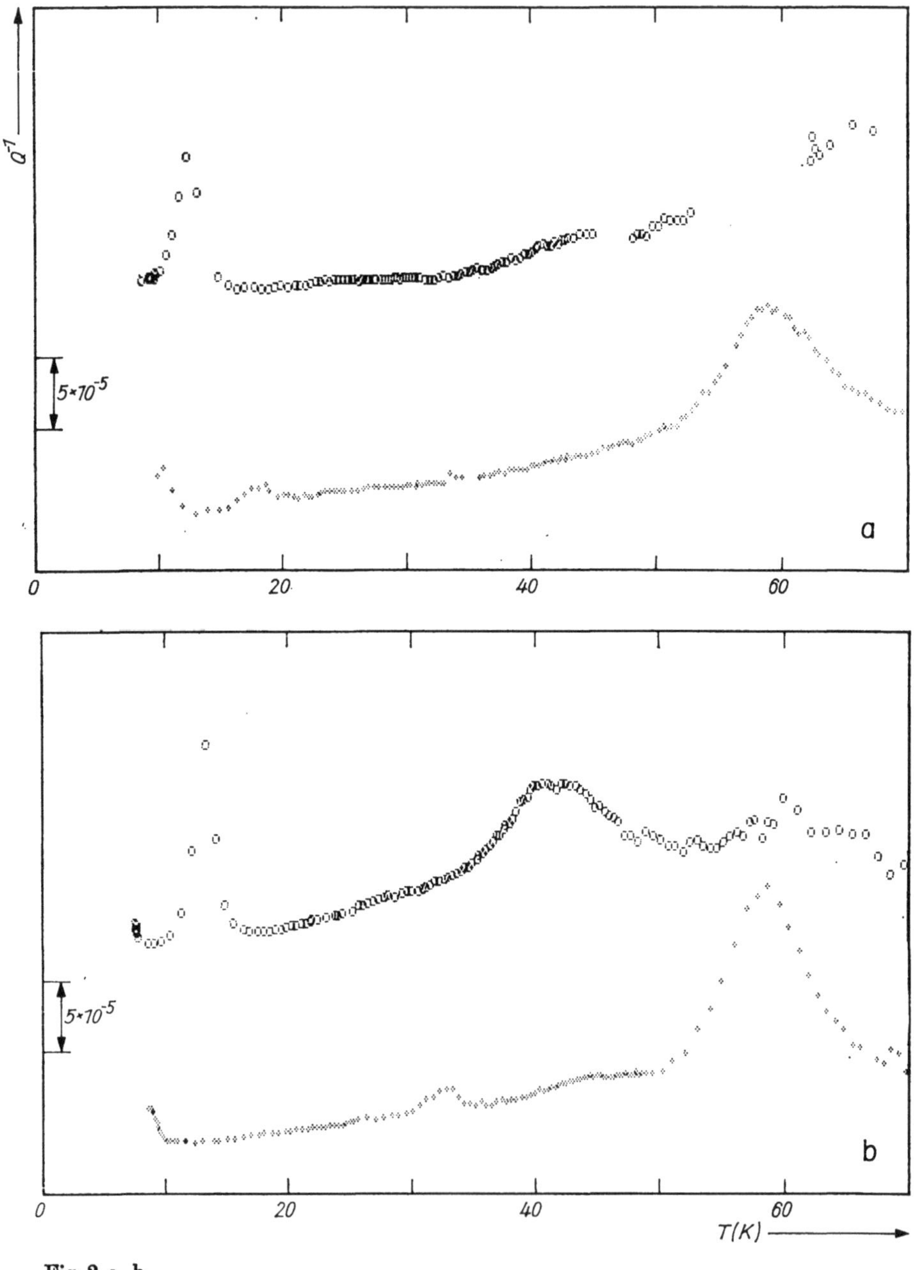


Fig. 2 a, b

at 55 to 60 K is due to the jump process between the configuration with an interstitial occupying a site of the ⟨111⟩ direction from a solute and the octahedral configuration after a warm-up to 55 to 60 K.

A further investigation is necessary to determine the configuration of the interstitial–solute complex with oversized solute in dilute alloys.

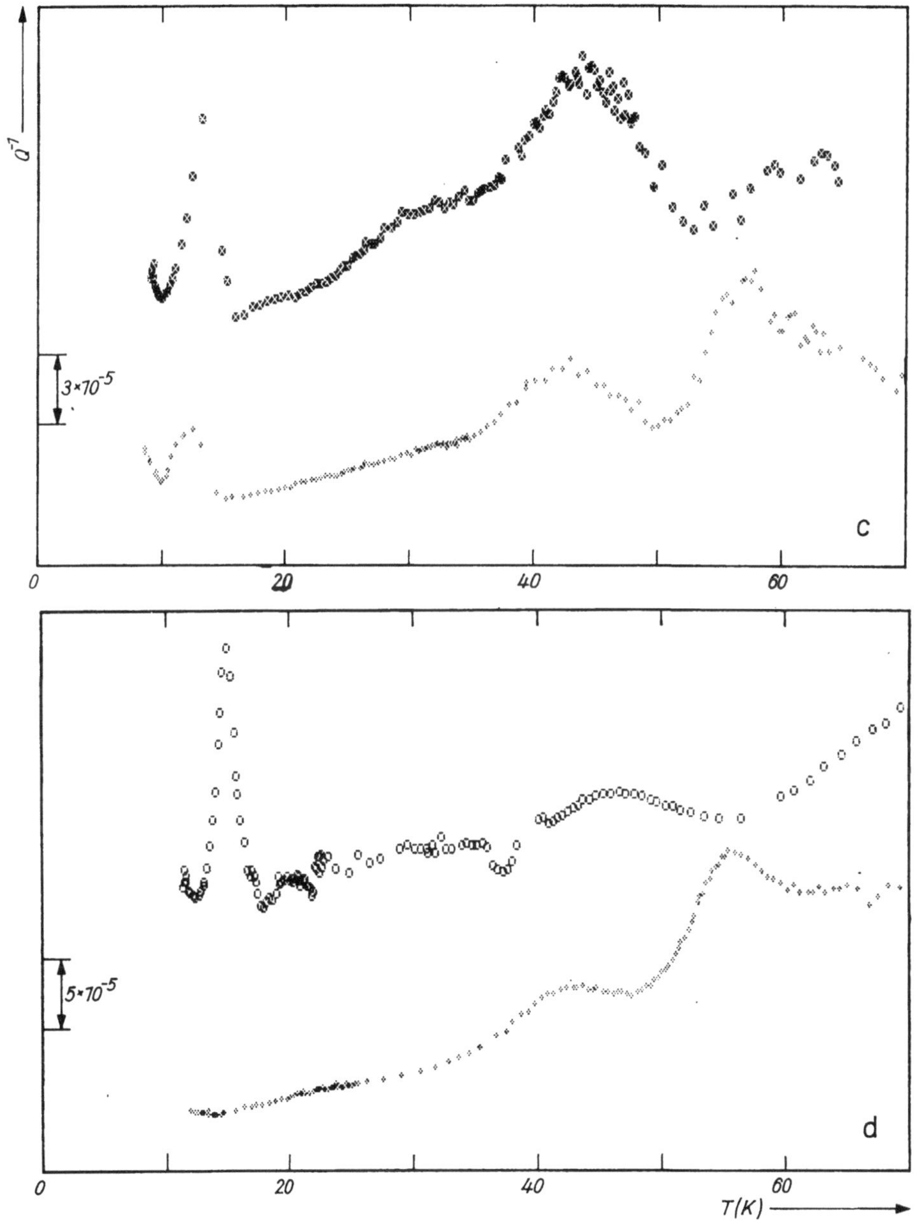


Fig. 2 c, d

Fig. 2. The same as Fig. 1 for a) Al–Zn, b) Al–Ge, c) Al–Sn, and d) Al–Pb alloys

## 4. Summary

Temperature dependence of internal friction was measured on single crystals with ⟨111⟩ and ⟨100⟩ direction in dilute Al–Si, Al–Zn, Al–Ge, Al–Sn, and Al–Pb alloys after 3 MeV electron irradiation. The spectra in these alloys are very similar. From the present internal friction experiment, all the solutes in these alloys seem to be oversized for interstitial–solute complexes. The two peaks at 10 to 12 and 55 to 60 K are attributed to the "frozen-free split" phenomenon. The peak at 10 to 12 K seems to be due to jumps to equivalent sites around a solute from the type b first-neighbour complex. When warming up to 50 to 60 K, part of the complex is moved to the position occupying a site of the ⟨111⟩ direction from a solute, the octahedral site. The peak at 55 to 60 K is considered to be caused by jumps between the equivalent octahedral sites around the solute.

***Acknowledgement***

The authors would like to thank Mr. H. Sunaga of Takasaki Research Establishment of JAERI for operation of electron irradiations.

## References

[1] A. V. Granato, J. Holder, K. L. Hulton, D. L. Johnson, G. G. Setser, P. Wallace, and H. Wong, Point Defects and Defect Interactions in Metals, Tokyo University Press, 1982 (p. 360).
[2] E. C. Johnson and A. V. Granato, Mater. Sci. Forum **15—18**, 581 (1987).
[3] S. Takamura and M. Kobiyama, Radiat. Eff. **91**, 21 (1985).
[4] M. Kobiyama and S. Takamura, phys. stat. sol. (a) **90**, 253 (1985).
[5] K. H. Robrock, see [1] (p. 353).
[6] G. Kollers, H. Jacques, L. E. Rehn, and K. H. Robrock, J. Physique **42**, C5–729 (1981).
[7] S. Takamura and M. Kobiyama, Radiat. Eff. **81**, 243 (1984).
[8] S. Takamura and M. Kobiyama, to be published.
[9] A. S. Nowick and B. S. Berry, Anelastic Relaxation in Crystalline Solids, Academic Press, New York 1972.
[10] P. H. Dederichs, C. L. Lehmann, H. R. Schober, A. Scholz, and R. Zeller, J. nuclear Mater. (Amsterdam) **69/70**, 176 (1978).
[11] S. Takamura, T. Aruga, K. Nakata, and M. Kobiyama, to be published.
[12] H. W. King, J. Mater. Sci. **1**, 79 (1966).

*(Received January 28, 1988)*

phys. stat. sol. (a) **108**, 225 (1988)

Subject classification: 65.70; S7.16

*Sektion Physik der Bergakademie Freiberg*[1]) *(a)*
*and Sektion Physik der Karl-Marx-Universität Leipzig*[2]) *(b)*

# Low-Temperature Thermal Expansion of $ZnSiAs_2$

By

P. DEUS (a), U. VOLAND (a), and H. NEUMANN (b)

Dedicated to Prof. Dr. E. GUTSCHE on the occasion of his 60th birthday

The thermal expansion coefficients $\alpha_a$ and $\alpha_c$ of the lattice parameters $a$ and $c$ of $ZnSiAs_2$ are measured in the temperature range from 30 to 300 K using the X-ray Bond method. Both $\alpha_a$ and $\alpha_c$ decrease with decreasing temperature and become negative at about 45 K. The results are compared with thermal expansion data for other chalcopyrite semiconductors and are discussed in terms of the principal Grüneisen parameters of these compounds.

Die thermischen Ausdehnungskoeffizienten $\alpha_a$ und $\alpha_c$ der Gitterparameter $a$ und $c$ von $ZnSiAs_2$ werden im Temperaturbereich von 30 bis 300 K mit Hilfe der Bond-Methode gemessen. Sowohl $\alpha_a$ als auch $\alpha_c$ fallen mit sinkender Temperatur und werden bei etwa 45 K negativ. Die Ergebnisse werden mit Daten zur thermischen Ausdehnung anderer Chalkopyrithalbleiter verglichen und mit Hilfe der prinzipiellen Grüneisenparameter dieser Verbindungen diskutiert.

## 1. Introduction

The thermal expansion behaviour of the ternary chalcopyrite semiconductor $ZnSiAs_2$ has been studied by X-ray diffraction measurements on powdered samples in the temperature range from 300 to about 900 K [1, 2]. In these investigations it has been found that the two principal linear thermal expansion coefficients $\alpha_a$ and $\alpha_c$ of the lattice parameters $a$ and $c$, respectively, are approximately constant in this temperature range and that always $\alpha_a > \alpha_c$. In the present work we report the results of an X-ray study of the anisotropic thermal expansion behaviour of $ZnSiAs_2$ for temperatures between 30 and 300 K. The experimental results obtained are compared with low-temperature thermal expansion measurements for some other ternary semiconductors with chalcopyrite structure and are discussed in terms of the principal Grüneisen functions of these compounds.

## 2. Experimental

In the present investigation the X-ray Bond method was used because it gives the highest precision in the determination of lattice spacings compared to other X-ray methods. A detailed description of the method can be found in [3], details of the experimental equipment used in the measurements have been reported elsewhere [4] and will not be repeated here. With the system used the temperature of the samples could be controlled with an accuracy of $\pm 0.1$ K in the temperatute range from 20 to 300 K. The relative error in the determination of the lattice spacings was always below $3 \times 10^{-5}$.

[1]) Bernhard-v. Cotta-Str. 4, DDR-9200 Freiberg, GDR.
[2]) Linnéstr. 5, DDR-7010 Leipzig, GDR.

Because of the axial symmetry of the chalcopyrite lattice the lattice parameters $a$ and $c$ vary independently with temperature. Therefore, in order to evaluate $a(T)$ and $c(T)$, at least two different lattice spacings must be measured using two types of single crystal specimens with different crystallographic orientations of the surface plane. The $ZnSiAs_2$ single crystals used in the experiments were grown by chemical transport with $ZnCl_2$ and $SiCl_4$ as transporting agents [5, 6]. The crystals had well formed and smooth (112) and (101) faces suitable for the experimental determination of the lattice spacings $d_{112}$ and $d_{101}$, respectively. Then, the lattice parameters $a$ and $c$ are given by

$$a = \left(\frac{2}{d_{101}^2} - \frac{1}{2d_{112}^2}\right)^{-1/2}, \qquad c = \left(\frac{1}{2d_{112}^2} - \frac{1}{d_{101}^2}\right)^{-1/2}. \tag{1}$$

The resulting relative error in the determination of $a$ and $c$ from $d_{112}$ and $d_{101}$ was found to be about $1 \times 10^{-4}$.

## 3. Results and Discussion

The measured temperature dependence of the lattice spacings $d_{112}$ and $d_{101}$ is shown in Fig. 1. In evaluating the lattice parameters $a$ and $c$ with (1) and the corresponding principal linear thermal expansion coefficients defined as

$$\alpha_a = \frac{1}{a(300\text{ K})}\frac{\mathrm{d}a}{\mathrm{d}T}, \qquad \alpha_c = \frac{1}{c(300\text{ K})}\frac{\mathrm{d}c}{\mathrm{d}T}, \tag{2}$$

we followed the method proposed in [7]. Accordingly, in order to determine $d_{112}$ and $d_{101}$ at a given temperature $T_i$ we fitted a polynomial of fourth order in the temperature $T$ to the data points of $d_{112}$ and $d_{101}$ in a limited temperature range around this temperature (usually $T_i \pm 25$ K). Then, using these polynomials we calculated the lattice spacings $d_{112}$ and $d_{101}$ and afterwards by means of (1) and (2) the lattice parameters $a$ and $c$ and their thermal expansion coefficients $\alpha_a$ and $\alpha_c$ at specific temperatures.

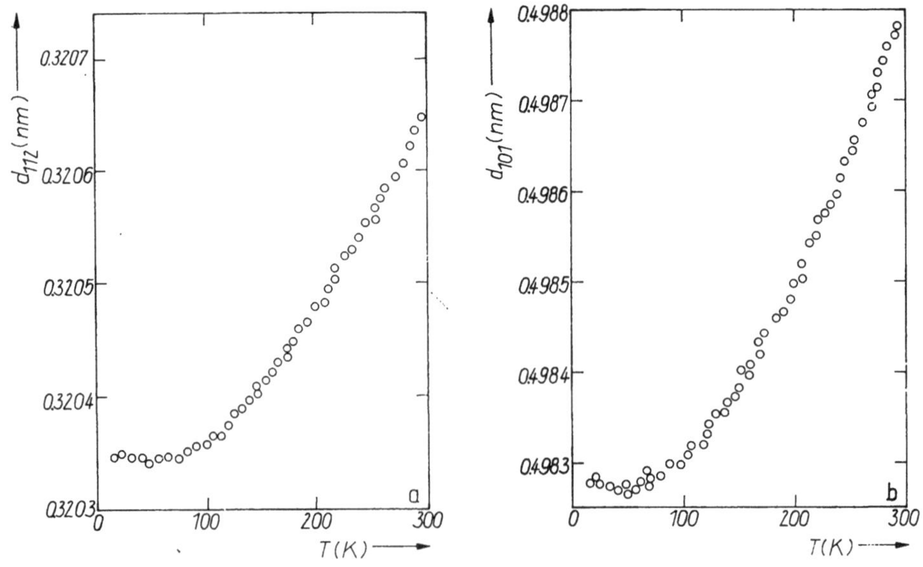


Fig. 1. Measured temperature dependence of the lattice spacings a) $d_{112}$ and b) $d_{101}$

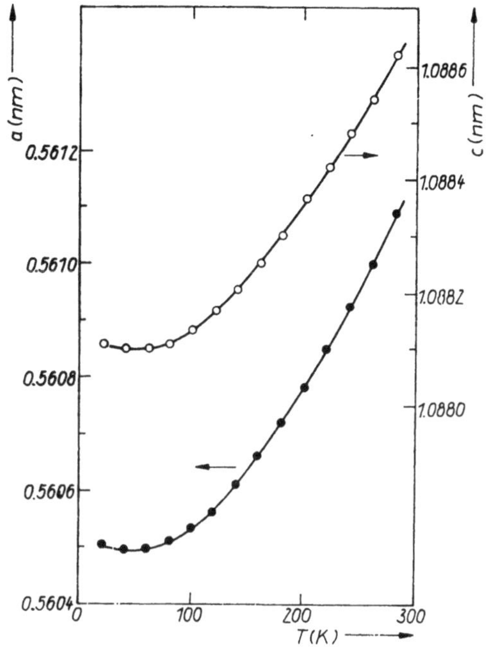


Fig. 2

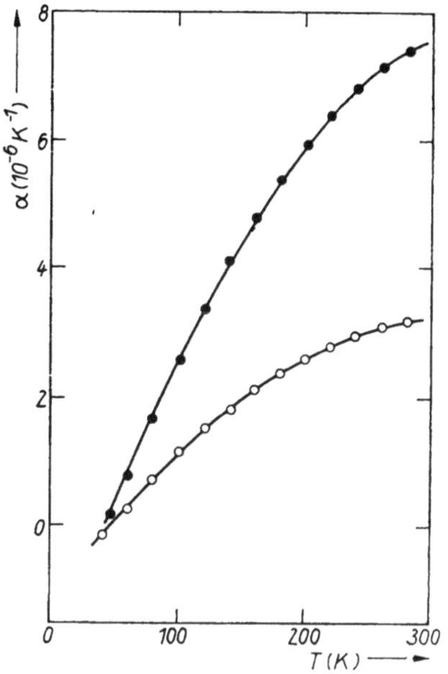


Fig. 3

Fig. 2. Temperature dependence of the lattice parameters $a$ (●) and $c$ (○)

Fig. 3. Temperature dependence of the principal linear thermal expansion coefficients $\alpha_a$ (●) and $\alpha_c$ (○)

The resulting temperature dependences of $a$, $c$, $\alpha_a$, and $\alpha_c$ are shown in Fig. 2 and 3, respectively. The room temperature values of these parameters are given in Tables 1 and 2, together with existing literature data compiled for comparison.

From Table 1 we see that within the given error limits our room temperature values for $a$ and $c$ are in very good agreement with those reported in [10, 11, 13]. The other literature values show slightly larger deviations from our values and those in [10, 11, 13], but the scatter in the data is generally small and remains in the range $a = (0.5610 \pm 0.0002)$ nm and $c = (1.0885 \pm 0.0007)$ nm. A similar scatter in the

Table 1

Room temperature lattice parameters of $ZnSiAs_2$

| $a$ (nm) | $c$ (nm) | ref. |
|---|---|---|
| 0.56117 ± 0.00006 | 1.08869 ± 0.00011 | this work |
| 0.5608 ± 0.0002 | 1.0891 ± 0.0002 | [8] |
| 0.5612 ± 0.0001 | 1.0878 ± 0.0002 | [9] |
| 0.5610 ± 0.0001 | 1.0884 ± 0.0002 | [10] |
| 0.5611 ± 0.0001 | 1.0885 ± 0.0002 | [11] |
| 0.560 ± 0.001 | 1.088 ± 0.001 | [12] |
| 0.5611 ± 0.0001 | 1.0886 ± 0.0002 | [13] |
| 0.56085 ± 0.00005 | 1.08920 ± 0.00020 | [1] |
| 0.56095 ± 0.00007 | 1.08812 ± 0.00016 | [14] |

Table 2

Room temperature values of $\alpha_a$ and $\alpha_c$ (in $10^{-6}$ K$^{-1}$) for $ZnSiAs_2$

| $\alpha_a$ | $\alpha_c$ | ref. |
|---|---|---|
| 7.6 ± 0.3 | 3.3 ± 0.2 | this work |
| 9.6 | 4.0 | [1] |
| 7.2 | 2.8 | [2] |
| 6.9 ± 1.7 | 2.3 ± 1.7 | [18] |
| 7.3 | 3.6 | [19] |

published lattice parameter values is also characteristic of other $A^{II}B^{IV}C_2^{V}$ compounds [13, 15 to 17] and is certainly caused by small systematic errors in the measurements reported by various authors.

With regard to the thermal expansion coefficients at 300 K we see from Table 2 that our values of $\alpha_a$ and $\alpha_c$ are in good agreement with those from [2] and by about 15% smaller than those given in [1]. The latter discrepancy is certainly due to relatively large errors in the lattice parameters at elevated temperatures reported in [1]. Indeed, using the experimental $a(T)$ and $c(T)$ data from [1] we estimate a relative error in the $\alpha_a$- and $\alpha_c$-values in [1] of about ±20%. We note that existing theoretical estimates of $\alpha_a$ and $\alpha_c$ for $ZnSiAs_2$ based on semi-empirical models for the thermal expansion behaviour of ternary chalcopyrite compounds [18, 19] agree with the experimental results, too (see Table 2).

Let us now consider the temperature dependence of the linear thermal expansion coefficients. As follows from Fig. 3 both $\alpha_a$ and $\alpha_c$ decrease below 300 K and become negative at temperatures below about 45 K. Below this temperature the determination of the thermal expansion coefficients becomes rather inaccurate because the change with temperature of both the lattice spacings (Fig. 1) and the lattice parameters (Fig. 2) is of the same order as the error in the determination of these parameters itself. However, the existence of a region with negative $\alpha_a$ and $\alpha_c$ below some critical temperature is clearly established.

This result is in complete analogy to the thermal expansion behaviour of the chalcopyrite compounds $ZnSiP_2$ [20], $CuInSe_2$ [21], and $CuInTe_2$ [22] for which also negative $\alpha_a$- and $\alpha_c$-values have been observed at sufficiently low temperatures. There are reasons to suppose that this type of low-temperature thermal expansion behaviour is characteristic of most, if not all, ternary $A^{II}B^{IV}C_2^{V}$ and $A^{I}B^{III}C_2^{VI}$ chalcopyrite semiconductors. Since thermal expansion is due to lattice anharmonicity it is usual to describe the thermal expansion properties of solids in terms of Grüneisen parameters [23, 24]. In the case of the uniaxial chalcopyrite lattice two independent principal Grüneisen parameters $\gamma_a$ and $\gamma_c$ must be introduced, which are related to the principal linear thermal expansion coefficients via

$$\gamma_a = \frac{V_m}{C_p} \left[(c_{11}^s + c_{12}^s)\,\alpha_a + c_{13}^s\alpha_c\right], \tag{3a}$$

$$\gamma_c = \frac{V_m}{C_p} \left[2c_{13}^s\alpha_a + c_{33}^s\alpha_c\right], \tag{3b}$$

where $V_m$ is the molar volume, $C_p$ the molar heat capacity at constant pressure, and the $c_{ij}^s$ are the adiabatic elastic stiffnesses [24]. So far, no experimental determinations of the $c_{ij}$ have been reported for $ZnSiAs_2$, $ZnSiP_2$, $CuInSe_2$, and $CuInTe_2$ which prevents a quantitative analysis of the temperature dependence of $\gamma_a$ and $\gamma_c$. However, some conclusions can be made regarding the relative magnitude and the sign of $\gamma_a$ and $\gamma_c$. Experimental elastic constant studies of the chalcopyrite semiconductors $AgGaS_2$ [25] and $CdGeAs_2$ [26] have shown that $c_{11} > c_{33} > c_{12} \approx c_{13} > 0$ in both compounds. Furthermore, it follows from theoretical estimates of the elastic constants of several $A^{II}B^{IV}C_2^{V}$ and $A^{I}B^{III}C_2^{VI}$ compounds in terms of a rigid-ion model [27] that the same relation between the elastic constants remains valid for all chalcopyrite semiconductors. Then, it is obvious from (3) and Fig. 3 that, firstly, $\gamma_a > \gamma_c$ over the whole temperature range considered and, secondly, both $\gamma_a$ and $\gamma_c$ become negative below about 40 K in $ZnSiAs_2$. Considering the low-temperature thermal expansion data of $ZnSiP_2$ [20], $CuInSe_2$ [21], and $CuInTe_2$ [22] we find the same trends in the behaviour of the principal Grüneisen parameters in these compounds, too, the only difference being the critical temperature for the transition to negative $\gamma_a$- and $\gamma_c$-values which differs from compound to compound. Now, since $\gamma_a$ and $\gamma_c$ are weighted averages of the respective principal mode Grüneisen parameters $\gamma_{ai}$ and $\gamma_{ci}$ in which the contribution of the $\gamma_{ji}$ of each normal mode to $\gamma_i$ is weighted by its contribution to the total heat capacity [24], the occurrence of negative values of $\gamma_a$ and $\gamma_c$ at low temperatures implies the existence of low-energy lattice vibrational modes with negative mode Grüneisen parameters.

At present, no direct experimental information about the mode Grüneisen parameters in $ZnSiAs_2$, $ZnSiP_2$, $CuInSe_2$, and $CuInTe_2$ is available, since the lattice vibration studies of these compounds reported in the literature ([28 to 31] and references cited therein) are limited to measurements at room temperature and ambient pressure. However, the existence of low-energy modes with negative Grüneisen parameters has been experimentally proved in several other ternary chalcopyrite compounds. Pressure-dependent Raman scattering measurements on $CdSiP_2$, $CuGaS_2$, and $AgGaS_2$ [32, 33] have shown that in these compounds the optical vibrational mode with lowest energy has a negative mode Grüneisen parameter. This mode with symmetry E in the chalcopyrite lattice is linked to the transverse acoustic mode at the point X of the Brillouin zone in the sphalerite-structure $A^{III}B^{V}$ and $A^{II}B^{VI}$ binary analogs of the ternary $A^{II}B^{IV}C_2^{V}$ and $A^{I}B^{III}C_2^{VI}$ compounds, and it has been argued in [32, 33] that the negative Grüneisen parameter of the low-energy E mode in the ternary compounds is closely related to the always negative Grüneisen parameter of the transverse acoustic mode at X in the binary analogs [34]. Furthermore, in analysing the pressure-dependent elastic constant measurements on $CdGeAs_2$ it has been established that there are long-wavelength transverse acoustic modes with negative mode Grüneisen parameters for several directions in the Brillouin zone [26]. Again, this result is in complete analogy to the situation in the sphalerite-structure $A^{III}B^{V}$ and $A^{II}B^{VI}$ binary analogs [34 to 36]. Then, if the analogy considerations used in [26, 32, 33] to explain the negative Grüneisen parameters of low-energy vibrational modes in $CdGeAs_2$, $CdSiP_2$, $CuGaS_2$, and $AgGaS_2$ in terms of corresponding mode Grüneisen parameter data in their binary analogs are extended to all ternary chalcopyrite compounds and if we take into account the behaviour of the Grüneisen parameters of the transverse acoustic modes in the binary analogs of these compounds [34 to 36], we come to the conclusion that the low-energy E mode and at least several of the transverse acoustic modes should have negative Grüneisen parameters in all chalcopyrite semiconductors. An additional argument in favour of this supposition is the experimental fact that many of the ternary chalcopyrite compounds undergo pressure-induced phase transitions to denser structures ([37 to 41] and references cited therein) which is also closely related to low-energy lattice vibrational modes with negative Grüneisen parameters. Thus, in complete analogy to the low-temperature thermal expansion behaviour observed in the $A^{III}B^{V}$ and $A^{II}B^{VI}$ semiconductors

([42] and references cited therein), it can be expected that the principal linear thermal expansion coefficients become both negative in all chalcopyrite compounds below some critical temperature. The results of the present study for $ZnSiAs_2$ and the previously reported thermal expansion data for $ZnSiP_2$ [20], $CuInSe_2$ [21], and $CuInTe_2$ [22] confirm this supposition. To get some more experimental evidence regarding the low-temperature thermal expansion behaviour of the chalcopyrite compounds it would be very interesting to extend the measurements to $AgGaS_2$ and $CuGaS_2$ for which quite extensive experimental data about mode Grüneisen parameters [32, 33] and, in the case of $AgGaS_2$, also elastic constants [25] have been reported in the literature.

*Acknowledgements*

The authors are grateful to Dr. E. Buhrig and Prof. K. Hein (Sektion Metallurgie und Werkstofftechnik, Bergakademie Freiberg) for providing the $ZnSiAs_2$ single crystals used in the present study.

## References

[1] I. I. KOZHINA, Vestnik Leningradskogo Univ., Ser. Fiz. i Khim. No. 3, 83 (1979).
[2] M. DE and S. P. SEN GUPTA, J. less-common Metals 86, 69 (1982).
[3] W. L. BOND, Acta cryst. 13, 814 (1960).
[4] P. DEUS, U. VOLAND, and H. OETTEL, Exper. Tech. Phys. 32, 433 (1984).
[5] K. WINKLER, K. HEIN, and K. LEIPNER, Kristall und Technik 9, 1223 (1974).
[6] K. WINKLER, E. BUHRIG, and K. HEIN, Freiberger Forsch.-H., Reihe B 207, 13 (1980).
[7] T. F. SMITH and G. K. WHITE, J. Phys. C 8, 2031 (1975).
[8] C. H. L. GOODMAN, Nature (London) 179, 828 (1957).
[9] K. MASUMOTO, T. ISOMURA, and W. GOTO, J. Phys. Chem. Solids 27, 1939 (1966).
[10] A. J. SPRING-THORPE and B. R. PAMPLIN, J. Crystal Growth 3/4, 313 (1968).
[11] A. A. VAIPOLIN, Fiz. tverd. Tela 15, 1430 (1973).
[12] M. D. LIND and R. W. GRANT, J. chem. Phys. 58, 357 (1973).
[13] N. A. GORYUNOVA and YU. A. VALOV, $A^{II}B^{IV}C_2^V$ Semiconductors, Izd. Sovetskoe Radio, Moscow 1974 (in Russian).
[14] A. A. VAIPOLIN, Crystal Res. Technol. 18, 415 (1983).
[15] J. L. SHAY and J. H. WERNICK, Ternary Chalcopyrite Compounds: Growth, Electronic Properties, and Applications, Pergamon Press, Oxford 1975.
[16] N. A. REDDY, P. KISTAIAH, and K. S. MURTHY, J. Phys. D 15, 2247 (1982).
[17] J. E. JAFFE and A. ZUNGER, Phys. Rev. B 29, 1882 (1984).
[18] H. NEUMANN, Kristall und Technik 15, 849 (1980).
[19] G. C. BHAR and L. K. SAMANTA, Phys. Rev. B 26, 2275 (1982).
[20] P. DEUS and H. A. SCHNEIDER, phys. stat. sol. (a) 79, 411 (1983).
[21] P. DEUS, H. NEUMANN, G. KÜHN, and B. HINTZE, phys. stat. sol. (a) 80, 205 (1983).
[22] H. NEUMANN, P. DEUS, R. D. TOMLINSON, G. KÜHN, and B. HINTZE, phys. stat. sol. (a) 84, 87 (1984).
[23] S. I. NOVIKOVA, Thermal Expansion of Solids, Izd. Nauka, Moscow 1974 (in Russian).
[24] T. H. K. BARRON, J. G. COLLINS, and G. K. WHITE, Adv. Phys. 29, 609 (1980).
[25] M. H. GRIMSDITCH and G. D. HOLAH, Phys. Rev. B 12, 4377 (1975).
[26] T. HAILING, G. A. SAUNDERS, W. A. LAMBSON, and R. S. FEIGELSON, J. Phys. C 15, 1399 (1982).
[27] A. S. POPLAVNOI, Izv. vuzov, Ser. Fiz. 29, No. 8, 5 (1986).
[28] W. H. KOSCHEL, F. SORGER, and J. BAARS, Solid State Commun. 15, 719 (1974).
[29] M. ATTORRESI, A. PINCZUK, and A. GAVINI, Proc. 12th Internat. Conf. Semicond. Phys., Teubner-Verlag, Stuttgart 1974 (p. 321).
[30] H. NEUMANN, R. D. TOMLINSON, W. KISSINGER, and N. AVGERINOS, phys. stat. sol. (b) 118 K51 (1983).

[31] G. D. Holah, A. A. Schenk, S. Perkowitz, and R. D. Tomlinson, Phys. Rev. B **23**, 6288 (1981).
[32] M. Bettini and W. B. Holzapfel, Solid State Commun. **16**, 27 (1975).
[33] C. Carlone, D. Olego, A. Jayaraman, and M. Cardona, Phys. Rev. B **22**, 3877 (1980).
[34] H.-Matsuo Kagaya and T. Soma, phys. stat. sol. (b) **142**, 97 (1987).
[35] T. Soma, J. Satoh, and H. Matsuo, Solid State Commun. **42**, 889 (1982).
[36] T. Soma, J. Satoh, and H. Matsuo, phys. stat. sol. (b) **109**, K47 (1982).
[37] H. Katzman, T. Donohue, W. F. Libby, H. L. Luo, and J. G. Huber, J. Phys. Chem. Solids **30**, 1609 (1969).
[38] H. Katzman, T. Donohue, W. F. Libby, and H. L. Luo, J. Phys. Chem. Solids **30**, 2795 (1969).
[39] A. Jayaraman, P. D. Dernier, H. M. Kasper, and R. G. Maines, High Temp.-High Pressures **9**, 97 (1977).
[40] A. Werner, H. D. Hochheimer, and A. Jayaraman, Phys. Rev. B **23**, 3836 (1981).
[41] A. Kraft, G. Kühn, and W. Möller, Z. anorg. allg. Chem. **504**, 155 (1983).
[42] H.-Matsuo Kagaya and T. Soma, phys. stat. sol. (b) **142**, 411 (1987).

*(Received February 1, 1988)*

phys. stat. sol. (a) 108, 233 (1988)

Subject classification: 68.55 and 73.60; S8

*Institute of Isotopic and Molecular Technology, Cluj-Napoca*[1])

# The Influence of the Film History on Some Electrophysical Properties of VE, CAD, and COD PbSe Films

By

RODICA M. CÂNDEA, L. P. BIRÓ, N. DĂDÂRLAT, G. BORODI, AL. DARABONT, P. FITORI, and RODICA TURCU

The influence of the "biography" of the PbSe films prepared by vacuum evaporation (VE) and chemical deposition (selenosulphate based procedure — CAD — and selenourea based procedure — COD) on their structure, phase composition, resistivity, and photoconductivity is studied. The considered processes perturbing the film–air system are ageing effects and different kinds of thermal treatments. It is evidenced that the number and the types of the active adsorption sites are the factors which differentiate between an "as-grown" film and a film with a certain history.

Es wird der Einfluß der „Biographie" von PbSe-Schichten, die mittels Vakuumaufdampfung (VE) und chemischem Niederschlag (Selenosulfat-Verfahren — CAD und Selenourea-Verfahren — COD) hergestellt werden, auf Struktur, Phasenzusammensetzung, Widerstand und Photoleitung untersucht. Die betrachteten Prozesse, die das Schicht–Luft-System stören, sind Alterungseinflüsse und verschiedene Arten thermischer Behandlung. Es werden Hinweise gefunden, daß die Zahl und Arten der aktiven Adsorptionsplätze die Faktoren sind, die zwischen einer ursprünglichen Schicht und einer Schicht mit gewisser Vergangenheit unterscheiden.

## 1. Introduction

An already commonly acknowledged fact is the high sensitivity to external factors of PbSe polycrystalline films, related to their spatial inhomogeneity [1]. The regions mostly affected by any external ambients and thermal disturbances are the intercrystalline regions that can be viewed as an amorphous matrix in which the PbSe crystallites are embedded [2, 3]. These regions being highly disordered and unstable will constitute the preferential places where the sorption–desorption processes and the forming–decomposing reactions of the new compounds will take place [4].

The sorption–desorption of active gases, as oxygen for PbSe, has a great influence on the band structure of the semiconductor and therefore on the electrophysical behaviour of the film [5 to 7]. From the point of view of these processes the film can be characterized by adsorption sites of different natures, the so-called "biographical" defects [5]. Related to this, in the present work, the influence of certain processes, as short- and long-term ageing effects, thermal pretreatments and annealings, the action of different environmental conditions and consecutive thermal cycle treatments in air on the structure, phase composition, resistivity, and photoconductivity of PbSe films is analysed.

[1]) P.O. Box 700, R-3400 Cluj-Napoca 5, Romania.

## 2. Experimental

The PbSe films used in this work were prepared either by vacuum evaporation in a standard manner at $10^{-3}$ Pa on glass supports heated at 200 °C (VE PbSe films) or by chemical deposition using the selenosulphate or selenourea based procedures (CAD or COD PbSe films) [8]. The thicknesses of films were in the range 0.7 to 1.1 μm.

A TUR-M61 diffractometer ($CuK_\alpha$ radiation) was used for the X-ray diffraction analysis.

The resistivity ($R$) versus temperature ($T$) dynamical behaviour and the photoconductivity (signal voltage $U_s$ and noise voltage $U_n$) were measured in a standard set-up [4, 9] at room temperature (RT).

## 3. Results and Discussion

The chemically deposited PbSe films, both CAD and COD, in the "as-grown" (AG) state exhibit a stable behaviour (see e.g. Fig. 1, curve a) with no significant short-term ageing effects. The $R$ and $U_s$ values measured on films stored in air at RT for months after the preparation are very close to those measured immediately after the films were deposited. However, we found that the thermal annealing at 350 °C in an open oven makes the films very sensitive to the interaction with air at RT. In Fig. 1 the variation of $R$ for typical COD films in the AG state, curve a, and annealed for 23 h, curve b, is represented versus the square root of the time elapsed from the moment when the sample was taken out from the oven. As can be seen several stages of interaction between the annealed PbSe film and air are evidenced. The existence of these stages is theoretically explained by the percolation model of Sukharev and Myasnikov [7] based on the fact that the kinetics of conductivity variations is determined by the charge exchange rate between semiconductor bands and surface levels.

The decrease of $R$ was accompanied by a decrease of $U_s$ and, surprisingly, by an increase of $U_n$. This latter effect should be attributed to the equilibrium of adsorbed molecular species. If the sample was placed in a low-pressure environment ($\leq 10^{-3}$ Pa) the value of the noise decreased by approximately one order of magnitude.

In Fig. 2 we represent the variation of the resistance versus the time elapsed from the annealing treatment for four CAD PbSe films. It can be seen that, depending on the annealing time, the samples (all being cut from the same film) exhibit different behaviour. This fact suggests that the duration of the thermal treatment might influence the nature and the number of active adsorption sites. We will refer to this later on.

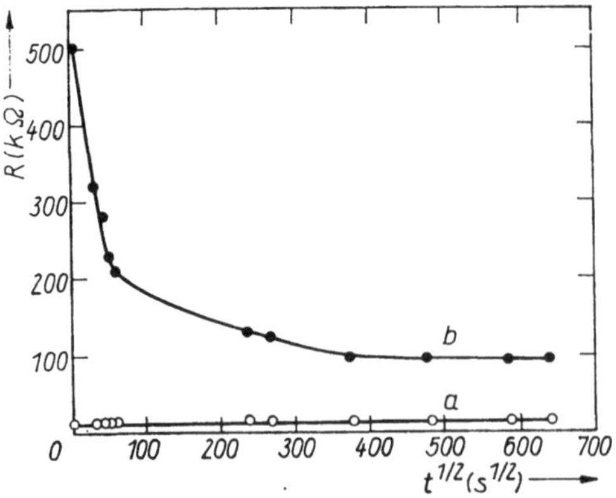


Fig. 1. The time dependence of the resistivity of a COD PbSe film (a) in the "as-grown" state and (b) subjected previously to a 23 h annealing in air at 350 °C

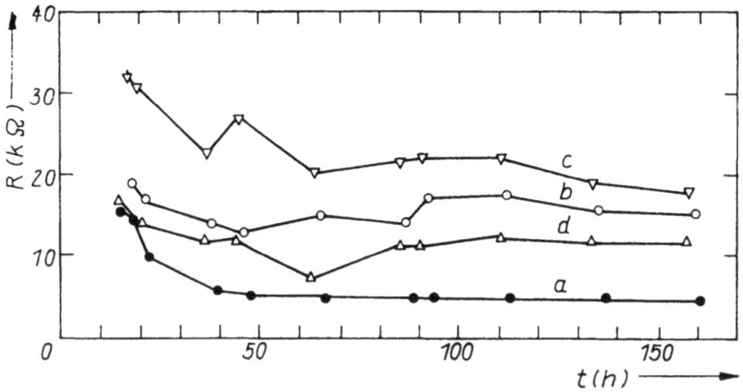


Fig. 2. The time dependence of the resistivity for CAD PbSe films after an annealing treatment in air at 350 °C for (a) $t = 21$, (b) 22, (c) 23, and (d) 24 h

Long-term ageing effects on the resistivity and photoconductivity were followed on a typical COD PbSe film pretreated for 16 h in an open oven and the results are given in Table 1. After 90 days a stable state is reached both for resistivity and photoconductivity. The deterioration of the performances is due to a decrease of $U_s$ and to an increase of $U_n$. $R$ decreases slightly. If a short baking in air at 350 °C is performed, the performance of the film in general can be completely restored.

Table 1

Long-term ageing effects on the resistivity and photoconductivity of a COD PbSe film, pretreated for 16 h in air at 350 °C

| time elapsed (d) | $R$ (kΩ) | $U_s$ (mV) | $U_n$ (μV) | $(1/R)(U_s/U_n)$ ($\Omega^{-1}$) |
|---|---|---|---|---|
| 0 | 19 | 96 | 6 | 0.8 |
| 90 | 10 | 24 | 7 | 0.34 |
| 94 | 12.4 | 36 | 10 | 0.29 |
| 96 | 12.3 | 39 | 10 | 0.31 |
| 102 | 12.7 | 40 | 10 | 0.31 |
| 139 | 11 | 37 | 12 | 0.28 |

Table 2

The temperature and pressure conditions of the thermal PTr of the VE PbSe films. The duration of the PTr was always 30 min. The phase composition was determined by X-ray diffractometry

| sample | $T_{PTr}$ (°C) | $P_{PTr}$ (Pa) | phase composition |
|---|---|---|---|
| VE-AG | | | PbSe |
| VE-a | 300 | $10^{-2}$ | PbSe + $PbSeO_3$ |
| VE-b | 250 | $10^{-1}$ | PbSe + $PbSeO_3$ + 2 PbO : $PbSeO_3$ (traces) |
| VE-c | 300 | $10^{-1}$ | PbSe + $PbSeO_3$ + 4 PbO : $PbSeO_3$ (traces) |

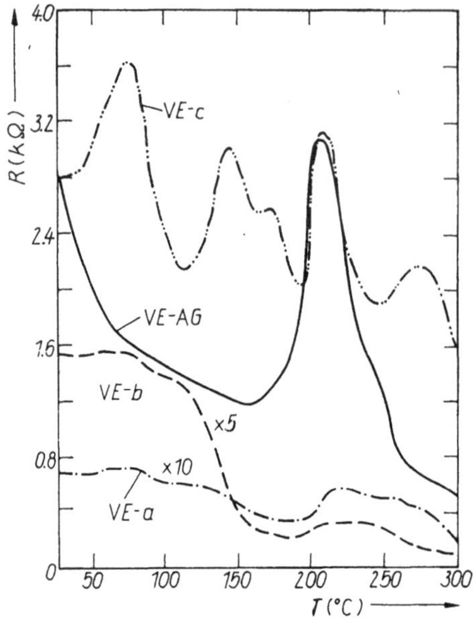


Fig. 3. The dynamical $R$–$T$ behaviour of VE PbSe films in the "as-grown" state (VE-AG) and pretreated as described in Table 2

Another way of studying the effect of the history of the film, and therefore of the "biographical" defects, was to subject samples from the same film to different pretreatments (PTr) and to compare their behaviour.

The thermal PTr of the VE PbSe films was done "in situ" immediately after the vacuum deposition of the films. The duration of the PTr was always 30 min. The temperature of the films and the pressure in the evaporation room during the PTr are given in Table 2. VE-AG is the PbSe film in the "as-grown" state. The access of air at atmospheric pressure in the evaporation room was allowed only after the films were cooled down to RT.

As was already mentioned in [4], a preferential orientation of the (100) crystallographic planes of the PbSe crystallites parallel to the plane of the film and crystallites several times larger in the [100] direction than along [110] were identified in the VE films.

In Fig. 3 the $R$–$T$ behaviour of films pretreated as described in Table 2 is presented for typical VE films.

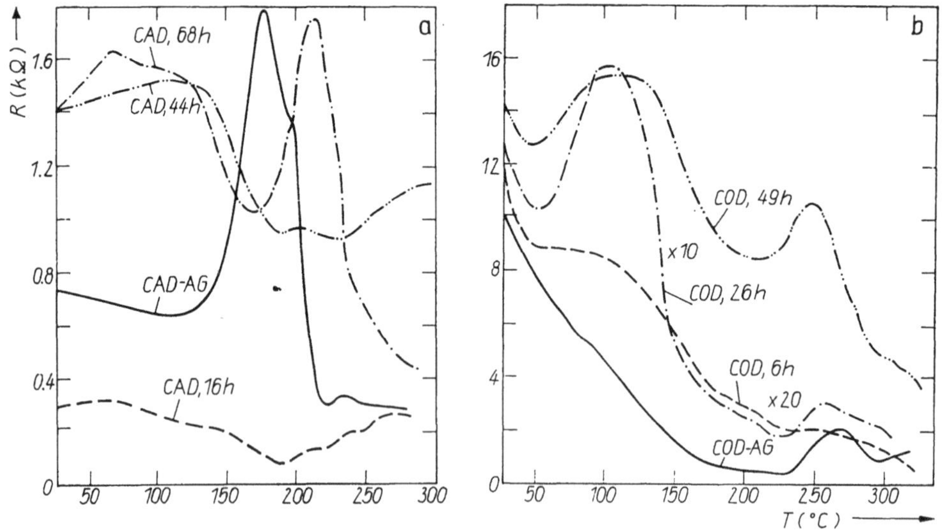


Fig. 4. The dynamical $R$–$T$ behaviour of a) CAD PbSe and b) COD PbSe films in the "as-grown" state and of films pretreated at 350 °C in air for a number of hours indicated by the figures following the CAD notation

The PTr of the CAD PbSe and COD PbSe films was performed after the deposition, in an open oven kept at 350 °C. In Fig. 4 a and b the *R–T* behaviour of typical CAD and COD PbSe films, respectively, is presented.

The presence and the origin of $PbSeO_3$ in films thermally treated in air was noticed and discussed in [4] where also the preferential orientation of the (100) planes of crystallites in CAD PbSe parallel with the plane of the film and an increase of the crystallites during the thermal treatment in the preferential plane were observed. This suggests that the penetration of oxygen is faster in the preferential plane. The preferential orientation was also found in the present work for COD films pretreated for more than 120 h and CAD films pretreated for more than 50 h.

From Fig. 3 and 4 the changes in the dynamical *R–T* behaviour due to the PTr are evident. In the AG state, as found already in [2, 4, 9] a typical "band" appears in the 180 to 270 °C range. PTr determines a more and more evident structure of "bands" in the *R–T* curves. The following final structure can be in general identified: $B_1$ in the 70 to 80 °C range, $B_2$ in the 120 to 140 °C range, $B_3$ in the 180 to 270 °C range, and $B_4$ in the 255 to 275 °C range. In the case of COD PbSe films the first two "bands" are probably superimposed and the last two appear to be shifted towards higher temperatures for some of the films, as shown in Fig. 4b. The $B_1$, $B_2$, $B_4$ "bands" appear only after the oxidation process and their amplitude and shape change as a function of the duration of the PTr. $B_3$ is characteristic of the AG state and it seems to be, at least partially, connected with the amorphous–crystalline transition of PbSe material in the intercrystalline regions [3].

No quantitative explanation for the "bands" in the *R–T* curves is tried at this moment. Qualitatively, two facts are clear. First, Petritz's theory of conductivity which models a polycrystalline film by a set of equally high and equidistant barriers cannot explain the complex *R–T* behaviour of the PbSe films. Second, the polycrystalline film should be viewed as a network of wide-spread grain barrier heights and the change of the conductivity of the film is determined by the change in the form of the grain barrier height distribution function [7].

The $O_2$ adsorption, due to the transfer of a fraction of chemisorbed O into the charged form leads to a variation of the energy barrier heights in the film conditioned by charge in built-in surface states which are in our case structural defects and impurity particles generated by the thermal pretreatment. The variation of the grain barrier heights occurring under $O_2$ adsorption leads to exponentially high barrier penetration [7] and consequently affects significantly the properties conditioned by transport effects as the *R–T* behaviour.

We also performed X-ray diffractometry on the CAD and COD films pretreated for different durations in order to find out if the change in the phase composition can explain the changes in the *R–T* behaviour in Fig. 4 a and b.

We determined the fraction $x_2$ of $PbSeO_3$ (in relative units) contained in the PbSe films treated for durations between 5 min and 120 h.

It is known that in the case of a two-component mixture one can write [10]

$$\frac{I_1}{I_{10}} = \frac{x_1 m_1}{x_2(m_1 - m_2) + m_1} \tag{1}$$

and

$$\frac{I_2}{I_{20}} = \frac{x_2 m_2}{x_2(m_2 - m_1) + m_1}, \tag{2}$$

where $I_1$, $I_2$ are the intensities of the diffraction lines for the two components, in our case PbSe and $PbSeO_3$, and $I_{10}$, $I_{20}$ are the intensities of the same diffraction lines for

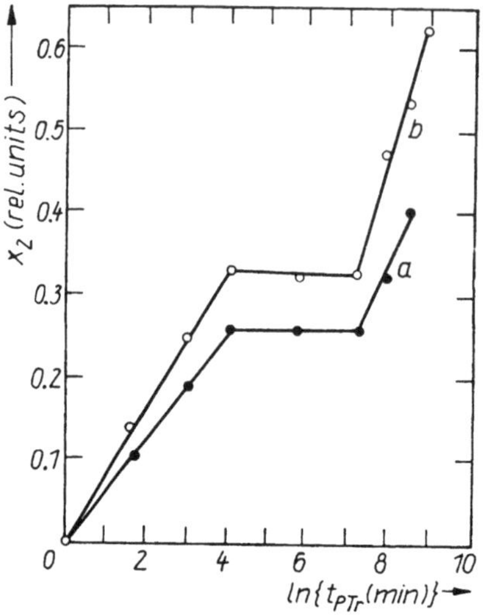


Fig. 5. The dependence of the relative amount of $PbSeO_3$ in (a) CAD and (b) COD films treated thermally in air at 350 °C, on the duration of the treatment

the components in the pure states. $x_1$ and $x_2$ represent the weight fractions and $m_1$, $m_2$ are the mass absorption coefficients for PbSe and $PbSeO_3$, respectively. From the ratio (1)/(2) and replacing $x_1 = 1 - x_2$ we obtain

$$\frac{I_{10}}{I_{20}}\frac{I_2}{I_1} = \frac{x_2^2(m_2^2 - m_1m_2) + x_2m_1m_2}{x_2^2(m_1^2 - m_1m_2) + x_2(m_1m_2 - 2m_1^2) + m_1^2} \tag{3}$$

and by measuring $I_2/I_1$ we can determine $x_2$ in relative units.

In order to minimize the effect of the preferential orientations on the value of $x_2$, instead of $I_2/I_1$ we considered the ratio of the sums of more diffraction lines, namely,

$$\frac{I_2}{I_1} = \frac{I_{3.43} + I_{3.32} + I_{3.17}}{I_{3.54} + I_{3.06}}, \tag{4}$$

where the corresponding interplanar spacings characterizing the diffraction lines are indicated.

Thus, for a certain $I_2/I_1$ ratio as defined by (4), corresponding to a film subjected to a certain PTr, $x_2$ is determined. In Fig. 5 the $x_2$ values are represented versus the duration of PTr. One can see that the amounts of $PbSeO_3$ found in the samples exhibit an exponential variation with three distinct stages. During the first hour of PTr the oxidation of the film surface occurs. This is also the period when the crystallization and the recrystallization of PbSe takes place [3]. The dependence of the average size of the PbSe crystallites being also of exponential type [3], it can be inferred that during this stage a thin $PbSeO_3$ layer is formed on the growing PbSe crystallites.

In the second stage, after the recrystallization is mostly accomplished, the diffusion of the "unoxidized" PbSe towards the surface is ended. The oxygen uptake continues, but now it is used for filling the Se vacancies which were created by a previous thermal elimination during the PTr. As Jones showed [11], the stable composition of p-type PbSe powder, after several cycles of oxidation and reducing, is $PbSe_{0.98}$. According to [2] the elimination of Se (after the crystallization of amorphous PbSe) is a slow process of diffusive nature. During the period when the Se vacancies are filled, the absorbed oxygen is used up by this process and the formation of new chemical compounds is interrupted. The oxygen filling the Se vacancies produces changes in the

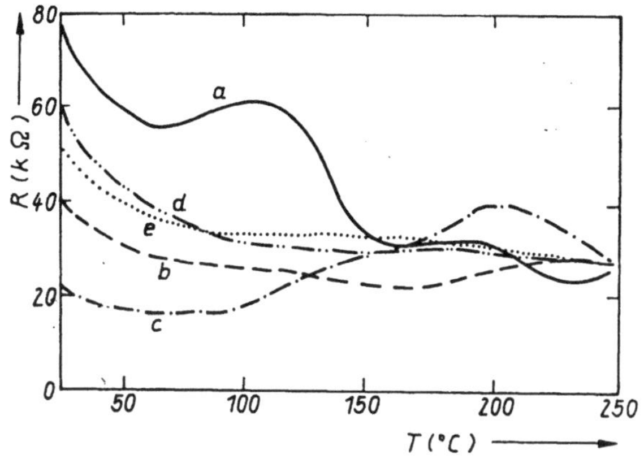


Fig. 6. The $R$–$T$ behaviour of a COD PbSe film pretreated for 16 h, during the following treatment: (a), (b) two consecutive heatings and coolings; 30 days storing; (c), (d) two consecutive heatings and coolings; (e) the 11-th heating cycle

resistance and photoconductive maxima as described also in [2, 12]. After the Se vacancies are filled the oxidation of PbSe starts again. During this third stage, as the higher slope in Fig. 5 suggests, the oxidation mechanism is probably of bulk type.

In order to find out if and when a reproducible equilibrium state is reached, the following experiment was performed. A COD PbSe film pretreated for 16 h at 350° in air was mounted in the resistivity chamber and it was subjected to a first heating (curve a in Fig. 6) and cooling cycle and then immediately to a second heating (curve b in Fig. 6) and cooling cycle. The treatment was done in air, the heating rate was 25 K/min and the cooling rate 10 K/min. Without removing the film, after 30 days, a new set of two heatings (curves c and d in Fig. 6) and coolings was performed. Curve e in Fig. 6 represents the $R$–$T$ behaviour of the film after eleven cycles and it is reproducible if the conditions of thermal cycling are not changed.

It is evident in Fig. 6 that the consecutive cycles tend to reduce the amplitude of the "bands" thus indicating that the film–atmosphere system reached an equilibrium state under the given temperature conditions. One can also notice that, after the period when the film was stored in air in the chamber, the $B_3$ "band" increased very much due to a time-allowed bulk diffusion process.

The influence of the PTr at 350 °C on the effects of a standard sensitization treatment ($T_{Tr} = 250$ °C, $T_{heating} = 42$ K/min, $T_{cooling} = 17$ K/min, $t_{Tr} = 0$) tested on COD PbSe films is depicted for a typical case in Table 3.

As can be seen the PTr reduces the effect of the Tr more and more as $t_{PTr}$ increases, because the state of the film approaches the equilibrium state.

Table 3

The changes induced by a standard sensitization treatment (Tr) in the performance of differently pretreated (PTr) COD PbSe films

| $t_{PTr}$ (h) | after PTr | | | after standard Tr | | | $\left[\frac{U_s}{RU_n}\right]^{Tr} / \left[\frac{U_s}{RU_n}\right]^{PTr}$ |
|---|---|---|---|---|---|---|---|
| | $R$ (kΩ) | $U_s$ (mV) | $U_n$ (μV) | $R$ (kΩ) | $U_s$ (mV) | $U_n$ (μV) | |
| 0 | 11.4 | 3.1 | 4.5 | 5.1 | 3 | 1.7 | 5.7 |
| 6 | 266 | 85 | 10 | 150 | 160 | 11 | 3.0 |
| 13 | 144 | 180 | 6.5 | 46 | 47 | 4 | 1.3 |
| 16 | 12 | 6.5 | 2 | 32 | 59 | 9 | 0.8 |
| 26 | 177 | 55 | 12.5 | 250 | 165 | 50 | 0.5 |

As in the case with no PTr, Tr produces an increase of $U_s$ but because of a simultaneous increase of $U_n$, the ratio $U_s/U_n$ does not always increase.

***Acknowledgement***

The authors express their gratitude to Dr. M. Ristoiu and to C. Neamtu for valuable technical assistance.

## References

[1] See for example: R. L. Petritz, Phys. Rev. **104**, 1508 (1956).
F. Briones, D. Golmayo, and C. Ortiz, Thin Solid Films 78, 385 (1981).
K. S. Bhat and V. D. Das, Phys. Rev. **32**, 6713 (1985).
[2] L. P. Biro, Al. Darabont, and P. Fitori, Europhys. Letters **4**, 691 (1987).
[3] L. P. Biro, Rodica M. Cândea, G. Borodi, Al. Darabont, P. Fitori, I. Bratu, and D. Dădârlat, Thin Solid Films, to be published.
[4] Rodica M. Cândea, Rodica Turcu, G. Borodi, and I. Bratu, phys. stat. sol. (a) **100**, 149 (1987).
[5] Th. Wolkenstein, Physico-Chimie de la Surface des Semiconducteurs, Izd. Mir, Moscow 1977.
[6] L. Šamaj, phys. stat. sol. (a) **100**, 157 (1987).
[7] V. Ya. Sukharev and I. A. Myasnikov, phys. stat. sol. (a) **100**, 277 (1987).
[8] Rodica M. Cândea, D. Dădârlat, P. Fitori, Rodica Turcu, E. Indrea, Al. Darabont, L. P. Biro, I. Bratu, and N. Aldea, St. Cerc. Fiz. **4**, 393 (1986).
[9] Rodica M. Cândea, Rodica Turcu, P. Mărgineanu, and D. Dădârlat, phys. stat. sol. (a) **96**, 337 (1986).
[10] H. P. Klug and L. E. Alexander, X-Ray Diffraction Procedure, Wiley, 1954 (p. 414).
[11] R. H. Jones, Proc. Phys. Soc. B **70**, 704 (1957).
[12] D. Dădârlat, Rodica M. Cândea, Rodica Turcu, I. Zasavitski, A. Shotov, and L. P. Biro, to be published.

*(Received January 5, 1988)*

phys. stat. sol. (a) **108**, 241 (1988)

Subject classification: 73.60; 72.15; S1.3

*Department of Physics, Faculty of Science, Tanta University*[1])

# Strain Coefficients of Electrical Resistance of Double-Layer Thin Metallic Films

By

F. KHATER and M. EL-HITI

For double-layer thin metallic films, general theoretical expressions for the longitudinal and transverse strain coefficients of electrical resistance are derived on the basis of the well-known Fuchs-Sondheimer theory. These expressions are expressed in terms of Poisson's ratio, temperature coefficient of resistance, conductivity, the electron scattering parameters at the film surfaces and interface and dimensions of the film layers, as well as, Poisson's ratio of the substrate. The variation in the strain coefficients of electrical resistance of Au/Ag double-layer film is studied numerically, as a function of the film thickness. The results show that these coefficients exhibit size effects.

Für dünne Metall-Doppelschichten werden allgemeine theoretische Ausdrücke für die longitudinalen und transversalen Spannungskoeffizienten des elektrischen Widerstands auf der Grundlage der wohlbekannten Fuchs-Sondheimer-Theorie abgeleitet. Diese Beziehungen werden mit dem Poissonverhältnis, dem Temperaturkoeffizienten des Widerstands, der Leitfähigkeit und den Elektronenstreuparametern an den Oberflächen und Grenzflächen der Schicht und den Dimensionen der Schichten sowie mit dem Poissonverhältnis des Substrats ausgedrückt. Die Änderung der Spannungskoeffizienten des Widerstands der Au/Ag-Doppelschicht wird numerisch als Funktion der Schichtdicke bestimmt. Die Ergebnisse zeigen, daß diese Koeffizienten Size-Effekte aufweisen.

## 1. Introduction

The changes in electrical resistance of thin metallic films, when they are deformed by applying stress along and perpendicular to the direction of the film surface, were studied theoretically [1 to 8] as well as experimentally [3 to 5, 9, 10]. In these studies the film was a single metallic layer deposited onto a non-metallic substrate, and the well-known Fuchs-Sondheimer (FS) [11, 12] conduction model was used. On the other hand, in the literature there are at present almost no detailed theoretical treatises on the strain coefficients of electrical resistance of double-layer thin metallic films.

The aim of this paper is to derive general expressions for the longitudinal and transverse strain coefficients of electrical resistance of double-layer thin metallic films, by taking into consideration the influence of reflection and refraction of the conduction electrons at the interface between the layers on the film resistivity in terms of FS theory. In our calculations we will not take any account of thermal expansion effects. We shall also study, numerically, the variations in the strain coefficients of electrical resistance when a thin gold overlayer is deposited onto a silver base layer.

## 2. Preliminary Definitions

For a firmly attached film-substrate system, a mechanical stress applied to the substrate in the length direction produces longitudinal $\mathrm{d}l_s/l_s$ and transverse $\mathrm{d}w_s/w_s$

[1]) Tanta, Egypt.

strains in the substrate of length $l_s$ and width $w_s$. The strains obey the relation

$$\frac{dw_s}{w_s} = -\mu_s \frac{dl_s}{l_s}, \tag{1}$$

where $\mu_s$ is the Poisson ratio of the substrate material. These strains in the substrate will be transmitted to the film producing total longitudinal $dl/l$ and transverse $dw/w$ strains given by the following equations:

$$\varepsilon_1 = \frac{dl}{l} = \varepsilon_L - \mu_f \varepsilon_T , \tag{2}$$

$$\varepsilon_2 = \frac{dw}{w} = -\mu_f \varepsilon_L + \varepsilon_T . \tag{3}$$

Hence

$$\varepsilon_L = (\varepsilon_1 + \mu_f \varepsilon_2)(1 - \mu_f^2)^{-1} , \tag{4}$$

$$\varepsilon_T = (\mu_f \varepsilon_1 + \varepsilon_2)(1 - \mu_f^2)^{-1} , \tag{5}$$

where $\mu_f$, $\varepsilon_L$, and $\varepsilon_T$ are the Poisson ratio, fictive longitudinal and transverse strains of the film, respectively.

However, when a longitudinal strain is applied to the substrate a transverse strain will occur. The transverse and longitudinal stresses are transmitted to the film but the thickness direction is stress free. Hence, under these assumptions it is now well established [13] that the elasticity formulae lead to the strain in the film thickness $t$, such that

$$\varepsilon_3 = \frac{dt}{t} = -\mu_f(\varepsilon_L + \varepsilon_T) . \tag{6}$$

Substituting (4) and (5) into (6), we have

$$\varepsilon_3 = -\mu_f \frac{1 - \mu_s}{1 - \mu_f} \frac{dl}{l} . \tag{7}$$

Since the longitudinal and transverse strain coefficients of film resistance, $\gamma_{fL}$ and $\gamma_{fT}$, are the relative changes in the film resistance $R_f$ with longitudinal and transverse mechanical strains applied to the film substrate [13], we can write

$$\gamma_{fL} = \frac{dR_f}{R_f} \Big/ \frac{dl}{l}, \tag{8}$$

$$\gamma_{fT} = \frac{dR_f}{R_f} \Big/ \frac{dw}{w}. \tag{9}$$

Similarly, the longitudinal and transverse strain coefficients of bulk conductivity, $\gamma_{0L}$ and $\gamma_{0T}$, (which can also be regarded as the strain dependence of the background scattering) are, respectively, given by [14]

$$\gamma_{0L} = -\frac{d\sigma_0}{\sigma_0} \Big/ \frac{dl}{l} = 1 + \gamma_{\lambda_0 L} , \tag{10}$$

$$\gamma_{0T} = -\frac{d\sigma_0}{\sigma_0} \Big/ \frac{dw}{w} = 1 + \gamma_{\lambda_0 T} , \tag{11}$$

where $\gamma_{\lambda_0 L}$ and $\gamma_{\lambda_0 T}$ are the longitudinal and transverse strain coefficients of the bulk mean free path $\lambda_0$. Here the indices s, f, and 0 refer to the substrate, film, and bulk materials, respectively. However, in the metals examined in the present study the variation of $\sigma_0$ with $l$ and $w$ may be entirely attributed to the alteration in the probability of the electron–phonon collisions due to the change in the amplitude of the thermal vibrations of the atoms produced by sample deformation. On this assumption $\gamma_{\lambda_0 L}$ and $\gamma_{\lambda_0 T}$ are given by $\gamma_{\lambda_0 L} \approx \gamma_{\lambda_0 T} \approx 2g(1 - 2\mu_f)$, where $g$ is the Grüneisen constant [14].

In this section we dealt with a single-layer thin metallic film, deposited onto a non-metallic substrate.

## 3. Strain Coefficients of Electrical Resistance of Double-Layer Thin Metallic Films

Let us consider a double-layer thin metallic film consisting of a base layer and a superimposed overlayer, made of two different metals, deposited onto a non-metallic substrate (see Fig. 1). The film is subjected to an external electric field parallel to its surface.

Using the well-known Fuchs-Sondheimer theory the electrical conductivity of the double-layer thin metallic film $\sigma_f$ was calculated [15], where the scattering of the conduction electrons at the interface between the layers was taken into consideration:

$$\sigma_f = \frac{t_1}{t_1 + t_2} \sigma_{01} F_1(K, P, Q) + \frac{t_2}{t_1 + t_2} \sigma_{02} F_2(K, P, Q) , \tag{12}$$

where the function $F_1(K, P, Q)$ is defined as follows:

$$F_1(K, P, Q) = 1 - \frac{3}{4K_1} \int\limits_0^1 dx_1 (x_1 - x_1^3) (1 - A) \times \times \{(1 - P_{10}) + D^{-1}(1 + P_{10}A) (X_1 + CQY_1)\} . \tag{13}$$

Here we have used the notations

$$X_1 = 1 - P_{12} + AP_{12}(1 - P_{10}) - B^2P_{20}(P_{21}(1 - P_{12}) + Q^2) + + AB^2P_{20}(Q^2 - P_{12}P_{21}) (1 - P_{10}) , \tag{14}$$

$$Y_1 = -(1 - B) (1 + BP_{20}) , \tag{15}$$

$$D = 1 - A^2P_{10}P_{12} - B^2P_{20}P_{21} + A^2B^2P_{10}P_{20}(P_{12}P_{21} - Q^2) , \tag{16}$$

$$A = \exp(-K_1/x_1) , \qquad B = \exp(-K_2/x_2) , \qquad C = \frac{\lambda_{02}m_1v_{1F}}{\lambda_{01}m_2v_{2F}} , \tag{17}$$

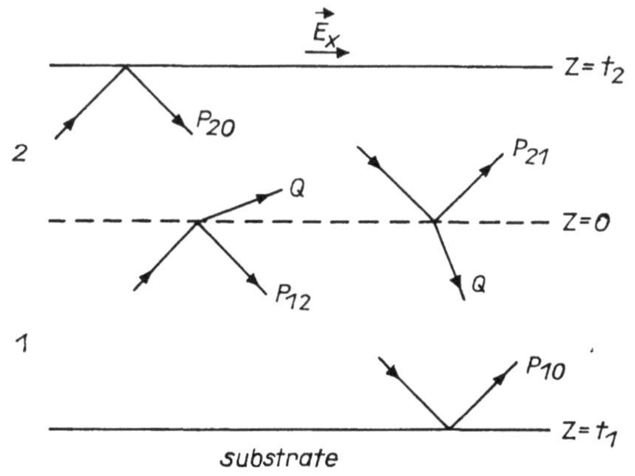


Fig. 1. The geometry of a double-layer thin metallic film deposited onto a non-metallic substrate. The arrows represent the reflection and refraction of the impinging electrons on the surfaces and interface

where

$$x_i = \cos\theta_i \quad \text{and} \quad K_i = \frac{t_i}{\lambda_{0i}}; \qquad i = 1, 2 .$$

Similarly we can define the function $F_2(K, P, Q)$ by changing in the integral (13) the index 1(2) by the index 2(1), the symbol $A(B)$ by the symbol $B(A)$, and $C$ by $C^{-1}$.

We have taken into consideration that the layers may be of different metals, so we consider different effective masses $m_1$, $m_2$, Fermi velocities $v_{1F}$, $v_{2F}$, and bulk mean free paths $\lambda_{01}, \lambda_{02}$. The parameters $P_{10}$, $P_{20}$, $P_{12}$, and $P_{21}$ are called the Fuchs specularity parameters, they characterize the probability of the reflection of the conduction electrons from the surfaces and the interface. Because of the roughness of the interface, $P_{12}$ need not be equal to $P_{21}$. The parameter $Q$ characterizes the probability that an electron is refracted on the interface, according to the law of refraction, where $Q_{12} = Q_{21} = Q$.

From (12), if the lengths and widths of the layers and substrate are equal, i.e. $l_1 = l_2 = l$ and $w_1 = w_2 = w$, we have

$$\frac{1}{R_f} = \frac{wt_1\sigma_{01}F_1(K, P, Q)}{l} + \frac{wt_2\sigma_{02}F_2(K, P, Q)}{l} \tag{18}$$

Logarithmic differentiation of (18) gives

$$-\mathrm{d}\ln R_f = \mathrm{d}\ln\left(\frac{wt_1\sigma_{01}F_1(K, P, Q)}{l}\right) + \mathrm{d}\ln\left(1 + \frac{t_2\sigma_{02}F_2(K, P, Q)}{t_1\sigma_{01}F_1(K, P, Q)}\right). \tag{19}$$

This leads to

$$\frac{\mathrm{d}R_f}{R_f} = -\left(\frac{\mathrm{d}t_1}{t_1} + \frac{\mathrm{d}\sigma_{01}}{\sigma_{01}} + \frac{\mathrm{d}F_1}{F_1}\right)\left(\frac{t_1\sigma_{01}F_1}{t_1\sigma_{01}F_1 + t_2\sigma_{02}F_2}\right) - \\ -\left(\frac{\mathrm{d}t_2}{t_2} + \frac{\mathrm{d}\sigma_{02}}{\sigma_{02}} + \frac{\mathrm{d}F_2}{F_2}\right)\left(\frac{t_2\sigma_{02}F_2}{t_1\sigma_{01}F_1 + t_2\sigma_{02}F_2}\right) - \left(\frac{\mathrm{d}w}{w} - \frac{\mathrm{d}l}{l}\right), \tag{20}$$

where $F_1$ and $F_2$ represent the functions $F_1(K, P, Q)$ and $F_2(K, P, Q)$, respectively.

When the two layers of the film are firmly attached to each other and to the substrate, we can suppose that the same mechanical state of the two layers is obtained under the action of the resulting strains $\mathrm{d}l/l$ and $\mathrm{d}w/w$ due to the substrate. Therefore,

$$\frac{\mathrm{d}t_1}{t_1} = -\mu_1 \frac{1 - \mu_s}{1 - \mu_1} \frac{\mathrm{d}l}{l}, \tag{21}$$

$$\frac{\mathrm{d}t_2}{t_2} = -\mu_2 \frac{1 - \mu_s}{1 - \mu_2} \frac{\mathrm{d}l}{l}. \tag{22}$$

### *3.1 Longitudinal strain coefficient of electrical resistance*

Dividing both sides of (20) on the longitudinal mechanical strain $\mathrm{d}l/l$ we obtain

$$\frac{\mathrm{d}R_f}{R_f}\Big/\frac{\mathrm{d}l}{l} = -\left(\frac{\mathrm{d}t_1}{t_1}\Big/\frac{\mathrm{d}l}{l} + \frac{\mathrm{d}\sigma_{01}}{\sigma_{01}}\Big/\frac{\mathrm{d}l}{l} + \frac{\mathrm{d}F_1}{F_1}\Big/\frac{\mathrm{d}l}{l}\right)\left(\frac{t_1\sigma_{01}F_1}{t_1\sigma_{01}F_1 + t_2\sigma_{02}F_2}\right) - \\ -\left(\frac{\mathrm{d}t_2}{t_2}\Big/\frac{\mathrm{d}l}{l} + \frac{\mathrm{d}\sigma_{02}}{\sigma_{02}}\Big/\frac{\mathrm{d}l}{l} + \frac{\mathrm{d}F_2}{F_2}\Big/\frac{\mathrm{d}l}{l}\right)\left(\frac{t_2\sigma_{02}F_2}{t_1\sigma_{01}F_1 + t_2\sigma_{02}F_2}\right) - \frac{\mathrm{d}w}{w}\Big/\frac{\mathrm{d}l}{l} + 1 . \tag{23}$$

Using (1), (8), (10), (11), (21), and (22), equation (23) can be rewritten in the form

$$\gamma_{fL} = \left(\gamma_{01L} + \mu_1 \frac{1-\mu_s}{1-\mu_1} - \frac{d\ln F_1}{d\ln l}\right)\left(\frac{t_1\sigma_{01}F_1}{t_1\sigma_{01}F_1 + t_2\sigma_{02}F_2}\right) + $$
$$+ \left(\gamma_{02L} + \mu_2 \frac{1-\mu_s}{1-\mu_2} - \frac{d\ln F_2}{d\ln l}\right)\left(\frac{t_2\sigma_{02}F_2}{t_1\sigma_{01}F_1 + t_2\sigma_{02}F_2}\right) + \mu_s + 1\,. \tag{24}$$

It is generally assumed [2, 3, 6, 16 to 19] that the change in the electronic specularity parameters $P$'s and the electronic refraction parameter $Q$ with strain or temperature is negligible. Consequently the derivatives $d\ln F_1/d\ln l$ and $d\ln F_2/d\ln l$ are

$$\frac{d\ln F_1}{d\ln l} = \frac{d\ln F_1}{d\ln K_1}\left(\gamma_{\lambda_{01}L} - \mu_1 \frac{1-\mu_s}{1-\mu_1}\right) + \frac{d\ln F_1}{d\ln K_1}\left(\gamma_{\lambda_{02}L} - \mu_2 \frac{1-\mu_s}{1-\mu_2}\right)\frac{d\ln K_1}{d\ln K_2}, \tag{25}$$

$$\frac{d\ln F_2}{d\ln l} = \frac{d\ln F_2}{d\ln K_2}\left(\gamma_{\lambda_{02}L} - \mu_2 \frac{1-\mu_s}{1-\mu_2}\right) + \frac{d\ln F_2}{d\ln K_2}\left(\gamma_{\lambda_{01}L} - \mu_1 \frac{1-\mu_s}{1-\mu_1}\right)\frac{d\ln K_2}{d\ln K_1}. \tag{26}$$

Assuming that the rigid band model of metals is valid and that the number of conduction electrons per unit volume is temperature independent in the experimental range, we obtain $\varrho_0\lambda_0 = \text{const.}$

Hence, the temperature coefficient of resistance $\beta_0$ of bulk material is given by

$$\beta_0 = -\frac{1}{\lambda_0}\frac{d\lambda_0}{dT}. \tag{27}$$

Using this definition, it can be easily found that

$$\frac{d\ln K_1}{d\ln K_2} = \frac{\beta_{01}}{\beta_{02}}. \tag{28}$$

Substituting (25), (26), and (28) into (24), then (24) can be rewritten in the form

$$\gamma_{fL} = \frac{t_1\sigma_{01}F_1}{t_1\sigma_{01}F_1 + t_2\sigma_{02}F_2}\left[\left(1 + \gamma_{\lambda_{01}L} + \mu_1\frac{1-\mu_s}{1-\mu_1}\right) + \right.$$
$$\left. + \frac{d\ln F_1}{d\ln K_1}\left\{\left(-\gamma_{\lambda_{01}L} + \mu_1\frac{1-\mu_s}{1-\mu_1}\right) + \left(-\gamma_{\lambda_{02}L} + \mu_2\frac{1-\mu_s}{1-\mu_2}\right)\frac{\beta_{01}}{\beta_{02}}\right\}\right] +$$
$$+ \frac{t_2\sigma_{02}F_2}{t_1\sigma_{01}F_1 + t_2\sigma_{02}F_2}\left[\left(1 + \gamma_{\lambda_{02}L} + \mu_2\frac{1-\mu_s}{1-\mu_2}\right) + \right.$$
$$\left. + \frac{d\ln F_2}{d\ln K_2}\left\{\left(-\gamma_{\lambda_{02}L} + \mu_2\frac{1-\mu_s}{1-\mu_2}\right) + \left(-\gamma_{\lambda_{01}L} + \mu_1\frac{1-\mu_s}{1-\mu_1}\right)\frac{\beta_{02}}{\beta_{01}}\right\}\right] + \mu_s + 1\,. \tag{29}$$

Based on the above assumptions, the temperature coefficient of electrical resistance of a doublelayer thin metallic film is given by [20]

$$\beta_f = \frac{t_1\sigma_{01}F_1\beta_{01}}{t_1\sigma_{01}F_1 + t_2\sigma_{02}F_2}\left(1 - \frac{d\ln F_1}{d\ln K_1}\right) + \frac{t_2\sigma_{02}F_2\beta_{02}}{t_1\sigma_{01}F_1 + t_2\sigma_{02}F_2}\left(1 - \frac{d\ln F_2}{d\ln K_2}\right). \tag{30}$$

This yields

$$\frac{\mathrm{d}\ln F_i}{\mathrm{d}\ln K_i} = 1 - \frac{\beta_\mathrm{f}}{\beta_{0i}}; \qquad i = 1, 2\,. \tag{31}$$

By considering (27) we obtain a general expression for the longitudinal strain coefficient of electrical resistance of a double-layer thin metallic film:

$$\gamma_{\mathrm{fL}} = \frac{t_1\sigma_{01}F_1}{t_1\sigma_{01}F_1 + t_2\sigma_{02}F_2}\left[\left(1 + \gamma_{\lambda_{01}\mathrm{L}} + \mu_1\frac{1-\mu_\mathrm{s}}{1-\mu_1}\right) + \right.$$
$$+\left(1 - \frac{\beta_\mathrm{f}}{\beta_{01}}\right)\left.\left\{\left(-\gamma_{\lambda_{01}\mathrm{L}} + \mu_1\frac{1-\mu_\mathrm{s}}{1-\mu_1}\right) + \left(-\gamma_{\lambda_{02}\mathrm{L}} + \mu_2\frac{1-\mu_\mathrm{s}}{1-\mu_2}\right)\frac{\beta_{01}}{\beta_{02}}\right\}\right] +$$
$$+ \frac{t_2\sigma_{02}F_2}{t_1\sigma_{01}F_1 + t_2\sigma_{02}F_2}\left[\left(1 + \gamma_{\lambda_{02}\mathrm{L}} + \mu_2\frac{1-\mu_\mathrm{s}}{1-\mu_2}\right) + \left(1 - \frac{\beta_\mathrm{f}}{\beta_{02}}\right)\times\right.$$
$$\times\left.\left\{\left(-\gamma_{\lambda_{02}\mathrm{L}} + \mu_2\frac{1-\mu_\mathrm{s}}{1-\mu_2}\right) + \left(-\gamma_{\lambda_{01}\mathrm{L}} + \mu_1\frac{1-\mu_\mathrm{s}}{1-\mu_1}\right)\frac{\beta_{02}}{\beta_{01}}\right\}\right] + \mu_\mathrm{s} + 1\,. \tag{32}$$

This expression may be reduced to

$$\gamma_{\mathrm{fL}} = \frac{t_1\sigma_{01}F_1}{t_1\sigma_{01}F_1 + t_2\sigma_{02}F_2}\left[\gamma_{\mathrm{f1L}} + \left(\frac{\beta_\mathrm{f}}{\beta_{01}} - 1\right)\left(\eta_2 - \mu_2\frac{1-\mu_\mathrm{s}}{1-\mu_2}\right)\frac{\beta_{01}}{\beta_{02}}\right] +$$
$$+ \frac{t_2\sigma_{02}F_2}{t_2\sigma_{02}F_2 + t_1\sigma_{01}F_1}\left[\gamma_{\mathrm{f2L}} + \left(\frac{\beta_\mathrm{f}}{\beta_{02}} - 1\right)\left(\eta_1 - \mu_1\frac{1-\mu_\mathrm{s}}{1-\mu_1}\right)\frac{\beta_{02}}{\beta_{01}}\right]. \tag{33}$$

### *3.2 Transverse strain coefficient of electrical resistance*

Similarly, when a transverse strain $\mathrm{d}w/w$ is applied, (20) leads to

$$\frac{\mathrm{d}R_\mathrm{f}}{R_\mathrm{f}}\Big/\frac{\mathrm{d}w}{w} = -\left(\frac{\mathrm{d}t_1}{t_1}\Big/\frac{\mathrm{d}w}{w} + \frac{\mathrm{d}\sigma_{01}}{\sigma_{01}}\Big/\frac{\mathrm{d}w}{w} + \frac{\mathrm{d}F_1}{F_1}\Big/\frac{\mathrm{d}w}{w}\right)\frac{t_1\sigma_{01}F_1}{t_1\sigma_{01}F_1 + t_2\sigma_{02}F_2} -$$
$$-\left(\frac{\mathrm{d}t_2}{t_2}\Big/\frac{\mathrm{d}w}{w} + \frac{\mathrm{d}\sigma_{02}}{\sigma_{02}}\Big/\frac{\mathrm{d}w}{w} + \frac{\mathrm{d}F_2}{F_2}\Big/\frac{\mathrm{d}w}{w}\right)\frac{t_2\sigma_{02}F_2}{t_1\sigma_{01}F_1 + t_2\sigma_{02}F_2} - 1 + \frac{\mathrm{d}l}{l}\Big/\frac{\mathrm{d}w}{w} \tag{34}$$

or

$$\gamma_{\mathrm{fT}} = \left(\gamma_{01\mathrm{T}} + \mu_1\frac{1-\mu_\mathrm{s}}{1-\mu_1} - \frac{\mathrm{d}\ln F_1}{\mathrm{d}\ln w}\right)\frac{t_1\sigma_{01}F_1}{t_1\sigma_{01}F_1 + t_2\sigma_{02}F_2} +$$
$$+\left(\gamma_{02\mathrm{T}} + \mu_2\frac{1-\mu_\mathrm{s}}{1-\mu_2} - \frac{\mathrm{d}\ln F_2}{\mathrm{d}\ln w}\right)\frac{t_2\sigma_{02}F_2}{t_1\sigma_{01}F_1 + t_2\sigma_{02}F_2} - 1 - \mu_\mathrm{s}\,. \tag{35}$$

Following the same procedure as in Section 3.1 the general expression for the transverse strain coefficient of electrical resistance of a double-layer thin metallic film can be expressed as

$$\gamma_{fT} = \frac{t_1\sigma_{01}F_1}{t_1\sigma_{01}F_1 + t_2\sigma_{02}F_2}\left[\left(1 + \gamma_{\lambda_{01}T} + \mu_1\frac{1-\mu_s}{1-\mu_1}\right) + \left(1 - \frac{\beta_f}{\beta_{01}}\right) \times \right.$$
$$\left. \times \left\{\left(-\gamma_{\lambda_{01}T} + \mu_1\frac{1-\mu_s}{1-\mu_1}\right) + \left(-\gamma_{\lambda_{02}T} + \mu_2\frac{1-\mu_s}{1-\mu_2}\right)\frac{\beta_{01}}{\beta_{02}}\right\}\right] +$$
$$+ \frac{t_2\sigma_{02}F_2}{t_1\sigma_{01}F_1 + t_2\sigma_{02}F_2}\left[\left(1 + \gamma_{\lambda_{02}T} + \mu_2\frac{1-\mu_s}{1-\mu_2}\right) + \left(1 - \frac{\beta_f}{\beta_{02}}\right) \times \right.$$
$$\left. \times \left\{\left(-\gamma_{\lambda_{02}T} + \mu_2\frac{1-\mu_s}{1-\mu_2}\right) + \left(-\gamma_{\lambda_{01}T} + \mu_1\frac{1-\mu_s}{1-\mu_1}\right)\frac{\beta_{02}}{\beta_{01}}\right\}\right] - \mu_s - 1\,, \quad (36)$$

which reduces to

$$\gamma_{fT} = \frac{t_1\sigma_{01}F_1}{t_1\sigma_{01}F_1 + t_2\sigma_{02}F_2}\left[\gamma_{f1T} + \left(\frac{\beta_f}{\beta_{01}} - 1\right)\left(\eta_2 - \mu_2\frac{1-\mu_s}{1-\mu_2}\right)\frac{\beta_{01}}{\beta_{02}}\right] +$$
$$+ \frac{t_2\sigma_{02}F_2}{t_2\sigma_{02}F_2 + t_1\sigma_{01}F_1}\left[\gamma_{f2T} + \left(\frac{\beta_f}{\beta_{02}} - 1\right)\left(\eta_1 - \mu_1\frac{1-\mu_s}{1-\mu_1}\right)\frac{\beta_{02}}{\beta_{01}}\right]. \quad (37)$$

Here $\gamma_{f1L}$, $\gamma_{f2L}$, $\gamma_{f1T}$, and $\gamma_{f2T}$ are the longitudinal and transverse strain coefficients of electrical resistance of single-layer thin metallic films of metals labeled by 1 and 2, respectively. Moreover, $\eta_1 \equiv \gamma_{\lambda_{01}L} = \gamma_{\lambda_{01}T}$ and $\eta_2 \equiv \gamma_{\lambda_{02}L} = \gamma_{\lambda_{02}T}$.

## 4. Numerical Results

We shall consider a Au/Ag bilayer model which can be viewed as a rough physical realization of the model proposed. For this model the potential energy step at the interface between the layers can be neglected. Therefore, the internal interface between the layers has no great influence on the scattering of the conduction electrons, i.e. the probability of diffuse scattering at this interface becomes very small. It means that the specularity parameters $P_{12}$ and $P_{21}$ are both equal to zero and the conduction electrons are totally transmitted through the interface, i.e. $Q$ equals unity. These values are reasonable, also, because Au and Ag have the same crystal structure. Moreover, for this model the parameters $P_{10}$ and $P_{20}$ are supposed to be totally specular and totally diffuse, i.e. $P_{10}$ equals unity and $P_{20}$ equals zero, respectively.

Making use of the fact that the Fermi velocity, electron concentration, and effective mass are nearly the same for both Au and Ag [21], the parameter $c = \lambda_{02}/\lambda_{01}$ can take the value 0.75.

For numerical calculations we need to know the values of some constants which can be determined (or estimated) from tables [13], namely,

$$\gamma_{01L} = 2.15\,, \qquad \eta_1 = 1.15\,, \qquad \mu_1 = 0.38\,, \quad \text{and} \quad \mu_2 = 0.42\,.$$

Since $\gamma_{0T}/\gamma_{0L} = 0.7$ [10], therefore, $\gamma_{01T} = 0.608$. Taking into account that the temperature coefficients of electrical resistance of Au and Ag are nearly the same [22], i.e. $\beta_{01} = \beta_{02}$, we can write $\gamma_{01L} \equiv \gamma_{02L}$ and $\gamma_{01T} \equiv \gamma_{02T}$. Using the above parameters, the dependences of the ratios of strain coefficients of electrical resistance $\gamma_{fL}/\gamma_{01L}$ and $\gamma_{fT}/\gamma_{01T}$ for unsupported Au/Ag bilayer model on the reduced thickness $K_1$ for some values of $K_2$ ($K_2 = 0$, 0.005, 0.05) were calculated using a computer. The results are shown in Fig. 2.

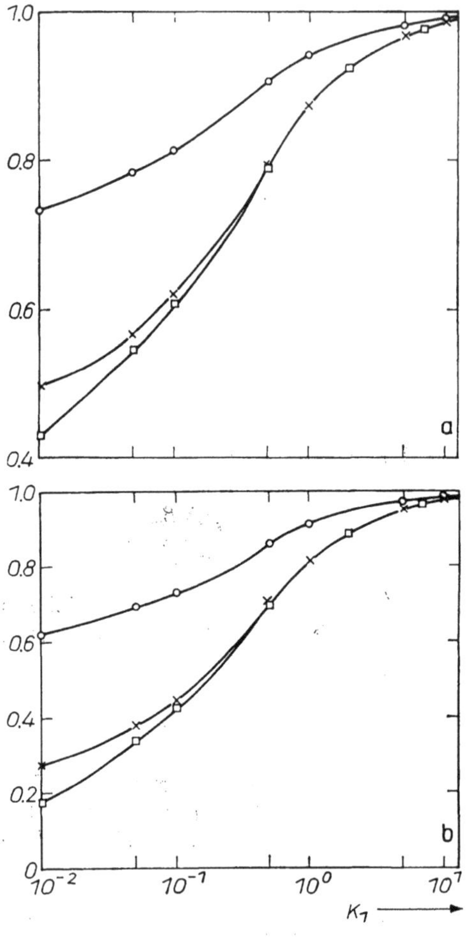


Fig. 2. The reduced a) longitudinal $\gamma_{fL}/\gamma_{01L}$ and b) transverse $\gamma_{fT}/\gamma_{01T}$ strain coefficients of electrical resistance as functions of the reduced thickness $K_1$ for various values of $K_2$ (○ $K_2 = 0$, × 0.005, □ 0.050)

The curves exhibit several features:

(i) An overlayer of Au onto a base layer of Ag will decrease its strain coefficient of electrical resistance. This decrease is markedly dependent on the thickness of the overlayer.

(ii) For large values of the reduced thickness, $K_1 > 10$, of the Ag base layer the size effect vanishes, i.e. the strain coefficients of electrical resistance ratio are independent of the reduced thicknesses $K_1$ and $K_2$.

(iii) The strain coefficients of electrical resistance of the Au/Ag bilayer model are dependent on the Ag overlayer thickness for reduced thicknesses <0.5 of the Ag base layer. On the other hand, for reduced thicknesses >0.5 of the Ag base layer the Au overlayer thickness has no great influence on the strain coefficients of electrical resistance of the Au/Ag bilayer model.

## 5. Conclusion

From the above analysis we have obtained general expressions for the strain coefficients of electrical resistance of double-layer thin metallic films. In addition numerical calculations are performed for the Au/Ag bilayer model.

## References


[1] Z. H. Meiksin and R. A. Hudzinski, J. appl. Phys. **38**, 4490 (1967).
[2] B. S. Verma and S. K. Sharma, Thin Solid Films **5**, R33, R47 (1970).
[3] C. Reale, Czech. J. Phys. **B21**, 662 (1971).
[4] B. S. Verma and G. C. Jain, Thin Solid Films **11**, 27 (1972).
[5] G. R. Witt, Thin Solid Films **22**, 133 (1974).
[6] C. R. Tellier, Thin Solid Films **48**, L9 (1978).
[7] B. K. Sharma and R. Srivastava, Thin Solid Films **113**, L19 (1984).
[8] C. R. Pichard, C. R. Tellier, and A. J. Tosser, J. Phys. D **13**, 1325 (1980).
[9] R. L. Parker and A. Krinsky, J. appl. Phys. **34**, 2700 (1963).
[10] B. S.Verma and H. J. Juretschke, J. appl. Phys. **41**, 4732 (1970).
[11] K. Fuchs, Proc. Cambridge Philos. Soc. **34**, 100 (1938).
[12] E. H. Sondheimer, Adv. Phys. **1**, 1 (1952).
[13] C. R. Tellier and A. J. Tosser, Size Effects in Thin Films, Chapt. 4, Elsevier, Amsterdam 1982.
[14] G. C. Kuczyhski, Phys. Rev. **94**, 61 (1954).
[15] F. Khater, Acta Phys. Slov. **33**, 43 (1983).
[16] B. S. Verma, Thin Solid Films **7**, 259 (1971).
[17] F. Warkusz, Thin Solid Films **41**, 261 (1977).
[18] F. Warkusz, Thin Solid Films **55**, 275 (1978).
[19] C. R. Tellier, C. R. Pichard, and A. J. Tosser, J. Mater. Sci. **16**, 2281 (1981).
[20] F. Khater and M. El-Hiti, to be published.
[21] C. Kittel, Introduction to Solid State Physics, 5th ed., Wiley, New York 1976 (p. 154).
[22] G. A. G. Bennet, Electricity and Modern Physics, 2nd ed., Edward Arnold, London 1983 (p. 23).


*(Received December 7, 1987)*

phys. stat. sol. (a) 108, 251 (1988)

Subject classification: 79.70; 68.20; S1; S4

*Department of Experimental Physics, Bashkir State University, Ufa*[1])

# Microroughness Field Emission Electron Energy Distribution and Electron-Phonon Interaction Spectroscopy Near the Metal Surface

By

R. Z. BAKHTIZIN and V. G. VALEEV

Field electron emission is considered from microtips on the metal surface. The possibility is shown of field emission spectroscopy of electron-phonon interaction near the emitting surface and the influence is studied of an increase in the surface potential barrier transparency by application of a weak high-frequency field altering the character of the field electron energy distribution in such a system.

Рассмотрен процесс автоэлектронной эмиссии из микронеоднородностей на поверхности металла. Показана возможность автоэмиссионной спектроскопии электрон-фононного взаимодействия вблизи эмиттируюшей поверхности. Изучено влияние роста прозрачности приповерхностного потенциального барьера под действием слабого высокочастотного поля на характер энергетического распределения автоэлектронов а такой системе.

## 1. Introduction

The present paper considers the energy distribution of the electrons field emitted from built-up tips. It has been shown by a number of authors [1 to 3] that in systems of this kind the current $J$, up to a value of $J \approx 100\ \mu A$, is determined totally by the emission from microtips, the characteristic size $l$ of which is $10^{-8}$ to $10^{-7}$ cm. It seems possible to single out microtips with $l$ shorter than the inelastic collision length $l_i$ (e.g. for tungsten at $T = 300$ K $l \ll 30$ nm).

Depending on the correlation of the microtip size and lengths of elastic ($l_i$) and inelastic ($l_e$) electron scattering, three modes of emission may be realized: ballistic ($l \ll l_e, l_i$), diffusional ($l_i \ll l \ll l_e$), and thermal ones ($l \gg l_e, l_i$). In the latter case the principal solution may be obtained by means of the hydrodynamic approximation [4]. Otherwise, in the case under consideration, the analysis should be based upon the direct solution of the kinetic equation for the electron distribution function. For metal microcontacts such a programme in the single-particle approximation is realized in [5]. In this case the energy scattering length $\lambda_e = (l_i l_e)^{1/2} \gg l$, thus heat discharges in a region of volume $\lambda_e^3$, exceeding considerably the region of formation of the microtip tunneling resistance ($\sim l^3$). According to [2], it is this particular phenomenon that causes the microtip thermal stability at greater values of the current density.

## 2. Theoretical Consideration

### 2.1 *The case of d-c field emission*

Proceeding from [5], let us build in the clean limit a microscopic theory of the energy distribution of the electron field emitted from microtips satisfying of the electron field

[1]) Ul. Frunze 32, SU-450074 Ufa, USSR.

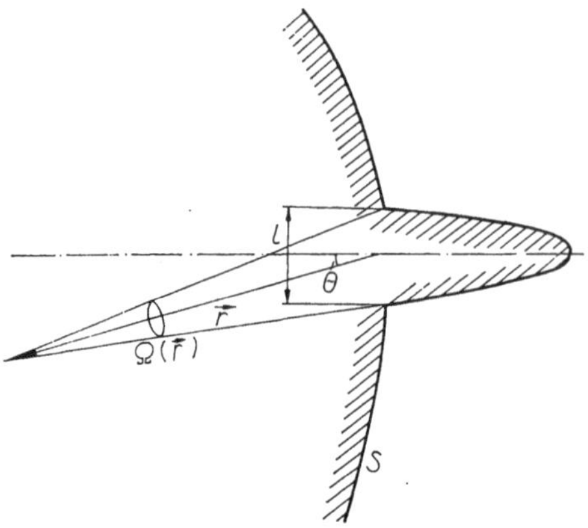


Fig. 1. Microtip on the surface $S$ of the field emitter; $\Omega(\boldsymbol{r})$ is the space angle inside the cathode, under point $\boldsymbol{r}$ the microtip base may be observed

emitted from microtips satisfying specific conditions (Fig. 1). The analysis will be carried out on the basis of the Boltzmann kinetic equation for the single-electron distribution function (which is possible for microroughness with $l \gg h/p_F \approx 10^{-8}$ cm),

$$\boldsymbol{V}\frac{\partial f}{\partial \boldsymbol{r}} + e\boldsymbol{E}\frac{\partial f}{\partial \boldsymbol{p}} = \hat{I}\{f\}\,, \tag{1}$$

where $\hat{I}\{f\}$ is the electron inelastic collision integral. As a source of energy relaxation let us consider electron–phonon scattering [6]. The electric field $\boldsymbol{E} = -\nabla\Phi(\boldsymbol{r})$ found in (1) should be obtained by solving the Maxwell equation with the charge and current densities corresponding to the distribution function $f$,

$$\varrho = 2e \sum_{\boldsymbol{p}} f(\boldsymbol{r}, \boldsymbol{p})\,, \tag{2}$$

$$\boldsymbol{j} = 2e \sum_{\boldsymbol{p}} \boldsymbol{V}\cdot f(\boldsymbol{r}, \boldsymbol{p})\,. \tag{3}$$

As is known in metals the equation for the electric field $E$ reduces to the condition of local electrical neutrality $\delta\varrho = 0$,

$$\sum_{\boldsymbol{p}} \left[f(\boldsymbol{r}, \boldsymbol{p}) - n_0\left(\frac{\mathcal{E}_p - \mu}{T}\right)\right] = 0\,. \tag{4}$$

Let us assume that emission occurs only from the microtip $z = 0$. At a distance from the cathode surface the current density and electric field are equal to zero while the distribution is equilibrium. With these assumptions made the boundary conditions to (1) and (4) acquire the form

$$f(\boldsymbol{r}, \boldsymbol{p})|_{z\to-\infty} = n_0\left(\frac{\mathcal{E}_p - \mathcal{E}_F}{T}\right) \equiv f_0(\mathcal{E}_p)\,, \tag{5}$$

$$\Phi(\boldsymbol{r})|_{z\to-\infty} = 0\,, \tag{6}$$

$$f(\boldsymbol{r}, \boldsymbol{p})|_{z=0,\,p_z<0} = [1 - D(\boldsymbol{p})]\cdot f(\boldsymbol{r}, \boldsymbol{p})|_{z=0,\,p_z>0}\,, \tag{7}$$

$$\boldsymbol{j}|_S = 0\,. \tag{8}$$

$D(\boldsymbol{p})$ in (7) is the surface potential barrier transparency. In the approximation of zero order in the parameter $l/l_e$ one has the collision-free Boltzmann equation

$$\boldsymbol{V} \frac{\partial f^{(0)}}{\partial \boldsymbol{r}} + e\boldsymbol{E}^{(0)} \frac{\partial f^{(0)}}{\partial \boldsymbol{p}} = \boldsymbol{0}\,. \tag{9}$$

Taking into account the boundary conditions (5) to (8), its solution will take the form

$$\begin{aligned} f^{(0)} &= f_0(\mathscr{E}_p + e\Phi^{(0)}(\boldsymbol{r})) \cdot \theta(\boldsymbol{V}\bar{\in}\Omega(\boldsymbol{r})) + \\ &+ (1 - D)\, f_0(\mathscr{E}_p + e\Phi^{(0)}(\boldsymbol{r})) \cdot \theta(\boldsymbol{V}\in\Omega(\boldsymbol{r}))\,. \end{aligned} \tag{10}$$

This solution is exact if we ignore scattered reflection of electrons from the sides of the microtip. Provided $\delta\varrho = 0$, we have for $\Phi^{(0)}$

$$\left(\frac{\partial n}{\partial \mu}\right) e\Phi^{(0)}(\boldsymbol{r}) = \sum_{\boldsymbol{p}} D(\boldsymbol{p})\, n_0\left(\frac{\mathscr{E}_p - e\Phi^{(0)}}{kT}\right) \cdot \theta(\boldsymbol{V}\in\Omega(\boldsymbol{r}))\,. \tag{11}$$

Let us confine ourselves to the WKB approximation for the barrier transparency. A simple analysis of (11) makes it possible to establish the character of the change of $\Phi^{(0)}(\boldsymbol{r})$ in the metal at large distances from the surface (see Fig. 1),

$$\Phi^{(0)}(\boldsymbol{r}) \approx \Phi_0 \begin{cases} \dfrac{1}{\sin 2\theta} \dfrac{l}{r}; & \dfrac{l}{r} \ll \max\{1, |\mathrm{tg} 2\theta|\}\,, \\ \left(\dfrac{l}{r}\right)^2; & |\mathrm{tg}\, 2\theta| \ll \dfrac{l}{r} \ll 1\,. \end{cases} \tag{12}$$

$\Phi_0$ is the potential within the microtip, it is constant throughout the length of the latter. It can be seen that the electric field penetrates the cathode to a depth of $\approx l \gg \lambda_{\text{Th-F}}$. The constant $\Phi_0$ is determined by the equation

$$y - \frac{1}{3\sqrt{\lambda}}\, \mathrm{e}^{-C} (\lambda + y)^{3/2}\, M\left[2, \frac{5}{2}, -\frac{3c}{2}(\lambda + y)\right] = 0\,, \tag{13}$$

where $C = (0.68\varphi^{3/2}/E_0)\, \theta(3.79E_0^{1/2}/\varphi)$, $\lambda = \mathscr{E}_F/\varphi_e$, $y = -e\Phi_0/\varphi$, $M(x, y, z)$ is the Whittaker function, $E_0$ the field near the microtip (in 10 V/nm), $\varphi$ the electron work function (eV), $\theta$ the Fowler-Nordheim function (see [7]). The solution of (13) for $\Phi_0$ in certain metals is given in Table 1. It can be seen that in zeroth approximation in the parameter $l_e/l$ a shift $e\Phi_0$ of the electron chemical potential appears near the microtip, which is conditioned by carrier heating by the external electric field. The shift $\Phi^{(0)}$ manifests itself in the corresponding shift of the maximum of the electron distribution function towards the high-energy region, bringing forth an increase of the emission total current by a factor $\exp(|e\Phi_0|/d)$, where $d^{-1} = (1.025\varphi^{3/2}/E_0)t(3.79E_0/\varphi)^{1/2}$ (see [7]).

To consider the effects of the electron inelastic relaxation we retain in (1) the terms of first order in the parameter $l/l_e$. In order to add $f^{(1)}(\boldsymbol{r}, \boldsymbol{p})$ to the electron distribution

Table 1

Values of the nonequilibrium contribution $|e\Phi_0|$ to the chemical potential for various cathode materials and different ac field amplitudes

| | $E_0 = 1.5$ V/nm | | | $E_0 = 1.6$ V/nm | | |
|---|---|---|---|---|---|---|
| metal | W | Zn | Al | W | Zn | Al |
| $\lvert e\Phi_0\rvert$ (eV) | 1.53 | 0.84 | 1.33 | 1.17 | 0.34 | 0.73 |

function we should have

$$\boldsymbol{V}\frac{\partial f^{(1)}}{\partial \boldsymbol{r}} + e\boldsymbol{E}^{(0)}\frac{\partial f^{(1)}}{\partial \boldsymbol{p}} = e\frac{\partial}{\partial \boldsymbol{r}}\Phi^{(1)}\frac{\partial f_0^{(0)}}{\partial \boldsymbol{p}} + \hat{I}_{\mathrm{ph}}^{(0)}(\boldsymbol{r}, \boldsymbol{p}), \tag{14}$$

where $\Phi^{(1)}$ is the addition to the electrostatic potential, satisfying the equation of neutrality, $\hat{I}_{\mathrm{ph}}^{(0)}$ the inelastic collision integral calculated for the distribution $f^{(0)}$. Taking into account the boundary conditions, we obtain for $f^{(1)}$

$$f^{(1)} = \{\theta(\boldsymbol{V}\bar{\in}\Omega(\boldsymbol{r})) + (1 - D)\,\theta(\boldsymbol{V}\in\Omega(\boldsymbol{r}))\} \times$$
$$\times \left\{e\Phi^{(1)}\frac{\partial n_0}{\partial \mathcal{E}_p} + \int_{-\infty}^{0} \mathrm{d}\tau\;\hat{I}^{(0)}((\boldsymbol{p}(\tau), \boldsymbol{r}(\tau))\right\}, \tag{15}$$

where $\tau$ is the time of electron motion along the classic trajectory at $\tau = -\infty$, $z = -\infty$. Taking into account the neutrality equation, we get from (15) for $\Phi^{(1)}$ to an accuracy up to terms $\sim D$

$$|e\Phi^{(1)}| = \frac{2\pi^2 h^3}{mp}\sum_{\boldsymbol{p}}\int_{-\infty}^{0} \mathrm{d}\tau\;\hat{I}_{\mathrm{ph}}^{(0)}(\boldsymbol{r}(\tau), \boldsymbol{p}(\tau))\left(\frac{1 - \Omega(\boldsymbol{r})}{2\pi}\right). \tag{16}$$

It is not difficult to write a corresponding expression for the correction to the function $\mathrm{d}j/\mathrm{d}\mathcal{E}$ (it is omitted here being unwieldy), also informative for the electron–phonon interaction.

The result (16) suggests that in principle a field-emission spectroscopy of the electron–phonon interaction near the metal surface is possible. The described experiment in principle allows simultaneous measurement of the two functions $\Phi^{(1)}$ and $\mathrm{d}j/\mathrm{d}\mathcal{E}$, demonstrating additional possibilities of such an approach as compared to the metal microcontact spectroscopy, based upon the analysis of the point contact current–voltage characteristic. The estimation $|e\Phi^{(1)}| \approx l_e/l\,|e\Phi_0| \approx (10^{-1}$ to $10^{-2})$ eV characterizes the necessary resolution of energy analyzers.

### *2.2 Introduction of a-c field stimulation*

Let us now consider the shift $e\Phi_{(0)}$ brought about by the tunnel emission, stimulated by the high-frequency field $\mathcal{E}\cos\Omega t$. As is shown in [8, 9], at $\Omega \gg \tau_s^{-1}$, where for $\mathcal{E} \ll E_0$, $\tau_s \equiv (2m\,|U|)^{1/2}/eE_0$ the time of subbarrier motion (here $U$ is the bound energy state of the electron in the metal), the external field amplitude determining the potential barrier transparency shows exponential amplification as compared to $\mathcal{E}$: $\mathcal{E}_{\mathrm{eff}} = \mathcal{E}\,\mathrm{sh}\,(\Omega\tau_s)$, thus entailing an increase of the tunneling probability $D$. Disregarding image forces at a random relation between $\mathcal{E}$ and $E_0$, the following relation holds:

$$D = (1 + \exp(A))^{-1}, \tag{17}$$

where

$$A = \frac{2\alpha^2}{\Omega}\left(\tilde{\Omega}\tau_0 - \frac{\mathrm{sh}\,(2\tilde{\Omega}\tau_0)}{2}\right) + \frac{8\alpha\beta}{\Omega^2}\,[\mathrm{sh}\,(\tilde{\Omega}\tau_0) - (\tilde{\Omega}\tau_0)\,\mathrm{ch}\,(\tilde{\Omega}\tau_0)] - \frac{4}{3}\beta^2\tau_0^2 + \tau_0,$$

$$\alpha = \frac{\mathcal{E}}{\tilde{\Omega}F_0}, \qquad \beta = \frac{E_0}{F_0}, \qquad \tilde{\Omega} = \frac{h\Omega}{|U|}, \tag{18}$$

$$F_0 = \frac{h^2\varkappa^3}{me}, \qquad \varkappa = \frac{\sqrt{2m\,|U|}}{h},$$

$$\tau_0 + \frac{\mathcal{E}\,\mathrm{sh}\,(\tilde{\Omega}\tau_0)}{E_0\tilde{\Omega}} = \frac{1}{2}\frac{F_0}{E_0}. \tag{19}$$

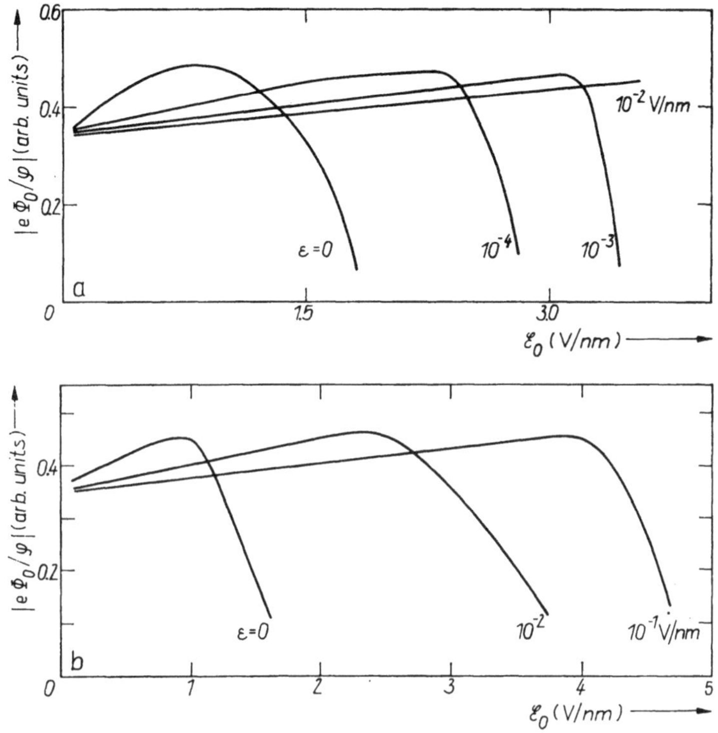


Fig. 2. Dependence of the nonequilibrium chemical potential term $|e\Phi_0|/\varphi$ on the dc electric field $E_0$ for a) tungsten and b) aluminium at different values of the amplitude of the ac field $\mathscr{E}$

The field electron energy distribution in an a.c. field undergoes nonuniform deformation along the energy axis, $U \to U(1 + 2\mathscr{E} \exp(\tilde{\Omega}\tau_0)/E_0)$.

Fig. 2 presents the results of the numerical analysis of (11) for tungsten (Fig. 2a) and aluminium (Fig. 2b), carried out allowing for (17). It can be seen that the application of an a.c. field $\mathscr{E}$ results in the shift of the curve maximum $|e\Phi(E_0)|/\varphi$ towards stronger fields.

## References


[1] V. G. Pavlov, A. A. Rabinovich, and V. N. Shrednik, Zh. tekh. Fiz. **45**, 2126 (1975).

[2] V. G. Pavlov and V. N. Shrednik, Proc. 7th Internat. Symp. Discharges and Electrical Insulation in Vacuum, Academy Sciences of the USSR, Siberian Branch, Novosibirsk 1976 (p. 213).

[3] I. D. Ventova and G. N. Fursey, Zh. tekh. Fiz. **47**, 1417 (1979).

[4] E. A. Litvinov and A. N. Starobinetz, in: Emissionnaya sylnotothnaya electronica, Izd. Nauka, Novosibirsk 1984 (in Russian).

[5] I. O. Kulik, A. N. Omelyanchuk, and R. I. Shechter, Fiz. nizkikh Temp. **3**, 1543 (1977).

[6] E. M. Liphshitz and L. P. Pitaevsky, Fizicheskaya kinetika, Izd. Nauka, Moscow 1979 (in Russian).

[7] R. E. Burgers, H. Kroemer, and J. M. Houston, Phys. Rev. **90**, 515 (1953).

[8] M. Buetticker and R. Landauer, Phys. Rev. Letters **49**, 1739 (1982).

[9] B. I. Ivlev and V. I. Melnikov, Zh. eksper. teor. Fiz., Pisma **41**, 116 (1986).


*(Received December 22, 1987)*

phys. stat. sol. (a) 108, 257 (1988)
Subject classification: 73.40; 68.55; 78.65; S7.1; S9.15

*Sektion Chemie*[1]) *(a) and Sektion Physik*[2]) *(b) der Karl-Marx-Universität Leipzig and Sektion Physik der Humboldt-Universität Berlin*[3]) *(c)*

# Multi-Quantum-Well Structures Prepared by the Flash Evaporation Technique

## Growth, Structural, and Optical Characterization

By
B. SCHUMANN (a), R. FLAGMEYER (b), and H. WEINERT (c)

Dedicated to Prof. Dr. E. GUTSCHE on the occasion of his 60th birthday

The flash evaporation technique under high-vacuum conditions is a possible method to produce multi-quantum-well structures and superlattices. Using this technique, epitaxial GaAs/GaP and, for the first time $GaAs/Ca_{0.43}Sr_{0.57}F_2$ multi-quantum-well samples are grown. The structural characterization is performed by reflection high energy electron diffraction and Rutherford back-scattering/channeling measurements. Optical investigations (absorption and light-induced grating measurements) show quantum confinement in both systems investigated and type II character for the $GaAs/Ca_{0.43}Sr_{0.57}F_2$ heterostructures.

Mit der Flash-Verdampfung im Hochvakuum können multi-quantum-well-Strukturen und Superlattices hergestellt werden. Dies wird am Beispiel von GaAs/GaP- und $GaAs/Ca_{0,43}Sr_{0,57}F_2$-multi-quantum-well-Systemen gezeigt. Die strukturelle Charakterisierung dieser Proben erfolgte mittels Beugung schneller Elektronen (Reflexion) und der Rutherford-Rückstreuung/Kanalisierung. Optische Messungen (Absorption und Methode der lichtinduzierten Gitter) ergeben ein quasi-zweidimensionales Verhalten der Ladungsträger in beiden Probenarten und den Typ-II-Charakter für die erstmals dargestellten und untersuchten $GaAs/Ca_{0,43}Sr_{0,57}F_2$-Heterostrukturen.

## 1. Introduction

After Esaki and Tsu's proposal [1] of semiconductor superlattices and early achievements in this field [2 to 4], there has been a continuously rising interest in these materials because of their importance for fundamental semiconductor physics as well as technological applications. The most common techniques for growing semiconductor superlattices (SL) and multi-quantum-well structures (MQW) by alternating thin layers of different semiconductors are clearly molecular beam epitaxy (MBE) and metal-organic chemical vapour phase epitaxy (MOCVD). Both of these growth techniques are able to produce high quality samples with sharp interfaces and excellent crystalline and doping properties. However, these methods are quite expensive. So it seems to be interesting to look for simpler and cheaper growing techniques for producing SL's and MQW's of sufficient quality.

The flash evaporation technique is a well established growing method for thin films of various materials [5, 6]. So far this technique under high-vacuum conditions has been used for the deposition of epitaxial single or double films of $A^{I}B^{III}C_2^{VI}$, $A^{III}B^{V}$,

[1]) Talstr. 35, Leipzig, DDR-7010, GDR.
[2]) Linnéstr. 5, Leipzig, DDR-7010, GDR.
[3]) Invalidenstr. 110, Berlin, DDR-1040, GDR.

and fluoride compounds [7 to 10]. The thickness of a single layer has been in the range of 100 nm.

In the present paper we report the first successful attempts to grow quantum-well heterostructures using the flash evaporation technique. We grew lattice-matched GaAs/$Ca_{0.43}Sr_{0.57}F_2$ and GaAs/GaP MQW structures with a thickness $t \gtrapprox 8$ nm for a single layer. Using different experimental techniques the structural properties of the samples were characterized. The results of optical measurements clearly show the quantum confinement of the excited carriers and the type II character of the GaAs/$Ca_{0.43}Sr_{0.57}F_2$ heterotransition.

## 2. Experimental

### *2.1 Sample preparation*

The structures were grown by the flash evaporation technique at pressures of about $2 \times 10^{-3}$ Pa during the evaporation [7 to 10]. In producing the multilayer samples the magazines of the feeder were alternately filled with desired amounts of the components. The powdered bulk material with grain sizes of about 150 μm was gravity fed to a resistance-heated tungsten boat used as evaporator. The substrate was located approximately 60 mm above the heater. The evaporator temperatures chosen in our experiments were around 1550 K, corresponding to the optimal conditions for the evaporation of the different film materials used. The growth rate of the layers amounted to 0.3 nm $s^{-1}$. The substrate temperature was varied from 670 to 870 K in steps of 25 K. The structures prepared contained up to 25 layers with a thickness $t \geqq 8$ nm for a single layer and consisted of GaAs or GaP substrates and GaAs/$Ca_{0.43}Sr_{0.57}F_2$ layers (lattice-matched at room temperature) or GaAs/GaP layers, respectively. Films of other materials, such as InAs, $CaF_2$, $SrF_2$, $BaF_2$, etc. were also deposited.

### *2.2 Sample characterization*

The structure of the samples was investigated by reflection high energy electron diffraction (RHEED) with an accelerating voltage of 50 kV and by Rutherford backscattering/channeling (RBS) using 1.2 MeV $^4He^+$ ions. Details of the analysis of GaAs–fluoride structures by the RBS technique were described in [11].

The optical properties of the GaAs/GaP MQW structures were investigated by conventional absorption measurements at $T = 77$ K.

For the lattice-matched GaAs/$Ca_{0.43}Sr_{0.57}F_2$ systems absorption measurements were also performed, but they gave no clear information about the quasi-two-dimensional character of the excited carriers. This was expected, because estimations of the absolute energy levels of the valence and conduction bands of both materials showed that these structures should be of type II [12], i.e. the holes should be confined to the GaAs layers, and the electrons to the $Ca_{0.43}Sr_{0.57}F_2$ layers. Therefore, for these samples the characterization of the recombination properties of the excited carriers was done with the experimental technique of light-induced gratings (for a recent review see [13]). The second harmonic radiation (532 nm) of a ns-YAG laser (pulse duration, 12 ns; repetition rate used here, 12 Hz) was split into two beams. These two beams were focussed at a definite angle onto the sample surface, which owing to interference effects gives a light-induced grating of a grating period $\Lambda$, which depends on the incident angle $\theta$ between the two beams. We measured the excitation intensity dependence of the first-order self-diffraction, which gives information about the recombination properties of the excited carriers (see [13]). In a second experiment we measured the time behaviour of the light-induced grating by probing the grating using a second, electronically delayed YAG laser beam (also the second harmonic radiation was used).

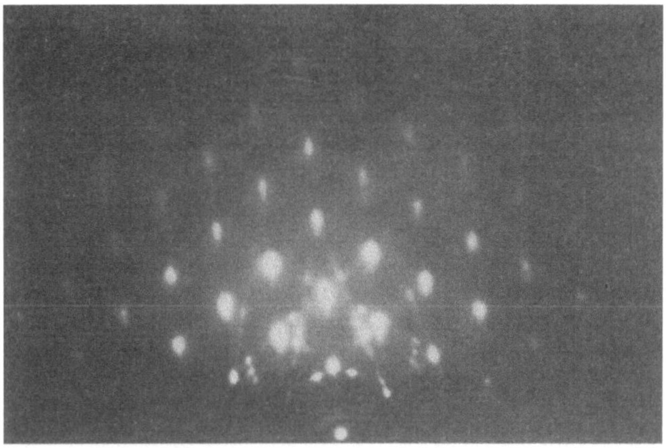

Fig. 1. RHEED diagram of a GaP/GaAs structure. Azimuth ⟨110⟩, twins around ⟨111⟩, deposition temperature 720 K, evaporator temperature 1520 K, substrate GaP (100), layers four GaP, 15 nm thick; three GaAs, 10 nm thick

This was performed for different grating periods and gave information about the predominant decay process of the light-induced grating. All these measurements were performed at room temperature.

## 3. Results and Discussion

### *3.1 RHEED measurements*

In the substrate temperature range from 720 to 790 K the films grew epitaxially on the substrates and were almost in all cases completely single-crystalline. It should be noted that with RHEED investigations only the uppermost layer is recorded, but epitaxial growth was found for all samples investigated (different layer materials and layer numbers up to 25). In Fig. 1 a RHEED diagram of a GaP/GaAs structure is shown. In addition to the matrix reflections, spots caused by twins around ⟨111⟩ in the uppermost GaP layer are seen. The RHEED pattern of a lattice-matched structure (Fig. 2) contains only the matrix reflections of the $Ca_{0.43}Sr_{0.57}F_2$ top layer.

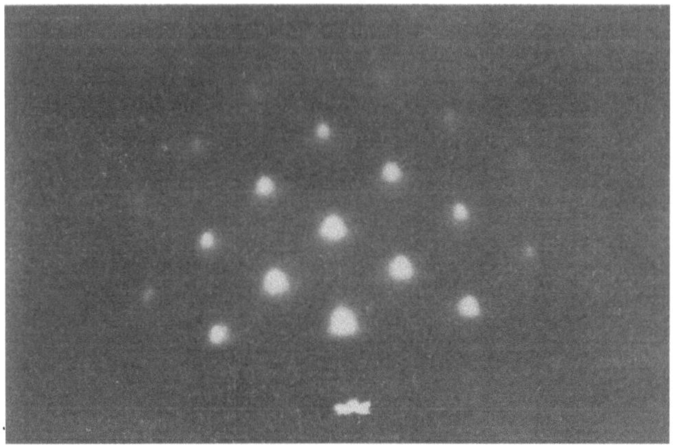

Fig. 2. RHEED diagram of a lattice-matched structure. Azimuth ⟨110⟩, deposition temperature 795 K, evaporator temperature 1570 K, substrate GaP (100), layers six GaAs, 8 nm thick; seven $Ca_{0.43}Sr_{0.57}F_2$, 25 nm thick

### *3.2 RBS measurements*

The structure of the heteroepitaxial system was determined from random and ⟨100⟩ aligned RBS spectra taken with 1.2 MeV $^4He^+$ ions. For normal incidence, shallow exit angles (cf. insets in Fig. 3) of the scattered ions were chosen so that the backscattering signals from the thin films were resolved as well as possible. As an example, Fig. 3a shows the random spectrum of a (Ca, Sr)$F_2$/GaAs/(Ca, Sr)$F_2$/GaAs (100) structure. The backscattering signal from the intermediate, thin GaAs layer is shifted to lower energies by about 30 keV with respect to the "gap" in the Sr backscattering yield and is superimposed on the Sr signal of the second fluoride layer. The area $A_{GaAs}$ (hatched) is related to a GaAs layer of (10 ± 1) nm thickness. Although the depth profile of the composition cannot be discussed, because of the limited depth resolution

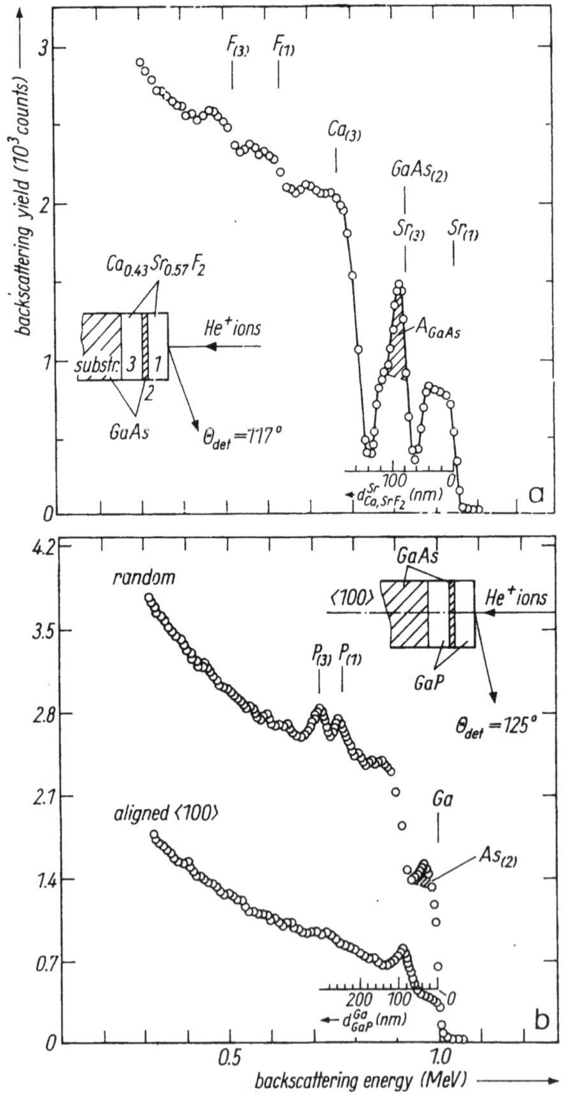


Fig. 3. a) RBS spectrum of a lattice-matched structure. Deposition temperature 795 K, substrate GaAs (100), layers one GaAs, 10 nm thick; two $Ca_{0.43}Sr_{0.57}F_2$, 60 nm thick. b) RBS/Channeling spectrum of a GaP/GaAs structure. Deposition temperature 720 K, substrate GaAs (100), layers one GaAs, 10 nm thick; two GaP, 45 nm thick

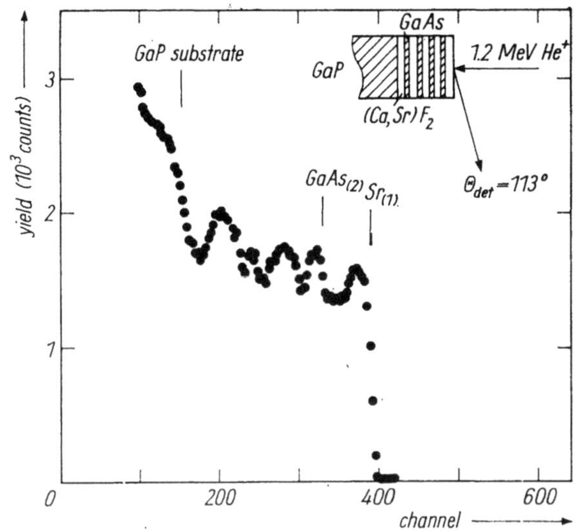


Fig. 4. RBS spectrum of a lattice-matched structure. Deposition temperature 795 K, substrate GaP (100), layers four GaAs, 20 nm thick; five $Ca_{0.43}Sr_{0.57}F_2$, 40 nm thick

(about 10 nm in this case), it can be concluded from the spectrum that the heterojunctions are sharp in the range of a few nm ($\leq$ 2 nm).

The thicknesses of the top and second insulating layers are 62 and 57 nm, respectively, giving a total thickness of the structure of 129 nm, as illustrated by the depth scale. The channeling minimum yield $\chi_{min}$ determined by the aligned spectrum is about 50%.

A somewhat different feature results from the ion backscattering of the GaP/GaAs/GaP/GaAs (100) structure (Fig. 3b). Here, the As signal from the intermediate GaAs layer is slightly shifted to higher energies, and the substrate As yield overlaps the Ga yield from the second GaP film. The As "peak" as well as the "gap" between the P signals (on top of the substrate yield) demonstrate the existence of a GaAs layer of some nm thickness, but an accurate estimation is impossible. The total thickness of the multilayer system amounts to 98 nm as expected from the quantity of evaporated material.

The channeling minimum yield $\chi_{min}$ 23% for the top GaP layer demonstrates their good epitaxial quality. Note, however, the backscattering peak in the aligned spectrum at about 0.92 MeV, which arises from extensive crystallographic defects at the interface between the layer structure and the substrate. Probably there is a dislocation network accommodating the lattice misfit.

From the RBS spectrum shown in Fig. 4 the total thickness of this multilayer structure appeared similar to that expected from the quantity of evaporated material. Optimizing the scattering geometry such that the Sr signals of the $(n-1)$-th layers and the GaAs signals of the $n$-th one coincided the alternation of five fluoride layers with four GaAs layers of thicknesses of 40 and 20 nm, respectively, was resolved. Obviously, since the thicknesses of the alternating layers in this sample are not exactly uniform, the spectrum becomes slightly blurred.

### *3.3 Optical measurements*

Fig. 5 shows a transmission spectrum obtained at $T = 77$ K for a GaAs/GaP structure. One can clearly see the step-like behaviour of the transmission, which is expected for any type I MQW structure (see [3]). For comparison the allowed $\Delta n = 0$ transi-

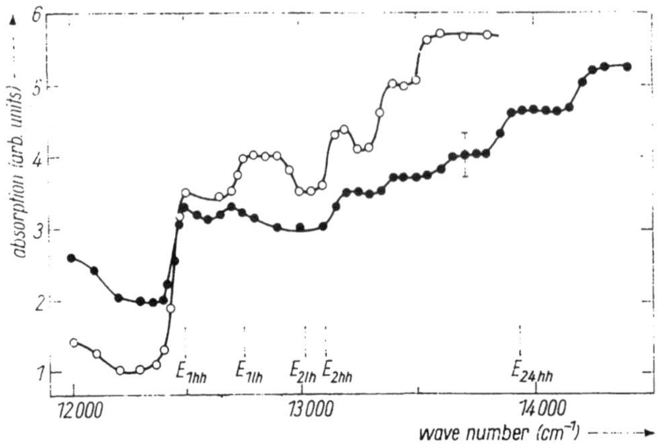


Fig. 5. Transmission spectra of GaP/GaAs/GaP structures at $T = 77$ K. Deposition temperature 720 K, substrates GaP (100), layers one GaAs, 8 nm thick (sample 1999), 15 nm thick (sample 2003), two GaP-layers, 40 nm thick. ● Sample 2003, ○ sample 1999. The marks show the calculated minimum energies for band-to-band transitions for sample 1999

tions between the different subbands of electrons and holes, expected for a square well structure, are indicated by marks. The subband energies were calculated using a simple Kronig-Penney model [14]. The following parameters were used for the calculation: Conduction band discontinuity $\Delta E_c = 30$ meV, valence band discontinuity $\Delta E_v = 801$ meV [15]. For the effective masses of electrons and holes in the GaP barriers we used $m_e^B = m_h^B = 0.54\ m_0$; for the effective masses in the GaAs quantum wells, $m_e = 0{,}067 m_0$, $m_{hh} = 0.48 m_0$, $m_{lh} = 0.083 m_0$ [16]. One can see that for the lowest-energy transitions the experimentally found steps agree well with the calculated energies, for higher energy transitions there are non-negligible deviations. These deviations are probably an effect of the interfaces, which are not sharp (i.e. extended) — the calculations were performed for an ideal square well structure. Nevertheless, the step-like character of the transmission spectrum is clear evidence for the quantum confinement of the electrons and holes in the GaAs layers. The absence of pronounced excitonic peaks, as are observed for GaAs/GaAlAs QW structures [3], may be related to a higher level of background doping.

Absorption and also photoluminescence measurements were performed on the type II lattice-matched structures obtained by alternate layers of GaAs and $Ca_{0.43}Sr_{0.57}F_2$. As expected the information obtained by these conventional methods was not easy to explain, and they gave no unambiguous evidence for a quasi-two-dimensional behaviour of the excited carriers. However, an excitation-dependent shift of the photoluminescence peak was observed.

In order to get more information about the relevant energy levels in these GaAs/$Ca_{0.43}Sr_{0.57}F_2$ MQW structures we used the light-induced grating technique [13]. At first we measured the dependence of the first-order self-diffraction efficiency $\eta_1$ on the excitation intensity. A typical result is shown in the inset in Fig. 6. The square dependence of $\eta_1$ on $I_{exc}$ obtained shows that linear recombination predominates [13]. This allows an easy interpretation of the results of time-resolved measurements shown in Fig. 6. There the decay kinetics of light-induced gratings with different periods (but identical $I_{exc}$) is shown. These results were obtained by measuring the first-order diffraction efficiency of an electronically delayed second YAG laser beam of low intensity. One can see that the diffraction intensity takes a surprisingly long time to reach

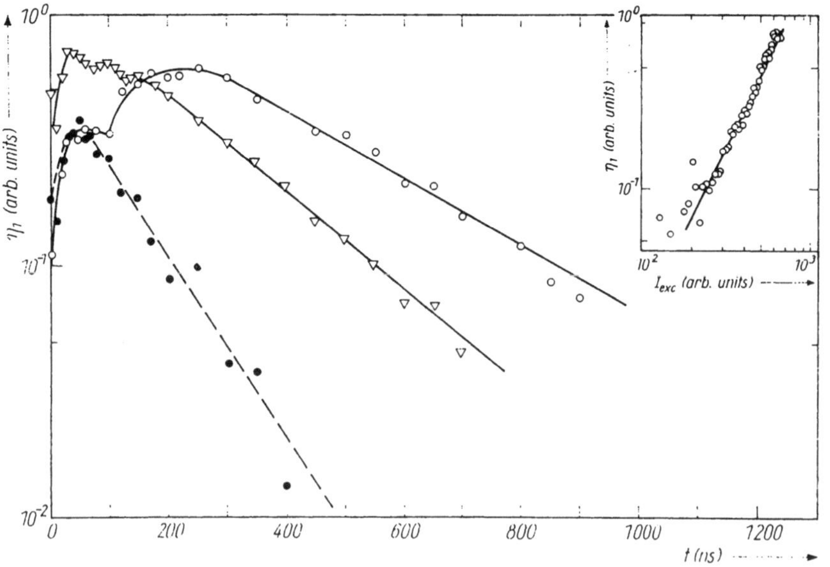


Fig. 6. First order diffraction efficiency (light-induced grating technique) of a lattice-mached structure (sample 1854). Deposition temperature 795 K, substrate GaP (100), layers two GaAs, 8 nm thick, three $Ca_{0.43}Sr_{0.57}F_2$, 50 nm thick. 532 nm radiation was used for the exciting and (electronically delayed) probing beams. The excitation intensity was constant, the different grating periods were obtained for different angles between the incident beams and normally measured using a microscope. $T = 300$ K; ● $\Lambda \approx 20.9, \tau = 290$; ▽ $\approx 25.6, 430$; ○ $\approx 29.7$ μm, 686 ns. Inset: Dependence of the first-order self-diffraction efficiency on the excitation intensity

its maximum value and that the rising kinetics are rather complicated. This rising behaviour of $\eta_1$ is not understood at present, but it is probably related to the fact that the 532 nm exciting laser beams generate both electrons and holes in the GaAs layers and it takes some time for the electrons to penetrate into the $Ca_{0.43}Sr_{0.57}F_2$ films. However, after $\eta_1$ reached its maximum value, we observed an exponential decay for every grating period $\Lambda$. The decay times thus obtained are given in the figure. Generally, if the predominant recombination processes are linear, a square dependence of the reciprocal decay time $1/\tau$ on the grating period $\Lambda$ is found, which gives information about the predominant process responsible for the decay of the light-induced grating [13, 17]. For the GaAs/$Ca_{0.43}Sr_{0.57}F_2$ structures investigated we found that the light-induced gratings decay predominantly by diffusion. This means that the recombination probability is so low that the contribution of recombination to the decay of the light-induced grating will be negligible compared with diffusion. This will be true in the time scale of the experiments, i.e., in a time scale of some 100 ns. Consequently, the results obtained by light-induced grating measurements are evidence that the GaAs/$Ca_{0.43}Sr_{0.57}F_2$ heterotransitions are indeed type II, i.e., that the electrons and holes are confined to different layers.

## 4. Conclusions

It has been shown that the flash evaporation technique under high-vacuum conditions is a possible method to produce multi-quantum-well structures and superlattices, both of lattice-matched and strained-layer types.

Using the flash evaporation technique GaAs/GaP and GaAs/$Ca_{0.43}Sr_{0.57}F_2$ multi-quantum-well structures were grown. To the authors' knowledge this paper is the first report on MQW structures with fluoride layers.

The samples were characterized by several experimental techniques. The epitaxial growth and the single-crystalline character were proved by RHEED measurements. RBS investigations showed the periodicity of the layer arrangement and the thicknesses of the films, which are approximately as expected from the growth conditions. The interfaces between alternate layers are probably extended in a scale of a few nm.

Optical measurements showed the quantum confinement of the carriers in both systems investigated and the type II character of the GaAs/$Ca_{0.43}Sr_{0.57}F_2$ heterostructures.

***Acknowledgements***

The authors wish to thank Prof. E. Gutsche, Prof. G. Kühn, and Prof. K. Unger, and Dr. A. Selle for stimulating discussions. We are grateful to Prof. A. Vaitkus and Dr. K. Jarašiunas for helpful discussions and enabling us to perform the light-induced grating measurements in the laboratories of Vilnius State University.

## References

[1] L. ESAKI and R. TSU, IBM J. Res. Developm. **14**, 61 (1970).
[2] H. L. STRÖMER, J. Phys. Soc. Japan **49**, Suppl. A., 1013 (1980).
[3] R. DINGLE, W. WIEGMANN, and C. J. HENRY, Phys. Rev. Letters **33**, 827 (1974).
[4] L. ESAKI, Proc. 17-th Conf. Physics of Semiconductors, San Francisco 1984, Ed. J. D. CHADI and W. A. HARRISON, Springer-Verlag, Berlin 1985 (p. 473).
[5] L. L. RICHARDS, The Use of Thin Films in Physical Investigations, Ed. J. C. ANDERSON, Academic Press, London 1966 (p. 71).
[6] R. GLANG, Handbook of Thin Film Technology, Ed. L. I. MAISSEL and R. GLANG, McGraw-Hill Publ. Co., New York 1970 (p. 1).
[7] B. SCHUMANN, A. TEMPEL, and G. KÜHN, Solar Cells **16**, 43 (1986).
[8] B. SCHUMANN, A. TEMPEL, and G. KÜHN, Crystal Res. Technol. **23**, 3 (1988).
[9] B. SCHUMAN, G. KÜHN, W. GASCH, G. WAGNER, R. FLAGMEYER, O. MÜLLER, and W. HAUFFE, J. Crystal. Growth **82**, 405 (1987).
[10] B. SCHUMANN, G. KÜHN, and G. WAGNER, Crystal Res. Technol. **21**, 1139 (1986).
[11] R. FLAGMEYER and B. SCHUMANN, Crystal Res. Technol. **22**, 893 (1987).
[12] K. UNGER, private communication.
[13] S. VAITKUS, K. JARAŠIUNAS, E. GAUBAS, L. JONIKAS, R. PRANAITIS, and L. SUBACIUS, IEEE Quantum Electronics **22**, 1298 (1986).
[14] A. CHOMETTE, B. DEVEAUD, M. BAUDET, P. AUVRAY, and A. REGRENY, J. appl. Phys. **59**, 3835 (1986).
[15] V. A. BUDYANU, S. N. CHECHUY, I. A. DAMASKIN, S. A. FEDOSEEV, A. A. NASAKIN, S. L. PYSHKIN, M. I. VALKOVSKAYA, and V. P. ZENCHENKO, phys. stat. sol. (a) **91**, 737 (1985).
[16] S. ADACHI, J. appl. Phys. **58**, R1 (1985).
[17] K. JARAŠIUNAS, private communication.

*(Received March 22, 1988)*

phys. stat. sol. (a) **108**, 265 (1988)

Subject classification: 73.60 and 79.20; S10.1

*Sektion Physik der Wilhelm-Pieck-Universität Rostock*[1])

# Secondary Electron Field Emission

By

H.-J. FITTING and D. HECHT

Dedicated to Prof. Dr. E. GUTSCHE on the occasion of his 60th birthday

Secondary electron field emission (SEFE) is based on Fowler-Nordheim tunneling from conducting substrates into thin insulating layers due to their strongly positive charging-up during electron bombardment. This self-consistent charging-up process is demonstrated by a computer simulation. Thereby the FN-injection current from the substrate is extremely field-sensitive, allowing a detection of less than $10^{10}$ elementary charges per $cm^2$ inside the top dielectric layer. Based on this high sensitivity an electron and hole trap spectroscopy by electron beam injection and thermal stimulation in thin insulating layers ($d < 150$ nm) is set up.

Sekundärelektronenfeldemission (SEFE) basiert auf Fowler-Nordheim-Tunnelung aus leitenden Substraten in dünne Isolatorschichten infolge ihrer starken positiven Aufladung während des Elektronenbeschusses. Dieser selbstkonsistente Aufladungsprozeß wird durch eine Computersimulation veranschaulicht. Dabei zeigt sich der FN-Injektionsstrom als extrem feldempfindlich und erlaubt einen Nachweis von weniger als $10^{10}$ Elementarladungen pro $cm^2$ in der dielektrischen Deckschicht. Begründet auf dieser hohen Empfindlichkeit wird eine Elektronen- und Löcher-Haftstellenspektroskopie mittels Elektronenstrahlinjektion und thermischer Stimulation in dünnen Isolatorschichten ($d < 150$ nm) entwickelt.

## 1. Introduction

In 1936 for the first time Malter [1] described the anomalous secondary emission from insulator–metal samples. This surprising phenomenon was characterized by post-emission of secondary electrons (SE) even after switching off the primary electron (PE) beam and is called Malter effect. The reason for that has to be seen in a general positive charging-up of the top insulating layer due to SE escape from that region and the successive field extraction of SE from the conducting substrate. Especially in case of porous MgO layers the Malter emission was observed and explained by field emission of electrons from the metal support and successive avalanche breakdown across the insulating layer maintaining the positive charge inside the latter one [2].

Though one did not succeed in building stable Malter cathodes, the interest in field enhanced SE emission from insulating layers has steadily increased, because of its applicability for electron beam storage screens, electron beam contacting, electrete techniques, and of course, for scanning electron microscopy and electron spectroscopy (AES, ELS) [3]. In 1979 the authors [6] set up a computer simulation of SEE and the associated self-consistent charge transport in insulating layers. Later this became the basis for the development of a charge trap spectroscopy in thin insulating layers [7, 8] based on the very field-sensitive tunneling effect of SE from the substrate.

---

[1]) Universitätsplatz 3, DDR-2500 Rostock, GDR.

## 2. Initial SEE from Uncharged Insulators

The charging-up of insulating emitter samples during electron emission is hardly avoidable. Consequently the resulting field distribution will affect the ratio of emitted electrons as well as their escape depth, angular and energy distributions.

The additional low energy electron exposure often used today was introduced by Salow [9] already in 1940 in order to fix the sample potential near the ground potential. Nevertheless, most of the data concerning emission efficiency $\sigma = n_{SE}/n_{PE}$ and escape depth $\lambda_e$ of SE from insulating materials were obtained by more or less theoretical or semiempirical estimations. Thus, e.g. a wide spread of data for escape depths is found varying between 5 and 32 nm [2, 10 to 12]. This is partly understandable because the effective escape depth $\lambda_{eff}$ will depend on the PE initial energy $E_0$ as we have shown in [12]. The SE generation is proportional to the PE energy loss $dE/dx$, i.e. their energy transfer to the target,

$$\lambda_{eff} = \frac{\int\limits_0^\infty x \frac{dE}{dx} \exp\left(-\frac{x}{\lambda_e}\right) dx}{\int\limits_0^\infty \frac{dE}{dx} \exp\left(-\frac{x}{\lambda_e}\right) dx}, \tag{1}$$

where $\lambda_e$ is the mean escape depth of a single SE. In [12] we estimated $\lambda_e$ by means of the Fröhlich theory [13] of electron–phonon interaction and have found $\lambda_e = 16$ nm for the insulator $SiO_2$. Because of the natural lack of inelastic electron–electron interaction in insulators this value is about one order of magnitude greater than for metals. Therefore, the variation of spatial SE generation proportional to $dE/dx(E_0, x)$ over SE attenuation $\exp(-x/\lambda_e)$ has to be taken into account according to (1). The result is presented in Fig. 1 and shows the wide spread of the effective escape depth $\lambda_{eff}$ towards low initial energies $E_0 < 2$ keV.

## 3. SEE under Charging-Up

The charging-up process in insulating layers during electron bombardment and SE emission comprises the following charge transport components:

a) penetration of *primary electrons* (PE) as a function of their initial energy $E_0$, elastic and inelastic scattering, and backscattering leading to a depth-depending

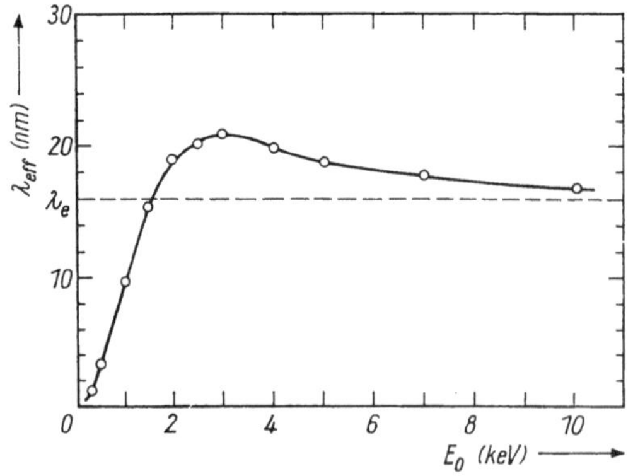


Fig. 1. Effective mean escape depth $\lambda_{eff}$ of SE's as a function of PE energy $E_0$ (in $SiO_2$) in comparison with the mean escape depth $\lambda_e$ of a single SE

current density,

$$j_{\mathrm{PE}}(x, E_0) \;;$$

b) generation, straggling, recombination, or emission of *secondary electrons* (SE). These electrons are created by PE's but have considerably lower energies ($E_{\mathrm{SE}} < 50$ eV) than the latter ones. Therefore, their straggling is affected by the electric field distribution $F(x')$ across the insulating layer,

$$j_{\mathrm{SE}}(x, E_0, F(x')) \;;$$

c) creation of *holes* as a consequence of SE generation from the valence band and their drift in internal fields,

$$j_{\mathrm{h}}(x, E_0, F) \;;$$

d) *Poole-Frenkel current* of field-enhanced thermal emission from traps with activation energy $E_{\mathrm{t}}$ at temperature $T$,

$$j_{\mathrm{PF}}(x, E_{\mathrm{t}}, T, F) \;;$$

e) *Fowler-Nordheim tunneling* from the substrate as a function of the field at the interface $x = d$,

$$j_{\mathrm{FN}}(x, F(d)) \;;$$

f) *retarding field current* of emitted SE but suppressed by an external electric field $F_{\mathrm{ex}}(x < 0)$ outside the emitter,

$$j_{\mathrm{R}}(x, F_{\mathrm{ex}}) \,.$$

Thus the total current through the depth $x$ is given by

$$j(x) = j_{\mathrm{PE}} + j_{\mathrm{SE}} + j_{\mathrm{h}} + j_{\mathrm{PF}} + j_{\mathrm{FN}} + j_{\mathrm{R}} \,. \tag{2}$$

The continuity equation in planar geometry,

$$\nabla \boldsymbol{j}^{+} = -\frac{\mathrm{d}}{\mathrm{d}x}\, j = -\frac{\mathrm{d}}{\mathrm{d}t}\, \varrho \,, \tag{3}$$

results in the charge distribution $\varrho(x)$, and with the Poisson equation

$$-\varepsilon_0 \varepsilon_{\mathrm{r}} \Delta U = \varepsilon_0 \varepsilon_{\mathrm{r}} \nabla \boldsymbol{F} = \varepsilon_0 \varepsilon_{\mathrm{r}} \frac{\mathrm{d}}{\mathrm{d}x} F = \varrho \tag{4}$$

we get the field distribution $F(x)$.

Substituting the charge density in (3) by (4) a partial differential equation

$$\varepsilon_0 \varepsilon_{\mathrm{r}} \frac{\partial}{\partial t} F(x) = j(x, t, F(x')) \tag{5}$$

for the field distribution $F(x, t)$ is obtained. Equations (2) to (5) are describing the self-consistent charging-up process in dielectric samples:

$$\begin{array}{ccccccc} \boldsymbol{j} & \rightarrow & \varrho & \rightarrow & \boldsymbol{F} & \rightarrow & U \\ \uparrow & & \downarrow & & \downarrow & & \\ \multicolumn{5}{c}{\rule{4em}{0.4pt}} & & \end{array} \tag{6}$$

where an especially strong feedback of the electric field to the original current is given.

Since different charge transport components contribute to (5), a very complicated partial integro-differential equation results, which cannot be solved analytically. Therefore, we solved (5) by a computer iteration technique [4 to 6],

$$F_n(x, t) = F(x, t - \Delta t) + \frac{\Delta t}{\varepsilon_0 \varepsilon_{\mathrm{r}}} \left[ j_n(x, F_{n-1}(x')) - j_n(x = 0, F_{n-1}(x')) \right] . \tag{7}$$

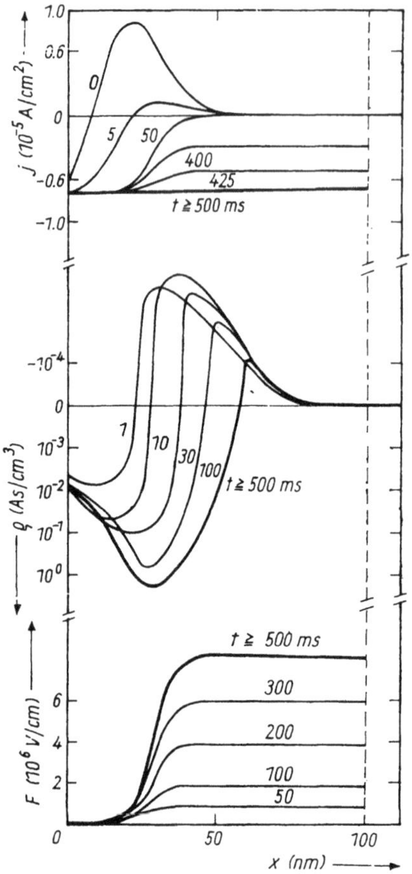


Fig. 2

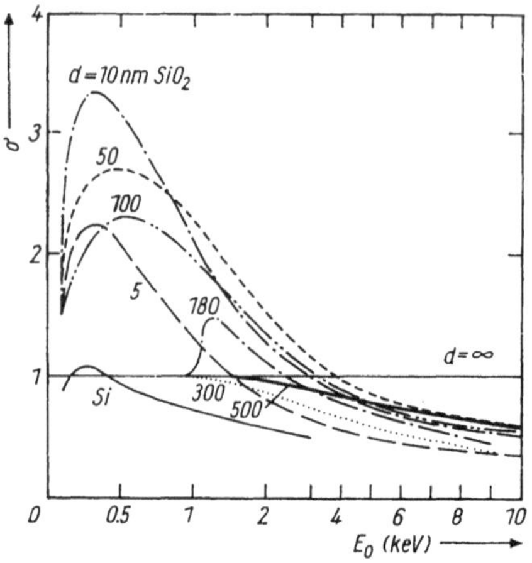


Fig. 3

Fig. 2. Depth distributions of current density $j$, charge density $\varrho$, and field strength $F$ within an insulating $SiO_2$ layer with thickness $d = 100$ nm on Si substrate during electron bombardment with energy $E_0 = 1$ keV, density $j_0 = 10^{-5}$ A/cm², and increasing time $t$

Fig. 3. Measurement of the total SE fraction $\sigma$ in dependence on PE energy $E_0$ and $SiO_2$ layer thickness $d$ on Si substrate

The second term in the brackets fulfils the boundary condition

$$F(x = 0, t) = 0 \tag{8}$$

at the surface of an open insulating layer on a conducting support. The iteration cycle $n$ is repeated till no further changes can be recognized, i.e. $F_n(x, t) = F(x, t)$.

Then $t$ is increased to $t + \Delta t$ and the iteration starts again. In case of no further time dependence the steady state $F(x, t) = F(x)$ is obtained. Results of this computer iteration for the current density $j(x, t)$, charge distribution $\varrho(x, t)$, and the internal electric field $F(x, t)$ within a 100 nm $SiO_2$ layer on Si substrate, when bombarded by electrons of energy $E_0 = 1$ keV and density $j_0 = 10^{-5}$ A/cm², are given in Fig. 2. It becomes obvious that at first, $t < 50$ ms, the SE's escape from surface-near regions $x < 40$ nm. With growing time the field towards the interface increases more and more and causes a Fowler-Nordheim injection of electrons into and through the dielectric layer. This phenomenon we will call secondary electron field emission — SEFE.

At about half a second the steady state condition $j(x, t) = \text{const}$ with no drains and sources, as well as $F(x, t) = F(x)$ is obtained. The total SEE fraction $\sigma$, i.e. the sum of true SE's and backscattered PE's is given by

$$\sigma(t) = 1 - \frac{j(x = 0, t)}{j_0}. \tag{9}$$

From Fig. 2 we find for the steady state condition of SEFE $\sigma_\infty \approx 1.75$ in good agreement with experimental values of Fig. 3. A comprehensive comparison between this computer simulation and experimental measurements is given in [6].

## 4. Field Emission Component of SEE

As already shown in Section 3, item e, and Fig. 2 the Fowler-Nordheim tunneling plays the dominant role in SEFE, when the SE fraction and the insulating layer thickness are $\sigma > 1$ and $d < 150$ nm, respectively. The tunnel current is given by the general expression

$$j_{FN} = AF^2 \exp\left(-\frac{B}{F}\right), \tag{10}$$

where $A$ and $B$ are constants depending on the material heterojunction at the interface and especially on the potential step between the conducting substrate and the top insulating layer. For a $Si$–$SiO_2$ junction the comparison of theoretical data by Lenzlinger and Snow [14] and experimental data by Powell [15] leads to appropriate values $A \approx 8.77 \times 10^{-8}$ A/V²; $B \approx 2.26 \times 10^8$ V/cm, see Fig. 4.

The expansion of (10) for $F = F_0 + \Delta F$ at $F_0 = 7 \times 10^6$ V/cm provides the approximation

$$j_{FN} \approx j_0(F_0) + A \exp\left(-\frac{B}{F_0}\right)(2F_0 + B)\,\Delta F \tag{11}$$

and the relative change in Fowler-Nordheim current,

$$\frac{\Delta j_{FN}}{j_0} = \frac{j_{FN} - j_0}{j_0} \approx \left(2 + \frac{B}{F_0}\right)\frac{\Delta F}{F_0}. \tag{12}$$

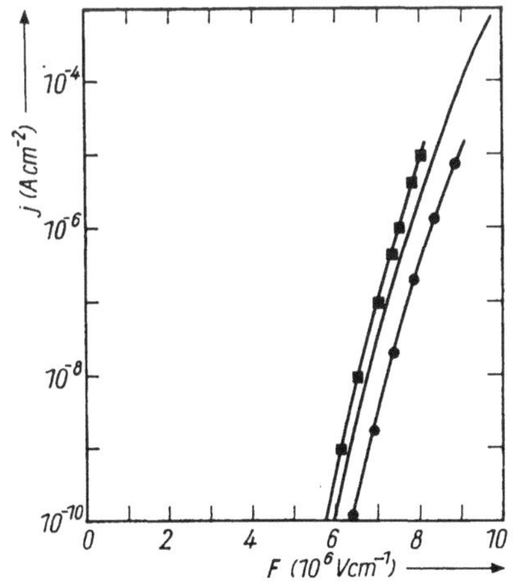


Fig. 4. Fowler-Nordheim tunneling current $j_{FN}$ at the $Si$–$SiO_2$ barrier in dependence on field strength $F$ according to Lenzlinger and Snow [14] (●), Powell [15] (■), as well as our approximation (solid curve). $j_{FN} = AF^2 \exp(-B/F)$ with $A = 8.77 \times 10^{-8}$ A/V² and $B = 2.26 \times 10^8$ V/cm

Substituting the constant $B$ for the Si–$SiO_2$ junction we get an estimation $\Delta j_{FN}/j_0 \approx \approx 34\Delta F/F_0$. This indicates a high field sensitivity of the Fowler-Nordheim current. According to the previous section the field is built up in a self-consistent process by the charge $\varrho(x)$ incorporated within the insulating layer. Consequently, the field strength at the interface $x = d$ is, according to (4),

$$F(d) = \frac{1}{\varepsilon_0 \varepsilon_r} \int_0^d \varrho(x) \, dx \,. \tag{13}$$

An estimation of a charge density $\Delta\varrho\Delta x \approx 8 \times 10^{-10}$ As/cm² in $SiO_2$ leads with (13) to a field change $\Delta F/F_0 \approx 3.5 \times 10^{-4}$ and with (12) to a current flux of $\Delta j_{FN}/j_0 \approx 1\%$. This detection sensitivity of about $5 \times 10^9$ elementary charges per cm² in the insulating layer is better than the possibilities of common capacity–voltage or photoinjection methods and favours the SEFE as a powerful tool in charge determination in thin dielectric layers.

## 5. Trap Spectroscopy in Thin Insulating Layers

First-order trapping kinetics of charges in localized levels of spatial concentration $n_{t0}$ with capture cross section $\sigma_t$ and thermal activation energy $E_t$ is described by a linear differential equation

$$\frac{dn_t}{dt} = \frac{j_i}{e_0} \sigma_t (n_{t0} - n_t) - f \exp\left(-\frac{E_t}{kT}\right) n_t \,, \tag{14}$$

where $n_t(t, T)$ is the actual concentration of occupied traps as a function of time $t$ and temperature $T$. $j_i/e_0$ means the injection density of charges and $f$ is a nearly constant, pre-exponential factor. The first term on the right-hand side of (14) describes the charge storage dominating at low temperatures and resulting in a simple exponential saturation law,

$$n_t = n_{t0}\left[1 - \exp\left(-\frac{j_i}{e_0}\sigma_t t\right)\right] = n_{t0}\left[1 - \exp\left(-\frac{t}{\tau}\right)\right], \tag{15}$$

with the mean filling time $\tau$.

On the other hand, absence or neglection of charge injection $j_i \to 0$ as well as a constant heating rate $\mu = dT/dt =$ const lead to the dominating role of the second term of (14) and to the decrease of occupied centers

$$n_t = n_{t0} \exp\left[-\frac{f}{\mu} \int_{T_0}^{T} \exp\left(-\frac{E_t}{kT}\right)\right] dT \,, \tag{16}$$

i.e. exhaustion of traps by carrier emission.

The negative slope of the second term of (14), describing carrier emission from traps,

$$S = -\frac{dn_t}{dT} = \frac{f}{\mu} \exp\left(-\frac{E_t}{kT}\right) n_t \tag{17}$$

is mainly ruled by the heating rate $\mu$. Applying different heating rates $\mu_1 \neq \mu_2$ we will get for equal trap filling $n_t(\mu_1, T_1) = n_t(\mu_2, T_2)$ the elimination of the actual $n_t$-value,

$$\frac{S_1(\mu_1, T_1)}{S_2(\mu_2, T_2)} = \frac{\mu_2}{\mu_1} \exp\left[\frac{E_t}{k}\left(\frac{1}{T_2} - \frac{1}{T_1}\right)\right], \tag{18}$$

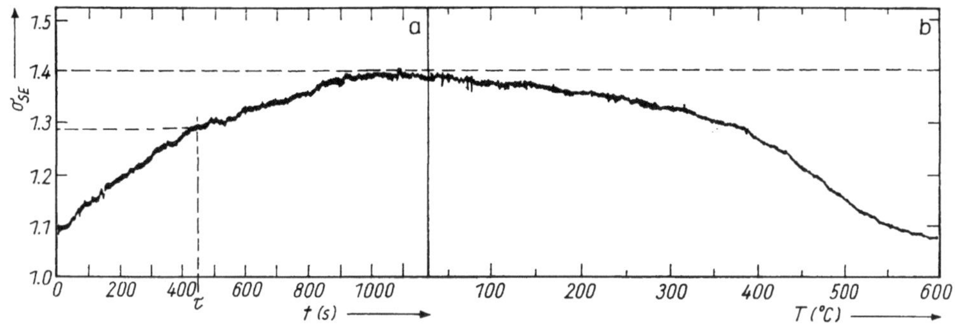


Fig. 5. a) Electron beam injection $E_0 = 1$ keV, $j_0 = 3 \times 10^{-6}$ A/cm² in a 60 nm $Ta_2O_5$ layer on n-Si substrate as well as b) thermal stimulation with a heating rate $\mu = 1.33$ K/s ($T = T_0 + \mu t$)

and immediately the trap activation energy

$$E_t = k \frac{T_1 T_2}{T_1 - T_2} \ln \frac{\mu_1 S_1(\mu_1, T_1)}{\mu_2 S_2(\mu_2, T_2)}. \tag{19}$$

The formalism described by (14) to (19) was checked by us for the SEFE method. Changes in trap occupation concentration $n_t$ were indicated by changes of SE current $j_{SE}$ according to (10) to (12) or the SE emission fraction $\sigma$ according to the definition in Section 3. We only have to keep the condition for Fowler-Nordheim field emission,

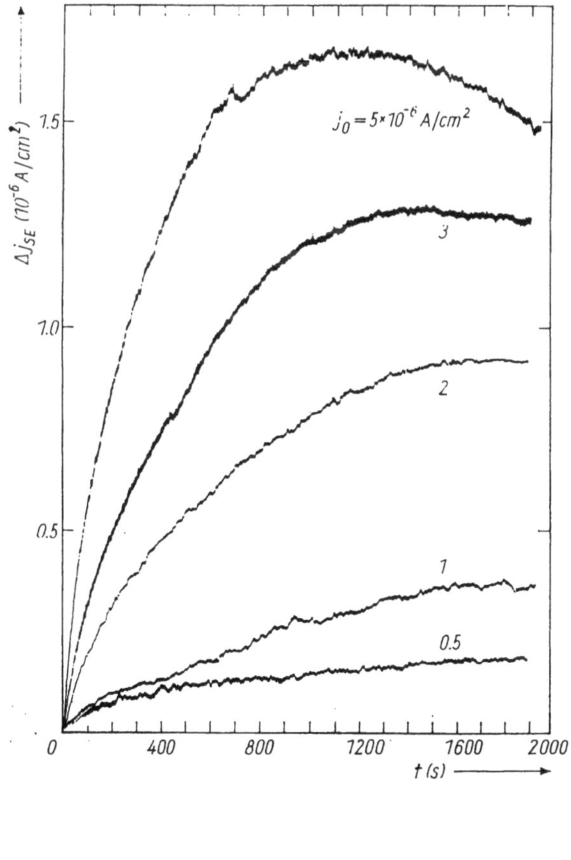


Fig. 6. Hole injection into $Ta_2O_5$ layers (as Fig. 5) at various PE current densities $j_0$

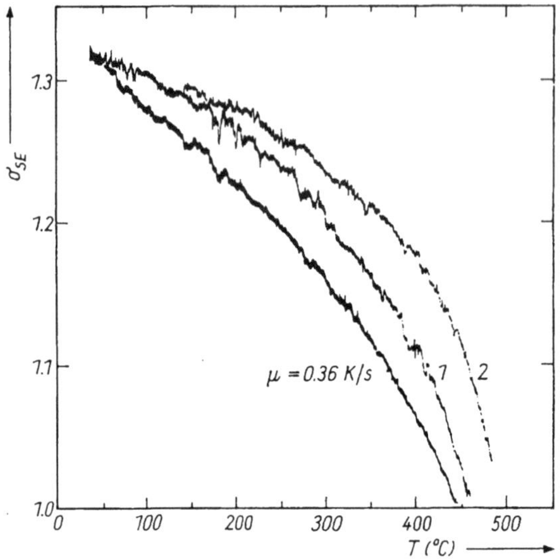


Fig. 7. Thermally stimulated release of holes in $Ta_2O_5$ layers (as Fig. 5) at various heating rates $\mu = dT/dt$

i.e. $\sigma > 1$ for energies $E_0 < 2$ keV and insulating layer thicknesses $d < 150$ nm. For anodically oxidized $Ta_2O_5$ layers with thickness $d = 60$ nm on n-Si substrate the full characteristic of charge storage and thermal stimulation is shown in Fig. 5. $Ta_2O_5$ is

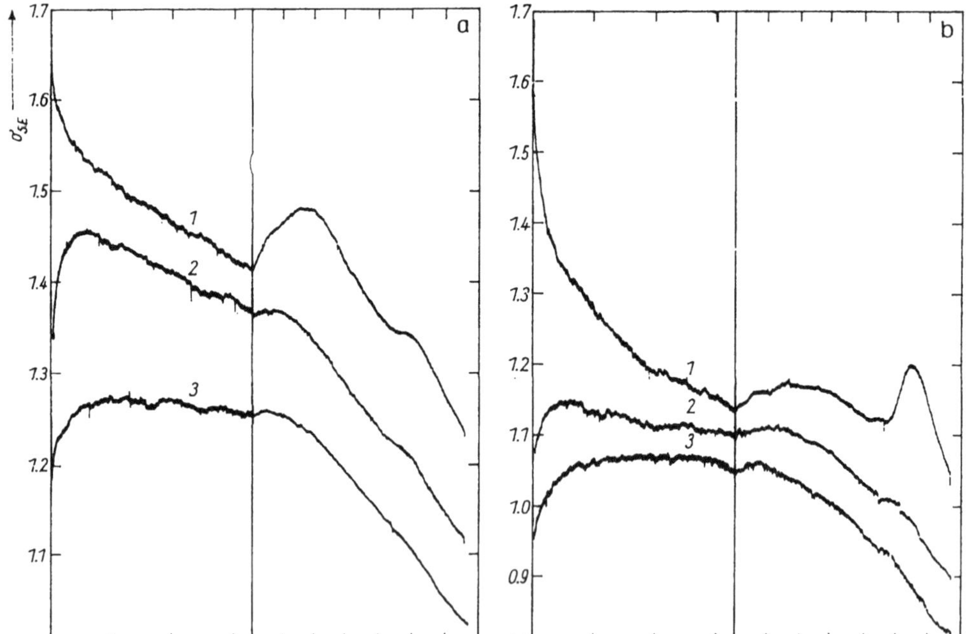


Fig. 8. Consecutive injection — stimulation cycles a) at 70 nm $Al_2O_3$ layers on n-Si and b) at $SiO_2$ layers on *n*-Si; $E_0 = 1$ keV, $j_0 = 10^{-5}$ A/cm², $\mu = 2$ K/s ((1), (2) and (3) are the first, second, and third injection — stimulation cycle, respectively)

known to have a relatively high hole trap concentration. Therefore, electron bombardment with energy $E_0 = 1$ keV and desity $j_0 = 3 \times 10^{-6}$ A/cm$^2$ predominantly leads to a storage of holes, indicated by the increase of SE emission fraction $\sigma$ with time in Fig. 5a. These holes are released again from their traps by thermal stimulation with a heating rate $\mu = 1.33$ K/s (Fig. 5b). In Fig. 6 the charge storage formula (15) was checked by variation of the injection density $j_i$. From the value $[1 - (1/e)] \Delta j_{sat}$, i.e. 63% of the saturation change $\Delta j_{sat}$ at mean saturation time $\tau$ we immediately find the capture cross section of carriers,

$$\sigma_t = \frac{e_0}{j_i \tau}. \tag{20}$$

For the present $Ta_2O_5$ layer the capture cross section of holes has been determined as $\sigma_t \approx (2 \pm 1) \times 10^{-17}$ cm$^2$. Application of the heating rate variation according to (18) and (19) in Fig. 7 results in hole activation energies $E_t \approx 0.2$ eV.

Fig. 8 shows charge storage and thermal stimulation of a 70 nm $Al_2O_3$ layer on n-Si. Here electron traps are dominating, shown by a decrease of $\sigma$ during injection and an increase of $\sigma$ for the successive thermal stimulation.

However, it becomes obvious that both kinds of charge carriers, electrons and holes, are stored as well as released in such an injection-stimulation cycle. The same phenomenon is observed in $SiO_2$ layers (Fig. 8b). In case of sufficiently different trap parameters: capture cross sections $\sigma_{t,i}$ and thermal activations energies $E_{t,i}$ it is possible to determine these parameters for electron and hole traps independently, as well as for two or more different traps of one and the same kind of charge carrier, electrons or holes. In a further paper we will present a more detailed evaluation formalism and special examples for trap spectroscopy in insulating layers by the SEFE method presented here.

## References


[1] L. Malter, Phys. Rev. 49, 478 (1936).
[2] H. Seiler, Z. angew. Phys. 22, 249 (1967).
[3] A. R. Shulman and S. A. Fridikhov, Secondary Emission Methods for Investigation of Solids, Izd. Nauka, Moscow 1977 (in Russian).
[4] M. Franke, Diplomarbeit, Rostock 1977.
[5] W. Müller, Diplomarbeit, Rostock 1978.
[6] H.-J. Fitting, H. Glaefeke, W. Wild, M. Franke, and W. Müller, Exper. Tech. Phys. 27, 13 (1979).
[7] H.-J. Fitting, Phys. Halbleiteroberfläche (Proc. Acad. Sci. GDR) 14, 183 (1983); 15, 277 (1984).
[8] H.-J. Fitting, Wirtschaftspatent DD 214690, G 01 N 23/22 (1984); Wirtschaftspatent DD 224941 A1, G 01 N 23/225, H01 L 21/76 (1985).
[9] H. Salow, Z. techn. Phys. 21, 8 (1940).
[10] A. J. Dekker, Solid State Phys. 6, 251 (1958).
[11] J. M. Bronshtein and B. S. Fraiman, Secondary Electron Emission, Izd. Nauka, Moscow 1969 (in Russian).
[12] H.-J. Fitting, J. Boyde, and J. Reinhardt, phys. stat. sol. (a) 81, 323 (1984).
[13] H. Fröhlich, Adv. Phys. 3, 325 (1954).
[14] M. Lenzlinger and E. H. Snow, J. appl. Phys. 40, 278 (1969).
[15] R. J. Powell, IEEE Trans. Nuclear Sci. 22, 2240 (1975).


*(Received December 3, 1987; in revised form April 4, 1988)*

phys. stat. sol. (a) **108**, 275 (1988)

Subject classification: 68.55; 78.30; S5.11

*Fachbereich Physik der Universität Kaiserslautern*[1])

# The Influence of Hydrogen Surface Reactions on the Growth of Evaporated a-Si:H Films[2])

By

S. ISELBORN, B. SCHRÖDER, and J. GEIGER†[3])

Thin a-Si:H films are deposited by a direct reaction of evaporated silicon with atomic hydrogen. It is shown that a definite ratio of the number of incoming Si and H atoms is necessary to obtain material without microstructure and with a clear dominance of diluted monohydride bonding configurations. An adequate hydrogen etching process has to take place at the surface of the growing film in order to produce high quality material by the evaporation technique. A steric model is presented which explains the selective etching of $SiH_2$ and $SiH_3$ bonding sites by atomic hydrogen.

Dünne a-Si:H-Schichten werden durch direkte Reaktion von aufgedampftem Silizium mit atomarem Wasserstoff hergestellt. Es wird gezeigt, daß ein bestimmtes Verhältnis der auf das Substrat auftreffenden Silizium- und Wasserstoffatome notwendig ist, um ein Material ohne Mikrostrukturen zu präparieren, in dem über das Volumen verteilte Monohydrid-Bindungskonfigurationen eindeutig dominieren. Nur wenn an der Oberfläche der wachsenden Schicht ein ausreichender Wasserstoffätzprozeß stattfindet, kann mit der Aufdampftechnik hochwertiges a-Si:H-Material abgeschieden werden. Ein sterisches Modell wird vorgestellt, welches die selektive Ätzung der $SiH_2$ bzw. der $SiH_3$-Bindungskonfigurationen durch atomaren Wasserstoff erklärt, die für ein inhomogenes Schichtwachstum verantwortlich sind.

## 1. Introduction

The preparation of amorphous silicon (a-Si) by evaporation is a very simple method and has frequently been used especially in the early stages of amorphous semiconductor physics. But the deposition of well-passivated, hydrogenated amorphous silicon by means of the evaporation technique (ev-a-Si:H) is considerably more difficult (see e.g. [1]) than the common glow discharge decomposition of silane (gd-a-Si:H). Since the electronic quality of ev-a-Si:H is not comparable to that of a-Si:H prepared by other techniques, e.g. gd-a-Si:H, the evaporation method became less interesting lately. Nevertheless this preparation technique can be very helpful for basic research on a-Si:H material and further improvements may enable the use of this method at least for some special applications.

Previous results [1] clearly proved that molecular hydrogen has only little influence on the electronic properties of evaporated amorphous silicon. As found by several authors (see [2]) atomic hydrogen is necessary for an efficient defect passivation. However, most of these studies on ev-a-Si:H mainly focus on the opto-electronic properties. The structural properties have not been extensively studied except for the investigations of Viturro and Weiser [2] concerning the influence of structural dis-

[1]) PF 3049, D-7650 Kaiserslautern, FRG.
[2]) This paper includes parts of the Thesis by S. Iselborn.
[3]) Prof. Dr. Jürgen Geiger died on 20th December 1985.

order on the band-edge absorption. Therefore, this study concentrates on structural investigations of a-Si:H films that have been grown by direct reaction between hydrogen atoms and silicon.

Knights [3] found a close correlation between the existence of microstructure in a-Si:H and its electronic properties. Although this conclusion does not seem to be fully clear (Moustakas [4]), it stresses the importance of structural investigations. In that context a better understanding of the physical processes which control the structure of the film might be helpful. Yet, such a knowledge is still rudimentary because of the complexity of processes in a glow discharge due to the large number of species involved and the consequently possible reaction mechanisms. Therefore, we have chosen an approach in which the silicon films can be investigated in an evaporation system independently of other processes.

## 2. Experiment

A scheme of the experimental setup for the film preparation is shown in Fig. 1. The main component is an UHV evaporation system consisting of two chambers for the electron beam evaporator and the substrate holder, respectively. During the deposition a hydrogen beam is partly dissociated and directed onto the substrate. The dissociation is accomplished by a newly developed microwave discharge system [5]. Compared to similar approaches (see e.g. [2]) the flux of hydrogen atoms could be enhanced in such a manner that deposited silicon atoms can be etched away to a considerable extent. The microwave discharge system consists of a tube of borosilicate

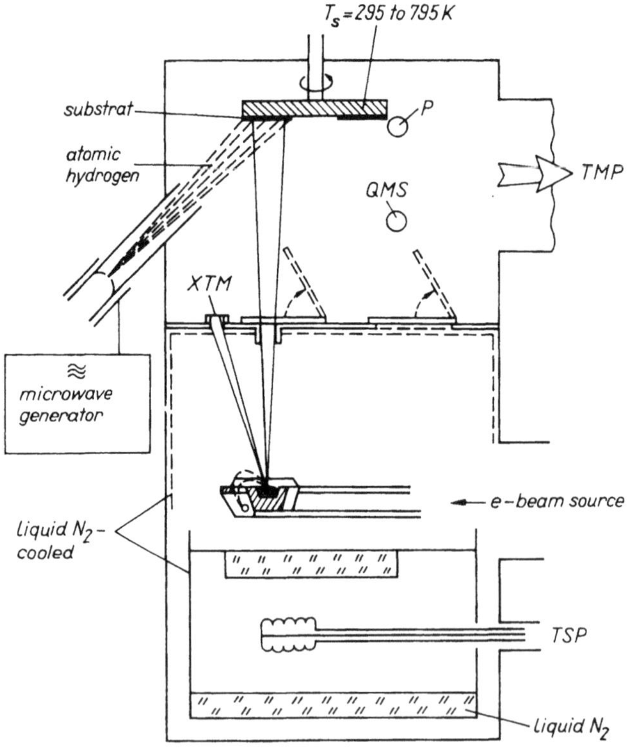


Fig. 1. Scheme of the deposition system: XTM crystal thickness monitor, QMS quadrupole mass spectrometer, P ionisation pressure meter, TMP turbo molecular pump, TSP titanium sublimation pump

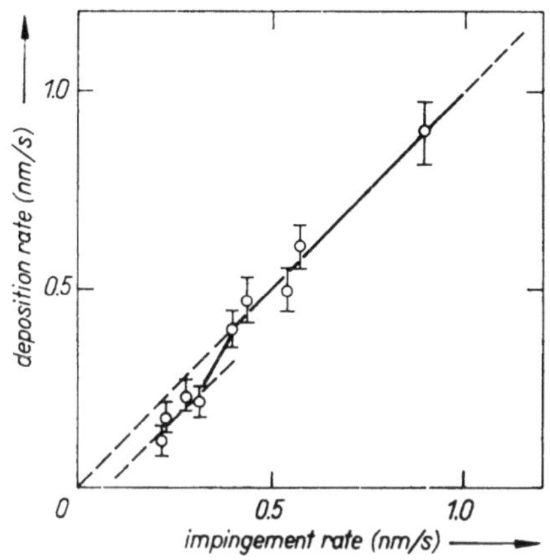


Fig. 2. Impingement rate vs. deposition rate ($T_s = 250$ to 320 °C, $p_H = 4 \times 10^{-3}$ Pa)

glass (pyrex) with an outlet nozzle of variable width. The diameter of the tube is 13 mm. The microwave power of typically 60 W is supplied by a power supply Microtron III of Electro Medical Supplies (England). The hydrogen pressure in the glass tube is held constant at 0.53 hPa to ensure constant and reproducible dissociation. The diameter of the nozzle controls the flux of hydrogen atoms onto the substrate and thus the hydrogen pressure in the deposition chamber. The hydrogen pressure is selected as the experimental parameter which defines the experimental situation concerning the hydrogen interaction. The determination of the flux or fraction of atomic hydrogen was not possible in our system. But from optical control of the microwave discharge and from measurements of Waibel [5] on a similar dissociation system we conclude that the directed hydrogen beam is dissociated to at least 30 to 50%. The silicon deposition rate and the film thickness can be measured in situ by an Inficon XTM quartz crystal monitoring system which was calibrated and steadily rechecked with optical methods. The calibration holds for almost all of the various deposition parameters. It becomes incorrect if the hydrogen particle current density is high enough to etch away a perceptible part of the film. In that case the quartz oscillator system measures a number which is proportional to the Si-impingement rate rather than to the Si-deposition rate. This is shown in Fig. 2 for the hydrogen pressure $p_H = 4 \times 10^{-3}$ Pa. At high deposition rates the deposition and impingement rates are equal. When the rate measured by the thickness monitor falls below 0.4 nm/s, etching is perceptible and the deposition rate becomes smaller than the impingement rate.

An extrapolation of the experimental data to the deposition rate $r_d = 0$ nm/s gives an etching rate $r_e = 0.1$ nm/s for $p_H = 4 \times 10^{-3}$ Pa and microwave dissociation as a rough estimate. Of course, the etching rate can depend on the impingement rate, thus this extrapolation has to be regarded with caution. In the next chapter it will be shown that the etching processes have considerable influence on the growth and structure of the films. The films were deposited on freshly cleaved NaCl crystals for electron microscopy (TEM Siemens Elmiskop) or on polished Si monocrystals for infrared spectroscopy (FTIR spectrometer Nicolet MX 1).

## 3. Results

### *3.1 Studies with the electron microscpe (TEM)*

The dependence of the morphology of a-Si and a-Si:H films on the experimental parameters is presented in Fig. 3. Fig. 3a shows the morphology of pure a-Si films. The appearance agrees with results reported earlier by other authors [6]. Fig. 3b to d

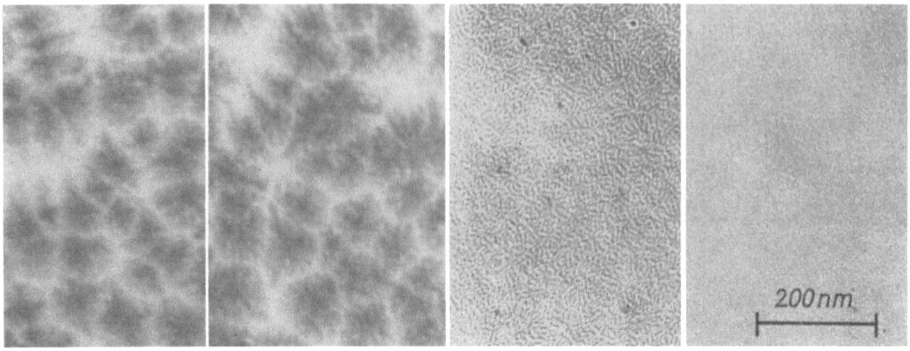


a b c d

Fig. 3. TEM micrographs of a) ev-a-Si and b) to d) different ev-a-Si:H films (for comments see text)

show micrographs of a-Si:H films which have been fabricated by reaction of atomic hydrogen with Si. The hydrogen content $c_H$, substrate temperature $T_s$, and growth rate $r_d$ are almost constant for all films ($c_H = 10$ at%, $T_s = 250$ to 300 °C, $r_d \approx \approx 0.3$ nm/s). If the hydrogen is decomposed by thermal dissociation (Fig. 3b), the film structure is characterized by dendritic growth giving rise to islands with diameters of 80 to 100 nm similar to pure a-Si. If the current density of atomic hydrogen at the substrate is increased by using the microwave dissociation system the structure of the films changes in a systematic manner: the contours smooth out (Fig. 3c, hydrogen pressure $p_H = 3 \times 10^{-4}$ Pa). The coarse island structure finally disappears in the case of surface etching by atomic hydrogen (Fig. 3d, $p_H = 4 \times 10^{-3}$ Pa) and the micrograph becomes similar to that of sp- or gd-a-Si:H material.

### 3.2 *Infrared absorbance*

In Fig. 4 a typical infrared transmission spectrum of evaporated a-Si:H is depicted. If no special care has been taken concerning etching reactions of atomic hydrogen at the growing surface (see below) five different absorption bands of localized silicon hydrogen vibrations will be observed. Quoting Brodsky et al. [7] these bands can be classified as valence vibrations between 1900 and 2200 $cm^{-1}$, scissors bending vibrations between 800 and 950 $cm^{-1}$, and rocking or wagging vibrations around 640 $cm^{-1}$.

Recently Beyer and Wagner [8] have reinterpreted the infrared spectra in terms of structure. They concluded that a-Si:H films show a vibrational absorption line at 2000 $cm^{-1}$ if the absorbing Si–H oscillator is incorporated in a compact bulk solid. On the other hand, a-Si:H absorbs infrared light at 2080 $cm^{-1}$ if H is bonded to (inner) surfaces. In the last case not only Si–H but also Si–$H_2$ bonding configurations are involved. This can be concluded from electron energy loss spectroscopy of hydrogen absorbed at crystalline silicon surfaces [8], as well as from the occurrence of weak absorption bands around 840 to 890 $cm^{-1}$ which have been associated with bending vibrations of the $SiH_2$ group [9]. The appearance of the 2080 $cm^{-1}$ stretching absorption band indicates an inhomogeneous a-Si:H material that usually does not possess device quality.

Following Shanks et al. [10] the total hydrogen content of a-Si:H films can be obtained by integrating the spectral line due to wagging absorption at 640 $cm^{-1}$ according to

$$c_H = A_w \int \frac{\alpha(\omega)}{\omega} \, d\omega \, . \tag{1}$$

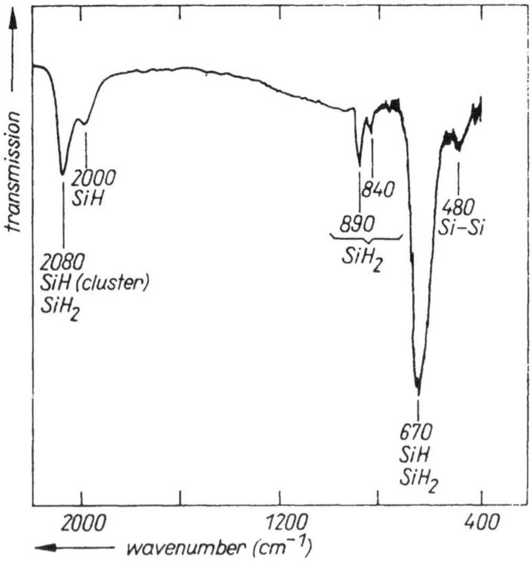


Fig. 4

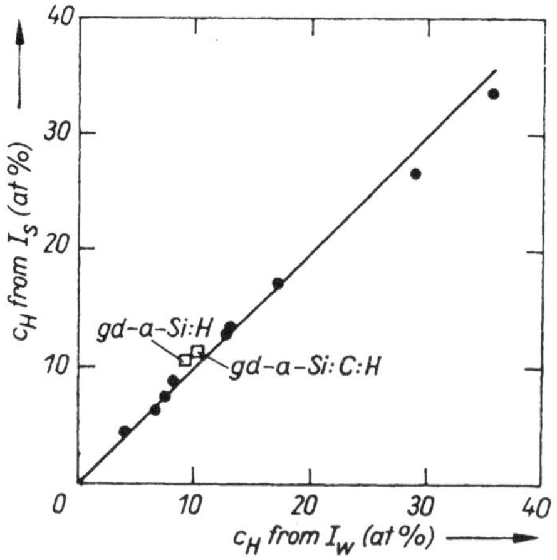


Fig. 5

Fig. 4. Typical IR transmission spectrum of evaporated a-Si:H ($T_s$ = 250 °C, $p_H = 3 \times 10^{-4}$ Pa, $r_d$ = 0.5 nm/s (read 640 instead of 670)

Fig. 5. Hydrogen concentration $c_H$ of evaporated a-Si:H samples calculated from the integrated absorption of the stretching vibration $I_s = I_s$ (2000) + $I_s$(2080) vs. $c_H$ from the integrated absorption of the wagging vibration $I_w = I_w$(640). $T_s$ = 300 to 570 K

Here $A_w = 1.6 \times 10^{19}$ cm$^{-2}$ is a semi-empirical proportionality constant [10]. Although some doubt was cast on this in the literature [9, 10], it is also possible to calculate the amount of hydrogen (at least for the evaporated and sputtered a-Si:H films) using one single proportionality constant $A_s = 1.3 \times 10^{20}$ cm$^{-2}$ for the stretching absorption spectral lines at 2000 and 2080 cm$^{-1}$ [11]. This is demonstrated in Fig. 5 which relates the hydrogen content $c_H$ of a-Si:H films calculated from the wagging mode to the hydrogen content obtained from the stretching mode. The squares in Fig. 5 are data on gd-a-Si:H and gd-a-Si:C:H films supporting the idea that this method can also be applied to samples that have been prepared by glow discharge decomposition of $SiH_4$. We will use the notation $c_H$ (640 cm$^{-1}$) = $c_H(I_w)$, $c_H$ (2000 cm$^{-1}$), $c_H$ (2080 cm$^{-1}$), and $c_H(I_s)$ = $c_H$ (2000 cm$^{-1}$) + $c_H$ (2080 cm$^{-1}$) to indicate from which spectral lines the hydrogen content has been calculated.

The influence of the deposition rate on the infrared spectra or the amount of hydrogen incorporated in different bonding configurations is shown in Fig. 6a and b for hydrogen pressures $p_H = 3 \times 10^{-4}$ and $4 \times 10^{-3}$ Pa, respectively. At the lower partial pressure, i.e. for smaller atomic hydrogen flux on the substrate surface (Fig. 6a), both stretching modes are visible for all deposition rates. The value for $c_H$ (2000 cm$^{-1}$) is independent of the deposition rate within a range of 5 at%, whereas the total hydrogen content $c_H$ (640 cm$^{-1}$) = $c_H$ (2000 cm$^{-1}$) + $c_H$ (2080 cm$^{-1}$) rises due to the increase of $c_H$ (2080 cm$^{-1}$). At the higher partial pressure (Fig. 6b) the total amount of hydrogen $c_H$ (640 cm$^{-1}$) almost remains constant ($c_H \approx 10$ at%). At deposition rates 0.4 nm/s $< r_d <$ 1.1 nm/s we obtain $c_H$ (2000 cm$^{-1}$) = 5 to 6 at%. Reducing the deposition rate below 0.4 nm/s results in a decrease of $c_H$ (2080 cm$^{-1}$) and an increase in $c_H$ (2000 cm$^{-1}$). At rates $r_d <$ 0.3 nm/s we find $c_H$ (2080 cm$^{-1}$) $\approx 0$ and $c_H$ (2000 cm$^{-1}$) = $c_H$ (640 cm$^{-1}$). Fig. 4 and 6 clearly show that for most of the experi-

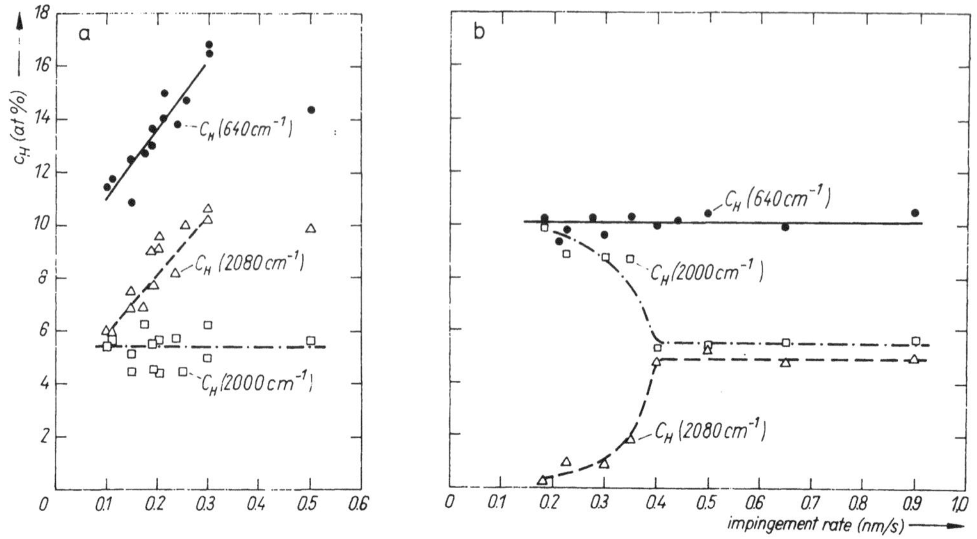


Fig. 6. Total (● $c_H$ (640)) as well as bulk (□ $c_H$ (2000)) and surface (△ $c_H$ (2080)) hydrogen concentration vs. deposition rate for a) small hydrogen bombardment ($p_H = 3 \times 10^{-4}$ Pa) and b) for large hydrogen bombardment ($p_H = 4 \times 10^{-3}$ Pa). $T_s$ = 250 to 320 °C

mental conditions H is distributed among both of the possible environments, i.e. diluted monohydride phase as well as SiH and $SiH_2$ bonding configurations at inner surfaces. In these circumstances the hydrogen in the deposition equipment has only little influence on the growth of the film. The role of hydrogen seems to be restricted to the passivation of dangling bonds. (Note that this passivation can be so efficient that drastic improvements of the electronic properties are observed [4].) If the experimental parameters are chosen in a way that perceptible etching occurs

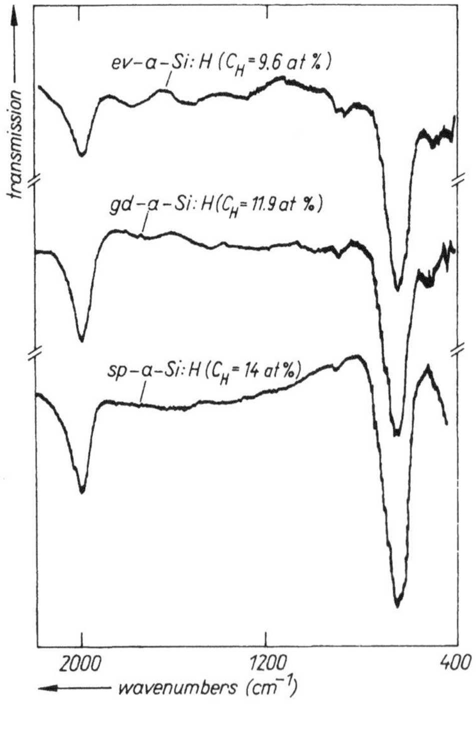


Fig. 7. Comparison of the IR absorption spectra of homogeneous ev-a-Si:H ($c_H$ = 9.6 at%), sp-a-Si:H ($c_H$ = 14 at%), and gd-a-S:H ($c_H$ = = 11.9 at%) films

($p_H = 4 \times 10^{-3}$ Pa, $r_d < 0.3$ nm/s in Fig. 6b), the infrared spectra change. Hydrogen is bonded to bulk solid sites only and the amount of hydrogen in this configuration increases to 10 at %. Obviously, etching reactions remove inner surface sites. In that way the formation of inner surfaces is suppressed and a homogeneous bulk solid grows. The infrared spectrum of such a film favourably compares with infrared spectra of "high electronic quality films" that have been prepared by glow discharge decomposition of silane or reactive magnetron sputtering (Fig. 7). Fig. 7 also shows that an asymmetry of the 2000 $cm^{-1}$ absorption band to higher wave numbers as well as small absorption features in the bending mode regime (840 to 890 $cm^{-1}$) are present in all spectra. These features indicate that a very small amount of $SiH_2$ bonding configurations remain even in high quality a-Si:H material.

The dependence of hydrogen incorporation on deposition rate and atomic hydrogen flux as well as the fact that a small number of $SiH_2$ configurations remain even in high quality a-Si:H material indicates that SiH bonding configurations on the growing surface are more stable against etching reactions by atomic hydrogen than $SiH_2$ bonding configurations. Therefore, a sufficient amount of atomic hydrogen striking the surface removes most of the $SiH_2$ sites and enables the growth of a homogeneous and dense a-Si:H network that contains only monohydride bonding configurations. If the amount of atomic hydrogen hitting the growing surface is insufficient, the remaining SiH bonding configurations are responsible for an inhomogeneous growth of the films (dendrites, voids, columnar structures) since $SiH_2$ sites cannot be incorporated in a

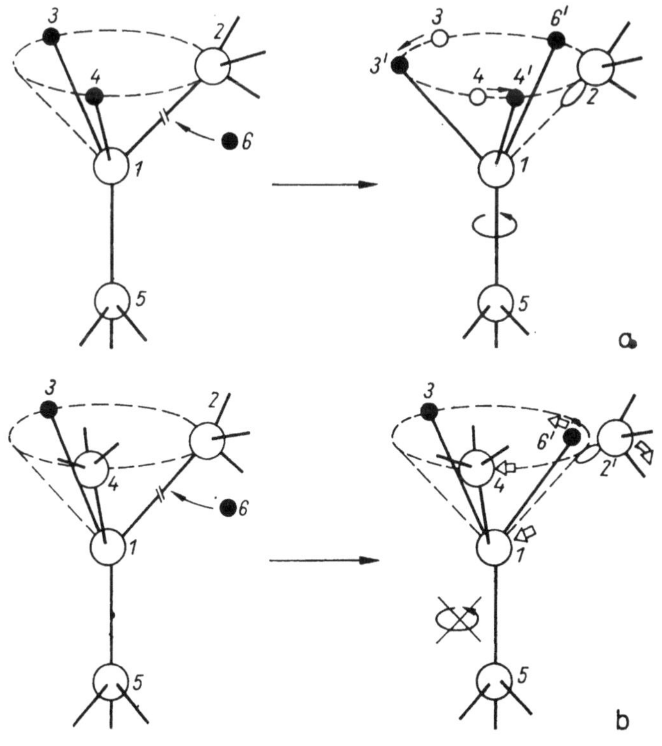


Fig. 8. Microscopic steric model of the selective hydrogenation a) $SiH_2 + H \rightarrow SiH_3$ and b) $SiH + H \rightarrow SiH_2$ (big open circles Si, small black circles H)

dense a-Si:H network. Of course it is not sufficient to argue that fewer steps are necessary to etch $SiH_2$ groups. There must be a reason why the reaction

$$Si:H(solid)\text{–}SiH + H \rightarrow Si:H(solid)\text{–}SiH_2 \tag{2}$$

is less likely than the reaction

$$Si:H(solid)\text{–}SiH_{2(3)} + H \rightarrow Si:H(solid)\text{–}SiH_{3(4)}\,. \tag{3}$$

The term Si:H(solid)–SiH in (2) means that there is a Si atom at the surface having a single hydrogen bond while the remaining three bonds are connected to Si atoms of the a-Si:H network (see Fig. 8).

## 4. A Steric Model for Hydrogen Etching Reactions

Differences between the two processes described by (2) and (3) can be suspected due to steric reasons. Geometric models of the structural units at the surface, at which the reactions occur, are shown in Fig. 8. The small black circles represent hydrogen atoms, the large balls are silicon atoms. All atoms are numbered to simplify the description. One main difference to the gas phase which must be taken into account is the fact that the hydrogenation step takes place at silicon atoms which are bonded to a solid. The site and the local tetrahedral surroundings are predetermined by the cluster before the reaction takes place. Only reactions which do not alter this topology are likely. Fig. 8a depicts reaction (3). The hydrogen atom 6 from the gas phase reacts with the surface in a way that the bond between the Si atoms 1 and 2 is broken and the H atom is attached to the silicon atom 1. But the Si atom 2 will remain with a dangling bond at its site because it is fixed to the solid by three other bonds. The repulsion between the involved orbitals will lead to a rotation (approximately 60°) of the whole $SiH_3$ group around the Si–Si bond (atoms 1 and 5). Thus a position is found for all atoms which is energetically favourable.

The analogous reaction (2) is less likely because of steric reasons. The repulsion between Si atom 2 and H atom 6 cannot be overcome by a similar mechanism because the atom $4 = 4'$ is also bonded to the network and cannot move easily. Therefore, a rotation of that structural unit will be hindered. The repulsion between the orbitals will lead to movements of the atoms. The movements are intimated by arrows in Fig. 8b. A simulation of these processes with a ball and stick model according to Polk [12] shows that strong local forces occur in the neighbourhood of the structural unit. To make the reaction possible, a considerable change or relaxation of the network influencing many bond lengths and angles would be necessary. Only if the number of hydrogen atoms arriving at the surface is so great that two bonds can be broken simultaneously, such a structural unit may be hydrogenated.

## 5. Discussion

The experimental data clearly show that the reason for the structural changes taking place in ev-a-Si:H films are etching reactions. These reactions suppress the development of dendrites already in the initial stages of the film growth. As dendrites have many dangling bonds and $SiH_2$ bonding configurations, they may easily be removed by H reactions. Compact regions of the film are more stable against etching. In that way the growth of homogeneous films is controlled by selective chemical reactions. In the case of glow discharge deposition or reactive sputtering of a-Si:H the mechanisms that control the homogeneity seem to be less clear. In his growth model, Knights [3] discusses that the glow discharge could enhance the surface mobility of silicon atoms. Enhanced surface mobility might improve the coalescence of islands to homogeneous films. Tsai et al. [13] argue that the film homogeneity and quality is correlated with the sticking coefficients $k$ of the species striking the growing film. For typical chemical vapour deposition (CVD) like glow discharge decomposition $k$ is close to zero (in contrast to typical physical vapour deposition (PVD) with $k \approx 1$). In agreement with that model hydrogen etching of our ev-a-Si:H films practically lowers the sticking coefficient of Si atoms and converts the evaporation (PVD) to a partial CVD process.

For the sputtering processes various groups have given different and even contradictory reasons for the improvement of structural homogeneity or electronic properties. Depending on the preparation conditions the bombardment of the substrate with Ar ions [14], energetic neutrals, or fast electrons [3], or ions and neutrals of moderate energies below the implantation threshold [15] have been argued to be helpful. However, the actual effect of bombardment is less clear. Columnar growth might be suppressed by resputtering of loosely bonded atoms or protrusions or/and by enhanced atomic diffusion [16]. On the other hand, atomic hydrogen will be produced during the sputtering process by dissociation of $H_2$ in the sputtering gas, too.

The hydrogen etching reaction of course influences the electronic properties of ev-a-Si:H. The ratio of photo-to-dark conductivity $\sigma_{Ph}(AM1)/\sigma_D$ that was pointed out to be a good measure of the photoelectronic quality for sputtered a-Si:H [17] reaches values above $10^4$ if the hydrogen etching produces a homogeneous ev-a-Si:H material. The density of states (DOS) near the Fermi level measured by the space charge limited current (SCLC) method results in values around $10^{17}$ cm$^{-3}$ eV$^{-1}$ for $E_c - E = 0.8$ eV ($p_H = 3 \times 10^{-4}$ Pa) [18] and $3 \times 10^{16}$ cm$^{-3}$ eV$^{-1}$ for $E_c - E = 0.6$ eV ($p_H = 4 \times 10^{-3}$ Pa) [19]. Using this homogeneous material simple Schottky-type solar cells have been fabricated showing an internal conversion efficiency of about 1% [19]. To our knowledge this is the highest value ever reported for evaporated material.

***Acknowledgements***

Helpful discussions with H. Rübel and U. Schneider are gratefully acknowledged. Thanks are due to I. Lorenz for preparational assistance and to H. Waibel for help and discussions concerning the microwave discharge. N. Kniffler provided us with the gd-a-Si:H and gd-a-Si:C:H samples. The work was supported by a grant of the Bundesministerium für Forschung und Technologie.

## References

[1] S. Iselborn, H. Rübel, B. Schröder, and J. Geiger, Phil. Mag. **B48**, 561 (1983).
[2] R. E. Viturro and K. Weiser, Phil. Mag. **B53**, 93 (1986) and references therein.
[3] C. Knights, J. non-crystall. Solids **35/36**, 159 (1980).
[4] T. D. Moustakas, Solar Energy Mater. 8, 187 (1982).
[5] H. Waibel, Thesis, Kaiserslautern 1987.
[6] A. Barna, P. B. Barna, G. Radnóczi, L. Tóth, and P. Thomas, phys. stat. sol. (a) **41**, 81 (1977).
[7] M. H. Brodsky, H. Cardona, and J. J. Cuomo, Phys. Rev. B **16**, 3556 (1977).
[8] W. Beyer and H. Wagner, J. non-crystall. Solids **59/60**, 161 (1983).
[9] M. Cardona, phys. stat. sol. (b) **118**, 463 (1983).
[10] H. Shanks, C. J. Fang, L. Ley, M. Cardona, F. J. Demond, and S. Kalbitzer, phys stat. sol. (b) **100**, 43 (1980).

[11] H. RÜBEL, S. ISELBORN, B. SCHRÖDER, and J. GEIGER, in: Proc. Mater. Res. Soc. Conf. Poly-Micro-Crystals and Amorphous Semiconductors Strasbourg 1984, Ed. P. PINARD and S. KALBITZER, Editions de Physique, Paris 1985 (p. 415).
[12] D. E. POLK, J. non-crystall. Solids **5**, 365 (1971).
[13] C. C. TSAI, J. C. SHAW, B. WACKER, and J. C. KNIGHTS, Proc. Mater. Res. Soc. Spring Meeting, Anaheim 1987, Vol. 95 (p. 219).
C. KNIGHTS, G. CHANG, and B. WACKER, J. appl. Phys. **59**, 2998 (1986).
[14] R. C. ROSS and R. MESSIER, J. appl. Phys. **52**, 5329 (1981).
[15] W. MÜLLER, Thesis, Kaiserslautern 1984.
[16] M. J. THOMPSON, in: The Physics of Hydrogenated Amorphous Silicon I, Ed. J. D. JOANNOPOULOS and G. LUCOVSKY, Topics appl. Phys., Vol. 55, Springer-Verlag Berlin 1984 (p. 119).
[17] H. RÜBEL, B. SCHRÖDER, and J. GEIGER, J. Vacuum Sci. Technol. A **4**, 1855 (1986).
[18] S. GANGOPADHYAY, S. ISELBORN, H. RÜBEL, B. SCHRÖDER, and J. GEIGER, Phil. Mag. Letters **51**, L33 (1985).
[19] S. ISELBORN, Thesis, Kaiserslautern 1985.

*(Received December 10, 1987; in revised form March 31, 1988)*

phys. stat. sol. (a) **108**, 285 (1988)

Subject classification: 68.55 and 73.60; 76.30; 78.65; S5.11

*Sektion Physik der Technischen Universität „Otto von Guericke" Magdeburg*[1])

# Photoconductive a-Si:H with Dominant Monohydride Bonding Prepared by DC-Magnetron Sputtering

By

T. DRÜSEDAU, M. ECKLER, and R. BINDEMANN

Dedicated to Prof. Dr. E. GUTSCHE on the occasion of his 60th birthday

A description is given of a dc-magnetron sputtering apparatus used for deposition of hydrogenated amorphous silicon (a-Si:H) by reactive sputtering in an argon/hydrogen discharge. The physical properties of the deposited a-Si:H films are determined by measuring the photo and dark conductivity, the Tauc gap, the IR absorption and by ESR and RBS measurements. The influence of the main deposition parameters (hydrogen partial pressure, dc-power) on the film properties is discussed. The films reveal a minimal value of dark conductivity of $\sigma_D = 10^{-11}$ $(\Omega cm)^{-1}$ and a four decade higher photoconductivity even for samples prepared at growth rates of 7 μm/h. In all films the hydrogen is predominantly bound in monohydride complexes (Si–H).

Es wird eine DC-Plasmatronbeschichtungsanlage zur Darstellung von hydrogeniertem amorphem Silicium (a-Si:H) mittels reaktivem Sputtern in Argon/Wasserstoffatmosphäre beschrieben. Die physikalischen Eigenschaften der abgeschiedenen a-Si:H Schichten werden durch die Messung der Photo- und Dunkelleitfähigkeit, der Fundamental- und IR-Absorption sowie durch ESR- und RBS-Untersuchungen charakterisiert. Es wird der Einfluß der wichtigsten Darstellungsparameter (Wasserstoffpartialdruck und DC-Leistung) auf die Schichteigenschaften diskutiert. Die Schichten weisen eine minimale Dunkelleitfähigkeit von $\sigma_D = 10^{-11}$ $(\Omega cm)^{-1}$ und eine um vier Größenordnungen höhere Photoleitfähigkeit auch für mit Aufwachsraten von 7 μm/h abgeschiedene Proben auf. Der Wasserstoff wird in allen Schichten vorwiegend in Monohydrid-Komplexen (Si–H) eingebaut.

## 1. Introduction

It is well known that a-Si: H prepared by plasma enhanced chemical vapour deposition (PCVD) of silane turns out to be a material with low concentration of defect states and excellent photoelectronic properties. Therefore, a variety of commercial applications of this material has been realized in the last few years [1, 2]. However, the main disadvantage of the PCVD is the low deposition rate being of importance for applications of a-Si: H requiring a large film thickness $d_s$ as for instance for electrophotographic photoreceptors ($d_s = 10$ to 20 μm). Recently, Savvides [3] and Tamahashi et al. [4] have shown that a-Si: H with low dark- and high photoconductivity can be deposited by dc- as well as rf-magnetron sputtering (MSP) at deposition rates in the range of 10 μm/h. The aim of this work is the optimization of preparation conditions for highly photosensitive a-Si: H by dc-magnetron sputtering at the mentioned high deposition rates. The sputtering apparatus and the magnetron source are described in Section 2. Section 3 gives an overlook of the physical properties of the prepared

[1]) PSF 124, DDR-3010 Magdeburg, GDR.

films showing a somewhat different behaviour in comparison with the results of previous works [3, 4]. In Section 4 the results will be discussed in order to point at some novel aspects in the conception of a-Si:H film deposition by sputtering.

## 2. Sputtering Apparatus

### *2.1 General aspects*

The sputtering apparatus is constructed on the base of a high vacuum evaporation equipment (B 55) and a planar magnetron source (PPS 5E) with a dc-power supply (HZP 53; Inst. v. Ardenne). Prior to sputtering the vacuum chamber was baked out and pumped down to a pressure of $p \approx 10^{-4}$ Pa. The following reduction of pumping speed led to a rise of the background pressure by a factor of about 3. The high purity (5N) gases hydrogen and argon were let in by needle valves. Partial pressures have been measured by a lin-log ionization gauge (IMG 060). For one deposition cycle six substrate sections were mounted on a rotatable substrate holder. Each substrate section contained six substrates of different materials in accordance to the characterization methods described below. Substrate temperature has been kept at $T_s = 200$ °C by a radiation heater and a Fe–Ko thermocouple pressed on the backside of the substrates. The target to substrate distance was fixed at the value of $d_{ts} = 55$ mm at an argon partial pressure of $p_{Ar} = 0.9$ Pa. Substrates have been cleaned by washing in organic solutions and by sputter-cleaning in an argon discharge ($p_{Ar} = 8$ Pa). The target was presputtered against a shutter for at least 15 min prior to each deposition cycle. Substrates were kept at ground potential. The thicknesses of the prepared films were in the range of 400 nm $< d_s <$ 550 nm.

### *2.2 Magnetron sputtering source*

The original planar magnetron was modified by means of a collar anode with an inner diameter of 120 mm and active pole shoes [5]. As targets we used slightly doped electronic grade crystalline silicon discs(125 mm in diameter) being metallurgically bonded onto a copper plate with a thickness of 5 mm and a resistivity of 12 to 20 Ω cm. The horizontal component of the magnetic induction on the target surface revealed a maximum of $B_{H\max} = 38$ mT at 28 mm distance from the centre. In general a magnetron can be described by its target voltage–power ($U_t$–$P$) characteristics which obeys the empirical formula [6, 7]

$$U_t = \left(\frac{P}{K}\right)^{1/n+1} \tag{1}$$

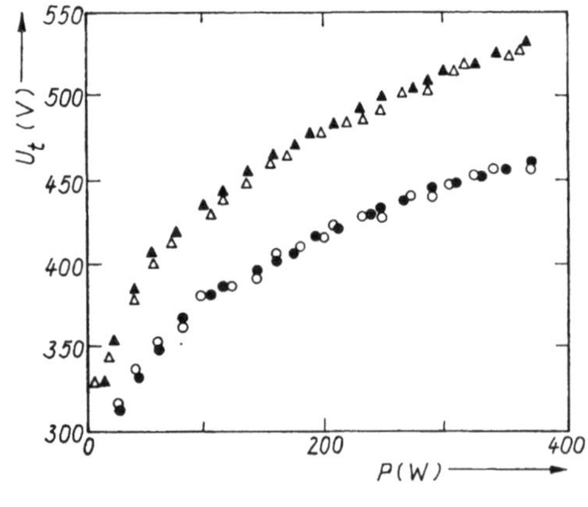


Fig. 1. Target voltage–dc-power characteristics $U_t(P)$ of the magnetron discharge obeying (1) with $n = 5.4$, $K = 3 \times 10^{-15}$ AV$^{-n}$ ($p_{Ar} = 0.9$ Pa, $p_{H_2} = 0.27$ Pa; triangles) and $n = 5.5$, $K = 10^{-15}$ AV$^{-n}$ ($p_{Ar} = 0.9$ Pa; no hydrogen, circles). Open symbols are taken with increasing and closed symbols with decreasing power

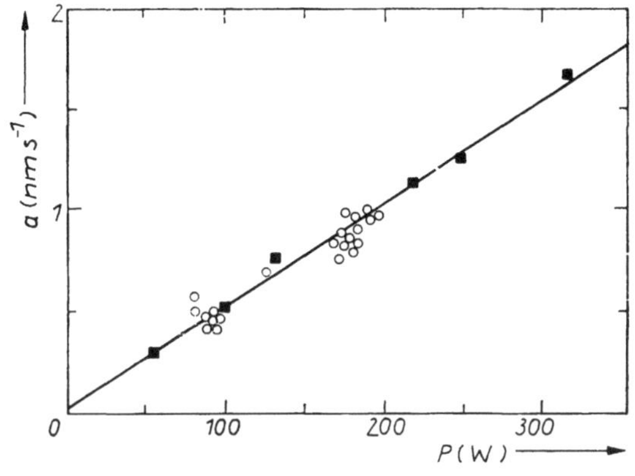


Fig. 2. Deposition rate a vs. dc-power $P$ (circles). The solid line represents $a = cP$ with $c = 5.1$ nm/kWs. Squares represent samples sputtered within one cycle at $p_{H_2} = 0.12$ Pa

with the empirical parameters $K$ and $n$. For the understanding of this relation two processes have to be taken into consideration: First, the coverage of the target by reactive gas atoms [6] and second, the charge transport within the plasma [7]. Fig. 1 shows $U_t$–$P$ characteristics at moderate conditions. It is evident that the target voltages used in the present work are low in comparison with values of other laboratories applying MSP (e.g. [3, 4, 8]) or typical target voltages of conventional sputtering [1, 2]. In contrast to the results of Savvides [3] the addition of hydrogen in our case causes an increase of target voltage. From one deposition cycle to another the target voltage at identical parameters (partial pressure, dc-power) can be reproduced only within an accuracy of $\pm 10$ V. This "uncertainty" of the $U_t$–$P$ characteristics seems to be the main reason for the limited reproducibility in the electrical parameters of a-Si:H (cf. Section 3.1) being a general behaviour for the deposition of a-Si:H by MSP [9, 10].

Savvides argued that the parameter $n$ can be used as an index for the overall efficiency of the magnetron discharge and concluded that the dc-magnetron is a much more efficient sputtering source than the rf-magnetron. By comparing the present results with those of Tamahashi et al. [4] who prepared a-Si:H by rf-MSP at very similar conditions the deposition rate for dc-MSP has been found to be higher by a factor of three. The deposition rate $a$, which was estimated by measuring the film thickness by means of optical transmission measurements in the NIR range and the deposition time, reveals a strong linear dependence on dc-power (cf. Fig. 2) being in agreement with the results of Savvides [3]. This dependence turns out to be a general feature of dc-MSP, whereas rf-MSP reveals a somewhat different $a$–$P$ characteristics [4].

## 3. Characterization of a-Si:H Films

### *3.1 Dark and photoconductivity*

The conductivity was measured in a coplanar electrode arrangement with evaporated Al electrodes of 10 mm length, interelectrode spacing of 1 mm and an electric field of 100 V/cm. The photoconductivity was excited by filtered light ($\lambda = 550$ nm) of a tungsten lamp with a radiation density of 0.8 mW/cm² at the film. For excluding the influence of adsorbates on the conductivity measurements the films have been heated up to a temperature of $T = 100$ °C under vacuum conditions. The reproducibility of dark and photoconductivity of samples prepared under identical conditions within one cycle was within a factor of two, whereas this from cycle to cycle was only within

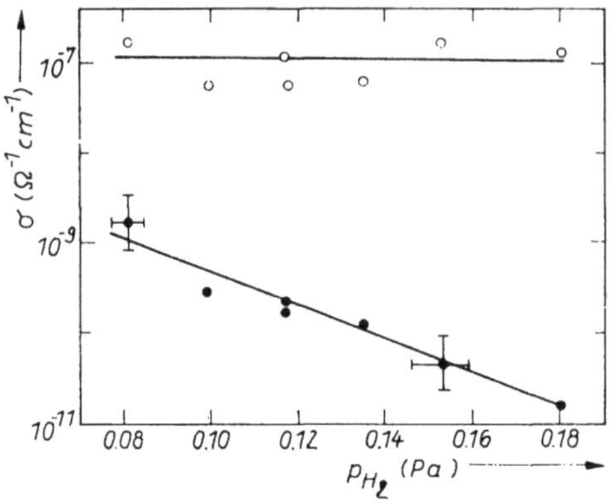


Fig. 3. Conductivity $\sigma$ in dependence on hydrogen partial pressure for a-Si:H films prepared at a dc-power of $P = 90$ W. Open circles photoconductivity (see text), closed circles dark conductivity

one order of magnitude, in agreement with findings of other authors [9, 10]. Fig. 3 shows the influence of hydrogen partial pressure on photo- and dark conductivity of different films. The decrease of dark conductivity results from a rising amount of incorporated hydrogen causing a widening of the mobility gap (cf. Section 3.2) and the energy distance between the conduction band mobility edge $E_c$ and the Fermi level $E_F$. The activation energies determined by measuring the thermally activated conductivity (measured by cooling down from 100 °C) of these samples have typical values of $E_A \approx 0.6$ eV and depend on the hydrogen partial pressure. The independence of the photoconductivity on hydrogen partial pressure can be interpreted in terms of unchanged concentration of Si dangling bonds acting as the main recombination centre in a-Si:H (cf. Section 3.5). Fig. 4 shows the influence of dc-power on the conductivity for two different hydrogen partial pressures. At medium dc-powers of about 150 W we find a minimum in the conductivity for the lower hydrogen pressure. At the higher pressure the conductivity is nearly independent on the dc-power in contrast to the results of Tamahashi et al. [4] and Webb et al. [11]. Defining the ratio of normalized photoconductivity ($\eta\mu\tau$-product) to dark conductivity as a measure for the photosensitivity of a-Si:H [12] a maximum value as high as $\eta\mu\tau/\sigma_d = 8 \times 10^3$ cm$^3$/A can be realized for the best films.

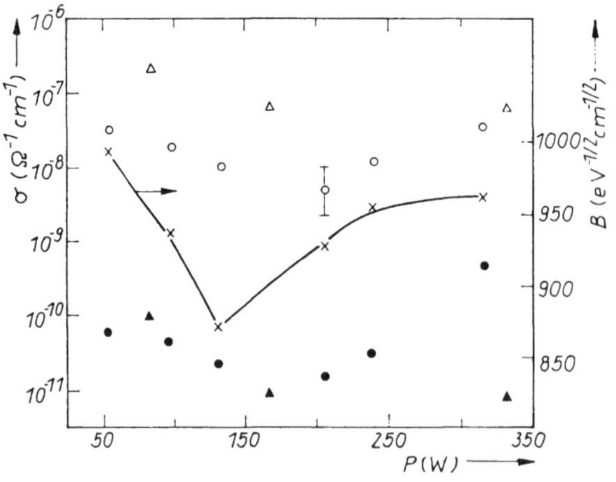


Fig. 4. Dark conductivity $\sigma_d$ ($p_{H_2} = 0.12$ Pa (circles) and $p_{H_2} = 0.18$ Pa (triangles)) and Tauc slope $B$ ($p_{H_2} = 0.12$ Pa (crosses); the solid curve is drawn to aid the eye) in dependence on dc-power $P$ at a different hydrogen partial pressures. Open symbols are to photoconductivity

### 3.2 Infrared-absorption spectroscopy

The infrared transmission spectra in the range of 300 to 4000 $cm^{-1}$ of a-Si:H films deposited on undoped polished crystalline silicon substrates were measured by means of a Perking-Elmer 5080 B spectrometer. The IR spectra were used to estimate the total hydrogen concentration $c_H$ of the films and to determine the microscopic incorporation of hydrogen, i.e. the percentage of mono- and dihydride groups. For simplicity we used the ratio $V$ of the absorption coefficients at 2090 $cm^{-1}$ ($Si-H_2$ stretching mode) to that at 2000 $cm^{-1}$ (Si–H stretching mode) as a measure for the relative concentration of dihydride bonding (cf. [8, 9, 13]). From Fig. 5 the dominance of monohydride bonding in our a-Si:H films prepared by dc-MSP is evident. The ratio $V$ for our samples generally increases with rising hydrogen pressure and decreases with increasing dc-power. For instance, films sputtered at $p_{H2} = 0.12$ Pa show a value of $V = 0.52$ at $P = 50$ W decreasing to $V = 0.23$ at $P = 250$ W. The dependence of total hydrogen concentration $c_H$ determined from the integral absorption of the Si–H wagging mode (as described in [13]) on dc-power reveals a maximum at medium powers $P = 150$ W being in coincidence with the minimum of the dark conductivity in Fig. 4. Fig. 6 shows the dependence of $c_H$ on hydrogen partial pressure. The experimental results are well described by the relation of Moustakas [2],

$$c_H = c_{HM} \left(1 - \exp\left(\frac{p_{H2} z \theta q}{a}\right)\right), \tag{2}$$

where $z = 8.3 \times 10^{18}$ $cm^{-2}$ $s^{-1}$ $Pa^{-1}$ is the hydrogen flux to the surface of the film, $q = 10^{-16}$ $cm^2$ is the cross section of a dangling bond for hydrogen and $\theta$ and $a$ are the sticking probability and the deposition rate in monolayers per second, respectively [2]. The solid line in Fig. 6 represents a fit to experimental data with $c_{HM} = 29.7$ at%

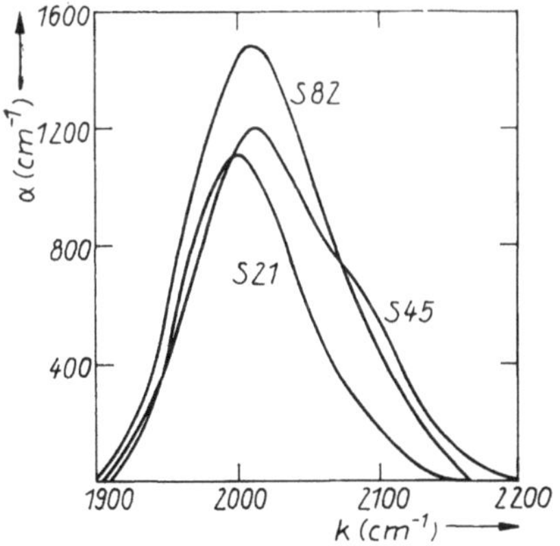


Fig. 5

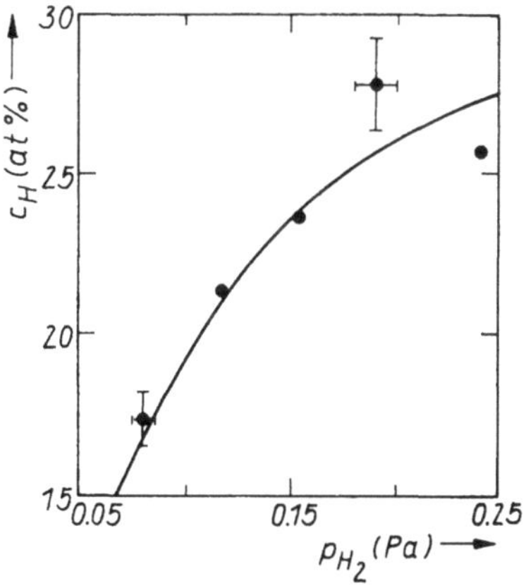


Fig. 6

Fig. 5. Absorption coefficient $\alpha$ in the region of the stretching mode in dependence on the wave number $k$ for three selected samples. S 21 and S 82 are the a-Si:H films with the lowest and the highest total hydrogen concentration from Fig. 6, respectively. S 45 is the film with the largest percentage of dihydride groups for the present work ($V = 0.52$, see text) sputtered at $p_{H_2} = 0.12$ Pa and $P = 50$ W.

Fig. 6. Hydrogen concentration $c_H$ determined from the integral absorption of the Si–H wagging mode vs. hydrogen partial pressure $p_{H_2}$ (see text). The solid line is given by (2) with $c_{HM} = 29.7$ at% and $z\theta q/a = 10.6$ $Pa^{-1}$

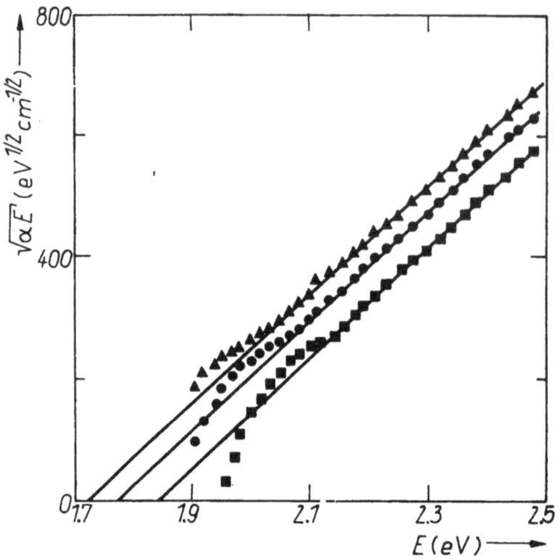


Fig. 7. Tauc plot of the absorption coefficient according to (3) for three samples containing various amounts of hydrogen: triangles $c_H$ = 17.3, circles 21.4, squares 25.8 at%. The shoulder at lower energies is due to the onset of interferences

and $\theta = 0.024$. The latter value is in good agreement with $\theta = 0.02$ resulting from Moustakas experimental data [2]. Further, we found that the dependence of the absorption coefficients $\alpha(2000\ cm^{-1})$ and $\alpha(2090\ cm^{-1})$ on the hydrogen partial pressure as an equivalent for the concentration of mono- and dihydride groups can also be described by (2). However, the corresponding values of the product $z\theta q$ for each curve are remarkably different.

### 3.3 Fundamental absorption

Optical absorption measurements were performed in the range of fundamental absorption of a-Si:H for wavelengths 500 nm $< \lambda <$ 650 nm. Assuming the existence of a square root density of states below and above the mobility edge of the valence and conduction band, respectively [1, 2], the absorption coefficient $\alpha(h\nu)$ obeys the relation

$$\sqrt{\alpha h\nu} = B(h\nu - E_g)\,, \tag{3}$$

where $h\nu$ is the photon energy, $B$ and $E_g$ are the slope of the Tauc plot and the optical gap, respectively. Fig. 7 shows experimental data and linear fits according to (3) for three samples prepared at different hydrogen partial pressures. The slope parameter $B$ is nearly independent of the hydrogen partial pressure and was estimated to be $B = 895\ (\text{eV cm})^{-1/2}$, whereas the optical gap is strongly influenced by the hydrogen concentration and can be described by the relation

$$E_g(\text{in eV}) = 1.47 + 0.014 c_H(\text{in at\%}) \tag{4}$$

being similar to the dependence given by Ley [1] for a-Si:H prepared by different methods. The influence of dc-power on the optical gap is negligible, whereas the slope parameter $B$ shows a minimum near the power of $P = 150$ W (cf. Fig. 4).

### 3.4 Rutherford backscattering

The RBS technique was applied to the a-Si:H films for determining the impurity concentration of heavy atoms in the material with special emphasis to argon (cf. [14]). The a-Si:H films deposited onto polished crystalline Si substrates were probed by a He ion beam with an energy of 1.8 MeV. Argon in-depth distribution profiles cal-

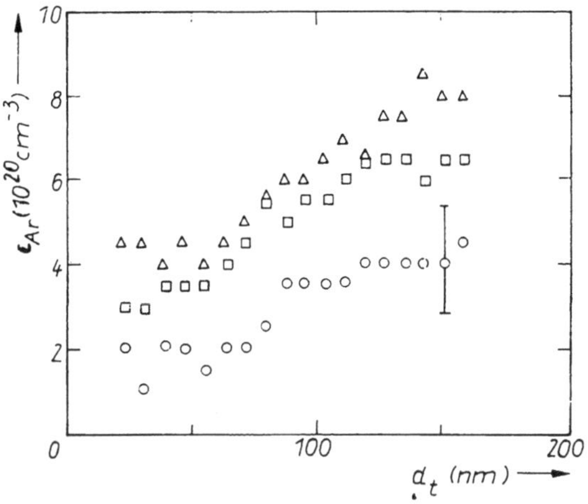


Fig. 8

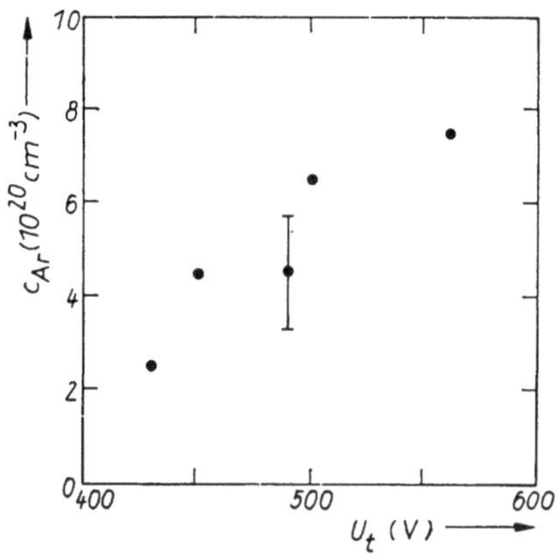


Fig. 9

Fig. 8. Argon concentration in-depth profile $c_{Ar}(d_t)$ for three samples calculated from RBS spectra

Fig. 9. Argon concentration $c_{Ar}$ in depth of $d_t = 160$ nm in dependence on the target voltage $U_t$

culated from the RBS spectra are shown in Fig. 8. The drastical decrease of Ar concentration near the film surface can be explained by the out-diffusion of argon atoms. In Fig. 9 the dependence of the Ar concentration on target voltage is demonstrated. A similar result was found by Martin and Pawlewicz [15] who prepared a-Si:H by dc-tetrode sputtering.

### *3.5 ESR measurements*

Electron spin resonance spectroscopy was performed by a Varian E 112 spectrometer. Unhydrogenated a-Si prepared by MSP showed a $g$-value of $g = 2.0056 \pm 0.0001$ and a spin density of $n_s = 5.6 \times 10^{19}$ cm$^{-3}$. Hydrogenation causes a decrease of spin densities of more than two orders of magnitude as shown in Fig. 10. The independence of spin density, i.e. of dangling bond concentration, on hydrogen content below a certain level agrees with the results of photoconductivity measurements. Müller [8] estimated the spin densities also to be in the range of $n_s \approx 10^{17}$ cm$^{-3}$, a typical value for MSP a-Si:H with film thickness of $d_s \approx 500$ nm. Further, we found the spin

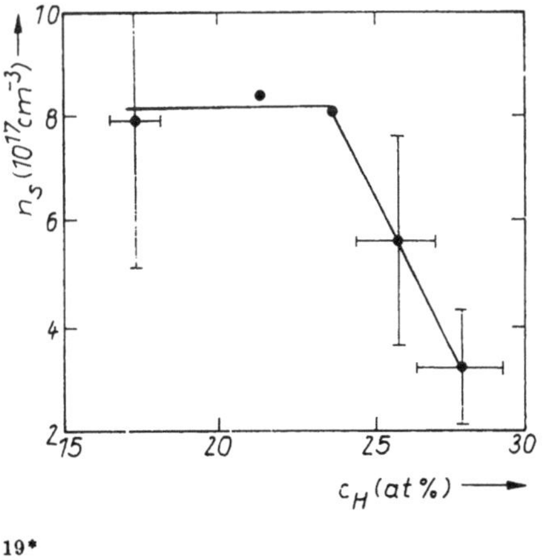


Fig. 10. Spin density $n_s$ as a function of hydrogen content $c_H$ for samples identical to Fig. 6 showing a $g$-values of $g = 2.0054 \pm 0.0003$

density to be proportional to the square root of the dc-power, i.e. $n_s \sim \sqrt{P}$, which agrees well with the results of Kottwitz et al. [16] for a-Si:H prepared by sputter-assisted PCVD in a magnetron discharge.

## 4. Discussion

The present results show that MSP a-Si:H because of its electric and photoelectric properties is suitable for some special applications, e.g. as photoreceptor in electrophotography. In spite of the low absolute value of the photoconductivity of our films in comparison with PCVD a-Si:H, their photosensitivity expressed in form of the ratio of normalized photoconductivity to dark conductivity is excellent. The realized deposition rate of maximum $a = 7$ μm/h exceeds typical values of PCVD by about one order of magnitude [1, 2, 16].

The main precondition for realizing high deposition rates using MSP is a low target to substrate distance of $d_{TS} \approx 50$ mm [3, 4]. On the other hand, Müller argued that a-Si:H with good electronic properties can be prepared only at a large distance $d_{TS} \approx 200$ mm because of the full thermalization of atoms hitting the growing film. However, our findings show that high quality a-Si:H can also be deposited at low distances $d_{TS}$. The film bombardment consists of three main components: (i) not fully thermalized Si atoms sputtered from the target, (ii) positively charged ions being directly accelerated from the plasma to the substrate, and (iii) ions recombining at the target with Auger electrons and being reflected towards the substrate. It is well known that a slight bombardment with Si atoms and Ar ions changes the structural and electronic properties of sputtered a-Si:H in a positive manner (Moustakas [2], Savvides [3], Rudder et al. [10], Ross and Messier [17]): a) columnar film growth is suppressed; b) mechanical film density increases; c) dark conductivity decreases, and d) the post-deposition oxygenation of a-Si:H films is lowered. Ferron et al. [18] have shown by a Monte-Carlo simulation of the growth process of sputtered a-Si:H film that the bombardment with $Ar^+$ ions turns out to be the main reason for a dense and non-porous film. However, the film bombardment causes a higher structural disorder of a-Si:H which for the present work is revealed by values of the Urbach energy $E_0$ [19] and of the incorporated argon both increasing with increasing target voltages. The latter can be explained by the mechanism (ii) because the rising target voltage results in a higher potential difference between plasma and substrate [17] causing a shift in the energy distribution of the $Ar^+$ ions hitting the a-Si:H layer surface with higher energies. Therefore, a larger number of $Ar^+$ ions possesses an energy above the threshold energy for implantation of $E_{th} \approx 20$ eV [1, 18]. The implantation of Ar atoms reflected from the target (cf. process (iii)) can be neglected because of the small number of reflected atoms with energy above the threshold value and their short mean free path being the reason for complete thermalization at the target to substrate distance and the pressure used [1] even for the atoms with the highest energies. Other interesting problems are the reason for the dominance of monohydride bonding of silicon atoms in the present layers and the validity of (2) for the concentration of both the mono- and dihydride complexes with different values of the product $\theta zq$, respectively. The answer to that question should also explain the reason for the increasing bonding of hydrogen in Si–H complexes with rising deposition power in our films in contrast to the findings of PCVD (e.g. [16]). For explaining the complex findings it is necessary to assume the existence of various simultaneously acting processes containing in detail: (i) the surface reaction of the growing layer with atomic hydrogen and $SiH_x$ radicals of the sputtering discharge [1]; (ii) the $H^+$ ions being directly accelerated from the plasma to the substrate and able to penetrate deep into the growing film for reacting with Si dangling bonds; (iii) the relaxation of the amorphous network consisting of reactions of Si dangling bonds with dihydride complexes, like

$$= SiH_2 + -Si \equiv \rightarrow = SiH\text{–}S \equiv + H\,. \tag{5}$$

The possibility of an analogous reaction induced by particle bombardment or radiation from the plasma of a monohydride complex with a dangling bond is much lower because of its strong pinning by three Si neighbour atoms, whereas a dihydride complex may easily turn around the axis of its two Si neighbours. Therefore, a certain amount of hydrogen bounded in dihydride complexes results in a remarkable lowering of the spin densities (cf. Fig. 10).

*Acknowledgements*

The authors are indebted to Dr. H. Sobotta and Dr. R. Böttcher (Karl-Marx-Universität Leipzig) for IR and ESR measurements, respectively. We thank Dr. A. Witzmann (Friedrich-Schiller-Universität Jena) for RBS spectroscopy as well as Prof. K. Schade and Dr. A. Kottwitz (Technische Universität Dresden) for stimulating discussions and providing technical help. We are grateful to Mr. W. Hempel (Inst. v. Ardenne Dresden) helping us to build up the plasmatron deposition apparatus.

## References

[1] J. D. Joannopoulos and G. Lucovsky (Ed.), Topics appl. Phys., Vol. 55, Springer, 1984.
[2] J. I. Pankove (Ed.), Semiconductors and Semimetals, Vol. 21, Academic Press (Orlando), 1984, Part A (p. 55); Part D (p. 1).
[3] N. Savvides, J. appl. Phys. 55, 4232 (1984).
[4] K. Tamahashi, M. Wakigi, T. Yamanaka, M. Murakawa, S. Onuma, and M. Chigasaki, Electrophotography 25, 141, 271 (1986).
[5] U. Heisig, K. Goedicke, P. Vetters, W. Hempel, J. Hartung, and G. Teschner, Proc. 6. Tagung Physik und Technik des Hochvakuums, Dresden 1978, Phys. Gesellsch. d. DDR (p. 48).
[6] K. Steenbeck, E. Steinbeiss, and K.-D. Ufert, Thin Solid Films 92, 371 (1982).
[7] W. D. Westwood, C. Maniv, and P. J. Scanlon, J. appl. Phys. 54, 6841 (1983).
[8] W. Müller, Dissertation, Universität Kaiserslautern, 1984.
[9] H. Rübel, B. Schröder, and J. Geiger, J. Vac. Sci. Technol. A 4, 1855 (1986).
[10] R. A. Rudder, J. W. Cook, J. F. Schetzina, and G. Lucovsky, J. Vacuum Sci. Technol. A 2, 326 (1984).
[11] J. B. Webb, J. appl. Phys. 53, 9043 (1983).
[12] N. Saito, H. Sannomiya, T. Yamaguchi, and H. Tanaka, Appl. Phys. A 35, 241 (1984).
[13] H. Shanks, C. J. Fang, L. Ley, M. Cardona, F. J. Demond, and S. Kalbitzer, phys. stat. sol. (b) **100**, 43 (1980).
[14] A. Turos, H. Frey, O. Meyer, W. Müller, and J. M. Pirrung, phys. stat. sol. (a) **83**, 437 (1985).
[15] P. M. Martin and W. T. Pawlewicz, AIP Conf. Proc. 52, 73 (1981).
[16] A. Kottwitz, M. Fuchs, K. Schade, R. Bindemann, T. Drüsedau, and W. Gerlach, J. non-crystall Solids 93, 230 (1987).
[17] R. C. Ross and R. Messier, J. appl. Phys. 52, 5329 (1981).
[18] J. Ferron and R. Arce, Phys. Rev. B 35, 7611 (1987).
[19] K. M. Eckler, I. Griessbach, T. Drüsedau, and R. Bindemann, to be published.

*(Received April 19, 1988)*

phys. stat. sol. (a) 108, 295 (1988)

Subject classification: 68.55 and 75.70; 76.80; S1.1

*Department of Metal Science and Technology, Kyoto University*[1])

# Structure and Magnetic Properties of Nonequilibrium Fe–Ru Alloys Produced by Vapor Quenching

By

M. HADA, K. SUMIYAMA, and Y. NAKAMURA

Sputter-deposited $Fe_{1-x}Ru_x$ alloy films are investigated by X-ray diffraction, magnetization, and Mössbauer effect measurements. The single b.c.c. phase is obtained for $0 \leqq x \leqq 0.12$, the b.c.c.–h.c.p. mixed phase for $0.12 < x \leqq 0.18$ and the single h.c.p. phase for $0.18 < x \leqq 1.0$. The lattice constants of sputter-deposited alloys are slightly larger than those of bulk alloys and increase with the increase in $x$. At low temperatures the b.c.c. alloys are ferromagnets, and their average magnetic moment decreases with the increase in $x$. Mössbauer effect measurements suggest that magnetic moments tend to align perpendicularly to the film plane in the b.c.c.–h.c.p. mixed alloy films.

Aufgestäubte $Fe_{1-x}Ru_x$-Legierungsschichten werden mittels Röntgenbeugung, Magnetisierung und Mößbauereffektmessungen untersucht. For $0 \leqq x \leqq 0{,}12$ wird eine einzige k.r.z. Phase erhalten, die gemischte k.r.z.–h.d.p.-Phase für $0{,}12 < x \leqq 0{,}18$ und die reine h.d.p.-Phase für $0{,}18 < x \leqq 1{,}0$. Die Gitterkonstanten der aufgestäubten Legierungen sind leicht größer als die der massiven Legierungen und nehmen mit wachsendem $x$ zu. Bei tiefen Temperaturen sind die k.r.z.-Legierungen ferromagnetisch und ihr mittleres magnetisches Moment nimmt mit der Zunahme von $x$ ab. Mößbauereffektuntersuchungen weisen darauf hin, daß die magnetischen Momente dazu neigen, sich senkrecht zur Schichtebene in den k.r.z.–h.d.p.-Mischlegierungsschichten anzuordnen.

## 1. Introduction

Vapor quenching method is powerful in producing a wide range of solid solutions. In sputter-deposited $Fe_{1-x}Mn_x$ alloys, for example, the single b.c.c. and α-Mn phases are obtained for $x \leqq 0.2$ and $> 0.3$, although they are stable only for $x \leqq 0.03$ and $\geqq 0.65$ in the equilibrium alloys at room temperature [1, 2].

In the equilibrium phase diagram of $Fe_{1-x}Ru_x$ alloys, the primary b.c.c. solid solution (α-Fe) is restricted to $x \leqq 0.05$, while the primary h.c.p. solid solution is obtained over a wide composition range of $0.24 \leqq x \leqq 1.0$ at low temperatures [3]. By the solid quenching method, the metastable b.c.c. phase is extended for $x \leqq 0.12$, while metastable f.c.c.–h.c.p. mixed phase is restricted to $0.12 < x \leqq 0.2$ due to the martensitic transformation [4].

This paper deals with the experimental results of X-ray diffraction, magnetization, and Mössbauer effect measurements on sputter-deposited $Fe_{1-x}Ru_x$ alloys. It also shows how widely single phase regions are extended in these sputter-deposited alloys.

## 2. Experimental Procedure

About 2 μm thick $Fe_{1-x}Ru_x$ films with $x = 0$ to 1.0 were prepared on polyimide film substrates by an rf sputtering method. The operational parameters of the sputtering are shown in Table 1. The compositions of sputter-deposited alloy films were determined within an experimental error of 1% by electron probe microanalysis (EPMA).

[1]) Sakyo, Kyoto 606, Japan.

Table 1
Sputtering parameters

| | |
|---|---|
| rf power | ≈170 W |
| rf frequency | 13.56 MHz |
| Ar pressure | 0.5 Pa |
| focussing magnetic field | $6.4 \times 10^3$ A/m |
| deposition rate | 8 to 12 nm/min |
| target–substrate distance | 40 mm |
| substrate | polyimide |

The crystal structure and the lattice constant of sputter-deposited alloy films were estimated by X-ray diffraction measurements using CoKα radiations and a graphite monochromator. A torsion balance magnetometer was used for magnetization measurements from 4.2 to 270 K. The Mössbauer effect measurements were made at room temperature and 4.2 K using a constant acceleration type spectrometer with a $^{57}$Co source doped in Cu. The magnetization curves of several deposited alloys were observed at 290 K using a vibrating sample magnetometer (VSM) in magnetic fields up to 15 kOe.

## 3. Experimental Results

Fig. 1 shows X-ray diffraction patterns of the sputter-deposited $Fe_{1-x}Ru_x$ alloys. Only b.c.c. peaks are observed for $x \leqq 0.12$, while h.c.p. peaks are dominant for $x \geqq 0.19$. For $0.14 \leqq x \leqq 0.18$, both b.c.c. and h.c.p. peaks are observed.

Fig. 2. shows lattice constants of sputter-deposited $Fe_{1-x}Ru_x$ alloy films estimated from the X-ray diffraction measurements. The lattice constants of both b.c.c. and h.c.p alloys increase monotonically with increasing $x$. The lattice constant of the b.c.c. phase and the $a$-values of the h.c.p. phase in sputter-deposited $Fe_{1-x}Ru_x$ alloys are slightly larger than those in bulk $Fe_{1-x}Ru_x$ alloys, while the $c$-values of sputter-deposited h.c.p. alloys are nearly the same to those of the bulk alloys. This lattice expansion in

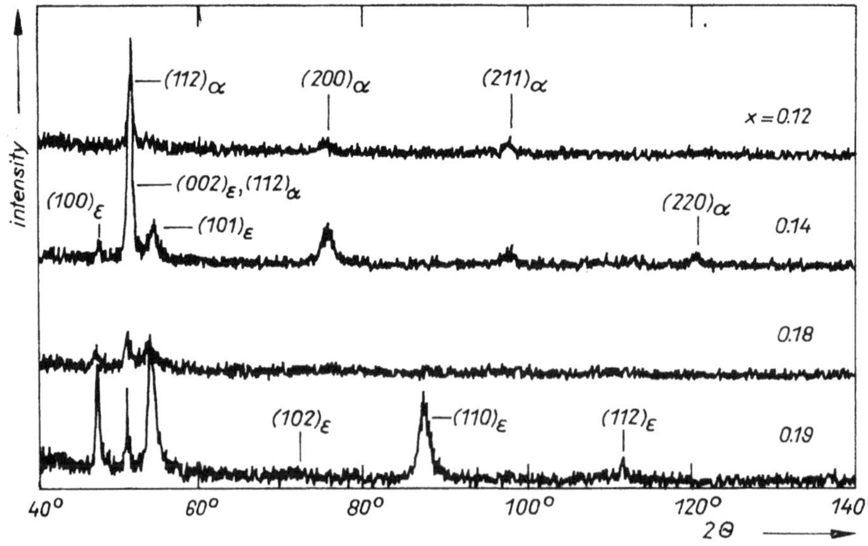


Fig. 1. X-ray diffraction patterns of sputter-deposited $Fe_{1-x}Ru_x$ alloy films

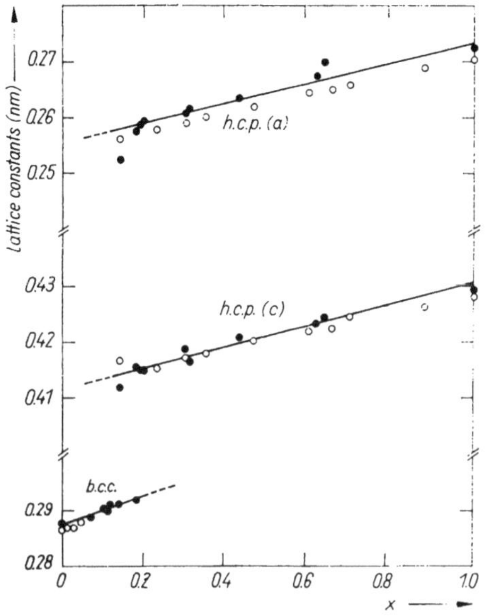


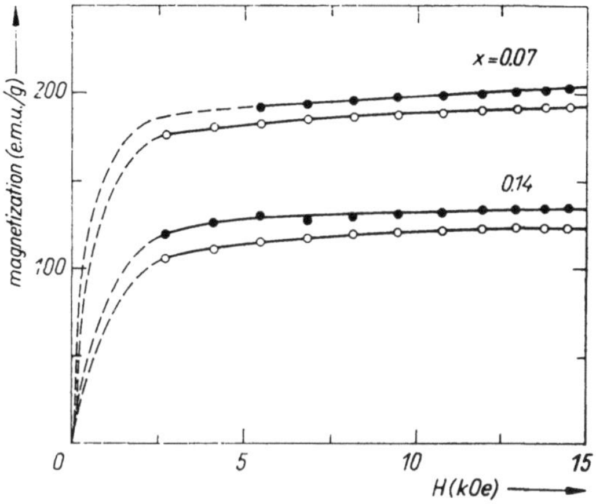


Fig. 2

Fig. 3

Fig. 2. Lattice constants of $Fe_{1-x}Ru_x$ alloys. The full circles indicate the results of the sputter-deposited alloys and the empty ones those of bulk alloys [8]

Fig. 3. Field dependence of magnetization of sputter-deposited $Fe_{1-x}Ru_x$ alloys. The full circles indicate the results at 4.2 K and empty ones those at 290 K

the sputter-deposited Fe–Ru alloys is attributable to a loosely packed structure due to an incorporation of Ar atoms and interstitial atoms caused by a peening effect in the sputter-deposited alloys (for example [5]).

Fig. 3 shows the field dependence of the magnetizations of sputter-deposited $Fe_{1-x}Ru_x$ alloys with $x = 0.07$ and 0.14 at 4.2 K and room temperature. The sputter-deposited $Fe_{1-x}Ru_x$ alloy films are ferromagnetic for $x \leqq 0.18$ and magnetizations almost saturate at magnetic fields beyond 2.7 kOe. The magnetizations at 4.2 K are a few percents larger than those at room temperature, indicating that their Curie temperatures are much higher than room temperature.

The average magnetic moment, $\bar{\mu}$, of sputter-deposited $Fe_{1-x}Ru_x$ alloys is estimated by linearly extrapolating the magnetization curve at high fields to zero field. As shown in Fig. 4, $\bar{\mu}$ of sputter-deposited $Fe_{1-x}Ru_x$ alloys is independent of $x$ up to around $x = 0.1$ and decreases rapidly when $x$ exceeds about 0.1 and becomes zero for $x > 0.2$. Using the results of the magnetization measurements and assuming that Ru atoms carry no magnetic moment, we have the average magnetic moment per Fe atom, $\overline{\mu_{Fe}}$, of sputter-deposited $Fe_{1-x}Ru_x$ alloys, as shown in Fig. 5: $\overline{\mu_{Fe}}$ of sputter-deposited $Fe_{1-x}Ru_x$ alloys initially increases slightly, reaches a maximum value around $x = 0.1$, and then decreases rapidly for $x > 0.12$ with the increase in $x$.

Fig. 6 shows Mössbauer spectra of sputter-deposited $Fe_{1-x}Ru_x$ alloys at room temperature. For $x \leqq 0.12$, a ferromagnetic sextet is observed. The line broadening is due to distributions of the magnetic hyperfine field and the electric field gradient [6]. For $x \geqq 0.19$, only a nonmagnetic single line is observed. For $0.14 \leqq x \leqq 0.18$, both the feromagnetic sextet and the nonmagnetic single line are observed. Fig. 7 shows the

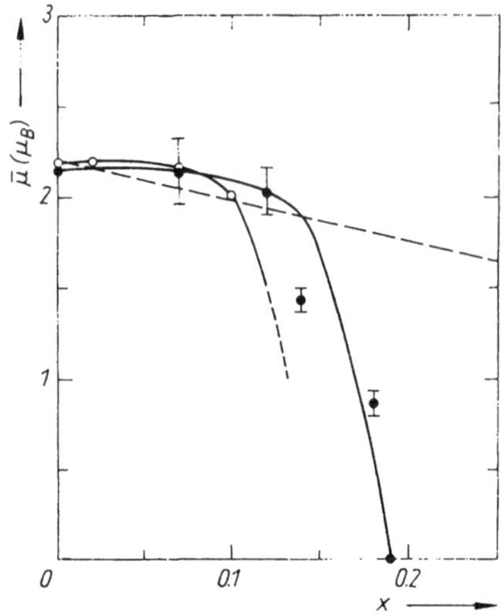


Fig. 4

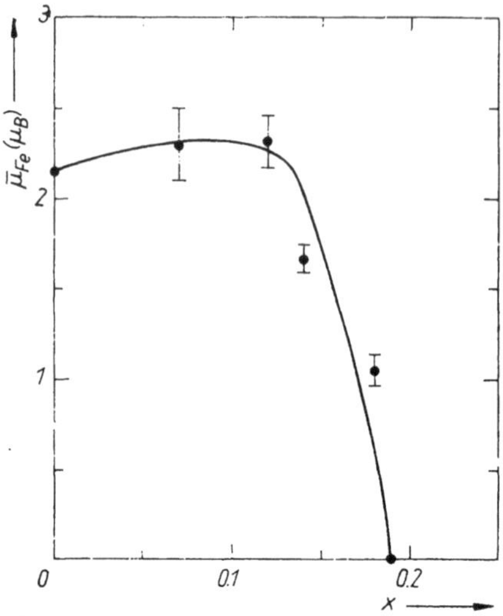


Fig. 5

Fig. 4. Concentration dependence of magnetic moment per atom, $\bar{\mu}$, at 4.2 K. Full circles indicate results of the present sputter-deposited alloys and empty ones those of bulk alloys [9]. The dashed line shows the simple dilution law on the assumption of no Ru moment

Fig. 5. Magnetic moment per Fe atom, $\overline{\mu_{Fe}}$, as a function of Ru concentration in $Fe_{1-x}Ru_x$ alloy

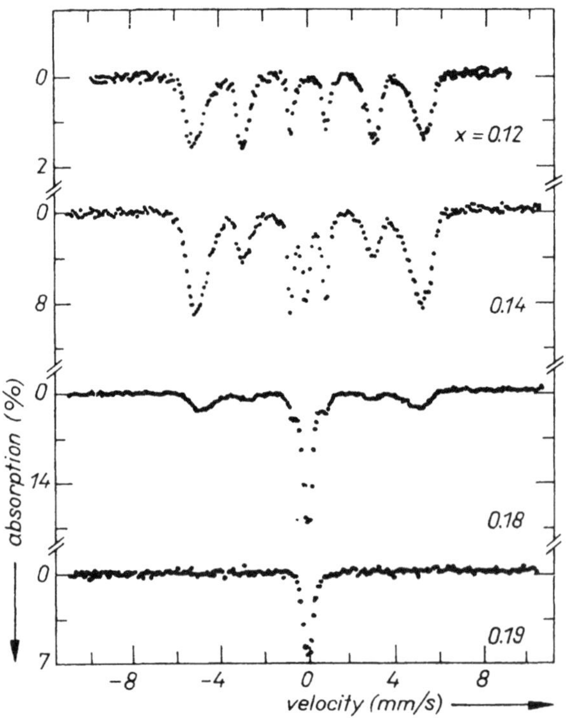


Fig. 6. Mössbauer spectra of sputterdeposited $Fe_{1-x}Ru_x$ alloys at 290 K

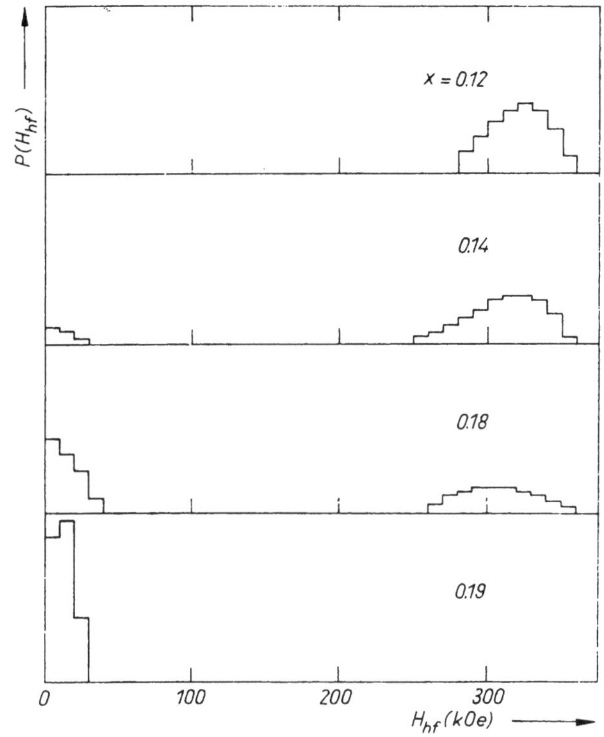


Fig. 7. Hyperfine field distribution, $P(H_{hf})$, for sputter-deposited $Fe_{1-x}Ru_x$ alloys at 290 K

magnetic hyperfine field distribution obtained by analyzing these Mössbauer spectra. Each spectrum was fitted to six Lorentzians with different Zeeman splittings assuming the halfwidth of a Lorentzian to be 0.25 mm/s. The isomer shift and the intensity ratios of the second and fifth peaks to the others were adjusted to get the best fitting. For $x = 0.12$, only the high field component around 300 kOe is observed. With further increase in $x$ for $x \geqq 0.14$, the high field component slightly shifts towards the lower field side, the width of the distribution increases, while the high field component disappears and the lower field component around 0 kOe becomes dominant for $x \geqq 0.19$. The average magnetic hyperfine field, $\overline{H_{hf}}$, is about 330 kOe for $x = 0.12$ at 4.2 K, which is slightly larger than that at room temperature, 315 kOe.

It is worthwhile noting the intensities of the second and fifth peaks of the sextet in Mössbauer spectra of $Fe_{1-x}Ru_x$ alloys with $x = 0.14$ and 0.18: the intensity ratios of the sextet are 3:1:1:1:1:3 for $x = 0.14$. This indicates that the average magnetic moment makes an angle of about 50° to the film plane of the sputter-deposited alloy films. As shown in Fig. 8, on the contrary, the magnetic hysteresis curves for $x = 0.14$ and 0.18 show that these alloys easily saturate in the in-plane direction but they do not saturate along a direction perpendicular to the film plane even in a magnetic field of 10 kOe. This indicates a contribution of the shape magnetic anisotropy of the films, $4\pi M$.

## 4. Discussion

Fig. 9a shows the nonequilibrium phase boundary diagram of vapor-quenched Fe–Ru alloys estimated from X-ray diffraction and Mössbauer effect measurements. The single b.c.c. phase is obtained for $0 \leqq x \leqq 0.12$, while the single h.c.p. phase for $x \geqq 0.19$.

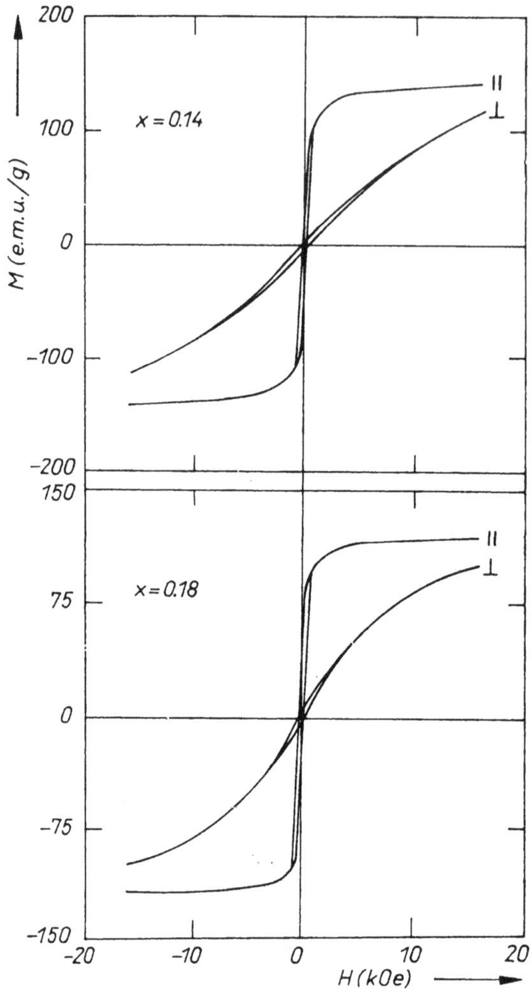


Fig. 8. Magnetic hysteresis curves of sputter-deposited $Fe_{1-x}Ru_x$ alloys at 290 K

The single phase regions of b.c.c. and h.c.p. are nearly the same as those in the solid quenched alloys (Fig. 9 b). However, in the two-phase regions, the h.c.p. and f.c.c. phases appear for the solid quenched alloys, while the b.c.c. and h.c.p. phases appear for the equilibrium alloys and the present vapor-quenched alloys. The bulk $Fe_{1-x}Ru_x$ alloys solid quenched from about 1450 °C undergo the martensitic transformation, i.e., the $\gamma$-phase (f.c.c.) transforms to the $\alpha$-phase (b.c.c.) for $x \leqq 0.12$, while it does to the $\varepsilon$-phase (h.c.p.) for $x \geqq 0.12$ [4].

Since the b.c.c. phase is ferromagnetic and the h.c.p. phase is paramagnetic at room temperature [4], the volume fraction of these phases can be roughly estimated as shown in Fig. 10 by comparing the integrated intensities of the sextet and the singlet in Fig. 6.

The concentration dependence of the magnetic moment generally follows an empirical rule in the Fe-based dilute alloys: in the iron matrix, $d\bar{\mu}/dc$ is positive for late transition metal impurities, while it is negative for early transition metal impurities [7], where $c$ indicates the impurity concentration. On the other hand, in the Fe-based alloys produced by vapor quenching, the volume expansion of Fe atoms origi-

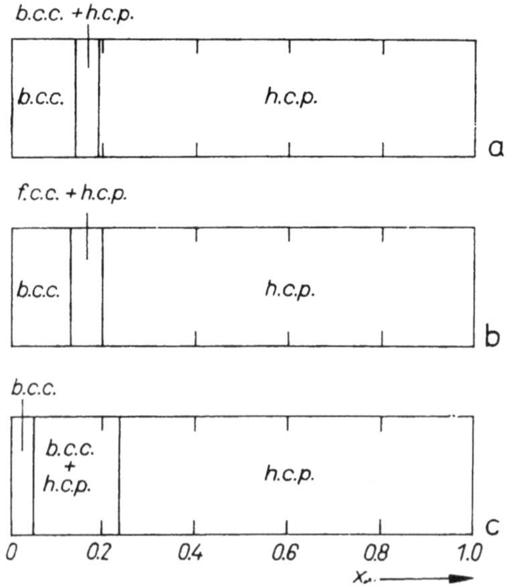


Fig. 9. a) Nonequilibrium phase boundary diagram of sputter-deposited $Fe_{1-x}Ru_x$ alloys at room temperature; b) nonequilibrium phase boundary diagram of solid quenched $Fe_{1-x}Ru_x$ alloys at room temperature [4]; c) equilibrium phase boundary diagram of $Fe_{1-x}Ru_x$ alloys at room temperature [3]

nated by the addition of an impurity element with a large atomic size often gives rise to an increase in the magnetic moment of Fe atoms. In this context, we expect an increase of the Fe moment in the Fe–Ru alloys, because the atomic radius of Ru is 0.134 nm, larger than that of Fe, 0.126 nm. Actually $\overline{\mu_{Fe}}$ of the sputter-deposited b.c.c. Fe–Ru alloys is slightly larger than 2.2 $\mu_B$ of pure b.c.c. Fe.

### *Acknowledgements*

The authors wish to thank Dr. M. Shiga, Dr. N. Kataoka, and Mr. R. Iehara for their experimental supports and valuable advices. The authors are indebted to Mr. T. Unesaki for making EPMA analysis.

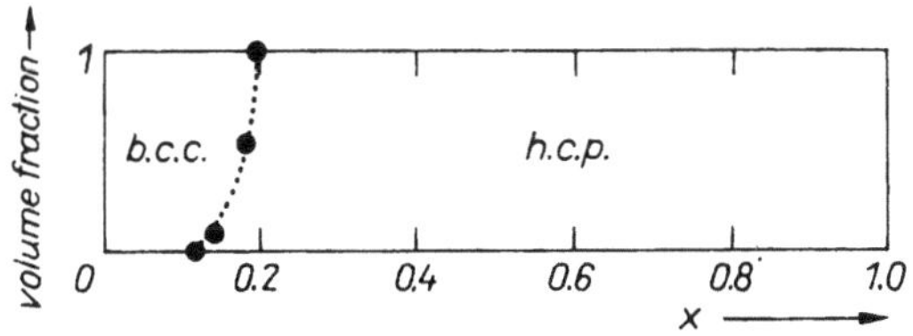


Fig. 10. Volume fractions of b.c.c. and f.c.c. phases in sputter-deposited $Fe_{1-x}Ru_x$ alloys estimated from Mössbauer effect measurements

## References

[1] K. SUMIYAMA, M. KADONO, and Y. NAKAMURA, Trans. Japan Inst. Metals **22**, 686 (1981).
[2] K. SUMIYAMA, N. OHSHIMA, and Y. NAKAMURA, phys. stat. sol. (a) **98**, 229 (1986).
[3] L. J. SWARTZENDRÜBER and B. SUNDMAN, Bull. Alloy Phase Diagrams 4 (2), Sep. (1983).
[4] H. FUJIMORI and H. SAITO, J. Phys. Soc. Japan **26**, 1115 (1969).
[5] J. J. CUOMO and R. J. GAMBINO, J. Vac. Sci. Technol. **14**, 152 (1977)-
[6] Y. SUMITOMO, T. MORIYA, H. INO, and F. E. FUJITA, J. Phys. Soc. Japan **35**, 461 (1973).
[7] M. F. COLLINS and G. G. LOW, Proc. Phys. Soc. **86**, 535 (1965).
[8] W. B. PEARSON, A Handbook of Lattice Spacings and Structures of Metals and Alloys, Vol. 2, Pergamon Press, 1967 (p. 928).
[9] S. CHIKAZUMI, K. OHTA, K. ADACHI, N. TSUYA, and K. ISHIHARA (Eds.) A Handbook of Magnetic Substance, Asakura-shoten, Tokyo 1975 (p. 305) (in Japanese).

*(Received April 7, 1988)*

phys. stat. sol. (a) **108**, 303 (1988)

Subject classification: 73.40; 72.40; 73.25; S7.13

*A. F. Ioffe Physico-Technical Institute, Academy of Sciences of the USSR, Leningrad*[1]) *(a) and Sektion Physik der Humboldt-Universität zu Berlin, Bereich Tieftemperatur-Festkörperphysik*[2]) *(b)*

# Investigation of Interface Peculiarities in Anodized p-InSb MOS Structures

By

V. A. BEREZOVETS (a), W. BRAUNE (b), N. KUBICKI (b), and A. O. SMIRNOV (a)

Dedicated to Prof. Dr. E. GUTSCHE on the occasion of his 60th birthday

$C$–$U$ measurements are made on p-InSb MOS structures in an extended frequency range (2 Hz to 50 kHz) at temperatures of 4.2 and 77 K. The effect of illumination with different intensities and also of stationary magnetic fields up to 5 T on the $C$–$U$ characteristic are analysed. The surface photoelectric electromotive force and its dependence of the illumination intensity are measured. The density of the localized surface states ($10^{12}$ eV$^{-1}$ cm$^{-2}$) and a state density peak ($10^{13}$ eV$^{-1}$ cm$^{-2}$) near the conduction band edge are determined from the measured dependences. The effect of magnetic field on the $C$–$U$ characteristic without illumination is explained qualitatively and quantitatively. The results of the surface photoelectric emf measurements are analysed qualitatively yielding to a special barrier mechanism for its formation.

Es werden Messungen der $C$–$U$-Charakteristick von p-InSb-MOS-Strukturen in einem breiten Frequenzbereich (2 Hz bis 50 kHz) bei Helium- und Stickstofftemperaturen durchgeführt. Der Einfluß von Beleuchtung verschiedener Intensität aber auch stationärer Magnetfelder bis 5 T auf die $C$–$U$-Charakteristiken wird untersucht. Bei Stickstofftemperaturen wird die Oberflächenfoto-EMK und ihre Abhängigkeit von der Lichtintensität gemessen. Aus den erhaltenen Abhängigkeiten werden bestimmt: Dichte der lokalisierten Oberflächenzustände ($10^{12}$ eV$^{-1}$ cm$^{-2}$), sowie ein Zustandsdichtepeak ($10^{13}$ eV$^{-1}$ cm$^{-2}$) nahe der Leitungsbandkante. Eine qualitative und quantitative Erklärung des Magnetfeldeinflusses auf die $C$–$U$-Charakteristik ohne Beleuchtung wird gegeben. Eine qualitative Analyse der Resultate der Oberflächenfoto-EMK Messungen wird durchgeführt und auf einen besonderen Barrierenmechanismus ihrer Formierung geschlußfolgert.

## 1. Introduction

At present there are many publications in the literature investigating InSb MOS structures. The interest in researching their properties results from the fact that InSb is a basic material for the infrared engineering. Therefore, the experimental studies are directed both to the perfection of InSb MOS structure preparation [1 to 5] and to the investigation of the physical properties of the interface between oxide and InSb [6 to 10].

Due to the peculiarities of the technological preparation procedures for InSb MOS structures there are essential differences in many results from several authors. In the present time general properties of the anodic oxide - InSb system are known, especially the hysteresis effect in the $C$–$U$ characteristics of the MOS structure at removing the

[1]) Ul. Politekhnicheskaya 26, SU-194 021 Leningrad, USSR.

[2]) Invalidenstraße 110, DDR-1040 Berlin, GDR.

accumulation state in inversion and vice versa by variation of gate voltage [9] and the existence of inserted charges in the oxide layer.

The present paper is aimed at the complex investigation of the interface properties between oxide and semiconductor of p-InSb MOS structures in a wide temperature range and under several external influences.

## 2. Experimental Method

An InSb single crystal of p-type having a carrier concentration of $p = 10^{13}$ cm$^{-3}$ at $T = 77$ K was used as basic material for the preparation of the MOS structures. The dielectric layer with a thickness of 100 nm on the (111)B-surface was produced by means of anodic oxidation [5]. Evaporated Au dots $(0.7 \times 0.7)$ mm$^2$ were used as gate electrodes. The capacitance–voltage $C$–$U$ characteristics were measured in order to study the interface properties of the system anodic oxide–InSb. The $C$–$U$ measurements were made by a quasi-stationary method using a slowly varying gate voltage and a sinusoidal test signal of small amplitude.

The frequency of the test signal and the gate voltage rate of change were varied between 2 Hz and 50 kHz and between 1 and 10 mV/s, respectively. The maximal gate voltage change amounted to $\pm 10$ V. The experiments were carried out at temperatures of 77 and 4.2 K. The effect of a magnetic field (up to 5 T) on the $C$–$U$ characteristics was studied at the temperature of 4.2 K. A GaAs-LED ($\lambda = 0.9$ μm) and a He–Ne laser ($\lambda = 0.699$ μm) were used for the stationary and modulated illumination of the structure.

## 3. Experimental Results and Discussion

### *3.1 Analysis of the C–U characteristics without illumination*

The results of the $C$–$U$ measurement at the studied MOS structures are shown in Fig. 1 for a wide range of frequency and temperature. A temperature decrease results

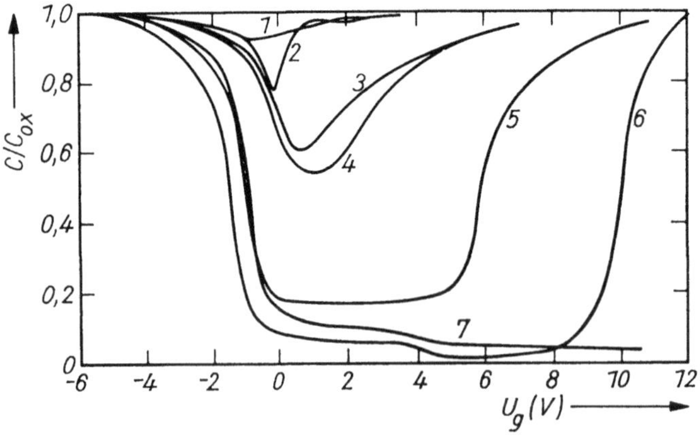


Fig. 1. Typical $C$–$U$ characteristics of p-InSb MOS structures at various temperatures and frequencies of the test pulse. (1) $f = 4$ Hz, $T = 77$ K; (2) 4 Hz, 61 K; (3) 40 kHz, 77 K; (4) 40 kHz, 61 K; (5) 2 Hz, 4.2 K; (6) 2 Hz, 1.6 K; (7) 2 kHz, 4.2K

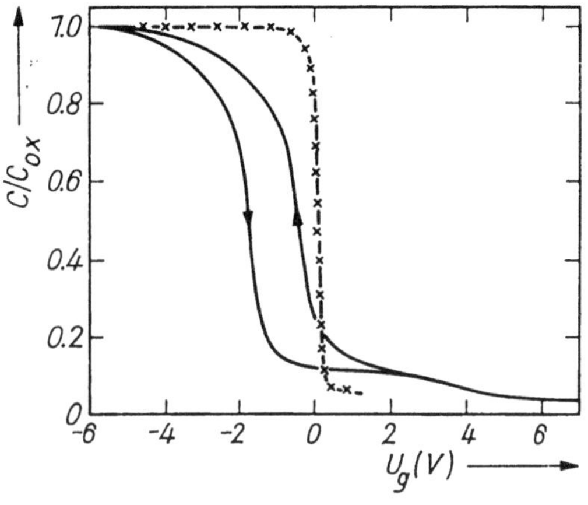


Fig. 2. Experimental (solid lines) and theoretical (dashed line with crosses) $C$–$U$ dependences at $T = 4.2$ K and $f = 2$ kHz

in a decrease of the recharge speed of the surface states, being represented by the appearance of the high frequency type $C$–$U$ charactetistics at low frequencies.

The $C$–$U$ hysteresis, appearing by changing the gate voltage from negative to positive values and vice versa, is a characteristic feature of all $C$–$U$ dependences. This effect can be explained by the polarization modification of the dielectric in electric fields due to the localized states near the semiconductor surface with charges opposite to the semiconductor interface states.

The $C$–$U$ characteristics at high frequency and the corresponding theoretical curve for an ideal MOS structure are shown in Fig. 2.

The voltage difference between the experimental and theoretical $C$–$U$ dependences in the flat-band regime allows us to evaluate the number of states recharged during a single voltage cycle,

$$N = \frac{C_{\text{ox}}}{q} \left(\Delta V_{\text{FB}}(U_{\text{acc.}\to\text{inv.}}) - \Delta V_{\text{FB}}(U_{\text{inv.}\to\text{acc.}})\right) , \tag{1}$$

where $C_{\text{ox}}$ is the geometrical oxide capacitance per unit area, $q$ the electron charge, and $\Delta V_{\text{FB}}$ the voltage difference between the experimental and theoretical $C$–$U$ dependences in the flat-band regime with voltage sweep from negative to positive values and vice versa [8]. Assuming $C_{\text{ox}} = 250$ pF we find the density of the recharged states $5 \times 10^{11}$ cm$^{-2}$ in the dielectric. By comparing the theoretical and the experimental $C$–$U$ characteristics a shift to larger negative voltage values is evident. This shift can be interpreted by the existence of inserted positive charges in the oxide layer. The increase of the initial voltage yields a larger displacement of the $C$–$U$ characteristics to the voltage axis without remarkable change of the hysteresis (Fig. 3). This is caused by the accumulation of positive charge carriers in the oxide as a result of the increased voltage swing in the cycle. The charge accumulation is maintained over a long time and vanishes by warming up to room temperature. This effect can be explained by the existence of traps with a very low generation–recombination velocity at low temperatures.

Using the $C$–$U$ dependences (Fig. 1) the surface state density and their energy distribution in the gap of InSb were calculated. The surface state density $N_{\text{ss}} \approx \approx 10^{12}$ eV$^{-1}$ cm$^{-2}$ in the midgap at 4.2 K was determined from the high-frequency $C$–$U$ curve ($f = 2$ kHz) using the Terman method [11].

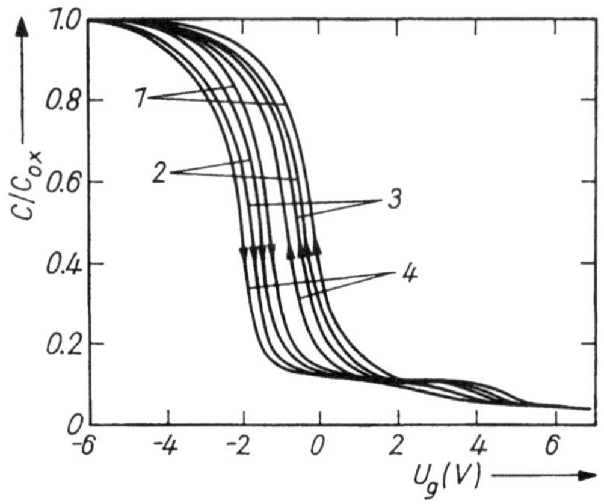


Fig. 3. Influence of the initial voltage on the charges incorporated into the oxide ($T = 4.2$ K, $f = 2$ kHz): (1) $U_0 = -4.5$; (2) $-5.0$; (3) $-5.5$; (4) $-6.0$ V

The spectrum of surface states at liquid nitrogen temperatures (61 to 77 K) was evaluated from the low-frequency $C$–$U$ characteristics ($f = 4$ Hz) by applying the Berglund method [12].

The results of calculation are shown in Fig. 4. The state density maximum near the conduction band edge points out the existence of monoenergetic surface states with increased state density in the background of a monotonous distribution. Similar results are found in [10] by investigations of the $C$–$U$ characteristics of n-InSb MOS structures without illumination. It is supposed that the formation of monoenergetic states can be caused by structural defects at the interface between anodic oxide and InSb, depending on the preparation technology of the MOS structure.

### 3.2 Magnetic field influence on the C–U characteristics

In Fig. 5 the $C$–$U$ characteristics as measured at low frequency and in strong magnetic fields perpendicular to the interface are shown. The experiments were carried out at 4.2 K and the frequency of 2 Hz. The influence of the magnetic fields yields an increased threshold voltage in the case of inversion, but in the accumulation range the $C$–$U$ characteristics remain essentially unchanged.

The measured influence of the magnetic field points out the change of the forbidden gap in the semiconductor ($\hbar\omega_c/2$) as a consequence of the magnetic splitting of the charge carrier energy spectrum. The influence of the magnetic field on the $C$–$U$ charac-

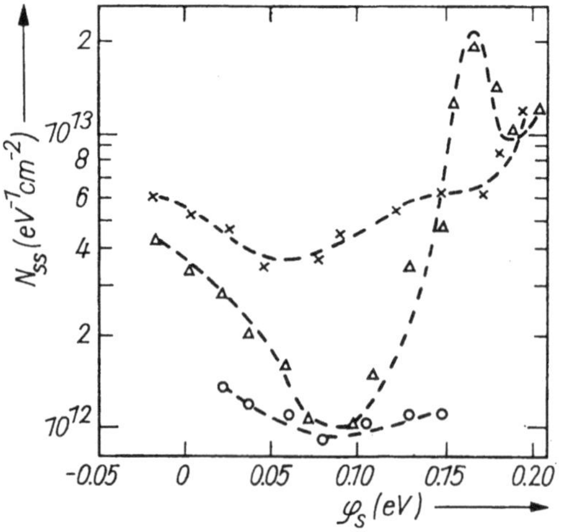


Fig. 4. Dependence of the surface state density on the band shift at the semiconductor surface. × $T = 77$; △ 61; ○ 4.2 K

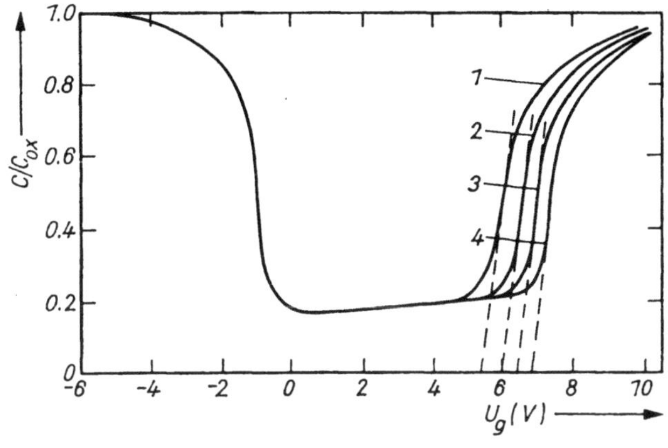


Fig. 5. Shift of the low-frequency $C$–$U$ characteristics in magnetic fields ($T$ = 4.2 K, $f$ = 2 kHz): (1) $B$ = 0; (2) 2.5; (3) 3.75; (4) 5 T

teristics is expected above all in the inversion range because the cyclotron masses of electrons and holes differ by more than one order of magnitude. For an estimation the following approximation formula of [13] was used. It connects the threshold voltage shift with the broadening of the semiconductor gap in the magnetic field,

$$\delta V_{\mathrm{T}}(H) = \left[\left(\frac{ed^2g_v^2}{4m^{*2}K_{\mathrm{r}}^2C^3}\right)^{1/3}\frac{H}{\gamma_0} + V_{\mathrm{T}}^{2/3}(0)\right]^{3/2} - V_{\mathrm{T}}(0)\,, \qquad (2)$$

where $K_{\mathrm{r}}$ is the dielectric constant ratio of oxide to semiconductor, $m^*$ the effective electron mass, $d$ the oxide thickness, $H$ the magnetic field strength, $g_v$ the degeneracy factor, and $V_{\mathrm{T}}(0)$ the threshold voltage of inversion at $H = 0$.

The threshold voltages are determined from the maximum ratio $\Delta C/\Delta V_{\mathrm{g}}$ in the inversion range (dashed curves in Fig. 5). With the parameters $m^* = 0.013$, $K_{\mathrm{r}} = 0.6$, $d = 100$ nm, and $\gamma_0 = 2.338$ (2) yields $\delta V_{\mathrm{T}}(H) = 1260$ meV at $B = 5$ T for the investigated MOS structures.

The experimental value of 1400 meV at this magnetic field is in good agreement with the calculated ones.

### *3.3 The illuminatiom influence on the interface properties of the system anodic oxide–InSb*

The measurements were carried out with constant illumination generated by a GaAs LED of wavelength $\lambda = 0.9$ μm, at the temperature of 4.2 K, and frequency of 2 kHz. The results are shown in Fig. 6. In the case of sufficiently high-intensity the high-fre-

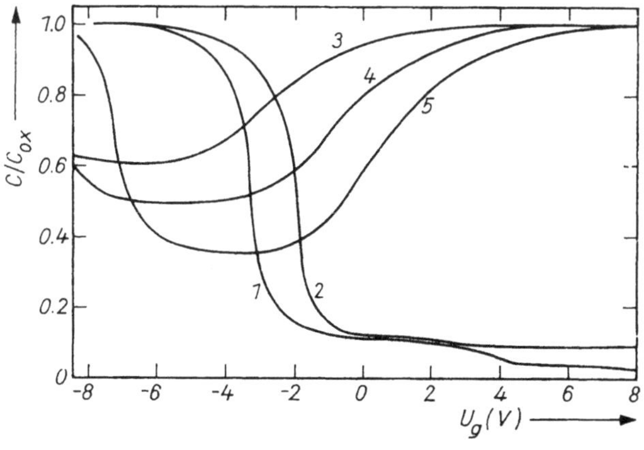


Fig. 6. Influence of stationary illumination ($\lambda = 0.9$ μm) of various intensity on the high-frequency $C$–$U$ characteristics ($T$ = 4.2 K, $f$ = 2 kHz): (1) $I = 0$; (2) $2.7 \times 10^{13}$; (3) $8.2 \times 10^{13}$; (4) $1.1 \times 10^{14}$; (5) $1.4 \times 10^{14}$ quanta/cm$^2$

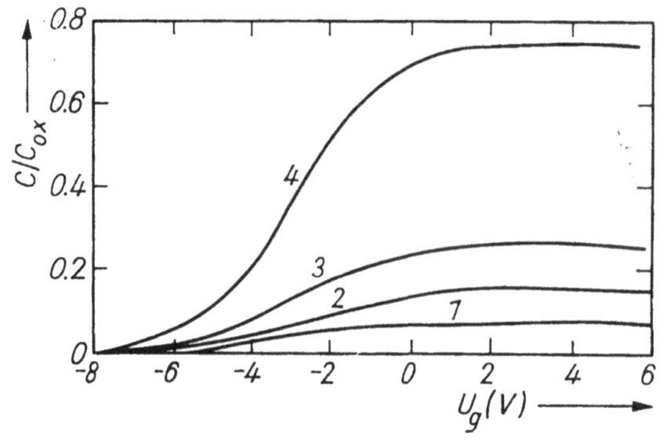


Fig. 7. Dependence of the photoelectric emf on the gate voltage at various illumination intensities ($T = 77$ K, $f = 80$ Hz): (1) $I_0 = 3 \times 10^{12}$ quanta/cm$^2$; (2) $2I_0$; (3) $4I_0$; (4) $12I_0$

quency $C$–$U$ dependence becomes similar to the low-frequency ones. The result can be interpreted in the following way: In the dark the semiconductor contains equilibrium charge carriers only. The charge of the inversion layer caused by electrons changes at the expense of generation–recombination processes of the space charge region. This process goes on slowly in comparison with the frequency of the test signal.

If the semiconductor is illuminated with frequencies $h\nu \geqq E_g$ non-equilibrium electron–hole pairs arise shielding the electric field near the interface. The charge of the inversion layer is already fixed by the non-equilibrium charge carriers in case of sufficiently high illumination intensity and the change takes place with the time constant of the diffusion–drift equilibrium setting (Maxwell relaxation time). The inversion layer charge responds to the high frequency signal and the $C$–$U$ characteristic is changed to the low-frequency type.

A decrease of the minimum in the characteristics is observed at increasing illumination intensity. The depth of the minimum depends on the alloy factor $\psi = p/n_i$, that means an increase of $\psi$ yields a decrease of the depth of the characteristics. The occurrence of non-equilibrium charge carriers caused by illumination gives rise to a non-equilibrium alloy factor $\psi' > \psi$.

As shown in [14] the $C$–$U$ characteristics in this case can be described by the same expression as in the dark case by replacing $\psi$ by $\psi'$. The large shift of the curves towards higher negative voltages with increasing illumination intensity is another characteristic peculiarity of the obtained results (Fig. 6). Such an effect is caused by excitation of non-equilibrium charge carriers from trap states in the oxide generated by the part of the exciting light having energy $h\nu \geqq E_g$. The collected charge is maintained after turning off the illumination and is vanishing after warming up to room temperature only.

The measurements of the surface photoelectric e.m.f. were carried out by means of frequency modulated light of wavelength $\lambda = 0.629$ µm. Using appropriate filters for this wavelength the calibrated illumination intensity was regulated. The results of the photoelectric e.m.f. measurement at 77 K and with different intensities are shown in Fig. 7. In earlier papers (for instance [15]) Gaussian-shaped dependences $U_{fe}(V_g)$ are found at InSb MOS structures and the photoelectric e.m.f. maximum is decreased by increasing illumination intensity. This effect is explained in [15] by an increase of the recombination centre concentration at the expense of defect formation at the interface between oxide and semiconductor due to increasing illumination intensity. The peculiarity of our structures is the saturation of the $U_{fe}(V_g)$ dependence at the band shift in inversion. Such a behaviour points to the fact that the photoelectric emf of the studied structures is determined by a barrier mechanism in the formation.

## References

[1] T. Sakurai, T. Suzuki, and Y. Noguchi, Japan. J. appl. Phys. **7**, 1491 (1968).
[2] L. C. Brennan and N. J. Giri, J. Electrochem. Soc. **117**, 391 (1970).
[3] M. L. Korwin-Pawlowski and E. L. Heasell, phys. stat. sol. (a) **24**, 649 (1974).
[4] O. V. Romanov, S. G. Sazonov, and N. V. Motaleva, Mikroelektronika **11**, 165 (1982).
[5] T. Jung, W. Braune, M. Schnürer, and J. Schulze, phys. stat. sol. (a) **81**, 463 (1984).
[6] R. K. Müller and R. L. Jacobson, J. appl. Phys. **35**, 1524 (1964).
[7] R. Y. Hung and E. T. Yon, J. appl. Phys. **41**, 2185 (1970).
[8] A. Etchelles and C. W. Fisher, J. appl. Phys. **47**, 4605 (1976).
[9] V. S. Vaviliv, A. F. Plotnikov, and V. E. Shubin, Fiz. Tekh. Poluprov. **5**, 2064 (1971).
[10] W. Braune, M. Schnürer, N. Kubicki, and R. Herrmann, phys. stat. sol. (a) **86**, 427 (1984).
[11] L. M. Terman, Solid State Electronics **5**, 285 (1962).
[12] C. N. Berglund, IEEE Trans. Electron Devices **13**, 701 (1966).
[13] M. Kaplit and J. N. Zemel, Phys. Rev. Letters **21**, 212 (1968).
[14] R. F. Pierret and C. T. Sah, Solid State Electronics **13**, 269 (1970).
[15] V. N. Davydov and E. A. Luskutova, Izv. vuzov Fizika **26**, 70 (1983).

*(Received February 18, 1988)*

phys. stat. sol. (a) **108**, 311 (1988)

Subject classification: 71.55; 61.70; S5.12

*Belorussian State University, Minsk*[1])

# Extrinsic Self-Trapping and Negative $U$ of Oxygen-Related Thermal Donors in Germanium

By

V. V. LITVINOV, G. V. PALCHIK, and V. I. URENEV

The occurrence of some types of configurationally bistable thermal donors with negative effective correlation energy ($U < 0$), that are formed at initial stages of heat treatment at $T \leqq 673$ K of the oxygen containing germanium crystals is presented. The study of the equilibrium electron occupation function of three transforming thermal donors types enables one to obtain the values of $U$ and the probability of double electron occupancy. By means of the kinetics of thermal donor transformation processes between two equilibrium configurations the activation energy and the cross-section for extrinsic self-trapping of charge carriers are determined. The results obtained are discussed in terms of very large lattice relaxation phenomena.

Показано существование в германии нескольких типов конфигурационно-бистабильных термодоноров с отрицательной эффективной корреляционной энергией ($U < 0$), которые образуются на начальных стадиях термообработки кислородосодержащих кристаллов германия при $T \leqq 673$ K. Исследования с помощью равновесной функции заполнения электронами трех типов перестраивающихся доноров позволили установить величину $U$ и уровни двухэлектронного заполнения. Благодаря кинетическому изучению процесса перестройки термодоноров между двумя равновесными конфигурациями определены энергия активации и сечение для несобственной автолокализации носителей тока. Полученные результаты обсуждаются в рамках эффекта большой релаксации решётки.

## 1. Introduction

The most fascinating effects of electron–phonon interaction in semiconductors are related with a new class of defects characterized by a large lattice relaxation [1]. Unlike ordinary defects, in which lattice relaxation is responsible for multiphonon trapping processes and Frank-Condon shift, these defects lead to new phenomena such as extrinsic self-trapping of charge carriers and a negative effective correlation energy $U$.

The idea of extrinsic self-trapping in semiconductor crystals was introduced for the first time by Lang and Logan [2] and later theoretically founded in Toyozava's work [3]. It was shown that the extrinsic self-trapping of charge carriers leads to a configurational bistability of defects or, in other words, to two minima of their adiabatic potential at the same charge state [4]. Such metastable centers are characterized by an activation dependence of the charge carrier trapping cross-section at low temperatures and they are the cause of the persistent conductivity of semiconductors [1 to 5].

---

[1]) Leninskii prospect 4, SU-220080 Minsk, USSR.

The negative-$U$ concept can be applied only to multiply charged centers and reflects the fact that the energy change of the defects, which is caused by the lattice relaxation at trapping of the second electron, exceeds the interelectron repulsion energy. Negative-$U$ properties in semiconductor crystals were proved for the vacancy and the boron interstitial in Si [6] and for the M-center in InP [7]. It has been recently discovered that the well-known oxygen-related thermal donors in Si and Ge [8, 9] are also negative-$U$ centers.

The results of our investigations on thermal donors in Ge, described in the present paper, show that extrinsic self-trapping and negative $U$ are features of these centers.

## 2. Experimental Procedure and Method of Analysis

We used Czochralski-grown n-Ge single crystals ($\varrho = 0.1$ to $40\ \Omega$ cm) containing oxygen with concentration (0.5 to 2) $\times 10^{17}$ cm$^{-3}$ and doped with antimony. The samples were heat treated at 623 K to introduce thermal donors (TD), which is related to agglomeration of oxygen dissolved in the lattice. The standard measurements of Hall effect, current conductivity, and spectral distribution of extrinsic photoconductivity were made both on initial and heat-treated crystals. The kinetics of slow conductivity relaxation of the heat-treated crystals was recorded at fixed ($\pm 0.1$ K) sample temperature. The free electron concentration ($n$) in the samples was calculated from Hall effect data taking into account the temperature change of the Hall factors [10].

The center parameters were determined from the analysis of the temperature dependence of $n$ by means of the neutrality equation using both the common method [11] and the differential one [12]. We took into account investigations of the electric properties [13] and far-IR absorption spectra of thermal oxygen donors in Ge [14], which provided evidence for their double-donor nature. Depending on the number of captured electrons ($j$) such centers can exist in three charge states ($++$, $+$, and 0): $TD^{++}(j = 0)$, $TD^{+}(j=1)$, and $TD^{0}(j = 2)$. The neutrality equation for the ionization of one double donor type with concentration $N$ in crystals, containing doped donors with concentration $N_d$ and acceptors with concentration $N_a$, can be presented as follows:

$$n + N_a^- = N_d^+ + N^+ + 2N^{++} + p\,, \tag{1}$$

where $p$ is the free hole concentration. The acceptors of the investigated crystals were completely occupied by electrons at all temperatures, i.e. $N_a = N_a^-$. The charged donor concentrations (see (1)) were found in terms of the partition function ($Z_j$) of a center with $j$ captured electrons [15, 16],

$$N_d^+ = N_d \left[1 + \lambda \frac{Z_{d1}}{Z_{d0}}\right]^{-1}, \tag{2}$$

$$N^+ = N \left[1 + \lambda \frac{Z_2}{Z_1} + \frac{1}{\lambda} \frac{Z_0}{Z_1}\right]^{-1}, \tag{3}$$

$$N^{++} = N \left[1 + \lambda \frac{Z_1}{Z_0} + \lambda^2 \frac{Z_2}{Z_1} \frac{Z_1}{Z_0}\right]^{-1}. \tag{4}$$

Here $\lambda \equiv \exp(E_F/kT)$, where $E_F$ is the Fermi level. The excited states were neglected and according to [17] it was supposed that the donor has a singly degenerate state, when it has captured two electrons or has no electrons, and the donor has a doubly degenerate state, when it has captured one electron. In this case the following relations hold [16]:

$$\frac{Z_{d1}}{Z_{d0}} = 2 \exp\left[-E_d(1, 0)/kT\right], \tag{5}$$

$$\frac{Z_1}{Z_0} = 2 \exp\left[-E(1, 0)/kT\right], \tag{6}$$

$$\frac{Z_2}{Z_1} = \frac{1}{2} \exp\left[-E(2, 1)/kT\right]. \tag{7}$$

Here $E_d(1, 0)$ and $E(1, 0)$, $E(2, 1)$ are the energy levels $E(j + 1, j)$ of doped and thermal donors as difference between energies of defects with $(j + 1)$ and $j$ captured electrons.

The above-mentioned relations do not depend on the sign of the effective correlation energy ($U$), determining the connection between double donor energy levels,

$$U = E(2, 1) - E(1, 0). \tag{8}$$

However, the statistics of centers with $U < 0$ has an essential peculiarity: the average number of double donors with one captured electron is zero under equilibrium conditions. This can be easily seen from the probability of a hole on a double donor,

$$f_p = \frac{N^+ + 2N^{++}}{N} = 2\left[1 + \frac{1 + 2n/n_2}{1 + 2n_1/n}\right]^{-1}. \tag{9}$$

We have replaced $\lambda$ by $(n/N_c) \exp(E_c/kT)$, where $E_c$ is the energy of the conduction band edge, and introduced the quantities [16]

$$n_1 = N_c \frac{Z_0}{Z_1 \exp(E_c/kT)}, \tag{10}$$

$$n_2 = N_c \frac{Z_1}{Z_2 \exp(E_c/kT)}. \tag{11}$$

According to (9) $f_p = 1$ if $n^2 = n_1 n_2$ or $2n_1 < n < n_2/2$. It is equivalent to the requirement on the Fermi level: $E_F = \frac{1}{2}[E(1, 0) + E(2, 1)]$ or $E(1, 0) < E_F < E(2, 1)$. The first condition here is irrelevant since it occurs only for one value of $E_F$. The second condition is not applicable for $U < 0$. The physics here is quite clear: at $E(2, 1) < < E(1, 0)$ it takes less energy to place two electrons on one center than one electron each on two centers. It means that if $U < 0$, the one-electron state is unstable and double thermal donors are either in the $TD^0$ state or in the $TD^{++}$ one under equilibrium conditions. Then $N^+ = 0$ and the hole occupation function is simplified to the following form if $|U| \gg kT$:

$$f_p \approx 2\left[1 + \exp\left(2\frac{E_F - E(2, 0)}{kT}\right)\right]^{-1} = 2\left[1 + \left(\frac{n}{N_c}\right)^2 \exp\left(2\frac{E_c - E(2,0)}{kT}\right)\right]^{-1}, \tag{12}$$

where

$$E(2, 0) = \frac{1}{2}[E(1, 0) + E(2, 1)]. \tag{13}$$

Here $E(2, 0)$ has the meaning of an "occupancy level", i.e. all defects are in the neutral state if $E_F > E(2, 0)$ and in the doubly ionized state if $E_F < E(2, 0)$.

Using the differential method of $n(T)$ analysis the energy levels and defect concentrations were calculated from the dependence of the derivative $X = kT\,dn/dE_F$ on $F = E_c - E_F$, which looks like spectral bands if $|U| \gg kT$ [12]. For ordinary defects ($U > 0$) the ordinate of the maximum is $X_m = 1/4N$, the abscissa of the maximum corresponds to the energy level position, and the half-width of each band is $\delta F = 3.5kT_m$ (where $T_m$ is the temperature corresponding to $F = F_m$). On the other hand, for double donors with $U < 0$, $X_m = N$, $F_m = E_c - E(2, 0)$, and $\delta F = 1.6kT_m$ [18].

## 3. Results and Discussion

### 3.1 Configurational bistability and negative U

Typical $n(T)$ dependences and corresponding dependences of $X = kT\,dn/dE_F$ on $F = E_c - E_F$ are shown in Fig. 1 both for the initial n-Ge crystal and that shortly heat-treated at 623 K. Curves 2 and 2′ were obtained from $n$ measurements when the sample was slowly cooled in the dark at $T = 300$ to 200 K, the equilibrium conduc-

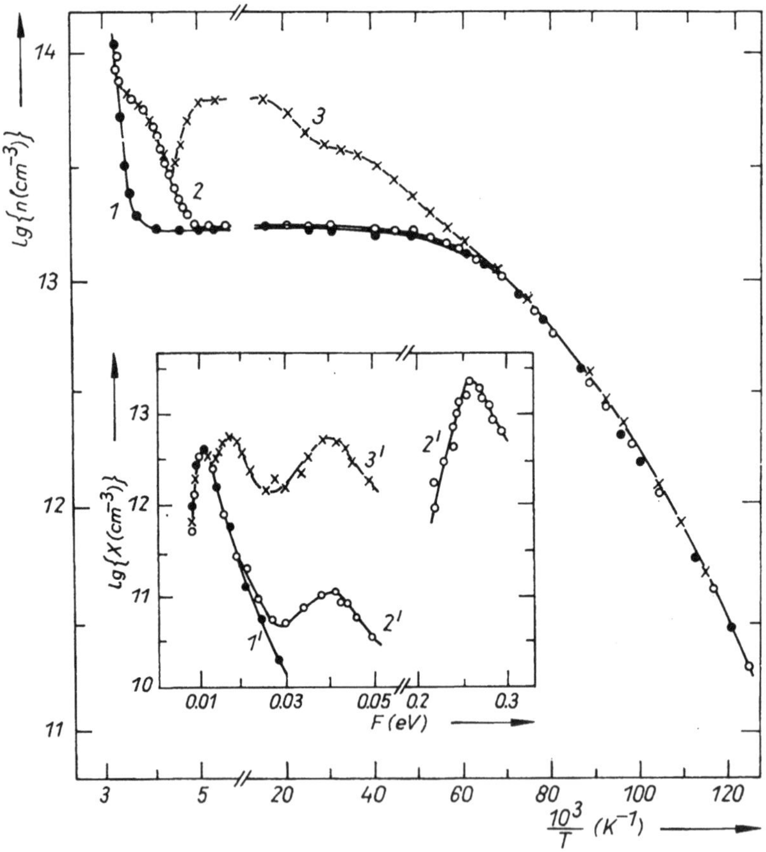


Fig. 1. Dependences of the electron concentration ($n$) on temperature and of $X$ on the Fermi level position $E_F$ ($F = E_c - E_F$) calculated from the $n(T)$ dependences for an n-Ge:Sb, O crystal (oxygen concentration $1.4 \times 10^{17}$ cm$^{-3}$) heat-treated at 623 K during 5 min. (1), (1′) Initial sample; (2), (2′), (3), (3′) after heat treatment and following slow cooling ((2), (2′)) or quenching to 150 K accompanied by illumination with white light ((3), (3′))

tivity was determined at every experimental point. Curves 3 and 3' are the results of $n(T)$ measurements during heating of the same sample which was previously cooled from room temperature to $T \leqq 150$ K either with white light illumination from a filament lamp or by quenching in the dark to liquid nitrogen. Such $n(T)$ dependence versus sample cooling regime illustrates the appearance of persistent conductivity in heat-treated crystals. The analysis of the curves in Fig. 1 at $T \leqq 150$ K (Table 1) shows that this persistent conductivity is caused by donors with two levels. The electron ionization energies from these levels to the conduction band are $\approx 40$ meV and $\approx 16$ meV, i.e. they are close to the binding energies for the first and second electrons of double thermal donors obtained from spectroscopic investigations [14]. Thus, the results of Fig. 1 show the instability of double thermal donors with energy levels $E(1, 0) \approx E_c - 40$ meV and $E(2, 1) \approx E_c - 16$ meV. In other words, there are two atomic configurations of thermal donors: A and B. The energy levels $E(1, 0)$ and $E(2, 1)$ correspond to the configuration A ($TD_A$) of thermal donors, and the B configuration is connected with the neutral thermal donor state ($TD_B^0$). So there are two ways of thermal donor recharging according to the following scheme:

$$TD_A^{++} \overset{a}{\rightleftarrows} TD_A^{+} \begin{matrix} \overset{b}{\rightleftarrows} TD_A^{0} \\ \underset{b'}{\rightleftarrows} TD_B^{0} \end{matrix} . \tag{14}$$

The sign change in scheme (14) indicates the participation of free charge carriers in the reaction, and the letter index change reflects the observed bistability of the neutral state of thermal donors. The defect ionization on the way a—b in (14) determines the levels $E(1, 0)$ and $E(2, 1)$ and causes the persistent conductivity of crystals at low temperatures (curve 3, Fig. 1). The energy levels $E(1, 0)$ and $E'(2, 1)$ correspond to defect recharging on the way a—b' in (14), however, these energy levels are not revealed in the temperature dependences of $n$ under equilibrium conditions. The change of the equilibrium electron concentration in the temperature range 200 to 300 K (curve 2, Fig. 1) indicates the reverse transformation of thermal donors between $TD_A^{++}$ and $TD_B^0$. It only means that the probability of thermal donors to be in the $TD_A^+$ state is close to zero under equilibrium conditions, and we observed the two-electron ionization at the same time. Such two-electron ionization is typical for negative-$U$ centers according to the statistical analysis in Section 2. The negative-$U$ nature of thermal donors is confirmed by the fact that the experimental values of the electron equilibrium concentration in the temperature range 200 to 300 K (curve 2, Fig. 1) are the same as the $n(T)$ dependence calculated by means of the neutrality equation (1) and the occupation function (12) if $E_c - E(2, 0) = (0.26 \pm 0.01)$ eV and $N = 2.4 \times 10^{13}$ cm$^{-3}$. It should be pointed out that an attempt to describe curve 2 by means of (1) leads to physically unrealistic values of degeneration multiplicity and ionization energy, when we take into account the ionization of an ordinary donor with double concentration ($2N$) [9]. On the other hand, the ionization of the double

Table 1

Analysis of $n(T)$ dependences (for $T \leqq 150$ K, Fig. 1) on the basis of (1) to (4)

| number of curve | $N_d$ ($10^{13}$ cm$^{-3}$) | $E_c - E_d$ (1, 0) (meV) | $N_a$ ($10^{11}$ cm$^{-3}$) | $N$ ($10^{13}$ cm$^{-3}$) | $E_c - E$ (2, 1) (meV) | $E_c - E$ (1, 0) (meV) |
|---|---|---|---|---|---|---|
| 1 | 1.72 | 10.2 | 7 | 0 | — | — |
| 2 | 1.7 | 10.2 | 10 | 0.04 | — | 40 |
| 3 | 1.7 | 10.2 | 12 | 2.34 | 16 | 40 |

donor with $U < 0$ is confirmed by the differential analysis of the $n(T)$ equilibrium dependence in the temperature range 200 to 300 K. Thus, the half-width of the maximum of the $X(F)$ dependence (curve 2′) is $1.6kT_m$ (at $T_m = 250$ K), i.e. approximately two times less than for ordinary centers. The values of $X_m$ and $F_m$ coincide with the thermal donor concentration ($N$) and the occupancy level $E(2, 0)$ calculated on the basis of the neutrality equation.

Taking into account the above-mentioned analysis of the data of Fig. 1 it can be concluded that in the initial stage of heat treatment the $n$ change is mainly determined by the first thermal donor type (TD-1), characterized by a configuration bistability and negative $U$. Fig. 2 shows the $n(T)$ and $X(F)$ equilibrium dependences in the temperature range 180 to 350 K for the subsequent stages of heat treatment of the same sample. The data presented indicate the successive generation of at least three thermal donor types (TD-1, TD-2, and TD-3) with negative $U$. The occupation levels $E(2, 0)$ obtained from $F_m$ essentially depend on the center type (Table 2). The existence of several types of double helium-like oxygen-related thermal donors in Ge crystals was also proved by spectroscopic studies [14]. Both the spectroscopic [14] and electric

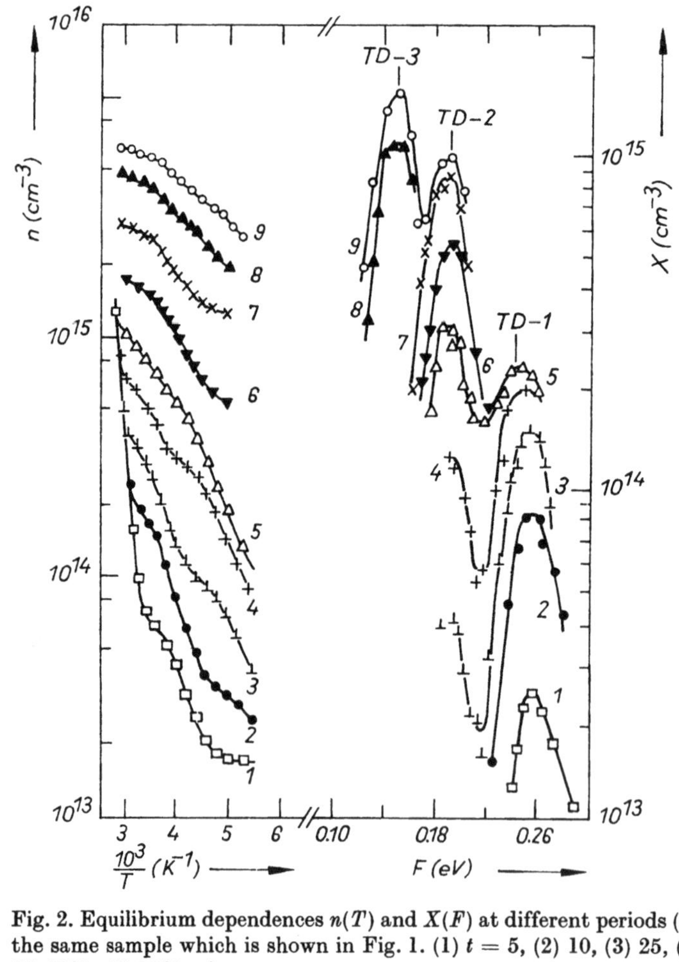


Fig. 2. Equilibrium dependences $n(T)$ and $X(F)$ at different periods ($t$) of heat treatment of one and the same sample which is shown in Fig. 1. (1) $t = 5$, (2) 10, (3) 25, (4) 60, (5) 120, (6) 240, (7) 480 (8) 1020, (9) 1860 min

Table 2
Occupation level positions and correlation energy ($U$) for three types of thermal donors in Ge

| type of thermal donor | $E_c - E(2,0)$ (eV) | $E_c - E(2,1)$ (eV) | $E_c - E(1,0)$ (eV) | $E_c - E'(1,0)$ (eV) | $U$ (eV) |
|---|---|---|---|---|---|
| TD-1 | 0.26 ± 0.01 | 0.016 ± 0.001 | 0.04 ± 0.01 | 0.48 ± 0.02 | −0.45 ± 0.03 |
| TD-2 | 0.19 ± 0.01 | — | — | 0.34 ± 0.02 | −0.31 ± 0.03 |
| TD-3 | 0.15 ± 0.01 | — | — | 0.26 ± 0.02 | −0.23 ± 0.03 |

[9, 13] investigations of the thermal donor energy spectrum at different heat treatment stages show that the energy levels $E(2,1)$ and $E(1,0)$ for all thermal donor types in the helium-like configuration A are located in a narrow energy interval (Table 2) contrary to the $E(2,0)$ occupation levels. On the basis of known $E(1,0)$ and $E(2,0)$ values we determined the $E'(2,1)$ level, which is the difference between the defect energies of $TD_B^0$ and $TD_A^+$ states. The obtained values of $E'(2,1) = 2E(2,0) - E(1,0)$ and $U = E'(2,1) - E(1,0)$ (Table 2) indicate the strong binding of electrons in the $TD_B^0$ configuration. Besides, the $E'(2,1)$ and $U$ values essentially decrease for center types originating at longer duration of heat treatment (Fig. 2). Since the generation kinetics of TD-1, TD-2, and TD-3 centers are interrelated with each other and satisfy the modified Kaiser model [19], the $U$ reduction can be related to the increase of the number of oxygen atoms in donor complexes.

### *3.2 Kinetic peculiarities of the transformation process of thermal donors*

In the previous section the configuration bistability and the negative $U$ of the oxygen-related thermal donors in Ge were shown on the basis of experimental results on the equilibrium function of electron occupation. Further properties of thermal donors were obtained by studying the slow conductivity relaxation kinetics of heat-treated crystals.

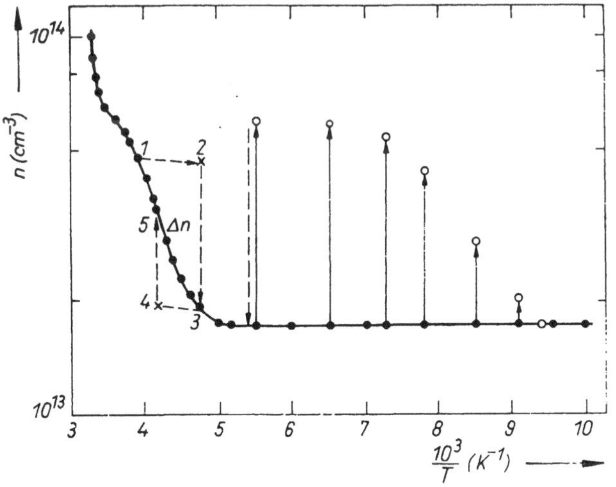


Fig. 3. Schematic representation of the excitation and relaxation process of non-equilibrium electrons in a sample with thermal donors. ● Equilibrium values of the electron concentration determined from curve 2 in Fig. 1, × values of the electron concentration $n$ at quick change of the sample temperature in the dark, ○ $n$ values after weak illumination by a filament lamp

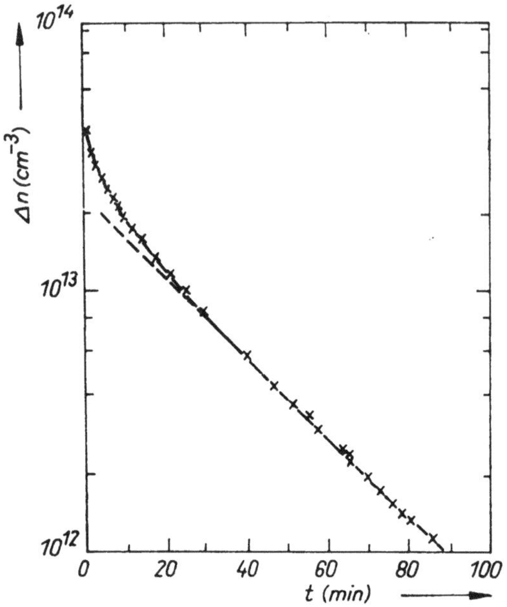


Fig. 4. Kinetics of $n$ change at 213 K after illumination of a sample containing the first type of thermal donors (TD-1)

The non-equilibrium conductivity in the samples can be caused both by white light illumination and quick cooling or quick heating in the dark in the temperature range 200 to 300 K. The change in electron concentration corresponding to the latter case is shown in Fig. 3 by the dashed curve (1 to 5). The influence of light from the intrinsic absorption region of Ge on the free electron concentration of a sample which was previously cooled to different temperatures, is also shown in Fig. 3. It should be pointed out that the same experiments with extrinsic light did not lead to a marked change of $n$, and light from the intrinsic absorption region of Ge does not cause persistent conductivity at temperatures below $\approx$ 100 K (Fig. 3). Such a thermal and photostimulated conductivity of heat-treated n-Ge crystals was discovered in [20] and was connected by the author with trapping of minority charge carriers by deep traps. The data obtained here show that these traps are bistable thermal donors in B configuration. According to the recharging scheme ((14), b′—a) the centers transform from the $TD_B^0$ state to the $TD_A^+$ one with change of their own configuration if they trap non-equilibrium holes generated by light. Since the Fermi quasi-levels for electrons are below the $E(1, 0)$ level, $TD_A^+$ releases the second electron to the conduction band and transforms to the $TD_A^{++}$ state. The velocity of achieving thermodynamic equilibrium of the reaction ((14), a—b′) is low and is controlled by the relaxation of the non-equilibrium electron concentration ($\Delta n$). The typical kinetics of such a $\Delta n$ relaxation at fixed temperature is shown in Fig. 4. One can see that the $\Delta n(t)$ dependence is described by first-order kinetics only at weak deviation from equilibrium ($\leqq 10\%$). The relaxation time constant ($\tau$) obtained in the quasi-equilibrium approximation does not depend on the kind of $\Delta n$ excitation presented in Fig. 3, but is a function of sample temperature, of the electron equilibrium concentration ($n_0$), and of the center type determining the non-equilibrium conductivity. The experimental $\tau^{-1}(T)$ dependences determining the recharging velocities of three thermal donor types at different positions of the equilibrium Fermi level are shown in Fig. 5. The curves consist of two parts characterized by different activation energies and can be described by

$$\tau^{-1} = A_1 \exp\left[-\frac{E_{a1}}{kT}\right] + n_0 A_2 \exp\left[-\frac{E_{a2}}{kT}\right]. \tag{15}$$

The corresponding values of the activation energies and the pre-exponential factors are given in Table 3. The observed activation energy change of the recharging process of thermal donors occurs when the Fermi level position corresponds to the occupancy

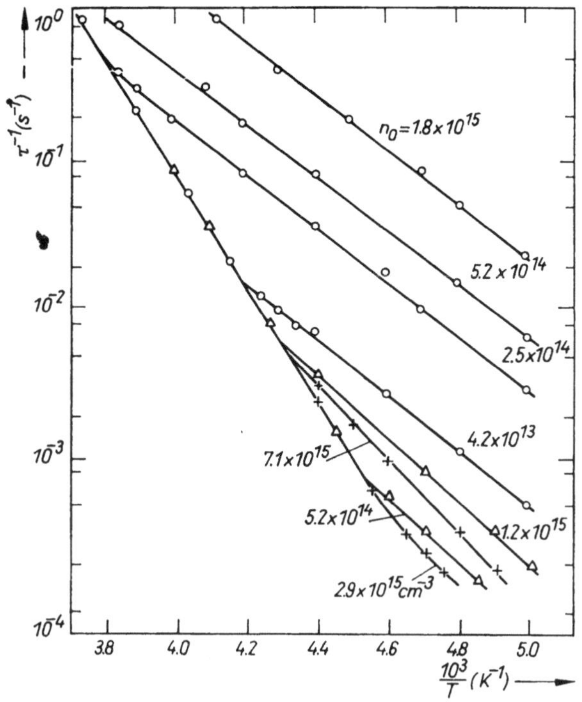


Fig. 5. Temperature dependence of the transformation rate of thermal donors at different values of the equilibrium electron concentration; ○ TD-1, △ TD-2, + TD-3

levels $E(2, 0)$. Meanwhile, according to [19], the first and second terms of the dependence (15) correspond to the emission rates of electrons (for $E_F \leqq E(2, 0)$) and holes (for $E_F \geqq E(2, 0)$) to the conduction and valence bands, respectively,

$$\tau_n^{-1} = \sigma_n^{(+)} v_n 2N_C \exp\left[-\frac{E_c - E'(2, 1)}{kT}\right] + \\ + n_0 \sigma_p^{(0)} v_p \frac{N_v}{N_c} \exp\left[-\frac{E_g - [E_c - E(2, 0)]}{kT}\right], \quad (16)$$

where $v_n$ and $v_p$ are the average thermal velocities of free electrons and holes, $\sigma_n^{(+)}$ and $\sigma_p^{(0)}$ are the cross-sections of electron and hole trapping by the $E'(2, 1)$ level, $N_v$ is the effective density of states in the valence band, $E_g$ is the energy gap.

A quantitative comparison of the experimental data of Fig. 5 (expression (15)) with the analytical expression (16) shows that the cross-sections of electron (hole)

Table 3

Experimental values of pre-exponential factors and activation energies of the transformation process of three types of thermal donors and parameters of electron and hole multiphonon trapping on the $E'$ (2, 1) level calculated according to the data of Fig. 5

| types of thermal donor | $A_1$ ($10^{14}$ s$^{-1}$) | $E_{a1}$ (eV) | $A_2$ ($10^{-8}$ cm$^3$ s$^{-1}$) | $E_{a2}$ (eV) | $\bar{\sigma}_n^{(+)}$ ($10^{-14}$ cm$^2$) | $E_n$ (eV) | $\bar{\sigma}_p^{(0)}$ ($10^{-17}$ cm$^2$) | $E_p$ (eV) |
|---|---|---|---|---|---|---|---|---|
| TD-1 | 2 | 0.76 | 4 | 0.39 | 5 | 0.25 | 2 | 0.1 |
| TD-2 | 2 | 0.76 | 4 | 0.43 | 5 | 0.38 | 2 | 0 |
| TD-3 | 2 | 0.76 | 4 | 0.49 | 5 | 0.47 | 2 | 0 |

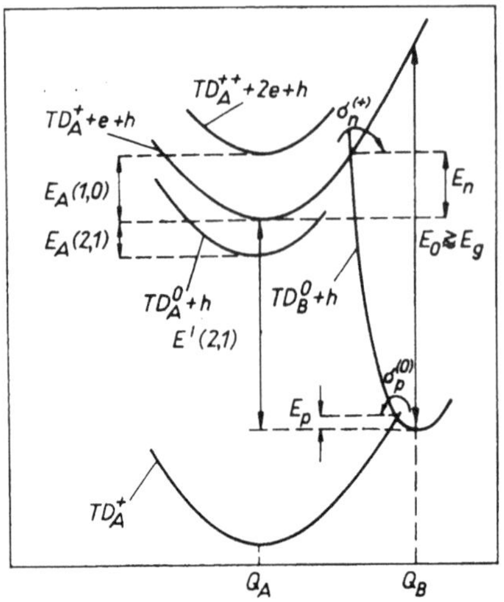


Fig. 6. Configuration-coordinate diagram for the bistable thermal donors. $E_0$ optical ionization energy of $TD_B^0$; e and h are free electrons and holes, respectively

trapping by the $E'(2, 1)$ level exponentially depends on temperature,

$$\sigma_{n(p)} = \bar{\sigma}_{n(p)} \exp\left[-\frac{E_{n(p)}}{kT}\right]. \tag{17}$$

The calculated values of "frequency terms" of cross-sections ($\bar{\sigma}_n$, $\bar{\sigma}_p$) and the thermal activation energies ($E_n$ and $E_p$) of this trapping are given in Table 3. It should be pointed out that the above-mentioned parameters were calculated taking into account the temperature dependences of $v_{n(p)}$, $N_{v(c)}$, and $E_g$ [21]. It can be seen from Table 3 that the values of $\bar{\sigma}_n^{(+)}$, $\bar{\sigma}_p^{(0)}$ for three thermal donor types are approximately the same, while the energy barrier ($E_n$) for electron trapping by the $E'(2, 1)$ level increases for centers formed at subsequent stages of heat treatment. Since the concentration of the latter centers dominates (Fig. 2), it explains the fact that thermal donors are observed only in the helium-like configuration A at long heat treatments [13, 22].

The activation dependence of the trapping cross-sections of carriers (17), observed by us at temperatures lower than the Debye temperature, cannot be explained within the ordinary multiphonon trapping model [5, 23], but it was experimentally and theoretically proved for multiphonon charge carrier trapping in terms of the extrinsic self-trapping mechanism. According to [1 to 5] we introduce the "external cross-section" of adiabatic potentials for $TD_A^+$ and $TD_B^0$ electron states and illustrate the configuration bistability of the neutral state of thermal donors (the configuration coordinates $Q_A$ and $Q_B$ in Fig. 6), the negative correlation energy $(E(1, 0) > E'(2, 1))$, and the large difference between the optical and thermal energies of $TD_B^0$. The latter correlates with the fact that light from the extrinsic absorption region did not affect the $TD_B^0$ state and the $E'(2, 1)$ levels were not observed in the extrinsic conductivity spectra [13]. The lacking overlap of vibrational states in the potential surface minima for the electron states $TD_A^{++}$ and $TD_B^0$ prevents charge carrier tunneling through the barriers $E_n$ and $E_p$, so both electron and hole trapping by the $E'(2, 1)$ level is also a thermal activation process at low temperatures. It causes the persistent conductivity of the heat-treated crystals and the possibility of observation of the metastable $TD_A^0$ and $TD_A^+$ states within the low-temperature range.

## 4. Conclusion

Up to now the oxygen-related thermal donors in Ge have been considered as double helium-like centers characterized by weak binding energies for the first ($\approx$ 40 meV) and the second ($\approx$ 16 meV) electrons. The data of this paper show that three types of thermal donors generated at the initial stages of heat treatment of Ge can also exist in other equilibrium configurations characterized by strongly bound electrons. The configuration transformation occurs in accordance with the extrinsic self-trapping mechanism and it is accompanied by so large lattice relaxation that both $TD_A^{++}$ and $TD_B^0$ equilibrium configurations form an Anderson negative-$U$ system. Thus, the well-known oxygen-related thermal donors in Ge can serve as objects to study a large number of effects of electron-phonon interaction. Moreover, the configurational bistability of thermal donors has to be taken into account in investigations of the accumulation kinetics and in developing a microscopic model of these centers in germanium.

## References

[1] J. M. Langer, Radiat. Eff. **72**, 55 (1983).
[2] D. V. Lang and R. A. Logan, Phys. Rev. Letters **39**, 635 (1977).
[3] Y. Toyozava, Physica **B116**, 7 (1983).
[4] M. Levinson, J. appl. Phys. **58**, 2628 (1985).
[5] V. A. Kovarskii, O. A. Sedlitskii, and L. V. Chernich, in: Kinetics Processes in Doped Semiconductors under External Fields, Ed. V. A. Kovarskii, Kishinev 1984 (in Russian).
[6] J. R. Troxell and G. D. Watkins, Phys. Rev. B **22**, 921 (1980).
[7] M. Tavola, M. Levinson, J. L. Benton, and L. C. Kimerling, Phys. Rev. B **30**, 832 (1984).
[8] L. F. Makarenko, V. P. Markevich, and L. I. Murin, Fiz. Tekh. Poluprov. **19**, 1935 (1985).
[9] V. V. Litvinov, G. V. Palchik, and V. I. Urenev, Fiz. Tekh. Poluprov. **19**, 1366 (1985).
[10] F. J. Morin and J. P. Maita, Phys. Rev. **94**, 1525 (1954).
[11] V. L. Bonch-Bruevich and S. G. Kalashnikov, Fizika poluprovodnikov, Izd. Nauka, Moskva 1977.
[12] H. J. Hoffman, Appl. Phys. A **16**, 307 (1979).
[13] V. V. Litvinov and V. I. Urenev, Fiz. Tekh. Poluprov. **17**, 1099 (1983).
[14] P. Clauws and J. Vennik, Phys. Rev. B **30**, 4837 (1984).
[15] P. T. Landsberg, Proc. Phys. Soc. B **69**, 1056 (1952).
[16] P. T. Landsberg and G. S. Kousik, J. appl. Phys. **56**, 1696 (1984).
[17] D. C. Look, Phys. Rev. B **24**, 5852 (1981).
[18] H. J. Hoffmann, Appl. Phys. A **27**, 39 (1928).
[19] V. V. Litvinov and V. I. Urenev, Dep. v VINITI 11. 06. 85, No. 4060-85, Tomsk (42 p).
[20] E. Adachi, J. Phys. Chem. Solids **28**, 1821 (1967).
[21] P. I. Baranskii, V. G. Klochkov, and I. V. Potikevich, Poluprovodnikovaya elektronika, Naukova Dumka, Kiev 1975.
[22] C. S. Fuller and F. H. Doleiden, J. Phys. Chem. Solids **19**, 251 (1961).
[23] V. N. Abakumov, I. A. Merkulov, V. I. Perel, and I. N. Yassievich, Zh. eksper. teor. Fiz. **89**, 1472 (1985).

*(Received October 6, 1987; in revised form March 7, 1988)*

phys. stat. sol. (a) **108**, 323 (1988)

Subject classification: 72.20 and 73.40; S5.11

*A. F. Ioffe, Physico-Technical Institute, Academy of Sciences of the USSR, Leningrad*[1]) *(a) and*
*Physico-Technical Institute, Academy of Sciences of the UzSSR, Tashkent (b)*

# Injection and Recombination Currents in a-Si:H Structures

By

O. A. Golikova (a), M. M. Mezdrogina (a),
N. A. Feoktistov (a), K. L. Sorokina (a), M. M. Kazanin (a),
P. M. Karageorgy-Alkalaev (b), and A. Yu. Leiderman (b)

Both, theoretical and experimental investigations of current–voltage characteristics of a-Si:H structures (Pt–i–n, p–i–n) are carried out. A choice in favour of the double-carrier injection mechanism is substantiated. A theory of G–R currents is applied to a semiconductor with a continuous density of localized states in the mobility gap.

Приведены результаты как теоретического, так и экспериментального исследования вольт–амперных характеристик структур на основе a-Si:H (Pt–i–n, p–i–n). Обоснован выбор в пользу механизма двойной инжекции. Теория генерационно-рекомбинационных токов, распространенная на случай материала с квазинепрерывным спектром локализованных состояний в щели подвижности, позволяет объяснить полученные результаты.

## 1. Introduction

Hydrogenated amorphous silicon (a-Si:H) being a promising material for applications in different microelectronic devices, the problem of a careful analysis of current–voltage ($IU$) characteristics for different structures has arisen.

First it concerns such basic structures as the p–i–n structure. Experimental data show a complicated behaviour of $I(U)$ [1, 2, 3]: $j = j_s \exp(qU/AkT)$ at low positive bias ($j_s = (10^{-8}$ to $10^{-11})A$ cm$^{-2}$, $A = 1$ to 2), and $j \sim U^m$ ($m = 3$ to 6) at $U > 1$ V. Schottky barrier structures appear to possess similar $I(U)$ characteristics and can be useful for process modelling. The question of the nature of the dark current has not yet been solved, though some theories of current transport have been applied for the explanation of experimental results [1, 2, 4].

The authors suggest to use the generation–recombination (G–R) current theory applied to a material with a continuous density of gap states (DOS). Our results of an experimental study of the dark current in both p–i–n and Pt–i–n structures are also presented to support this point of view.

## 2. Theory

The presence of a continuous DOS introduces some peculiarities of injection phenomena. The most important of them are the reduction of injection modulation of the volume of the semiconductor [5] and the intensive G–R current in the space charge region (SCR) [2, 6, 7], leading to a variation of the injection coefficient.

[1]) Politekhnicheskaya 26, SU-194021 Leningrad, USSR.

The system of equations used for the steady state is as follows:

$$\frac{1}{q}\,j_{\rm n} = \mu_{\rm n} nE + D_{\rm n}\,\frac{{\rm d}n}{{\rm d}x}, \tag{1}$$

$$\frac{1}{q}\,j_{\rm p} = \mu_{\rm p} pE + D_{\rm p}\,\frac{{\rm d}p}{{\rm d}x}, \tag{2}$$

$$\frac{1}{q}\,\frac{{\rm d}j_{\rm n}}{{\rm d}x} = R\,, \tag{3}$$

$$\frac{1}{q}\,\frac{{\rm d}j_{\rm p}}{{\rm d}x} = R\,, \tag{4}$$

$$j_{\rm n} + j_{\rm p} = j\,, \tag{5}$$

$$\frac{\varepsilon}{4\pi q}\,\frac{{\rm d}E}{{\rm d}x} = n - p - N_{\rm d} + n_{\rm R} - p_{\rm R}\,, \tag{6}$$

where $q$ is the electron charge, $E$ the electric field, $j_{\rm n}(j_{\rm p})$ the electron (hole) current density, $\mu_{\rm n}(\mu_{\rm p})$ the free electron (hole) mobility, $n(p)$ the free electron (hole) concentration, $n_{\rm R}(p_{\rm R})$ the trapped electron (hole) concentration, $N_{\rm d}$ the donor concentration, $R$ the recombination rate, and $D_{\rm n}(D_{\rm p})$ the electron (hole) diffusion coefficient.

$n_{\rm R}$ and $p_{\rm R}$ depend especially on the DOS shape, $g(E_{\rm R})$,

$$n_{\rm R} = \int\limits_{E_{\rm v}}^{E_{\rm c}} g(E_{\rm R})\, f_{\rm R}(E_{\rm R}, F_{\rm n}, F_{\rm p}, F_0)\,{\rm d}E_{\rm R}\,, \tag{7}$$

$$p_{\rm R} = \int\limits_{E_{\rm v}}^{E_{\rm c}} g(E_{\rm R})\, f_{\rm Rp}(E_{\rm R}, F_{\rm n}, F_{\rm p}, F_0)\,{\rm d}E_{\rm R}\,. \tag{8}$$

$E_{\rm c}(E_{\rm v})$ being the energy of the c- (v-) band edge, $F_{\rm n}(F_{\rm p})$ the quasi-Fermi energy for electrons (holes), $F_0$ the equilibrium Fermi energy, $f_{\rm R}(f_{\rm Rp})$ the electron (hole) statistical function.

Using these equations one can obtain different $I(U)$ characteristics for various conditions.

Both single- and double-carrier injection in a-Si:H alloys have been found recently [8]. In this case the evidence for the latter will be shown below; we consider the i-base volume to be quasineutral,

$$\varrho = \frac{\varepsilon}{4\pi}\left|\frac{{\rm d}E}{{\rm d}x}\right| < q(n + n_{\rm R})\,, \quad q(p + N_{\rm d} + p_{\rm R})\,. \tag{9}$$

Now some simplifications can be made in the case of temperature and voltage variations in a rather narrow scale:

1. The density of states near the quasi-Fermi level can be approximated by an exponential function

$$g(E_{\rm R}) = g_{F_0} \exp\,(-\alpha|E_{\rm R} - F_0|)\,, \tag{10}$$

where $g_{F_0}$ is the DOS at $E_{\rm R} = F_0$ and $\alpha$ a parameter characterizing the trap distribution.

2. The variation of the lifetime due to injection is negligible and recombination is mono-molecular with constant recombination time for electrons ($\tau_{\rm n}$) and holes ($\tau_{\rm p}$),

$$R = \frac{p - p_{\rm n}}{\tau_{\rm p}} = \frac{n - n_{\rm n}}{\tau_{\rm n}}\,. \tag{11}$$

$n_{\rm n}(p_{\rm n})$ being the equilibrum electron (hole) concentration.

3. According to [5, 9] the dependence of the trapped hole concentration on the free one is weaker than in the case of monoenergetic traps[2]),

$$p_R(p) \approx \gamma p^{\alpha kT}, \tag{12}$$

where

$$\gamma = \frac{g_{F_0}}{\alpha} p_n^{-\alpha kT}.$$

4. If most of the positive charge is trapped, the quasi-neutrality condition (9) will be

$$n(p) \approx p_R(p) \gg p, \quad \frac{\varepsilon}{4\pi q} \left|\frac{dE}{dx}\right|. \tag{13}$$

Double-carrier injection in the semiconductor with a continuous DOS leads to the quasi-neutral excess free carrier distribution

$$D_a \frac{d^2p}{dx^2} - \frac{p - p_n}{\tau_p} = 0, \tag{14}$$

where diffusion dominates and $D_a$ is the ambipolar diffusion coefficient. Under condition (13) $D_a$ is almost independent of the shape of the DOS, $D_a \approx D_p$. Therefore, $p(x)$ does not feel the broadening of $g(E_R)$.

However, the sublinear relationship between $n$ and $p$, typical for the continuous DOS, subjects $E$ to the variations

$$E = \frac{\frac{j}{q} - \left(D_n \frac{dn}{dx} - D_p \frac{dp}{dx}\right)}{\mu_n n + \mu_p p} \tag{15}$$

both in the drift $E_j \approx j/q\mu_n \gamma p^{\alpha kT}$ and in the Dember $E_D \approx -\alpha kT(kT/q)(1/p)(dp/dx)$ components. Therefore, the voltage drop

$$V = \int_0^{d_i} E\,dx = \int_0^{d_i} E_j\,dx + \int_0^{d_i} E_D\,dx = V_j + V_D \tag{16}$$

will be sensitive enough to the DOS shape ($d_i$ is the i-layer thickness). So the influence of the DOS shape on the final $I(U)$ characteristics can be predicted.

The spatial distribution of excess carriers as a solution of (14) is

$$p(x) = B_1 \exp(x/L_p) + B_2 \exp(-x/L_p) + p_n, \tag{17}$$

where

$$B_1 = \frac{p(W) - p(0) \exp(d_i/L_p)}{2\mathrm{sh}(d_i/L_p)}; \qquad B_2 = \frac{p(0) \exp(d_i/L_p) - p(W)}{2\mathrm{sh}(d_i/L_p)},$$

$L_p$ is the minority carrier diffusion length, $p(0)$ and $p(W)$ are the boundary concentrations at the ends of the i-base, Fig. 1.

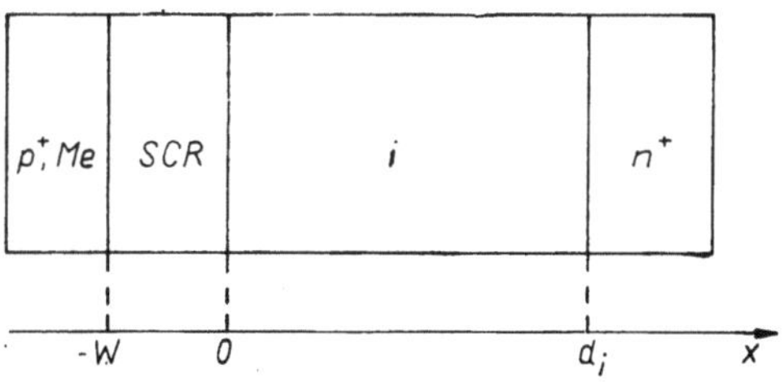


Fig. 1. p–i–n structure scheme

[2]) The levels in the upper half of the mobility gap are donor-like.

If the i-base is thick enough so that $d_i/L_p \gg 1$, the base voltage drop will be the determining factor for $I(U)$: $U \approx U_f$. To find the magnitude and the current dependence of $U_f$ it is necessary to formulate the boundary conditions.

For a-Si:H diode structures the current in the SCR of the p–n junction is controlled by the G–R processes

$$j(x = 0) = j_s \exp\left(qU_{p\text{-}n}/AkT\right), \tag{18}$$

$U_{p\text{-}n}$ being the voltage drop through the SCR. According to [10], the concentration at $x = 0$ (Fig. 1) is

$$p(0) \approx p_n j^A / j_s^A . \tag{19}$$

The factor $A$ is determined by the G–R current in the SCR and the saturation current $j_s$ is

$$j_s \approx q\sqrt{C_n C_p}\, \frac{n_i}{2}\, g_{\text{eff}}(E_R)\, \frac{kT}{q}\, \frac{1}{E_{sc}},$$

where $g_{\text{eff}}(E_R)$ is the effective DOS [6], $n_i$ the "intrinsic" carrier concentration, $C_n(C_p)$ the effective electron (hole) capture cross section, $E_{sc}$ the effective field in the SCR.

The theory of G–R currents in the semiconductor with a continuous DOS predicts for $A$ the values 1, 4/3, 2 with rising applied voltage, which were confirmed experimentally [9].

The back i–n$^+$ contact at the plane $x = d_i$, Fig. 1, is supposed to provide ideal injection so that according to [10]

$$j(d_i) = qD_\gamma \left(\left.\frac{dn}{dx}\right|_{x=d_i} + \frac{n(d_i)}{p(d_i)} \left.\frac{dp}{dx}\right|_{x=d_i}\right) \tag{20}$$

and correspondingly

$$p(d_i) = \left(\frac{L_p}{(\gamma + 1)\,\alpha kT}\right)^{1/\alpha kT} j^{1/\alpha kT} . \tag{21}$$

Formulae (19), (21) enable to obtain $I(U)$.

The concrete shape of $I(U)$ depends on the injected carrier distribution in the i-base and the ratios of $U_f$, $U_{p\text{-}n}$, $U_{i\text{-}n^+}$, $U_D$ ($U_{i\text{-}n^+}$ being the voltage drop through the i–n$^+$ contact).

In the simplest case $p(x)$ drops monotonously from the p–n junction to the i–n$^+$ contact and $U \approx U_f \gg U_{p\text{-}n}$, $U_{i\text{-}n^+}$, $U_D$; then

$$j = \left(\frac{q\mu_n g_{F_0}}{j_s^{A\alpha kT} L_p}\right)^m U^m , \tag{22}$$

where

$$m = 1/(1 - A\alpha kT) . \tag{23}$$

At $T =$ const $m$ increases owing to the rise of $A$ with rising voltage. If $T \neq$ const in the part of $I(U)$ where $A =$ const, $m$ will increase with rising temperature. However, the variation of temperature can change the length of different parts of the G–R current along the $U$-axis, which can lead to lower magnitudes of $A$. Then $m$ can happen to decrease with rising temperature.

### 3. Sample Preparation

The samples were prepared from the following layers: dielectric substrate, Ni (NiCr), P-doped a-Si:H, undoped a-Si:H, B-doped a-Si:H, and NiCr or Pt (Pd) for p–i–n structures or Schottky diodes, respectively. The a-Si:H layers were prepared by the glow discharge method. The thickness of i-, $n^+$-, and $p^+$-layer was 200 to 2500 nm, 50, and 10 to 20 nm, respectively.

### 4. Experimental Results

The dark $I(U)$ curves demonstrate three regions: 1. at negative bias the current varies weakly, 2. at low positive bias with an exponential dependence, 3. at high positive bias with a superlinear dependence.

For the second region $j = j_s \exp(qU/AkT)$, $j_s$ and $A$ depending on the DOS in the i-base (information about the DOS has been drawn from capacitance measurements; $j_s$ appeared to vary directly with $g(E_R)$). The magnitudes of $j_s$ and $A$ vary also with the sort of metal in the Schottky barrier or with the $p^+$-layer thickness in p–i–n structures. Nevertheless some common features can be derived for both types of structures.

At $T \geqq 300$ K there is one region with the constant slope in lg $j$–$U$ coordinates. But at $T < 300$ K there are two regions with different slopes, Fig. 2, the transition voltage depending on temperature.

Fig. 3 presents temperature dependences for $j_s$ and for the current in the low-voltage region, $j|_{U=0.05\text{V}}$. At high temperature $j_s$ and $j|_{U=0.05\text{V}}$ coincide but at low temperature they diverge; the activation energy of the latter (0.55 to 0.7 eV) is equal to the dark conductivity one within 0.05 eV. Fig. 4 shows $A$ versus temperature for the same p–i–n and Schottky structures. The structures are different in the magnitude of $A$ but not in the shape of the temperature dependences.

When the voltage exceeds the initial band bending the transition to a superlinear function $j \sim U^m$ appears. Fig. 5 demonstrates that this transition in dependence on the value of the DOS occurs the earlier, the larger the density of states is (curves 1, 3); the transition voltage depends also on the thickness of the i-layer (Fig. 5, curves 1, 2). For thin i-layer structures with low DOS the superlinear dependence is not observed because of the current restriction by the contact resistance. The superlinear part of the $I(U)$ is rather complicated: as usual, several superlinear regions with different $m$ can be distinguished for both p–i–n structures and Schottky diodes (as an example, $I(U)$ for the Pt–i–n structure is shown in Fig. 6; p-i-n structures possess analogous characteristics).

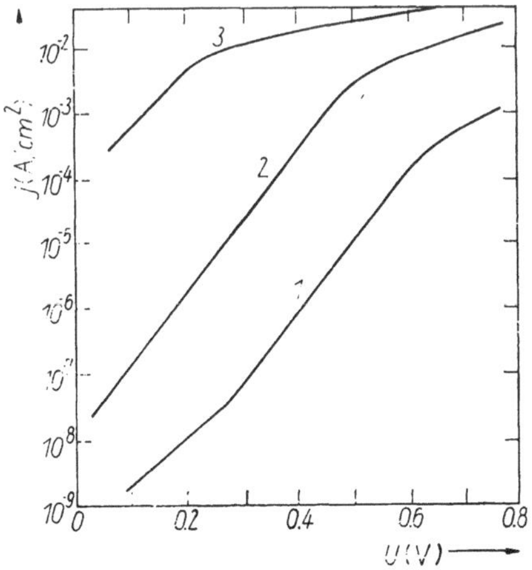


Fig. 2. Dark $I(U)$ of Pt–i–n structure at low positive bias; (1) $T = 243$, (2) 318, (3) 435 K

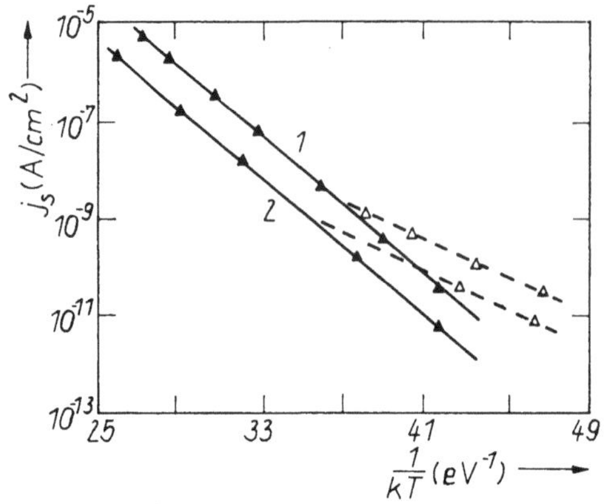


Fig. 3

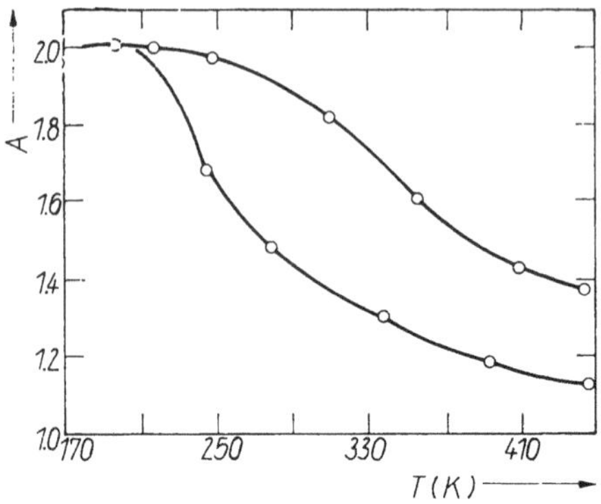


Fig. 4

Fig. 3. Temperature dependences of saturation current $j_s$ (solid curve) and $j|_{U=0.05\,V}$ (dashed curve) for Pt–i–n (1) and p–i–n (2) structures

Fig. 4. Temperature dependences of factor $A$ for Pt–i–n (lower curve) and p–i–n (upper curve) structures

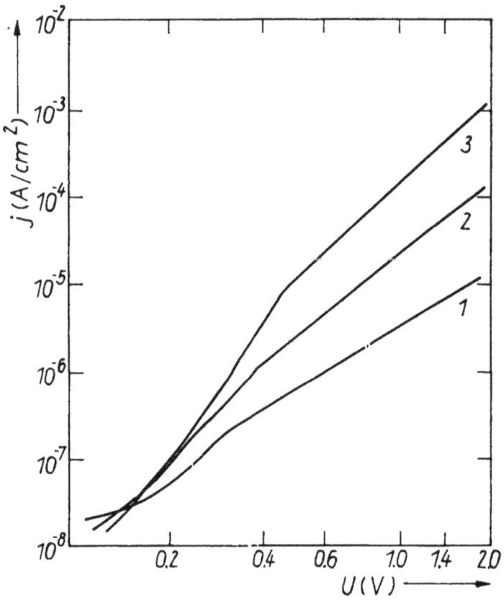


Fig. 5

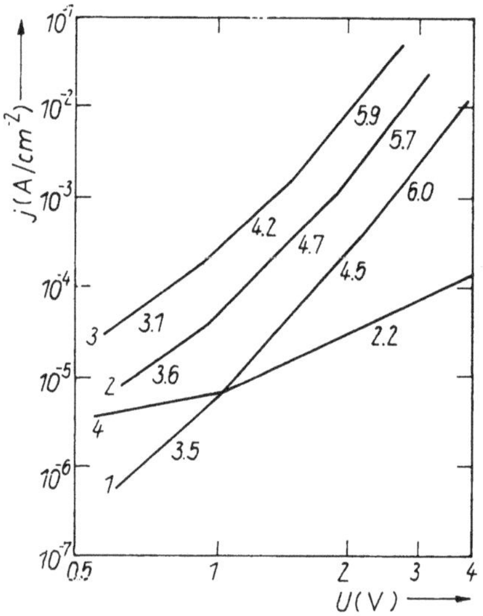


Fig. 6

Fig. 5. Dark $I(U)$ of Pt–i–n structures with various thickness of i-layer $d_i$ and density of states $\bar{g}$ (average DOS near $F_0$); (1) $d_i = 100$ nm, $\bar{g} = 10^{18}$ eV$^{-1}$ cm$^{-3}$; (2) $d_i = 500$ nm, $\bar{g} = 10^{18}$ eV$^{-1}$ cm$^{-3}$; (3) $d_i = 1000$ nm, $\bar{g} =$ (5 to 8) $10^{16}$ eV$^{-1}$ cm$^{-3}$; $T = 293$ K

Fig. 6. Dark $I(U)$ of Pt–i–n structure ($\bar{g} =$ (5 to 8) $10^{13}$ eV$^{-1}$ cm$^{-3}$, $d_i = 1500$ nm); (1) $T = 253$, (2) 272, (3) 295 K (figures near curves indicate values of $m$); (4) $I(U)$ of n–i–n structure ($\bar{g} =$ (5 to 8) $10^{16}$ eV$^{-1}$ cm$^{-3}$, $d_i = 1500$ nm, $T = 293$ K)

## 5. Discussion

The theory of G–R currents in the SCR has been developed in [6] and compared with the experimental data for Schottky diodes in [2]. Now p–i–n structures have been shown to possess similar $I(U)$ characteristics.

Indeed, at low positive bias the dependences $j_s(T)$ and $A(T)$ are similar for both types of structures, Fig. 3, 4, and the magnitudes of $j_s$ are also close to one another. The main difference between Schottky diode and p–i–n structure is the larger experimental $A$ for the latter. The reason is the stronger band bending which shifts the G–R current saturation towards high voltages. The temperature dependence of $A$ is due to a shift of the exponential region of the $I(U)$ towards higher voltage with decreasing temperature. Besides, the length of the regions with different $A$ depends on temperature. For instance, the voltage at the transition from the $A = 4/3$ region to the $A = 2$ one is

$$U_{\text{tr}} = \frac{4kT}{q} \ln \left[ \frac{j_s(2)}{j_s(4/3)} \right]. \tag{24}$$

Thus, $A$ falls with rising temperature at $U = \text{const}$, which has been found experimentally.

As to the superlinear $I(U)$, Fig. 6, such characteristics have been ascribed lately to the mechanism of space charge limited currents (SCLC) [4] but then double-carrier injection has been supposed to be responsible for superlinear dependences in the case of p–i–n structures [8]. In our case the latter point of view is confirmed by some estimations: both the Debye length and the hole diffusion length $L_p$ are found to be smaller than $d_i$ ($L_p/d_i \approx 0.1$ [10]). A decisive argument for the double-carrier injection mechanism could be the experimental increase of $m$ with rising temperature[3]) but this is not observed, Fig. 6. It seems to be due to a change of the length of different parts of the G–R current (see Section 2). In addition, the data concerning $I(U)$ for n–i–n structures have also been considered, Fig. 6 (curve 4), when the SCLC mechanism has been reliably identified. Fig. 6 demonstrates quite different shapes of the $I(U)$ curves for the two cases compared. The presence of $I(U)$ regions with different $m$ (Fig. 6, curves 1 to 3) is due to the $A$ variation with rising voltage. According to the theory of G–R currents [6, 7] the value of $A$ runs through a consequent set of values 1, 4/3, 2 with rising voltage. So, as (23) predicts that the transition to the next value of $A$ is accompanied by a change of $m$, $m$ will increase with rising voltage at $T = \text{const}$.

## References


[1] P. E. Carlson, C. R. Wronsky, J. I. Pankove, P. J. Zancchi, and P. L. Stabler, RCA Rev. **38**, 211 (1977).

[2] A. A. Andreev, O. A. Golikova, P. M. Karageorgy-Alkalaev, A. Yu. Leiderman, M. M. Mezdrogina, V. S. Rubin, and N. A. Feoktistov, Fiz. Tekh. Poluprov. **18**, 373 (1984).

[3] W. E. Spear, Adv. Phys. **26**, 811 (1977).

[4] M. K. Han, W. A. Anderson, P. Sung, and R. Lahri, phys. stat. sol. (a) **75**, 283 (1983).

[5] A. Yu. Leiderman, phys. stat. sol. (a) **55**, 661 (1979).

[6] A. Yu. Leiderman, phys. stat. sol. (a) **87**, 363 (1985).

[7] S. A. Asimov, P. M. Karageorgy-Alkalaev, A. Yu. Leiderman, and V. S. Rubin, phys. stat. sol. (a) **83**, K81 (1984).

[8] M. Hack and W. den Boer, J. appl. Phys. **58**, 1554 (1985).

[9] A. Yu. Leiderman, in: Physical Phenomena in Semiconductor Structures with Deep Levels and Opto-Electronics, Izd. FAN, Tashkent 1977 (p. 98), in Russian.

[10] E. I. Adirovich, P. M. Karageorgy-Alkalaev, and A. Yu. Leiderman, Double-Carrier Currents in Semiconductors, Izd. "Sovetskoe Radio", Moscow 1978 (p. 320), in Russian.


*(Received September 1, 1987)*

---

[3]) Single-carrier injection theory also predicts the $j \sim U^m$ dependence but with $m$ decreasing when temperature rises.

phys. stat. sol. (a) **108**, 331 (1988)

Subject classification: 72.80; S12

*Institute of Chemistry, Pedagogical University, Opole*[1])

# An Explanation of Electrical Conductivity Data for Copper Phthalocyanine-Acetylene Carbon Black Mixtures

By

R. PIASECKI, M. ZĄBKOWSKA-WACŁAWEK, and W. WACŁAWEK

The macroscopic aspects of the conducting properties of copper phthalocyanine-acetylene carbon black mixtures are investigated. The experimental conductivity data are explained for mixtures with carbon black weight fraction $x < 0.1$ by using linear approximations and by the simplest effective medium approximation for $x > 0.1$. Remarkable agreement with experiment is obtained.

Die makroskopischen Aspekte der Leitfähigkeit der Kupferphthalozyaninemischungen mit Azethylenruß werden untersucht. Es werden die Untersuchungsergebnisse der Leitfähigkeit der Mischungen bei einem Rußgewichtsanteil $x < 0,1$ unter Benutzung der linearen Näherung erklärt, und für $x > 0,1$ wird die einfachste Effektivmediumnäherung benutzt. Es wird gute Übereinstimmung mit dem Experiment erreicht.

## 1. Introduction

There are plenty of papers on conducting mixtures e.g. [1 to 7]. Some authors describe experimental facts, others use theoretical treatment. In many papers an S-like shape of the electrical conductivity as a function of content of the highly conducting component in the mixture was observed. This behaviour is usually interpreted in terms of percolation theory. Only few authors e.g. [8, 9] discovered a linear dependence of the activation energy of conduction on the amount of the conductive (nonconductive) component, respectively. Paper [8] describes the electrical conductivity of the copper phthalocyanine-acetylene carbon black, (CuPc)C, mixtures. (CuPc)C is macroscopically inhomogeneous since its inhomogeneity is on a scale, much larger than any of the important microscopic lengths of the material (static inhomogeneity).

In the present paper we are concerned with purely macroscopic aspects of the conducting properties of (CuPc)C. The aim is to explain a characteristic dependence of the overall conductivity $\sigma$ as a function of the acetylene carbon black weight fraction $x$ described in [8].

## 2. Samples

Samples investigated in [8] were mixtures of copper phthalocyanine and acetylene carbon black of different compositions. They were prepared by grinding in an agate mortar and compressing (at pressure of $10^8$ Pa) to make pellets of 8 and 10 mm diameter.

[1]) Ul. Oleska 48, PL 45-052 Opole, Poland.

## 3. Electrical Conductivity

The (CuPc)C conductivity is seen to increase by more than eight orders of magnitude as $x$ varied from 0 to 0.3. For samples with $x < 0.1$, $\sigma(x, T)$ decreases as the temperature $T$ is reduced. For $x > 0.1$ the electrical conductivity is almost frequency and temperature independent. In our case such a behaviour of $\sigma(x, T)$ suggests a transition from predominantly thermally assisted tunneling between the highly conducting particles of carbon black to predominantly percolative conduction at $x_c = 0.1$. From the mathematical point of view the experimental $\sigma$ curve has a point of inflexion at $x_c$ and that will be utilized later (see Fig. 1 in [8]). Physically, the $x_c$ point may be connected with a so-called threshold concentration and coincides with the formation of the first continuous highly conducting filament through the (CuPc)C body.

Now we propose that the dc conductivity $\sigma(x, T)$ of the (CuPc)C composite may essentially be described by considering it to be a random binary mixture of acetylene carbon black powder (good conductor) in copper phthalocyanine (poor conductor). We assume the simple activated form of $\sigma(x, T)$,

$$\sigma(x, T) = \sigma_0(x) \exp\left[-\frac{E(x)}{kT}\right], \tag{1}$$

where $E(x)$ is the apparent activation energy of the composite.

### *3.1 Linear approximation for $x < x_c$ (qualitative model)*

For $x \in \mathrm{I}$, $\mathrm{I} = \{0 \leqq x < x_c\}$ from the experimental data taken from [8] the activation energy $E_\mathrm{I}(x)$ decreases linearly according to formula

$$E_\mathrm{I}(x) = (0.885 \pm 0.005) - (8.75 \pm 0.08)\, x \tag{2}$$

with the correlation coefficient 0.9998. Next, when the experimental $\sigma$ curve is redrawn in coordinates $(x, \lg \sigma)$ the first part of this curve (for $x \in \mathrm{I}$) is nearly linear except in the immediate neighbourhood of $x_c$. Thus, according to (1) and (2) we obtain

$$\lg \sigma_{0,\mathrm{I}} = (-1.8 \pm 0.2) + (10.4 \pm 0.3)\, E_\mathrm{I}\,. \tag{3}$$

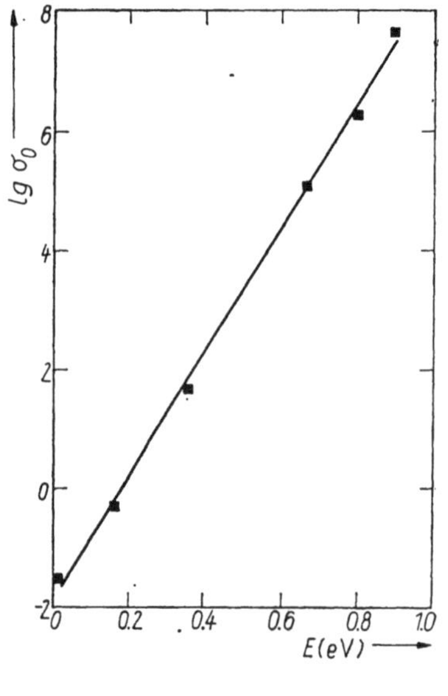


Fig. 1. Compensation effect for (CuPc)C mixture ($\sigma_{0,\mathrm{I}}$ in S m$^{-1}$)

This linear dependence of $\lg \sigma_{0,\mathrm{I}}$ on $E_\mathrm{I}$ is called the compensation rule. Equation (3) has a correlation coefficient as high as 0.998 (see Fig. 1). In our case the compensation effect results from the linearity of activation energy $E_\mathrm{I}$ as a function of weight fraction $x$. This behaviour suggests the following heurestic formula for the apparent activation energy $E_\mathrm{I}$:

$$E_\mathrm{I}(x) = p(x)\, E_\mathrm{cb} + \bigl(1 - p(x)\bigr)\, E_\mathrm{Pc}\,, \tag{4}$$

where $E_{cb}$, $E_{Pc}$ are the activation energies for acetylene carbon black and pure CuPc, respectively, with linear dependence of $p(x)$ on $x$. The $p(x)$ function must now be specified subject to the following constraints: $p(0) = 0$ and $p(x_c) = 1$. Hence we obtain $p(x) = x/x_c$. On the other hand, one ought to expect that the choice of (4) is based on the concept of the volume fraction $x_v$ of acetylene carbon black dependence of $E_I$ rather than weight fraction $x$. To convert weight fraction $x$ to volume fraction $x_v$ the following relationship is used:

$$x_v = \frac{x}{\alpha + (1 - \alpha)\, x}, \tag{5}$$

where $\alpha = (\varrho_{cb}/\varrho_{Pc}) \approx 0.54$, $\varrho_{cb}$ and $\varrho_{Pc}$ are the densities of acetylene carbon black and copper phthalocyanine in the sample, respectively. For $x \in$ I we can use (5) linearized with respect to $x$, namely $x_v \approx x/\alpha$ which gives $x_v$ values with an inaccuracy $\Delta x_v$ less than 8.5%, for instance $\Delta x_v = 4.3\%$ for $x = 0.05$. Therefore, it is safe to say that (4) is roughly fulfilled for $x_v$, too. Such a linear behaviour of the activation energy of the composite has been also discussed in [9] for ionic conductors containing a dispersed insulating second phase.

### *3.2 Effective medium approximation for $x > x_c$ (quantitative model)*

For $x \in$ II, II $= \{x_c < x < 1\}$ $E_{II}(x)$ is very low and highly sensitive to the kind of sample (some of them exhibit the positive thermal coefficient of resistivity), [8]. Therefore, (1) becomes inapplicable because the activation energy $E_{II}(x)$ is unknown. To consider the second part of the experimental $\sigma$ curve (for $x \in$ II) we intend to use the simplest analytic theory for percolative transport since the percolation always dominates $\sigma(x, T)$ when $x$ exceeds $x_c$. As we have already noted, the critical weight fraction $x_c$ or, equivalently by (5), the critical volume fraction $x_{v,c}$ is the point at which a continuous chain of macroscopic length of carbon black particles appears first.

For the static properties of homogeneous media one usually studies equivalent random networks of conductors and insulators, while for an inhomogeneous random mixture, such networks represent the original systems only. Various methods have been used to study these networks, for example, the position space renormalization methods [10, 11], scaling arguments [12], critical path analysis [13], and the effective medium approximation (EMA) [14]. Besides the last one the above-mentioned methods are applicable only in the immediate vicinity of the percolation threshold. The EMA is valid over a wide range of compositions except in close proximity to $x_c$. Within the single-bond EMA the inhomogeneous medium is replaced by a uniform medium with effective conductors chosen in such a way that the average effect of one of the real components embedded in the effective medium is zero. The effective conductivity $\sigma_{EMA}(x, T)$ of the random binary mixture can be calculated by representing it as a regular lattice of random resistors [14]. In a notation suitable to our problem we have

$$\sigma_{EMA}(x, T) = \sigma_{Pc} \frac{(A + \sqrt{A^2 + 8\gamma})}{4}, \tag{6}$$

where $A = (2 - \gamma) + 3x_v(\gamma - 1)/0.52$, $\gamma = \sigma_{cb}(T)/\sigma_{Pc}(T) \gg 1$ is the ratio of acetylene carbon black to copper phthalocyanine conductivities at any given temperature, and $x_v$ is determined by (5).

It is interesting to note that the lg $\sigma_{EMA}$ curve as well as the experimental lg $\sigma$ curve has a point of inflexion for $A = 0$; hence the critical volume fraction $x_{v,c} = 0.173$. It

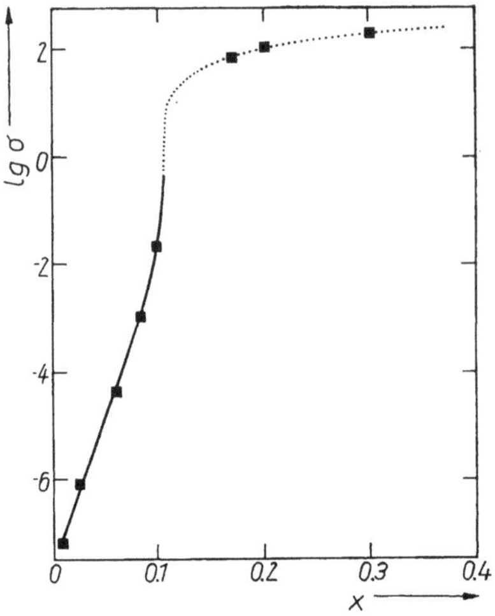


Fig. 2. Room-temperature electrical conductivity ($\sigma(x, T)$ in S $m^{-1}$) as a function of weight fraction of acetylene carbon black for the (CuPc)C system. The calculated $\sigma_{EMA}(x, T)$ (the dotted line through the data) is obtained from EMA for $x > x_c$

is a value near to the one observed in [15] for carbon black in an insulating resin equal to 0.175. From the $x_{v,c}$ value using (5) we get the critical weight fraction, $x_c = 0.10$, which is in remarkable agreement with the experiment [8].

Let us compare the $\sigma_{EMA}$ curve with the experimental data (see Fig. 2). We take the conductivities of each pure component to be thermally activated, i.e. $\sigma_i = \sigma_{0,i} \exp(-E_i/kT)$, where $i$ = cb and Pc, respectively. For the CuPc we use values $E_{Pc} = 0.885$ eV and $\sigma_{0,Pc} = 4.63 \times 10^7$ S $m^{-1}$ as measured. For acetylene carbon black contained in the samples, $\sigma_{cb}$ has been estimated by fitting the $\sigma_{EMA}$ curve to the data. The best fit obtained is $\sigma_{cb} = 221$ S $m^{-1}$ (see the dotted line in Fig. 2). At length, it should be noted that the dependence of the apparent activation energy of the composite on concentration $x$, may be used for a precise determination of the value $x_c$.

## 4. Conclusions

A macroscopic description of the dependence of the observed conductivity of (CuPc)C on the concentration of acetylene carbon black has been presented: It rests on the assumption that a linear approximations is valid only for small concentrations ($x \ll x_c$) indicating that clustering of conducting particles is negligible. For $x > x_c$ the simplest analytic theory for percolative transport, i.e. single-bond effective medium approximation, agrees remarkably well with the experimental data [8].

***Acknowledgements***

The authors are grateful to Mr. M. Antas, M.Sc. for performing numerical computations. This work was partly supported by Technical University of Wrocław under the contract C.P.B.P. 01.02.V.10.

## References

[1] H. J. MAIR and S. ROTH, Elektrisch leitende Kunststoffe, Carl Hanser Verlag, München/Wien, 1986.
[2] D. STAUFFER, Introduction to Percolation, Taylor & Francis, London 1985.
[3] Electrical Transport and Optical Properties of Inhomogeneous Media, Ed. J. C. GARLAND and D. B. TANNER, Amer. Inst. Physics, New York 1978.

[4] R. D. Sherman, L. M. Middleman, and S. M. Jacobs, Polymer Engng. Sci. **23**, 36 (1983).
[5] B. Lundberg and B. Sundqvist, J. appl. Phys. **60**, 1074 (1986).
[6] J. V. Mantese, W. A. Curtin, and W. W. Weeb, Phys. Rev. B **33**, 7897 (1986).
[7] D. S. McLachlan, J. Phys. C **19**, 1339 (1986).
[8] W. Wacławek, M. Kulesza, and M. Ząbkowska, J. Physique **44**, C3-673 (1983).
[9] Chen Li-Quan, Zhao Zong-Yuan, Wang Chao-Ying, and Li Zi-Rong, Acta phys. Sin. **34**, 1027 (1985).
[10] R. B. Stinchcombe and B. P. Watson, J. Phys. C **9**, 3221 (1976).
[11] S. Kirkpatrick, Phys. Rev. B **15**, 1533 (1977).
[12] D. Stauffer, Phys. Rep. **54**, 1 (1979).
[13] J. Bernasconi, Phys. Rev. B **1**, 2252 (1973).
[14] S. Kirkpatrick, Rev. mod. Phys. **45**, 574 (1973).
[15] G. Fug, R. Canet, and P. Delhaes, C.R. Acad. Sci. (France) B **5**, 287 (1978).

*(Received December 3, 1987)*

phys. stat. sol. (a) **108**, 337 (1988)

Subject classification: 74.70; 75.20; S10.15

*Department of Physics, Jilin University, Changchun*[1])

# Magnetic and Superconducting Properties of $REBa_2Cu_3O_{9-y}$ (RE = Sm, Eu, Gd, Dy, Ho, Er, Tm)[2])

By

Changyong Sun, Wenhui Su, Hongjian Liu, Jianshi Zhou, Fatang Wang, and Yupu Zhang

Some high-$T_c$ superconductors $REBa_2Cu_3O_{9-y}$ (RE = Sm, Eu, Gd, Dy, Ho, Er, Tm) are prepared and their zero-resistance temperatures $T_{ce}$ are about 90 K. The effect of magnetic rare earth ions on the transition temperatures is not very great but their magnetic behaviours are different from one another. The magnetization loops of these compounds are measured at 77 K. The coexistence of superconductivity with magnetic ordering (COSM) occurs in the samples of RE = Gd, Dy, Ho, Er, Tm, which predictes that there are two possible ways of coexistence in the samples. The experiments show that the relation of paramagnetism to the atomic number of rare earth elements for these superconductors is similar to that of rare earth ions in some paramagnetic salt crystals, which indicates that the magnetism of the samples originates from the local magnetic moments of rare earth ions. Some samples show antiferromagnetic ordering at low temperature, and the effective Bohr magneton numbers are consistence with that of free rare earth ions.

Es werden Hoch-$T_c$-Supraleiter $REBa_2Cu_3O_{9-y}$ (Re = Sm, Eu, Gd, Dy, Ho, Er, Tm) mit Nullwiderstandstemperaturen $T_{ce}$ von etwa 90 K präpariert. Der Einfluß der magnetischen Seltenerdionen auf die Übergangstemperaturen ist nicht sehr groß, jedoch unterscheidet sich ihr magnetisches Verhalten voneinander. Die Magnetisierungsschleifen dieser Verbindungen werden bei 77 K gemessen. Koexistenz der Supraleitung mit magnetischer Ordnung (COSM) tritt in den Proben von RE = Gd, Dy, Ho, Er, Tm auf, was zeigt, daß es zwei mögliche Wege der Koexistenz in den Proben gibt. Die Experimente zeigen, daß die Beziehung des Paramagnetismus zur Atomzahl der Seltenerdelemente für diese Supraleiter ähnlich der der Seltenerdionen in einigen paramagnetischen Kristallen ist, was zeigt, daß der Magnetismus der Proben aus den lokalen magnetischen Momenten der Seltenerdionen stammt. Einige Proben zeigen antiferromagnetische Ordnung bei tiefen Temperaturen, und die effektiven Bohrschen-Magneton-Zahlen sind mit denen der freien Seltenerdionen konsistent.

## 1. Introduction

Since the discovery of the new high $T_c$ superconducting materials $YBa_2Cu_3O_{9-y}$ [1], a number of substitutions for Y have been made with various rare earth atoms, and no apparent change in their $T_c$'s has been found due to these substitutions. As most of the rare earth ions have large magnetic moments, someone suggested that the magnetic rare earth ions are being shielded in the $REBa_2Cu_3O_{9-y}$ [2, 3]. In order to understand the superconducting mechanism of the new kind of superconductors, it is useful to systematically investigate the role of rare earth ions, their magnetic states, and the COSM phenomena in these compounds. In this paper, we will discuss the magnetic behaviour of rare earth ions and their effects on the superconductivity by

[1]) Changchun, People's Republic of China.

[2]) A part of the results of the present paper has been reported at the 1st Chinese Youth Symposium on Material Science, Oct. 16 to 19, 1987, Shenyang, People's Republic of China.

measuring the hysteresis loops, susceptibilities, and the temperature dependence of resistance in an applied magnetic field.

## 2. Experiments

The samples were prepared by mixing CuO(sp), $BaCO_3$(sp) with high-purity $Sm_2O_3$, $Eu_2O_3$, $Gd_2O_3$, $Dy_2O_3$, $Ho_2O_3$, $Er_2O_3$, $Tm_2O_3$ powders, respectively, then sintering at 920 °C for 10 h in air after grinding and shaping; regrinding and pressing into pellets of 13 × (1 to 2) mm²; annealing at 920 °C for 8 h and cooling down to room temperature by switching off the current. The four-terminal technique was employed to measure the resistance. The zero-resistance temperatures of $REBa_2Cu_3O_{9-y}$ (RE = = Sm, Eu, Gd, Dy, Ho, Er, Tm) were about 86, 80, 89, 90, 91, 91, 90 K, respectively. The samples have been examined by X-ray diffraction. The *M–H* and *M–T* curves have been measured using the vibrating sample magnetometer JDM-1, and the *R–T* curves in an applied magnetic field have been checked, too.

## 3. Results and Discussions

1. The magnetization loops of $REBa_2Cu_3O_{9-y}$ (RE = Sm, Eu, Gd, Dy, Ho, Er, Tm; Y) at 77 K are shown in Fig. 1. The lower critical fields of all of them are in a range of about 10 mT. For the samples of $REBa_2Cu_3O_{9-y}$ (RE = Gd, Dy, Ho, Er, Tm) the magnetizations in the diamagnetic state increase with increasing the magnetic field at first, then decrease and become zero at the points $\mu_0H_p$, i.e. the magnetizations change from negative to positive across the points. This means that the paramagnetism appears. The values of $\mu_0H_p$ are in the range from 300 mT for $ErBa_2Cu_3O_{9-y}$ to 500 mT for $TmBa_2Cu_3O_{9-y}$. The relations between resistance and temperature of the five samples have been measured in magnetic field beyond the values of $\mu_0H_p$. It is found that there still exists evident superconductivity in them and the zero-resistance temperatures of $ErBa_2Cu_3O_{9-y}$ and $GdBa_2Cu_3O_{9-y}$ are still above the boiling point of liquid nitrogen. While the zero-resistance temperatures of the other three compounds fall into the range of below 77 K. Remarkable changes have not been observed for the mid-transition temperatures of these specimens, which implies that the upper critical fields are very high. On increasing the magnetic field, the magnetizations increase along the positive direction. These facts suggest that the magnetism of the samples results from the superposition of the diamagnetism due to the Meissner effect and the magnetism originating from the rare earth ions, i.e., there is a COSM in the compounds.

In order to probe the magnetic properties, the temperature dependence of magnetic susceptibilities $\chi$ were measured at normal states in $\mu_0H = 315$ mT, and the relations of $\chi^{-1}$ versus temperature are shown in Fig. 2. The Curie-Weiss law is correctly fulfilled except for the sample with RE = Er whose magnetic behaviour is complicated, and the samples show obviously antiferromagnetic ordering. The asymptotic Curie temperatures $\theta_p$ for the samples of RE = Gd, Dy, Tm are of −18, −9, −12 K (magnetic moments per ion are 8.3, 9.9, 7.7$\mu_B$, close to the values of free rare earth ions), respectively. There is no anomaly at normal state for the function $\chi(T)$, which implies that the Néel temperatures $\theta_N$ of the samples are lower than their superconducting transition temperatures. The $\theta_p$'s are lower, indicating that there exist very weak antiferromagnetic correlations among the $Gd^{3+}$ ions in dilute concentration in the structure, and the ordering Néel temperature will also be lower. It can be speculated that in the temperature range of liquid nitrogen the samples are in the paramagnetic state above $\theta_N$. So the coexistence discussed above is the coexistence of superconductivity with paramagnetism.

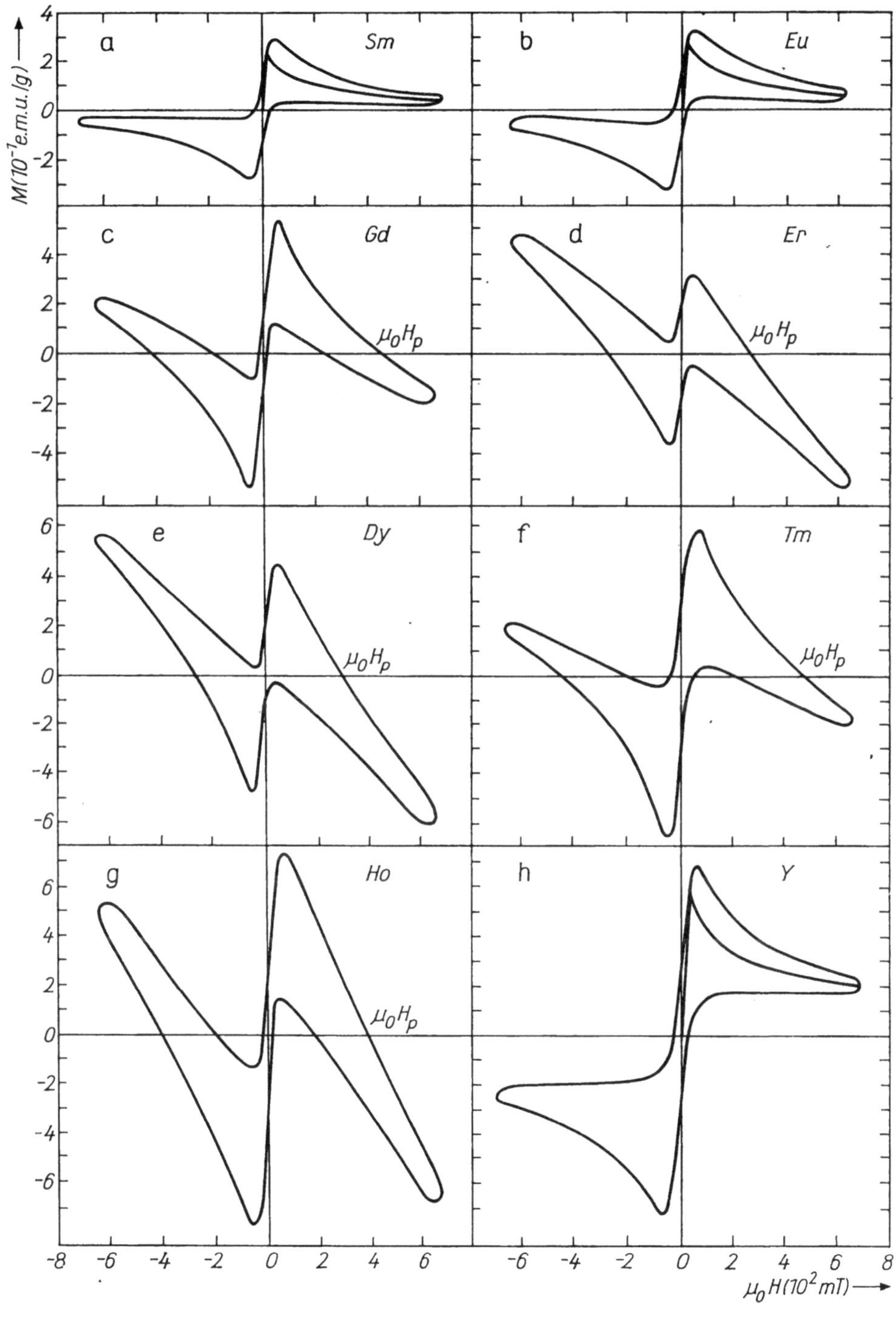


Fig. 1. Magnetization loops of $REBa_2Cu_3O_{9-y}$ measured at 77 K. The reversed curves of the samples RE = Sm, Eu, and Y are given in Fig. 1 a, b, and h, respectively.

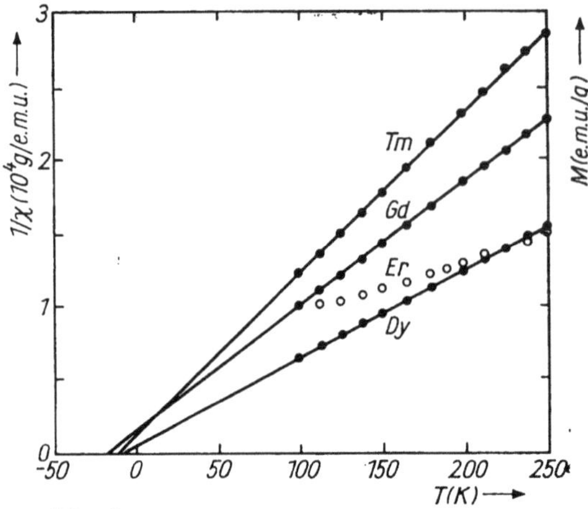


Fig. 2

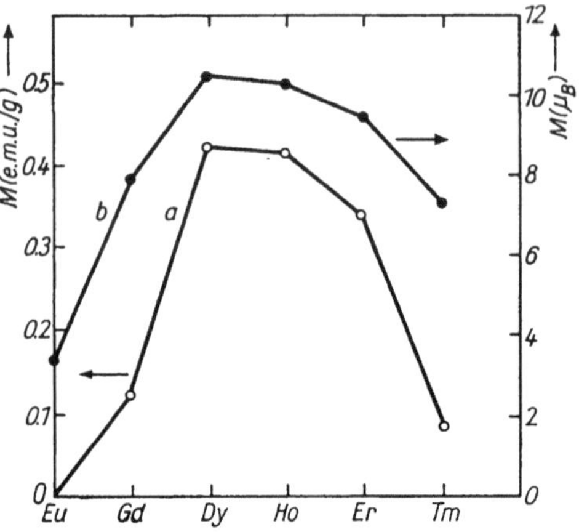


Fig. 3

Fig. 2. $\chi^{-1}$ as a function of temperature, with the measurement performed in a field of 315 mT

Fig. 3. (a) Paramagnetism for samples of RE = Eu, Gd, Dy, Ho, Er, and Tm (in a field of 650 mT). (b) Magnetic moments of $RE^{3+}$ in some paramagnetic salts

The magnitude of the magnetization in the paramagnetic state of the five samples have been measured and calculated in the range of magnetic field of 650 mT, being higher than $\mu_0 H_p$, i.e. in which the magnetization from rare earth ions has surpassed the magnetization due to the superconductivity. The results are shown in Fig. 3, curve a. It is evident that the paramagnetism of $DyBa_2Cu_3O_{9-y}$ and $HoBa_2Cu_3O_{9-y}$ is the most intensive in the compounds; the rest become weaker in order, while for the samples of RE = Sm, Eu the paramagnetism is not exhibited in the magnetic field range above given. This regularity is consistent with the behaviour of rare earth ions in other paramagnetic salt crystals (Fig. 3, curve b). It can be seen that the magnetism of the samples comes from the local magnetic moments of rare earth ions in the lattice. From X-ray diffraction patterns it is shown that all the samples possess the same orthorhombic structure of $YBa_2Cu_3O_{9-y}$. Therefore, the magnetic moments of rare earth ions in this layered perovskite structure are similar to that in other paramagnetic salts containing rare earth elements, and all are not shielded by Cu–O layers. That is to say, under the condition that the magnetic moments of $RE^{3+}$ are all not shielded, the superconductivity of the compounds would also appear, and this is not in agreement with the hypothesis stated in [2, 3]. In a weak magnetic field, the diamagnetism of the supercurrent is dominant. With increasing magnetic field however, the magnetization of rare earth ions increases gradually, and the total magnetization will also be increased positively, consequently it results in the coexistence of superconductivity with the magnetism of $RE^{3+}$.

The appearance of the COSM predicts that there are two possible ways of coexistence in the $REBa_2Cu_3O_{9-y}$ (RE = Gd, Dy, Ho, Er, Tm): (i) there are two phases in the samples, one is responsible for superconductivity, the other for paramagnetism. The superconducting phase is able to prevent or resist the disturbance of magnetism coming from the magnetic phase and pertains itself superconducting. The competition between them results in the COSM, (ii) In the layered perovskite structure of $REBa_2Cu_3O_{9-y}$ some Cu–O atomic layers with one- or two-dimensional superconductivity are sandwiched between two magnetic $RE^{3+}$ layers whose magnetism is not shielded, thus it bring about the COSM. The Y(RE)–Ba–Cu–O systems which are single-phase structures according to X-ray powder diffraction, in fact, are usually poly-phase structures in their microstructures. Consequently both the above-mentioned ways of coexistence are possible. However, either of the two ways is different in concept from the COSM for the superconducting alloy. To further study the COSM is an important problem for understanding the superconductivity mechanism of the new kind of ceramic superconductors. Also, to further study on the magnetic structures of various magnetic rare earth ions in $REBa_2Cu_3O_{9-y}$ is another valuable work.

2. From Fig. 1 it is clear that the magnetization loops are hysteretic. The residual magnetization of $TmBa_2Cu_3O_{9-y}$ is the most intensive of all the samples, and it is expected to have a relatively high critical current [4]. This property is different from that of the La–Ba–Cu–O and La–Sr–Cu–O systems whose hystereses are very small [5]. The hysteresis results from pinning which is due to the inhomogeneity and the boundaries between grains in the samples. Although both the two grains connected by the boundary are made from the same materials, the crystallite orientations are different, so the superconductivity is anisotropic, which makes the boundary pin the magnetic flux on the pinning centres. Besides, the defects such as dislocations, holes, etc. can also pin the flux. On the basis of the above-mentioned discussion the substitution of Tm for Y in the $REBa_2Cu_3O_{9-y}$ would increase the pinning, too, which might be of benefit for the increase of $J_c$.

3. The samples are type II superconductors and weak pinning systems because the area of the magnetic hystereses are small. Therefore, the reversed magnetization curves can be obtained from $M_{rev} = (1/2)(M^+ + M^-)$ (Fig. 1a, b, h), where $M^+$ and $M^-$ are the magnetizations in the same magnetic field for increasing and decreasing field. In relatively lower magnetic fields, $H_{c_1} \ll H \ll H_{c_2}$, the Ginzburg-Landau (GL) result [6] might be used, and hence

$$-M = H_{c_1} \frac{\ln\left[\frac{\alpha'}{\xi}\left(\frac{2\Phi_0}{3B}\right)^{1/2}\right]}{\ln k},$$

from any two points on the reversed magnetization curves, we have

$$k = \left(\frac{B_2}{B_1}\right)^{H_{c_1}/2(M_2-M_1)}, \tag{1}$$

where $k$, $\xi$, $\Phi_0$, and $\alpha'$ are GL parameter, coherence length, flux quantum, and constant, respectively. From (1) and the data of Fig. 1a, b, h we obtained GL parameter $k = 81, 156$, and 43 for samples of RE = Sm, Eu, and Y, respectively. Because $k \gg 1$,

Table 1

Superconductivity parameters derived from the reversed magnetization curves shown in Fig. 1

| | $\mu_0 H_{c_1}$ (mT) | $\mu_0 H_c$ (T) | $\mu_0 H_{c_2}$ (T) | $\xi$ (nm) | $\lambda$ (nm) | $k$ |
|---|---|---|---|---|---|---|
| $YBa_2Cu_3O_{9-y}$ | 8 | 0.13 | 8 | 6.5 | 280 | 43 |
| $SmBa_2Cu_3O_{9-y}$ | 6 | 0.16 | 18 | 4.3 | 350 | 81 |
| $EuBa_2Cu_3O_{9-y}$ | 7.5 | 0.33 | 72 | 2.1 | 330 | 156 |

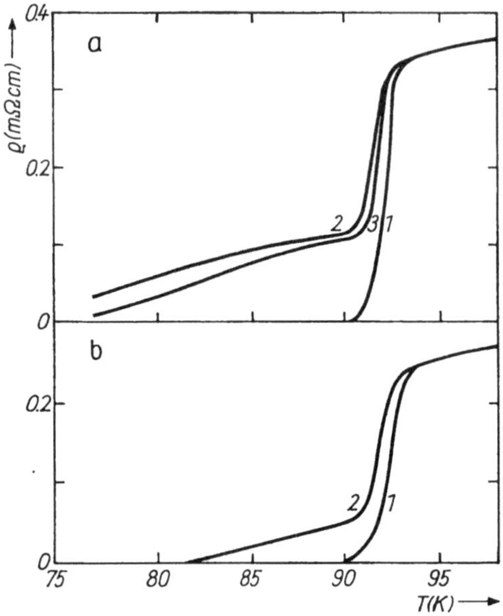


Fig. 4. Temperature dependence of resistivities for samples RE = a) Dy and b) Tm measured in magnetic fields of (1) 0; (2) 500; (3) 170 mT

the GL formula

$$H_{c_1} = H_c \frac{\ln k}{\sqrt{2}k},$$

$$H_{c_2} = \sqrt{2}\, k H_c,$$

$$\mu_0 H_{c_2} = \frac{\Phi_0}{2\pi\xi^2}$$

can be used to obtain the upper critical field $H_{c_2}$, thermodynamic critical field $H_c$, $\xi$, and penetration depth $\lambda$ as shown in Table 1. The parameters for other samples have not been determined because of the appearance of magnetic ordering.

4. Fig. 4 shows the $\varrho$–$T$ curves for $DyBa_2Cu_3O_{9-y}$ and $TmBa_2Cu_3O_{9-y}$ measured in magnetic fields both below and above $\mu_0 H_p$. On cooling the samples, a slight shoulder below the large resistance drop has been observed for both samples. This can be ascribed to the boundary effect. When the temperature decreases up to the vicinity of $T_c$, the granules are turned from normal into superconducting state, and the resistivity decreases rapidly. On further cooling, the coupling between the granules becomes weaker due to the existence of magnetic field. So the Josephson junctions are destroyed, and only at very lower temperature they can be transformed into the superconducting state. By then, a slowly variable shoulder will occur in the $\varrho$–$T$ curves.

## References


[1] M. K. WU, J. R. ASHBURN, C. J. TORNG, P. H. HOR, R. L. MENG, L. GAO, Z. J. HUANG, Y. Q. WANG, and C. W. CHU, Phys. Rev. Letters **58**, 908 (1987).

[2] I. K. SCHALLER, D. G. HINKS, M. A. BENO, D. W. CAPONE II, L. SODERHOLM, J. P. LOCQUET, Y. BRUYNSERAEDE, C. U. SEGRE, and K. ZHANG, Solid State Commun. **63**, 385 (1987).

[3] P. H. HOR, R. L. MENG, Y. Q. WANG, L. GAO, Z. J. HUANG, J. BECHTOLD, K. FORSTER, and C. W. CHU, Phys. Rev. Letters **58**, 1891 (1987).

[4] J. J. NEUMEIER, Y. DALICHAOUCH, J. M. FERREIRA, R. R. HAKE, B. W. LEE, M. B. MAPLE, M. S. TORIKACHVILI, K. N. YANG, and H. ZHOU, Appl. Phys. Letters **51**, 371 (1987).

[5] D. K. FINNEMORE, R. N. SHELTON, J. R. CLEM, R. W. MCCALLUM, H. C. KU, R. E. MCCARLEY, S. C. CHEN, P. KLAVINS, and V. KOGAN, Phys. Rev. **35**, 5319 (1987).

[6] D. SAINT-JAMES, G. SARMA, and E. J. THOMAS, Type II Superconductivity, Pergamon Press Ltd., Oxford 1969 (pp. 69).


*(Received December 28, 1987)*

phys. stat. sol. (a) 108, 343 (1988)

Subject classification: 72.15; S1.2

*Department of Physics and Center for Advanced Material Processing, Clarkson University, Potsdam*[1])

# Electrical Resistivity of Some Directionally Solidified Nickel-Base Eutectic Composites

By

S. Arajs and T. A. Jenkins

The electrical resistivity of directionallysolidified Ni-base eutectic composites is measured as a function of temperature between 77 and 300 K. These measurements are made with electric current being parallel and perpendicular to the fibers. Various theories are applied to the measured longitudinal and transverse electrical resistivities to obtain the resistivities of the components.

Der elektrische Widerstand von richtungsverfestigten eutektischen Zusammensetzungen auf Ni-Basis wird als Funktion der Temperatur zwischen 77 und 300 K gemessen. Diese Messungen werden mit einem elektrischen Strom parallel und senkrecht zur Faser durchgeführt. Verschiedene Theorien werden für die gemessenen longitudinalen und transversalen elektrischen Leitfähigkeiten herangezogen, um die Widerstände der Komponenten zu erhalten.

## 1. Introduction

The central problem in understanding of composite materials is the prediction of the overall properties in terms of the individual properties of the components, their morphology, and their concentration. From a general viewpoint, this is a relatively unexplored area. For this reason we have initiated extensive experimental and theoretical studies on different types of multiphase solid systems. In this paper we describe experimental electrical resistivity studies on Ni-base eutectic alloys from 77 to 300 K and discuss their significance.

## 2. Experimental Considerations

The directionally solidified eutectic alloys used in this study were provided by Dr. E. R. Thompson of the United Technologies Research Center, East Hartford, Connecticut, USA. The composition, type of structure, and directional solidification rates of the alloys studied are given in Table 1.

Table 1

Eutectic alloys studied

| material | composition (wt%) | structure | solidification rate (cm/h) |
|---|---|---|---|
| $Ni–Ni_3Nb$ | 77.6 Ni, 22.4 Nb | lamellar | 20 |
| $Ni(Cr)/Ni_3Al–Ni_3Nb$ | 71.8 Ni, 19.7 Nb 6.0 Cr, 2.5 Al | lamellar | 3 |

[1]) Potsdam, New York 13676, USA.

The Ni–Nb system forms a eutectic at 23.3 wt% Nb (16.2 at%) and 1281 °C [1]. When directionally solidified, this alloy has a lamellar microstructure consisting of alternating aligned lamellae of $Ni_3Nb$ in Ni–Nb solid solution. At the eutectic temperature the composite consists of approximately 24 vol% $Ni_3Nb$. At lower temperatures the vol% of $Ni_3Nb$ is higher since the solubility of Nb in Ni decreases with decreasing temperature, and at room temperature it is about 32%. Since the solubility of Nb in Ni at the eutectic temperature is 14 at%, this change in vol% yields a value of 12.7 at% Nb in the Ni matrix at room temperature. This is much larger than the equilibrium value of 3 at%. In making the alloy studied, 22.4 wt% Nb was used since Ni is lost during melting, and thus the resulting sample had approximately the eutectic composition. The bivariant pseudo-binary eutectic Ni(Cr)/$Ni_3Al$–$Ni_3Nb$($\gamma/\gamma'$–$\delta$) is related to the Ni–$Ni_3Nb$ ($\gamma$–$\delta$) and $Ni_3Al$–$Ni_3Nb$ ($\gamma'$–$\delta$) systems. For the sample studied, the liquidus temperature is 1260 °C and the solidus temperature is 1243 °C [2]. Near the solidus temperature the two phases are a Ni–Cr–Al solid solution and $Ni_3Nb$. The $Ni_3Al$ solution temperature is between 1150 and 1175 °C. On cooling below this temperature the aluminum precipitates out of solution in the form of $Ni_3Al$. The microstructure consists of alternating lamellae of Ni–Nb and Ni–Cr solid solution with $Ni_3Al$ precipitates. The lamellae are continuous in the longitudinal direction but are faulted in the transverse direction. The $Ni_3Al$ precipitates are globular in shape. The $Ni_3Nb$ phase occupies about 40 vol%. The interlamellar spacing is 4 μm.

The eutectic alloys were produced as cylindrical ingots approximately 10 cm long and 0.5 to 1 cm in diameter. Transverse and longitudinal samples in the form of rectangular parallelepipeds were then cut from the ingots. The length of the transverse samples was limited by the diameter of the composite ingot. The longitudinal samples were typically 3 to 4 cm long and of approximately square cross section.

The electrical resistivity was measured using the standard four-point method. Potential and current leads of 0.25 mm Cu wires were spot-welded to the samples. The temperature was held constant to ±0.1 K with the apparatus described by Arajs and Conroy [3]. The temperature was measured with a calibrated Pt resistance thermometer to an accuracy of ±0.1 K. The constant current was supplied by a North Hills Electronics Precision Current source (Model CS-12). The current was constant to ±0.01%. The current was chosen so as to provide a sufficiently high potential difference for accurate readings (>10 μV at 77 K) yet not to cause unnecessary resistive heating effects. The potential difference was measured with a nanovolt potentiometer (Guildline Type 9176-G) accurate to ±0.01 μV. Provision was made to reverse the current through the sample and galvanometer to correct the thermal voltages.

The accuracy of the absolute resistivity was 5 to 12%, due mostly to a large uncertainty in the geometrical form factor. The error in the normalized resistivity, $\varrho/\varrho_{298}$, was less than 0.5%.

The longitudinal electrical resistivity, $\varrho_L$, of a composite material consisting of parallel fibers of resistivity $\varrho_f$ and volume fraction $c_f$ embedded in a matrix of resistivity $\varrho_m$ and volume fraction $c_m$ is, neglecting any interface effects,

$$\frac{1}{\varrho} = \frac{c_f}{\varrho_f} + \frac{c_m}{\varrho_m} . \tag{1}$$

The transverse resistivity (the electric current is perpendicular to the fibers), $\varrho_T$, cannot be found exactly. The results of different theoretical approximations are as follows:

$$\varrho_T = c_f\varrho_f + c_m\varrho_m , \tag{2}$$

$$\varrho_T = \frac{2\varrho_f\varrho_m^2}{2\varrho_f\varrho_m + c_f(\varrho_m^2 - \varrho_f^2)} , \tag{3}$$

$$\varrho_T = \frac{[2\varrho_m + c_f(\varrho_f - \varrho_m)]\,\varrho_f}{2\varrho_m - c_f(\varrho_f - \varrho_m)}\,, \tag{4}$$

$$\varrho_T^2 = \frac{c_f\varrho_f + c_m\varrho_m}{(c_f/\varrho_f) + (c_m/\varrho_m)}\,, \tag{5}$$

$$\frac{c_f(\varrho_T - \varrho_f)}{\varrho_T + \varrho_f} + \frac{c_m(\varrho_T - \varrho_m)}{\varrho_T + \varrho_m} = 0\,. \tag{6}$$

Equation (2) gives the upper bound [4] for the transverse electrical resistivity of a lamellae-aligned composite. Equations (3) to (5) result from the general formulae [5] for the effective electrical conductivity of a two-component composite using different types of approximations for the electrical field ratios. Equation (6) comes from the effective medium theory [6, 7] and its extensions [8, 9].

For anisotropic composites, the equations for the longitudinal and transverse resistivities give two independent equations in $\varrho_f$ and $\varrho_m$ that can be solved simultaneously to give $\varrho_f$ and $\varrho_m$ as functions of $\varrho_L$ and $\varrho_T$. Solving the equations gives quadratic or higher order equations for $\varrho_f$ and $\varrho_m$. It is then necessary to use the appropriate pairs of roots depending on whether $\varrho_f > \varrho_m$ or $\varrho_f < \varrho_m$. Solving (1) for the resistivity of the components gives rise to lower limits for the component resistivities in terms of the longitudinal resistivity

$$\varrho_f > c_f\varrho_L \tag{7}$$

and

$$\varrho_m < c_m\varrho_L\,. \tag{8}$$

If, for example, $\varrho_f \approx c_f\varrho_L$, then the calculated values for $\varrho_m$ are very sensitive to experimental errors in the measurement of $c_f$, $\varrho_f$, or $\varrho_L$.

## 3. Results and Discussion

### *3.1 Ni–Ni$_3$Nb*

The experimental data in the form of normalized electrical resistivity to 298 K for the Ni–Ni$_3$Nb lamellar eutectic alloy are shown in Fig. 1. The measured room temperature longitudinal and transverse resistivities for the Ni–Ni$_3$Nb composite at room temperature are $\varrho_L = (89.2 \pm 0.2)\ \mu\Omega$cm and $\varrho_T = (97.8 \pm 0.4)\ \mu\Omega$cm, respectively. The temperature dependence can be seen from the log–log plot shown in Fig. 2. The resistivity ratio $\varrho_T/\varrho_L$ shows a marked temperature dependence as can be seen in Fig. 3. This result is expected in a lamellar composite because of enhanced electron scattering with the phase boundaries in the transverse direction.

Fig. 4 shows the calculated values for the resistivity of the Ni (12.7 at% Nb) matrix using the theoretical predictions described above. The increase in transverse resistivity with decreasing temperature has been subtracted out, assuming it was due entirely to interface effects. Arajs [10] found that Matthiessen's rule is not obeyed at any temperature for Ni-rich Ni–Nb alloys. The electrical resistivity for such an alloy can be written as

$$\varrho = \varrho_0 + \varrho_i(T) + \Delta\varrho_i(T, c)\,, \tag{9}$$

where $\varrho_0$ is the residual resistivity, $\varrho_i(T)$ the intrinsic contribution, and the additional term $\Delta\varrho_i(T, c)$ is the change in resistivity as a result of alloying. By extrapolating Arajs' data, the Curie temperature of the 12.7 at% Nb alloy is found to be 30 K.

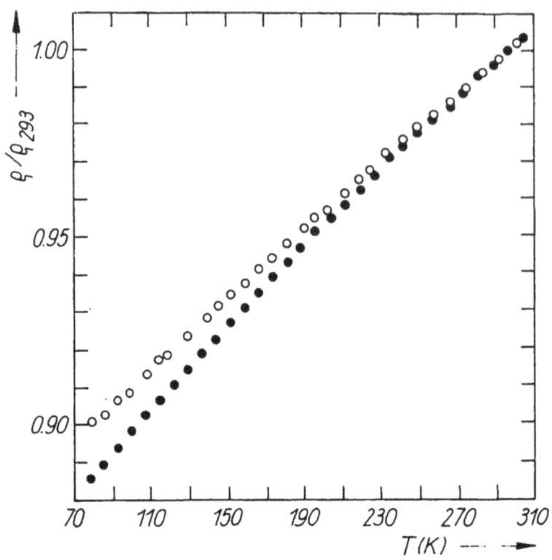


Fig. 1

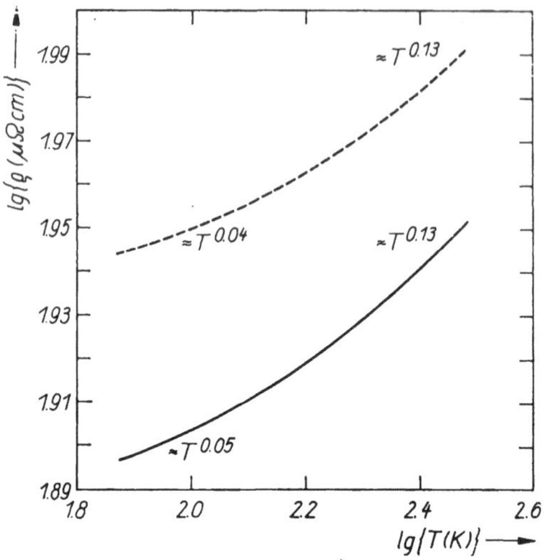


Fig. 2

Fig. 1. $\varrho/\varrho_{298}$ vs. $T$ of Ni–$Ni_3Nb$ composite. ● longitudinal, ○ transverse

Fig. 2. lg $\varrho$ vs. lg $T$ of Ni–$Ni_3Nb$ composite. ——— longitudinal, — — — transverse

Thus, the alloy is paramagnetic over the entire temperature range studied. The quantity $\Delta\varrho_i(T, c)$ was found to be negative in the paramagnetic region and its magnitude to decrease with increasing temperature. The residual resistivity is 5.0 μΩcm/(at% Nb).

The two effective medium theories for semialigned lamellae, (4) and (5), give reasonable agreement with the data for the above matrix calculated using (9). The theories for aligned lamellae (2) and dilute dispersions of lamellae (3) give results that

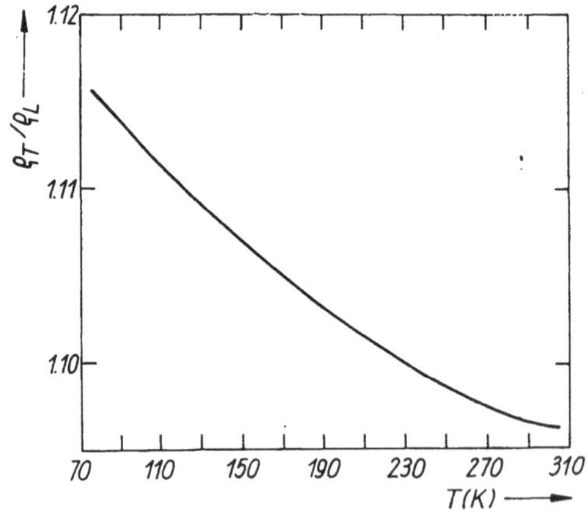


Fig. 3

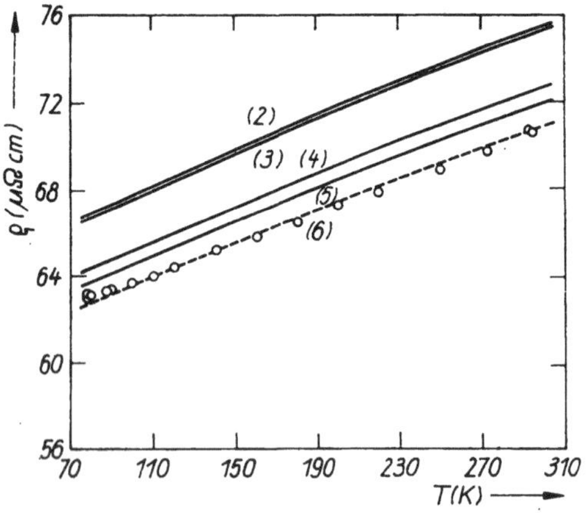


Fig. 4

Fig. 3. $\varrho_T/\varrho_L$ vs. $T$ of Ni–$Ni_3Nb$ composite

Fig. 4. $\varrho$ vs. $T$ of Ni (12.7 at% Nb) in Ni–$Ni_3Nb$. Lines calculated from $\varrho_L$ and $\varrho_T$ data using different theoretical equations. Points calculated from $\varrho$ of Ni using (9) and including corrections for interface effects (number of the equation used is marked at the curves)

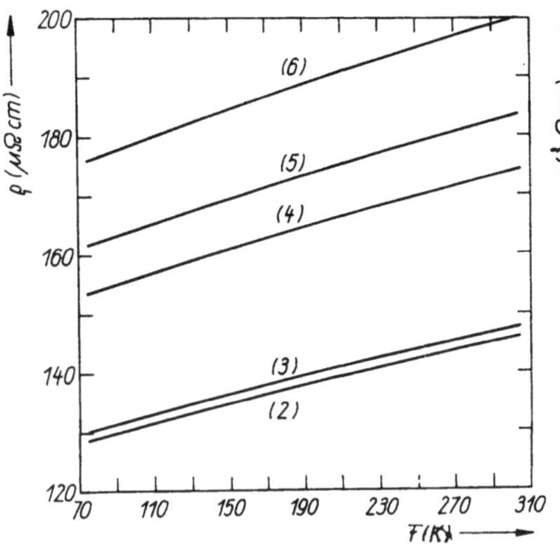


Fig. 5

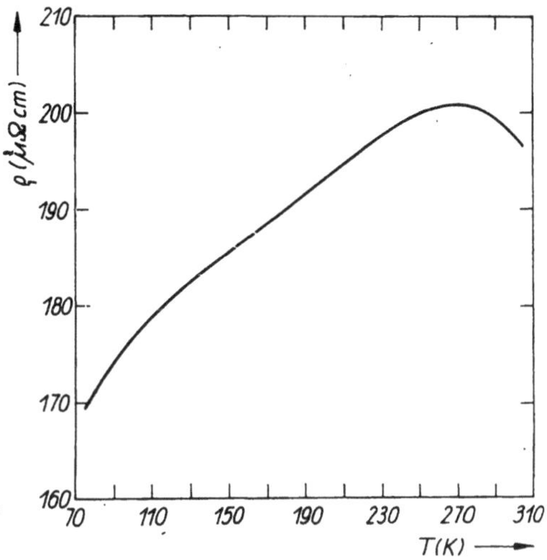


Fig. 6

Fig. 5. $\varrho$ vs. $T$ of $Ni_3Nb$ lamellae in Ni–$Ni_3Nb$ calculated from $\varrho_L$ and $\varrho_T$ data using different theoretical predictions and including interface effects (number of the equation used is marked at the curves)

Fig. 6. $\varrho$ vs. $T$ of $Ni_3Nb$ lamellae in Ni–$Ni_3Nb$ calculated from $\varrho_L$ and $\varrho_T$ of Ni data

are too high. It is expected that the theory for dilute dispersions would be inappropriate since the volume fraction is only 0.32. It is particularly interesting that the effective mean field theory for fibers (6) gives very good agreement with the calculated points. This implies that the physical assumptions associated with the effective mean theory are much more important than exact details of the geometry of the phases.

The resistivity of the $Ni_3Nb$ lamella calculated from $\varrho_L$ and $\varrho_T$ data, including the increase in the resistivity due to the size effect, is shown in Fig. 5. Fig. 6 shows the resistivity evaluated from $\varrho_L$ and $\varrho_T$ of Ni data. These calculations give results that are much greater than those reported in the literature, i.e., 45 to 50 μΩcm [11].

### 3.2 *Ni(Cr)/Ni$_3$Al–Ni$_3$Nb*

The experimental data for the Ni(Cr)/$Ni_3Al$–$Ni_3Nb$ directionally solidified eutectic are shown in Fig. 7. The measured longitudinal and transverse resistivities at room temperature of this lamellar composite are $(105.5 \pm 0.3)$ μΩcm and $(141.5 \pm 0.8)$ μΩcm, respectively. The temperature dependence of the resistivities can be seen from the log–log plot shown in Fig. 8. As in the Ni–$Ni_3Nb$ composite, the exponent in $T^n$ is much less than unity and increases with increasing temperature. In this composite, however, there is a large difference in the temperature dependences of the longitudinal and transverse resistivities.

The same type of analysis was used on the data for this system as for Ni–$Ni_3Nb$. The results for the Ni(Cr)/$Ni_3Al$ matrix are shown in Fig. 9a and those for the $Ni_3Nb$ lamellae in Fig. 9b. The resistivity of the $Ni_3Nb$ lamellae is much greater in this composite than in the Ni–$Ni_3Nb$ composite. This probably results from interface effects which are particularly pronounced in this composite as can be seen from the resistivity ratio $\varrho d/\varrho d$ (Fig. 10).

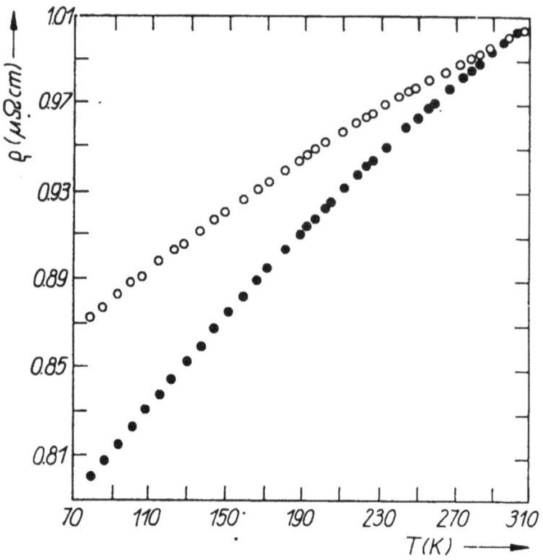


Fig. 7

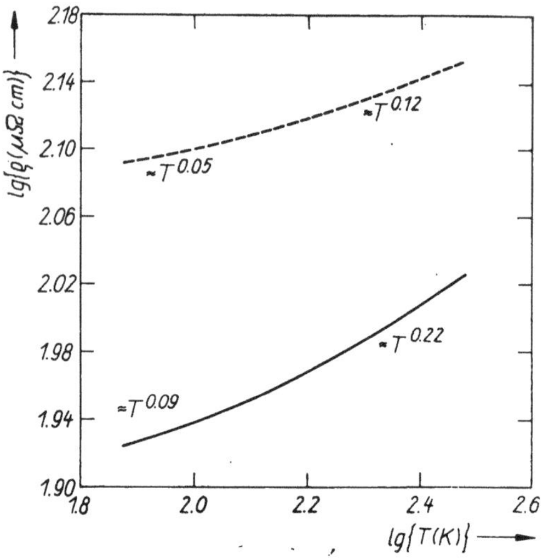


Fig. 8

Fig. 7. $\varrho/\varrho_{298}$ vs. $T$ of Ni(Cr)/$Ni_3Al$–$Ni_3Nb$ composite. ● longitudinal, ○ transverse

Fig. 8. lg $\varrho$ vs. lg $T$ of Ni(Cr)/$Ni_3Al$–$Ni_3Nb$ composite. ——— longitudinal, – – – transverse

The matrix in this composite material may also be considered as a composite material consisting of $Ni_3Al$ particles in a Ni(Cr) matrix. This means that the composite as a whole should be considered a ternary composite. To treat this in detail, it would be necessary to combine the theories for particulate and lamellar composites, and

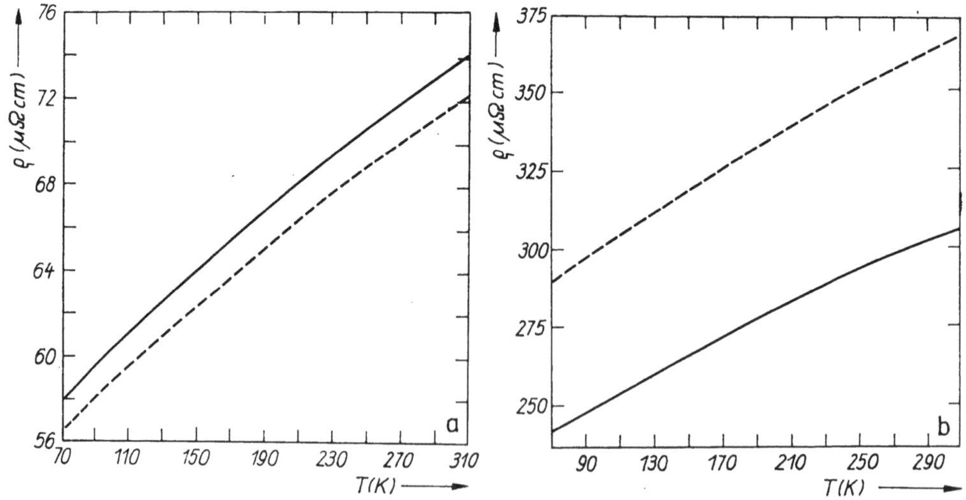


Fig. 9. $\varrho$ vs. $T$ a) of Ni(Cr)/$Ni_3Al$–$Ni_3Nb$ composite and b) of $Ni_3Nb$ lamellae in Ni(Cr)/$Ni_3Al$ calculated from $\varrho_L$ and $\varrho_T$ data (solid line using (4), dashed line (5))

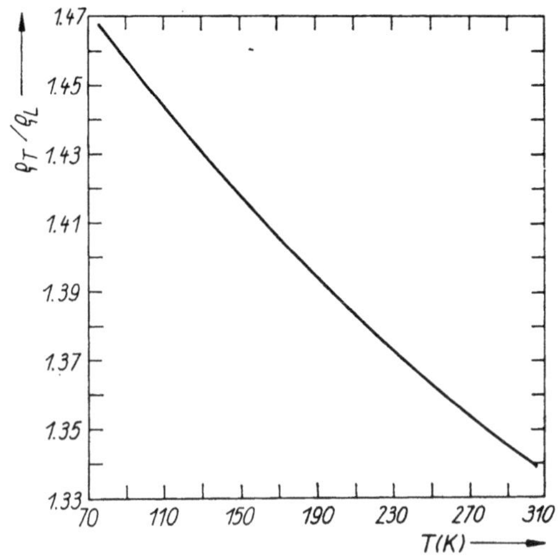


Fig. 10. $\varrho_L/\varrho_T$ vs. $T$ of $Ni(Cr)/Ni_3Al–Ni_3Nb$ composite

the resistivities of the phases would have to be known in order to compare the experimental results with the theory. At the present time this is impossible to do.

## References


[1] M. Hansen, Constitution of Binary Alloys, 2nd ed., McGraw-Hill Publ. Co., New York 1958.
[2] F. D. Lemkey and E. R. Thompson, Proc. Conf. In-Situ Composites, Vol. II, Xerox Individualized Publishing, Lexington 1973 (p. 105).
[3] S. Arajs and J. W. Conroy, Japan. J. appl. Phys. 8, 614 (1969).
[4] O. Wiener, Abh. math.-phys. Kl. Sächs. Akad. Wiss. **32**, 509 (1912).
[5] J. A. Reynolds and J. M. Hough, Proc. Phys. Soc. **B70**, 769 (1957).
[6] D. A. G. Bruggeman, Ann. Phys. (Leipzig) **24**, 636 (1936).
[7] D. A. G. Bruggeman, Ann. Phys. (Leipzig) **25**, 645 (1936).
[8] D. J. Bergman, Physics Rep. **43**, 379 (1978).
[9] W. E. A. Davies, J. Phys. D 7, 120 (1973).
[10] S. Arajs, J. appl. Phys. **32**, 97 (1961).
[11] N. V. Petrushin, A. V. Logunov, A. I. Kovalev, A. F. Zverev, V. M. Toropov, and N. N. Fedotov, Teplofiz. vysokikh Temp. **14**, 649 (1976).


*(Received January 14, 1988)*

phys. stat. sol. (a) **108**, 351 (1988)

Subject classification: 72.40; 71.55; S8.16

*Dipartimento di Scienze Fisiche, Centro Interuniversitario di Struttura della Materia, G.N.S.M./C.N.R., Università di Cagliari*[1]) *(a) and*
*Laboratoire de Spectroscopie et d'Optique du Corps Solide, Unité associée au C.N.R.S. no 232, Université Louis Pasteur, Strasbourg*[2]) *(b)*

# On the Photoconductivity Relaxation in $ZnIn_2S_4$

By

A. SERPI (a) and J. P. ZIELINGER (b)

The relaxation process of photoconductivity in $ZnIn_2S_4$ is investigated together with optical quenching and thermally stimulated current. Photocurrent decays are processed by four-gate technique of photoinduced current transient spectroscopy (PICTS). A non-exponential behaviour of the decays, due to a non-linear multichannel recombination mechanism, leads to "anomalous" PICTS spectra that in the temperature range 200 to 300 K allow to evidence a thermal emission from a centre with a repulsive potential barrier. The photoconductivity relaxation in the 80 to 200 K range is attributed to two competitive recombination channels. Models for the relaxation through both radiative and non-radiative recombination channels are proposed on the basis of optical quenching effect. In any case the recombination mechanism is dominated by energy levels exponentially distributed below the conduction band with a "slope" of 33 $eV^{-1}$.

Le processus de relaxation de la photoconductivité dans $ZnIn_2S_4$ est étudié en connexion avec le quenching optique et le courant stimulé thermiquement. Le déclin en fonction du temps du photocourant a été analysé par une technique quadrupole-porte des transitoires de courant photoinduits (PICTS). Une allure nonexponentielle du déclin du photocourant par suite de l'intervention d'un mécanisme de recombinaison multicanal nonlinéaire produit des spectres PICTS „anomaux" qui dans l'interval de temperature 200 à 300 K permettent de mettre en évidence une émission thermique par un centre avec une barrière de potentiel répulsive. La relaxation du photocourant dans l'interval 80 à 200 K est attribuée à deux canaux de recombinaison qui sont en compétition. Des modèles pour la relaxation par les deux recombinaisons radiative et non-radiative sont proposés sur la base de l'effet du quenching optique. Dans tous les cases le mécanisme de recombinaison est dominé par une distribution exponentielle de niveaux d'énergie sous la bande de conduction avec une „pente" de 33 $eV^{-1}$.

## 1. Introduction

$ZnIn_2S_4$ is characterized, like other semiconductor compounds of the $AB_2X_4$ family, by a high number of "intrinsic" lattice defects (vacancies, antisite defects, etc.) which are responsible for the presence of several levels inside the energy gap. The material shows, in spite of that, a very high resistivity ($10^{10}$ to $10^{12}$ $\Omega$ cm) due to strong self-compensation.

From photoconductivity (PC) studies it resulted in particular [1 to 3] that in this compound there are several, simultaneously active recombination centres, and an exponential distribution of trapping levels. Photoelectrical measurements indicated also switching behaviour [4], negative resistance [5], and the presence of a negatively charged repulsive centre connected with PC quenching [6]. These multiple impurity levels can account for the complicated kinetics of the relaxation process of the photo-

[1]) Via Ospedale 72, I-09100 Cagliari, Italy.
[2]) 5, rue de l'Université, F-67084 Strasbourg, Cédex, France.

response. The broad luminescence band centred at 1.7 eV at 80 K has been first attributed to electron transitions from the trap distribution to an acceptor level [7]. A recent model [8] proposed that the emission is more likely due to electron transitions from a single donor level to the acceptor, leaving out any role of the trapping levels in the luminescence process. More recently [9], from PC measurements it has been concluded that the density of trapping states decays more slowly with increasing energy depth than close to the conduction band and it may even increase towards the middle of the band gap.

The aim of this work is to clarify the role of the trap distribution in the relaxation process of PC and to obtain if possible information about the mechanism of radiative recombination and the electronic transitions which are responsible for it. To this purpose we shall analyse quenched and unquenched PC spectra, thermally stimulated current (TSC) curves, and PC transients for times longer than 0.3 ms. The decay analysis will be ruled out by means of the recently proposed technique of photoinduced current transient spectroscopy (PICTS) [10] that generally allows to plot a global spectrum of the trapping levels which take part in the PC relaxation process.

## 2. Experimental

$ZnIn_2S_4$ single crystals were grown by the well-known method of vapour phase transport reactions [11, 12] using gaseous iodine as transporter. Indium contacts in a coplanar configuration were evaporated on a natural surface. The contacts were ohmic in the explored temperature range. All the samples were n-type with a resistivity of about $10^{10}\ \Omega$ cm. Both the quenched and unquenched PC spectra were obtained using a dc technique of measurement. A halogen lamp coupled with a McPherson monochromator was employed as primary light source. A small He–Ne laser (5 mW) was used as secondary source. The quenching spectrum was obtained by means of the halogen lamp equipped with an interferential filter (405 nm) (primary source) and another halogen lamp coupled with the monochromator (secondary source). As the value of the forbidden band gap of $ZnIn_2S_4$ at 91 K is 2.99 eV [13], it corresponds to intrinsic excitation from the primary source.

Thermally stimulated current measurements were achieved by exciting the sample for 10 min at a fixed temperature by means of the halogen lamp coupled with the 405 nm interferential filter; then the temperature was lowered down to 70 K keeping the sample in the dark. Finally, while temperature was linearly raised (about 0.5 K/s), the dark current flowing in the sample was measured.

Photocurrent transients, excited by the halogen lamp coupled with the 405 nm interferential filter, were obtained by means of an electronic shutter (0.3 ms closing time). The voltage drop across a low resistor, in series with the sample and a low dc voltage supply, was amplified by means of a PAR 113 amplifier operated in dc mode. The output signal was fed, via an ac converter, to an acquisition system (PC XT3 IBM). The computer checked first for saturation of the PC signal, then operated the shutter closing. Sampling frequency was properly set via software according to the transient length (maximum frequency $2 \times 10^4$ Hz). In order to enhance the signal-to-noise ratio several transients (typically 8) were averaged. The entire decay was recorded at various temperatures (about every 2 K) during a very slow thermal scan and the results are stored on a hard disk with all the parameters of the measurement for subsequent analysis. Quenched transients were obtained in the same way, but keeping the sample continuously excited by the He–Ne laser radiation.

## 3. Results

### *3.1 Photoconductivity spectra*

In Fig. 1 (curve a) the PC spectrum at 80 K is shown. It has been normalized for equal incident energy to take into account the spectral response of the light source and the optical system. Two extrinsic peaks are present at 440 and 520 nm, according to the fact that $ZnIn_2S_4$ exhibits strong self-compensation: all the acceptor states are ionized at low temperature and therefore it is possible to excite electrons from the acceptor level to the conduction band.

Curve b of Fig. 1 is the PC spectrum quenched by the 633 nm line of the He–Ne laser (secondary light source). As may be seen, the quenching radiation erases the extrinsic PC.

The quenching spectrum at 80 K, shown in Fig. 2, was obtained by exciting the sample with the primary radiation at 405 nm (intrinsic PC) and varying the wavelength of the secondary radiation. The quenching rate is defined as usual as

$$Q = (I_1 - I_2)/I_0 , \tag{1}$$

where $I_1$ is the PC signal due to the primary source only, $I_2$ the signal due to both the primary and secondary light, and $I_0$ a normalization function which takes into account the spectral transmission of the optical system. It is to be noted that a quenching effect is observed only below 200 K. According to other measurements [1, 6], a maximum quenching is obtained at about 700 nm.

### *3.2 Thermally stimulated current*

In Fig. 3 we show TSC curves corresponding to various excitation temperatures (curves a, c, d, e). The excitation radiation was fixed at 405 nm. The TSC peak shifts to the right and its height decreases with increasing excitation temperature. If we plot the logarithm of the TSC signal as a function of $1/T$ for the different excitation temperatures, the initial part of each curve up to its maximum yields a straight line

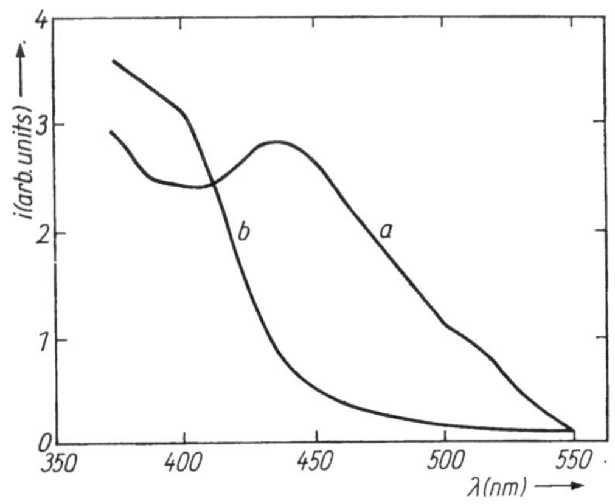


Fig. 1

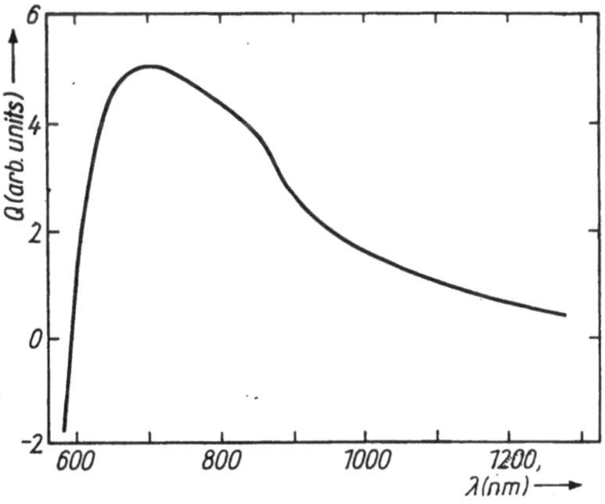


Fig. 2

Fig. 1. PC excitation spectra measured at 80 K: (a) unquenched spectrum, (b) quenched spectrum (additional illumination with 633 nm radiation)

Fig. 2. Quenching spectrum of PC measured at 80 K (see (1)). The primary radiation was fixed at 405 nm

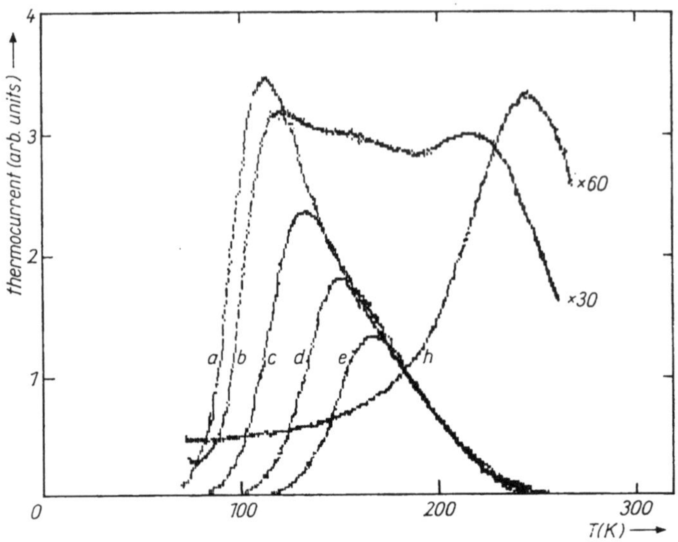


Fig. 3. TSC curves. Curves (a), (c), (d), (e) refer to excitation from 405 nm radiation at different temperatures: (a) 83, (c) 110, (d) 130, (e) 150 K. Curves (b) and (h) refer to different excitation conditions at 83 K: (b) excitation at the same time from both 405 and 633 nm radiations, (h) excitation first from 405 nm, then from 633 nm radiation

whose slope furnishes an "activation energy" $E_t$ which is experimentally related to the excitation temperature $T_0$ by

$$E_t = 12 k_B T_0 , \tag{2}$$

where $k_B$ is the Boltzmann constant.

Fig. 3 (curves b, h) shows also the effect of the secondary light radiation (633 nm of He–Ne laser) on the TSC curve at the excitation temperature $T = 83$ K. Curve b refers to the sample excited at the same time both by the primary and secondary light sources. Curve h is obtained by exciting the sample first with the primary light, then with the quenching radiation. As can be seen, the quenching radiation kills the TSC peak, keeping a low signal in the high-temperature range. Taking into account the scale factors, this low signal may be hidden in the unquenched curves under the low-temperature TSC peak.

### 3.3 PC transients

In Fig. 4 PC decays at different temperatures are represented for the case of intrinsic primary pulsed excitation. It can be seen that the time dependence of the transient in the whole explored temperature range is of the type

$$i = (a + bt)^{-1} . \tag{3}$$

Decays obtained while keeping the sample continuously illuminated by the quenching radiation, are plotted in Fig. 5 in a double logarithmic representation. Quenching radiation clearly affects both the response time and the behaviour of the transient signal. In the low-temperature range the response time is reduced by a factor 20. From Fig. 5 a behaviour of the type

$$i \sim t^{-\gamma} \tag{4}$$

may be inferred, especially for the final part of the decay.

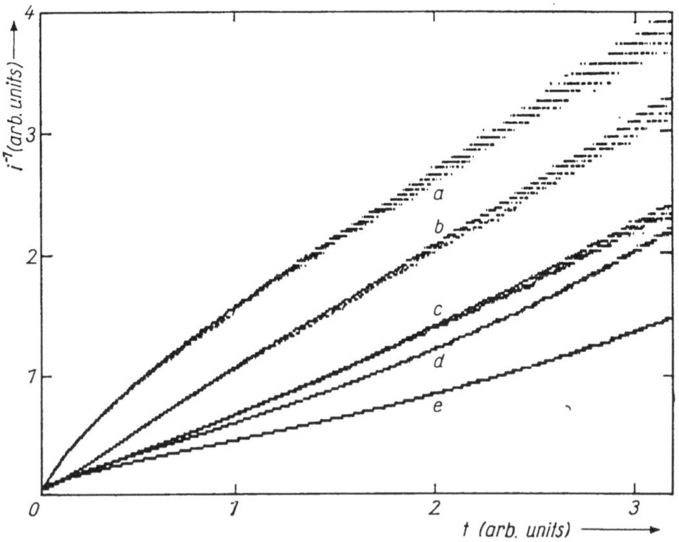


Fig. 4. Unquenched decays. The inverse of the photocurrent as a function of $t$ for different excitation temperatures: (a) 293, (b) 250, (c) 200, (d) 150, (e) 100 K

## 4. Transient Data Processing

In order to apply the PICTS technique to our PC decays we must keep in mind that we are concerned with a complicated non-exponential relaxation process, as shown in the preceding section. So we found it convenient to apply the four-gate method of

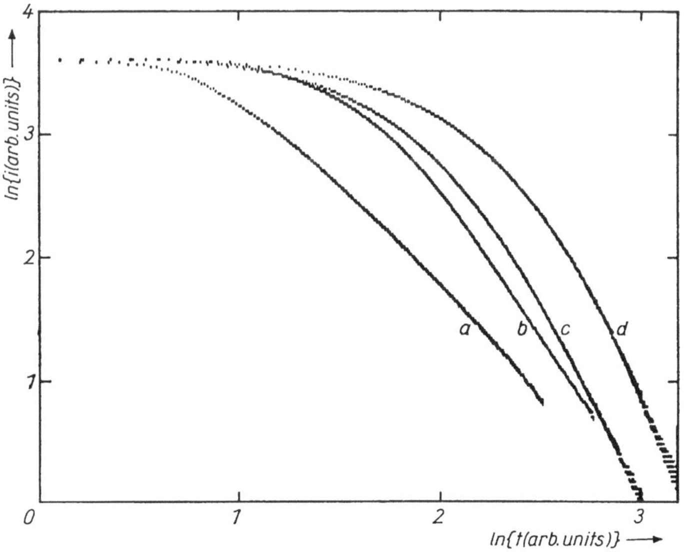


Fig. 5. Quenched decays. ln of the photocurrent as a function of ln $t$ for different excitation temperatures: (a) 293, (b) 250, (c) 200, (d) 100 K

PICTS [10] which consists in plotting the ratio

$$Y(T) = \frac{i(t_1) - i(t_2)}{i(t_0) - i(t_3)} \tag{5}$$

as a function of temperature, where $i(t)$ is the amplitude of the signal at temperature $T$ and at instant $t$ after the shutter has been closed; $t_0$, $t_1$, $t_2$, and $t_3$ are four reading times chosen in a convenient manner. Let us assume a purely exponential decay with a time constant $\tau(T)$ (single trap case). Then it may be shown [10] that, if $t_3 \gg t_2 > > t_1 > t_0$, $Y(T)$ presents a maximum at a temperature $T_m$, where

$$\tau(T_m) = \frac{t_2 - t_1}{\ln\left(\dfrac{t_2 - t_0}{t_1 - t_0}\right)} = (S_t v N_t)^{-1} \exp(E_t/k_B T_m)\,. \tag{6}$$

Here $S_t$ is the apparent capture cross section for free electrons of the trap, $N_t$ and $E_t$, respectively, its density and activation energy, and $v$ the thermal velocity. Here, of course, a classical mechanism of thermal emission is assumed for the PC relaxation. The height of the peak $Y(T)$ depends on the choice of $t_0$, $t_1$, and $t_2$. In particular for $t_2 = 3t_0$ and $t_1 = 2t_0$ a value of 0.25 is expected for the maximum, independently of the $t_0$ value. This fact can be used as a direct check of the postulated physical model, that is, if the maxima of $Y(T)$ corresponding to various $t_0$ have the same amplitude, close to the expected value, one may conclude that an exponential trap relaxation followed by a quasi-linear recombination of emitted carriers is involved in the decay. At this point two remarks are to be made:

1. The four-gate method loses any information about the initial trapped electron density, so, if we have two trapping centres with different concentrations, two peaks with the same amplitude are expected in the PICTS spectra, provided of course that their thermal emissions do not significantly overlap.

2. In principle the PICTS spectrum yields the complete spectrum of the various trapping levels that are active in a given temperature range. In the case of electronic levels distributed continuously within the forbidden energy gap, the four-gate method gives a broad, flat response at a level that can fall considerably below 0.25 [10].

For plotting our spectra we chose $t_2 = 3t_0$ and $t_1 = 2t_0$. In Fig. 6 we show a few curves obtained in the case of unquenched transients. In the high-temperature range (small $t_0$) a well defined peak is present, the position of which shifts as $t_0$ is changed. For larger $t_0$ values a broadening of the peak is observed but its amplitude remains quite constant. Therefore, we have to deal with PICTS spectra that are somewhat "anomalous", that is they are not referable to the classical kinetics of thermal emission-recombination. The low-temperature, relatively flat part of the spectrum, for large $t_0$ values, may be ascribed to some role played in the relaxation by a continuous trap distribution, while the high-temperature part, for small $t_0$ values, involves more likely the presence of a single trapping level. The "anomaly" consists in the fact that one would expect a flat PICTS signal due to the trap distribution also for small $t_0$ values in the low-temperature region, if the traps were acting as thermal emitting centres. In conclusion the main or at least the unique role played by the trap distribution is not that of thermal emission.

In Fig. 7a we show the Arrhenius plot obtained on the basis of (6) from the PICTS peaks of Fig. 6 in the high-temperature range. This plot provides an activation energy $E_t = 0.32$ eV and a cross section $S_t = 10^{-22}$ cm$^2$.

At this point it must be noticed that the PICTS technique can reveal the presence of a continuous trap distribution, but is unable to furnish the characteristic parameters

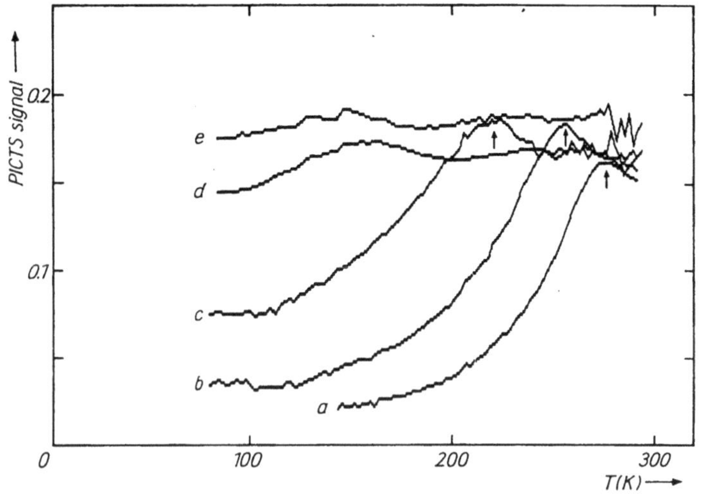


Fig. 6. PICTS spectra of unquenched PC decays for different values of $t_0$: (a) 1.95, (b) 7.81, (c) 62.5 ms, (d) 1, (e) 2 s

of the trap distribution. This must be derived from a different treatment of experimental data. We have already observed that unquenched decays show a time dependence of the type (3). This law can be easily related with bimolecular kinetics of recombination,

$$\frac{dn}{dt} = -vS_r mn\,, \tag{7}$$

where $n$ and $m$ are the densities of free electrons in the conduction band and of holes in the recombination centre, respectively, $S_r$ is the capture cross section of the recombination centre for free electrons. If we assume $m = n$, by integration we obtain

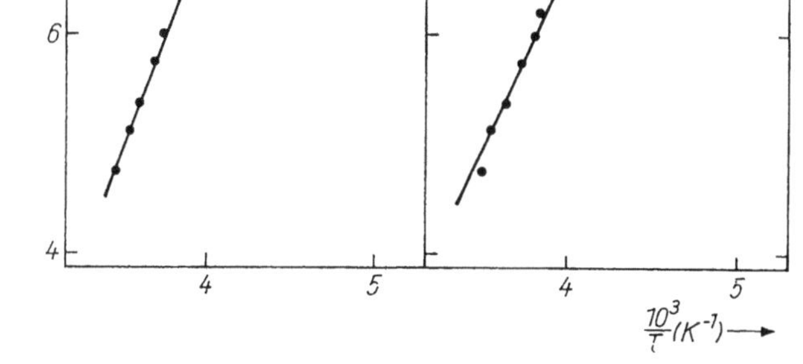


Fig. 7. Arrhenius plots of PICTS peaks: a) unquenched decays (Fig. 6), b) quenched decays (Fig. 9)

for the PC signal $I(t)$ normalized to the initial value $i(0)$,

$$I(t) = \frac{i(t)}{i(0)} = (1 + vS_t n_0 t)^{-1} = (1 + Bt)^{-1} , \tag{8}$$

where $n_0$ is the free electron density at the beginning of the recombination process, and $B = vS_t n_0$. The parameter $B$ can be obtained from the transients. To do it we used the statistical covariance method [14], that is, putting $y = I(t)$ and $x = t$, the $B$ value results as

$$B = \frac{\text{cov}\,(x, y)}{v(x)} = \frac{\text{cov}\,(x, y)}{v(y)} , \tag{9}$$

where cov $(x, y)$, $v(x)$, and $v(y)$ are the covariance of $x$ and $y$, the variance of $x$, and the variance of $y$, respectively. The reliability of (8) can be evaluated from the correlation coefficient $\sigma$,

$$\sigma = \frac{\text{cov}\,(x, y)}{v(x)\, v(y)} , \tag{10}$$

that should approach unity. $B$ has been evaluated on the basis of (8) by using the whole decay (that is 3000 digitalized values). The correlation coefficient $\sigma$ always resulted greater than 0.99.

$B$ values derived from unquenched decays are plotted in Fig. 8 versus $T$. In the hypothesis $S_t$ = const, the initial density of free electrons is seen to change exponentially between 100 and 260 K according to a function of the form

$$n_0(T) = N_0 \exp{(\beta T)} . \tag{11}$$

Fig. 8 provides $\beta = 3.2 \times 10^{-2}\ \text{K}^{-1}$.

Fig. 9 shows some four-gate PICTS spectra relative to quenched decays. In this case also there is a peak in the high-temperature range for small values of $t_0$ and the temperature of the maximum changes as $t_0$ is varied. In the low-temperature range, for large $t_0$, a signal quite independent of temperature is obtained, whose amplitude

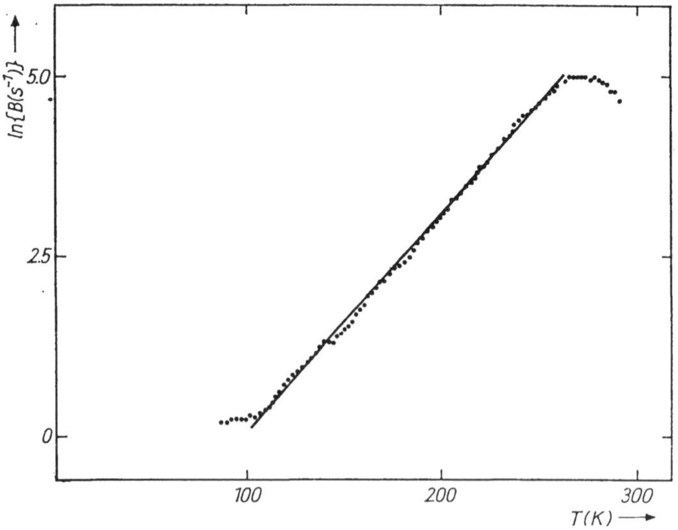


Fig. 8. ln $B$ vs. $T$ (see (8)) as derived from unquenched decays

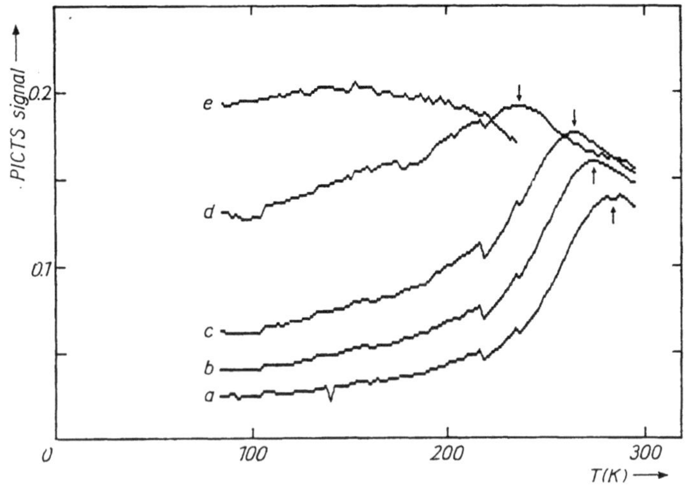


Fig. 9. PICTS spectra of quenched PC decays for different values of $t_0$: (a) 0.97, (b) 1.95, (c) 3.90, (d) 7.91, (e) 62.5 ms

approaches 0.20. These spectra are not very different from those related to the unquenched transients, so that the same comments are valid. Fig. 7b is the Arrhenius plot corresponding to the peaks of Fig. 9. An activation energy $E_t = 0.28$ eV and a cross section $S_t = 2 \times 10^{-23}$ cm$^2$ can be derived.

Again the parameters characteristic of the trap distribution may be obtained from a different treatment of the decays. As remarked before, the time dependence of quenched decays is of type (4), especially for its final part. It is known [15 to 17] that this behaviour, which is common in amorphous semiconductors, may originate from a thermalization process of the trapped carriers. In this case the value $\gamma$ is expected to show a linear dependence on temperature. $\gamma$ has been obtained by applying the covariance method to the final part of the quenched decay (the last 2500 points out of 3000). In Fig. 10 $\gamma$ is shown as a function of temperature. As may be seen in the range 80 to 200 K $\gamma$ changes linearly with temperature with a slope of $3.0 \times 10^{-3}$ K$^{-1}$.

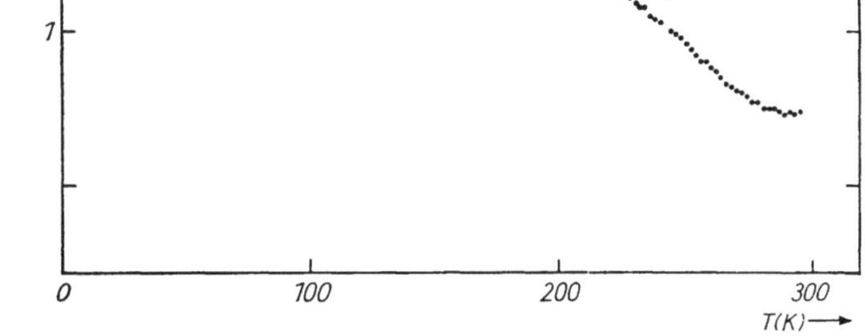


Fig. 10. $\gamma$ vs. $T$ (see (4)) as derived from quenched decays

## 5. Discussion

First of all it must be pointed out that the observation of a quenching effect implies the presence of two different recombination paths in the PC relaxation process of $ZnIn_2S_4$. Depending on temperature, one of the two recombination mechanisms may predominate, or both may be active at the same time. One of the processes is, without doubt, the radiative recombination of free electrons with holes in the acceptor level and has been ascribed to a kind of donor-acceptor pair transition [7, 8]. In our case quenching manifests itself not only by the extinction of the extrinsic PC at low temperature, but also by a considerable reduction of the TSC peak. This fact implies that quenching irradiation favours considerably the radiative recombination at the expense of the other recombination channel. This is in agreement with the experimentally observed enhancement of the luminescence intensity accompanying PC quenching [7]. Then we can establish the following points:

1. The decay of PC below 200 K implies two types of relaxations, one radiative, due to recombination of electrons from some level below the conduction band with holes in the acceptor level; the other, likely non-radiative, due to recombination of free electrons with holes in a deeper level.

2. Quenching radiation fills the non-radiative recombination centre with electrons and the acceptor centre with holes: therefore, the non-radiative path is stopped and the radiative recombination is enhanced. From the quenching spectrum (Fig. 2) we can infer that the non-radiative centre lies 2 eV above the valence band. The decrease of the response time of the low-temperature PC, induced by the quenching radiation, may be accounted for by a considerable increase of the hole density at the acceptors, allowing thus faster recombination.

The problem that remains now is to know the exact nature of the radiative recombination, that is, whether the electrons which recombine with holes in the acceptors come from some single level or from the trap distribution. This question cannot be answered directly from the PC properties, but we can derive indirectly information about the radiative transition. We may suppose that the radiative transition is faster than the non-radiative one. We further assume that immediately after the sample excitation with intrinsic radiation a fraction of the photoexcited electrons recombine radiatively through the trap distribution with holes in the acceptor centres. The remaining part of photoelectrons recombine slowly with the non-radiative centre. It is this latter process that produces the unquenched PC decay with bimolecular kinetics. We can evaluate the fraction of the photoexcited electrons that take part in this latter process. This fraction depends on the rate of filling of the trap distribution during the photoexcitation, that is on the quasi-Fermi level position. We can write

$$E_L = E_F - E_0 , \tag{12}$$

where $E_L$ is the quasi-Fermi level energy for electrons when the sample is excited by light, $E_F$ the Fermi level energy in the dark, and $E_0$ the Fermi level variation due to the light excitation. Assuming $E_F = E_t$ and $E_0$ dependent only on the excitation intensity, we have on the basis of (2),

$$E_L = 12k_BT - E_0 . \tag{13}$$

The fraction of the photoexcited electrons that are in the conduction band at the beginning of PC decay may be written as

$$n_0/N_0 = g(0)/g(E_L) , \tag{14}$$

where $n_0$ is the free electron density at the beginning of the decay, $N_0$ the total density of the photogenerated electrons, $g(0)$ and $g(E_L)$ the densities of trap states per unit energy immediately below the conduction band and at the energy $E_L$ below the conduction band, respectively. It is to be noted that in (14) the Boltzmann factor $\exp(-E_L/k_BT)$ is absent: indeed we deal with a process indirectly controlled by temperature through the density of states $g(E)$, so the trap centres do not act as thermal emission centres. If we assume an exponential distribution, the density of states can be expressed as

$$g(E) = A \exp(-\alpha E) . \tag{15}$$

It follows

$$n_0/N_0 = [g(0)/A] \exp(12\alpha k_B T - E_0) . \tag{16}$$

It is to be noted that $A$ is not, in general, the actual value of $g(E)$ at $E = 0$. Thus the initial number of free electrons depends exponentially on temperature. This fact may be used to interpret the data of Fig. 8. A value $\alpha = 31\ eV^{-1}$ is derived, in good agreement with other measurements [1, 3, 9, 18].

Now we shall examine the case of quenched transients. In this case the non-radiative recombination channel is blocked by the quenching radiation. The whole recombination traffic takes now place through the radiative centre, involving the trap distribution. The PC decay is now ascribed to a dispersive transport due to thermalization of the carriers in the trapping centres. It is known [17] that in a dispersive transport process without recombination we may write for the fraction of photogenerated electrons that are mobile

$$n/N_0 = [g(0)/g(E_d)] \exp(-E_d/k_BT) \tag{17}$$

with the same meanings as before. $E_d$ is a demarcation level defined so that $\nu(E_d)\, t = 1$, where $\nu(E)$ denotes the rate at which electrons are released from states of energy $E$. This means that states at energy $E_d$ are just coming into thermal equilibrium at time $t$. Since

$$\nu(E) = \nu_0 \exp(-E/k_BT) , \tag{18}$$

then

$$E_d = k_BT \ln(\nu_0 t) , \tag{19}$$

where $\nu_0$ is the usual prefactor of order $10^{12}\ s^{-1}$. Of course, $E_d$ shifts away from the bottom of the conduction band as time increases. Assuming an exponential trap distribution for $g(E)$ as in (15), we get

$$n/N_0 = [g(0)/A]\, (\nu_0 t)^{-1+\alpha k_BT} . \tag{20}$$

It must be pointed out, however, that in our case dispersive transport includes recombination of trapped electrons with holes in the acceptor centres (radiative recombination). On the other hand, $E_d$ is also the energy for which the release to the conduction band has the same probability as radiative recombination [19]. Therefore, we can write

$$n/N_0 = [g(0)/g(E_d)] \exp(-2E_d/k_BT) = [g(0)/A]\, (\nu_0 t)^{-2+\alpha k_BT} . \tag{21}$$

The experimental values of $\gamma$ reported in Fig. 10 are in good overall agreement with (21). We find $\alpha = 35\ eV^{-1}$ for the slope of the trap distribution in good agreement with the previous value obtained from the unquenched transients. From this analysis one could conclude that the trap distribution plays a direct role in the radiative process. However, this conclusion has been obtained in an indirect manner. It needs further confirmation by performing simultaneous PC and photoluminescence measurements.

At this point it is to be noted that we performed a simulation of four-gate PICTS spectra on the basis of decays of the type (3) and (4). A PICTS signal quite constant over a wide temperature range was obtained in both the cases according to our interpretation of the PICTS spectra of Fig. 6 and 9 for large values of $t_0$.

As regards PC in the temperature range 230 to 300 K we can point out that in this range there is no quenching of PC, while a thermal quenching of luminescence is observed [7]. On the other hand, PICTS spectra of both quenched and unquenched decays show the presence of a single trapping centre with a very small capture cross section. This means that it is a centre surrounded by a repulsive potential barrier. Its activation energy, about 0.30 eV, has thus to be considered as "apparent": its actual position in the energy gap may be much closer to the conduction band.

## 6. Conclusions

Our investigation of the relaxation processes of PC in $ZnIn_2S_4$ suggests that two competitive recombination channels are active in the temperature range 80 to 200 K. The non-radiative channel, which dominates the unquenched PC, constitutes the slower process and is due to recombination of free photoexcited electrons with holes in a centre lying 2 eV above the valence band. The radiative path, which prevails in the quenched PC, is faster and may be attributed to electron-hole pair transitions from the trapping states of an exponential distribution to the acceptor level. However, the continuous trap distribution affects both recombination channels and plays an essential role in the whole PC process. A correct analysis of both quenched and unquenched transients provides the same slope of the tail edge states.

In the high-temperature range (230 to 300 K), the PC decay is dominated by thermal emission from a repulsive centre with an "apparent" activation energy of 0.30 eV.

Finally, it is to be noticed that the PICTS technique, although unable to furnish the characteristic of the trap distribution, provides useful information about the mechanism of the relaxation process, allowing thus to set the most suitable analysis for the observed mechanism.

***Acknowledgement***

The authors would like to thank Prof. F. Raga for valuable comments on this report.

## References

[1] A. SERPI, J. Phys. D **9**, 1881 (1976).
[2] A. CINGOLANI, M. FERRARA, A. MINAFRA, F. ADDUCI, and P. TANTALO, phys. stat. sol. (a) **23**, 367 (1974).
[3] L. HERNANDEZ, O. VIGIL, and F. GONZALEZ, phys. stat. sol. (a) **36**, 33 (1976).
[4] S. MORA, C. PAORICI, and N. ROMEO, J. appl. Phys. **42**, 2064 (1971).
[5] W. LUDWIG and G. VOGT, phys. stat. sol. **24**, K161 (1967).
[6] N. ROMEO and O. VIGIL, phys. stat. sol. (a) **10**, 447 (1972).
[7] E. GRILLI, M. GUZZI, E. CAMERLENGHI, and F. PIO, phys. stat. sol. (a) **90**, 691 (1985).
[8] M. COLOCCI, F. FERMI, R. QUERZOLI, L. TARRICONE, and A. VINATTIERI, Proc. 7th Conf. Ternary and Multinary Comp., Material Research Society, 1987 (p. 271).
[9] D. CRANDLES and E. FORTIN, phys. stat. sol. (a) **95**, K47 (1986).
[10] J. C. BALLAND, J. P. ZIELINGER, M. TAPIERO, J. G. GROSS, and C. NOGUET, J. Phys. D **19**, 71 (1986).
[11] H. SCHAFER, Z. anorg. allg. Chem. **286**, 27 (1956).
[12] H. SCHAFER, Z. anorg. allg. Chem. **286**, 42 (1956).
[13] S. SHIONOYA and A. EBINA, J. Phys. Soc. Japan **19**, 1150 (1964).
[14] S. GOLDBERG, Probability, Prentice-Hall Inc., New Jersey 1962 (p. 232).
[15] M. POLLAK, Phil. Mag. **36**, 1157 (1977).
[16] T. TIEDJE and A. ROSE, Solid State Commun. **37**, 48 (1981).
[17] J. ORESTEIN and M. A. KASTNER, Phys. Rev. Letters **46**, 1421 (1981).
[18] M. GUZZI and G. BALDINI, J. Lum. **9**, 514 (1975).
[19] R. A. STREET, Adv. Phys. **30**, 593 (1981).

*(Received January 22, 1988)*

phys. stat. sol. (a) **108**, 363 (1988)

Subject classification: 72.10; 72.15; S1.4

*Institute for Low Temperature and Structure Research, Polish Academy of Sciences, Wrocław*[1])

# Resistance of Some Lanthanum and Lutetium Compounds of the Type (La, Lu)$Me_3$

By
Z. Kletowski

Temperature dependences for four nonmagnetic compounds of the type (La, Lu)$Me_3$ are measured and compared with those calculated according to Grüneisen-Bloch formula. The results show that the lanthanum compounds exhibit more clearly a d character of the electrical resistivity than the lutetium ones. Additionally the $LaSn_3$ shows a poor similarity to the simple metal.

Ein Vergleich der gemessenen mit den aus der Grüneisen-Bloch-Relation berechneten Werten des Widerstandes wird für vier nichtmagnetische Verbindungen des Typs (La, Lu)$Me_3$ durchgeführt. Die Ergebnisse weisen darauf hin, daß die Lanthanum-Verbindungen einen ausgeprägteren d-Charakter des elektrischen Widerstandes als die Lutetium-Verbindungen haben. Die $LaSn_3$-Verbindung weist nur geringe Ähnlichkeit zum einfachen Metall auf.

## 1. Introduction

Intermetallic compounds with stoichiometry $REMe_3$ (where RE is a rare earth and Me = Sn, In, or Ga) have been intensively investigated recently because of their 4f shell character and consequent effects such as a crystal field, mixed valence, Kondo behaviour, or complicated phase transitions observed. Among RE elements only two, i.e. lanthanum and lutetium, do not possess the well-defined f electrons and therefore compounds containing either element are often treated as reference standards for the magnetic $REMe_3$ compounds and are sometimes called "nonmagnetic" compounds. In this paper we present an analysis of the temperature dependence of the resistivity $(\varrho(T))$ for four of such nonmagnetic compounds, i.e. for $LaSn_3$, $LaIn_3$, $LuIn_3$, and $LuGa_3$. This analysis is based on a comparison of the measured $\varrho(T)$ dependences with those calculated according to the Grüneisen-Bloch (GB) relation. To our best knowledge such an analysis has not been performed for (La, Lu)$Me_3$ compounds so far, although transport properties of $LaSn_3$ and $LaIn_3$ are relatively well known [1, 2, 3]. Resistivity data for both lutetium compounds are known fragmentary only [4].

The investigated compounds crystallize with a $AuCu_3$ crystallographic type of structure and display metallic character of conductivity. Electronic structure calculations [5, 6] performed for $LaSn_3$ and $LaIn_3$ showed that the Fermi level lies at the vicinity of a sharp peak in the density of states and the conduction band consists of s and p states. However, the lanthanum d states can hybridize with Sn p states modifying this p band. A lack of the 4f magnetic electrons causes the total resistance observed in these compounds to result from two scattering mechanisms only: electron–phonon and electron–residual crystal imperfection scattering. Since the conduction band possesses predominantly s and p character, the GB relation seems to be applicable to describe the phonon contribution. Deviations from this law would indicate the appearance of additional types of scattering mechanisms apart from that assumed in the GB approximation.

[1]) Pl. Katedralny 1, PL-50-950 Wrocław, Poland.

## 2. Experimental and Results

The samples used in these investigations were cut out from a bulk single crystal grown by slow cooling of the metal. More details concerning the single crystal preparation are available in a previous paper [7]. The samples had the form of rectangular prisms with dimensions of 0.3 × 0.3 × 4.0 mm³. The length of the $LuGa_3$ sample was shorter and equal to 2 mm. Metallographic and X-ray examinations confirmed the single crystal form and the lattice constants of the samples. A conventional four probe dc method was used to measured the resistivity.

Temperature dependences of the resistivity are shown in Fig. 1. Curves are similar to each other and display the typical metallic character of resistivity. Both compounds with the lighter RE element (lanthanum) exhibit higher resistivity than those with the heavier one (lutetium). A comparison of the observed and theoretically calculated lattice resistances is given in Fig. 2. The resistance in the GB approximation was calculated according to the well-known formula

$$\varrho(T) = \frac{C}{M\theta_\mathrm{R}} \left(\frac{T}{\theta_\mathrm{R}}\right)^5 \int_0^{\theta_\mathrm{R}/T} \frac{z^5\,\mathrm{d}z}{(\mathrm{e}^z - 1)(1 - \mathrm{e}^{-z})}\,, \qquad (1)$$

where $M$ is the atomic weight and $C$ a constant. $\theta_R$ is a temperature characteristic of the metal lattice resistivity analogous to the Debye characteristic temperature $\theta_\mathrm{D}$ charactetistic of the solid lattice specific heat. A horizontal straight line in this figure corresponds to zero deviation from the GB law. The curves show the least deviations of the experimentally found values $\varrho(T)$ from those calculated from the GB law for different $\theta_R$ parameter. The least-squares procedure was used to obtain such fits. Values of $\theta_R$ obtained in such a way are listed in Table 1. The best fitting is observed at higher temperatures, where the temperature dependence of $\varrho(T)$ is nearly linear. As one could expect the fitting becomes rapidly worse for all compounds at temperatures below $\theta_R/5$. Both lutetium compounds show less discrepancies than the lanthanum compounds. Especially, for $LaSn_3$ the resistivity deviates strongly from the GB relation in the whole investigated temperature range. In the low temperature region the com-

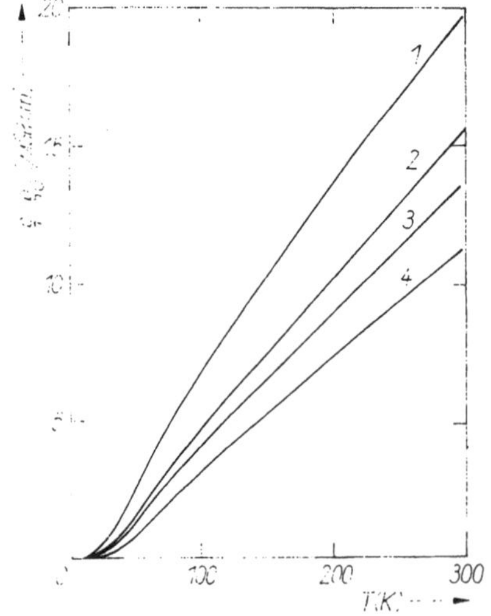


Fig. 1. Temperature dependences of the restistivity of the compounds investigated. The residual part is subtracted. (1) $LaSn_3$, (2) $LaIn_3$, (3) $LuIn_3$, (4) $LuGa_3$

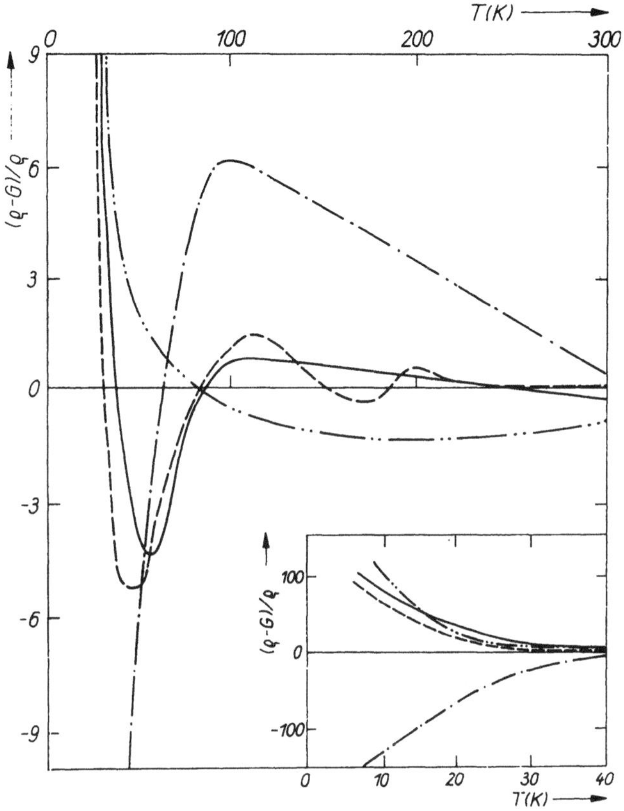


Fig. 2. Temperature variations of the least deviations from the Grüneisen-Bloch relation (in %), $\varrho$ stands for experimental value of the resistivity and $G$ represents the resistance calculated according to (1). For $\theta_R$ the values listed in Table 1 were used. — — — $LuIn_3$, ——— $LuGa_3$, —·—·— $LaSn_3$, —··— $LaIn_3$

pounds investigated exhibit a parabolic change of $\varrho$ with temperature. The values of the power exponents found experimentally range between 2.2 and 3.6 (see Fig. 3 and Table 1).

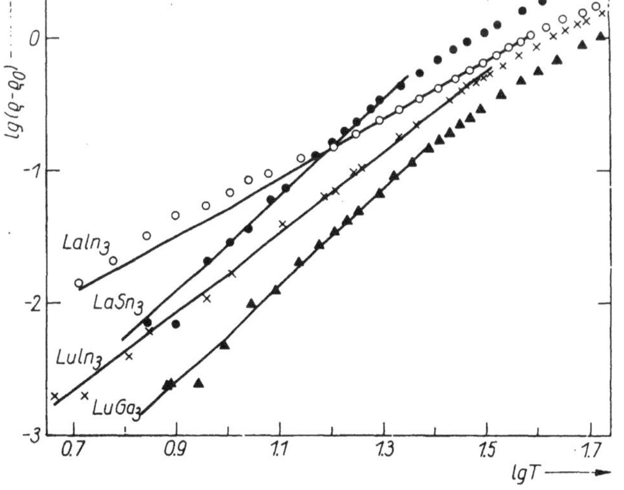


Fig. 3. A lg–lg plot of resistivity vs. temperature at low temperatures. Estimated power exponents: $LaSn_3$ (●) 3.6; $LaIn_3$ (○) 2.2; $LuIn_3$ (×) 3.0; $LuGa_3$ (▲) 3.7

Table 1

| compound | $\varrho_{300}$ (μΩcm) | $\varrho_{300}/\varrho_{4.2}$ | low temperature exponent | mean square deviation (%) | $\theta_R$ (K) | $\theta_D$ (K) | $\theta_R^*$ (K) |
|---|---|---|---|---|---|---|---|
| $LaSn_3$ | 19.5 | 31 | 3.6 | 1.59 | 82[a]) | | 178 |
| $LaIn_3$ | 15.7 | 83 | 2.2 | 0.64 | 153 | 170 | 137 |
| $LuIn_3$ | 13.6 | 37 | 3.0 | 0.26 | 151 | | 151 |
| $LuGa_3$ | 11.2 | 20 | 3.7 | 0.22 | 187 | | 171 |

$\theta_D$ is the Debye temperature obtained in specific heat measurements [9].
[a]) See comment in text.

## 3. Conclusions

We interpret the discrepancies observed between experimental and calculated data of the resistivity for the nonmagnetic $REMe_3$-type compounds as the consequence of a contribution of the d electrons to the scattering processes. Our result indicates that this contribution is much more effective in the lanthanum compounds and manifests itself in stronger deviations from the GB relation. Extremely high deviations are observed for $LaSn_3$ where a negative curvature of $\varrho(T)$ is visible at temperatures $T > \theta_R$ as well. In our opinion it suggests the existence of an additional scattering process in $LaSn_3$. Most probably this is an s–d interband scattering (Mott scattering).

The values of $\theta_R$ obtained in our measurements are lower than those obtained in specific heat studies. The $\theta_R$ values seem to be reasonable except for $LaSn_3$. The value of $\theta_R$ estimated for this compound has only formal meaning. According to a simple formula given by Meaden [8],

$$\theta_R^* = 48.2 \left(\frac{\varrho_{295}}{\varrho_{20}}\right)^{1/4}, \tag{2}$$

one would rather expect a value of about 180 K for $LaSn_3$.

It is also worth mentioning that two other experimental facts become evident for the $LaIn_3$ compound. Firstly, as is seen in Fig. 3, the experimental points deviate from the straight line below 15 K. It coincides with an anomalous dependence of the thermal expansion coefficient observed lately for $LaIn_3$ at temperatures below 20 K [10]. That fact was interpreted as a possible effect of the hybridization of d or f states with the conduction band. Moreover, we would like to point out that the results of both resistivity and specific heat measurements were carried out on samples originating from the same single crystal and an influence of impurities cannot be excluded.

Secondly, the value of 2.2 of the power exponent found in this work for $LaIn_3$ differs from the value of 3 obtained previously for a polycrystalline sample [1]. Such a low exponent is seldom observed even for transition metals. It is well known from theory [11] that the exponent 2 is expected for the s–d electron–electron Coulomb-type scattering process. This would be an explanation of the observed low power exponent if current carriers with large effective masses are present in $LaIn_3$. Although the electron band structure calculations [6] indicate this possibility in the sixth Brillouin zone, the available de Haas-van Alphen investigations [12] because of their incompleteness do not solve this question.

## References

[1] W. D. GROBMAN, Phys. Rev. B **5**, 2924 (1972).
[2] B. STALIŃSKI, Z. KLETOWSKI, and Z. HENKIE, phys. stat. sol. (a) **19**, K165 (1973).
[3] Z. KLETOWSKI, P. J. MARKOWSKI, and B. STALIŃSKI, Solid State Commun. **57**, 293 (1986).
[4] Z. KLETOWSKI and P. J. MARKOWSKI, Solid State Commun. **62**, 299 (1987).
[5] A. HSAEGAWA, J. Phys. Soc. Japan **50**, 3313 (1981).
[6] A. HASEGAWA, Crystalline Electric Field Effects in f-Electron Magnetism, Ed. K. P. GUERTIN, W. SUSKI, and Z. ZOŁNIEREK, Plenum Press, New York 1982 (p. 201).
[7] Z. KLETOWSKI, N. ILIEV, Z. HENKIE, and B. STALIŃSKI, J. less-common Metals **110**, 235 (1985).
[8] G. T. MEADEN, Electrical Resistance in Metals, Heywood Books, London 1966 (p. 107).
[9] A. M. VAN DIEPEN, R. S. CRAIG, and W. E. WALLACE, J. Phys. Chem. Solids **32**, 1867 (1971).
[10] A. CZOPNIK, R. POTT, H. MÄDGE, N. ILIEV, and B. STALIŃSKI, phys. stat. sol. (a) **93**, K137 (1986).
[11] W. G. BEBER, Proc. Roy. Soc. **A158**, 383 (1937).
[12] H. KITAZAWA, Q. Z. GAO, H. SHIDA, T. SUZUKI, H. HASEGAWA, and T. KASUYA, J. Magnetism magnetic Mater. **52**, 286 (1985).

*(Received January 19, 1988)*

phys. stat. sol. (a) **108**, 369 (1988)

Subject classification: 72.15; 75.80; S1.1

*Department of Physics and Center for Advanced Materials Processing, Clarkson University, Potsdam*[1])

# Stress Dependence of Magnetoresistance in Crystalline and Amorphous Ferromagnets

By

A. RAOUFI and S. ARAJS

A detailed study of the influence of tensile stress on the electrical magnetoresistivity of polycrystalline Fe of high purity and glassy alloys Metglas 2605SC, 2605CO, 2826MB, 2705M, and 2705X is reported at 300 K. The stress dependence of magnetoresistance can be understood by the domain theory in both, crystalline and amorphous ferromagnets. While there exists an equivalence between magnetic field and stress in Fe samples, the glassy alloys, however, do not obey this equivalence principle. In addition, the magnetoresistance of zero magnetostrictive alloys is insensitive to the applied stress. Furthermore, the ferromagnetic anisotropy of resistivity (FAR) is found to be approximately stress-independent in both, Fe samples as well as in amorphous metallic alloys.

Eine ausführliche Untersuchung des Einflusses von Dehnungsspannung auf den elektrischen Magnetowiderstand von hochreinem polykristallinem Eisen und glasartigen Legierungen, Metglas 2605SC, 2605CO, 2826MB, 2705M und 2705X wird bei 300 K durchgeführt. Die Spannungsabhängigkeit des Magnetowiderstandes läßt sich durch die Domänentheorie sowohl in kristallinen als auch in amorphen Ferromagneten verstehen. Während eine Äquivalenz zwischen Magnetfeld und Spannung in Fe-Proben existiert, befolgen die glasartigen Legierungen dieses Äquivalenzprinzip nicht. Überdies ist der Magnetowiderstand von Legierungen mit verschwindender Magnetostriktion unempfindlich gegenüber der angelegten Spannung. Darüberhinaus wird gefunden, daß die ferromagnetische Anisotropie des Widerstands (FAR) näherungsweise spannungsunabhängig sowohl in Fe-Proben als auch in den amorphen Metallegierungen ist.

## 1. Introduction

Recently there has been a strong interest in the effects of magnetic field and tensile stress on the electrical resistivity of amorphous alloys [1 to 4]. However, detailed studies of stress-dependence of magnetoresistance in amorphous in comparison with similar studies in crystalline materials are not very extensive. For this reason, we initiated a new investigation of the effects of magnetic field and tensile stress on the electrical resistivity of some metallic glasses and crystalline Fe samples at 300 K.

## 2. Experimental Considerations

The amorphous metallic systems used in this study were obtained from Allied Chemical Corporation, Morristown, New Jersey. Table 1 gives their trade names and nominal compositions. These materials, produced by melt spinning, were in the form of thin ribbons. Specifically, metglas 2705X were approximately 2.54 cm wide and 0.0038 cm thick. The other composition ribbons had typically 0.3 to 0.5 cm width and 0.001 cm thickness.

[1]) Potsdam, NY 13676, USA.

Table 1
Compositions of glassy metals used in this study

| metglas | nominal composition |
|---|---|
| 2826MB | $Fe_{40}Ni_{38}Mo_4B_{18}$ |
| 2605SC | $Fe_{81}B_{13.5}Si_{3.5}C_2$ |
| 2605CO | $Fe_{66}Co_{18}B_{15}Si_1$ |
| 2705X | $(Co_{.925}Fe_{.075})_{78}B_{15}Mo_2Si_5$ |
| 2705M | $Co_{69}Fe_4Ni_1Mo_2B_{12}Si_{12}$ |

The Fe samples were in the form of wires, approximately 0.1 cm in diameter and with different lengths. They were of high purity (99.99%) and were prepared by an oxidation-zone refining technique [5].

The experimental data, with the long axis of ribbons perpendicular or parallel to the magnetic field, were obtained using an automated direct current comparator resistance bridge, Model 9975, Guildline Instrument, Inc. This bridge compares four terminal resistors by measuring the current ratio corresponding to voltage drop equality. It reads the ratio of the resistance under the test to a standard resistance directly. Therefore, this technique greatly eliminates the problems associated with a conventional potentiometric method and increases the precision in electrical resistivity measurements up to eight digits. The above facility has been automated by means of a DSP-1 microcomputer (S-100 Bus) with a main frame manufactured by Ithaca Intersystem, Inc. It consists of a power supply, a mother board, front panel switches, and a 4 MHz Z-100 microprocessor board. A combination of two memory boards (total of 56 K) manufactured by Godbout Electronics was used. Both boards have been designed to run full speed at 4 MHz. The microcomputer was interfaced with two Keithley multimeters.

## 3. Results and Discussion

Fig. 1a shows the relative change in the electrical resistivity, $(\Delta\varrho/\varrho)_H$, of metglas 2605CO as a function of longitudinal magnetic field, $H$, for different tensile stresses. The quantity $(\Delta\varrho/\varrho)$ is defined by

$$\left(\frac{\Delta\varrho}{\varrho}\right)_H = \frac{\varrho(H, \sigma) - \varrho(0, \sigma)}{\varrho(0, \sigma)},$$

where $H$ (either $H_{\parallel}$ or $H_{\perp}$) and $\sigma$ represent the magnitude of the magnetic field and stress, respectively. A similar plot for polycrystalline Fe is given in Fig. 1b. Both of these figures indicate that $(\Delta\varrho/\varrho)_{H_{\parallel}}$ increases with $H_{\parallel}$ reaching a saturation value at higher values of $H_{\parallel}$. Similar plots for the above materials with the magnetic field $H_{\perp}$ perpendicular to the long axis of the ribbon (parallel to the ribbon plane) are shown as Fig. 2 a and b. In these last cases the quantity $(\Delta\varrho/\varrho)_{H_{\perp}}$ is negative. It also reaches a saturation value but at higher values of $H_{\perp}$. Quantitatively, the behavior of other ferromagnetic metglas samples were found to be similar to the results for MG 2605CO presented above.

Since, according to the domain theory the electrical resistivity of a ferromagnet depends on the angle between the magnetization and the electric current, in the longitudinal field both processes of domain wall displacements and rotations occur

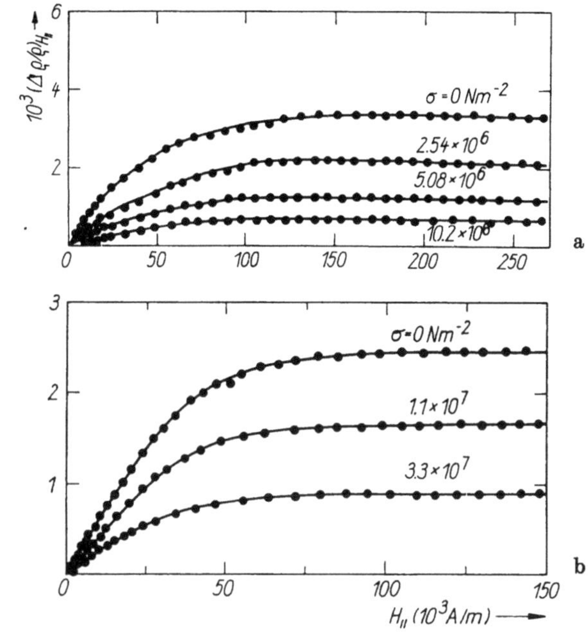


Fig. 1. Relative change in the electrical resistivity of a) metglas 2605CO and b) Fe sample as a function of longitudinal magnetic field for different tensile stresses

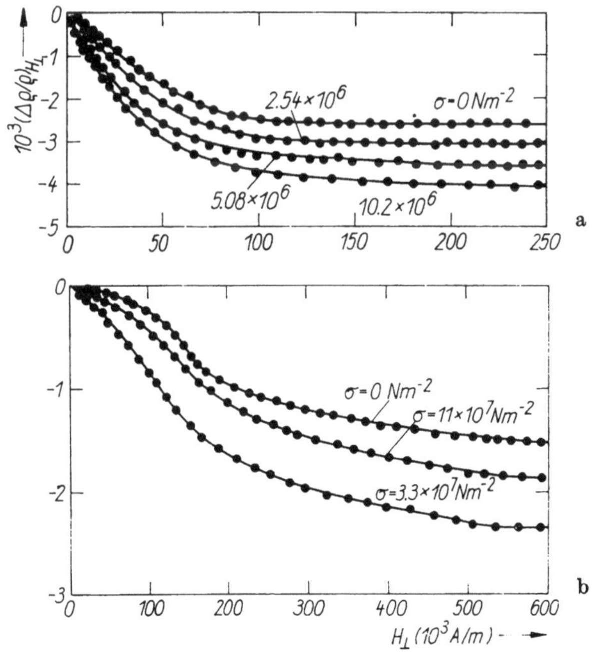


Fig. 2. Relative change in the electrical resistivity of a) metglas 2605CO and b) Fe sample as a function of transverse magnetic field for different tensile stresses

simultaneously. In particular, the domain rotation saturates in a lower field than that needed for the saturation of the magnetization. Therefore, the longitudinal magnetoresistivity, which depends on magnetization, reaches its saturation value in a rather lower field. In the transverse field the domain wall displacements, which are not effective in a magnetoresistivity change, prevail at lower fields, whereas the domain rotation becomes dominant at higher fields. Therefore, the magnetoresistivity reaches its saturation value at higher fields.

Fig. 1 and 2 also show that the application of tensile stress causes the saturation value of $\Delta\varrho/\varrho$ to decrease both in the longitudinal and transverse magnetic fields. This implies that the effect of stress on the magnetoresistance of these ferromagnets also can be interpreted by the domain theory. In fact, a tensile stress parallel to the long axis of the ribbon, or more exactly in the direction of the electric current, induces a preferred domain orientation parallel to the stress axis. When a magnetic field is applied thereafter to the stressed ribbon, the change in magnetoresistivity will increase or decrease depending on the field direction. In the longitudinal orientation the saturation value of $(\Delta\varrho/\varrho)_{\parallel}$ decreases with increasing value of $\sigma$. This results because the tension already has produced a rotation of the domains toward the current direction. Therefore, there is a decrease in the average angle between the domain magnetization vectors and the electric current, leading to an increase in magnetization in the remanent state at $H_{\parallel} = 0$. On the other hand, a transverse field $H$ competes with the tensile stress in regard to domain alignment. In this case the stress tends to align the domain magnetization parallel to the current while the magnetic field tends to turn the domain perpendicular to the current, resulting in a decrease in $(\Delta\varrho/\varrho)_{\perp}$.

In the case of amorphous alloys with zero magnetostriction, such as metglas 2705X and 2705M, it was found that the magnetoresistance is independent of the applied stress as can be seen in Fig. 3, where the longitudinal and transverse magnetoresistivities are shown as a function of tensile stress for MG 2705X. Within the experimental error, the quantity $(\Delta\varrho/\varrho)_\sigma$ in both cases is zero. The quantity $(\Delta\varrho/\varrho)_\sigma$ is defined as

$$\left(\frac{\Delta\varrho}{\varrho}\right)_\sigma = \frac{\varrho(H,\sigma) - \varrho(H,0)}{\varrho(H,0)},$$

where, as before, $H$ can be $H_{\parallel}$ or $H_{\perp}$. A similar behavior to that shown in Fig. 3 was also observed for MG 2705M. In order to clearly see the effects of the tensile stress on the ferromagnets, the samples were exposed to selected constant magnetic fields and the relative changes in the electrical resistivity due to tensile stress $\sigma$ were measured. Fig. 4 a and b show $(\Delta\varrho/\varrho)_H$ versus $\sigma$ for metglas 26050CO in different constant magnetic fields. These figures clearly indicate that, first $(\Delta\varrho/\varrho)_H$ increases linearly with $\sigma$ in both longitudinal and transverse magnetic fields, and, second, that the presence of the applied magnetic field plays a minor role in determining the resistivity increase due to $\sigma$. On the other hand, Fig. 5 a and b show $(\Delta\varrho/\varrho)_\sigma$ versus $\sigma$ for an Fe sample where $(\Delta\varrho/\varrho)_\sigma$ both in the longitudinal and transverse magnetic fields increases with increasing stress and eventually reaches a saturation value. This behavior, which also

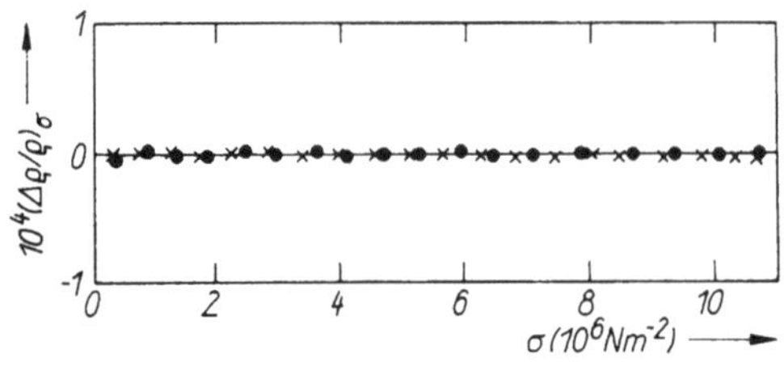


Fig. 3. The longitudinal (●) and transverse (×) magnetoresistivities of zero magnetostrictive alloy metglas 2705X as a function of tensile stress

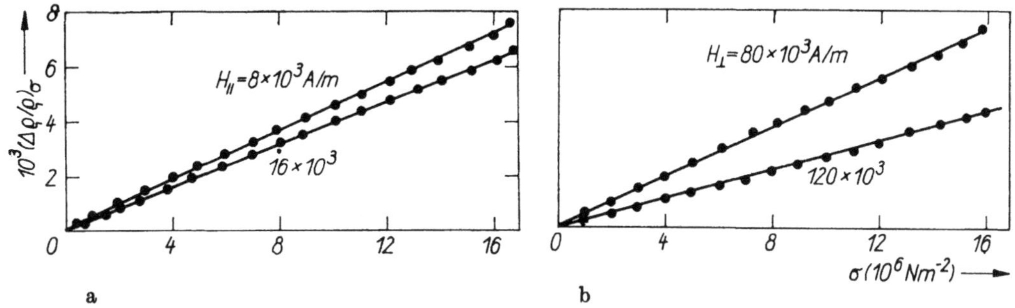


Fig. 4. Relative change in the a) longitudinal and b) transverse magnetoresistivity of metglas 2605CO as a function of tensile stress in constant magnetic fields

exists, for example, in crystalline Fe–Ni alloys [6] is known as the equivalence principle between the electrical resistivity changes resulting from the magnetic fields and tension.

From the data obtained in the present work, the values of ferromagnetic anisotropy of resistivity (FAR) have been determined. This quantity is defined by $(\varrho_{\parallel} - \varrho_{\perp})/\varrho_0$, where $\varrho_0$ is the resistivity in zero internal magnetic field $H_{\text{int}}$, $\varrho_{\parallel}$ and $\varrho_{\perp}$ are the longitudinal and transverse magnetoresistivities extrapolated to $H_{\text{int}} = 0$. Typical values for FAR in glassy alloys and Fe samples were found to be $5.7 \times 10^{-3}$ and $4.4 \times 10^{-3}$, respectively. Furthermore, FAR also seemed to be approximately independent of the applied stress.

## 4. Conclusions

From our experimental results, it is concluded that the behavior of the electrical magnetoresistivity of ferromagnetic metallic glasses and that of crystalline Fe samples can be best interpreted by the domain theory. From the comparison of the magneto-

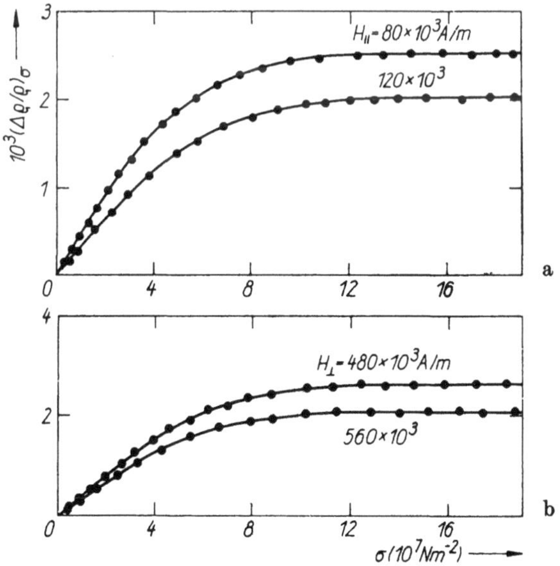


Fig. 5. Relative change in the a) longitudinal and b) transverse magnetoresistivity of Fe sample as a function of tensile stress in constant magnetic fields

resistivity and its stress dependence, it is concluded that both the amorphous ferromagnetic glassy alloys and crystalline Fe samples are analogous to each other in regard to their magnetoresistance behavior. This results because the magnetic energies, responsible for the domain structure of amorphous ferromagnets, are essentially independent of their amorphous character. In fact, the major difference between the electric and magnetic properties of amorphous and crystalline ferromagnets comes from the change in the electronic environment around each atom caused by the metalloid but not by the structure disorder [7]. The striking difference between the two systems is the fact that the equivalence between magnetic field and tensile stress, which exists for crystalline Fe, is not satisfied by the amorphous ferromagnets. This might be due to an anisotropy in the magnetostriction in the amorphous alloys with positive magnetostriction.

Owing to their zero magnetostriction, the ferromagnetic glassy alloys metglas 2705M and 2705X have smaller saturation values of magnetoresistivity as compared to those of positive magnetostrictive alloys and Fe samples. These smaller saturation values imply that magnetic anisotropy is minimum in these alloys. Furthermore, they were also found to be independent of applied tensile stress in both the longitudinal and transverse magnetic fields. This observation has an interesting implication. The stress insensitivity of these alloys, coupled with the small magnetic anisotropy, make them candidates for excellent soft magnetic devices.

Finally, from the comparison of the data of FAR of Fe and glassy samples, it is observed that the FAR values are much smaller in glassy alloys than those of Fe samples. This is an additional evidence that the magnetic anisotropies are considerably smaller in amorphous alloys than in crystalline materials. Therefore, it is concluded that studies of anisotropic magnetoresistance effect in amorphous alloys can be used as a supplementary method for determining possible magnetic anisotropies in these systems. Furthermore, FAR was found to be approximately independent of the applied stress in both the crystalline Fe and amorphous samples.

## References

[1] K. KERN and V. GONSER, J. Magnetism magnetic Mater. **21**, 182 (1980).
[2] R. AIDUN, S. ARAJS, E. K. DOKYI, and M. C. MARTIN, J. appl. Phys. **53**, 8243 (1982).
[3] A. RAOUFI and S. ARAJS, phys. stat. sol. (a) **95**, 193 (1986).
[4] A. RAOUFI and S. ARAJS, Physica Scripta **35**, 62 (1986).
[5] B. F. OLIVER, F. GAROFALO, R. PRIESTNER, and S. ARAJS, Proc. Symp. Micrometallurgy, National Metallurgical Laboratory, Jamshedpur (India) 1968 (p. 45).
[6] R. M. BOZORTH, Ferromagnetism, van Nostrand Co, 1955 (p. 745).
[7] F. F. LUBORSKY, Ferromagnetic Materials, Ed. E. P. WOHLFARTH, North-Holland Publ. Co., 1980 (p. 451).

*(Received March 18, 1988)*

phys. stat. sol. (a) **108**, 375 (1988)

Subject classification: 75.60; 75.80; S1.1; S1.4

*Institute of Molecular Physics, Polish Academy of Sciences, Poznań*[1])

# Properties of Amorphous $Sm_xFe_{80-x}B_{20}$ Ribbons

By

B. IDZIKOWSKI and A. WRZECIONO

Properties of amorphous $Sm_xFe_{80-x}B_{20}$ ribbons are investigated. Saturation magnetization and Curie temperature decrease whereas the crystallization temperature increases with increasing samarium content. Saturation magnetostriction varies from $37 \times 10^{-6}$ for $x = 1$ to $12 \times 10^{-6}$ for $x = 9$. The coercivity decreases with decreasing magnetostriction. The domain structure is also studied.

Es werden die Eigenschaften der amorphen Legierung $Sm_xFe_{80-x}B_{20}$ untersucht. Die Sättigungsmagnetisierung und Curie-Temperatur fallen und die Kristallisationstemperatur wächst mit $x$. Die Sättigungsmagnetostriktion variiert von $37 \times 10^{-6}$ für $x = 1$ bis $12 \times 10^{-6}$ für $x = 9$. Die Koerzitivkraft verringert sich, wenn die Magnetostriktion kleiner wird. Es wird auch die Bereichstruktur untersucht.

## 1. Introduction

Amorphous magnetic alloys containing rare earth elements are of scientific and technological interest for several reasons.

In crystalline alloys the rare earth ions produce large anisotropies and high magnetostrictive strains. Since amorphous alloys are nearly isotropic, the average value of the anisotropy should be very low. It may therefore be possible to create magnetically soft amorphous alloys with isotropic magnetostriction.

The melt spinning technique can be used to produce permanent magnetic materials of the R–Fe–B type which show a microcrystalline structure having excellent hard magnetic properties [1, 2].

## 2. Experimental

Amorphous alloys of $Sm_xFe_{80-x}B_{20}$ were prepared over the range $0 \leqq x \leqq 6$. The purity of the starting materials was Sm 99.9, Fe 99.998, and B 99.8%. Sm–Fe–B ternary alloys were obtained by melting appropriate amounts of the constituents in a high-frequency induction furnace in a cold crucible to a homogeneous button in argon atmosphere. The molten alloy was then rapidly quenched using a Cu disk. Argon ejection pressure of 7 to 10 kPa and a substrate surface speed of 40 m/s were employed. The gap width of the quartz crucible was 270 μm. The ribbons were 30 to 40 μm thick and approximately 1.5 mm wide.

The amorphous nature of the samples was established by X-ray diffraction using CuKα radiation. Saturation magnetization was measured at 77 and 300 K. The Curie temperatures were determined from the curves of squared magnetization $\sigma^2$ versus temperature $T$, as well as from the diagrams of differential thermal analysis (DTA) at heating rates of 7.5 and 11 K/min, respectively. Saturation magnetostriction was measured by the three-terminal capacitance method at room temperature.

[1]) Smoluchowskiego 17, PL-60-179 Poznań, Poland.

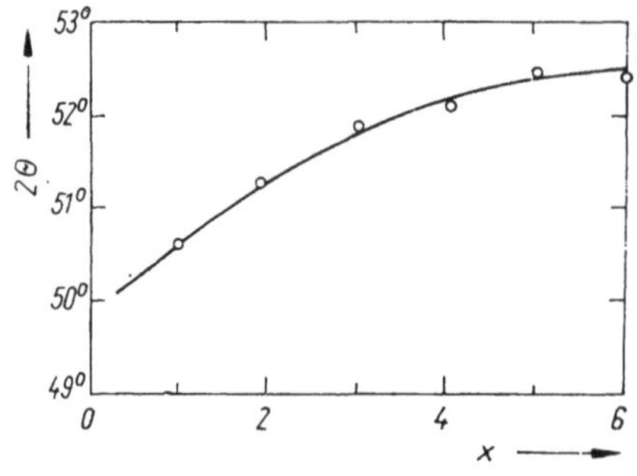


Fig. 1. The "top angle", $2\theta$, of the diffused broad maximum of amorphous $Sm_xFe_{80-x}B_{20}$ alloys as a function of Sm content $x$

Quasi-static hysteresis loops were obtained by means of a loop tracer. The frequency of magnetic reversal was 0.02 Hz. Coercivity was measured before and after annealing of the ribbons at a temperature of 300 °C.

The domain structure was observed on the surfaces of the as-quenched ribbons by the powder method. The surfaces were not polished mechanically or electrolytically.

## 3. Results

### *3.1 X-ray analysis*

X-ray diffraction patterns of Sm–Fe–B ribbons in as-quenched state exhibit a diffused broad maximum, which is typical for amorphous solids. With an increase of samarium concentration a shift of this maximum to higher values was observed (Fig. 1). Fully amorphous alloys $Sm_xFe_{100-x}$ were not obtained for $x < 17$ [3].

Fig. 2 shows the X-ray diffraction pattern for sample $x = 6$ after being annealed for one week at 900 °C in argon atmosphere. As seen from the figure, the annealed ribbon showed sharp lines corresponding to $Sm_2Fe_{17}$, $Fe_3B$, and α-Fe.

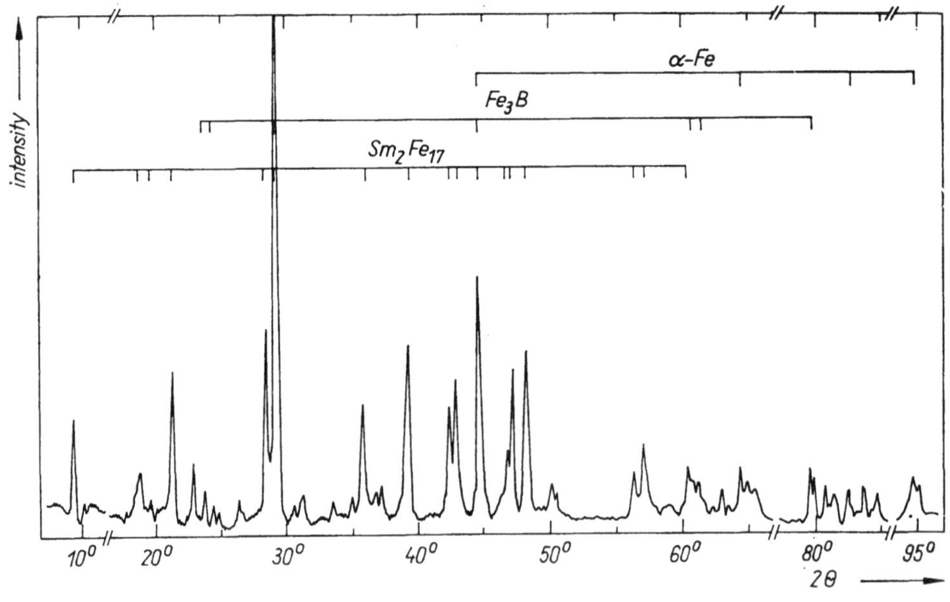


Fig. 2. CuKα X-ray diffraction pattern for annealed (900 °C, one week) amorphous $Sm_6Fe_{74}B_{20}$ alloy

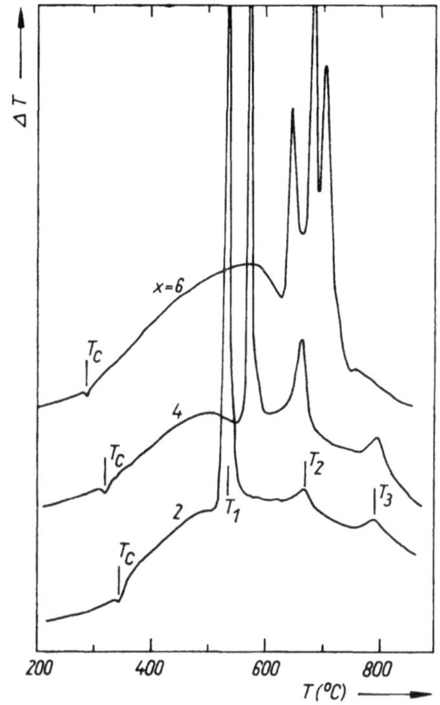


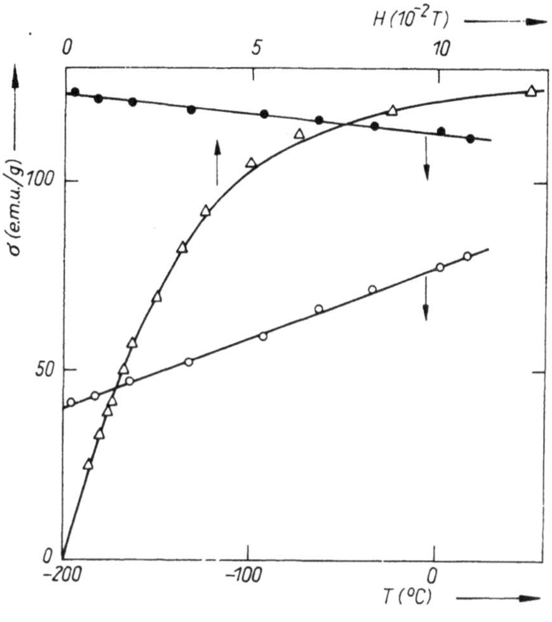


Fig. 3 Fig. 4

Fig. 3. DTA traces of amorphous $Sm_xFe_{80-x}B_{20}$ alloys; $x = 2$, 4, and 6

Fig. 4. Magnetization of $Sm_5Fe_{75}B_{20}$ at 0.0125 (○) and 0.125 T (●) as a function of temperature and magnetization at 77 K as a function of external field (△)

### *3.2 DTA diagrams*

In general three exothermic heat effects were observed in the DTA traces. The first exothermic peak temperature in the DTA traces has been interpreted as crystallization temperature, $T_{cr}$. Some DTA diagrams are shown in Fig. 3 with a heating rate of 11 K/min. The results of DTA studies are summarized in Table 1.

In all cases the three step crystallization behaviour ($T_1$, $T_2$, and $T_3$) is preserved. The distance between the three thermal events is seen to become smaller as the samarium concentration is increased.

Table 1

The Curie $T_C$ and transformation temperatures $T_1$, $T_2$, $T_3$ (all in °C) observed in DTA experiments (heating rate 11 K/min) for the Sm–Fe–B system

| alloy | $T_C$ | exothermic | | |
|---|---|---|---|---|
| | | $T_1 = T_{cr}$ | $T_2$ | $T_3$ |
| $Sm_1Fe_{79}B_{20}$ | 372 | 518 | 672 | 824 |
| $Sm_2Fe_{78}B_{20}$ | 350 | 549 | 681 | 796 |
| $Sm_3Fe_{77}B_{20}$ | 334 | 570 | 680 | 801 |
| $Sm_4Fe_{76}B_{20}$ | 319 | 584 | 673 | 792 |
| $Sm_5Fe_{75}B_{20}$ | 296 | 625 | 693 | 776 |
| $Sm_6Fe_{74}B_{20}$ | 278 | 655 | 693 | 733 |

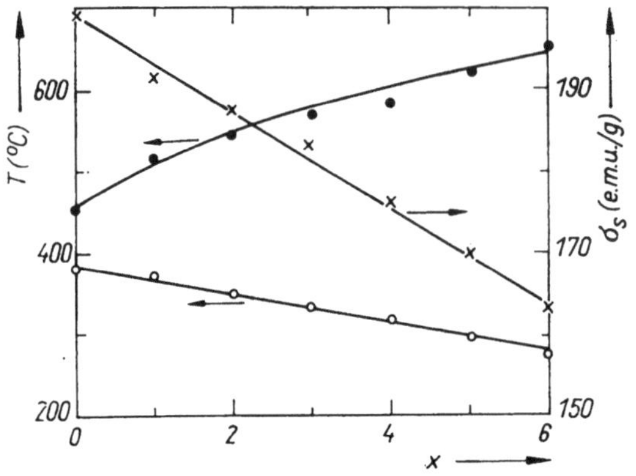


Fig. 5. Curie (○) and crystallization (●) temperatures and saturation magnetization (×) (at room temperature) as functions of Sm content of amorphous Sm–Fe–B alloys

### *3.3 Magnetization, crystallization temperature*

Fig. 4 shows the temperature dependence of magnetization measured in an external field of 0.0125 T and in a field ten times higher. In the field of 0.0125 T the magnetization increases with increasing temperature, in the field of 0.125 T magnetization decreases with increasing temperature. This result is analogous to the behaviour of Nd–Fe amorphous alloys [4].

Fig. 5 shows the concentration dependences of Curie and crystallization temperature and the concentration dependence of saturation magnetization.

### *3.4 Hysteresis loops, coercivity, magnetostriction*

The hysteresis curves were obtained for all samples in the as-quenched state and after annealing at 300 °C. The annealing time was 3 and 5 h. Fig. 6 shows an example of hysteresis loops before and after annealing. All results are summarized in Table 2.

The coercive force is very structure sensitive. Annealing causes a decrease of $H_c$, which is related to the stress relaxation.

The saturation magnetostriction decreases from $37 \times 10^{-6}$ for $Sm_1Fe_{79}B_{20}$ to $12 \times \times 10^{-6}$ for $Sm_9Fe_{71}B_{20}$.

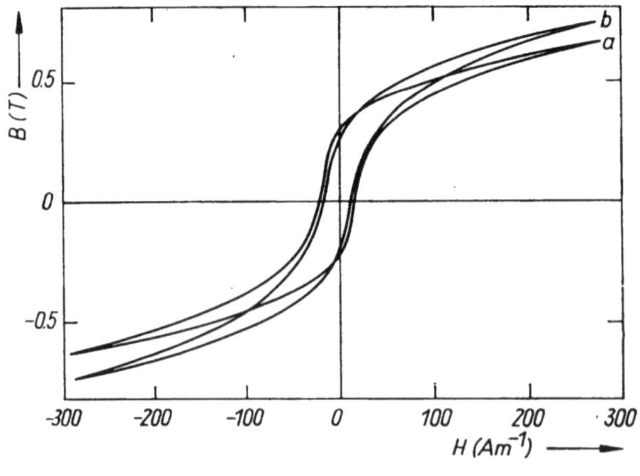


Fig. 6. Hysteresis loops of $Sm_6Fe_{74}B_{20}$ (a) before and (b) after annealing (5 h, 300 °C) at room temperature

Table 2

Coercivity $H_c$ (in A/m) of as-quenched and of annealed amorphous Sm–Fe–B alloys

| alloy | as-quenched | after annealing at 300 °C | |
|---|---|---|---|
| | | 3 h | 5 h |
| $Fe_{80}B_{20}$ | 5.5 | — | — |
| $Sm_1Fe_{79}B_{20}$ | 32.5 | 27.7 | 25.2 |
| $Sm_2Fe_{78}B_{20}$ | 28.6 | 25.5 | 23.2 |
| $Sm_3Fe_{77}B_{20}$ | 24.9 | 23.2 | 22.0 |
| $Sm_4Fe_{76}B_{20}$ | 22.5 | 21.4 | 20.1 |
| $Sm_5Fe_{75}B_{20}$ | 20.2 | 19.8 | 18.5 |
| $Sm_6Fe_{74}B_{20}$ | 18.5 | 16.1 | 14.9 |

The room temperature magnetostriction of the crystalline compound $SmFe_2$ amounts to $-1560 \times 10^{-6}$ [5]. Assuming this large magnetoelastic coupling to occur also in the dilute Sm alloys, the addition of 1% Sm to $Fe_{80}B_{20}$ is expected to decrease the magnetostriction nearly to zero, so an alloy containing 6% Sm may have $\lambda_s = -150 \times 10^{-6}$.

The experimental value of magnetostriction is, however, an order of magnitude smaller than the expected value. This result is analogous to the influence of Tb on the magnetostriction of the $(Fe_{82}B_{18})_{0.9}La_5Tb_5$ alloy [6].

The coercivity decreases with decreasing magnetostriction (Fig. 7) as in a variety of amorphous alloy systems.

### *3.5 Domain structure*

The domain structure of the investigated amorphous ribbons is composed of two characteristic types of domain patterns: wide laminae, sometimes forming star-like arrangements and narrow laminae forming patches of domains 2 to 3 μm wide. Within wide domains the magnetic easy axis is oriented parallel to the surface of the ribbon. Narrow laminae correspond to a closure structure of domains arranged perpendicular to the ribbon. Within the closure domains the spontaneous magnetization lies parallel

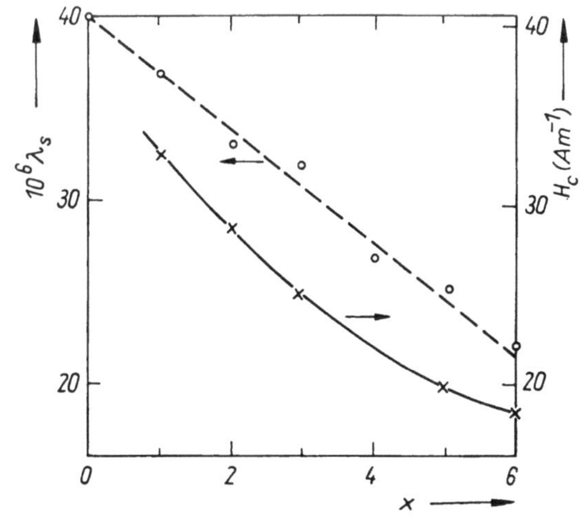


Fig. 7. Variation of magnetostriction $\lambda_s$ and coercivity with Sm content for Sm–Fe–B alloys at room temperature

to the surface, whereas below the closure domains the easy direction lies perpendicular to the ribbon surface.

The local magnetic anisotropies cannot be measured by a global method like a torque measurement since the different local anisotropies cancel each other.

The narrow domains vanish if the applied in-plane field corresponds to the crystal field [7],

$$H_k = \frac{2K_c}{M_s}, \tag{1}$$

where $K_c$ is the local anisotropy constant within the narrow domains and $M_s$ the saturation magnetization per unit volume. With the measured magnetic fields $H_k$ and the measured values of $M_s$ the $K_c$-values were obtained from (1).

Using the relations for the closure domain structure of Landau-Lifshits type the specific domain wall energy, $\gamma$, and the wall width, $\delta$, have been calculated for many places in the as-quenched ribbons. $H_k$ varies from 2.4 to 4.8 kA/m, $K_c$ from 3 to 7 kJ/m$^3$, $\gamma$ from 0.3 to 1.0 mJ/m$^2$, and $\delta$ from 0.06 to 0.14 $\mu$m.

## 4. Conclusions

Saturation magnetization and Curie temperature of amorphous $Sm_xFe_{80-x}B_{20}$ alloys decrease, whereas the crystallization temperature increases with increasing samarium content. During crystallization these amorphous ribbons transform into the phases $Sm_2Fe_{17}$, $Fe_3B$, and $\alpha$-Fe. The magnetoelastic coupling of Sm seems to be smaller in dilute alloys than in the compound $SmFe_2$.

***Acknowledgement***

This work was supported by the Polish Academy of Sciences from the program RPBP 01.8.

### References

[1] J. J. Croat, J. F. Herbat, R. W. Lee, and F. E. Pinkerton, J. appl. Phys. 55, 2078 (1984).
[2] G. C. Hadjipanayis, C. P. Wong, and Y. F. Tao, Rapidly Quenched Metals, Vol. 2, Ed. S. Steeb and H. Warlimont, North-Holland Publ. Co., 1985 (p. 1307).
[3] Yang Xingbo, T. Miyazaki, K. Takakura, K. Hisatake, T. Hattori, and M. Takahashi, J. Magnetism magnetic Mater. 62, 293 (1986).
[4] R. C. Taylor, T. R. McGuire, J. M. D. Coey, and A. Gangulee, J. appl. Phys. 49, 2885 (1978).
[5] A. Clark, AIP Conf. Proc. 18, 1015 (1974).
[6] E. du Tremolet de Lacheisserie and A. Chamberod, J. Physique Lettres 44, L-1013 (1983).
[7] G. Schroeder, R. Schäfer, and H. Kronmüller, phys. stat. sol. (a) 50, 475 (1978).

*(Received January 18, 1988)*

phys. stat. sol. (a) **108**, 381 (1988)

Subject classification: 75.60; S1.2

*Department of Physics, Faculty of Science, University of Cairo*[1])

# On the Influence of Copper Atoms on Domain Wall Energy and Pinning Force in Ni–Cu Alloys

By

M. A. FAHIM and A. R. ALI

The maximum magnetic susceptibility $\chi_{max}$, the critical magnetic field $H_{cr}$, and the energy loss $E$ are used to study the effect of grain boundaries on the behaviour of domain wall displacements in pure Ni and Ni–Cu alloys. The observed dependence of $\chi_{max}$, $H_{cr}$, and $E$ on the grain size is used to calculate the magnetic domain wall surface energy and the boundary pinning force, and leads to the confirmation of the theoretical predictions made by Guyot and Globus for Ni–Cu alloys. The existence of Cu-solute atoms in the Ni-matrix is found to decrease both, the magnetic domain wall energy and the boundary pinning force, which is assumed to be due to the expected interaction and the pinning action of Cu solute atoms that appear in the field of the matrix.

Die maximale magnetische Suszeptibilität $\chi_{max}$, das kritische Magnetfeld $H_{cr}$ und der Energieverlust $E$ werden benutzt, um den Einfluß von Korngrenzen auf das Verhalten von Domänenwandverschiebungen in reinem Ni und Ni–Cu-Legierungen zu untersuchen. Die beobachtete Abhängigkeit von $\chi_{max}$, $H_{cr}$ und $E$ von der Korngröße wird benutzt, um die Oberflächenenergie der magnetischen Domänenwand und die Grenzenanhaftkraft zu berechnen, und führt zu der Bestätigung der theoretischen Vorhersagen von Guyot und Globus für Ni–Cu-Legierungen. Die Existenz von Cu-Lösungsatomen in der Ni-Matrix vergrößert sowohl die magnetische Domänenwandenergie als auch die Grenzenanhaftkraft, von der angenommen wird, daß sie von der erwarteten Wechselwirkung und Anhaftwirkung der Cu-Lösungsatome herrührt, die im Feld der Matrix auftreten.

## 1. Introduction

Recently attention has been focussed on the effect of impurity atoms on the behaviour of magnetic domain wall motion in pure Ni and Ni-alloys [1, 2]. It has been found that a small amount of impurities can have a strong influence on the maximum magnetic susceptibility $\chi_{max}$, the critical magnetic field $H_{cr}$, and the energy loss $E$. Attempts have been made to explain this in terms of the interaction of lattice imperfections with the magnetic domain walls and hence effectively limiting the reversible and irreversible displacement of magnetic domain walls [3, 4]. On the other hand, Globus [5, 6] came to the conclusion that the magnetic susceptibility of polycrystalline materials is attributed to the interaction of a domain wall with the grain boundaries. It was assumed that the wall portion, enclosed in the grain, formed the responsible parameter for the relation between magnetic susceptibility and grain diameter. Based on the Guyot-Globus model [7, 8] in connection with the magnetization mechanisms in ferrimagnetic material important specific properties characterizing a domain wall, such as domain wall surface energy $\gamma$ and the boundary pinning force per unit length $f$, are determined.

From these considerations, the present work reports experimental results on the interaction behaviour of grain boundaries with migrating domain walls which con-

[1]) Cairo, Giza, Egypt.

firm the Guyot and Globus model in ferromagnetic material. Also, it is aimed here principally to study the influence of copper atoms on domain wall energy and the pinning force in Ni–Cu alloys. Associated changes in maximum magnetic susceptibility, critical magnetic field, and energy loss with the grain size of pure Ni and Ni–Cu alloys were taken as a tool of study in the present work.

## 2. Experimental

Samples of Ni–Cu alloys were prepared from high-purity Ni and Cu by induction melting and suitable homogenization at 1200 °C for 24 h. The material was then shaped by extrusion into rods of 3 mm diameter. The rods were then subjected to a reduction in cross-sectional area by swaging at room temperature to wires of 1 mm diameter. The atomic absorption method was used in order to determine the composition of the alloys. Samples of different grain size were prepared by annealing for 5 h at different high temperatures. The mean grain size of the samples was measured using an ordinary microscope.

The cathode ray technique was employed to get the room temperature hysteresis loop, the area of each loop is equal to the energy $E$ lost per unit volume during one cycle [8]. The magnetic susceptibility was obtained from the relation $\chi = B/H$, which characterizes the magnetization of both reversible and irreversible domain wall motion [1]. The relative remanent magnetization ($m_r$) was obtained from the relation $m_r = M_r/M_s$, where $M_r$ is the remanent magnetization and $M_s$ the saturation magnetization.

## 3. Results and Discussions

### 3.1 *Effect of grain size on $\chi_{max}$ and $H_{cr}$*

The magnetic susceptibility $\chi$ measured at room temperature as a function of the magnetizing field $H$ for pure Ni, Ni–1.5 at% Cu, and Ni–4.5 at% Cu of different mean grain sizes is presented in Fig. 1. The curves are characterized by pronounced peak values in magnetic susceptibility $\chi_{max}$ which are found to shift their positions with the magnetic field $H_{cr}$ with increasing mean grain size of the samples. The dependence of both the maximum susceptibility $\chi_{max}$ and the critical magnetic field $H_{cr}$ on the mean grain size is illustrated in Fig. 2a, b. It is clear that the maximum susceptibility is proportional to the mean grain size $D$. On the contrary, the critical magnetic field is obviously proportional to $1/D$ (see Fig. 2b). Based on the Globus theory [5, 6], the bulging and subsequent displacement of magnetic domain walls bounded by the grain boundaries formed the responsible parameter for the relation between magnetic susceptibility and the grain diameter. Therefore, the values of the domain wall surface energy $\gamma$ and the boundary pinning force $f$ could be determined using the Globus equation,

$$\gamma = \frac{3DM_s^2}{16\chi_{max}} \tag{1}$$

and

$$f = \tfrac{1}{2} H_{cr} M_s D\,, \tag{2}$$

where $M_s$ is the saturation magnetization.

In the present work, the saturation magnetization $M_s$ of the different samples was determined and found to be 0.0485, 0.0468, and 0.0460 T for pure Ni, Ni–1.5 at% Cu, and Ni–4.5 at% Cu, respectively. The values of the domain wall surface energy and

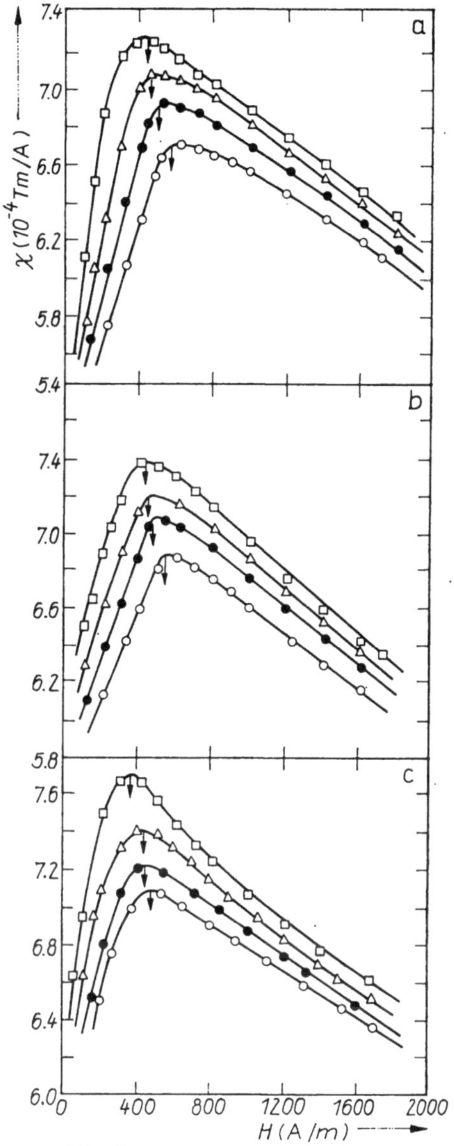


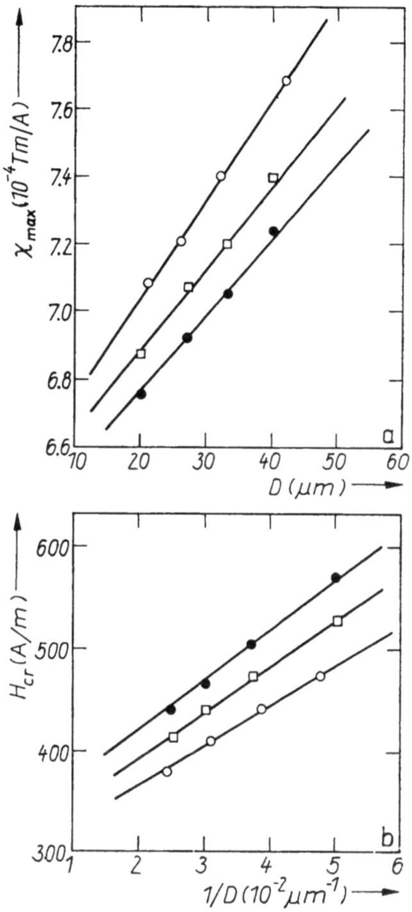


Fig. 1

Fig. 2

Fig. 1. The dependence of magnetic susceptibility on the magnetic field of a) pure Ni, b) Ni–1.5 at% Cu, and c) Ni–4.5 at% Cu with different mean grain sizes (a), b) □ $D = 40$, △ 33, ● 27, ○ 20 μm; c) □ $D = 42$, △ 32, ● 26, ○ 21 μm). The arrows show the values of critical magnetic field $H_{cr}$

Fig. 2. The dependence of a) $\chi_{max}$ on the mean grain size $D$, and b) $H_{cr}$ on the reciprocal grain size $1/D$ for pure Ni (●), Ni–1.5 at% Cu (□), and Ni–4.5 at% Cu (○) (see Table 1)

the boundary pinning force were calculated from the total susceptibility curves for the different samples and are given in Fig. 2a, b. Table 1 summarizes the present results on the domain wall surface energy and boundary pinning force as calculated

Table 1

| material | domain wall energy $\gamma$ ($10^{-4}$ J/m$^2$) | boundary pinning force $f$ ($10^{-4}$ J/m$^2$) | $f/\gamma$ |
|---|---|---|---|
| pure Ni (99.999%) | 1.96 | 1.15 | 0.59 |
| Ni–1.5 at% Cu | 1.68 | 1.04 | 0.62 |
| Ni–4.5 at%Cu | 1.32 | 0.86 | 0.65 |

from total susceptibility curves for pure Ni and Ni Cu alloys. From these results, the general trend is apparent that both the domain wall surface energy and the boundary pinning force decrease with increasing Cu solute concentration in Ni matrix.

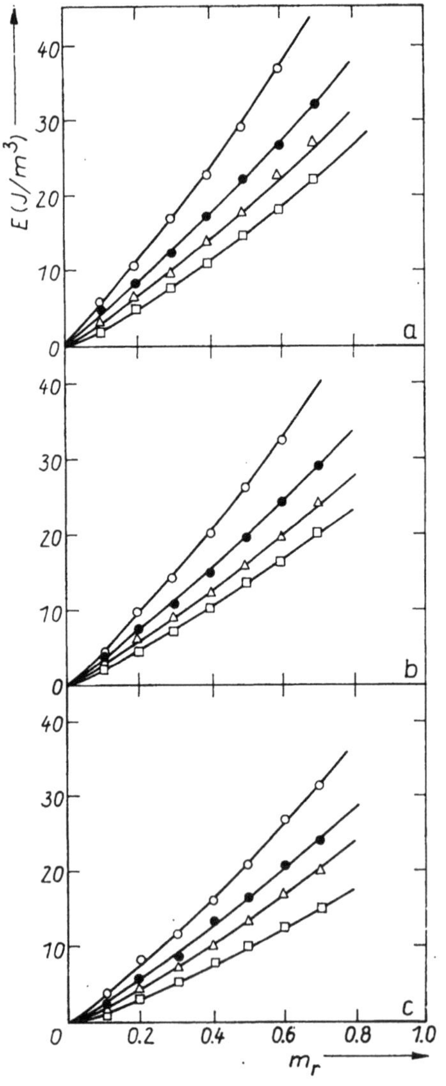


Fig. 3. Variation of the energy loss $E$ as a function of the relative remanence $m_r = M_r/M_s$ for a) pure Ni, b) Ni–1.5 at% Cu, and c) Ni–4.5 at% Cu with different mean grain sizes (for symbols see Fig. 1)

### *3.2 Effect of grain size on the energy loss*

The dependence of the energy loss $E$ per unit volume on the relative remanent magnetization $m_r$ of pure Ni and Ni–Cu alloys o different mean grain size $D$ is shown in Fig. 3. It is evident that, for a given value of $m_r$, the energy loss $E$ is inversely proportional to the mean grain size for pure Ni and Ni–Cu alloys (see Fig. 4). Therefore, according to the theoretical prediction made by Guyot and Globus [7, 8] in connection with the magnetization in ferrimagnetic material, the domain wall surface energy $\gamma$ and the boundary pinning force $f$ could be determined using the equation

$$\frac{ED}{8m_r} = f + \frac{m_r}{3}. \tag{3}$$

In the present work, plotting $(ED/8m_r)$ as a function of the relative remanent magnetization $m_r$, straight lines were obtained for pure Ni and Ni–Cu alloys (see Fig. 5). From these lines, which are independent of the grain size, it is possible to determine the

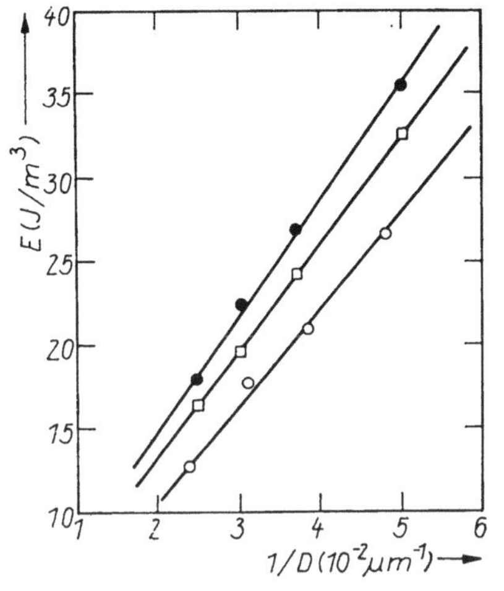


Fig. 4

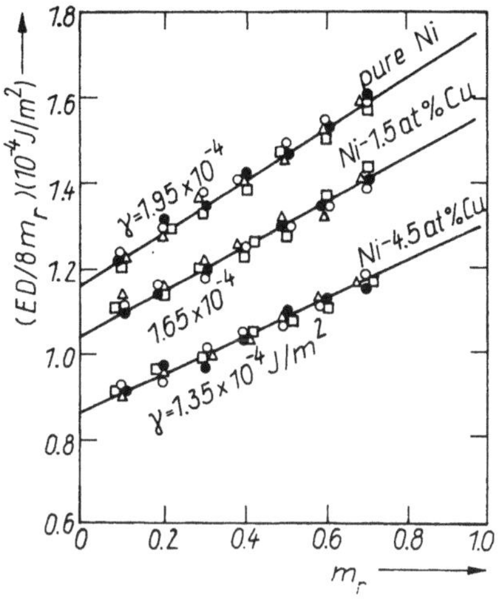


Fig. 5

Fig. 4. The dependence of the energy loss $E$ on the reciprocal grain size $1/D$ as deduced from Fig. 3 for a given $m_r = 0.6$ for different samples (for symbols see Fig. 2)

Fig. 5. Experimental values of domain wall surface energy $\gamma$ and boundary pinning force $f$ as calculated from the energy loss measurements for different samples. For pure Ni ($f = 0.59$), for Ni–1.5 at% Cu ($f = 0.63$), and for Ni–4.5 at% Cu ($f = 0.64$). ○ $D = 20$; ● 27; △ 33; □ 40 µm (for pure Ni and Ni–1.5 at% Cu), ○ $D = 21$; ● 26; △ 32; □ 42 µm (for Ni–4.5 at%Cu)

parameters $\gamma$ and $f$ by using (3). The values of $\gamma$ are given on each line in Fig. 5, in good agreement with the present values directly obtained from the maximum susceptibility $\chi_{max}$ and critical magnetic field $H_{cr}$ measurements (see Table 1). Hence, these results lead to the confirmation of the Guyot-Globus model in annealed ferromagnetic polycrystalline material.

The observed dependence of maximum susceptibility, critical magnetic field, and energy loss on grain size for pure Ni and Ni–Cu alloys might be attributed to the bowing and the propagation of magnetic domain walls. This process is governed by the existence of structural defects producing a configuration of potential energy controlling the location of magnetic domain walls. The maximum susceptibility was found to exhibit a significant dependence on the grain size (see Fig. 2a). This can be easily understood by remarking that $\chi_{max}$ reflects the magnetization of both the short-range reversible displacement and irreversible wall displacement of magnetic domain. The reversible displacement of magnetic domain walls was found to be practically independent of the grain size [10]. On the contrary, the irreversible displacement of domain walls is obviously sensitive to the grain size [9, 10]. In fact, the influence of the grain boundaries might be hidden during the reversible displacement of magnetic domain walls by the pinning effect arising from dislocations, which are weak but very numerous obstacles, and the existence of Cu solute atoms in the matrix, until the density of grain boundaries becomes high enough to act effectively as pinning centres to the domain walls. The situation is different for the effect of grain size on the critical magnetic field $H_{cr}$ and the energy loss $E$, which reflects the amplitude of irreversible displacement of magnetic domain walls. On the other hand, the boundary pinning force $f$ was found to be a definite fraction of the domain wall surface energy $\gamma$ (see Table 1), which supports in another way the idea that the break-off mechanism of the magnetic domain walls is quantitatively related not to the grain boundary behaviour but to the domain wall behaviour. Therefore, the observed decrease in both the domain wall surface energy and boundary pinning force with increasing solute content of Cu atoms, could be reasonably attributed to the expected interaction and pinning action of Cu solute atoms in the field of the matrix.

## References


[1] A. R. ALI, S. A. MAHMOUD, and G. SAID, phys. stat. sol. (a) **84**, K67 (1984).
[2] H. KRONMÜLLER, Z. angew. Phys. **30**, 9 (1970).
[3] A. R. ALI and G. SAID, Physica (Utrecht) **B112**, 241 (1982).
[4] J. DEGAUQUE and B. ASTIE, phys. stat. sol. (a) **74**, 201 (1982).
[5] A. GLOBUS, J. Physique **38**, C1-1 (1977).
[6] A. GLOBUS, Proc. Soft Magnetic Materials Conf., Cardiff 1975.
[7] A. GLOBUS, P. DUPLEX, and M. GUYOT, IEEE Trans. Magnetics **7**, 617 (1971).
A. GLOBUS and M. GUYOT, phys. stat. sol. (b) **52**, 427 (1972).
[8] M. GUYOT and A. GLOBUS, phys. stat. sol. (b) **59**, 447 (1973); J. Physique **38**, C1-157 (1977).
[9] J. DEGAUQUE, B. ASTIE, J. L. PORTESEIL, and R. VERGNE, J. Magnetism magnetic Mater. **26**, 261 (1981).
[10] B. ASTIE, J. DEGAUQUE, J. L. PORTESEIL, and R. VERGNE, J. Magnetism magnetic Mater. **28**, 149 (1982).


*(Received December 29, 1987)*

phys. stat. sol. (a) 108, 387 (1988)

Subject classification: 75.60; S1.1

*Institut für Angewandte Physik der Universität Münster*[1])

# Observation of a New Domain Nucleation Process in Soft Magnetic Materials

By

U. HARTMANN[2])

**A critical bowing of 180° Bloch walls during the magnetization reversal process leads to a structural decomposition of these walls into a periodic arrangement of 90° transverse domains. The nucleation process is accompanied by a considerable decrease of the susceptibility and provides a source of irreversibilities in the overall magnetization process even under the influence of external fields which are small compared to the magnetocrystalline anisotropy field. The process is investigated on isolated subdivided 180° Bloch walls in iron whisker samples by means of X-ray topography and magneto-optical observation techniques. The nucleation field distribution and the critical inclination of the walls are deduced from inductive measurements.**

**Eine kritische Biegung der 180°-Blochwände während des Magnetisierungsumkehrprozesses führt zu einer strukturellen Zerlegung dieser Wände in eine periodische Anordnung von transversalen 90°-Domänen. Der Keimbildungsprozeß ist von einem beträchtlichen Rückgang der Susceptibilität begleitet und liefert eine Quelle für Irreversibilitäten im Gesamtmagnetisierungsprozeß selbst unter dem Einfluß äußerer Felder, die im Vergleich zum magnetokristallinen Anisotropiefeld klein sind. Der Prozeß wird an isolierten unterteilten 180°-Blochwänden in Eisenwhiskerproben mittels Röntgentopografie und magnetooptischer Beobachtungstechniken untersucht. Die Verteilung des Keimbildungsfeldes und die kritische Neigung der Wände werden aus Induktionsmessungen abgeleitet.**

## 1. Introduction

The magnetization reversal process in soft magnetic materials is largely governed by the displacement of 180° Bloch walls. Recent theoretical as well as experimental results have confirmed that these walls are generally of a complex structure, at least in the near-surface region of bulk crystals [1 to 3]. The demagnetizing fields formed at the intersection of a Bloch wall with the crystal surface cause the wall to divide into a periodic configuration of oppositely polarized wall segments [4, 5]. At the junction of two successive wall segments a Néel line (commonly called a Bloch line also, but it is preferable to reserve this second terminology for the analogous situation of a singularity in a Néel wall [6]) is located which provides a quasiabrupt change of the chirality, i.e., the handedness, of the wall [3]. The thermodynamic equilibrium configuration of the wall, i.e., the actual periodicity, is fixed by the balance between the self-magnetostatic energy and the energy due to the presence of the Néel lines [4, 5].

Subdivided 180° Bloch walls have been observed on Perminvar specimens [7], silicon iron transformer sheets [8], yttrium iron garnet (YIG) single crystals [9], Permalloy elements [10], and on iron whisker crystals [1 to 3, 11]. Due to their

---

[1]) Corrensstr. 2/4, D-4400 Münster, FRG.

[2]) Permanent address: Institut für Angewandte Physik, Universität Gießen, Heinrich-Buff-Ring 16, D-6300 Gießen, FRG.

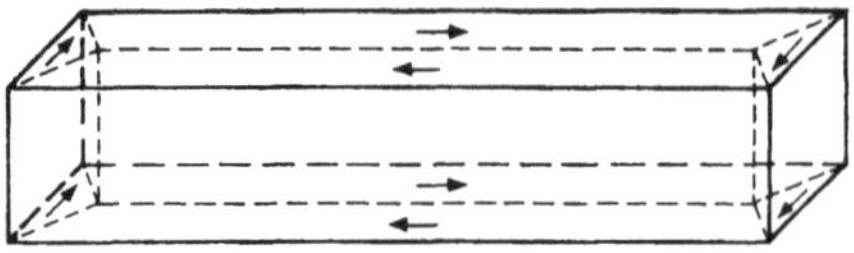

Fig. 1. Schematic sketch of a ⟨100⟩–{100}-oriented iron whisker exhibiting a Landau domain structure

naturally formed shape and their virtually ideal crystallographic structure the latter specimens are ideal sample materials for an experimental analysis of fundamental magnetization reversal mechanisms in soft magnetic materials.

Iron whiskers grown in a ⟨100⟩ direction and bounded by {100} crystallographic faces exhibit generally the simple and transparent Landau domain structure consisting of two opposite domains in axial direction and triangular flux closure domains at the ends of the whisker (see Fig. 1). The arrangement and internal structure of the subdivided 180° Bloch wall separating the two long antiparallel domains are shown in Fig. 2.

The effect of structural material defects, such as grain boundaries, non-magnetic inclusions, stresses, and dislocation arrangements on the magnetization process has been studied for a very long time and in great detail. An excellent review and extension of this general subject has been given by Träuble [12]. However, until the present time, no detailed analysis has been performed, concerning the effect of the internal Bloch wall structure on the overall magnetization process. The present paper is focussed on an experimental investigation of the influence of the internal wall configuration on the magnetization reversal process in iron whiskers.

## 2. Experimental Equipment

Iron whiskers were grown by hydrogen reduction of ferrous chloride with a small addition of carbon black according to a modified Brenner technique [13, 14]. The whiskers suitable for the investigations are grown in a ⟨100⟩ direction and bounded by optically perfect {100} crystallographic faces. They have nearly square cross sections, 10 to 200 μm on a side, and a length ranging from 5 to 20 mm. The cut end of the whisker was inserted in a capillary tube and the naturally terminated end available for observation.

In situ observations of the domain structure during the magnetization reversal process have been performed employing the X-ray topography [15, 16]. A suitable wavelength of 0.070926 nm is provided by the $MoK_{\alpha1}$ radiation. The surface mag-

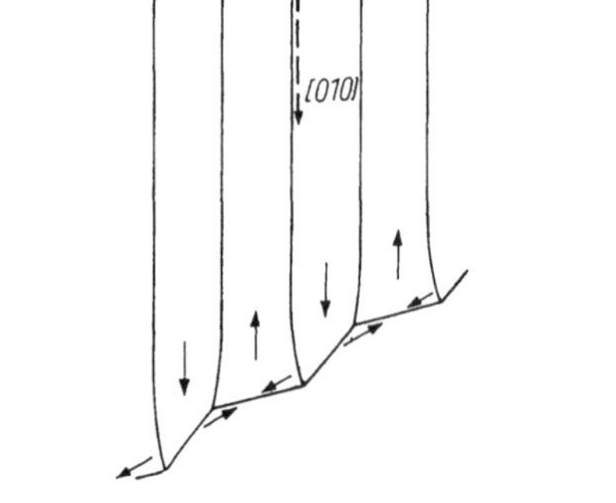


Fig. 2. Internal structure of a subdivided 180° Bloch wall in a whisker. The slight surface tilting of the individual wall segments causes a flux exchange between the wall and the adjacent domains [3]

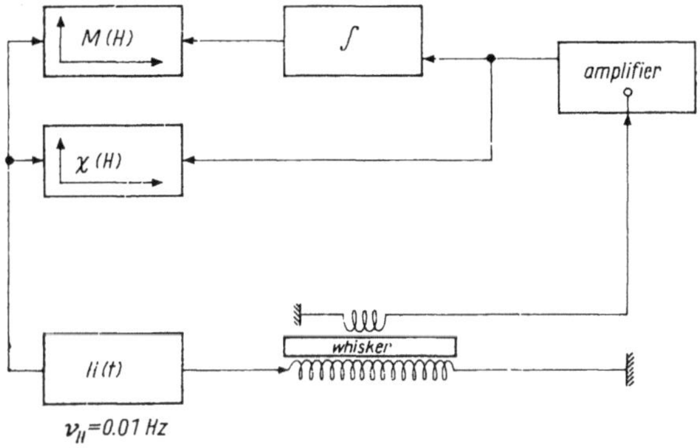


Fig. 3. Scheme of the experimental arrangement for quasistatic inductive measurements

netization of the specimens up to a depth of about 10 nm has been analyzed by means of high-resolution Kerr microscopy. An enhancement of the magneto-optical effect was obtained by the application of a $\lambda/4$ zinc sulfide layer [17]. Additionally, fine structures of the domain walls at the surface of the specimens were observed at high spatial resolution using the recently developed interference contrast colloid (ICC) technique [1, 18, 19].

The net magnetization of a whisker as a function of an externally applied homogeneous magnetic field in $\langle 100 \rangle$ direction (see Fig. 1) has been measured under quasistatic conditions employing an inductive hysteresis loop tracer. A schematic sketch of the apparatus is given in Fig. 3. The pick-up coil of 0.25 mm inner diameter, 2.8 mm length, and 2800 windings is compensated for the air flux and coupled tightly to the whisker. The surrounding long solenoid is driven with a rate of field change of 0.01 Hz. The pick-up coil is connected to a low-noise amplifier. The data displayed with respect to the applied field yield directly the susceptibility diagram of the magnetization process. The analog integration leads to the quasistatic hysteresis loop of the whisker.

### 3. Nucleation of Transverse Domains

The increase of the net magnetization of a whisker under the influence of an external homogeneous magnetic field applied in longitudinal direction is predominantly determined by the bowing of the long 180° wall separating the axial domains [20]. It has been shown in a preliminary publication [3] that the overall bowing of a subdivided 180° Bloch wall causes characteristic modifications of the internal wall structure. The regions in which the wall slope is partly compensated by the segment tilting, indicated in Fig. 2, grow at the expense of regions with enhanced segment tilting. In the vertex region of the bowed wall the slope is nearly zero and the segmentation remains largely unaffected.

Fig. 4 shows a sequence of X-ray topographs taken in an increasing external field. Up to a field strength of $H = 1.21$ kA/m the magneto-elastic lattice distortion is not sufficient to cause a visible X-ray contrast. Only the 90° walls bounding the triangular flux closure domain at the tip of the whisker yield a clear constrast. The topography for $H = 1.67$ kA/m exhibits a considerable modification of the overall domain structure. The strong broken contrast appearing along the bowed wall indicates the existence of 90° walls incorporated directly in the 180° wall. Additional 90° walls are running from the bowed wall to the edge of the whisker. For further increasing longitudinal field the 90° transverse domains grow and proceed from the tip to the vertex region of the bowed wall.

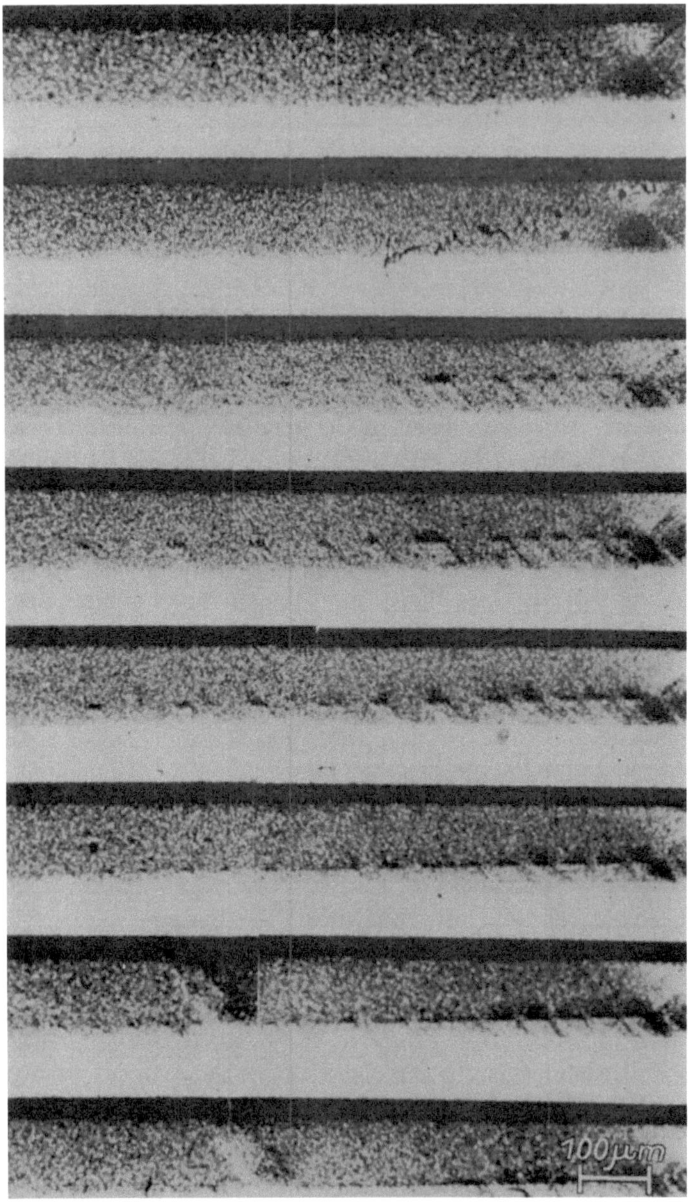


Fig. 4. Sequence of (200) $MoK_{\alpha 1}$ X-ray topographs showing changes in the domain wall structure during the magnetization reversal process of an iron whisker

For a closer investigation of the transverse domain nucleation process the near-surface magnetization has been observed employing the high-resolution Kerr microscope. Fig. 5 reveals the abrupt change in the wall structure appearing between $H = 1.65$ and $1.66$ kA/m. The simultaneous observation of all surfaces of the whisker confirms that the shallow transverse domains are restricted to the near-surface region

Fig. 5. High-resolution Kerr micrograph showing the wall-induced nucleation of transverse domains. The critical slope of the subdivided 180° Bloch wall is reached for about $H = 1.65$ kA/m. The appearing transverse domains nucleate at the most strongly tilted segments of the complex wall. The bounding 90° walls cause the broken contrast lines obtained in the X-ray topographs of Fig. 4

of the top and bottom faces of the whisker. A schematic sketch of the resulting surface magnetization distribution is given in Fig. 6. It has been shown previously [3] that the most strongly tilted segments of the bowed subdivided 180° Bloch wall serve as nucleation centers for the creation of transverse domains. The resulting, energetically more favorable, structure provides an intensified flux exchange between the magnetically charged wall and the adjacent domains [3].

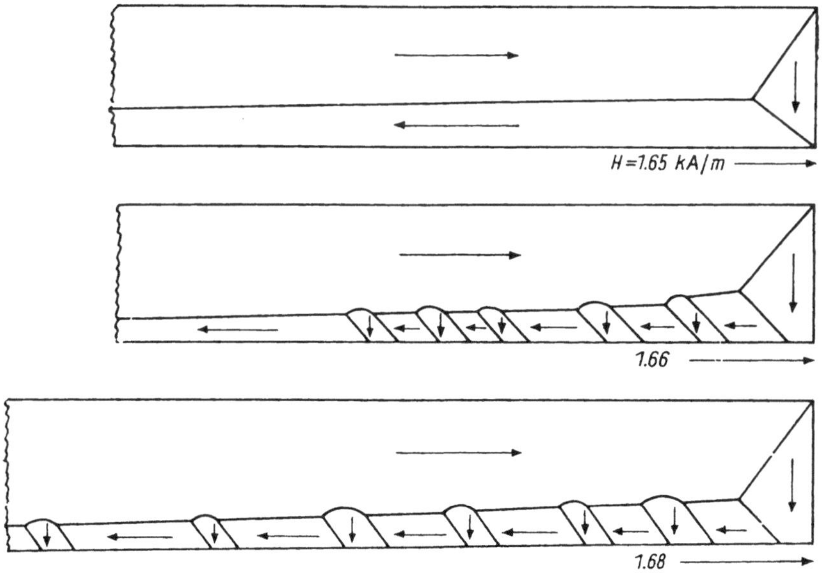


Fig. 6. Surface magnetization distribution of the multi-domain structure shown in Fig. 5. The 90° transverse domains incorporated in the long 180° wall are restricted to the near-surface region

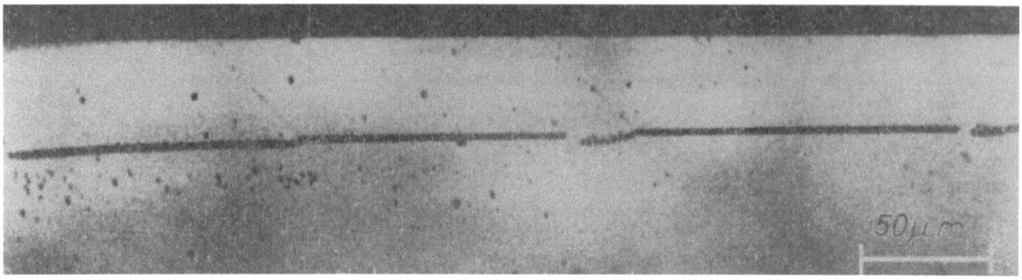


Fig. 7. High-resolution ICC micrograph showing the structural decomposition of the 90° walls incorporated in the long 180° wall. The symmetrical inclination of the wall segments causes a slightly curved course of the walls and a serrated appearance of the ICC pattern

A particular detail of the transverse domains is shown by the ICC micrograph in Fig. 7. The slightly serrated appearance of the colloid deposit at the 90° walls incorporated in the 180° wall indicates a structural decomposition of the former. It has been suggested that a symmetrical inclination with respect to the crystallographic {110} planes, i.e. a zig-zag shape of the 90° walls, might lower the total wall energy [21, 22].

A sequence of X-ray topographs taken in an increasing external field oriented antiparallel with respect to Fig. 4 is given in Fig. 8. Again an abrupt nucleation of transverse domains is observed between $H = 1.21$ and 1.67 kA/m. Contrary to the domain development shown in Fig. 4, a second flux closure domain, i.e., an edge domain, is nucleated at about $H = 2.67$ kA/m. The bounding 180° Bloch wall exhibits an initial structural decomposition with an incorporation of transverse domains.

It should be noted that the Bloch-wall-induced nucleation of transverse domains has already been observed by Scott and Coleman [23] using the conventional colloid technique and by Fowler et al. [24] employing the Kerr magneto-optic effect. However, in both cases no detailed information has been obtained concerning the origin and nature of the occurring discontinuous modifications of the overall domain structure.

## 4. Wall-Induced Discontinuities in the Magnetization Curve

The influence of the wall-induced nucleation of a multi-domain structure on the field-dependent actual susceptibility of the whisker is directly revealed by the susceptibility diagram shown in Fig. 9. The linear displacement of the long 180° Bloch wall separating the axial domains leads to a constant quasistatic susceptibility in the central parts of the curve [20]. Only a few weak Barkhausen jumps occur. At a field strength of about $H = \pm 1.65$ kA/m the susceptibility exhibits a discontinuous decrease and then shows a gradual decrease with further increasing field. The abrupt changes of the quasistatic susceptibility correspond clearly to the domain-wall-induced nucleation of transverse domains observed in Fig. 4, 5, and 8. In comparison to the simple Landau structure (see Fig. 1), which corresponds to the high-susceptibility parts of the curve in Fig. 9, the nucleation of the multi-domain structure leads to a considerable lowering of the overall susceptibility.

The finite slope at the turning points of the susceptibility curve indicates that the whisker is not totally saturated even in the maximum applied fields. The slow susceptibility decrease during the asymptotic approach to saturation results from the

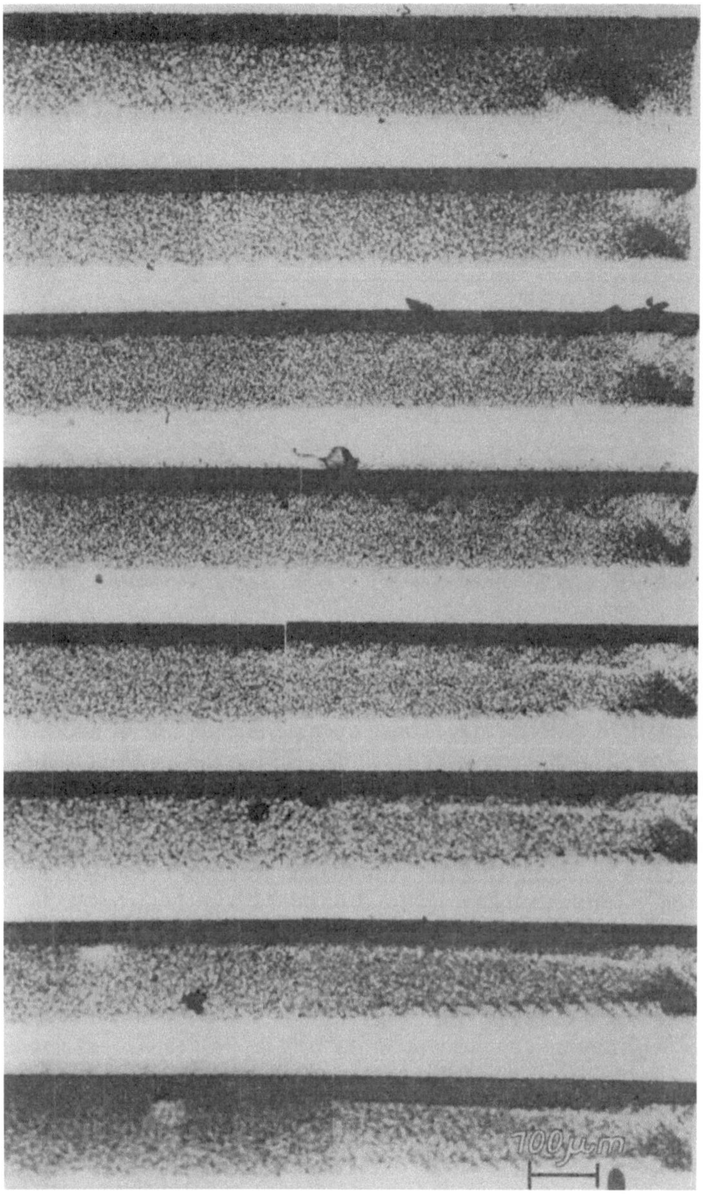


Fig. 8. Sequence of X-ray topographs taken in an increasing external field oriented oppositely to that in Fig. 4. The wall-induced nucleation of transverse domains appears again between $H = 1.21$ and 1.67 kA/m

extremely high local demagnetizing fields created around the sharp corners of the whisker. As a result, small flux closure domains are locally stabilized even in strongest external fields [24, 25]. Hence, starting from the quasi-saturated state, the magnetization reversal in a decreasing field does not take place by a true micromagnetic nucleation process in the Brownian sense [26, 27], but proceeds by a gradual expansion of

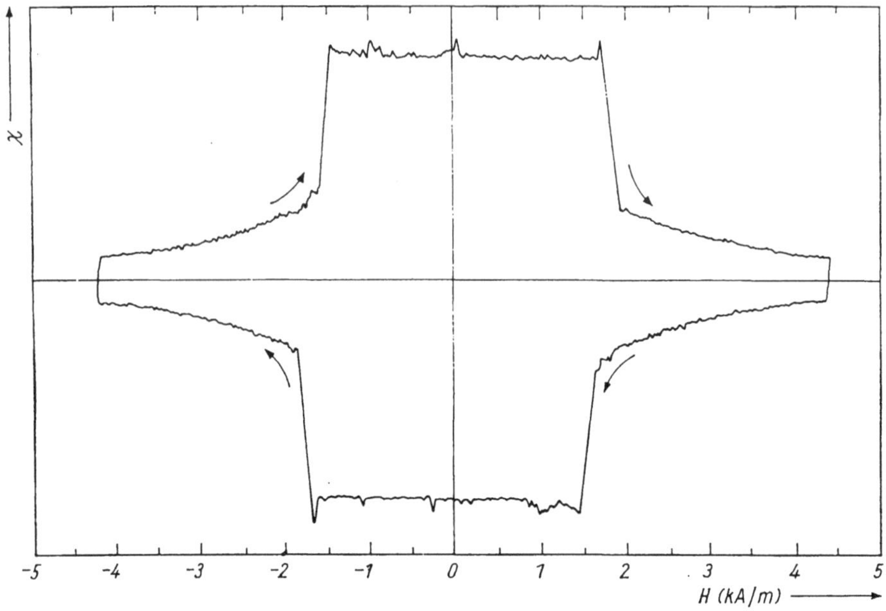


Fig. 9. Diagram taken from the middle region of the whisker showing the quasistatic susceptibility $\chi$ depending on the externally applied field $H$ during a complete magnetization cycle

the persistent closure domains. Consequently, these ever persistent nuclei make the Brownian coercivity paradox [27] a somewhat fictitious problem: since the magnetization reversal during the existence of the Landau domain structure is reversible [20] the coercive field of the whisker is only determined by the small difference between the departure field and the nucleation field of the Landau configuration. In Fig. 9 this difference amounts to about 45 A/m which leads to a coercive field much smaller than the magnetocrystalline anisotropy field of 44.8 kA/m for iron at 25 °C [26].

Additional evidence as to the validity of the interpretation of these observations could probably be obtained by the study of the effect of shape on the coercive force, as by gradually etching the sharp edges of the whiskers.

The integration of the field-dependent susceptibility in Fig. 9 over the externally applied homogeneous field yields the quasistatic hysteresis loop of the whisker shown in Fig. 10. The curve confirms clearly the linear change of the magnetization during the existence of the Landau configuration. The transition to the transverse domain phase introduced by the structural decomposition of the subdivided 180 ° Bloch wall causes a considerable and abrupt decrease of the previously constant slope of the curve. It is clearly seen that both, coercive field and remanence, are determined by the difference between the departure field and the nucleation field of the Landau phase. The gradually decreasing closure domains in high applied fields cause an asymptotically vanishing slope of the magnetization curve during the approach to saturation.

Recent investigations [2, 3] have confirmed that the internal subdivision of a complex 180° Bloch wall is modified by a homogeneous magnetic field applied vertically to the surface of the specimen, i.e. in the direction corresponding to the magnetization axis of the antiparallel wall segments. Under the influence of an increasing vertical

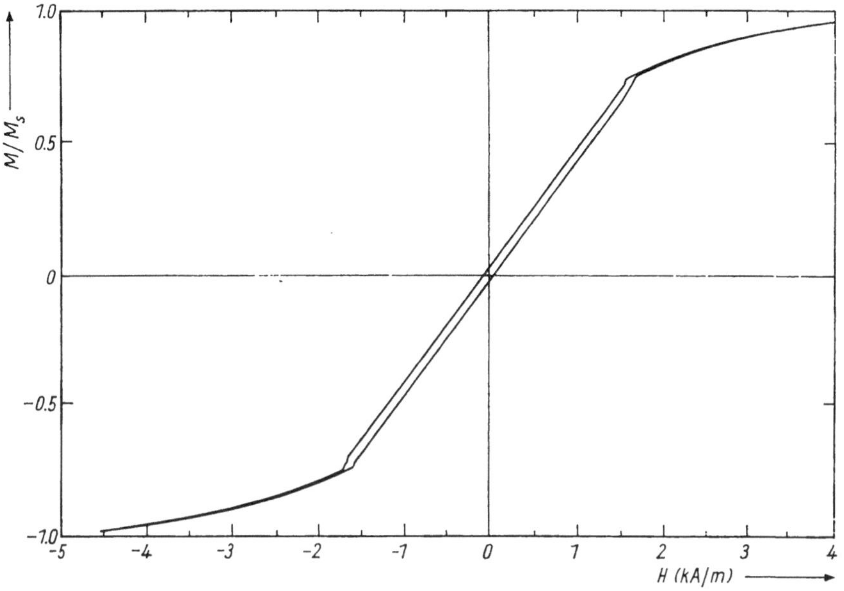


**Fig. 10.** Quasistatic hysteresis loop of the whisker showing the dependence of the reduced net magnetization $M/M_s$ on the applied field $H$ in the central part of the whisker

field the wall segments with the polarization oriented parallel to the external field grow at the expense of the adjacent antiparallel segments. The proceeding polarization reversal of the wall is governed by the motion of the Néel lines localized at the transition region of two successive wall segments. The recording of a complete polarization cycle of the subdivided wall permits a determination of the actual wall saturation field, the wall coercivity and remanence, and of the effective Bloch wall width [3].

Through the direct effect on the internal subdivision of the complex 180° wall, the vertical field has a significant bearing upon the actual bowing of the wall under the influence of a given longitudinal field. Consequently, the shape of the hysteresis loop of the whisker ought to depend critically upon the value of an applied vertical bias field.

The experimental confirmation of this effect is given in Fig. 11. The magnetization curves have been recorded near the tip of the whisker, corresponding to a region of positive slope of the bowed wall in a positive longitudinal field (see also Fig. 12). The curves marked by the thick lines are taken without the application of a bias field. The hysteresis loop marked by the thin line in Fig. 11a shows the influence of a positive vertical field of $H_\perp = 8$ kA/m. This field is oriented parallel to those wall segments which exhibit a sense of inclination partly compensating the overall slope of the bowed wall (see Fig. 2 and 12).

In an increasing positive longitudinal field the modified magnetization curve exhibits a somewhat higher value of the initial susceptibility during the existence of the Landau phase, i.e., in the linear response region. The structural decomposition of the bowed wall, and thus the transition of the Landau phase to the multidomain configuration, is shifted to a reduced departure field. The wall-induced nucleation of 90° transverse domains is again accompanied by an abrupt decrease of the susceptibility. Upon decreasing the longitudinal external field from positive saturation,

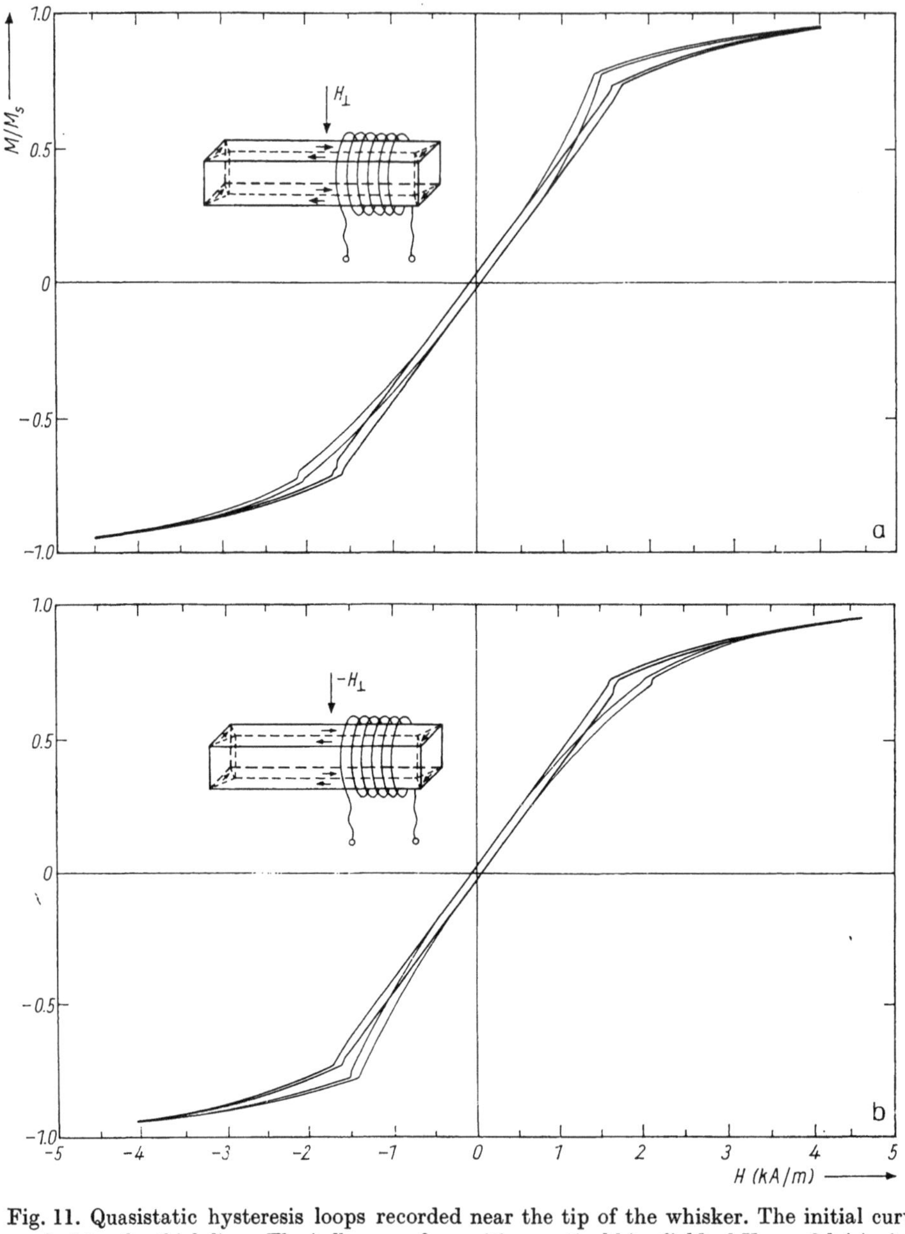


Fig. 11. Quasistatic hysteresis loops recorded near the tip of the whisker. The initial curves are marked by the thick lines. The influence of a positive vertical bias field of $H_\perp = 8$ kA/m is shown in a), whereas a negative bias field of $H_\perp = -8$ kA/m has been applied in b)

the nucleation field of the Landau phase appears to be symmetrically shifted with respect to the departure field.

A completely different situation is obtained during the application of an increasing negative longitudinal field. The bowing of the long 180° wall is now oppositely to that indicated in Fig. 12. Hence, the measuring section corresponds to a region of

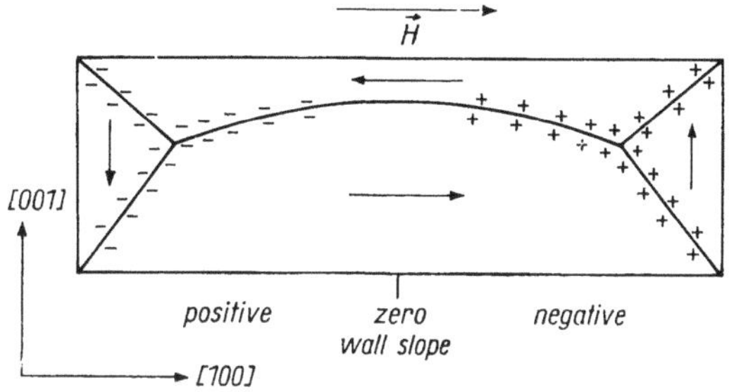


Fig. 12. Schematic illustration of the bowing of a long 180° Bloch wall anchored between 90° closure domains at the tips of the whisker under the influence of a longitudinal field

negative wall slope. Consequently, the bias field is oriented parallel to those segments of the complex wall which exhibit a sense of inclination that enhances the local slope of the wall. This leads to a somewhat lower susceptibility of the modified hysteresis loop with respect to the initial curve. As a result, the structural subdivision of the complex wall is shifted to a more negative value of the critical field. Upon decreasing the negative longitudinal field, the nucleation field of the Landau phase exhibits again a symmetrical shift with respect to the departure field.

The analogous case of a negative vertical bias field of $H_\perp = -8$ kA/m is shown in Fig. 11b. A comparison to Fig. 11a confirms that the inversion of the bias field leads to an exchange of the roles between the positive and the negative longitudinal driving fields.

If the magnetization curve is recorded from the central part of the whisker, i.e. from a region of nearly vanishing wall slope (see Fig. 12), any applied vertical field has no influence on the slope of the curve. Hence, the latter corresponds always to the hysteresis loop shown in Fig. 10.

It ought to be emphazised that, due to the arising high geometrical demagnetizing factor, the application of a vertical bias field has no direct effect on the actual domain configuration. Consequently, the modifications of the magnetization curves occurring during the application of vertical fields provide a striking experimental confirmation to the suggested effect of the actual domain wall structure upon the overall magnetization reversal process of the specimens [3].

## 5. Distribution of the Nucleation Fields

The fact that the mode of development of domains during the structural decomposition of subdivided 180° Bloch walls is the same for all whiskers observed, indicates that this nucleation mechanism is not connected with crystalline imperfections or surface irregularities, but rather with the macroscopic shape of the specimens. Since the actual slope of the bowed wall is determined by the balance between an externally applied longitudinal field and the arising overall demagnetizing field, the field strength causing a critical inclination — and thus a structural decomposition of the bowed wall — should only depend on the actual geometry of the specimens.

Both, theoretical [28 to 30] and experimental [31 to 33] investigations have clearly confirmed that the bowed 180° Bloch wall exhibits a symmetrical parabolic curvature along the axis of the whisker. This is schematically shown in Fig. 13. Under the influence of a moderate longitudinal field $H$ the displacement $y_0(H)$ of the vertex changes linearly with field. The latter can be concluded directly from the inductive measurements shown in Fig. 9 and 10 which yield a constant susceptibility during the existence of the Landau phase. Hence, the actual displacement $y(x, H)$ at an axial distance $x$

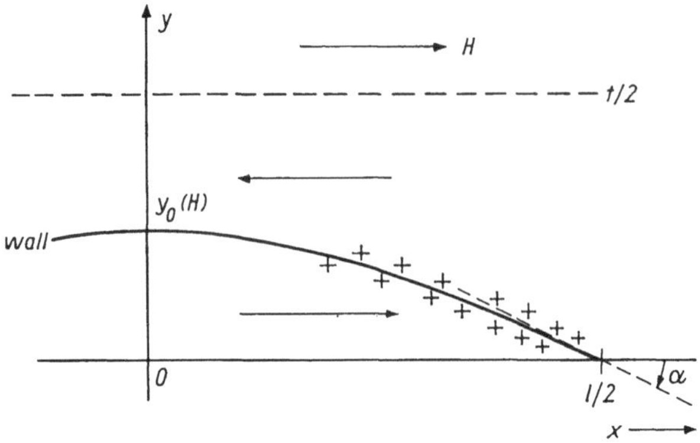


Fig. 13. Geometrical relationships used for the calculation of the actual slope of a bowed 180° Bloch with a length $l$ and a free transverse path length $t$ corresponding to the length and the thickness of a whisker sample, respectively

from the midpoint of the wall is given by

$$y(x, H) = y_0(H) \left\{1 - \left(\frac{2}{l} x\right)^2\right\}, \tag{1}$$

where $l$ denotes the length of the 180° Bloch wall. The local slope of the bowed wall is then determined by

$$\frac{\partial}{\partial x} y(x, H) = -\frac{8}{l^2} y_0(H) x . \tag{2}$$

The maximum slope of the wall occurs at the tips of the whisker for $x = \pm 1/2$ and is given by

$$\left(\frac{\partial y}{\partial x}\right)_{\max} = \mp \frac{4}{l} y_0(H) . \tag{3}$$

Assuming a moderate bowing of the wall, the angle of inclination is locally determined by

$$\tan \alpha(x, H) \approx \alpha(x, H) = \frac{\partial}{\partial x} y(x, H) , \tag{4}$$

and with (3),

$$\alpha_{\max} = \mp \frac{4}{l} y_0(H) . \tag{5}$$

In a highly perfect material, the initial susceptibility $\chi_0$ should only depend on the actual geometrical demagnetization factor $N$ [34],

$$\chi_0 = N^{-1} . \tag{6}$$

Hence, the magnetization in the vertex region is given by

$$M(H) = \frac{H}{N} . \tag{7}$$

On the other hand, the geometrical relationships in Fig. 13 yield a reduced magnetization of

$$\frac{M_0(H)}{M_s} = \frac{y_0(H)}{t/2} , \tag{8}$$

with the thickness of the whisker given by $t$. Combining (7) and (8) leads to

$$y_0(H) = \frac{H}{M_s} \frac{t}{2N}. \tag{9}$$

Thus, the maximum angle of inclination is given by

$$\alpha_{max} = \mp 2 \frac{H}{M_s} \frac{t}{lN}. \tag{10}$$

A boundary condition for the structural decomposition is provided by the fact that the critical angle of inclination has to be reached before the wall touches the edges of the whisker. The latter process is an alternative mechanism also causing the formation of edge domains [33]. Using (9) the boundary condition

$$y_0(H) < \frac{t}{2} \tag{11}$$

leads to the fundamental constraint

$$\frac{H}{NM_s} < 1. \tag{12}$$

In a systematic investigation the critical field strength for both mechanisms, the structural decomposition of the wall and the contact of the wall with the edge of the crystal, has been determined inductively for whiskers of various geometry factors $l/t$. The appropriate demagnetizing factor has been deduced according to (7) from the recorded magnetization curves. The result is shown in Fig. 14. The discrimination of both alternative processes has been performed by means of optical observations employing the ICC technique. The open and solid circles indicate the measured result of the nucleation field for the structural decomposition and the contact with the crystal edge, respectively. The line for small values of $l/t$ is a least-squares computer fit which has been done using (10) and the approximate expression

$$\lim_{\chi_0 \to \infty} N = \frac{4.02 \lg (l/t) - 0.92}{2(l/t)^2}, \tag{13}$$

that has been derived by Neumann and Warmuth [35] for long cylindrical specimens. The obtained fit to the experimental results yields a critical angle of wall inclination of 2°. This agrees satisfactorily with the optical observations (see Fig. 5). The solid line corresponding to the "edge contact effect" has been obtained combining the

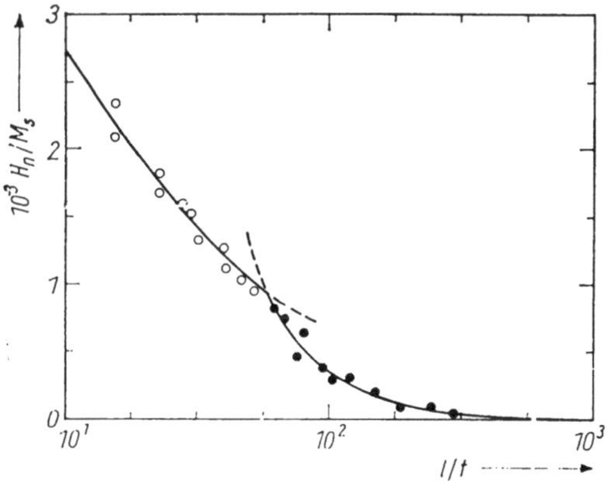


Fig. 14. Distribution of the nucleation field $H_n$ for the two alternative mechanisms causing the formation of closure domains as a function of the geometry factor of the specimens. ○ Structural decomposition; ● contact with the crystal edge

criterion given in (12) with (13),

$$\frac{H}{NM_s} = 1 \,. \tag{14}$$

Again, there is a good agreement between the experimental points and the theoretically obtained result.

It should be noted that the experimentally observed thermodynamic equilibrium inclination of the successive segments of a subdivided 180° Bloch wall in the absence of an external field amounts to about 2° to 3° [3]. This is just below the critical angle of inclination causing the structural decomposition of the wall.

Finally, using the experimental result $\alpha_{crit} = 2°$, the actual nucleation field $H_n$ depending on the geometry of the specimen is given by (10),

$$\frac{H_n}{M_s} = \frac{\pi}{180} \frac{l}{t} N \,. \tag{15}$$

Combining this result with (14), the demarcation line between the two alternative nucleation processes is given by

$$\frac{\pi}{180} \frac{l}{t} = 1 \,. \tag{16}$$

Fig. 15 indicates the resulting occurrence of both nucleation mechanisms depending on the actual geometry of the specimens. Additionally, the experimentally observable area has been indicated, which results from natural restrictions concerning the dimensions of iron whiskers.

## 6. Concluding Remarks

Observations on several iron whiskers confirm the substantial influence of the actual internal configuration of complex 180° Bloch walls on the overall domain structure and thus on the net magnetization of a specimen. Particularly the abrupt structural decomposition caused by a critical inclination of a bowed wall leads through the formation of flux closure domains to discontinuous and irreversible modifications of the whole domain population. The occurring 90° transverse domains provide the origin of hysteresis effects and a considerable decrease in the susceptibility of the magnetization curve. The relevant nucleation field depends only on the actual demagnetizing factor acting on the wall and is always small compared to the mag-

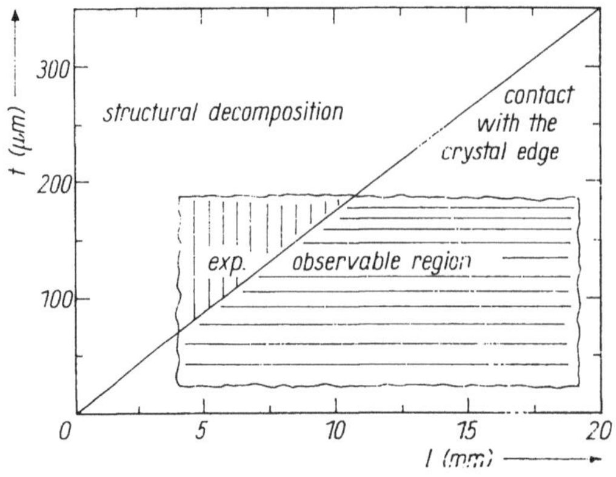


Fig. 15. Demarcation line between the two alternative nucleation processes. The experimentally observable area results from the restrictions given by the nature shape of iron whisker samples

netocrystalline anisotropy field required for a coherent spin rotation. It should be noted that an extensive derivation of the underlying micromagnetics and magnetostatics has been performed by Bloomberg and Arrott [36] and Templeton and Arrott [37].

The important role of the actual domain wall sub-configuration within the overall magnetization process is directly revealed by the effect of a vertical bias field on the longitudinal magnetization curve of the specimens. Using appropriate bias fields the shape of the hysteresis loop can be modified in a wide range even including a complete smoothing of the curve by a suppression of the structural wall decomposition. This fact might be of importance with respect to the various technical applications using the magnetization reversal of soft magnetic materials.

## References

[1] U. Hartmann and H. H. Mende, J. Phys. D **18**, 2285 (1985).
[2] U. Hartmann and H. H. Mende, Phys. Rev. B **33**, 4777 (1986).
[3] U. Hartmann and H. H. Mende, J. appl. Phys. **59**, 4123 (1986).
[4] S. Shtrikman and D. Treves, J. appl. Phys. **31**, 147S (1960).
[5] S. Shtrikman and D. Treves, J. appl. Phys. **31**, 1304 (1960).
[6] M. Kléman, Phys. Rev. B **13**, 3091 (1976).
[7] H. J. Williams and M. Goertz, J. appl. Phys. **23**, 316 (1952).
[8] L. Schön and U. Buchenau, Internat. J. Magnetism **3**, 145 (1972).
[9] M. Labrune, J. Miltat, and M. Kléman, J. appl. Phys. **49**, 2013 (1978).
[10] I. Khayer, J. appl. Phys. **49**, 1906 (1978).
[11] R. W. De Blois and C. D. Graham, Jr., J. appl. Phys. **29**, 932 (1958).
[12] H. Träuble, in: Probleme der Metallphysik, Ed. A. Seeger, Springer-Verlag, Berlin/Heidelberg/New York 1966 (p. 157).
[13] S. S. Brenner, Acta metall. **4**, 62 (1956).
[14] S. Kittaka and T. Kaneko, Japan. J. appl. Phys. **8**, 860 (1969).
[15] H. H. Mende and H. Galinski, Appl. Phys. **5**, 211 (1974).
[16] H. Galinski and H. H. Mende, phys. stat. sol. (a) **27**, 35 (1975).
[17] Th. Kleinefeld and H. H. Mende, phys. stat. sol. (a) **51**, 487 (1979).
[18] U. Hartmann and H. H. Mende, J. Physique **46**, C6–279 (1985).
[19] U. Hartmann and H. H. Mende, Z. Phys. B **61**, 29 (1985).
[20] U. Hartmann and H. H. Mende, Proc. 7th Internat. Conf. Soft Magnetic Materials, Blackpool (U.K.) 1985 (paper 2 to 8).
[21] R. W. De Blois and C. D. Graham, Jr., J. appl. Phys. **29**, 528 (1958).
[22] M. Labrune, J. Physique **37**, 1033 (1976).
[23] G. G. Scott and R. V. Coleman, J. appl. Phys. **28**, 1512 (1957).
[24] C. A. Fowler, E. M. Fryer, and D. Treves, J. appl. Phys. **31**, 2267 (1960).
[25] S. Shtrikman and D. Tevers, J. appl. Phys. **31**, 72S (1960).
[26] W. F. Brown, Jr., Rev. mod. Phys. **17**, 15 (1945).
[27] W. F. Brown, Jr., Micromagnetics, Wiley New York/London 1963.
[28] A. S. Arott and B. Heinrich, Canad. J. Phys. **50**, 710 (1972).
[29] A. S. Arott and B. Heinrich, IEEE Trans. Magnetics 8, 352 (1972).
[30] A. S. Arott, B. Heinrich, and D. S. Bloomberg, AIP Conf. Proc. **10**, 941 (1972).
[31] W. Hagedorn and H. H. Mende, Z. angew. Phys. **30**, 68 (1970).
[32] H. H. Mende and Th. Kleinefeld, Appl. Phys. **20**, 159 (1979).
[33] W. Hagedorn and H. H. Mende, IEEE Trans. Magnetics **17**, 2946 (1981).
[34] R. M. Bozorth, Ferromagnetism, 2nd ed., Van Nostrand Co., Toronto 1953.
[35] K. Warmuth, Arch. Elektrotech. **30**, 761 (1936).
[36] D. S. Bloomberg and A. S. Arrott, Canad. J. Phys. **53**, 1454 (1975).
[37] T. L. Templeton and A. S. Arrott, J. appl. Phys. **59**, 4123 (1984).

*(Received February 25, 1988)*

phys. stat. sol. (a) 108, 403 (1988)

Subject classification; 75.30; 75.50; 75.60; S1.2

*Institut für Physik, Max-Planck-Institut für Metallforschung, Stuttgart*[1]) *(a) and Krupp Widia GmbH, Essen*[2]) *(b)*

# Investigations of the Magnetic Properties and Demagnetization Processes of an Extremely High Coercive Sm(Co, Cu, Fe, Zr)$_{7.6}$ Permanent Magnet

## I. Determination of Intrinsic Magnetic Material Parameters

By

K.-D. DURST[3]) (a), H. KRONMÜLLER (a), and W. ERVENS (b)

The intrinsic magnetic material parameters of the 2:17 type matrix phase and the 1:5 type precipitation phase in sintered multiphase $Sm(Co_{0.68}Cu_{0.08}Fe_{0.22}Zr_{0.02})_{7.6}$ permanent magnets are determined in the whole ferromagnetic temperature regime between 10 K and $T_C = 1070$ K. Magnetization curves measured parallel and perpendicular to the alignment axis are analysed taking into account the volume fractions of ferromagnetic phases, their magnetostatic interactions, and the angular distribution of grain axes. The influence of annealing treatments on the intrinsic magnetic properties of the relevant ferromagnetic phases are investigated.

Die intrinsischen magnetischen Materialparameter der 2:17-Matrixphase und der 1:5-Ausscheidungsphase in gesinterten Mehrphasen-$Sm(Co_{0,68}Cu_{0,08}Fe_{0,22}Zr_{0,02})_{7,6}$-Permanentmagneten werden im gesamten ferromagnetischen Temperaturbereich zwischen 10 K und $T_C = 1070$ K bestimmt. Die parallel und senkrecht zur Ausrichtungsachse gemessenen Magnetisierungskurven werden unter Berücksichtigung der ferromagnetischen Volumenanteile, ihrer magnetostatischen Wechselwirkungen und der Winkelverteilung der Kornachsen analysiert. Der Einfluß von Temperungen auf die intrinsischen magnetischen Eigenschaften der relevanten ferromagnetischen Phasen wird untersucht.

## 1. Introduction

Sintered magnets with the composition $Sm(Co_{0.68}Cu_{0.08}Fe_{02.2}Zr_{0.02})_{7.6}$ belong to permanent magnet materials with the highest coercive fields attainable today. Coercive fields of 3.5 T at room temperature can be obtained. The most important advantage of this magnet material, compared with the new class of Fe–Nd–B magnets [1 to 3], is their high Curie temperature and the excellent temperature stability of the coercive field at elevated temperatures. The mean pre-requisite for high coercive fields in sintered $Sm(Co_{0.68}Cu_{0.08}Fe_{0.22}Zr_{0.02})_{7.6}$ permanent magnets is a complex annealing treatment which leads to a complicated precipitation structure which acts as a very strong pinning centre for domain walls. TEM photographs parallel and perpendicular to the easy axis of magnetization ($c$-axis) of the "precipitation hardened" alloy $Sm(Co_{0.68}Cu_{0.08}Fe_{0.22}Zr_{0.02})_{7.6}$ are shown in Fig. 1 a and b. The rhombic structure parallel to the $c$-axis (Fig. 1a) and the hexagonal network in the basal plane (Fig. 1b) indicate that the precipitation phase lies on pyramidal surfaces.

[1]) Heisenbergstr. 1, D-7000 Stuttgart 80, FRG.

[2]) Münchner Str. 90, D-4300 Essen, FRG.

[3]) Present address: Porsche Research and Development Center, Porschestr., D-7251 Weisach, FRG.

a b

Fig. 1. $Sm(Co_{0.68}Cu_{0.08}Fe_{0.22}Zr_{0.02})_{7.6}$: structure a) parallel, b) perpendicular to the *c*-axis

Various investigations [e.g. 4 to 10] showed that the interior of rhombic cells (region A of Fig. 1a) has the rhombohedral $Sm_2Co_{17}$ crystal structure. These cells are separated from each other by a coherent cell boundary phase (region B of Fig. 1a) with the hexagonal $SmCo_5$ crystal structure. The very thin lines perpendicular to the *c*-axis (region C of Fig. 1a) result from an extremely thin, modified hexagonal $Sm_2Co_{17}$ phase.

As the result of the known phase diagrams of Sm–Co–Cu [11, 12], Sm–Co–Fe [13, 14], Sm–Co–Zr [15] and results from STEM investigations of $Sm(Co, Cu, Fe, Zr)_{7.6}$ permanent magnets it can be concluded that in the magnetically hardened state Cu should be soluble almost exclusively in the hexagonal 1:5 type precipitation phase, Fe in the rhombohedral 2:17 type matrix phase and Zr in the extremely thin hexagonal 2:17 phase. A review of the complex metallurgical behaviour of the five-component system Sm–Co–Cu–Fe–Zr has been given by Ray [16, 17].

The extremely high coercive field of sintered $Sm(Co_{0.68}Cu_{0.08}Fe_{0.22}Zr_{0.02})_{7.6}$ alloys can be obtained by a homogenization at 1160 °C for 1 h, followed by an isothermal annealing treatment at 840 °C for 24 h with subsequent slow cooling down to 400 °C in 3 h. The result of the homogenization heat treatment at 1160 °C is a one-phase hexagonal 2:17 structure which during the 840 °C annealing treatment transforms into the rhombohedral 2:17 structure with subsequent growth of the 1:5 precipitation structure. At the end of the 840 °C annealing treatment the cellular morphology is fully developed, but the resulting coercive field still is very low ($\mu_0 H_c = 0.02$ T at room temperature [20]). Only by means of a subsequent slow cooling process down to 400 °C the coercive field is increased substantially ($\mu_0 H_c = 3.5$ T at room temperature). Because this drastic increase of $H_c$ is not correlated with observable variations of the microstructure [21] it is assumed that both, changes of the chemical composition of the matrix and precipitation phase and modifications of the direct phase boundary matrix/precipitation, should be responsible for this effect [20, 22].

A pre-requisite for a precise analysis of the magnetic hardening mechanism in $Sm(Co_{0.68}Cu_{0.08}Fe_{0.22}Zr_{0.02})_{7.6}$ magnets is the accurate knowledge of the intrinsic magnetic material parameters of the real appearing phases for various characteristic points of the annealing treatment required to obtain high $H_c$ values. Accordingly we have determined the intrinsic magnetic material parameters, such as the spontaneous magnetization, $M_s$, magnetocrystalline anisotropy constant, $K_1$, and Curie temperature, $T_C$, for the relevant ferromagnetic phases in the whole ferromagnetic temperature region from 10 to 1070 K for $Sm(Co_{0.68}Cu_{0.08}Fe_{02.2}Zr_{0.02})_{7.6}$ sintered magnets after the following treatments: (i) 1160 °C homogenization heat treatment, (ii) 840 °C isothermal annealing treatment, and (iii) slow cooling down to 400 °C. This final state of the magnet will be denoted as the processed state. These results allow conclusions about phase compositions and comparison of experimental $H_c$ values with theoretical predictions for pinning fields of different magnetic phases.

## 2. Experimental

The magnetic properties of sintered magnets with the nominal composition $Sm(Co_{0.68}Cu_{0.08}Fe_{0.22}Zr_{0.02})_{7.6}$ were studied in the temperature range between 10 and 1100 K in field strengths up to 11 T using a vibrating sample magnetometer. The magnetic measurements were performed on cubes of size $3 \times 3 \times 3$ mm$^3$ whose demagnetizing factor could be approximated by that of a sphere (1/3).

The grain alignment was determined by observations of the orientation of magnetic domains by means of the magneto-optical Kerr effect [23].

The heat treatment of the sintered samples which were wrapped in Ta foils was given in quartz ampoules evacuated to $10^{-4}$ Pa and refilled with pure argon.

## 3. Experimental Results and Discussion

### 3.1 $Sm(Co_{0.68}Cu_{0.08}Fe_{0.22}Zr_{0.02})_{7.6}$ *magnets in the processed state*

#### 3.1.1 *Thermomagnetic analysis*

Fig. 2 shows the temperature dependence of the magnetic polarization of a Sm(Co, Cu, Fe, Zr)$_{7.6}$ sample. The polarization was measured with an applied field of 0.5 T parallel to the alignment direction. The Curie temperature of the sample is 1070 K. Sm(Co, Cu, Fe, Zr)$_{7.6}$ magnets in the processed state do not show two different $T_C$ values as expected for magnetic two-phase materials. The reasons for this

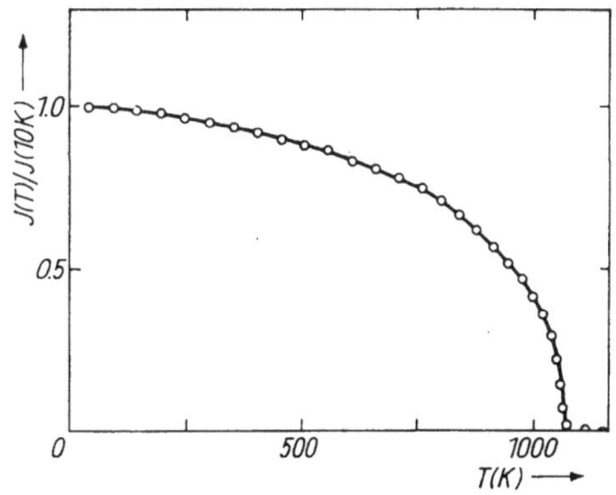


Fig. 2. Temperature dependence of polarization of a $Sm(Co_{0.68}Cu_{0.08}Fe_{0.22}Zr_{0.02})_{7.6}$ permanent magnet in the processed state. The applied field of 0.5 T was parallel to the alignment direction

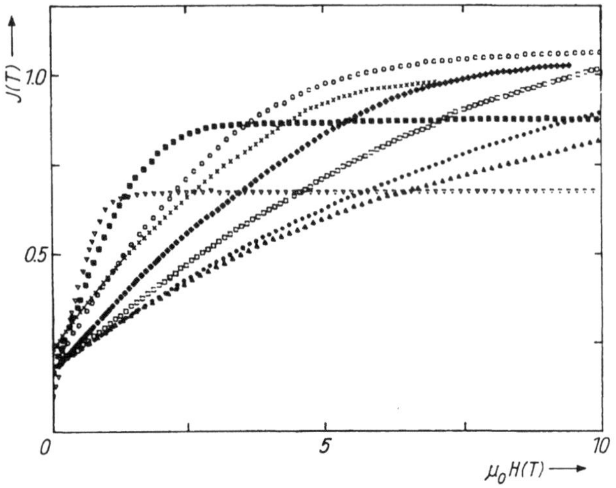


Fig. 3. Reversible recoil curves at selected temperatures (▲ 10, ● 220, □ 285, ◆ 400, × 465, ○ 525, ■ 735, ▽ 925 K) measured with H applied perpendicular to the alignment axis of as intered $Sm(Co_{0.68}Cu_{0.08}Fe_{0.22}Zr_{0.02})_{7.6}$ permanent magnet in the processed state

behaviour could be:

a) one phase has a continuous concentration gradient of one of its components (perhaps Cu) which would lead to a continuous distribution of $T_C$-values;

b) the $T_C$-values of both phases are identical;

c) the magnetostatic stray field of the ferromagnetic phase with the higher $T_C$ (presumably the matrix phase of 2:17 type) smears out the magnetic phase transition of the phase with the lower $T_C$ (presumably 1:5 type precipitation phase).

### 3.1.2 *Hard direction magnetization curves*

Fig. 3 shows the recoil curves with $H$ applied perpendicular to the alignment direction. As shown in [23, 24] for these curves magnetization takes place only by reversible rotation procesess and therefore can be analyzed.

### 3.1.3 *Determination of the anisotropy field and saturation magnetization*

From the magnetization curves measured perpendicular to the alignment direction (Fig. 3) the anisotropy fields of the 2:17 matrix phase and the 1:5 precipitation phase ($H_a^{2:17}$, $H_a^{1:5}$) should be determinable. The most convenient method is to plot the

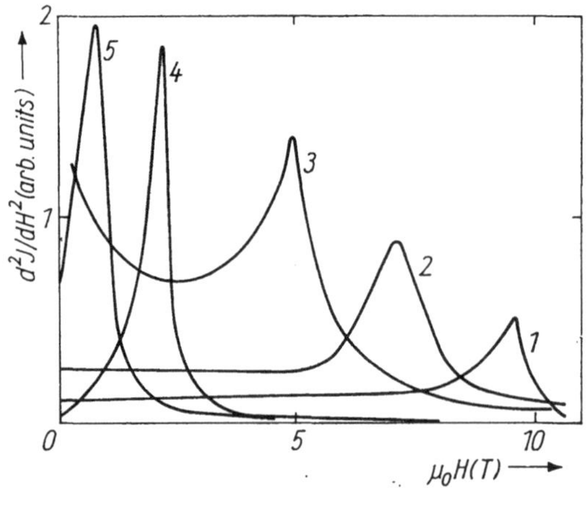


Fig. 4. Plot of $d^2J/dH^2$ vs. $H$ at various temperatures ((1) 285, (2) 400, (3) 520, (4) 735, (5) 925 K) for a $Sm(Co_{0.68}Cu_{0.08}Fe_{0.22}Zr_{0.02})_{7.6}$ permanent magnet in the processed state

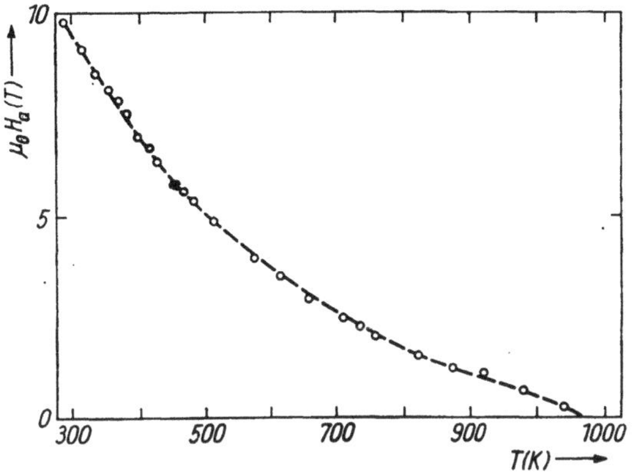


Fig. 5. Temperature dependence of the experimentally determined anisotropy field of $Sm(Co_{0.68}Cu_{0.08}Fe_{0.22}Zr_{0.02})_{7.6}$ in the processed state

second field derivative of polarization in the hard direction as a function of the field [25]. The anisotropy field is the field strength where $d^2J/dH^2$ shows a peak. Fig. 4 shows for some selected temperatures that at each temperature only one peak could be determined. This means that only one anisotropy field in the polarization curves perpendicular to the alignment axis for multiphase Sm(Co, Cu, Fe, Zr)$_{7.6}$ magnets could be determined definitely. Fig. 5 shows the temperature dependence of the experimentally determined anisotropy field, $H_a$. The experimental values for $H_a$ are restricted to temperatures above 270 K because at lower temperatures $H_a$ exceeds our maximum field of 11 T. The reason for the definite determination of only one anisotropy field in the polarization curves perpendicular to the alignment direction presumably is due to the small product $J_s v$ of the 1:5 type precipitation phase with its small volume fraction, $v$ ($\approx 25\%$). The experimentally determined values of $H_a$ therefore are ascribed to the 2:17 type matrix phase.

Identical results to the SPD technique (Fig. 4) can be obtained by plotting $J$ in the hard direction versus the inverse square of the field [23, 26].

For the determination of the saturation polarization, $J_s$, of the processed Sm(Co, Cu, Fe, Zr)$_{7.6}$ magnets the polarization curves measured parallel to the alignment direction can be used. By plotting $J$ versus $1/H^2$ and extrapolation to $H \to \infty$ with consideration of the so-called paraprocess [27] which is effective at high fields [26] the temperature dependence of $J_s$ for processed Sm(Co, Cu, Fe, Zr)$_{7.6}$ shown in Fig. 6 can be determined.

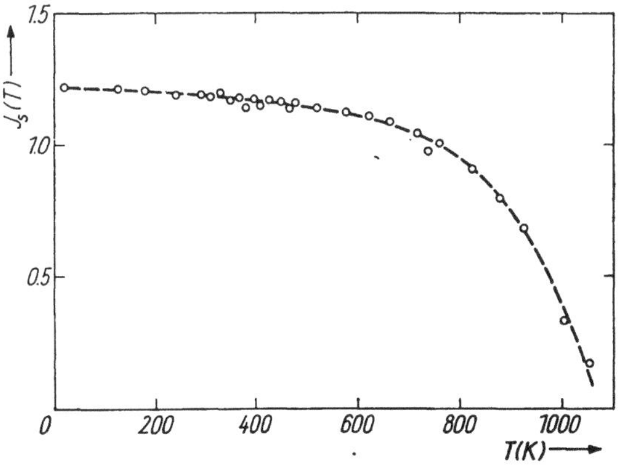


Fig. 6. Temperature dependence of the experimentally determined saturation polarization of processed $Sm(Co_{0.68}Cu_{0.08}Fe_{0.22}Zr_{0.02})_{7.6}$ magnets

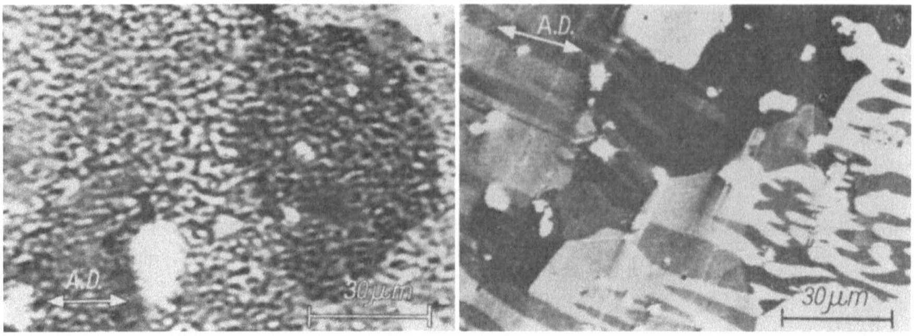


a b

Fig. 7. Domain structures of a plane parallel to the alignment direction (A.D.) for a) a processed and b) a 1160 °C-homogenized $Sm(Co_{0.68}Cu_{0.08}Fe_{0.22}Zr_{0.02})_{7.6}$ magnet in the thermally demagnetized state

### *3.1.4 Determination of the distribution of grain orientations*

For a processed $Sm(Co, Cu, Fe, Zr)_{7.6}$ magnet in the thermally demagnetized state Fig. 7 a shows a micrograph of a plane parallel to the alignment direction taken with the magneto-optical Kerr effect. Fig. 7a shows a rather narrow and more or less irregular laminar domain structure which is the consequence of the cellular precipitation microstructure (Fig. 1) of the processed magnet. It is obvious that determination of grain orientations by observation of the orientation of magnetic domains is not possible for the processed state (Fig. 7a). If the $Sm(Co, Cu, Fe, Zr)_{7.6}$ magnet is only homogenized at 1160 °C without subsequent annealing treatment the cellular precipitation microstructure has not yet developed (compare Section 1) and Fig. 7b shows that in this state domains are large and rather regularly shaped. Therefore, the homogenized state of $Sm(Co, Cu, Fe, Zr)_{7.6}$ is appropriate to determine the distribution of grain orientations. Fig. 8 shows the histogram of the angle $\varphi_0$ which gives the probability $p(\varphi_0)$ to find a grain with a deviation $\varphi_0$ from the ideal alignment axis. $p(\varphi_0)$ is approximately a Gaussian $\exp(-\varphi_0^2/2\sigma^2)$ with $\sigma = 14°$.

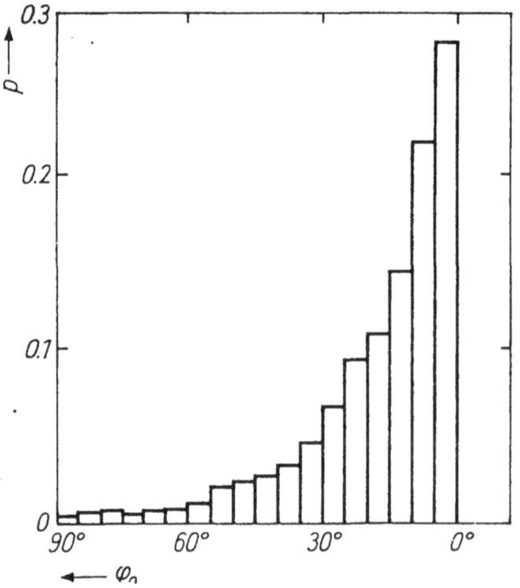


Fig. 8. Histogram of the angle $\varphi_0$ for an aligned $Sm(Co_{0.68}Cu_{0.08}Fe_{0.22}Zr_{0.02})_{7.6}$ magnet

### *3.1.5 Determination of the intrinsic magnetic material parameters of the 2:17 type matrix and 1:5 precipitation phase*

In [23] the authors showed that in nearly oriented sinter magnets consisting of one magnetic phase the intrinsic magnet material parameters of this magnetic phase can be determined with high accuracy. Now these results are extended to polycrystalline magnets consisting of two main magnetic phases as essentially is the case for processed Sm(Co, Cu, Fe, Zr)$_{7.6}$ magnets.

The total free energy, $\varphi_t$, of uniaxial Sm(Co, Cu, Fe, Zr)$_{7.6}$ magnets consisting of the 2:17 type matrix (phase 1) and 1:5 type precipitations (phase 2) with their $c$-axis oriented parallel is composed of the sum of crystalline anisotropy energy, magnetostatic energy, and stray field energy of both phases. If we neglect the stray field energy due to internal interfaces (phase boundaries between 2:17 and 1:5 type phases) the total free energy per unit volume for one grain composed of 2:17 and of 1:5 type phases is given by

$$\varphi_t = v_1 K_{1,1} \sin^2 \varphi_1 + v_1 K_{2,1} \sin^4 \varphi_1 + v_2 K_{1,2} \sin^2 \varphi_2 + v_2 K_{2,2} \sin^4 \varphi_2 - - v_1 H M_{s,1} \cos (\varphi_0 - \varphi_1) - v_2 H M_{s,2} \cos (\varphi_0 - \varphi_2) , \quad (1)$$

where $v_1(v_2)$ means the volume fraction of phase 1(2), $K_{1,1}(K_{1,2})$ the first magnetic anisotropy coefficient of phase 1(2), $K_{2,1}(K_{2,2})$ the second magnetic anisotropy coefficient of phase 1(2), $M_{s,1}(M_{s,2})$ the saturation magnetization of phase 1(2), $\varphi_1(\varphi_2)$ the angle between $M_{s,1}(M_{s,2})$ and the $c$-axis, and $\varphi_0$ the angle between the applied magnetic field $\boldsymbol{H}$ and the $c$-axis (Fig. 9). If a spectrum of grain orientations is considered, (1) is transformed to

$$\varphi_t' = v_1 K_{1,1} \sin^2 \varphi_1 + v_1 K_{2,1} \sin^4 \varphi_1 + v_2 K_{1,2} \sin^2 \varphi_2 + v_2 K_{2,2} \sin^4 \varphi_2 - - v_1 H M_{s,1} \sum_{i=1}^{N} \cos (\varphi_0^i - \varphi_1)\, p(\varphi_0^i) - v_2 H M_{s,2} \sum_{i=1}^{N} \cos (\varphi_0^i - \varphi_2)\, p(\varphi_0^i) , \quad (2)$$

where the summation has been approximated by $N$ angle intervals between 0° and 90°. $\varphi_0^i$ is the angle between the applied magnetic field and the $c$-axis in the angle interval $i$, and $p(\varphi_0^i)$ is the constant probability for a misorientation angle in the angle interval $i$. Variations with respect to the angles $\varphi_1$ and $\varphi_2$ gives for phase 1

$$H = \frac{2K_{1,1} \sin \varphi_1 \cos \varphi_1 + 4K_{2,1} \sin^3 \varphi_1 \cos \varphi_1}{M_{s,1} \sum_{i=1}^{N} \sin (\varphi_0^i - \varphi_1)\, p(\varphi_0^i)} \quad (3)$$

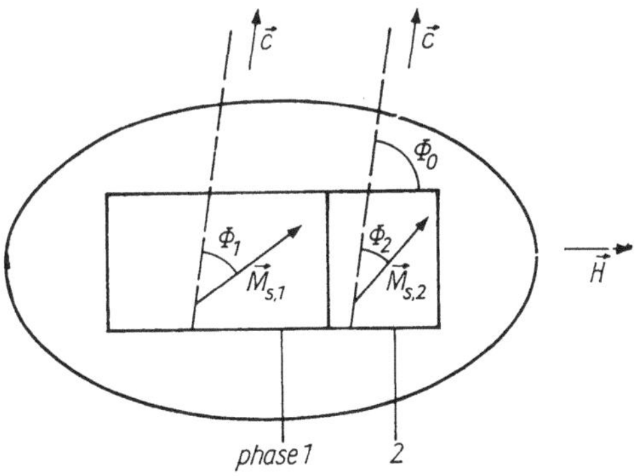


Fig. 9. Two ferromagnetic phases with applied external field

and for phase 2

$$H = \frac{2K_{1,2} \sin \varphi_2 \cos \varphi_2 + 4K_{2,2} \sin^3 \varphi_2 \cos \varphi_2}{M_{s,2} \sum_{i=1}^{N} \sin (\varphi_0^i - \varphi_2)\, p(\varphi_0^i)} . \tag{4}$$

With (3), (4) and the relation for the component of the total magnetization, $M$, in direction of the applied field,

$$M = v_1 M_{s,1} \sum_{i=1}^{N} \cos (\varphi_0^i - \varphi_1)\, p(\varphi_0^i) + v_2 M_{s,2} \sum_{i=1}^{N} \cos (\varphi_0^i - \varphi_2)\, p(\varphi_0^i) , \tag{5}$$

the magnetization curve $M(H)$ of both phases can be calculated in principle. The following relations between the experimentally determinable parameters and the intrinsic magnetic material parameters exist:

anisotropy field of phase 1 $$H_{a,1} = \frac{2K_{1,1} + 4K_{2,1}}{M_{s,1}} , \tag{6}$$

anisotropy field of phase 2 $$H_{a,2} = \frac{2K_{1,2} + 4K_{2,2}}{M_{s,2}} , \tag{7}$$

total saturation magnetization $$M_s = v_1 M_{s,1} + v_2 M_{s,2} , \tag{8}$$

effective anisotropy constant $$K_{eff} = v_1(K_{1,1} + K_{2,1}) + v_2(K_{2,1} + K_{2,2}) . \tag{9}$$

The effective anisotropy constant can be determined experimentally from the area between the reversible polarization curves in the easy and hard direction. Thus the number of six unknown material parameters ($K_{1,1}$, $K_{1,2}$, $K_{2,1}$, $K_{2,2}$, $M_{s,1}$, $M_{s,2}$) in (3) to (5) can be reduced to two (e.g. $K_{1,1}$ and $K_{1,2}$). If the phase volumes ($v_1$, $v_2$) and the grain orientations are known, a determination of all six parameters is possible by variation of the values of two parameters in (3) to (5) until correspondence between experimental and theoretical magnetization curves is achieved.

In principle the magnetizations of both phases are not independent of each other as assumed so far. They are coupled by stray fields which result from magnetic surface poles due to the discontinuity of magnetization at phase boundaries. Estimations of relevant stray field contributions to the total free energy in processed $Sm(Co, Cu, Fe, Zr)_{7.6}$ magnets lead to the following relations [26]:

$$\begin{aligned} H = \{ & 2K_{1,1} \sin \varphi_1 \cos \varphi_1 + 4K_{2,1} \sin^3 \varphi_1 \cos \varphi_1 + \pi[M_{s,1}^2 v_1[\sin (120^\circ - 2\varphi_1) + \\ & + \sin (240^\circ - 2\varphi_1)] - 2M_{s,1} M_{s,2} v_2 \sin (120^\circ - \varphi_1) \cos (120^\circ - \varphi_2) - \\ & - 2M_{s,1} M_{s,2} v_2 \sin (60^\circ - \varphi_1) \cos (60^\circ - \varphi_2)]\} / [M_{s,1} \sum_{i=1}^{N} \sin (\varphi_1 - \varphi_0^i)\, p(\varphi_0^i)] \end{aligned} \tag{10}$$

for phase 1, and to the relation

$$\begin{aligned} H = \{ & 2K_{1,2} \sin \varphi_2 \cos \varphi_2 + 4K_{2,2} \sin^3 \varphi_2 \cos \varphi_2 - \pi[M_{s,2}^2 v_2 (\sin (120^\circ - 2\varphi_2) + \\ & + \sin (240^\circ - 2\varphi_2)) + 2M_{s,1} M_{s,2} v_1 \sin (60^\circ - \varphi_2) \cos (60^\circ - \varphi_1) + \\ & + 2M_{s,1} M_{s,2} v_2 \sin (120^\circ - \varphi_2) \cos (120^\circ - \varphi_1)]\} / [M_{s,2} \sum_{i=1}^{N} \sin (\varphi_2 - \varphi_0^i)\, p(\varphi_0^i)] \end{aligned} \tag{11}$$

for phase 2.

Thus for the determination of the intrinsic magnetic material parameters of the 2:17 type matrix (phase 1) and the 1:5 type precipitation (phase 2) (5), (10), (11) together with (6) to (9) have to be solved.

Fitting of experimental magnetization curves by means of (5), (10), (11) requires the experimental determination of $H_{a,1}$, $H_{a,2}$, $M_s$, $K_{eff}$ (6) to (9) and the phase volumes $v_1$, $v_2$. By means of a quantitative microstructure analysis [26] from TEM photographs (e.g. Fig. 1) the following phase volumes of the 2:17 type matrix (phase 1) and the 1:5 type precipitation (phase 2) were obtained:

$$v_1 = 0.75\,, \qquad v_2 = 0.25\,. \tag{12}$$

In Section 3.1.3 it was shown that only one anisotropy field in the experimental magnetization curve could be determined. This experimentally determinable anisotropy field was ascribed to the 2:17 type matrix (phase 1). The effective anisotropy constant, $K_{eff}$, in principle is another experimentally determinable parameter (9). Because of grain misorientations in the investigated highly anisotropic processed Sm(Co, Cu, Fe, Zr)$_{7.6}$ magnets an experimental determination of $K_{eff}$, however is not possible at temperatures below about 700 K. The following four parameters may be determined:

$$v_1, v_2, H_{a,1}, J_s\,, \tag{13}$$

whereas the magnetic material parameters, $H_{a,2}$ and $K_{eff}$, are lacking. However, a determination of the parameters of both phases is possible if the second magnetic anisotropy coefficients ($K_{2,1}$ and $K_{2,2}$) for both phases are neglected. As known from literature, e.g. [29 to 33], the second magnetic anisotropy coefficient of Sm–Co based magnets is of minor importance.

With $v_1 = 0.75$ and $v_2 = 0.25$ (12), the grain orientation distribution of Fig. 8 and the experimentally determined parameters at 285 K (Fig. 5, 6),

$$J_s = 1.21\ \mathrm{T}\,, \qquad \mu_0 H_{a,1} = 9.60\ \mathrm{T}\,, \tag{14}$$

it is possible to fit the experimental polarization curve at 285 K (Fig. 3) by means of (5), (10), (11) with the following fit parameters:

$$K_{1,1} = 5.0 \times 10^6\ \mathrm{J/m^3}\,, \qquad K_{1,2} = 1.9 \times 10^6\ \mathrm{J/m^3}\,. \tag{15}$$

With (6) to (8) it follows:

$$J_{s,1} = 1.32\ \mathrm{T}\,, \qquad J_{s,2} = 0.80\ \mathrm{T}\,, \qquad \mu_0 H_{a,2} = 6.0\ \mathrm{T}\,. \tag{16}$$

Fig. 10 shows the calculated polarization curves (5), (10), (11) at 285 K (with the material parameters (14) to (16)) of the 2:17 type matrix (phase 1) and 1:5 type

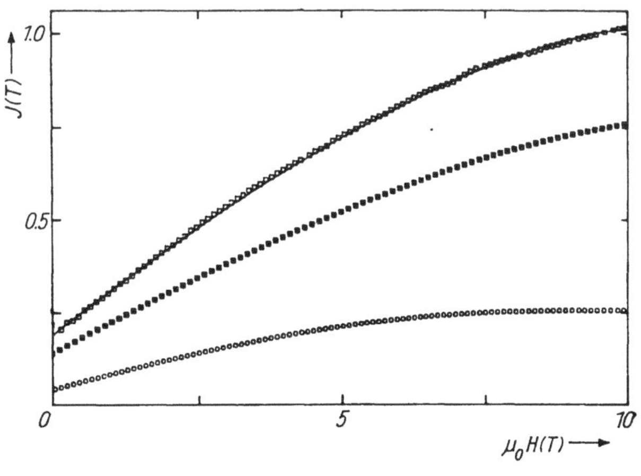


Fig. 10. Experimental (□) and calculated (solid line) polarization curves in the hard direction at $T = 285$ K for a processed $Sm(Co_{0.68}Cu_{0.08}Fe_{0.22}Zr_{0.02})_{7.6}$ permanent magnet (■ 2:17, phase 1; ○ 1:5, phase 2)

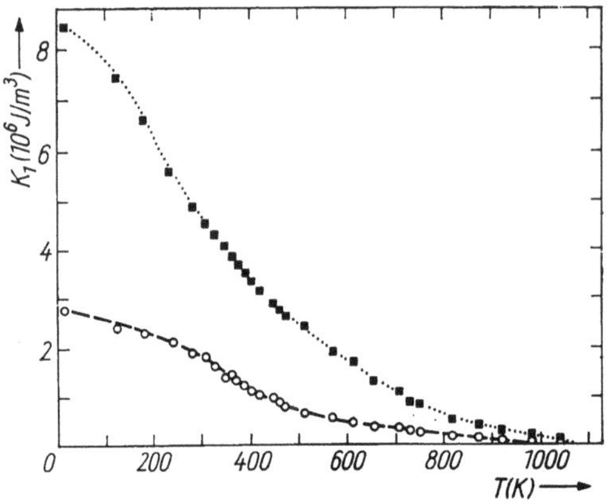


Fig. 11. Temperature dependence of crystal anisotropy constant $K_1$ of the 2:17 type matrix (■) and 1:5 type phase (○) of processed $Sm(Co_{0.68}Cu_{0.08}Fe_{0.22}Zr_{0.02})_{7.6}$ magnets

precipitation (phase 2) whose superposition is in very good correspondence with the experimentally determined magnetization curve in the hard direction of the processed $Sm(Co, Cu, Fe, Zr)_{7.6}$ magnet. The temperature dependence of the fit parameters $K_{1,1}$ and $K_{1,2}$ is shown in Fig. 11, and the temperature dependences of the parameters $J_{s,1}$, $J_{s,2}$, and $H_{a,2}$ (calculated from (6) to (8)) are shown in Fig. 12. From Fig. 12a the Curie temperatures of both phases can be determined to

$$T_{C,1} \approx 1070\ \mathrm{K}\,, \qquad T_{C,2} \approx 950\ \mathrm{K}\,. \tag{17}$$

It should be noted that neglecting the stray field coupling between both ferromagnetic phases (3) to (5) leads to values for the material parameters of both phases which differ only about 5% from the more precise results (14) to (16).

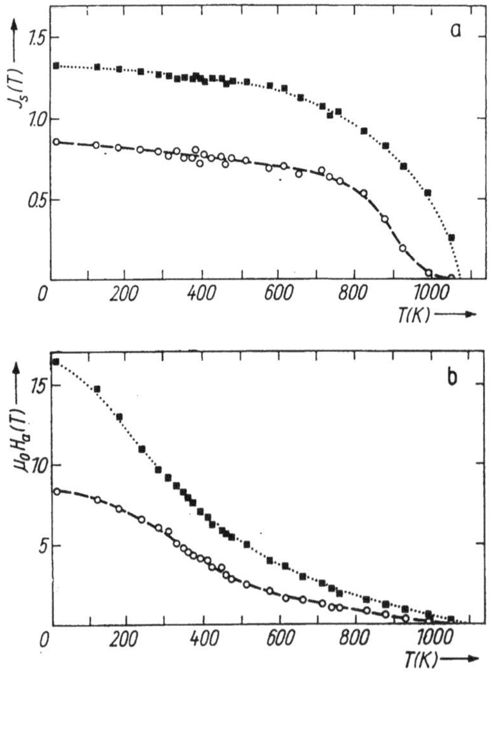


Fig. 12. Temperature dependence a) of saturation polarization and b) of anisotropy fields of 2:17 (■) and 1:5 (○) phases of processed $Sm(_{0.68}Cu_{0.08}Fe_{0.22}Zr_{0.02})_{7.6}$ magnets

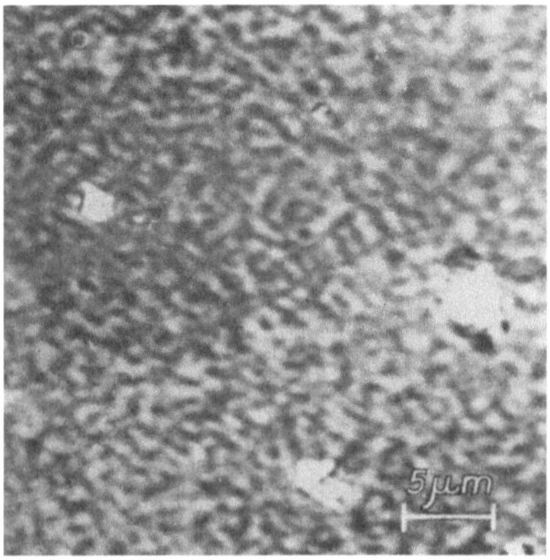


Fig. 13. Domain structure on a plane perpendicular to the alignment axis for a processed $Sm(Co_{0.68}Cu_{0.08}Fe_{0.22}Zr_{0.02})_{7.6}$ magnet in the thermally demagnetized state

A precise direct determination of the specific domain wall energies of both phases $\gamma_1$, $\gamma_2$, the exchange constants $A_1$, $A_2$, and the domain wall thicknesses $\delta_{B,1}$, $\delta_{B,2}$ is not possible. However, we can give estimations for these parameters. From the closure domains of a processed Sm(Co, Cu, Fe, Zr)$_{7.6}$ magnet in the thermally demagnetized state (Fig. 13) only an "effective" wall energy $\gamma_{eff}$ can be obtained. Using the specific procedure of Bodenberger and Hubert [34] we obtain at room temperature

$$\gamma_{eff} = 3.1 \times 10^{-6}\ \text{J/cm}^2 . \tag{18}$$

From the relation for the effective exchange constant

$$A_{eff} = \frac{\gamma_{eff}^2}{16 K_{eff}} \tag{19}$$

and $K_{eff} = 4.2 \times 10^6\ \text{J/m}^3$ (9) at room temperature follows:

$$A_{eff} = 1.4 \times 10^{-13}\ \text{J/cm} . \tag{20}$$

If we use for the exchange constants of both phases (with their very similar $T_C$-values (17)) this value of $A_{eff}$ as a first approximation, we get from the relation $\gamma = 4\sqrt{AK_1}$ for 180° Bloch walls

$$\gamma_1 = 3.4 \times 10^{-6}\ \text{J/cm}^2 , \qquad \gamma_2 = 2.0 \times 10^{-6}\ \text{J/cm}^2 , \tag{21}$$

and from the relation $\delta_B = \pi \sqrt{A/K_1}$ the domain wall thicknesses

$$\delta_{B,1} = 5.2\ \text{nm} , \qquad \delta_{B,2} = 8.5\ \text{nm} . \tag{22}$$

In summary for processed Sm(Co, Cu, Fe, Zr)$_{7.6}$ magnets the following material parameters were obtained:

2:17 type matrix:
$$J_s = 1.32\ \text{T} , \quad K_1 = 5.0 \times 10^6\ \text{J/m}^3 ,$$
$$\mu_0 H_a = 9.6\ \text{T} ,$$
$$\gamma \approx 3.4 \times 10^{-6}\ \text{J/cm}^2 , \quad \delta_B \approx 5.2\ \text{nm (RT-values)} ,$$
$$T_C = 1070\ \text{K}; \tag{23a}$$

1:5 type precipitation:
$$J_s = 0.80\ \text{T} , \quad K_1 = 1.9 \times 10^6\ \text{J/m}^3 ,$$
$$\mu_0 H_a = 6.0\ \text{T} ,$$
$$\gamma \approx 2.0 \times 10^{-6}\ \text{J/cm}^2 , \quad \delta_B \approx 8.5\ \text{nm (RT-values)} ,$$
$$T_C = 950\ \text{K} . \tag{23b}$$

### *3.1.6 Why do the 1:5 type precipitations act as a "soft" magnetic phase?*

If the values for the material parameters of the 2:17 type matrix and 1:5 type precipitation phase (23) of processed $Sm(Co, Cu, Fe, Zr)_{7.6}$ are compared to the corresponding values of the pure binaries $SmCo_5$ and $Sm_2Co_{17}$, e.g. [35 to 38], it can be concluded that in processed $Sm(Co, Cu, Fe, Zr)_{7.6}$ magnets the 2:17 matrix has better hard magnetic properties than pure $Sm_2Co_{17}$ ($J_s$, $K_1$, $H_a$ are higher, $T_C$ lower), whereas the 1:5 type precipitation phase shows worse hard magnetic properties than the pure $SmCo_5$ phase ($J_s$, $K_1$, $H_a$, $T_C$ lower). The enhancement of $K_1$ and $J_s$ of the 2:17 type matrix as compared to pure $Sm_2Co_{17}$ can be explained by substitution of Co by Fe and Zr on sites with planar anisotropy in the rhombohedral $Sm_2Co_{17}$ structure (e.g. [39]). The reduction of $J_s$ and $T_C$ of the 1:5 type phase as compared to pure $SmCo_5$ can be explained by the magnetic dilution effect due to Cu atoms on Co sites. According to our results the crystal anisotropy of the Cu-enriched 1:5 type precipitation phase is drastically reduced by a factor of about 7 as compared to pure $SmCo_5$. If we assume that in the processed $Sm(Co, Cu, Fe, Zr)_{7.6}$ magnets all Cu is dissolved in the 1:5 type precipitations (with a volume fraction of about 25%) we get for the 1:5 type phase the composition $Sm(Co_{0.65}Cu_{0.35})_5$ under the assumption of a Sm: (Co + Cu) stoichiometry of 1:5. If we compare the results (23b) for the 1:5 type phase with results from literature for $Sm(Co, Cu)_x$ phases (with $x = 5$ to 6) [40 to 43] it can be concluded that the 1:5 type phase should not have the Sm: (Co + Cu) stoichiometry of 1:5 but rather of 1:6. Various investigations [11, 12, 41] have shown that the Sm: (Co + Cu) ratio can be $<1/5$ although the structure still is of 1:5 type. In such an overstoichiometric 1:5 precipitation phase therefore not only Cu atoms (which reduce crystal anisotropy) are important but also excess Co atoms. These Co atoms substitute Sm atoms in the $SmCo_5$ structure preferentially on so-called "dumbbell" places in the form of Co pairs [44, 45] by means of which an approximation of the $Sm_2Co_{17}$ structure with its much smaller crystal anisotropy is achieved. The drastic reduction of the crystal anisotropy of the precipitation phase in processed $Sm(Co, Cu, Fe, Zr)_{7.6}$ magnets thus could be explained in principle.

### ***3.2 Influence of the annealing treatment on intrinsic magnetic properties***

All former investigations have been realized with processed $Sm(Co, Cu, Fe, Zr)_{7.6}$ magnets which have been homogenized at 1160 °C, subsequently isothermally annealed at 840 °C for 24 h and then slowly cooled down to 400 °C in 3 h. Important for the interpretation of the magnetic hardening mechanism of processed $Sm(Co, Cu, Fe, Zr)_{7.6}$ magnets is not only to know the intrinsic material parameters of the phases in the processed state, but also after the 1160 °C homogenization treatment and after the subsequent 840 °C isothermal heat treatment. For these characteristic temperatures of the annealing treatment we just report the results because the procedure is the same as described in the foregoing sections.

For the one-phase hexagonal 2:17 type structure after the *1160 °C homogenization treatment* the following magnetic parameters have been determined [26]:

$$\text{2:17 phase:}\quad J_s = 1.21\ \mathrm{T}\,,\quad K_1 = 3.9 \times 10^6\ \mathrm{J/m^3}\,,\quad \mu_0 H_a = 8.2\ \mathrm{T},\ \text{(RT-values)},\quad T_C = 1025\ \mathrm{K}\,. \tag{24}$$

After the 840 *°C annealing treatment* which leads to the characteristic precipitation structure (Fig. 1) the following magnetic parameters have been determined [26]:

$$\begin{aligned}&\text{2:17 type matrix:} && J_s = 1.30\ \mathrm{T}\,,\quad K_1 = 4.7 \times 10^6\ \mathrm{J/m^3}\,,\quad \mu_0 H_a = 9.1\ \mathrm{T}\,,\\ & && \text{(RT-values)}\,,\quad T_C = 1065\ \mathrm{K};\\ &\text{1:5 type precipitation:} && J_s = 0.94\ \mathrm{T}\,,\quad K_1 = 2.1 \times 10^6\ \mathrm{J/m^3}\,,\quad \mu_0 H_a = 5.6\ \mathrm{T}\,,\\ & && \text{(RT-values)}\,.\end{aligned} \tag{25}$$

The enhancement of $J_s$, $K_1$, $H_a$, and $T_C$ of the 2:17 type phase in the state after the 840 °C annealing treatment (25) compared to the state after the 1160 °C homogenization treatment (24) can be explained by diffusion of Cu into the new 1:5 type precipitation phase and Fe enrichment in the 17:2 phase. Comparison of material parameters of the 2:17 and 1:5 phases in the state after the 840 °C annealing treatment (25) and in the state after slow cooling down to 400 °C (23) shows only small changes. As a consequence of the slow cooling process from 840 to 400 °C the 2:17 type matrix becomes only a little bit magnetically "harder" and the 1:5 type precipitation phase a little bit magnetically "softer" but nevertheless the coercive field is increased enormously. This result will be discussed separately in Part 2 of this paper by the same authors.

## 4. Summary

From analysis of magnetization curves, domain structures, and TEM photographs the intrinsic magnetic material parameters of the real appearing ferromagnetic phases of multiphase sintered $Sm(Co_{0.68}Cu_{0.08}Fe_{0.22}Zr_{0.02})_{7.6}$ magnets have been determined in the whole ferromagnetic temperature range from 10 to 1070 K. The variations of intrinsic magnetic properties of the relevant 2:17 type matrix phase and of the 1:5 type precipitation phase due to annealing treatments have been determined. For the processed state of $Sm(Co_{0.68}Cu_{0.08}Fe_{0.22}Zr_{0.02})_{7.6}$ permanent magnets we obtained the results that the Cu rich 1:5 type precipitation phase is magnetically softer than the 2:17 type matrix phase and that the 1:5 type should not have the Sm: (Co + Cu) stoichiometry of 1:5 but rather of 1:6.

## References

[1] M. Sagawa, S. Fujimura, M. Tagawa, H. Yamamoto, and Y. Matsuura, J. appl. Phys. **55**, 2083 (1984).
[2] J. J. Croat, J. F. Herbst, R. W. Lee, and F. E. Pinkerton, J. appl. Phys. 55, 2078 (1984).
[3] K.-D. Durst, H. Kronmüller, and G. Schneider, 5-th Internat. Symposium on Magnetic Anisotropy and Coercivity in Rare Earth-Transition Metal Alloys, Ed. C. Herget, H. Kronmüller, and R. Poerschke, Bad Soden 1987, DPG-GmbH, Bad Honnef 1987 (p. 209).
[4] J. Fidler and P. Skalicky, J. Magnetism magnetic Mater. 27, 127 (1982).
[5] G. C. Hadjipanayis, Proc. 6-th Internat. Workshop on RE–Co Permanent Magnets, Ed. J. Fidler, Techn. Univ., Vienna 1982 (p. 609).
[6] L. Rabenberg, R. K. Mishra, and G. Thomas, J. appl. Phys. **53**, 2389 (1982).
[7] J. D. Livingston and D. L. Martin, J. appl. Phys. **48**, 1350 (1977).
[8] R. W. Lee, J. appl. Phys. **52**, 2549 (1981).
[9] E. Potenziani, K. S. Choe, and D. I. Paul, IEEE Trans. Magnetics **18**, 1457 (1982).
[10] J. Fidler, J. Magnetism magnetic Mater. **30**, 58 (1982).
[11] R. Glardon and W. Kurz, Z. Metallk. **70**, 386 (1979).
[12] A. J. Perry, J. Less Common Metals **51**, 153 (1977).
[13] G. Schneider, E.-Th. Henig, H. L. Lukas, and G. Petzow, J. less-common Metals **110**, 159 (1985).
[14] K. H. J. Buschow, J. less-common Metals **31**, 359 (1973).
[15] C. H. Allibert, S. Derkaoui, F. Delannay, and J. Laforest, 9-th Internat. Workshop on Rare-Earth Magnets, Ed. C. Herget, H. Kronmüller, and R. Poerschke, Bad Soden 1987, DPG-GmbH, Bad Honnef 1987 (p. 363).
[16] A. E. Ray, IEEE Trans. Magnetics **20**, 1614 (1984).
[17] A. E. Ray, J. appl. Phys. **55**, 2094 (1984).
[18] L. Rabenberg, R. K. Mishra, and G. Thomas, see [5] (p. 599).
[19] Chou Sochen, Wang Run, and Sun Guan fei, see [5] (p. 693).

[20] H. KRONMÜLLER, K.-D. DURST, W. ERVENS, and W. FERNENGEL, IEEE Trans. Magnetics **20**, 1569 (1984).
[21] J. FIDLER, private communications.
[22] K.-D. DURST, H. KRONMÜLLER, F. T. PARKER, and H. OESTERREICHER, phys. stat. sol. (a) **95** 213 (1986).
[23] K.-D. DURST and H. KRONMÜLLER, J. Magnetism magnetic Mater. **59**, 86 (1986).
[24] K.-D. DURST and H. KRONMÜLLER, Proc. 8-th Internat. Workshop on RE Magnets, Ed. K. J. STRNAT, Univ. of Dayton, 1985 (p. 725).
[25] G. ASTI and S. RINALDI, Phys. Rev. Letters **28**, 1584 (1972).
[26] K.-D. DURST, Thesis, Univ. Stuttgart 1986.
[27] T. HOLSTEIN and H. PRIMAKOFF, Phys. Rev. **58**, 1098 (1940).
[28] K.-D. DURST, Diploma Thesis, Univ. Stuttgart 1983.
[29] M. G. BENZ and D. L. MARTIN, J. appl. Phys. **55**, 4733 (1984).
[30] M. JURCZYK and A. WRZECIONO, IEEE Trans. Magnetics **20**, 1578 (1984).
[31] R. KÜTTERER, Thesis, Univ. Stuttgart 1978 .
[32] W. I. KHAN and D. MELVILLE, phys. stat. sol. (a) **48**, 209 (1978).
[33] A. S. ERMOLENKO and E. V. SHCHERBAKOVA, Fiz. Metallov i Metallovedenie **48**, 275 (1979).
[34] R. BODENBERGER and H. HUBERT, phys. stat. sol. (a) **44**, K7 (1977).
[35] R. KÜTTERER, H. R. HILZINGER, and H. KRONMÜLLER, J. Magnetism magnetic Mater. **4**, 1 (1977).
[36] P. GAUNT and C. K. MYLVAGANAM, Phil. Mag. **B44**, 569 (1981).
[37] W. E. WALLACE, J. Magnetism magnetic Mater. **40**, 32 (1984).
[38] K. H. J. BUSCHOW, Rep. Progr. Phys. **40**, 1149 (1977).
[39] M. V. SATYANARAYANA, H. FUJII, and W. E. WALLACE, J. appl. Phys. **53**, 2374 (1982).
[40] B. BARBARA, Proc. 2-nd Internat. Symp. Coercivity and Anisotropy of Rare Earth-Transition Metal Alloys, Ed. K. J. STRNAT, Univ. of California, 1978 (p. 137).
[41] H. NAGEL and A. MENTH, IEEE Trans. Magnetics **14**, 67 (1978).
[42] H. NAGEL, J. appl. Phys. **50**, 1026 (1979).
[43] K. N. MELTON and H. NAGEL, J. appl. Phys. **48**, 2608 (1977).
[44] G. DWORSCHAK and Y. KHAN, J. Phys. Chem. Solids **35**, 1021 (1974).
[45] K. H. J. BUSCHOW and A. S. VAN DER GROOT, J. less-common Metals **14**, 323 (1968).

*(Received February 18, 1988)*

phys. stat. sol. (a) **108**, 417 (1988)

Subject classification: 75.60; 62.20; S1.1

*Center for Non Destructive Evaluation, Iowa State University, Ames*[1])

# Variation of the Magnetic Properties of AISI 4140 Steels with Plastic Strain

By

D. C. JILES

The magnetic properties of five different microstructures of AISI 4140 steel are measured as a function of compressive plastic deformation. Coercivity, hysteresis loss, remanence, permeability, differential permeability, and anhysteretic magnetisation are determined for strains of up to 10%. Coercivity, hysteresis loss, and initial permeability are found to be dependent on residual plastic strain and hardness. Remanence is only weakly dependent on plastic strain, while maximum differential permeability is insensitive to plastic strain over the range 2 to 10%. The results are interpreted in terms of changes in the number density of dislocations caused by straining beyond the yield point. Correlations are found between coercivity and hysteresis loss and between initial anhysteretic permeability and maximum differential permeability.

Die magnetischen Eigenschaften von fünf unterschiedlichen Mikrostrukturen von AISI 4140-Stahl werden als Funktion einer plastischen Kompressionsdeformation gemessen. Koerzitivkraft, Hysteresisverluste, Remanenz, Permeabilität, differentielle Permeabilität und anhysteretische Magnetisierung werden für Dehnungen bis zu 10% bestimmt. Es wird gefunden, daß die Koerzitivkraft, Hysteresisverlust und Anfangspermeabilität von den plastischen Restspannungen und Härte abhängig sind. Die Remanenz ist nur schwach von der plastischen Dehnung abhängig, während die maximale differentielle Permeabilität im Bereich 2 bis 10% unempfindlich gegenüber plastischer Dehnung ist. Die Ergebnisse werden mit Änderung der Dichte der Versetzungen interpretiert, die durch Dehnung über den Fließpunkt erzeugt werden. Korrelationen zwischen Koerzitivkraft und Hystereseverluste und zwischen anhysteretischer Anfangspermeabilität und maximaler differentieller Permeabilität werden gefunden.

## 1. Introduction

This paper reports on an investigation of the changes in bulk magnetic properties of some AISI 4140 steels with microstructure and plastic deformation. It is well known that the magnetic properties of steels are affected by the mechanical treatment of the material. This means conversely that magnetic measurements can be used to determine the mechanical state of the steel. The magnetic properties are affected differently by differences in the mechanical treatment, for example the properties are altered in distinct ways by an applied stress causing elastic strain, or by residual strain (plastic deformation) so that these two effects can be distinguished from one another using bulk magnetic measurements.

The objective of the present work was to make a basic study of magnetic properties of several different microstructures of 4140 steel as a function of compressive plastic deformation. The magnetic properties of steels under elastic stress have been the subject of a number of investigations over the years including Ewing [1], Bozorth and Williams [2], Lliboutry [3], Brugel and Rimet [4], Craik and Wood [5], Birss,

[1]) Ames, Iowa 50011, USA.

Faunce and Isaac [6] and Schneider and Richardson [7]. By comparison the effect of plastic deformation on magnetic properties has received less attention, although in principle the interpretation of the results should be simpler, since it is not necessary in this case to distinguish between stress applied before the magnetic field is changed and stress applied after.

Ewing [1] investigated the effect of straining iron beyond its elastic limit. The results showed that tensile plastic deformation increased hysteresis loss and coercivity while the differential permeability was reduced by the compressive residual stress at all points on the hysteresis loop. In the light of present knowledge of magnetic hysteresis these results can be fairly easily interpreted since the plastic strain increases the density of dislocations which impede the movement of domain walls. This leads to a reduction in permeability and an increase in coercivity. Bozorth [8] also investigated the effect of plastic deformation on various magnetic materials. It was found that on application of tensile stress to iron the relative permeability at a field strength of 240 A/m increased from 2600 to 3300 in the elastic region. Once plastic deformation began the relative permeability was found to decrease reaching a value of 1600 at a stress of $2.8 \times 10^8$ N $m^{-2}$. When the applied load was removed a further decrease in relative permeability to 400 was observed although no further plastic deformation arose. These results seem to indicate that tensile stress increases the permeability of iron in the low field region (where the magnetostriction coefficient is positive) but that plastic deformation resulting from large tensile stresses reduces permeability.

There have been a number of studies of the effect of plastic deformation on the magnetic properties of iron, steel, and nickel by Cullity [9 to 13] which provide probably the only extensive investigation of the subject in recent years. In these investigations the magnetic properties such as hysteresis loss, initial permeability, and magnetostriction were measured on specimens that had been subject to tensile plastic deformation. The residual stress in the materials was then measured using X-ray diffraction [9] and the results correlated with the magnetic measurements. Cullity [10] was able to explain the seemingly contradictory results obtained when measuring residual stress by X-ray diffraction and mechanical methods. In plastically deformed metals the residual stress determined from X-ray diffraction results indicates for example that a specimen stretched beyond its yield point in tension and then unloaded has a residual compressive stress. This compressive stress can not be detected by mechanical relaxation. Therefore a number of authors have labelled it "fictitious" stress.

The magnetic hysteresis loss was shown to increase after tensile plastic deformation indicating that some microscopic mechanical changes had occured within the metal. It was suggested [10] that these results could be explained if most of the volume of the metal was strained in compression but that a small volume fraction was highly strained in tension, giving rise to a net zero stress in the specimen as a whole. The small volumes, which contain the large residual tensile stresses, act as pinning sites for domain walls thereby increasing hysteresis loss. In this way it was possible to resolve contradictions and explain both the magnetic and X-ray results. The regions with high residual stresses were identified as groups of dislocations forming sub-grain boundary walls while the regions with low compressive residual stresses were identified as the interiors of the sub-grains.

This interpretation of the effect of tensile plastic deformation was used to explain the observed changes in magnetostriction of nickel [11] which depended on the existence of a residual compressive stress throughout most of the volume. Further work on bulk magnetoresistance of nickel [12] served to confirm these conclusions.

The initial permeability of iron was measured as a function of bulk plastic deformation by Rusnak and Cullity [13]. They showed that not only the irreversible but also the reversible component of permeability in the Rayleigh region was dependent on plastic deformation. The permeability of the normal magnetisation curve can be written in the Rayleigh form

$$\mu(H) = \mu_n' + vH \,, \tag{1}$$

where $\mu_n'$ is the initial normal differential permeability which is the reversible component. An analysis of their results for the present work found $\mu_n'$ to vary with plastic deformation $\varepsilon$ according to the relation

$$\mu_n'(\varepsilon)/\mu_n'(0) = 1/(1 + c\varepsilon^{1/2}) \,, \tag{2}$$

where $c$ was a constant of value 17.3 for pure iron. This is an interesting and somewhat surprising result which supports the idea of a correlation between initial permeability and coercivity as observed by Jiles and Utrata [14] and Jiles [15]. Rusnak and Cullity were also able to relate their observed changes in initial permeability to dislocation density $N$ and dislocation spacing $D$, showing that the initial permeability was proportional to dislocation spacing where $D = (N)^{1/2}$ and $N$ was proportional to the plastic deformation $\varepsilon$. Further analysis of their results leads to the following relation between $N$ and $\varepsilon$:

$$N = N_0 + k\varepsilon \,, \tag{3}$$

where $N_0 = 4.7 \times 10^7\ \text{cm}^{-2}$ is the dislocation density before plastic deformation and $k = 3.2 \times 10^{10}$ per unit strain. The relation between initial permeability and dislocation density is then

$$\mu_n'(\varepsilon)/\mu_n'(0) = 1/\bigl(1 + \beta(N - N_0)^{1/2}\bigr) \,, \tag{4}$$

where $\beta = 1.02 \times 10^{-4}$.

The effects of stress and strain on magnetic properties are closely related, but the two effects can be separated. Langman [16] and Jiles and Atherton [17] have recently shown that coercivity is virtually independent of applied stress, while the maximum differential permeability, measured at the coercive point, does vary with stress. For iron and steel, the maximum differential permeability, and hence the differential permeability of the anhysteretic at the origin, increases slightly under tension and decreases markedly under compression [18]. Coercivity is strongly affected by plastic deformation as is indicated in results reported by Ewing [1], Kersten [19] and more recently by Jiles and Utrata [14, 15]. However the maximum differential permeability and the anhysteretic differential permeability at the origin were found to be virtually independent of plastic deformation [14].

In view of the strong effects of mechanical treatment on magnetic properties there have been suggestions that magnetic measurements be used for stress analysis in steels. Hoselitz ([20], p. 180) has suggested that coercive force, initial permeability and remanence should be employed for this. Studies were undertaken by Kersten [21] and Stanley [22].

It seems clear that information about the mechanical condition of a ferromagnetic material can be derived from the bulk magnetic properties. However restriction to measuring only coercivity, initial permeability and remanence is likely to exclude information contained in parameters such as hysteresis loss, anhysteretic magnetisation and maximum differential permeability. Therefore in this work we have measured a wide range of magnetic parameters in order to investigate correlations of these magnetic properties with plastic strain. The results show clear differences between the various microstructures in the magnetic properties as a function of strain. Hysteresis loss and coercivity were found to be closely related to mechanical hardness. Both were found to be dependent on plastic strain, as was the initial permeability. Cross correlations between maximum differential permeability and initial anhysteretic permeability, and between coercivity and hysteresis loss were found.

## 2. Experimental Procedure

Five specimens of 2.54 cm diameter AISI 4140 steel rod were subjected to different heat treatments indicated in Table 1. The specimens had nominal composition

Table 1
Heat treatment of specimen

| specimen | heat treatment | microstructure |
|---|---|---|
| A | hold at 860 °C for 2 h then cool in furnace | coarse pearlite |
| B | hold at 860 °C for 2 h then cool in air | fine pearlite |
| C | hold at 860 °C for 2 h then quench in salt pot at 500 °C hold for 16 h then air cool | bainite |
| D | hold at 860 °C for 2 h then quench in salt pot at 500 °C hold for 1 h then air cool | bainite |
| E | hold at 860 °C for 2 h then quench in oil to room temperature, temper at 200 °C for 1 h | martensite |

97 wt% Fe, 0.4 wt% C, 0.9 wt% Mn, 0.035 wt% P, 0.04 wt% S, 0.2 wt% Si, 1.0 wt% Cr, and 0.2 wt% Mo. All specimens were heated into the high temperature austenitic phase at 860 °C as shown in Fig. 1. They were then subjected to different heat treatments. The pearlite specimens were cooled slowly from the high temperature phase. The bainite specimens were quenched into a salt pot at 500 °C. Finally the martensite was rapidly quenched to room temperature in oil and then tempered at 200 °C.

Optical micrographs were made of the surfaces of the specimens in order to examine the microstructure. Hardness measurements were made using a Rockwell hardness

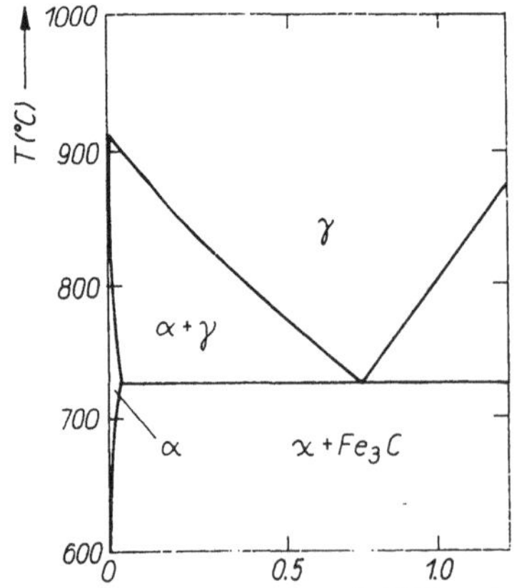


Fig. 1. Phase diagram of the iron carbon system

Table 2
Microstructure, grain size and hardness of the various specimen

| specimen | microstructure | grain size (μm) | hardness (multiples of $R_c$) |
|---|---|---|---|
| A | coarse pearlite | 15 | 22 |
| B | fine pearlite | 8 | 43 |
| C | bainite | — | 42 |
| D | bainite | — | 48 |
| E | martensite | — | 57 |

tester. Results of these examinations are shown in Table 2. Specimens were subjected to compressive stress and the residual strain was measured after the load was removed. Measurements were made on specimens at strain intervals of 2% up to a limit of 10% plastic deformation.

Magnetic measurements were made using a custom designed, computer controlled magnetic hysteresisgraph. Magnetic induction was measured on a Walker MF-3A fluxmeter and magnetic field on an MG-3A field indicator. Hard copy of the results were output on a chart recorder, while numerical values of $B$ and $H$ at 697 points on the $B$, $H$ plane were stored on floppy disc.

Specimens were carefully demagnetised under computer control using a software routine for demagnetisation. Thereafter the magnetic hysteresis characteristics of the material were measured, including initial permeability $\mu_n$ $(=\mu_n')$, remanence $B_R$, coercivity $H_c$, maximum differential permeability $\mu'_{max}$ and hysteresis loss $W_H$. The specimens were then demagnetised again, the initial magnetisation curve was measured, followed by a further demagnetisation and an anhysteretic magnetisation measurement.

From the results six important empirical magnetic parameters were determined using a software data analysis package. These were the five parameters indicated above which were used to define the hysteresis curve together with the initial anhysteretic permeability $\mu_a$ $(=\mu_a')$. The variation of these six magnetic quantities with such factors as hardness and plastic deformation were then studied. It should be noted in passing that in general we have found differential permeabilities to be rather more useful that permeabilities when relating magnetic properties to mechanical properties of ferromagnetic materials.

## 3. Results

The magnetic properties of the various specimens after 0 to 10% compressive plastic deformation are shown in Fig. 4 to 9. Generally it was observed that the pearlite specimens had the highest initial relative permeabilities (62 to 66 at 0% strain) and the lowest coercivities (1600 to 2160 A/m at 0% strain) and hysteresis losses (8.7043 to 10.769 J/m$^3$ at 0% strain) while the martensite had the lowest initial relative permeability (55 at 0% strain) and the highest coercivity (3280 A/m at 0% strain) and hysteresis loss (15.673 J/m$^3$ at 0% strain). The bainites were found to have properties which were intermediate between pearlite and martensite.

### *3.1 Correlation between magnetic properties and hardness*

The hardness of steels other than martensite is determined principally by the size and distribution of the second phase carbide precipitates. Generally the steel is made harder by the presence of more and finer precipitates and is made softer by fewer and larger precipitate particles. The maximum attainable hardness is a simple function of the carbon content and has been given by Digges, Rosenberg and Geil [23]. The reason for the dependence of hardness on carbide precipitates is that the strength of the metal is determined by the production and motion of dislocations. In a precipitation hardened alloy such as steel the increased strength is caused by interaction of dislocations with the precipitate phase which reduces the mobility of the dislocations and consequently leads to hardening ([24], p. 405).

In martensite there is an additional strengthening due to the shear transformation from austenite to martensite. This leads to an increase in hardness far beyond that which would be expected solely on the basis of precipitation hardening ([25], p. 227). The high strength of the martensite is due to the strong barriers to dislocation motion which are due mainly to solid solution hardening via the localised strains produced by the individual carbon atoms, and to precipitation hardening ([24], p. 531).

The factors which contribute to hardening of the steels and hence to the increase of yield strength are also factors which affect the motion of magnetic domain walls. That is localised microstrains and inhomogeneous second phase particles. It is therefore not surprising that a correlation was observed between some of the structure sensitive magnetic properties and the hardness. The main interest was therefore to find which magnetic quantities were dependent on hardness and which were not.

The coercivity $H_c$ varied with hardness as shown in Fig. 2. Such a relationship has been noted before in studies by Jiles and Verhoeven [26], and Jiles and Utrata [14]. Since the hardness can be related to the yield strength $\sigma_y$ of the steel as indicated by Dieter ([25], p. 228) and in the formula proposed by Shaw [27],

$$P/A = 3\sigma_y\ , \tag{5}$$

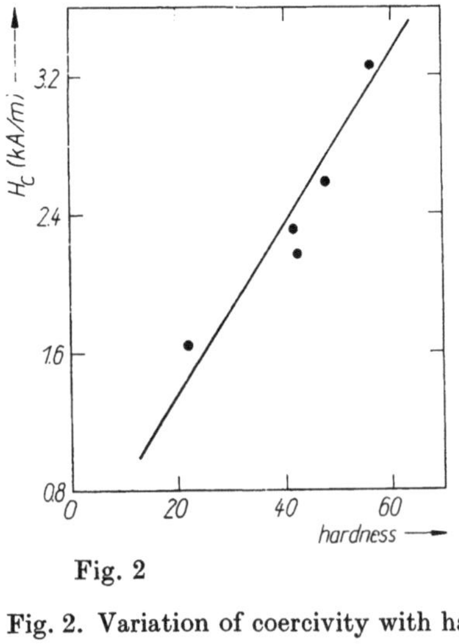


Fig. 2

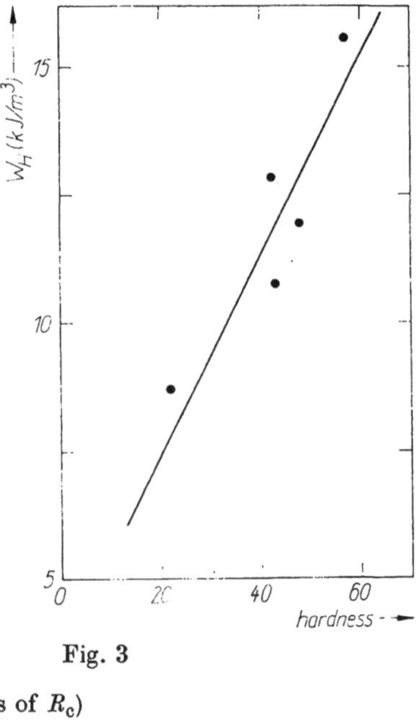


Fig. 3

Fig. 2. Variation of coercivity with hardness (in units of $R_c$)

Fig. 3. Variation of hysteresis loss with hardness (in units of $R_c$)

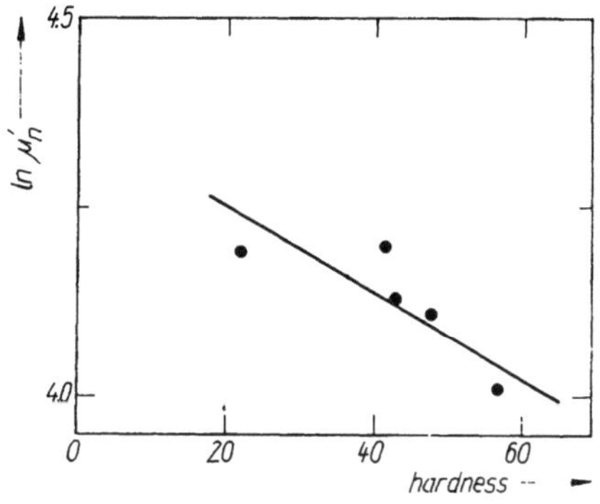


Fig. 4. Variation of initial relative permeability with hardness (in units of $R_c$)

the coercivity can be used as a measure of yield strength of the steel, which gives a useful non destructive test of an important mechanical property. The hysteresis loss $W_H$, which is closely related to coercivity $H_c$ since both are caused by impedances to domain wall motion, also revealed a simple dependence on hardness as shown in Fig. 3.

The initial permeability $\mu_n$, which is primarily determined by a reversible process, is known to be affected by a number of factors including the presence of inclusions and dislocations. These affect the reversible translation and bending of domain walls as discussed by Hoselitz ([20], p. 47), Kersten [28] and Becker [29]. They have shown that as the level of internal strains increases so the initial permeability decreases. Similarly as the number of magnetic inclusions increases the initial permeability decreases. Results of the present investigation show that the initial permeability decreases with hardness as indicated in Fig. 4, which is consistent with this interpretation.

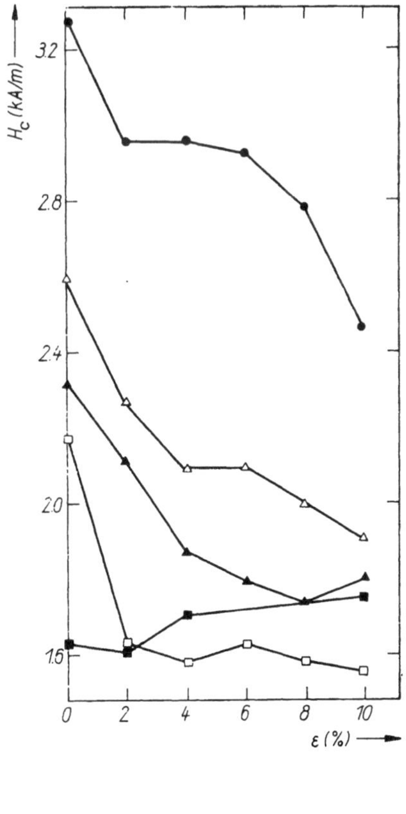


Fig. 5. Dependence of coercivity on plastic deformation (■ specimen A, □ B, ▲ C, △ D, ● E)

### *3.2 Variation of magnetic properties with residual plastic deformation*

As the material is deformed the number density of dislocations increases linearly with plastic deformation [30] and hence the amount of residual microstrain increases. Since the magnetic domain walls interact with these localised regions of strain as shown by Seeger, Kronmüller, Rieger and Trauble [31], Hilzinger and Kronmüller [32] and Scherpereel, Kazmerski and Allen [33], changes in the number density of the dislocations can be expected to alter structure sensitive magnetic properties.

The coercivity $H_c$ was affected by plastic deformation rather differently depending on the microstructure of the material, as shown in Fig. 5. As expected the pearlite specimens had the lowest coercivity and hence the lowest level of residual microstrains. The coercivity of the pearlites was not greatly affected by the level of macroscopic plastic strain induced. In the case of the furnace cooled coarse pearlite which had the lowest level of microstrain, the coercivity was found to increase by typically 16 A/m per % strain. The fine pearlite showed a slight reduction in coercivity over the range of strains used. The bainitic specimens which had somewhat higher levels of microstrain to begin with, had higher coercivities than the pearlites. After plastic deformation however the coercivity was considerably reduced, by 60 A/m. per % strain, a result that was somewhat unexpected in view of the recent results on 4130 steels [15] in which the coercivity of the bainites was almost independent of strain.

The martensite showed a large reduction in coercivity with plastic deformation which was of the order of 80 A/m per % strain. This reduction was also observed in 4130 steels and was attributed to a sweeping out of dislocations from some volumes of the martensite by the applied stresses during macroscopic plastic deformation. The martensitic transformation which occurs during the rapid quench from 860 °C to room temperature causes a marked expansion of the material and leads to a very high density of dislocations in the material. This accounts for its high hardness and for the coercivity of 3280 A/m. As the martensite is plastically deformed the stress moves some of the dislocations, thereby reducing their number density in some volumes of the material. These dislocations become pinned elsewhere and hence do not return to their original location when the stress is removed. The net result is that the material experiences a reduction in coercivity as observed both in the present investigation and in the study of 4130 steels.

The behaviour of the hysteresis loss $W_H$ as a function of plastic deformation as indicated in Fig. 6 was similar to the behaviour of the coercivity. The initial permeability varied with plastic deformation in just the opposite way from the coercivity. For the coarse pearlite initial permeability decreased with plastic deformation strain while

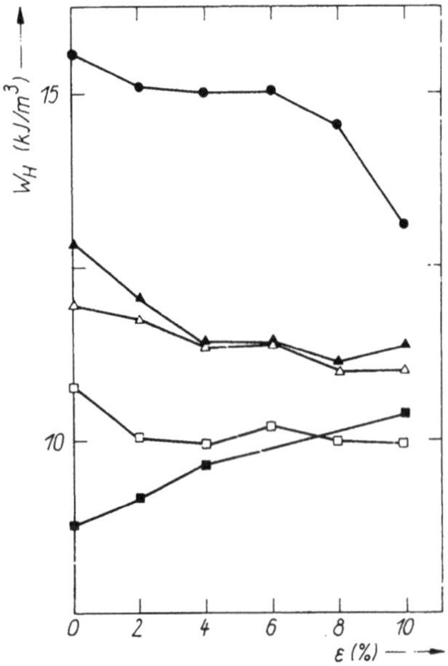


Fig. 6. Dependence of hysteresis loss on plastic deformation (for symbols see Fig. 5)

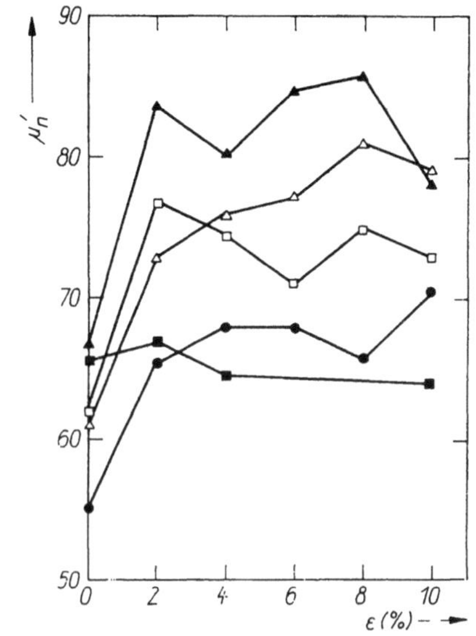


Fig. 7

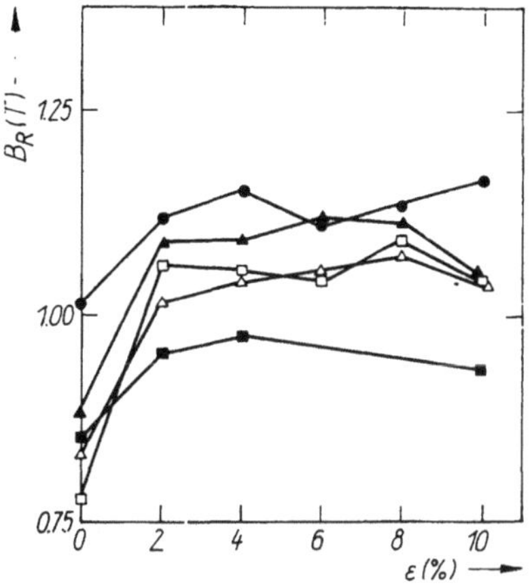


Fig. 8

Fig. 7. Dependence of initial relative permeability on plastic deformation (for symbols see Fig. 5)

Fig. 8. Dependence of remanence on plastic deformation (for symbols see Fig. 5)

for the martensite and the bainites it increased with strain as shown in Fig. 7. Remanence was found to increase with plastic deformation in Fig. 8.

The maximum differential permeability $\mu'_{max}$ measured for these materials at the coercive point, is a significant parameter for the determination of stress, as was indicated in recently by Jiles and Utrata [14]. In this case the variation of $\mu'_{max}$ with plastic deformation over the range 2 to 10%. This is in complete agreement with recent results on a single microstructure (heat treatment) of 4140 steel after tensile plastic deformation [14].

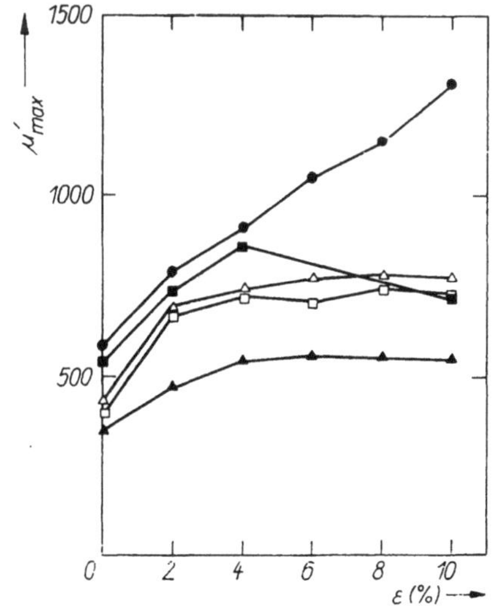


Fig. 9. Dependence of maximum differential relative permeability on plastic deformation (for symbols see Fig. 5)

In both cases some slight changes in $\mu'_{max}$ were observed over the first 2% of deformation. Under compression this was a slight increase while under tension a slight decrease was observed. These results can be interpreted on the basis of changes in the uniaxial magnetic anisotropy with residual stress caused by plastic deformation as reported by Krause and Cullity [34]. Most of the volume of the material will be under tensile residual strain once the load is removed. This couples with the positive magnetostriction coefficient of iron at low inductions to increase the initial anhysteretic permeability under tension (after compressive plastic deformation) and to reduce the initial anhysteretic permeability under compression (after tensile plastic deformation).

Krause and Cullity made measurements on specimens with plastic deformations of up to 15%, but rapid changes in anisotropy were only observed over the first 25% of plastic deformation and thereafter either remained constant or decreased slightly. This was interpreted by speculating that uniaxial anisotropy is created by the formation of a subgrain structure, but that the subgrains in nickel start to decrease in size beyond 2.5% plastic deformation. This increases the volume fraction of subgrain walls (which are composed of a high density of dislocations) while reducing the volume fraction of subgrains, which then prevents much further change in stress anisotropy, since although the residual stress level in the subgrains will rise with further plastic deformation the volume that they occupy will decrease. The explanation appears to account for present observations on AISI 4140 steels as well as those on 4130 steels [15] and 4140 steels after tensile plastic deformation.

The behaviour at the coercive point is known to lead under certain conditions to a maximum differential permeability that in theory is identical to the differential permeability of the anhysteretic at the origin $\mu'_a$. This equivalence has also been noted independently by Craik [35]. The initial anhysteretic permeability is known to be dependent on applied stress through an induced stress anisotropy, but to be independent of the number density of impediments to domain wall motion. Present results seem to confirm this.

### *3.3 Covariance of bulk magnetic properties*

The bulk magnetic properties of ferromagnetic materials, such as coercivity, hysteresis loss, initial permeability, maximum differential permeability, initial anhysteretic permeability and remanence are interrelated, although the exact nature of these interrelations are not always clear. For example the equality of initial anhysteretic permeability and maximum differential permeability under certain conditions emerges from the theory of hysteresis [36], and coercivity and hysteresis loss are caused by a quasi-frictional force, which opposes changes in magnetisation. Nevertheless the exact relationship between initial permeability and coercivity is not yet clear although experimental results seem to indicate that they are related, (initial permeability decreases as coercivity increases) even though the initial permeability is primarily a reversible process and coercivity is entirely dependent on the existence of irreversibility.

The relationship between coercivity and hysteresis loss is much easier to understand. Although it is probably only a first approximation, a linear relationship first suggested by Yensen and Ziegler [37], and this was found to fit the present results reasonably well as shown in Fig. 10. Similar behaviour has been observed in previous results [14, 15]. In this case the results for each of the microstructures indicated a single relationship between the quantities, much as was found for the 1000 series steels [15] in which the microstructures remained the same but the carbon content varied. Some divergences from a single relationship were observed in results on 4130 steels which suggests that the relationship between coercivity and hysteresis loss is weakly microstructure dependent.

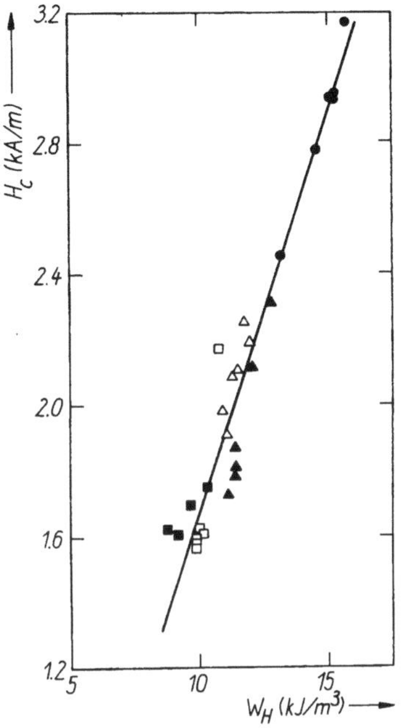


Fig. 10

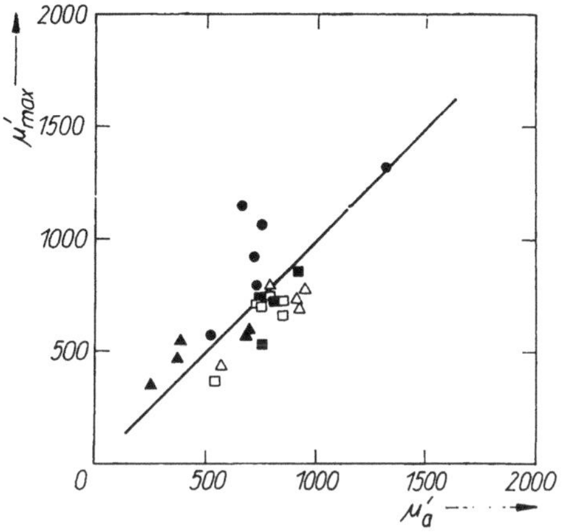


Fig. 11

Fig. 10. Correlation between coercivity and hysteresis loss (for symbols see Fig. 5)

Fig. 11. Correlation between initial anhysteretic relative permeability and maximum differential relative permeability (for symbols see Fig. 5)

The equivalence of initial anhysteretic permeability $\mu_a$ and the maximum differential permeability $\mu'_{max}$ in the present results is shown in Fig. 11. There is some scatter in the results which is to be expected from the errors incurred in determining the slopes of the curves at the relevant points, but a definite trend is discernible which is scattered symmetrically about the line $\mu_a = \mu'_{max}$. The equality of these parameters has an important role to play in the evaluation of applied stress (elastic strains) since the maximum differential permeability is much easier to measure in practice than the initial anhysteretic permeability.

## 4. Analysis

It has been shown elsewhere [38] that the stress in a ferromagnetic material is related to the maximum differential susceptibility $\chi'_{max}$ by the relation

$$3b\sigma/\mu_0 = \left(1/\chi'_{max}(0)\right) - \left(1/\chi'_{max}(\sigma)\right), \tag{6}$$

where $b = \lambda/M^2$ where $\lambda$ is the magnetostriction and $b$ has a value of $2.5 \times 10^{-15}$ $(A/m)^{-2}$. Substituting the values of maximum differential susceptibility of specimen $A$ at zero plastic deformation, $\chi'_{max} = 42.5$, and at 4% plastic deformation, $\chi'_{max} = 68.7$, leads to a value of residual stress of $\sigma = 1.2 \times 10^6$ $Nm^{-2}$ in tension. This compares with a value of $2.15 \times 10^6$ $Nm^{-2}$ obtained after 6% plastic deformation of a specimen of steel of similar composition by Abuku [39].

## 5. Conclusions

The work presented here was an investigation into the dependence of magnetic properties of several heat treatments of AISI 4140 steel on plastic deformation. A wide range of magnetic properties were measured including coercivity, hysteresis loss, remanence, initial permeability of the normal magnetisation curve, initial anhysteretic permeability and maximum differential permeability.

Some of the properties were found to be dependent on plastic deformation, notably $H_c$, $W_H$, $\mu'_n$, and $B_R$. Others were not, for example $\mu'_a$ and $\mu'_{max}$. The dependences were interpreted in terms of changes in dislocation densities within the metal caused by the plastic deformation, under the assumption that we could neglect any changes due to the effect of magnetic inclusions with plastic strain in comparison with the changes in dislocation densities.

For very pure metals it would be expected that $H_c$ and $W_H$ increase with plastic deformation while $\mu'_n$ decreases. In the present investigation it was found that for hard steels such as martensite, $H_c$ and $W_H$ decreased with plastic deformation while $\mu'_n$ increased.

Correlations were noticed between initial normal permeability, hysteresis loss and coercivity and the mechanical hardness of the steel. It is suggested that these measurements could be used for estimating the yield strength of the metal non-destructively. The residual stress in the material was calculated from changes in the maximum differential susceptibility and the bulk magnetostriction.

## References

[1] J. A. EWING, Magnetic Induction in Iron and Other Metals, The Electrician, London 1900.
[2] R. M. BOZORTH and H. J. WILLIAMS, Rev. mod. Phys. **17**, 72 (1945).
[3] L. LLIBOUTRY, Ann. Phys. (France) **6**, 731 (1951).
[4] L. BRUGEL and G. RIMET, J. Phys. Radium **27**, 589 (1966).
[5] D. J. CRAIK and M. J. WOOD, J. Phys. D **3**, 1009 (1970).
[6] R. R. BIRSS, C. A. FAUNCE, and E. D. ISAAC, J. Phys. D **4**, 1040 (1971).
[7] C. S. SCHNEIDER and J. M. RICHARDSON, J. appl. Phys. **53**, 8136 (1982).
[8] R. M. BOZORTH, Ferromagnetism, Van Nostrand, New York 1951.
[9] B. D. CULLITY, J. appl. Phys. **35**, 1915 (1964).
[10] B. D. CULLITY, Trans. MS AIME **227**, 356 (1963).
[11] B. D. CULLITY and O. P. PURI, Trans. MS AIME **227**, 359 (1963).
[12] D. K. BAGCHI and B. D. CULLITY, J. appl. Phys. **38**, 999 (1967).
[13] R. M. RUSNAK and B. D. CULLITY, J. appl. Phys. **40**, 1581 (1969).
[14] D. C. JILES and D. UTRATA, J. non-destructive Evaluation, in the press (1988).
[15] D. C. JILES, J. Phys. F (1988) (in press).
[16] R. LANGMAN, Non Destr. Test. Internat. **14**, 255 (1981).
[17] D. C. JILES and D. L. ATHERTON, J. Phys. D **17**, 1265 (1984).
[18] L. G. DOBRANSKI, D. C. JILES, and D. L. ATHERTON, J. appl. Phys. **57**, 4229 (1985).
[19] M. KERSTEN in: Probleme der technischen Magnetisierungskurve, Ed. R. BECKER, Springer, Berlin 1938.
[20] K. HOSELITZ, Ferromagnetic Properties of Metals and Alloys, Oxford University Press, London, Oxford 1951.
[21] M. KERSTEN, Z. Phys. **82**, 723 (1933).
[22] J. K. STANLEY, Trans. AIME **162**, 106 (1945).
[23] T. G. DIGGES, S. J. ROSENTHAL, and G. W. GEIL, Heat Treatment and Properties of Iron and Steel, NBS Monograph No. 88, November 1966.
[24] J. D. VERHOEVEN, Fundamentals of Physical Metallurgy, Wiley, New York 1975.
[25] G. E. DIETER, Mechanical Metallurgy, 3rd Ed., McGraw-Hill, New York 1986.
[26] D. C. JILES, and J. D. VERHOEVEN, Review of Progress in Quantitative NDE, Vol. 6B, Ed. D. O. THOMPSON and D. E. CHIMENTI, Plenum Press, New York 1986 (p. 1681).

[27] M. C. Shaw, Mechanical Behaviour of Materials, Ed. F. A. McClintock and A. S. Argon, Addison/Wesley, Reading, Mass., USA 1966.
[28] M. Kersten, Phys. Z. **39**, 860 (1938).
[29] R. Becker, Z. Phys. **64**, 660 (1930).
[30] D. Hull and D. J. Bacon, Introduction to Dislocations, 3rd Ed. Pergamon Press, New York 1984.
[31] A. Seeger, H. Kronmüller, H. Rieger, and H. Träuble, J. appl. Phys. **35**, 740 (1964).
[32] H. R. Hilzinger and H. Kronmüller, J. Magnetism magnetic Mater. **2**, 11 (1976).
[33] D. E. Scherpereel, L. L. Kazmerski, and C. W. Allen, Metallurg. Trans. **1**, 517 (1970).
[34] R. F. Krause and B. D. Cullity, J. appl. Phys. **39**, 5532 (1968).
[35] D. J. Craik, Structure and Properties of Magnetic Materials, Pion, London 1971.
[36] D. C. Jiles and D. L. Atherton, J. Magnetism magnetic Mater. **61**, 48 (1986).
[37] T. D. Yensen and N. A. Ziegler, Trans. ASM **23**, 556 (1935).
[38] D. C. Jiles, J. Phys. D. 1988 (in press).
[39] S. Abuku, Japan. J. appl. Phys. **16**, 1161 (1977).

*(Received February 11, 1988; in revised form April 21, 1988)*

phys. stat. sol. (a) 108, 431 (1988)

Subject classification: 78.60; S7.1

*Institute of Solid State and Semiconductor Physics, Academy of Sciences of the Belorussian SSR, Minsk*[1]) *(a) and V. I. Lenin Belorussian State University, Minsk*[2]) *(b)*

# Influence of High Pressure on Cathodoluminescence of Cubic Boron Nitride

By

V. B. SHIPILO (a), E. M. SHISHONOK (a), A. M. ZAITSEV (b), A. A. MELNIKOV (b), and A. I. OLEKHNOVICH (a)

It is shown that the pressure treatment of boron nitride ($BN_{sph}$) polycrystals at room temperature results in the change of broad band parameters (A, B, C) and zero-phonon lines of cathodoluminescent (CL) spectra, and in the appearance of RC centers. In the pressure treated ceramic samples $BN_{sph}$ the PC-1 (2.84 eV), PC-2 (2.325 eV), PC-3 (1.79 eV) centers of presumably interstitial nature are found. The comparative analysis of the dependence of the broad bands parameters and zero-phonon lines of the GC centers in the CL spectra of boron nitride on pressure makes it possible to connect the appearance of the A and B bands in the energy range 1.85 to 3.25 eV with electronic states in the band gap due to multi vacancy complexes based on boron and nitrogen vacancies, respectively, while the appearance of the C band (1.55 to 1.85 eV) with the plastic deformation of $BN_{sph}$.

Es wird gezeigt, daß eine Druckbehandlung von Bornitrid ($BN_{Sph}$)-Polykristallen bei Zimmertemperatur zu einer Änderung der Parameter der breiten Banden (A, B, C) und der Null-Phononenlinien der Kathodolumineszenz (CL)-Spektren und dem Auftreten von RC-Zentren führt. In den druckbehandelten keramischen $BN_{Sph}$-Proben werden PC-1 (2,84 eV)-, PC-2 (2,325 eV)- und PC-3 (1,79 eV)-Zentren mit wahrscheinlich Zwischengitternatur gefunden. Die vergleichende Analyse der Druckabhängigkeit der Parameter der breiten Banden und der Null-Phononenlinien der GC-Zentren im CL-Spektrum von Bornitrid ermöglicht, das Auftreten der A- und B-Banden im Energiebereich 1,85 bis 3,25 eV mit elektronischen Zuständen in der Bandlücke infolge Mehrfach-Vakanzenkomplexe auf der Basis von Bor- bzw. Stickstoff-Leerstellen zu verknüpfen, während das Auftreten der C-Bande (1,55 bis 1,85 eV) mit der plastischen Deformation von $BN_{Sph}$ verbunden ist.

## 1. Introduction

In the luminescent spectra of cubic boron nitride ($BN_{sph}$) broad bands with maxima near 2.3 and 2.65 eV, zero-phonon lines of optically active GC centers at 1.76, 1.63, 1.55, 1.44 eV, as well as lines of radiative RC centers at 2.23, 2.15, 1.99 eV have been observed [1 to 3]. The GC-1 and GC-2 centers, sensitive to elastic microstress and the perfection of the $BN_{sph}$ structure, are assumed [3] to be impurity–vacancy complexes based on nitrogen and boron vacancies.

The influence of pressure on the cathodoluminescent (CL) spectra of $BN_{sph}$ polycrystals has been investigated in this paper for elucidating the nature of the broad luminescent bands.

[1]) Ul. P. Brovki 17, SU-220726 Minsk, USSR.

[2]) Leninskii prospekt 4, SU-220080 Minsk, USSR.

## 2. Experimental

Polycrystals with different structure perfection served as samples of cubic boron nitride $BN_{sph}^{1p}$ and $BN_{sph}^{2p}$, synthesized from high-purity pyrolytic boron nitride at different parameters by direct phase transition $BN_{h}^{p}$–$BN_{sph}^{p}$ and $BN_{sph}^{c}$ ceramics, obtained by sintering $BN_{sph}$ single crystals (10 to 100 μm in size) grown in a system with nitrogen excess at high pressures and temperatures. In order to avoid diffusion of impurities from the container material (lithographic stone) and the heater, the samples were placed in tantalum capsules which, as evidenced by the activation analysis, did not practically enter the $BN_{sph}$ lattice during the synthesis.

The samples, polycrystalline $BN_{sph}$, cut in the form of parallelepipeds $2 \times 2 \times 3\ mm^3$ in size with polished facets, placed in graphite-like boron nitride containers, were subjected to pressure up to 9.5 GPa at room temperature. The pressure increased up to the given values in steps of 3 MPa/s, the exposure under pressure being 1 h. After a complete cycle of the pressure treatment (up to 9.5 GPa) of the samples some insignificant changes in the lattice parameters and of the thin crystal structure of the material have been observed.

The CL spectra of the initial and the pressure-treated samples were recorded at 80 and 300 K. The energy of the exciting electrons was 15 keV at a current density of $6\ mA/mm^2$.

## 3. Results

The CL spectra of the initial $BN_{sph}^{1p}$ and $BN_{sph}^{2p}$ revealed a broad band consisting of two bands A and B (Fig. 1 a, b) with maxima at 2.2 and 2.45 eV, a more narrow band C in the range 1.55 to 1.85 eV, as well as the zero-phonon lines of the GC-1 center interacting with the phonons 49 and 64 meV and the GC-2 center with a weak phonon replica. A structure (at 2.84, 2.95, 3.13 eV) was noted in the high-energy section of band B. The microhardness (63.5 GPa) of $BN_{sph}^{2p}$ with more intensive band C in the CL spectra was higher than that (48.5 GPa) of the $BN_{sph}^{1p}$ samples.

In the CL spectra of the initial $BN_{sph}^{c}$ the broad bands were observed with band A being dominant (Fig. 1c). The pressure treatment of the $BN_{sph}$ samples leads to a drastic change in the intensity of bands A, B, C as well as in the widths and intensities of the zero-phonon lines (Fig. 2, 3). In this case the centers RC-1 and RC-2, observed earlier [2] only after radiation treatment, have been recorded in the $BN_{sph}^{1p}$ spectra, while in the $BN_{sph}^{c}$ spectra the PC-1 (at 2.84 eV) and PC-2 (at 2.325 eV) centers and the PC-3 (at 1.79 eV) center equidistantly located at steps of 0.53 eV. A narrow zero-phonon line split into a doublet of the PC-3 center has been observed after pressure treatment of the $BN_{sph}^{c}$ sample at 9.5 GPa. The PC centers have been first recorded in this investigation.

In the CL spectra of the investigated samples the C band ($BN_{sph}^{1p}$) is dominant with respect to intensity, and the zero-phonon lines of the GC and PC-1 centers, with the latter being recorded at 70 K, have been displayed at room temperature. The zero-phonon lines of the RC and PC-2 centers have been observed at room temperature (Fig. 1a, 2c, 3a).

From the dependences of the intensities of the A, B, and C bands (Fig. 4a) and half-widths of the zero-phonon lines of the GC-1, GC-2, and RC-2 centers (Fig. 4 b, c) on the pressure applied to BN it is seen that the most pronounced change of the parameters of the bands and lines has been noted at pressure 3.5 GPa, where, from the data given in [4], plastic deformation of cubic boron nitride ($BN_{sph}$) at room temperature takes place. Up to 3.5 GPa the narrowing of the zero-phonon lines of the GC

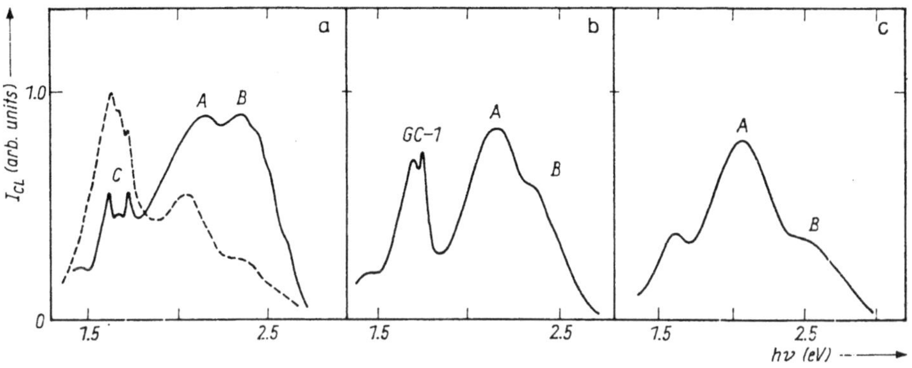


Fig. 1. CL spectra of initial samples at 70 (solid line) and 300 K (dashed line). a) $BN^{1p}_{sph}$, b) $BN^{2p}_{sph}$, c) $BN^{c}_{sph}$

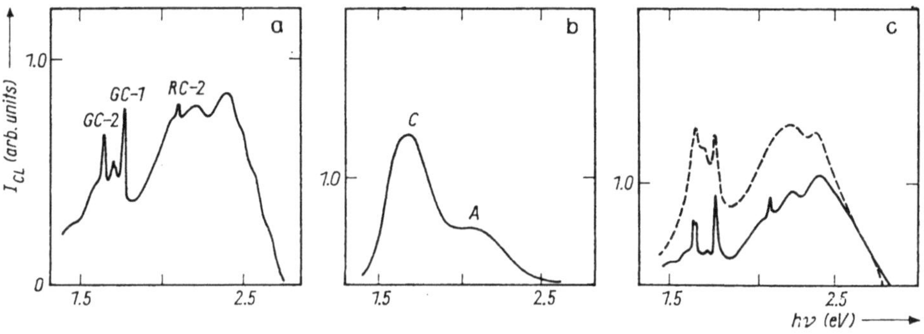


Fig. 2. CL spectra of $BN^{1p}_{sph}$ polycrystal at 70 (solid lines) and 300 K (dashed line) subjected to various pressures. a) 2.5, b) 3.5, c) 9.5 GPa

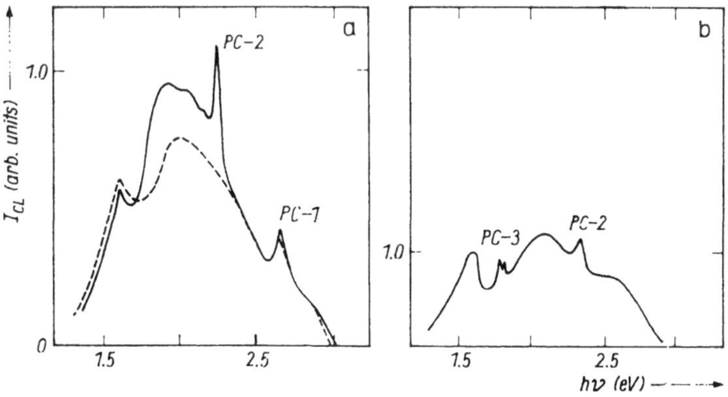


Fig. 3. CL spectra of $BN^{c}_{sph}$ ceramics at 70 (solid lines) and 300 K (dashed line) subjected to various pressures. a) 2.5, b) 9.5 GPa

centers and steady enhancement of the intensity of the C band have been observed. The intensity of the C band attains its maximum at $P = 3.5$ GPa (Fig. 2b, 4a), the intensity of the bands A and B drops, and the zero-phonon lines disappear. Then the intensity of the C band decreases rapidly to zero and the zero-phonon lines, though somewhat broad, reappear. Further enhancement of pressure up to 9.5 GPa results in increasing intensity of the A, B bands and the zero-phonon lines of the GC and RC centers, in the splitting of the zero-phonon lines of the GC-2 center (Fig. 2c, 4b, c).

The dependence of the intensities of the A and B bands in the CL spectra of $BN^{c}_{sph}$ on the pressure treatment is about the same as that for $BN^{1p}_{sph}$. However, the intensity minimum has been recorded at higher pressure (4.5 GPa). The parameters of the zero-phonon lines of the PC-1 and PC-2 centers are less sensitive to changes of pressure and plastic deformation of the samples than those of the GC and RC centers. They

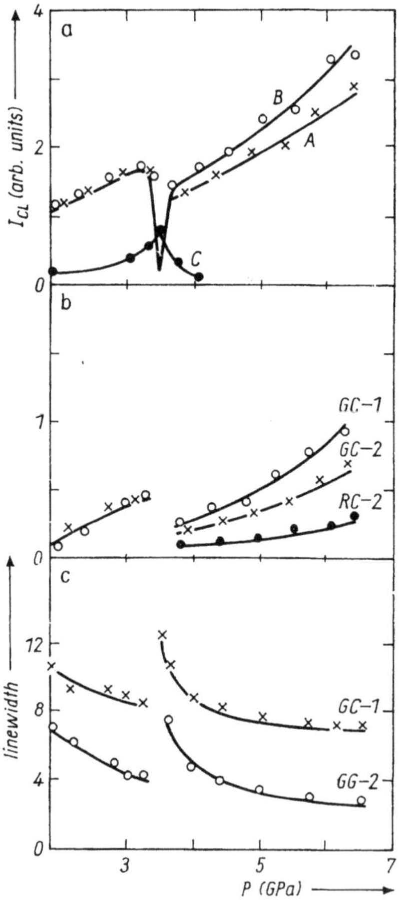


Fig. 4

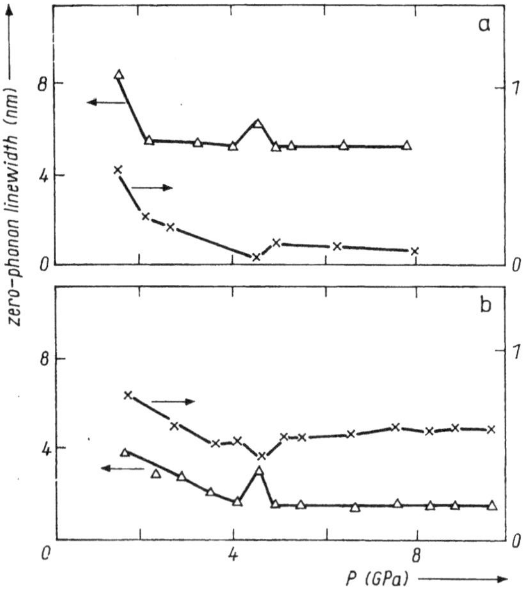


Fig. 5

Fig. 4. Intensities, half-widths of the lines A, B, C, and zero-phonon lines of GC and RC centers of $BN^{1p}_{spn}$ against pressure treatment

Fig. 5. Intensity (×) and half-width (△) of a) PC-1 and b) PC-2 zero-phonon lines against pressure treatment remain practically unchanged when pressure above 4.5 GPa is applied to $BN_{sph}^{c}$. At pressure above 8 GPa the zero-phonon lines of the PC-1 center have not been noted (Fig. 5 a, b).

## 4. Discussion

The sensitivity of the C band to the change of dislocation density (microhardness) in the initial $BN_{sph}^{1p}$ and $BN_{sph}^{2p}$, the intensity growth in the range of plastic deformation at room temperature, allows one to make two assumptions. First, the radiative recombination of cubic boron nitride $BN_{sph}$ in the range 1.55 to 1.85 eV can be associated with the electronic states in the band gap of $BN_{sph}$ which form the dislocation band. Second, it is possible that the dislocations due to the plastic deformation distort the GC centers to a great extent to produce inhomogeneous broadening, while the appearance of the substantial background (C band) may be connected with that of strong interaction with long-wave lattice acoustic vibrations, the possibility of the interaction with which increases when a distortion of the structure of the defect occurs. Such low-energy phonons will be effectively excited with the rise of temperature, which manifests itself in the increase of intensity of the general background of the GC spectra and the emergence of a pronounced anti-Stokes wing.

It is worthwhile to notice that the change in the intensity of bands A and B, the lines of the GC and RC centers, when $BN_{sph}^{1p}$ is subjected to pressure, is of general character. The optically active defect centers GC-1 and GC-2 are most simple impurity–vacancy complexes resistant to temperature which contain nitrogen and boron vacancies and the RC centers are also vacancy-type intrinsic defects [3]. Thus one can assume that the appearance of bands A and B in the CL spectra of cubic boron nitride is connected with the presence of multivacancy defects sensitive to elastic microstress. The high stability of the bands allows one to suppose that the impurity atoms, connecting the vacancies into stable complexes are included in their composition. On the other hand, from the correspondence observed between the growth of the intensity of band B (Fig. 3a) and the zero-phonon line of the GC-1 center (Fig. 3b) dominant with respect to band A and the zero-phonon line of the GC-2 center in $BN_{sph}^{1p}$ subjected to pressure above 5.0 GPa one can assume that the vacancies due to boron and nitrogen, respectively, form part of the complexes which give rise to the appearance of bands A and B and that the formation of vacancies due to nitrogen and its complexes mainly takes place when the cubic boron nitride is subjected to pressure.

The continuity of bands A and B in the range from 1.85 to 3.25 eV and their sensitivity to temperature make it possible to assume the presence of a rather broad band with a quasicontinuous set of levels in the band gap of $BN_{sph}$. The thermoactive and recombination processes depending on the lifetime of nonequilibrium carriers are characteristic of these levels.

The enhancement of the intensity of the zero-phonon lines of the GC centers and their narrowing during the pressure treatment of polycrystals above 4.5 GPa testify the decrease of microstress in the $BN_{sph}$ sample in proportion to the pressure applied. The decrease of microstresses is likely to be due to the processes of annihilation of dislocations owing to the increase in their mobility at rising pressure.

As mentioned above the parameters of zero-phonon lines of the PC-1 and PC-2 centers are practically insensitive to pressure after the pressure treatment of the polycrystals beyond the range of their plastic deformation in contrast to those of the GC centers. The energy of their interaction with quasilocal oscillations (100 meV) exceeds greatly that of similar oscillations of the defects of vacancy type (56.64 meV). Consequently, the composition of the above-mentioned centers does not include vacancies but most likely interstitial atoms.

## 5. Conclusions

1. The pressure treatment of $BN_{sph}$ polycrystals up to 3.4 to 4.5 GPa results in plastic deformation of the material.

2. In the band gap of $BN_{sph}$ the existence of electronic states forming the dislocation band in the range 1.55 to 1.85 eV is possible, while in the range from 1.85 to 3.25 eV there is a broad band with a quasicontinuous set of levels associated with the presence of multivacancy defects in the material.

3. The optically active PC defects which possibly have interstitial nature have first been recorded in this investigation.

## References

[1] V. D. TKACHEV, V. B. SHIPILO, and A. M. ZAITSEV, Fiz. Tekh. Poluprov. **19**, 795 (1985).

[2] V. B. SHIPILO, A. M. ZAITSEV, E. M. SHISHONOK, and A. A. MELNIKOV, Zh. prikl. Spektroskopii **45**, 601 (1986).

[3] A. M. ZAITSEV, A. A. MELNIKOV, V. B. SHIPILO, and E. M. SHISHONOK, phys. stat. sol. (a) **94**, K125 (1986).

[4] N. V. NOVIKOV, V. P. BONDARENKO, YU. A. KOCHESZHINSKII, A. B. BELYANKINA, V. K. GUERASIMENKO, E. V. STOLYAROV, and V. M. TOVSTOGAN, Sverkhtverdye mater. **2**, 17 (1985).

*(Received September 18, 1987; in revised form February 12, 1988)*

phys. stat. sol. (a) **108**, 437 (1988)

Subject classification: 78.50; S11.1

*A. V. Shubnikov Institute of Crystallography, Academy of Sciences of the USSR, Moscow*[1]) *and Moscow State University*

# X-Ray and UV Influence on the Optical Absorption Spectra of the Non-Photorefractive Lithium Niobate

By

T. R. VOLK and N. M. RUBININA

A comparative study is made of optical properties in $LiNbO_3$:Fe, $LiNbO_3$:Mg, and $LiNbO_3$:Fe:Mg. The value of the photorefraction in $LiNbO_3$:Fe:Mg decreases not less than by $10^3$ as compared with $LiNbO_3$:Fe. X-ray and UV-induced change of the absorption spectra reveal a change in the type of the radiation-induced centre in Mg-doped crystals. X-ray induced extra $Fe^{2+}$ is not revealed in $LiNbO_3$:Fe:Mg indicating the alteration of the electron acceptor centre due to Mg incorporation.

Проведено сравнительное исследование оптических свойств кристаллов $LiNbO_3$:Fe, $LiNbO_3$:Mg, $LiNbO_3$:Fe:Mg. Величина фоторефракции в кристаллах $LiNbO_3$:Fe:Mg падает не менее, чем на три порядка по сравнению с кристаллами $LiNbO_3$:Fe. Исследование спектров поглощения кристаллов, облученных рентгеновским и УФ-излучением, выявило изменение характера радиационного центра в кристаллах легированных Mg. Избыточное $Fe^{2+}$, образующееся при облучении кристаллов $LiNbO_3$:Fe, в $LiNbO_3$:Fe:Mg не возникает, что указывает на изменение типа акцептора электронов при введении Mg.

## 1. Introduction

Lithium niobate is a well-known electrooptic crystal, but its application is limited by the photorefractive effect. Magnesium doping was reported to reduce the value of the photorefraction (e.g. [1, 2]). It was shown [2] that the improved performance of the "non-photorefractive" $LiNbO_3$:Mg was due to a significant increase of the photoconductivity while the value of the Glass constant was not affected by the Mg doping.

For Mg concentrations higher than some threshold (4.5 mol%, i.e. 1.23 wt% [1]) the increase of the photoconductivity by a factor of $10^2$ as compared to the pure $LiNbO_3$ [1] leads to the corresponding decrease of the photorefraction. Thus Mg doping is likely to alter the type of the photoelectric processes in $LiNbO_3$. The photoconductivity of $LiNbO_3$:Fe (and apparently of $LiNbO_3$) depends on the $[Fe^{2+}]/[Fe^{3+}]$ ratio, $Fe^{2+}$ being a donor and $Fe^{3+}$ an acceptor of electrons [3]. The ESR investigations in $LiNbO_3$:Mg [2] give evidence of the change in the $Fe^{3+}$ location in the $LiNbO_3$ lattice owing to the Mg incorporation. The increase of the photoconductivity in $LiNbO_3$:Mg was supposed to be due to a decrease of the capture cross-section of electrons by $Fe^{3+}$ [4].

The origin of the influence of magnesium on the photoelectric and photorefractive centres in $LiNbO_3$ is not clear. The data of the X-ray influence were shown [5, 6] to be useful for the investigation of the photovoltaic and photorefractive centres.

---

[1]) Leninskii prospekt 59, SU-117333 Moscow, USSR.

In the present paper we report on comparative investigations of the X-ray and UV influence on the optical absorption spectra in $LiNbO_3$ doped with Mg and/or Fe. The results indicate the alteration of the radiation-induced centres in $LiNbO_3$:Mg as compared to $LiNbO_3$.

## 2. Experimental Results

The $LiNbO_3$ crystals used in the present work were grown by the Czochralski technique from the congruent melt with $Fe_2O_3$ (0.01 and 0.03 wt%) and/or MgO (1 to 2.5 wt%)[2]) added to the melt (according to [1, 2] the sharp alteration of the photorefractive properties takes place for Mg concentrations exceeding ≈1.23 wt% MgO). The concentration of contaminations in pure $LiNbO_3$ is less than $10^{-4}$ wt%. All the crystals were grown under identical conditions from the same starting materials which provides the possibility of comparisons.

Non-polarized optical absorption spectra in the range from 300 to 800 nm have been obtained with the spectrophotometer Specord M40 on (001) cuts of crystals with thicknesses of 0.1 cm. The source of X-radiation was an X-ray tube with Mo target ($MoK_\alpha$, 20 keV) with intensities of $10^4$ R/min (absorbed doses of approximately $10^5$ Gy). The source of the UV light was a 500 W high-pressure mercury lamp. All the results were obtained at room temperature.

First of all we have measured the photorefraction in all crystals under investigation. The measurements in magnesium-doped $LiNbO_3$ did confirm the sharp decrease of photorefraction due to Mg doping. In our undoped $LiNbO_3$ crystals the maximum value of photorefraction is $\delta\Delta n \approx 2 \times 10^{-4}$ (Ar-ion laser: 0.488 μm, 500 W/cm²). In $LiNbO_3$: 1% Mg and $LiNbO_3$: 2% Mg crystals no photorefraction was detected that means it did not exceed $2 \times 10^{-6}$ (the sensitivity of our optical compensator technique).

Measurements of the photorefractive effect in $LiNbO_3$:Fe:Mg have not been performed before. Our investigations have shown that Mg doping of $LiNbO_3$:Fe leads to the same results as in pure $LiNbO_3$, namely to the sharp reduction of the photorefraction. The maximum values of the photorefraction in our $LiNbO_3$: 0.01% Fe and $LiNbO_3$: 0.03% Fe crystals were $2 \times 10^{-4}$ and $1.3 \times 10^{-3}$, respectively, while in $LiNbO_3$: 0.01% Fe: 2.4% Mg and $LiNbO_3$: 0.03% Fe: 2.4% Mg thc photorefraction was not observed. That means that its value as compared to $LiNbO_3$:Fe is reduced at least by a factor of $10^3$.

Fig. 1a, b demonstrate the X-ray induced change in the optical transmission spectra of $LiNbO_3$:Fe and $LiNbO_3$. The difference curves 3 were obtained by subtracting the initial spectra (curves 1) from the spectra of X-irradiated crystals (curves 2). The difference curves reveal two X-ray induced absorption bands at 21000 and 26500 cm$^{-1}$ (475 and 375 nm, respectively). These bands are stable revealing a slight relaxation in time during 20 to 40 h in darkness [6] and can be optically erased.

The X-ray influence on the optical absorption spectra of Mg-doped crystals (Fig. 2a to d) sharply differs from the above results. First of all in $LiNbO_3$:Mg and $LiNbO_3$:Fe: Mg we have observed not only an X-ray-induced but similarly a UV-induced (by $\lambda \leq 300$ nm) change in the optical absorption spectra. The spectral characteristics and the maximum values of the changes in the optical absorption coefficients are the same

[2]) Further on the investigated crystals are denoted as $LiNbO_3$:1% Mg, $LiNbO_3$:2% Mg, $LiNbO_3$:0.01% Fe:2.4% Mg, and $LiNbO_3$:0.03% Fe:2.4% Mg.

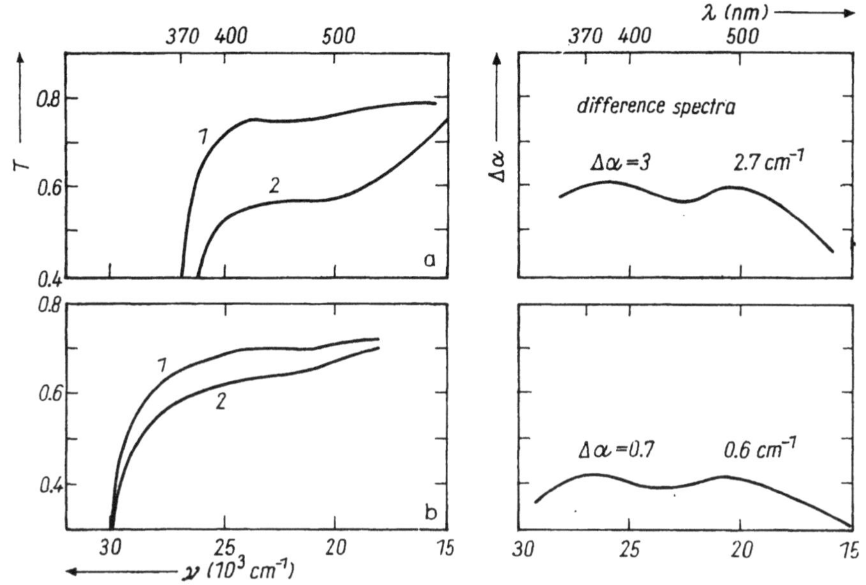


Fig. 1. X-ray-induced change of the optical transmission spectra in $LiNbO_3$ crystals. a) $LiNbO_3$:0.001% Fe, b) $LiNbO_3$. (1) initial spectra, (2) after X-irradiation with a dose of $10^5$ Gy. Difference absorption spectra are obtained by subtracting curves 1 from 2

for the X-ray- and UV-induced absorption spectra, so the given curves (Fig. 2) refer to the results for the both types of irradiation.

Fig. 2a to d demonstrate X-ray and UV-irradiation influence on the optical transmission spectra of $LiNbO_3$:1% Mg, $LiNbO_3$: 2% Mg, $LiNbO_3$: 0.01% Fe: 2.4% Mg, $LiNbO_3$: 0.03% Fe: 2.4% Mg crystals. The difference curves 3 were obtained by subtracting the initial spectra (curves 1) from the spectra of the irradiated crystals (curves 2). If comparing Fig. 1 and 2 one can come to the following conclusions:

1. In the case of Mg-doped crystals X-ray and UV-irradiation induces a single absorption band 375 nm (26500 $cm^{-1}$) in the visible region. The spectra of the irradiated Mg-doped crystals do not reveal the absorption band 475 nm (the initial absorption band at ≈480 nm does not increase in $LiNbO_3$:Fe:Mg).

2. The X-ray and UV-induced absorption band 375 nm in Mg-doped crystals with [Mg] $\geqq$ 2% is metastable fading completely during 1 to 2 h in the darkness (at room temperature) while the X-ray induced bands in $LiNbO_3$ and $LiNbO_3$:Fe (Fig. 1) reveal only a slight decrease for 20 to 40 h [6].

3. The spectra of irradiated $LiNbO_3$: 1% Mg (i.e. with Mg doping lower than the "threshold") behave in another way than those of $LiNbO_3$ with higher Mg concentration. In the former X-ray and UV-induced weak absorption band in the broad region 350 to 500 nm (Fig. 2a) is stable and does not decrease in darkness.

4. In all the Mg-doped crystals the X-ray and UV-induced band 375 nm is bleached by the visible light 350 nm $< \lambda <$ 600 nm, while illumination by UV light with $\lambda \leqq$ 300 nm on the contrary increases this band to saturation. In $LiNbO_3$ and $LiNbO_3$: Fe both X-ray induced bands are erased by light of the whole spectral region $\lambda <$ $<$ 600 nm.

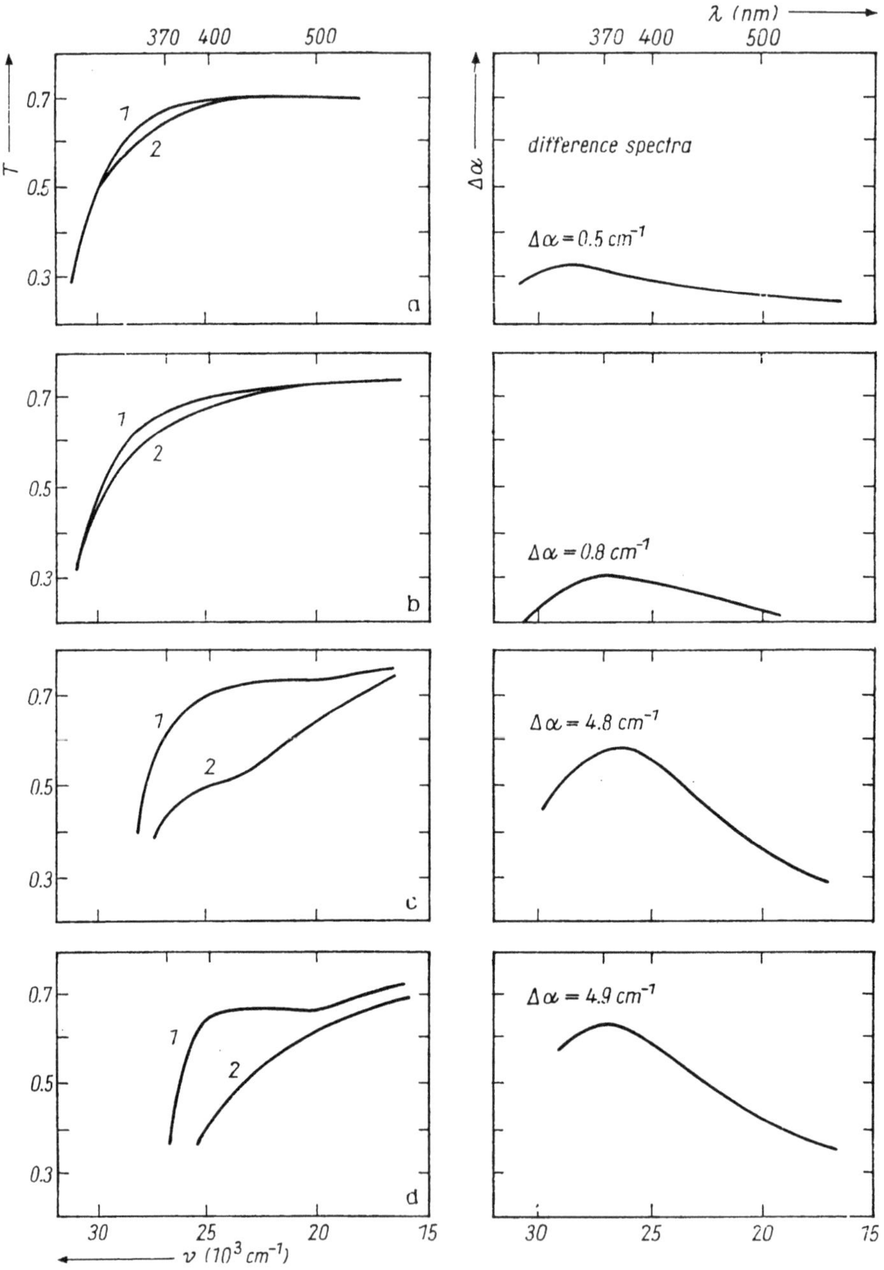


Fig. 2. X-ray- and UV-induced change of the optical transmission spectra in Mg-doped $LiNbO_3$ crystals. a) $LiNbO_3$:1% Mg, b) $LiNbO_3$:2% Mg, c) $LiNbO_3$:0.01% Fe:2.4% Mg, d) $LiNbO_3$:0.03% Fe:2.4% Mg. (1) initial spectra, (2) after X-rays or UV-irradiation. Difference absorption spectra are obtained by subtracting curves 1 from 2

### 3. Discussion

The difference in the optical absorption spectra of the irradiated Mg-doped (Fig. 2) and not Mg-doped (Fig. 1) $LiNbO_3$ crystals indicates different types of radiation-induced centres in these two groups of crystals.

First of all we discuss the origin of the X-ray induced absorption bands in $LiNbO_3$: Fe and $LiNbO_3$ (Fig. 1) basing on the current interpretation of their absorption spectra [7]. The X-ray (i.e. band-to-band) excitation of electron–hole pairs involves the capture of electrons and holes. IR and ESR investigations [6] have shown that X-irradiation of $LiNbO_3$:Fe crystals leads to the formation of "extra" $Fe^{2+}$ due to the capture of electrons by the acceptor centres $Fe^{3+}$. Previously $Fe^{2+}$ was proved to be a hole acceptor in $LiNbO_3$:Fe [3]. However, in the case of X-irradiation the increase of the $Fe^{2+}$ concentration was shown to exceed significantly the corresponding decrease of the $Fe^{3+}$ concentration [6], which requires the existence of another hole acceptor centre. Thus with regard to this scheme in $LiNbO_3$:Fe the X-ray induced band 475 nm is due to the capture of electrons by $Fe^{3+}$ (that means $[Fe^{2+}]$ increase), while the band 375 nm is connected with hole trapping by some hole acceptor centre discussed below. As for undoped $LiNbO_3$ (Fig. 1b) the origin of the X-ray induced band $\approx$480 nm is not evident and in some works [8, 9] it is explained by contaminations. In this connection the induced band at 375 nm might be in any case due to hole capture.

Now we discuss the X-ray (and UV) induced spectra in Mg-doped crystals (Fig. 2). The main distinction from those in the not Mg-doped crystals (Fig. 2) is the absence of the induced band at 475 nm both in $LiNbO_3$:Mg and $LiNbO_3$:Fe:Mg. Corresponding to the considered donor–acceptor scheme this means that X-ray (and UV) irradiation of $LiNbO_3$:Mg and $LiNbO_3$:Fe:Mg does not lead to the formation of extra $Fe^{2+}$. Consequently in Mg-doped $LiNbO_3$ crystals the $Fe^{3+}$ centres are no more electron acceptors.

The obtained results directly confirm the a priori suggested [2, 4] alteration of the $Fe^{3+}$ acceptor properties due to Mg incorporation. The effect is sharply manifested in the $LiNbO_3$ crystals with Mg concentration exceeding the threshold value (2 and 2.4%) while the spectrum of the irradiated, low-doped crystal (1% Mg) behaves in an "intermediate" way.

The absorption spectra of the $LiNbO_3$: 2% Mg crystals, X-irradiated at low temperatures, reveal an X-ray induced band at 1200 nm [2]. The authors [2] interpret its appearance as electron capture at the acceptor $Mg^{2+}$ on $Li^+$ site.

Thus our results together with [2] lead to the conclusion that Mg doping alters the type of the electron acceptor centre in $LiNbO_3$: Fe. Consequently the radiation-induced $Fe^{2+}$ centre which is characteristic of $LiNbO_3$:Fe crystals, does not appear in the case of $LiNbO_3$: Fe:Mg.

Note that the spectral position of the X-ray and UV-induced absorption band at 375 nm in Mg-doped crystals coincides with that of the second absorption band in the X-irradiated crystals without Mg (Fig. 1). So it is reasonable to suppose a general origin of the X-ray and UV-induced absorption band 375 nm in all the $LiNbO_3$ crystals and to relate it to the capture of a photoexcited hole by some intrinsic hole acceptor centre.

The origin of hole-trapping centres in $LiNbO_3$ was widely discussed. For instance, the authors [9] have observed similar absorption bands in $LiNbO_3$ and $LiNbO_3$:Cu, X-irradiated at low temperatures (in the second case the band does not disappear up to 300 K) and associated it with holes trapped at Li vacancies. However, our data are not in agreement with this conclusion. The "sensitivity" to X-ray (and UV) irradiation both in Mg-doped and not Mg-doped crystals reveals the sharp dependence on the presence of even very low iron concentrations (see Fig. 1, 2 and [6]). For instance, the maximum irradiation-induced changes in the absorption coefficient at $\lambda = 375$ nm for $LiNbO_3$: 2% Mg and $LiNbO_3$:0.01% Fe: 2.4% Mg are $\Delta\alpha_{max} \approx$ $\approx 0.3$ to 0.5 and $\approx$10 to 12 $cm^{-1}$, respectively, the initial value of $\alpha$ being similar in both crystals ($\alpha_0 \approx 0.3$ to 0.8 $cm^{-1}$).

Thus iron acts as a "trigger" of the hole-trapping centre. The acceptor properties of the Li vacancies are unlikely to depend so sharply on the low Fe doping. The radiation-induced band at 375 nm which is characteristic of all the investigated crystals seems to be attributed to a $V^-$-type radiation-induced hole centre ($O^-$ ion with the adjacent defect) which is well known for alkaline-earth oxides [10].

## 4. Conclusion

1. Mg doping of $LiNbO_3$:Fe crystals reduces (as in pure $LiNbO_3$) the value of photorefraction at least by a factor of $10^3$.

2. Radiation-induced centres in Mg-doped and not Mg-doped $LiNbO_3$ crystals are different. In $LiNbO_3$:Fe X-irradiation leads to the formation of extra $Fe^{2+}$ due to the electron capture at $Fe^{3+}$, while in $LiNbO_3$:Fe:Mg extra $Fe^{2+}$ does not arise under X-ray and UV-irradiation. That means the electron acceptor type changes due to Mg incorporation and consequently the photorefractive properties change.

## References

[1] D. A. BRYAN, R. GERSON, and H. E. TOMASCHKE, Appl. Phys. Letters **44**, 847 (1984).
[2] K. L. SWEENEY, L. E. HALLIBURTON, D. A. BRYAN, R. R. RICE, and R. GERSON, J. appl. Phys. 57, 1036 (1985).
[3] E. KRÄTZIG and H. KURZ, Optica Acta **24**, 475 (1977).
[4] R. GERSON, J. E. KIRCHHOFF, L. E. HALLIBURTON, and D. BRYAN, J. appl. Phys. **60**, 3553 (1986).
[5] T. R. VOLK, S. SHRAMCHENKO, and L. A. SHUVALOV, Soviet Phys. — Solid State **27**, 2334 (1985).
[6] T. R. VOLK, M. A. IVANOV, and M. L. MEILMAN, Soviet Phys. — Solid State **29**, 871 (1987).
[7] H. KURZ, E. KRÄTZIG, W. KEUNE, H. ENGELMANN, U. GONSER, and B. DISCHLER, Appl. Phys. **12**, 355 (1977).
[8] O. F. SHIRMER and D. VON DER LINDE, Appl. Phys. Letters **33**, 35 (1978).
[9] L. ARIZMENDI, Y. M. CABRERA, and F. AGULLÓ-LÓPEZ, J. Phys. C **17**, 515 (1984).
[10] B. HENDERSON and J. E. WERTZ, Defects in Alkaline Earth Crystal Oxides, Taylor and Francis, London 1977 (Sect. 4.2).

*(Received January 11, 1988)*

phys. stat. sol. (a) 108, 443 (1988)

Subject classification: 78.20; S8.13

*Sektion Physik der Humboldt-Universität zu Berlin, Bereich Halbleiteroptik*[1])

# Photoacoustic Spectroscopy of CdTe and (Hg, Cd)Te

By

O. GOEDE, W. HEIMBRODT, and TH. KÖPP

Dedicated to Prof. Dr. E. GUTSCHE on the occasion of his 60th birthday

The photoacoustic (PA) spectroscopy using a piezoelectric PZT transducer attached to the sample is shown to be successfully employable also for small gap II–VI semiconductor bulk crystals and epilayers both in the low- and high-absorption region. The PA measurements of CdTe and (Hg, Cd)Te mixed crystals carried out at room temperature yield information about the quality and composition of the samples, respectively. Especially, the PA spectroscopy can be used to study the exciton absorption region of a layer also in the case of a substrate with smaller energy gap, as demonstrated for CdTe epilayers on GaAs as an example.

Es wird gezeigt, daß die photoakustische (PA) Spektroskopie mittels eines an die Probe geklebten piezoelektrischen PZT-Transducers auch bei schmallückigen II–VI-Halbleiter-Kompaktkristallen und -Epitaxieschichten sowohl im Bereich niedriger als auch hoher Absorption erfolgreich anwendbar ist. Die bei Zimmertemperatur durchgeführten PA-Messungen an CdTe und (Hg, Cd)Te-Mischkristallen liefern Informationen über Qualität bzw. Zusammensetzung der Proben. Die PA-Spektroskopie kann insbesondere benutzt werden, um den Exzitonen-Absorptionsbereich von Schichten auch im Fall von Substraten mit kleinerer Energielücke zu untersuchen, wie am Beispiel von CdTe-Epitaxieschichten auf GaAs demonstriert wird.

## 1. Introduction

The photoacoustic (PA) spectroscopy [1 to 3] can be applied profitably also for optical investigations of semiconductors because of its specific advantages in comparison with the conventional absorption spectroscopy. Especially, the PA spectroscopy with a piezoelectric transducer directly attached to the sample has often been used (see, e.g. [4 to 10]). In this method of measurements the transducer converts the thermal strains in the sample induced by the absorption of modulated light into a measurable voltage.

In a recent paper [5] we have shown for various broad-gap II–VI semiconductors as examples, that PA spectroscopy with a piezoelectric transducer can be applied just in the $\alpha d$ regions being complementary to the range of products of absorption coefficient ($\alpha$) and sample thickness ($d$) about $\alpha d = 1$, which is accessible for the conventional absorption spectroscopy. In the low-absorption (LA) region ($\alpha d \ll 10^{-1}$) there is a linear dependence between the measured PA signal and the absorption coefficient $\alpha(E)$ of the sample. In the high-absorption (HA) region ($\alpha d > 10^2$) at photon energies near the gap energy, this relation is not known quantitatively but is monotonous and, therefore, allows at least the determination of exciton peak positions.

In the present paper results are reported on the PA spectroscopy of the smaller-gap II–VI semiconductors CdTe and (Hg, Cd)Te which are important materials for IR detector devices. Information concerning the quality and composition of bulk crystals

---

[1]) Invalidenstr. 110, DDR-1040 Berlin, GDR.

and epilayers, respectively, can be obtained by room temperature measurements, in spite of some open questions with respect to the theoretical interpretation of the method.

## 2. Experimental

The investigated CdTe bulk crystals were grown by the travelling-heater and Bridgman methods. The (111) CdTe epilayers on (100) GaAs with a thickness $d = 10\,\mu m$ were obtained using hot-wall epitaxy. The (Hg, Cd)Te bulk crystals were grown also by the travelling-heater method, the composition was determined by X-ray fluorescence measurements.

The cylindrical lead zirconium titanate (PZT) transducer having 3 mm diameter and 0.5 mm height was attached to the sample using conducting silver paste. The optical arrangement for the PA spectroscopy consisted of 150 W halogen lamp, grating or prism monochromator, and light chopper with a chopper frequency of 12.5 Hz. The PA signal was measured using preamplifier (impedance $> 100\,M\Omega$), selective nanovoltmeter, and recorder. Further experimental information is given in [5].

## 3. Experimental Results and Discussion

### *3.1 CdTe bulk crystals and epilayers on GaAs substrate*

In Fig. 1a the typical PA spectrum for pure CdTe samples is shown. In the LA region, at photon energies $E$ sufficiently below the band gap, the measured PA signal increases with increasing $E$ due to growing $\alpha(E)$ up to $\alpha d \approx 10^{-1}$. Then a steep decrease of the PA signal down to zero is observed, probably caused by a compensation of the absorption-induced homogeneous expansion of the sample by bending effects as discussed in more detail in [4, 5]. Therefore, a maximum appears in the LA region of the

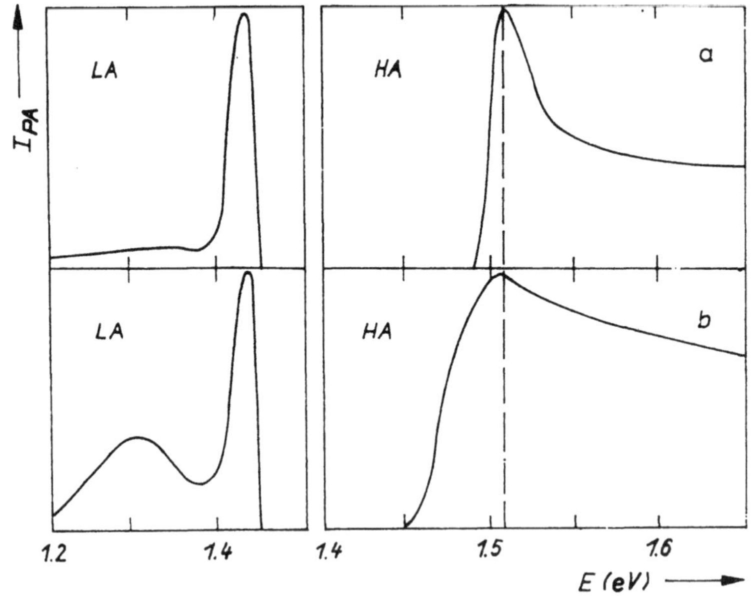


Fig. 1. a) Typical (normalized) PA spectra of high purity CdTe samples in the LA and HA region, respectively. b) Examples for PA spectra of lower-quality CdTe samples. $T = 300$ K

PA signal at a sample-thickness-dependent position $E_{LA}^{max}$ given by $\alpha(E_{LA}^{max}) \approx 10^{-1}/d$. As demonstrated in Fig. 1b for a lower-quality CdTe sample, in this LA region of the PA spectrum an additional peak at about 1.3 eV can be observed which corresponds to a real impurity absorption band with an estimated peak absorption of about $10^{-1}$ $cm^{-1}$. The nature of these impurities is not known at present.

For sufficiently small distances between light spot and transducer on the sample also in the HA region at photon energies near or above the gap energy a PA signal can be detected being phase shifted by about 180° with respect to the PA signal in the LA region. The position of the observed peak in the HA region corresponds to the free exciton energy as could be shown experimentally in [5] for various broad-gap II–VI semiconductors. For an explanation of *structured* PA spectra in the HA region, in spite of the absorption saturation for $\alpha d \gg 1$, the radiationless deexcitation processes of excitons near the sample surface[2]) and the appearance of acoustic surface oscillations must be taken into account [5]. As can be seen in Fig. 1a also for CdTe a distinct peak is found in the HA region at the exciton energy $E_{exc} = 1.51$ eV at 300 K [11]. The halfwidth of this peak increases considerably for lower-quality samples (see Fig. 1b).

A very interesting advantage of the PA spectroscopy with piezoelectric transducer is the possibility of absorption studies in the exciton region of thin semiconductor layers deposited on a substrate having a smaller energy gap. This is demonstrated in Fig. 2 by the PA spectrum measured for a CdTe epilayer on a GaAs substrate with the PZT transducer attached to GaAs. In the same measurement the LA and HA parts of the PA spectrum were obtained both for GaAs and CdTe as can be seen in Fig. 2. The HA peak for the GaAs substrate agrees well with the corresponding exciton energy $E_{exc}$ (GaAs, 300 K) = 1.41 eV [12]. In comparison to a CdTe bulk crystal the LA peak position $E_{LA}^{max}$ is shifted to higher energies because of the smaller thickness of the layer. On the basis of these experimental results the application of the PA spectroscopy with piezoelectric transducer can be suggested also for absorption studies in the case of multi-quantum well structures or superlattices on substrates with smaller energy gap.

### 3.2 (Hg, Cd)Te mixed crystals

In Fig. 3 the measured PA spectra are shown for (Hg, Cd)Te bulk mixed crystals with various compositions, again consisting of the LA and HA regions as in the case of CdTe. With increasing Hg content both the HA and LA peaks strongly shift to lower

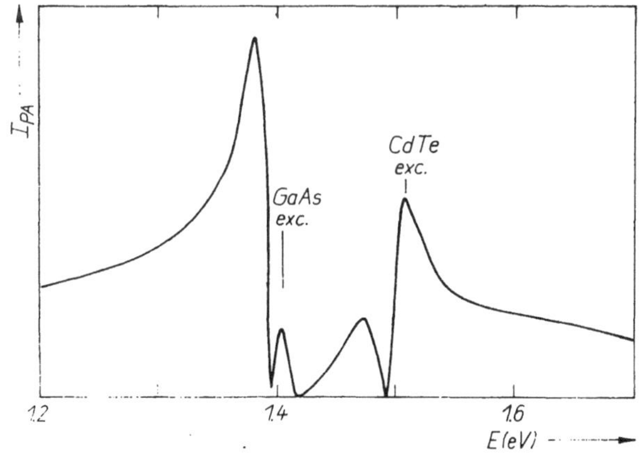


Fig. 2. PA spectrum of a CdTe epilayer on GaAs substrate measured with transducer on GaAs. Layer thickness $d = 10$ µm. $T = 300$ K

[2]) A comparable situation exists in the case of luminescence or photoconductivity excitation spectroscopy for $\alpha d \gg 1$.

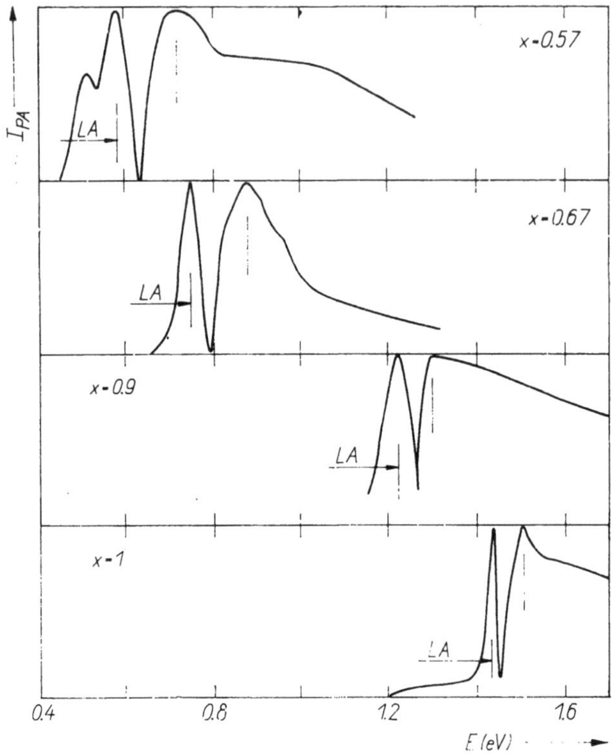


Fig. 3. Normalized PA spectra of $Hg_{1-x}Cd_xTe$ crystals with the same thickness ($d \approx 1$ mm) and various compositions $x$. The marks denote exciton peak positions, respectively. $T = 300$ K

energies. The peak position of the HA peak which corresponds to the exciton energy is plotted in Fig. 4 as a function of the Hg concentration. A good agreement can be established with the empirically obtained relation for $E_g(x, 300\text{ K})$ given in [13][3]. Therefore, the PA spectroscopy can successfully be applied to determine the composition of (Hg, Cd) Te mixed crystals at room temperature with an accuracy of about $\pm$ 0.003.

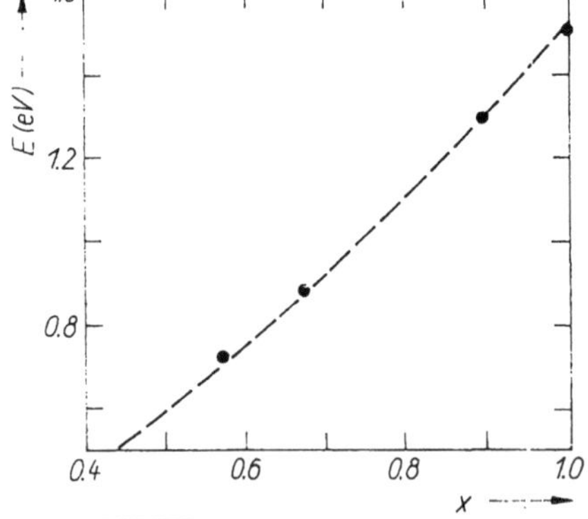


Fig. 4. Exciton peak positions from the PA spectra of $Hg_{1-x}Cd_xTe$ after Fig. 3 (experimental points). Dashed line: Energy gap as a function of composition $x$ for $T = 300$ K after [13]

[3]) After correcting an obvious misprint in the second term this relation reads $E_g(x, T) = -0.303 + 1.985x - 0.655x^2 + 0.579x^3 + (5.5 - 13.92x + 5.84x^2) \times 10^{-4}T$. Taking into account the large halfwidths at 300 K, in the present context the difference between $E_g$ and the exciton energy ($\leqq$10 meV) can be neglected.

The observed broadening of both the HA and LA peak and the increase of the peak distance $E_{exc} - E_{LA}^{max}$ with increasing Hg concentration are caused by potential-fluctuation-induced flattening of the absorption edge being well known from broad-gap II–VI semiconductor mixed crystals. It should be pointed out, however, that this effect is surprisingly less pronounced for (Hg, Cd)Te than for the anionically mixed systems Cd(S, Se) and Cd(S, Te) studied in [5]. This different behaviour may be related to the comparatively very small deviations between the covalent radii of Cd and Hg in comparison to the corresponding values for the chalcogen anions. Therefore, smaller *structural* disorder and lower density-of-states tails can be expected in the case of (Hg, Cd)Te.

The additional lower-energy peak in the LA region of the (Hg, Cd)Te crystal with $x = 0.57$ can be attributed to an (unidentified) impurity absorption and indicates the applicability of the PA spectroscopy for an impurity detection also in small-gap II–VI semiconductor mixed crystals.

### *Acknowledgement*

We would like to thank our collegues of the crystal growth department of the Sektion Physik for supplying the crystals and epilayers, respectively.

### References

[1] A. Rosencwaig, Adv. Electronics and Electron Phys. **46**, 207 (1978).
[2] A. C. Tam, Rev. mod. Phys. **58**, 381 (1986).
[3] C. S. Sunandana, phys. stat. sol. (a) **105**, 11 (1988).
[4] W. Jackson and N. M. Amer, J. appl. Phys. **51**, 3343 (1980).
[5] O. Goede, W. Heimbrodt, and F. Sittel, phys. stat. sol. (a) **93**, 277 (1986).
[6] J. Morimoto, T. Ito, and T. Miyakawa, Japan. J. appl. Phys. **21** 101 (1982).
[7] H. Hordvik and H. Schlossberg, Appl. Optics **16**. 101 (1977).
[8] K. Wasa, K. Tsubouchi, and N. Mikoshiba, Japan. J. appl. Phys. **19**, L475 (1980).
[9] T. Hata, Y. Sato, and M. Kurebayashi, Japan. J. appl. Phys. **22**, 205 (1983).
[10] T. Hata, Y. Sato, Y. Nagai, and T. Hada, Japan. J. appl. Phys. **23**, 75 (1984).
[11] N. C. Giles-Taylor, R. N. Bicknell, D. K. Blanks, T. H. Myers, and J. F. Schetzina, J. Vacuum Sci. Technol. **A3**, 76 (1985).
[12] S. Adachi, J. appl. Phys. **58**, R1 (1985).
[13] R. Legros and R. Triboulet, J. Crystal Growth **72**, 264 (1985).

*(Received April 8, 1988)*

phys. stat. sol. (a) 108, 449 (1988)

Subject classification: 78.65; S11.1

*Fachbereich Physik der Universität Osnabrück*[1])

# Extraordinary versus Ordinary Refractive Index Change in Planar $LiNbO_3$:Ti Waveguides

By

S. FRIES, P. HERTEL, and H. P. MENZLER

Various planar waveguides are fabricated by indiffussion of one or two titanium layers into y-cut $LiNbO_3$ substrates. The effective indices of guided TE- and TM-modes are measured for red, green, and blue light. The refractive index profiles are reconstructed by a new reliable inverse WKB procedure. Despite of widely differing profiles there is, for each wavelength, a one to one relationship between the changes $\Delta n_o$ of the ordinary and $\Delta n_e$ of the extraordinary refractive index. For low and for high refractive index changes $\Delta n_e$ and $\Delta n_o$ grow proportionally, but their ratios differ appreciably. The results support recent findings that the $LiNbO_3$:Ti system exists in two distinct phases.

Verschiedene planare Wellenleiter werden durch Eindiffusion von einer oder zwei Titanschichten in y-geschnittene $LiNbO_3$-Substrate hergestellt. Die effektiven Indizes der geführten TE- und TM-Moden werden für rotes, grünes und blaues Licht gemessen. Mit einem neuen zuverlässigen inversen WKB-Verfahren werden die Brechungsindexprofile rekonstruiert. Trotz stark unterschiedlicher Profile existiert für jede Wellenlänge eine eins-zu-eins-Beziehung zwischen den Änderungen $\Delta n_o$ des ordentlichen und $\Delta n_e$ des außerordentlichen Brechungsindexes. Für niedrige und hohe Brechungsindexänderungen wachsen $\Delta n_o$ und $\Delta n_e$ proportional, jedoch unterscheiden sich ihre Verhältnisse beträchtlich. Die Ergebnisse bestätigen die kürzlich gemachte Entdeckung, daß das $LiNbO_3$:Ti-System in zwei unterschiedlichen Phasen existiert.

## 1. Introduction

Already in 1974 Schmidt and Kaminow [1] reported on a new waveguide which had been fabricated by indiffusion of titanium into a $LiNbO_3$ single domain substrate. Today such waveguides are indispensible in the growing field of integrated optics. Although everybody knows how to make them, the underlying physics is still badly understood.

The initial stages of the indiffusion process have been studied in great detail by De Sario et al. [2]. How titanium ions migrate is not known, but where they settle down has been investigated recently by Skeath et al. [3]. If the concentration is higher than a critical value of $5 \times 10^{20}\,cm^{-3}$ titanium ions sit on off-axis sites. Skeath et al. argue why the extraordinary and the ordinary refractive indices are modified differently in the low and high density phase.

The changes $\Delta n_o$ of the ordinary refractive index and $\Delta n_e$ of the extraordinary refractive index do not depend linearly on the titanium concentration $c$. This fact is known since long.

Minakata et al. [4] noted that $\Delta n_o$ was larger than $\Delta n_e$ in one case and smaller in another, it therefore is impossible that both grow in proportion with the number of titanium ions per unit volume.

[1]) Barbarastraße 7, D-4500 Osnabrück, FRG.

Lüdtke et al. [5] measured titanium concentration profiles $c = c(x)$ by secondary ion mass spectroscopy (SIMS). They obtained Gaussian distributions characteristic of a normal diffusion process. A local relationship between refractive index changes and titanium concentration — linear for $\Delta n_e$, power law for $\Delta n_o$ — could explain the data. Their analysis was, however, restricted to fairly low titanium concentration.

Normal diffusion behaviour was assumed by Vollmer et al. [6]. Since the reconstructed refractive index profiles were found to be of Gaussian shape, a local power law relationship $\Delta n \sim c^\eta$ was postulated. Satisfactory agreement could be reached with $\eta_0 = 0.62$ for the ordinary and $\eta_e = 0.88$ for the extraordinary index change (at $\lambda = 633$ nm).

Čtyroký et al. [7] assumed refractive index profiles of Gaussian shape and fitted the measured propagation constants accordingly. Their findings agreed reasonably well with earlier measurements.

Fouchet et al. [8] use an electron microprobe to scan the surface of a 1° bevel. Although the power law relationship $\Delta n = Ac^\eta$ is confirmed, the proportionality constant $A$ now depends on $c_0$, the surface titanium concentration. They claim that there is no local relationship between $\Delta n_o$ or $\Delta n_e$, respectively, and the distribution of titanium ions.

We undertook this investigation with the following idea in mind. If both $\Delta n_o$ and $\Delta n_e$ depend locally on $c$ then, for each wavelength, a universal function $f$ should exist such that

$$\Delta n_e(x) = f(\Delta n_o(x)) \tag{1}$$

holds true. This can be tested without measuring the titanium concentration which is a difficult and error prone task. Instead we concentrated on reconstructing the refractive index profiles as accurately as possible. Waveguides which differ by titanium layer thickness or diffusion time or which have been treated twice were investigated. If their $\Delta n_e$ versus $\Delta n_o$ graphs do not coincide, then (1) cannot be true, and the relationship between titanium concentration and refractive index change is non-local. Conversely, if (1) is supported by experiment, we know that $\Delta n_o$ and $\Delta n_e$ depend locally on some variable describing the $LiNbO_3$:Ti state. This might be, but must not be, the titanium concentration $c$.

Now, (1) turned out to hold true, and we shall discuss some implications in the last section. In contrast to earlier results the $\Delta n_e$ versus $\Delta n_o$ graph consists of two straight lines with substantially different slopes. Our results agree with the findings of Skeath et al. that two different mechanisms are responsible for the refractive index change.

## 2. Experimental Procedure

We used y-cut $LiNbO_3$ substrates of 1 mm thickness which came from the same boule and were of congruent composition. The latter fact was established by measuring the phase matching temperature for second harmonic generation. Samples of $25 \times 5$ mm$^2$ were cut such that the optical axis is parallel to the short side. After polishing and cleaning titanium layers of thickness $d_1$ were deposited by electron gun evaporation. The samples were wrapped in a platinum foil and heated in argon atmosphere. Having been kept at 1000 °C for a certain time $t_1$ they were cooled down to room temperature in an oxygen atmosphere. Some samples were treated twice in this way. After depositing and indiffusing a first titanium layer a second layer of thickness $d_2$ was deposited and indiffused for a time $t_2$ at 1000 °C. We here report on eight waveguides:

In some cases it was necessary to gently polish the surface once more in order to improve the waveguide quality.

Table 1

Thickness $d_1$ of first and $d_2$ of second Ti layer as well as diffusion times $t_1$ and $t_2$ of first and second treatment. Waveguides 2, 5, and 8 have been treated twice

| waveguide | $d_1$ (nm) | $t_1$ (h) | $d_2$ (nm) | $t_2$ (h) |
|---|---|---|---|---|
| 1 | 180 | 24 | — | — |
| 2 | 180 | 26 | 60 | 11 |
| 3 | 180 | 27 | — | — |
| 4 | 60 | 8 | — | — |
| 5 | 60 | 8 | 80 | 20 |
| 6 | 60 | 27 | — | — |
| 7 | 80 | 16 | — | — |
| 8 | 80 | 16 | 50 | 11 |

Effective indices $N_{\text{eff}}$ of guided TE- and TM-modes were determined by the well-know prism coupling method [9]. We estimate the accuracy to be better than $\pm 5 \times \times 10^{-5}$. Mode spectra were measured for $\lambda = 633$, 514, and 458 nm.

## 3. Reconstructing Refractive Index Profiles

It is easy to calculate effective indices from a given refractive index profile. The inverse problem is much more difficult. Without plausibility constraints there cannot be a unique solution, there are infinitely many profiles for the same set of effective indices.

Many authors still use the so-called inverse WKB procedure of White and Heidrich [10]. The refractive index profile is approximated by a continuous, piecewise linear function which passes through the measured effective indices at certain turning points. One has to guess the surface refractive index, and the algorithm then allows to determine the turning points step by step.

Chiang [11] has considerably improved this procedure. He interpolates by a polynomial between measured effective indices such that $N_{\text{eff}}(m)$ is available for real $m$, not for mode labels $m = 0, 1, \ldots$ only. The WKB equation then serves to compute the refractive index profile as a smooth function. A reconstructed profile has vanishing — as a square root — derivative at the waveguide surface and falls off exponentially at large depth. Moreover, the bulk refractive index cannot be used as an input, it is returned.

We remedied these shortcomings [12]. The functional behaviour of $\Delta n(x)$ at $x = 0$ (surface) and at $x \to \infty$ (bulk) may be enforced along with the bulk value $n_{\text{sub}} = n(\infty)$ itself. Our method proves to be much more accurate and numerically more stable than Chiang's.

One particular point must be mentioned. We strongly restrict the refractive index profile by prescribing its behaviour near the surface and at large depth. As a consequence, solutions exist for a certain range of the bulk refractive index only. Likewise, a $10^{-3}$ error for one effective index may already prevent a solution. However, if a profile is returned at all, one may trust it.

## 4. Results

In Fig. 1 we contrast the refractive index profiles of samples 5 and 6 at $\lambda = 514$ nm. The former has been treated twice, it carries eight TE modes and six TM modes. The latter has been treated once, it carries five TE and five TM modes. All profiles

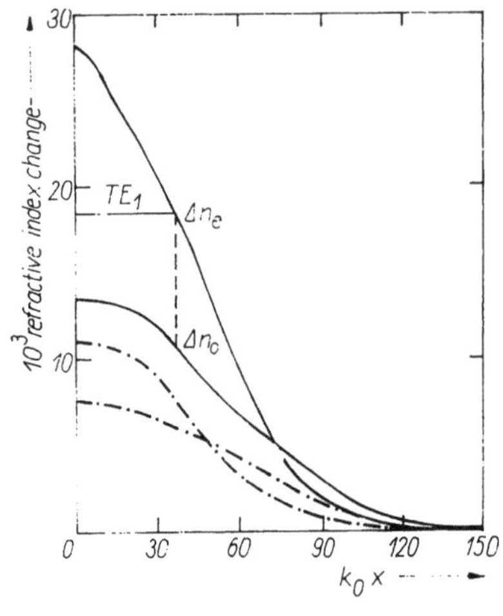


Fig. 1. Extraordinary and ordinary refractive index change vs. dimensionless depth $k_0x$ for waveguides 5 and 6 at $\lambda = 514$ nm. Waveguide 5 (solid lines) was treated twice: 60 nm Ti indiffused for 8 h, then 80 nm Ti indiffused for 20 h. Waveguide 6 (dash-dotted lines) has been treated once: 50 nm Ti indiffused for 27 h. For an effective index (horizontal line) a pair of $\Delta n_o$ and $\Delta n_e$ values are shown which refer to the same depth

were reconstructed with a surface behaviour

$$\Delta n(x) = \Delta n(0) - ax^2 + \dots \tag{2}$$

and

$$\lg \Delta n(x) \to -bx^2 \tag{3}$$

at large depth.

The titanium concentration gradient will vanish at $x = 0$ since all samples have been diffused for a relatively long time. Whatever the (local) relationship between titanium concentration and refractive index may be, its slope then vanishes at the surface. This explains (2). Likewise, whatever the exact diffusion equation will be, we expect a Gaussian concentration profile at large depth. We furthermore assume that the relationship between index changes and titanium concentration be linear for sufficiently low concentration. Therefore (3) was enforced. By the way, most mode spectra could not be fitted at all if a linear decrease was assumed instead of (2).

We have depicted one of the TE modes of waveguide 5 by a horizontal line. It intersects the extraordinary profile at a certain depth $d$. At that depth the ordinary refractive index change is $\Delta n_o$ and the extraordinary index change is $\Delta n_e$. This pair $(\Delta n_o, \Delta n_e)$ will represent the mode in the following figures.

Fig. 2 a to c show such pairs of refractive index changes for three different wavelengths. For each mode there is an entry $(\Delta n_o, \Delta n_e)$ which refers to index changes at the same depth. Either $\Delta n_o$ or $\Delta n_e$ corresponds to an effective index. Neighbouring points from different waveguides represent values at different depths. Data from eight differently fabricated waveguides are merged into these figures.

Obviously the correlation between $\Delta n_o$ and $\Delta n_e$ values is very strong. If the ordinary and the extraordinary refractive index change depended locally on the titanium concentration, then — ideally — all points would lie on one and the same graph. There is some scattering, small for blue and larger for red light, but it should be attributed to reconstruction inaccuracies. The same refractive index profile looks broader for shorter wavelengths and will carry more modes. The short wavelength refractive index profiles — since reconstructed from more input data — are therefore more reliable and should give rise to less scattering.

The large refractive index changes clearly lie on a straight line. Its slope $d\Delta n_e/d\Delta n_o$ is 2.2 for $\lambda = 458$ nm, 2.3 for 514, and 2.4 for 633 nm. This straight line would intersect the $\Delta n_e = 0$ axis at $\Delta n_o = 0.0029$, 0.0027, and 0.0024, respectively.

The lower part ($\Delta n_e < 0.004$ for all three wavelengths) looks like another straight line with slope 0.8. There the data can also be fitted by a power law (as implied by [5] and [6]). A third possibility is that the relatively large scattering may point to a non-local relationship.

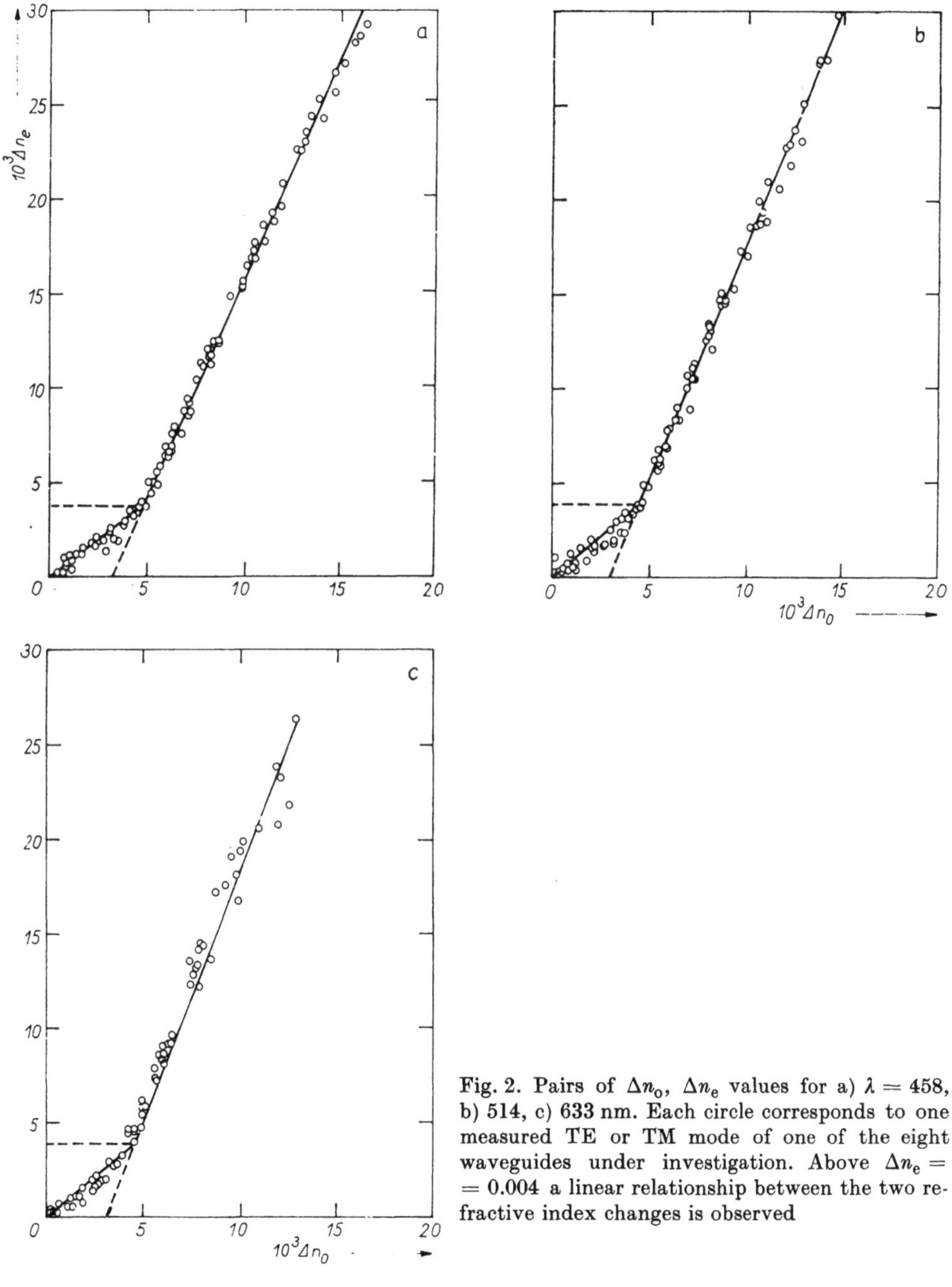


Fig. 2. Pairs of $\Delta n_o$, $\Delta n_e$ values for a) $\lambda = 458$, b) 514, c) 633 nm. Each circle corresponds to one measured TE or TM mode of one of the eight waveguides under investigation. Above $\Delta n_e =$ $= 0.004$ a linear relationship between the two refractive index changes is observed

## 5. Discussion

We have presented evidence that the ordinary and the extraordinary refractive index changes depend locally on some variable describing the $LiNbO_3$:Ti state. This state variable may be, but must not be, the titanium concentration. Internal strain is one of many other possibilities. Moreover, the $\Delta n_e$ versus $\Delta n_o$ graph is a straight line for high refractive indices. This linearity regime sets on at $\Delta n_e = 0.004$.

Note that [5, 6, 8] report a power law relationship between refractive index changes and titanium concentration. This in turn implies a power law relationship between $\Delta n_o$ and $\Delta n_e$ which is definitely ruled out by our new data. Previous investigations relied on less sophisticated methods to reconstruct profiles, and a smoothened version of our curves was to be found. Furthermore, by indiffusing two titanium layers we have to account for a considerably wider range of refractive index changes.

There is a rather sharply defined onset of linearity. The high value part of the universal $\Delta n_e$ versus $\Delta n_o$ curve, when extrapolated to $\Delta n_e = 0$, misses the origin by a significant amount. Therefore, it seems very likely that different mechanisms are responsible for the index changes at low and high titanium concentrations.

Assume that it is the internal strain which causes the refractive indices to change. After all, the ionic radius of $Ti^{4+}$ is smaller than that of $Nb^{5+}$, and a titanium distribution will cause internal strain [13]. Since we have planar waveguides and y-cut substrates the strain tensor has one non-vanishing element only, namely $S_{22}$. The elasto-optic index changes are

$$\Delta n_o = -2n_o^3 p_{1122} S_{22} \quad \text{and} \quad \Delta n_e = -2n_e^3 p_{3322} S_{22} \,. \tag{4}$$

See [14] for the notation and a table of photoelastic strain coefficients. With $p_{1122} = 0.08$ and $p_{3322} = 0.17$ a $\Delta n_e : \Delta n_e$ ratio of 2.1 results. This is compatible with the high $\Delta n$ part. But what, then, explains the low $\Delta n$ data?

It is much more likely that the $LiNbO_3$:Ti system exists in two distinct phases. After all, the kick in Fig. 2 is rather marked. Skeath et al. have put forward convincing evidence that the titanium ion sits on the $LiNbO_3$ symmetry axis at low concentration and is displaced off the axis at high concentration. Its contribution to the respective refractive index changes therefore will be different below and above the critical density $5 \times 10^{20}$ cm$^{-3}$. Note that this concentration corresponds to $\Delta n_e = 4 \times 10^{-3}$ according to [6] and others.

One may also speculate that above a certain critical density titanium ions tend to cluster. A simple pairing model, however, cannot explain the distinct onset of linearity. Assume that titanium ions cluster to pairs. With $c_1$ as density of single ions and $c_2$ as density of pairs one would write

$$2n \, \Delta n = c_1 \alpha_1 + c_2 \alpha_2 \tag{5}$$

for the refractive index change. $\alpha_1$ and $\alpha_2$ are atomic polarizabilities for single ions and pairs, respectively. During the indiffusion process the balance equation

$$c_1^2 = K c_2 \tag{6}$$

will hold, and together with $c_1 + c_2/2 = c$ one may calculate the $\Delta n_e$ versus $\Delta n_o$ curve. If $\alpha_{1,e}$ and $\alpha_{1,o}$ for the ordinary and extraordinary index change differ sufficiently from $\alpha_{2,e}$ and $\alpha_{2,o}$, then different slopes at low and high $\Delta n_o$ can be fitted. The transition between the two regions, however, turns out to be too smooth to fit the data.

### *Acknowledgements*

This work was performed within the program of SFB 225. We gratefully acknowledge financial support from Deutsche Forschungsgemeinschaft. Thanks are also due to K. Betzler and N. Schmidt for the SHG measurements.

### References

[1] R. V. Schmidt and I. P. Kaminov, Appl. Phys. Letters **25**, 458 (1974).
[2] M. De Sario, M. N. Armenise, C. Canali, A. Carnera, P. Mazzoldi, and G. Gelotti, J. appl. Phys. **57**, 1482 (1985).
[3] P. Skeath, W. T. Elam, W. K. Burns, F. A. Stevie, and T. H. Briggs, Phys. Rev. Letters **59**, 1950 (1987).
[4] M. Minakata, S. Saito, M. Shibata, and S. Miyazawa, J. appl. Phys. **49**, 4677 (1978).
[5] H. Lüdtke, W. Sohler, and H. Suche, in: Digest of Workshop on Integrated Optics, Berlin 1980 (p. 121).
[6] J. Vollmer, J. P. Nisius, P. Hertel, and E. Krätzig, Appl. Phys. A **32**, 125 (1983).
[7] J. Čtyroký, M. Hofman, J. Janta, and J. Schröfel, IEEE J. Quantum Electronics **20**, 400 (1984).
[8] S. Fouchet, A. Carenco, C. Daguet, R. Guglielmi, and L. Riviere, J. Lightwave Techn. **LT—5**, 700 (1987).
[9] P. K. Tien and R. Ullrich, J. Opt. Soc. Amer. **60**, 1325 (1970).
[10] J. M. White and P. F. Heidrich, Appl. Optics **15**, 151 (1976).
[11] K. S. Chiang, J. Lightwave Tech. **LT—3**, 385 (1985).
[12] P. Hertel and H. P. Menzler, Appl. Phys. B **44**, 75 (1987).
[13] K. Sugii, M. Fukuma, and H. Iwasaki, J. Mater. Sci. **13**, 523 (1978).
[14] R. S. Weis and T. K. Gaylord, Appl. Phys. A **37**, 191 (1985).

*(Received January 11, 1988)*

phys. stat. sol. (a) **108**, 457 (1988)

Subject classification: 77.20 and 77.40; S9.12

*Department of Physics, Indian Institute of Technology, Kharagpur*[1])

# Dielectric Properties of $CaF_2$ Single Crystals Irradiated with X-Rays under a High Electric Field or γ-Rays

By

A. DE and K. V. RAO

The dielectric constant ($K$), loss ($\tan \delta$) and hence the dielectric conductivity ($\sigma$) of $CaF_2$ single crystals under different conditions are measured in the frequency region $10^2$ to $10^7$ Hz and in the temperature range 30 to 400 °C. The different conditions are when the crystals are (i) quenched from 800 °C, (ii) X-ray irradiated under an electric field of 10 kV/cm (AC or DC), (iii) γ-ray irradiated or (iv) subjected to a combination of these treatments. The dielectric properties of $CaF_2$ are significantly affected due to the charged defects produced in it by these treatments. An attempt is made to understand the results and it is concluded that quenched $CaF_2$ crystals X-ray irradiated under a DC field 10 kV/cm, contain the largest concentration of charged defects.

Die Dielektrizitätskonstante ($K$), der Verlustwinkel ($\tan \delta$) und somit die dielektrische Leitfähigkeit ($\sigma$) von $CaF_2$-Einkristallen unter verschiedenen Bedingungen werden im Frequenzbereich $10^2$ bis $10^7$ Hz und Temperaturbereich 30 bis 400 °C gemessen. Diese verschiedenen Bedingungen sind: (i) die Kristalle werden von 800 °C abgeschreckt, (ii) die Kristalle werden in einem elektrischen Feld von 10 kV cm (Gleich- oder Wechselfeld) röntgenbestrahlt, (iii) γ-bestrahlt oder (iv) einer Kombination dieser Behandlungen ausgesetzt. Die dielektrischen Eigenschaften von $CaF_2$ werden signifikant durch geladene Defekte beeinflußt, die bei diesen Behandlungen erzeugt werden. Es wird versucht, die Ergebnisse zu verstehen, und es wird angenommen, daß abgeschreckte $CaF_2$-Kristalle, die unter einem Gleichspannungsfeld von 10 kV/cm röntgenbestrahlt werden, die größte Konzentration geladener Defekte enthalten.

## 1. Introduction

A study of the colour centre phenomena and dielectric properties of alkali halide crystals X-ray irradiated under high electric fields yielded valuable information regarding the defect processes occurring in them [1 to 3]. Similar studies carried out on the optical absorption and thermoluminescence of alkaline earth fluorides — particularly $CaF_2$ — also gave useful data [4, 5]. It is the aim of this paper to report the data of our measurements on the dielectric properties of $CaF_2$ single crystals taken under these conditions. Dielectric constant ($K$) and loss ($\tan \delta$) of $CaF_2$ have been measured — and hence conductivity $\sigma$ calculated — in the frequency region $10^2$ to $10^7$ Hz and in the temperature range 30 to 400 °C. These measurements have been carried out on quenched $CaF_2$ (from 800 °C), when these crystals are later irradiated with X-rays under a high electric field (ac or dc) or when they are irradiated with γ-rays.

## 2. Experimental Methods

The $CaF_2$ crystals used in the present work are a gift from Crystal Physics Laboratory, M.I.T. Cambridge, U.S.A. Samples were cleaved from big boules, ground, and optically polished. The final dimensions of the samples were about $1 \times 1 \times 0.1\ cm^3$.

---

[1]) Kharagpur 721302, India.

Flame photometric and chemical analyses of these samples showed impurity concentrations less than 2 ppm. Some of the impurities looked for were strontium, barium, manganese, cadmium, sodium, and potassium. X-ray irradiation was done using a Norelco unit run at 35 kV, 10 mA for 1.5 h — half the time on either side of the crystal. The details of X-ray irradiation under high ac or dc fields were reported earlier [6, 7]; care was taken to see that the same face of the crystal was towards the same polarity of the dc field while X-ray irradiation was being carried out. $\gamma$-ray irradiation was carried out with a $Co^{60}$ source of about $10^3$ Ci giving a dose of 7.8 MGy/h.

The dielectric measurements were taken on a GR 716 type capacitance bridge in the frequency range $10^2$ to $10^5$ Hz [8] and on a Marconi circuit magnification meter type TF 329G in the range $10^5$ to $10^7$ Hz using the resonance curve principle [9]. Silver paint was applied on either side of the samples as electrodes. The accuracy of the measurement of the dielectric constant ($K$) is 2% and of the loss ($\tan \delta$) about 5%. At least two samples were studied and the agreement was within the limits mentioned.

### 3. Results

Fig. 1 presents K and $\tan \delta$ at room temperature ($\approx$ 30 °C) as a function of frequency for $CaF_2$ crystal under different conditions. K of $CaF_2$ is 6.8 and is frequency independent; $\tan \delta$ has a value below the detection level of the instrument ($\tan \delta < < 1 \times 10^{-4}$). Quenching, X-ray irradiation under an electric field of 10 kV/cm (ac or dc), or $\gamma$-ray irradiation is found to increase $K$ and $\tan \delta$ values appreciably at low frequencies; for example, $K$ at $10^2$ Hz is increased by about 35% (compared to the as-cleaved one) when quenched $CaF_2$ is X-ray irradiated under a dc field of 10 kV/cm and $\tan \delta$ increases by three decades.

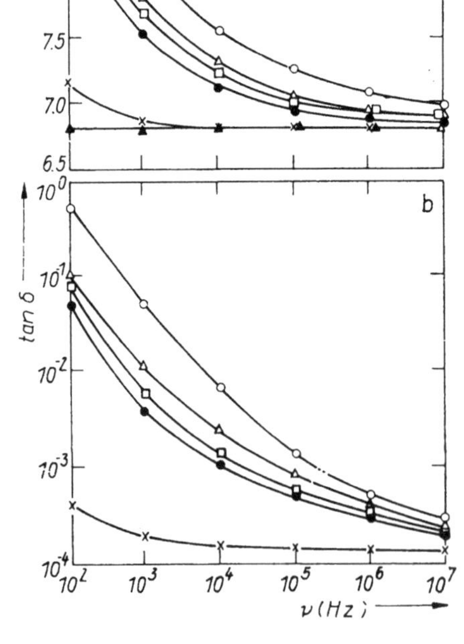


Fig. 1. a) Dielectric constant ($K$) and b) loss ($\tan \delta$) as functions of frequency at 30 °C for $CaF_2$ crystals under different conditions mentioned. ▲ As-cleaved, × quenched, ● quenched + X-rayed, □ quenched + X-rayed under 10 kV/cm ac, △ quenched + $\gamma$-rayed for 50 h, ○ quenched + X-rayed under 10 kV/cm dc

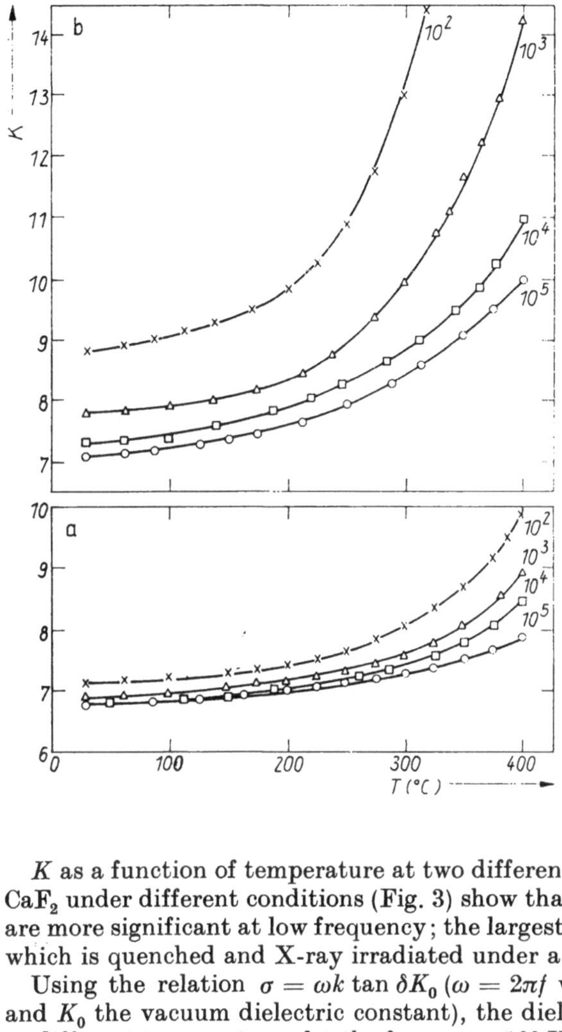


Fig. 2. Variation of dielectric constant ($K$) with temperature at different frequencies (given at the curves in Hz) for $CaF_2$ crystals. a) Quenched, b) quenched + γ-rayed for 50 h

The variation of dielectric constant with temperature at different frequencies for quenched $CaF_2$ and later γ-ray irradiated, is shown in Fig. 2. We notice that $K$ increases slowly up to about 200 °C beyond which its increase is larger and more frequency dependent; $K$ has larger values at lower frequencies at any temperature.

$K$ as a function of temperature at two different frequencies e.g. $10^2$ and $10^5$ Hz, for $CaF_2$ under different conditions (Fig. 3) show that the changes in $K$ with temperature are more significant at low frequency; the largest changes have been observed in $CaF_2$ which is quenched and X-ray irradiated under a dc field of 10 kV/cm.

Using the relation $\sigma = \omega k \tan \delta K_0$ ($\omega = 2\pi f$ where $f$ is the measuring frequency and $K_0$ the vacuum dielectric constant), the dielectric conductivity $\sigma$ of the samples at different temperatures for the frequency $10^2$ Hz has been calculated and plotted as a function of $1/T$ (Fig. 4) for the crystals under different conditions. The conductivity curves exhibit the usual extrinsic and intrinsic conductivity regions; the value of activation energy for conduction in the intrinsic region (which is 1.4 eV for as-cleaved crystals [10]) is found to decrease with various treatments being minimum for the crystals which have been quenched and X-ray irradiated under a dc field of 10 kV/cm.

Pertinent data of the present work are summarised in Table 1.

## 4. Discussion

The dielectric constant of a material consists of four contributions: electronic, ionic, dipolar, and space charge polarisations. All these may be active at low frequencies. In fact, the nature of the variation of the dielectric constant with frequency indicates which contributions are present.

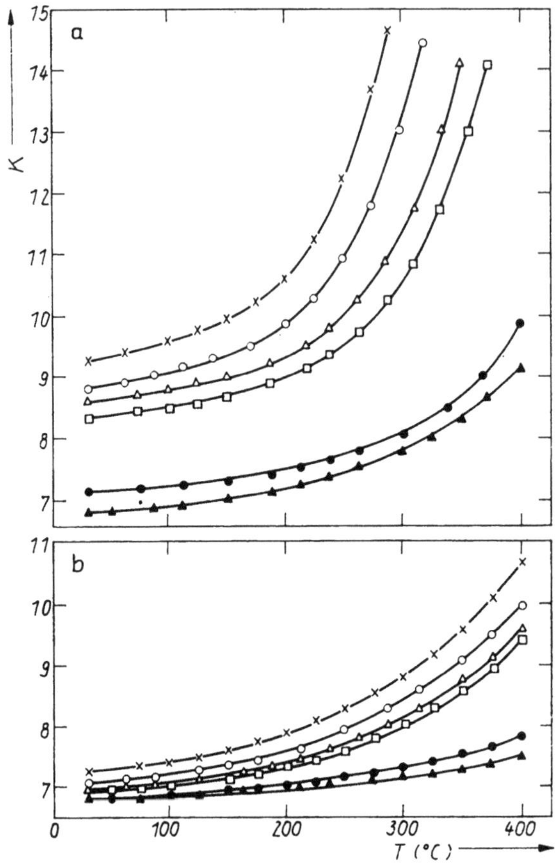


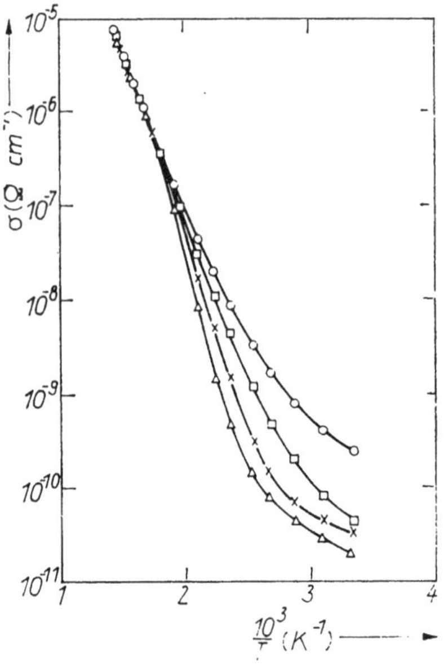


Fig. 3 Fig. 4

Fig. 3. Variation of dielectric constant ($K$) with temperature at two frequencies (a) $10^2$ and b) $10^5$ Hz) for $CaF_2$ crystals under different conditions mentioned. ▲ As-cleaved, ● quenched, □ quenched + X-rayed, △ quenched + X-rayed under 10 kV/cm ac, ○ quenched + γ-rayed for 50 h, × quenched + X-rayed under 10 kV/cm dc

Fig. 4. Conductivity ($\sigma$) as a function of $T^{-1}$ at $10^2$ Hz for $CaF_2$ crystals under different conditions˙ △ Quenched + X-rayed, × quenched + X-rayed under 10 kV/cm ac, □ quenched + γ-rayed for 50 h, ○ quenched + X-rayed under 10 kV/cm dc

The space charge contribution will depend on the purity and perfection of the crystals. Its influence is negligible at very low temperatures and is noticeable in the low-frequency region. The dipolar orientational effect can sometimes be seen in some materials up to $10^{10}$ Hz. Ionic and electronic polarisations always exist below $10^{13}$ Hz.

Quenching, γ-ray irradiation, or X-ray irradiation of solids under high electric fields is known to produce charged defects like vacancies [1, 2]. These defects, under the influence of the electric field used to measure the dielectric properties, will contribute to an appreciable space charge polarisation which increases the low-frequency $K$ and $\tan\delta$ values as observed in the present measurements. The present results indicate that the concentration of such defects is largest in crystals which are quenched and X-ray irradiated under a dc field of 10 kV/cm. This conclusion is supported by

Table 1

Summary of the data on the dielectric properties of $CaF_2$ single crystals

| sample condition | K at 30 °C | | $10^4$ tan δ at 30 °C | | activation energy for conduction in the intrinsic region (eV) |
|---|---|---|---|---|---|
| | $10^2$ Hz | $10^6$ Hz | $10^2$ Hz | $10^6$ Hz | |
| as-cleaved | 6.8 | 6.8 | <1 | – | 1.4 |
| quenched | 7.16 | 6.8 | 4 | 1 | 1.2 |
| quenched + X-rayed | 8.35 | 6.9 | 500 | 2.0 | 1.1 |
| quenched + X-rayed under 10 kV/cm ac | 8.62 | 6.95 | 800 | 2.2 | 1.0 |
| quenched + γ-rayed | 8.80 | 6.95 | 1000 | 2.5 | 1.0 |
| quenched + X-rayed under 10 kV/cm dc | 9.25 | 7.08 | 5000 | 3 | 0.9 |

the fact that the rate of increase of $K$ with temperature ($\mathrm{d}K/\mathrm{d}T$) at low frequency (viz. $10^2$ Hz) and at high temperatures — which can be considered as a measure of the concentration of the defects [11] — has the largest value under these conditions of the crystals.

The decrease in activation energy for conduction in the intrinsic region may be attributed to the increase in charge carrier concentration. Again we find that the quenched $CaF_2$ crystals X-ray irradiated under a dc field of 10 kV/cm exhibit the lowest activation energy indicating that these samples contain the largest concentration of defects.

It may be mentioned here that the changes in dielectric constant and loss of quenched $CaF_2$ crystals subjected to various treatments, are more pronounced compared to those in as-cleaved $CaF_2$ subjected to the same treatments, as such, data are presented on quenched crystals only.

In summary it may be mentioned that a considerable concentration of lattice defects like vacancies can be produced in $CaF_2$ crystals by subjecting them to various treatments like quenching, γ-ray irradiation, or X-ray irradiating them under high electric fields or a combination of these treatments; such defects have been found to affect the dielectric properties of the crystals appreciably.

*Acknowledgement*

The authors are grateful to Prof. A. Smakula, erstwhile Director, Crystal Physics Laboratory, M.I.T. Cambridge, U.S.A. for kindly supplying the samples used in the present work.

## References

[1] A. Subrahmanyam, and K. V. Rao, phys. stat. sol. (a) **52**, K147 (1979).
[2] P. K. Chakraborty and K. V. Rao, phys. stat sol. (b) **120**, K107 (1983).
[3] P. K. Chakraborty, Ph. D. Thesis, Indian Institute of Technology, Kharagpur 1984.
[4] A. De and K. V. Rao, phys. stat. sol. (a) **105**, 297 (1988).
[5] A. De and K. V. Rao, J. Mater. Sci., in the press.
[6] S. Govinda and K. V. Rao, Solid State Commun. **16**, 297 (1975).
[7] P. K. Chakraborty and K. V. Rao, Crystal Lattice Defects **10**, 67 (1983).
[8] W. B. Westphal, Techn. Rep. No. 182, Lab. Insulating Research, M.I.T., Cambridge (USA) 1963.
[9] K. V. Rao, J. Phys. Chem. Solids **20**, 193 (1961).
[10] K. V. Rao and A. Smakula, J. appl. Phys. **37**, 319 (1966).
[11] G. Prasad and K. V. Rao, phys. stat. sol. (a) **97**, 455 (1986).

*(Received December 16, 1987)*

phys. stat. sol. (a) 108, 463 (1988)
Subject classification: 78.60; S9.11

*Faculty of Technology, Kanazawa University*[1])

# Decrease of the TL Glow Peak Temperature by Partial Cleaning

By

Y. NAKAGAWA, S. NAKAMURA, and N. TAKEUCHI[2])

An interesting phenomenon is observed for a thermoluminescence (TL) glow peak of an additively colored and X-irradiated KCl:Cu, Cs single crystal. The peak temperature decreases by performing the partial cleaning. This is out of our common sense, because it is expected that a lower temperature TL peak should decay earlier than higher ones when they are overlapping. As the result, the composite peak should shift to the higher temperature side. A tentative calculation is performed to explain the effect assuming the transfer of electrons from one type of traps to the other during the TL.

Es wird ein interessantes Phänomen für ein Thermolumineszenz (TL)-Glowmaximum in einem additiv verfärbten und röntgenbestrahlten KCl:Cu, Cs-Einkristall gefunden. Die Glowmaximumtemperatur nimmt mit der partiellen Ausheizung der Glowkurve ab. Dies ist eine unerwartete Erscheinung, da erwartet wird, daß ein TL-Maximum mit niedrigerer Temperatur eher als eins mit höherer Temperatur bei sich überlappenden Maxima verschwinden sollte. Als Ergebnis sollte sich das zusammengesetzte Maximum zu höheren Temperaturen verschieben. Eine Berechnung wird versuchsweise durchgeführt, um den Effekt zu erklären, wobei Transfer von Elektronen von einem Typ von Haftstellen zu einem anderen während der TL angenommen wird.

## 1. Introduction

Thermoluminescence (TL) is often observed when irradiated solids are heated. This effect is used as a powerful tool to examine the nature of trapped electrons and/or holes in irradiated materials. For example, color centers in alkali halide crystals have largely been investigated by this method.

In investigating a TL glow peak which is composed of several overlapping peaks, a partial cleaning method is usually applied to remove the peak located at the lowest temperature of the composite peak region. By repeating the procedure from one lowest peak to the next, the composite peak will generally shift to higher temperatures step by step. The following experiment may be related to this shift of the TL glow peak. However, the result is quite out of our general sense.

It should be confessed that the first experimental result concerning this study was obtained fortuitously with an additively colored KCl: Cu, Cs single crystal which was X-irradiated at RT for several hours to ascertain the kinetics order of the TL. It was expected that the unbalancing between the trapped holes and the trapped electrons caused by the additive coloration should induce a change on the kinetics order of the TL [1]. The idea of co-doping of Cs with Cu was based on the low work function of the former, which might easier release electrons from colloidal metals. However, the peculiarity of the data that a shift of the main TL glow peak to lower temperatures was caused by a partial cleaning of the TL as explained below attracted our interest. This is the motivation of the present work.

[1]) Kodatsuno 2-40-20, Kanazawa 920, Japan.
[2]) To whom the correspondence should be addressed.

## 2. Experimental

### 2.1 Sample preparation

Single crystals of KCl:Cu, Cs were grown from the melt of reagent grade KCl powder doped with 0.3 mol% CuCl and 1.0 mol% CsCl by the Kyropoulos method in air. They were additively colored by heating at 650 °C in potassium vapor for 5 h and cooled to RT. Several pieces of crystals of about $5 \times 5 \times 1$ mm³ in size were cleaved from the same ingot and used for the measurements. They were irradiated by X-rays from a Cu-target operated at 40 kV, 25 mA for 30 min before the TL measurement.

### 2.2 Optical absorption spectra

Optical absorption spectra were recorded by a Hitachi 624-type spectrometer at RT. Fig. 1, curves 1 and 2, show the optical absorption spectra of an additively colored sample before and after X-irradiation, respectively. It is seen that the F-band intensity observed at 2.21 eV (560 nm) is stronger after the X-irradiation than before indicating the formation of new F-centers by the irradiation in addition to the original F-centers formed by the additive coloration. The broad optical absorption region extending to the higher energy side of the F-band may partially be related to the colloidal Cu centers [2].

### 2.3 Thermoluminescence

Thermoluminescence (TL) glow curves were measured by a home-made apparatus with a constant heating rate of 60 K/min. Fig. 2 shows the TL glow curve for a KCl:Cu, Cs crystal additively colored and subsequently X-irradiated. Several glow peaks are observed, among which an asymmetric peak at 195 °C is the strongest. The geometrical factor, $\mu_g = (T_m - T_1)/(T_2 - T_1)$, of a TL glow peak defined by Chen [3], where $T_m$, $T_1$, and $T_2$ are the temperatures of peak maximum, half intensities at lower and higher energy sides of the peak, respectively, is 0.29 for the 195 °C peak in Fig. 2. This value is extremely small compared with the normal values of 0.42 for the first-order kinetics and 0.52 for the second-order kinetics. Therefore, it was doubted that this peak might be composed of two or more overlapping peaks.

In order to isolate the overlapping TL peaks, a partial cleaning method is usually applied, in which the TL glow curve measurement is performed up to a defined tem-

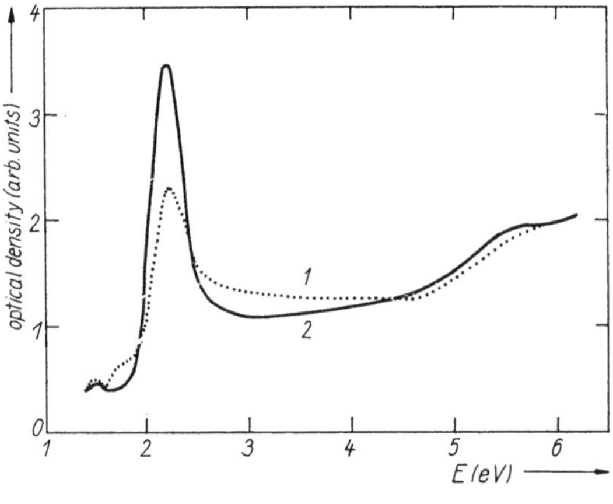


Fig. 1. Optical absorption spectra of an additively colored KCl:Cu, Cs crystal measured at RT. (1) and (2) show the spectra before and after X-irradiation at RT, respectively

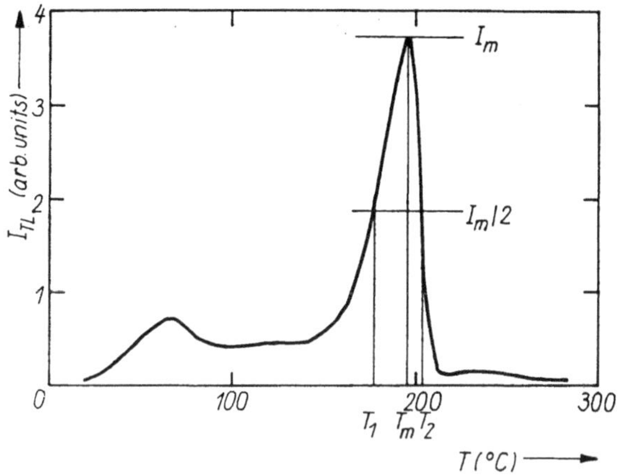


**Fig. 2.** TL glow curve for an KCl:Cu, Cs crystal additively colored and subsequently X-irradiated. Heating rate was 60 K/min. Extraordinary asymmetry of the TL peak at about 195 °C is observed

perature and returned to RT, and successively TL glow curve measurements are performed to higher and higher temperatures in this way. Then, the overlapped TL glow peaks are removed step by step from peaks located at lower temperatures. As a result, the composite peak position will shift to higher temperatures according to the process of cleaning. However, the result was different, as indicated in Fig. 3, which shows the partial cleaning of the TL glow curve for an additively colored KCl: Cu, Cs X-irradiated for 30 min. The curve 0 shows the TL before the partial cleaning. The defined temperatures are successively increased from curves 1 to 5, i.e., 165, 172, 174, 176, and 182 °C, respectively. It is observed that the peak temperature decreases with the increase of the cleaning temperature. This is quite out of our common sense. The peak should shift to the higher temperature side, if a component of the composite peak was removed from the lower temperature side.

#### *2.4 TL emission spectra*

The spectral distribution of the light emitted during a TL measurement at different temperatures are shown in Fig. 4. A JASCO CT-50 grating monochromator equipped with R-446 photomultiplier was used. Since the intensity of the glow peak varied

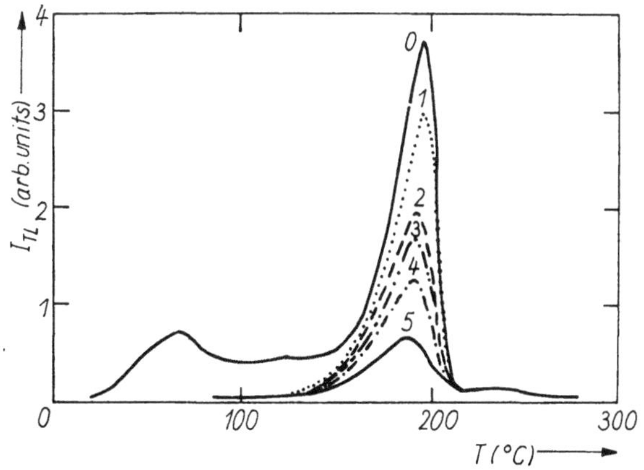


Fig. 3. TL glow curves for an additively colored and X-irradiated KCl:Cu, Cs crystal before (curve 0) and after (1) to (5) the partial cleaning. (1) to (5) are the results after heating to 165, 172, 174, 176, and 182 °C, respectively. Heating rate was 60 K/min and after reaching the indicated temperature, the sample was kept there for 30 s each. The peak shifts to lower temperatures by each step

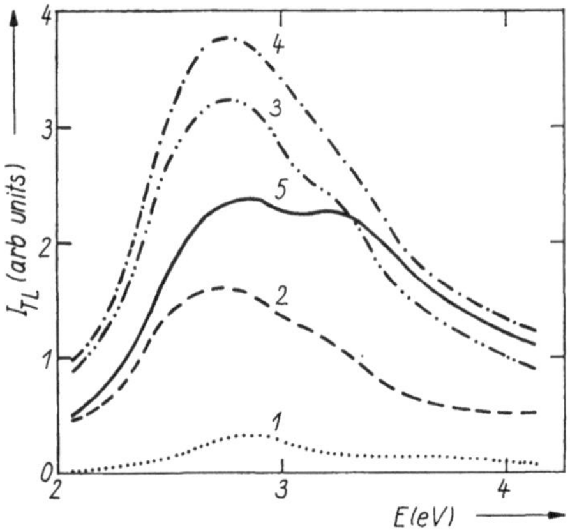


Fig. 4. Emission spectra for an additively colored and X-irradiated KCl:Cu, Cs crystal. (1) to (5) are measured at 94, 156, 165, 174, and 182 °C, respectively

during the scanning of the spectrometer, the average was taken with the two spectra obtained by scanning from 300 to 600 nm and the next one from 600 to 300 nm. Curves 1 to 5 in Fig. 4 show the results for an additively colored and subsequently X-irradiated KCl: Cu, Cs sample measured at 94, 156, 165, 174, and 182 °C, respectively. It is seen that curves 2 to 4 are nearly the same in their spectral distribution, though their relative intensities are different. However, curve 5 is different from those three. This suggests that the centers which emit TL light at 182 °C and 156 to 174 °C are of different nature. (Curve 1 is also different from the others indicating that the TL centers are different. However, emission at 94 °C is out of our present interest.) Therefore, the noted asymmetry of the high temperature TL peak is again considered due to the overlapping of the two different peaks which emit different spectra of light.

### 2.5 *Effect of partial optical bleaching on the glow curve*

It has been reported that KCl single crystals doped with divalent impurities shift their TL glow peak to lower temperatures when they are illuminated in the F-band before the TL measurements [4]. It was discussed that the effect was caused by the formation of new color centers which decay at lower temperatures than the normal F-centers.

Fig. 5 shows the effect of such F-band illumination for an additively colored KCl: : Cu, Cs sample X-irradiated for 30 min. Curves 1 and 2 are the TL glow curves direct-

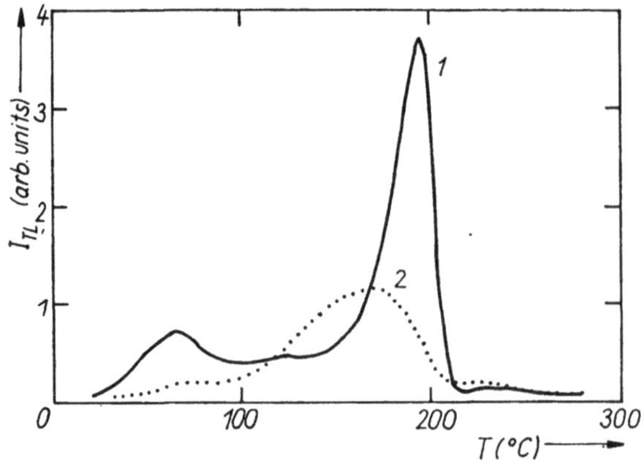


Fig. 5. TL glow curves (1) before and (2) after F-light illumination for an additively colored and X-irradiated KCl: Cu, Cs crystal

ly after the X-irradiation and after the subsequent F-light irradiation for 60 s from an 500 W xenon lamp through a JASCO CT-50 monochromator, respectively. It is observed that the 195 °C peak decayed largely and a new peak at about 170 °C with a shoulder at about 130 °C appeared suggesting that the F-centers had changed into other color centers which decay at lower temperatures.

## 3. Discussion

As has been shown in Fig. 3, the fact that the 195 °C TL peak shifts to lower temperatures by the partial cleaning is an interesting phenomenon, because if this TL peak is composed of two or more overlapping peaks, the lower one should thermally be more unstable and the composite peak should shift to higher temperatures by the partial cleaning. In order to analyze the effect, we make the simplifying assumption that there are two trapping states for electrons, which are called traps 1 and 2, respectively, as shown in Fig. 6. Trap 2 may indicate an F-center and F-light illumination will transfer the trapped electron from trap 2 to trap 1, though the structure of the latter is unknown. The reason why we assume this transformation of centers is the experimental result shown in Fig. 5, where the TL peak at 195 °C shifts to one at 170 °C by the F-light illumination. We assume that the 195 °C peak is related to trap 2 and the 170 °C peak to trap 1. As shown in Fig. 4, light emission during the TL may cover the tail of the F-band, which may effectively be bleached by absorbing the light emitted by the TL. Therefore, the result should be very similar to Fig. 5, which is optically bleached from the outside of the sample to induce the shift of the TL glow peak to lower temperatures. The effect presently considered may proceed inside the sample and the TL light is partially consumed in the sample, i.e., it may be a special type of energy transfer between the centers.

The set of differential equations governing the reactions related to Fig. 6 are as follows:

$$\frac{dn_c}{dt} = n_1 p_1 + n_2 p_2 - A_r n_h n_c = 0 \,, \tag{1}$$

$$\frac{dn_1}{dt} = -n_1 p_1 + A_m n_2 \,, \tag{2}$$

$$\frac{dn_2}{dt} = -n_2 p_2 - A_m n_2 \,, \tag{3}$$

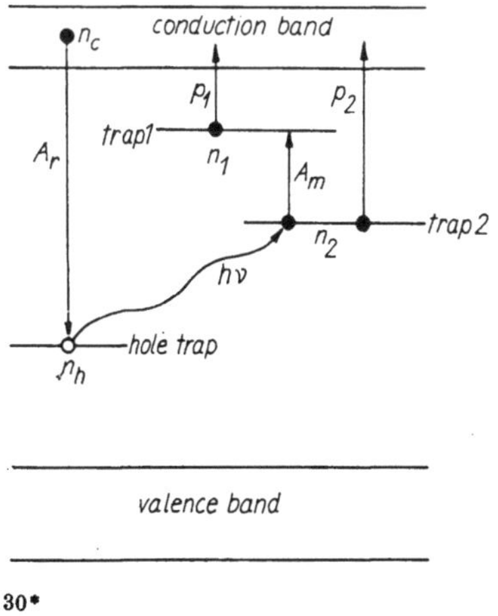


$$I = -\frac{dn_h}{dt} = A_r n_h n_c = -\left(\frac{dn_1}{dt} + \frac{dn_2}{dt}\right), \tag{4}$$

$$p_i = s_i \exp\left(\frac{-E_i}{kT}\right), \tag{5}$$

$$A_m = aI \,, \tag{6}$$

Fig. 6. Schematic mechanism of the transfer of trapped electrons from traps 1 to 2 by absorbing TL light. The recombination energy is partly dissipated by phonons

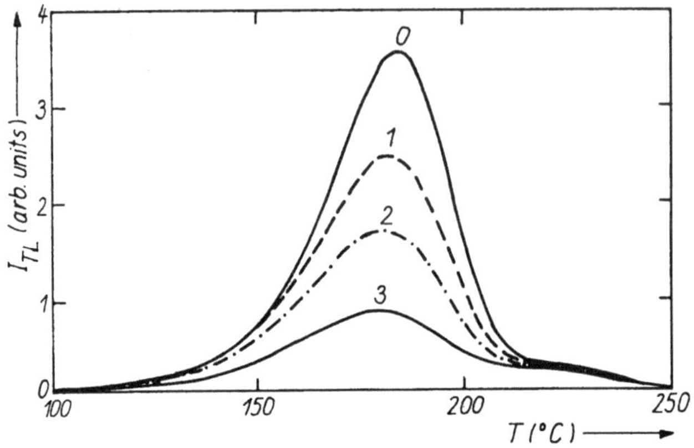


Fig. 7. Calculated results for the TL glow curves by partial cleaning. (0) is before the cleaning, and (1) to (3) are after heating to 150, 160, and 170 °C, respectively. The same conditions as in the experiment were assumed for the calculation. The equations and parameters are written in the text

where the symbols $n_1$, $n_2$, and $n_c$ are the concentrations of electrons at trap 1, trap 2, and in the conduction band, respectively. The symbol $n_h$ refers to the concentration of trapped holes which will recombine with thermally released electrons and give rise to the TL light, $h\nu$, $p$ is the probability, $s$ the frequency factor, $A_m$, and $A_r$ the transition coefficients referring to Fig. 6, respectively; $I$ is the TL intensity and $a$ the intermediate transition coefficient, suffixes, $i = 1$ or 2, refer to the type of traps, respectively.

The numerical calculation was performed by using a microcomputer NEC PC-9801 model-XL following the Runge-Kutta-Gill method. The results are shown in Fig. 7, where the initial values of concentrations are chosen tentatively as $n_{1,0} = 0.5 \times 10^{16}$ and $n_{2,0} = 1.0 \times 10^{16}$. The frequency factors, $s_1$ and $s_2$, are $1 \times 10^{10}\,s^{-1}$ for the two processes and the activation energies $E_1 = 1.0$ eV and $E_2 = 1.1$ eV. The intermediate transition coefficient $a = 1.6$ was assumed. As the heating rate 60 K/min was used and after heating to the defined temperature, the sample was kept at each temperature for 30 s before cooling to RT. Other parameters in (1) to (6) are determined self-consistently. Then, curve 0 in Fig. 7 shows the TL glow curve before the partial cleaning. Curves 1 to 3 show it after successive partial cleaning at 150, 160, and 170 °C, respectively.

It is seen from Fig. 7 that the tentative calculation assuming two types of electron traps, between which electrons are transferred by absorbing the TL light, gives the observed shift of the main TL peak to lower temperatures by the successive partial cleanings of the TL, though the agreement with the experimental results is not yet satisfactory enough. It should also be noticed that the geometry factor of 0.42 observed after the partial cleaning at high temperatures of 176 or 182 °C as seen in Fig. 3 is in good agreement with the calculated result in Fig. 7.

## References

[1] N. TAKEUCHI, phys. stat. sol. (a) 78, K93 (1983).
[2] N. TAKEUCHI, K. ASAMI, and M. ISHIGURO, Japan. J. appl. Phys. 9, 1095 (1970).
[3] R. CHEN, J. appl. Phys. 40, 570 (1969).
[4] K. INABE, N. TAKEUCHI, and M. ISHIGURO, phys. stat. sol. (a) 24, K7 (1974).

*(Received January 25, 1988)*

phys. stat. sol. (a) **108**, 469 (1988)

Subject classification: 73.40 and 85

*VEB Werk für Fernsehelektronik, Berlin*[1]) *(a),*
*Sektion Physik der Humboldt-Universität Berlin*[2]) *(b),*
*Zentralinstitut für Elektronenphysik*
*der Akademie der Wissenschaften der DDR, Berlin*[3]) *(c),*
*and Sektion Informationswissenschaft der Technischen Universität Dresden*[4]) *(d)*

# Temperature Distribution in Square-Shaped Semiconductor Chips

By

U. HENNIGER (a), B. SCHMIDT (b), R. TEMPEL (c), and H. MOELLER (d)

Dedicated to Prof. Dr. E. GUTSCHE on the occasion of his 60th birthday

Advantages and disadvantages of both, the eigenvalue and the Green's function methods for the evaluation of temperature distributions in square-shaped homogeneous and hetero-layer semiconductor chips are analyzed. Combining the eigenvalue and Green's function techniques in order to overcome the particular disadvantages of both the methods a new procedure is proposed which allows a fast three-dimensional temperature modelling of integrated circuits. The new mixed method is applied to the calculation of temperature profiles of a power GaAs MESFET.

Es werden Vor- und Nachteile sowohl der Eigenwert- als auch der Greenfunktionsmethoden für die Berechnung der Temperaturverteilung in quaderförmigen homogenen bzw. Mehrschichthalbleiterchips analysiert. Ein neues Verfahren wird vorgeschlagen, welches die Eigenwert- und Greenfunktionstechniken miteinander kombiniert, um die spezifischen Nachteile beider Methoden zu überwinden, und welches schnelle Modellierungsrechnungen dreidimensionaler Temperaturverteilungen in integrierten Schaltungen gestattet. Diese neue, gemischte Methode wird zur Berechnung von Temperaturprofilen in einem Leistungs-MESFET auf GaAs-Basis angewandt.

## 1. Introduction

The problems of heat transport and temperature distribution in semiconductor chips have been investigated for decades now. Initially the interest in these problems resulted from the characteristics of single high-performance devices constructed from homogeneous semiconductor chips. Following the tendency of monolithic integration of high-performance functional and control units [1], theoretical calculations of the temperature distribution under stationary and dynamical conditions turns out to be more and more important for the design of the corresponding semiconductor circuitry. At least thermal overheating and swell effects in highly integrated circuits are serious problems. Therefore, the optimization of devices must include the distribution of heat sources, spatial parameters of the chip, measures for heat draining, and related phenomena.

For modelling the temperature distribution in semiconductor chips with square or cylindrical symmetry various mathematical methods have so far been applied. Espe-

[1]) Ostendstr. 1/5, DDR-1060 Berlin, GDR.
[2]) Invalidenstr. 110, DDR-1040 Berlin, GDR.
[3]) Rudower Chaussee 5, DDR-1199 Berlin, GDR.
[4]) Mommsenstr. 13, DDR-8027 Dresden, GDR.

cially concerning cases where two-dimensional analysis is sufficient due to symmetry reasons, a remarkably large variety of procedures has been developed. The most multivalently applicable procedure is the solution of the heat conduction equation using the method of finite elements (for principles, see [2]). This procedure allows any shape and distribution of heat sources, temperature- and space-dependent heat conductivities, as well as arbitrary choice of boundaries and boundary conditions. Problems arise, however, regarding the demand for calculating memory space and time which are considerably high even for two-dimensional calculations if heat sources are small compared with the chip size which is valid for the majority of integrated circuits concerned. Finite-difference procedure causes similar problems.

In the case of simple chip geometries as, for example, square or cylindrical plates, considerable significance was gained by analytical procedures such as the eigenvalue method, function-theoretical approaches based on conformal representation, as well as the Green's function method. Having defined sufficiently simple boundary conditions, the advantage of the eigenvalue method consists in the ability to calculate the exact solution of the heat conduction problem as a series, the chip being homogeneous or a sandwich of planar layers with different heat conductivities (Section 2.1). In the case of very small heat sources the eigenvalue method becomes problematic for the same reasons as the purely numeric procedures: too many terms of the solution series must be calculated which results in serious imprecisions. The Green's function method (Section 2.2) avoids such problems. Heat-emitting chip surfaces, however, may not be included in this approach, a fact which is caused by the limited analytic complication that can be practically handled. Green's functions for multi-layer systems are highly complicated analytic expressions [3] the application of which necessitates expensive numerical integration over all the heat sources.

In Section 2.3 a procedure is proposed which combines the eigenvalue and Green's function methods striving to eliminate in this way the most striking disadvantages of both methods.

Function-theroretical techniques are definitely limited to two-dimensionally describable problems. They are only applied to the investigation of very simple structures. In the present work which is mainly concerned with the three-dimensional aspect of heat conduction they are not regarded.

## 2. Theoretical Foundation

### *2.1 Mathematical model*

Calculation of the temperature distribution $T(\boldsymbol{r}, t)$ in a semiconductor chip starts from heat conduction equation

$$c(\boldsymbol{r})\,\varrho(\boldsymbol{r})\,\frac{\mathrm{d}T}{\mathrm{d}t} = \nabla_{\boldsymbol{r}}(k(\boldsymbol{r})\,\nabla_{\boldsymbol{r}}T) + P(\boldsymbol{r}, t) \tag{1}$$

which satisfies the boundary conditions

$$\left.\frac{\partial T}{\partial \nu} + \beta(\boldsymbol{r})\,(T - T_\mathrm{a})\right|_{\boldsymbol{r}\,\in\,\mathrm{surface}} \equiv 0\,, \tag{2}$$

where $c(\boldsymbol{r})$ is the local heat capacity, $\varrho(\boldsymbol{r})$ the mass density, and $k(\boldsymbol{r})$ the heat conductivity. The function $k(\boldsymbol{r})$ is space-dependent not only explicitly but also implicitly due to its dependence on the local temperature. The slight temperature dependence of $c(\boldsymbol{r})$ and $\varrho(\boldsymbol{r})$ may be omitted. Regarding semiconductor devices relevant for microelectronics, the temperature dependence of heat conductivity in the range of 200 K $<$ $< T <$ 500 K may be described by [4]

$$k(\boldsymbol{r}) = K(\boldsymbol{r}, T_a) \left(\frac{T(\boldsymbol{r})}{T_a}\right)^w , \tag{3}$$

where $T_a$ is the ambient temperature which can be taken for constant, and $K(\boldsymbol{r}, T_a)$ contains only the explicit space dependence if this is concerned at all. In the present work anisotropy effects are neglected. The exponent $w$ amounts to $-4/3$ for Si according to Ioffe's law [5] and $-5/4$ for GaAs [6].

The function $P(\boldsymbol{r}, t)$ in (1) is the heating performance density, $\beta(\boldsymbol{r})$ the Newton emission coefficient. The latter is taken for constant over each of the square surfaces. In the case of hetero-layer structures and structured surfaces $\beta$ represents suitable mean values.

Heat sources are assumed to be square rods or plates parallel to the chip surfaces, having a homogeneous distribution of heating performance. If dynamic behaviour is investigated, these heat sources are switched on at a time $t_{on}$ and switched off at a time $t_{off}$. It is suitable to further assume that any heat source is located within one of the semiconductor layers. More complicated heat sources may be constructed from several of the simple heat sources described.

The boundary value problem (1), (2) is non-linear due to (3). In most of the practically relevant cases, however, temperature differences within a chip are not greater than maximum 200 K. Thus it is justified to substitute (3) by

$$k(\boldsymbol{r}) \approx K^*(\boldsymbol{r}, T_a, T_M) \left(\frac{T(\boldsymbol{r})}{T_a}\right)^{-1} , \tag{3'}$$

where the modified heat conductivity $K^*$ slightly depends on the maximum temperature $T_M$. For $T_M$ it is sufficient to evaluate the maximum expected temperature. Using

$$T(\boldsymbol{r}) = T_a \exp\left(\frac{\vartheta(\boldsymbol{r})}{T_a}\right) \tag{4}$$

the space operator in (1) is linearized.

Literature references also use the Kirchhoff transformation [4],

$$\vartheta(\boldsymbol{r}) = T_a + \int_{T_a}^{T} \mathrm{d}T \left(\frac{T}{T_a}\right)^w . \tag{4'}$$

The difference of this transformation to (4) may be neglected. If cooling is concerned along with heating, (4) turns out to be numerically more reliable.

For dynamical calculations the coefficient $c(\boldsymbol{r})\, \varrho(\boldsymbol{r})$ in (1) is replaced by a suitable mean value in order to simplify the mathematical problem. Thus the linearized heat conduction equation is expressed by

$$\frac{1}{\omega} \frac{\partial \vartheta}{\partial t} = \nabla_{\boldsymbol{r}}\left(K^*(\boldsymbol{r}, T_a, T_M)\, \nabla_{\boldsymbol{r}} \vartheta\right) + P(\boldsymbol{r}, t) , \tag{5}$$

where

$$\frac{1}{\omega} = \left\langle c(\boldsymbol{r})\, \varrho(\boldsymbol{r}) \exp\left(\frac{\vartheta(\boldsymbol{r})}{T_a}\right)\right\rangle_{\text{whole chip}} . \tag{6}$$

The boundary condition is now

$$\frac{\partial \vartheta}{\partial \nu} + \beta(\boldsymbol{r})\, T_a \left[1 - \exp\left(-\frac{\vartheta}{T_a}\right)\right] \approx \frac{\partial \vartheta}{\partial \nu} + \beta(\boldsymbol{r})\, \vartheta \Big|_{\boldsymbol{r} \in \text{surface}} \equiv 0\,, \tag{7}$$

because $|\vartheta/T_a| < 1$ is valid at an ambient temperature of about 300 K in the majority of relevant cases.

Starting from the linearized boundary value problem (5) and (7) the eigenvalue approach will be applied in Section 2.2, while Section 2.3. presents the Green's function approach.

### 2.2 Eigenvalue method

The general chip structure is presented in Fig. 1. The $z$-axis is counted from top to bottom perpendicularly to any of the layers which, in turn, are homogeneous each and contact each other abruptly.

Thus the $\boldsymbol{r}$-dependence of $K^*(\boldsymbol{r}, T_a, T_M)$ is reduced to $z$-dependence. Determination of eigenvalues and eigenfunctions of the linearized heat-conduction operator is carried out as usual by means of the Fourier method of separation of variables [7]. The total space eigenfunction is

$$\vartheta_{\mu\nu\varkappa^*}(\boldsymbol{r}) = \varphi_\mu(x)\, \psi_\nu(y)\, \chi_{\varkappa^*}(z) \tag{8}$$

with the normalized partial eigenfunctions

$$\varphi_\mu(x) = \frac{\mu \cos(\mu x) + \beta_s \sin(\mu x)}{\sqrt{\frac{a}{2}(\mu^2 + \beta_s^2) + \beta_s}}\,, \tag{9a}$$

$$\psi_\nu(y) = \frac{\nu \cos(\nu y) + \beta_s \sin(\nu y)}{\sqrt{\frac{b}{2}(\nu^2 + \beta_s^2) + \beta_s}}\,, \tag{9b}$$

$$\chi_{i,\varkappa^*}(z) = \frac{\Gamma_{i-1,\varkappa^*}}{A_{\varkappa^*}} \cos\left(\varkappa_i(c_{i-1} - z)\right) - \frac{\Gamma'_{i-1,\varkappa^*}}{\varkappa_i A_{\varkappa^*}} \sin\left(\varkappa_i(c_{i-1} - z)\right). \tag{9c}$$

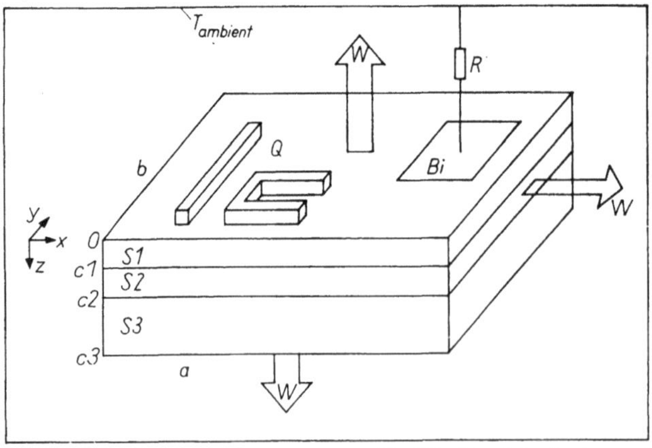


Fig. 1. General hetero-layer system S1, S2, S3 with the layer boundaries c1, c2, c3 = c. The edges of sources $Q$ are parallel to the chip edges $a$, $b$, $c$. A bond area Bi is connected with the environment $U$ via the thermal resistance $R$. Heat flows $W$ pass through the chip surface

Index $i$ runs through all the layers from 1 to $N$. The parameter $\varkappa^*$ is related to the $z$ eigenvalue of the maximum-conductivity layer having $\hat{K}(T_a, T_M)$. Moreover, it holds

$$\varkappa_i^2 = \frac{\hat{K}}{K_i}\varkappa^{*2} + \left(\frac{\hat{K}}{K_i} - 1\right)(\mu^2 + \nu^2)\,, \tag{10}$$

$$A_{\varkappa^*}^2 = \sum_{i=1}^{N}\left\{\frac{\Gamma_{i-1,\varkappa^*}^2}{2}\left[d_i + \frac{\sin(2\varkappa_i d_i)}{2\varkappa_i}\right] + \frac{\Gamma_{i-1,\varkappa^*}'^2}{2\varkappa_i^2}\left[d_i - \frac{\sin(2\varkappa_i d_i)}{2\varkappa_i}\right] + \right.$$
$$\left. + \Gamma_{i-1,\varkappa^*}\Gamma'_{i-1,\varkappa^*}\frac{\sin^2(\varkappa_i d_i)}{\varkappa_i^2}\right\}, \tag{11}$$

as well as the recursion formulae

$$\begin{pmatrix}\Gamma_{i,\varkappa^*}\\ \Gamma'_{i,\varkappa^*}\end{pmatrix} = \begin{pmatrix}\cos\varkappa_i d_i & -\dfrac{\sin\varkappa_i d_i}{\varkappa_i}\\ \varkappa_i\sin\varkappa_i d_i & \cos\varkappa_i d_i\end{pmatrix}\begin{pmatrix}\Gamma_{i-1,\varkappa^*}\\ \Gamma'_{i-1,\varkappa^*}\end{pmatrix} \tag{12}$$

with the starting values $\Gamma_{0,\varkappa^*} = 1$ and $\Gamma'_{0,\varkappa^*} = -\beta_u$.

The eigenvalues satisfy the following equations:

$$\tan(\mu a) = \frac{2\mu\beta_s}{\mu^2 - \beta_s^2}, \qquad (\mu > 0)\,, \tag{13a}$$

$$\tan(\nu b) = \frac{2\nu\beta_s}{\nu^2 - \beta_s^2}, \qquad (\nu > 0)\,, \tag{13b}$$

$$\Gamma_{N-1,\varkappa^*}[\beta_l\cos(\varkappa_N d_N) - \varkappa_N\sin(\varkappa_N d_N)] =$$
$$= \Gamma'_{N-1,\varkappa^*}\left[\cos(\varkappa_N d_N) + \beta_l\frac{\sin(\varkappa_N d_N)}{\varkappa_N}\right]. \tag{13c}$$

Parameters $\beta_s$, $\beta_u$, and $\beta_l$ denote the (mean) emission coefficients of the side, upper, and lower surfaces of the chip and the substrate, respectively. The rigorous solution of the boundary value problem (5), (7) is

$$\vartheta(\mathbf{r}, t) = \sum_{h,\mu,\nu,\varkappa^*}\tau_{h,\lambda}(t)\, M_{h,\mu,\nu,\varkappa^*}\varphi_\mu(x)\,\psi_\nu(y)\,\chi_{\varkappa^*}(z)\,, \tag{14}$$

where $h$ counts the heat sources and $\lambda$ denotes a single total eigenvalue,

$$\lambda^2 = \hat{K}(\mu^2 + \nu^2 + \varkappa^{*2})\,. \tag{15}$$

Furthermore,

$$\tau_{h,\lambda}(t) = \begin{cases}0\,; & t \leqq t_{h,\mathrm{on}}\,,\\ 1 - \exp[-\lambda\omega(t - t_{h,\mathrm{on}})]\,; & t_{h,\mathrm{on}} < t \leqq t_{h,\mathrm{off}}\,,\\ \exp[-\lambda\omega(t - t_{h,\mathrm{off}})] - \\ \quad - \exp[-\lambda\omega(t - t_{h,\mathrm{on}})]\,; & t_{h,\mathrm{off}} < t\,,\end{cases} \tag{16}$$

$$M_{h,\mu,\nu,\varkappa^*} = \frac{P_h}{\lambda}\int_{V_h} \mathrm{d}^3\mathbf{r}'\,\varphi_\mu(x')\,\psi_\nu(y')\,\chi_{\varkappa^*}(z')\,, \tag{17}$$

using $P_h$ the heating performance density and $V_h$ the volume of the $h$-th heat source.

The particular profit of the eigenvalue method is the ability, first, to process mixed boundary conditions (7), and, second, to handle multi-layer systems without serious analytical complications. Moreover, the expansion coefficients $M_{h,\mu,\nu,\varkappa^*}$ may be integrated simply analytically in case of square-shaped heat sources parallel to the chip surfaces (see (17)). The main disadvantage of the eigenvalue approach consists in the poor resolving capacity which is caused by the extreme increase in the number of eigenfunctions to be included if heat sources are small compared with the chip size.

### 2.3 Green's function approach

If the Green's function $G(\boldsymbol{r}, \boldsymbol{r}', t - t')$ of the boundary value problem (5), (7) is granted, the solution of the latter may be calculated by

$$\vartheta(\boldsymbol{r}, t) = \int_{\text{volume}} \mathrm{d}^3\boldsymbol{r}' \int_0^t \mathrm{d}t'\, G(\boldsymbol{r}, \boldsymbol{r}', t - t')\, P(\boldsymbol{r}', t') \,. \tag{18}$$

The construction of the Green's function for a homogeneous, square-shaped region is carried out by means of the reflection method, the boundary condition (7) being reduced to the two extremal cases of ideally reflecting surface $\vartheta(\boldsymbol{r}, t)|_{\boldsymbol{r} \in \text{surface}} \equiv 0$ and ideally open surface $\partial\vartheta(\boldsymbol{r}, t)/\partial v|_{\boldsymbol{r} \in \text{surface}} \equiv 0$. Generally some of the square surfaces may be reflective while the others are open.

For a square with the lengths $a$, $b$, $c$ and reflecting surfaces (Dirichlet problem) the stationary Green's function for the Laplace operator including multiple reflection amounts to

$$G(\boldsymbol{r}, \boldsymbol{r}') = \frac{1}{4\pi} \sum_{l,m,n=-\infty}^{\infty} \sum_{u,v,w=-1}^{1} \frac{uvw}{\sqrt{Z}}, \tag{19}$$

using

$$Z = (x - 2la - ux')^2 + (y - 2mb - vy')^2 + (z - 2nc - wz')^2 \,.$$

Reflection procedures for open surfaces and reflecting surfaces, respectively, are identical except that heat performances of the reflected heat sources are to be taken as the positive original heat performance in the open case and the negative one in the reflecting case.

The Green's function (19) is the core of the solution for the stationary heat conduction problem for a homogeneous square space. The Green's function for the corresponding dynamical problem is

$$G(\boldsymbol{r}_0, \boldsymbol{r}', t - t') = \sum_{l,m,n=-\infty}^{\infty} \sum_{u,v,w=-1}^{1} \frac{\theta(t - t')\,\omega}{[2\sqrt{\pi\omega k(t - t')}]^3} \times$$
$$\times \exp\left\{\frac{-[(x_0 - 2la - ux')^2 + (y_0 - 2mb - vy')^2 + (z_0 - 2nc - wz')^2]}{4\omega k(t - t')}\right\}. \tag{20}$$

In spite of the advantage that heat sources may have any spatial form (especially important for extremely small sources) the Green's function method includes the serious implication that boundary conditions are restricted to either Dirichlet or Neumann conditions which may be an approximation too rough for applications where the chip is embedded in an inert and partly cooling medium. Furthermore, there may be cases where analytical calculations are to be done for device structures included within a semiconducting substrate where a heat current should be assumed leading through the surfaces of the substructure embedded into the other substrate regions. Another disadvantage of the Green's function technique is the striking increase in numerical expense if multi-layer structures are investigated. Additionally, in some cases it is necessary to include more than five shells of reflected heat sources in the calculation which may result in high numerical expense.

### *2.4 Mixed procedure*

The aim of the mixed procedure is the combination of Green's function and eigenvalue techniques in order to overcome the disadvantages of both methods.

For beginning the chip is assumed to consist of a homogeneous medium. If some of the heat sources contained are too small (i.e. if the period of the highest-rank eigenfunction included does not fit in any space direction at least once into the corresponding length of the source), the eigenvalue method will yield significant imprecision. This is avoided by covering the small source $P_h$ by a sufficiently large source $P'_h$ using a special construction procedure which tends to optimally adjust both sources centredly. Naturally, the large source must be fully contained within the original chip. For the enlarged source the eigenvalue method is applied leading to sufficiently accurate results. The error caused by adding a source $P'_h$ is compensated by applying the Green's function method to the problem which contains the original source $P_h$ minus the additionally constructed enlarged source $P'_h$ (the latter is added with negative heating performance). Symbolically written

$$\vartheta(\boldsymbol{r}, t) = \{EV\} * P' + \{GF\} * (P - P') , \tag{21}$$

where $EV$ is the eigenvalue operator and $GF$ the Green's function operator. The result is the temperature distribution which fits to the exact heat conduction problem within and near the small source, and to the exact boundary conditions in the farther surroundings of the source. Even if the sources are situated fairly near or at the surface (which is true for most of the semiconductor devices) the influence of the actual boundary conditions (Dirichlet, Neumann, or mixed) on the Green's functional result is more or less suppressed because of the compensation of heating performances. The only alternative to be decided in this case is whether the surface must be considered reflecting or open.

For investigating hetero-layer systems by means of this method without constructing sophisticated exact Green's functions the chip will be "homogenized", i.e. the layers are distorted such as to result in a homogeneous heat conductivity. Necessarily the side surfaces are structures in this approximation. In most practical applications, however, the distance of the heat sources from the side surfaces is high enough to justify the neglect of such structuring effects in the following calculation. Homogenization for a suitable mean heat conductivity $K$ is carried out in a way that the total thickness $c$ (length in $z$-direction perpendicular to the layers) remains constant,

$$c = \sum_{i=1}^{N} d_i = \sum_{i=1}^{N} d_i^* , \tag{22}$$

where $d_i^*$ are the $z$ lengths of the transformed layers. For the distortion factors $(\lambda_i = d_i/d_i^*)$,

$$\lambda_i = \frac{1}{c} \sum_{j=1}^{N} d_j \sqrt{\frac{k_i}{k_j}} \tag{23}$$

is valid. The mean heat conductivity is calculated according to

$$\overline{K} = \left(\frac{1}{c} \sum_{i=1}^{N} \frac{d_i}{\sqrt{k_i}}\right)^{-2} . \tag{24}$$

The new point coordinates of the sources or their calculation origins in the Green's function are transformed by

$$x^* = \frac{a}{2} + \frac{1}{\lambda_i}\left(x - \frac{a}{2}\right),$$
$$y^* = \frac{b}{2} + \frac{1}{\lambda_i}\left(y - \frac{b}{2}\right), \tag{25}$$
$$z^* = c_{i-1}^* + \frac{1}{\lambda_i}(z - c_{i-1}),$$

using $c_i^* = \sum_{j=1}^{i} d_j^*$; $c_0 = c_0^* = 0$ and $P \in i$-th layer.

In cases of practical relevance there are generally thin hetero-layers applied on a relatively thick substrate layer, and heat sources are sufficiently far from the side surface (except semiconductor laser diodes). Therefore, the influence of the side-surface structuring from distortion procedure on the results of the Green's function technique using compensated heat sources is negligible. The most important effect of local heating of a small heat source is fully represented by the Green's function results, while the specific properties of the real hetero-layer construction are treated by the eigenvalue method.

### 3. Example

Fig. 2 displays the distribution of heat sources and bonding areas of a power MESFET on GaAs semiconductor chip. The dimensions of the chip are $a = 500$ µm, $b = 400$ µm, and $c = 100$ µm. Heat sources are the active transistor channels, especially those between gate and drain. They have the dimensions 0.5 µm × 100 µm × 0.2 µm and are located symmetrically to the chip centre in a parallel way having distances of 20 µm. Situated 0.1 µm under the upper chip surface they have a performance of 0.26 W each. Size and location of the bonding areas are shown in Fig. 2. Their thermal resistances are approximated to 4000 K/W, that of the cooling lower chip surface to 25 K/W. It is physically justified to assume Neumann-type boundary conditions for all the chip surfaces.

The areas for bonding and cooling are formally described as negative heat sources which simulate the physical effect of heat draining. The corresponding heat performances are determined self-consistently from all the other thermal performances and thermal transition resistances. Using a thermal conductivity of $K = 45$ W/Km at 300 K for GaAs the temperature profiles shown in Fig. 3 and 4 were calculated.

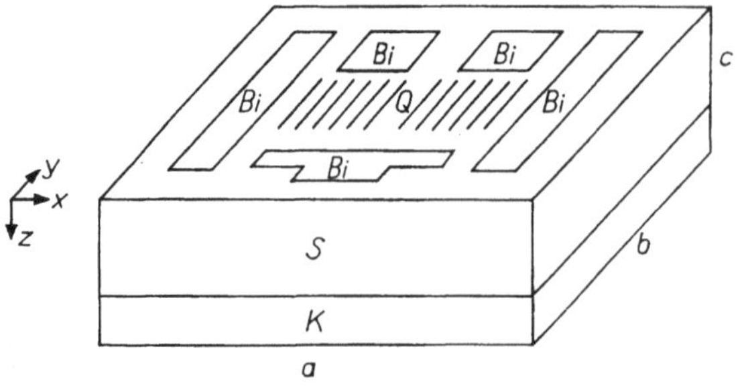


Fig. 2. Diagram of a power MESFET with ten heat sources $Q$ and five bond areas Bi. The chip S has the dimensions $a$, $b$, $c$ and is mounted on a sink K

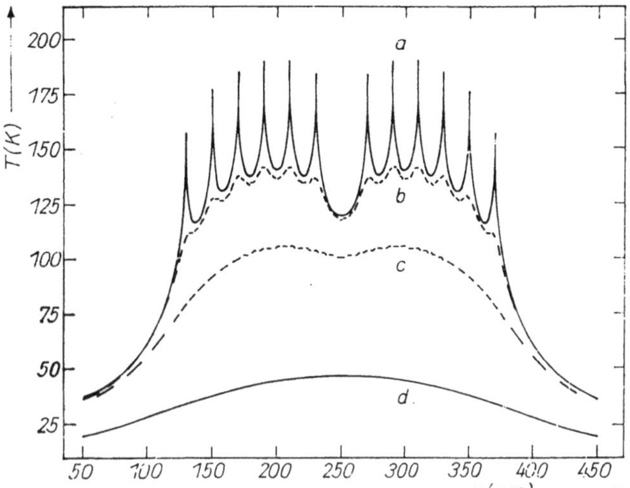


Fig. 3. Temperature profiles parallel to the $x$-axis at $y = b/2$ and at different $z$ depths: (a) $z = 0$; (b) 5; (c) 25; (d) 100 μm

Fig. 3 displays the temperature differences from the ambient temperature (300 K) for different $z$ depths at $y = b/2$ parallel to the $x$-axis. There are distinct peaks of extreme temperature exposure within the heat sources caused by the relatively small diameter of the heating rods. If limits from technology and material parameters for thermal exposure are given, it is possible to theoretically derive approximate limits for the maximum power loss.

Moreover, our example shows distinct differences of more than 10% in the peak temperatures. Changes in the design of the chip, as e.g. different performance distributions or distances between the heat sources, may result in a homogenization of thermal exposure.

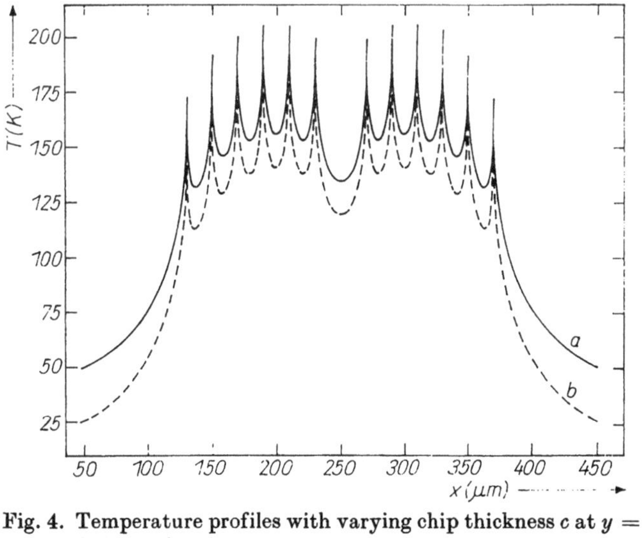


Fig. 4. Temperature profiles with varying chip thickness $c$ at $y = b/2$ along the surface ($z = 0$ μm). (a) $c = 100$; (b) 25 μm

Fig. 4 shows temperature profiles at the chip surface parallel to the $x$-axis at $y = b/2$ calculated for identically designed chips with two different $z$ diameters ($c = 100\,\mu m$, $c = 25\,\mu m$). Most remarkable is the small influence of $c$ (differing by 75%) on the peak temperatures within the heat sources, which results from the very high ratio $c/L_{z,\text{source}}$. One of the conclusions is that despite of significant reduction of the chip thickness the thermal exposure peaks are reduced only slightly while serious mechanical and technological problems may arise. Therefore, this measure can only be part of a practically economic compromise.

## 4. Conclusion

The mixed eigenvalue–Green's function method for the evaluation of the stationary and dynamical temperature distribution in square-shaped homogeneous and heterolayer semiconductor chips overcomes the particular disadvantages of both the pure eigenvalue and the pure Green's function techniques and is favourably applied to fast modelling and design of semiconductor devices. Because of the high calculation speed and the sufficient precision compared with experimental measurements the method proposed is suited, additionally, to analyze given temperature profiles measured for re-determination of real heat source parameters (dimensions, heat performance).

## References

[1] P. R. GRAY, IEEE J. Solid State Circuits, 7, 474 (1972).
[2] P. A. MARKOWICH, The Stationary Semiconductor Equations, Springer-Verlag, Wien/New York 1985.
[3] F. BECHSTEDT and R. ENDERLEIN, phys. stat. sol. (b) **131**, 53 (1985).
[4] M. S. CARSLAW and J. C. JAEGER, Conduction of Heat in Solids, Clarendon Press, Oxford 1959 (p. 11).
[5] A. F. IOFFE, Fizika poluprovodnikov, Izd. Nauka, Moskva/Leningrad 1957.
[6] W. NAKWASKI, IEEE Trans. Electron Devices, **33**, 889 (1986).
[7] W. S. WLADIMIROW, Gleichungen der mathematischen Physik, Deutscher Verlag der Wissenschaften, Berlin 1972 (p. 240).

*(Received April 4, 1988)*

See Short Note by J. PIQUERAS et al.

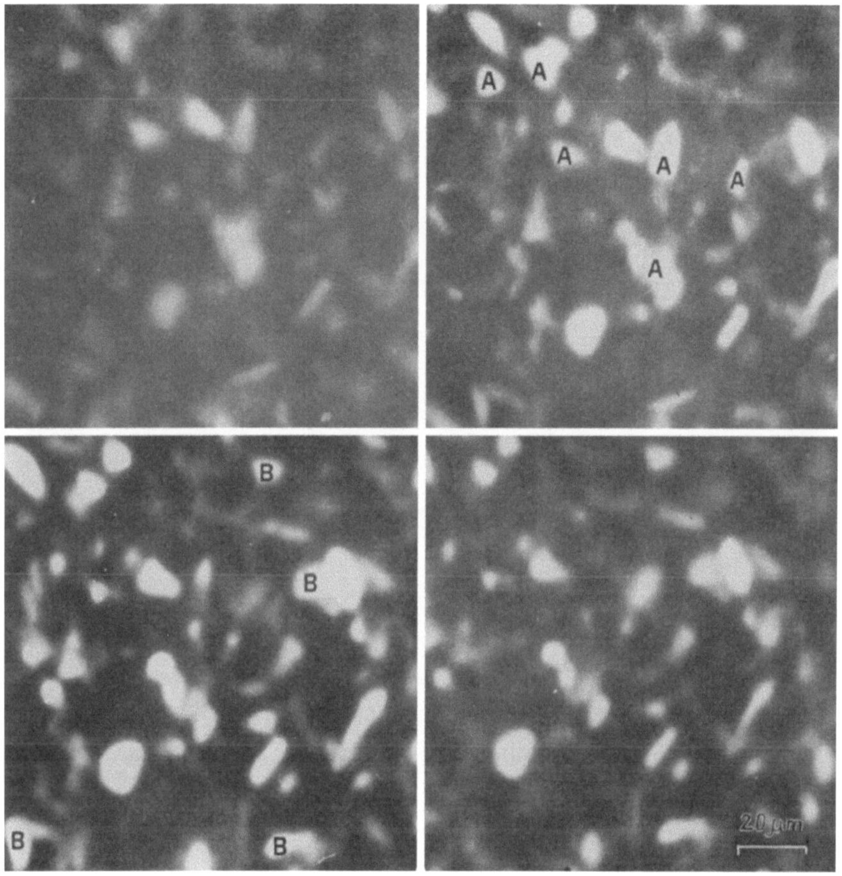


Fig. 1 CL images of the same sample region at a) room temperature, b) 215 K, c) 140 K, d) 100 K

# *Short Notes*

phys. stat. sol. (a) 108, K1 (1988)
Subject classification: 61.70 and 62.20; S4

Belorussian State University, Minsk[1)]

## The Loss of Strength of Al Alloy after $Sb^+$ Implantation

By
V. P. GOLTSEV and S. I. ZHUKOVA

Introduction The advantages of ion-bombardment make this technique rather promising for the modification of surface-controlled physical properties of metals and alloys. The possibility of a purposeful modification of such material characteristics as durability, fatigue strength, hardness is demonstrated for a great number of materials /1/, exhibiting a favourable effect only for some ion-target combinations and implantation conditions. A great deal of problems are involved in the interpretation of the changes observed in each particular case because there is no single mechanism of the implantation effect on the mechanical properties of materials.

Experimental techniques The present note deals with the antimony ion-bombardment effect on the steady-state creep and microhardness of the aluminium alloy D16 (4.5% Cu, 0.3% Fe, 0.4% Mn, 1.6% Mg, < 0.1% Si, Al is the remainder). The samples were mechanically treated by removing the plated aluminium layer and then polishing with diamond paste. Then the samples were annealed in vacuum at T = 720 K to exclude heating effects during bombardment and relaxation of stresses due to treatment. The annealed samples were chemically polished.

Antimony ions of 35 keV energy and $9 \mu A/cm^2$ current density were implanted up to the dose $10^{17}$ $Sb^+/cm^2$. Following /2/, the range and straggling of antimony ions are 22.7 and 6.75 nm, respectively.

The steady-state creep was studied at temperatures between 600 and 700 K and stresses from 2 to 15 MPa using four-point bending of the samples. Microhardness and its depth distribution was measured with indenter loadings 0.1 to 2 N.

---

1) Lenin Avenue 4, SU-220080 Minsk, USSR.

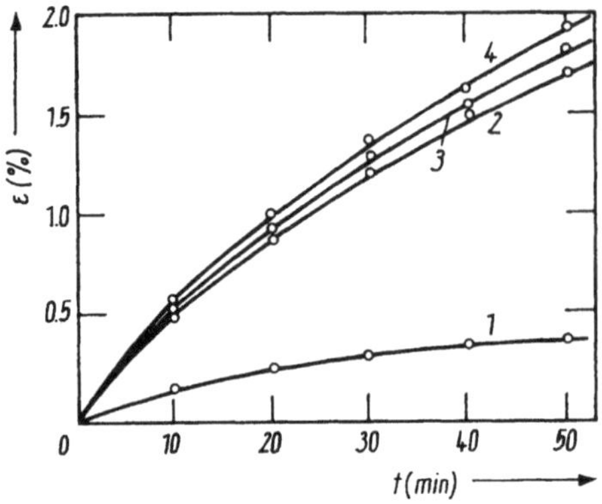


Fig. 1. Creep curves at T = 660 K and $\tau$ = 10 MPa for the alloy D16 before and after implantation; (1) 0, (2) $2x10^{16}$, (3) $6x10^{16}$, (4) $10^{17}$ $Sb^+/cm^2$

Results and discussion Fig. 1 shows the time dependence of the alloy D16 deformation at 660 K before and after implantation. The reported data imply that after antimony implantation the steady-state creep rate increases; the deformation stored at the developed stage for the bombarded samples is about an order greater than for initial ones. For the chosen range of temperatures and stresses the steady-state creep rate ε can appropriately be described by the Weertman equation:

$$\dot{\varepsilon} = A\tau^n \exp(-U/kT) ,$$

where U is the activation energy of steady-state creep, n is the creep rate response to the stress $\tau$.

The resulting values of the steady-state creep parameters are presented in Table 1.

The increase in the creep rate (for example, at $\tau$ = 10 MPa the ratio $\dot{\varepsilon}_{imp}/\dot{\varepsilon}_{nonimp}$ is equal to 4) and in the creep rate response (parameter n) as well as the decrease in the activation energy U evidence the creep resistance reduction after antimony ion bombardment.

Table 1

Characteristics of the steady-state creep

| dose ($10^{16}$ $Sb^+/cm^2$) | U (eV) | n |
|---|---|---|
| 0 | 2.34 ± 0.02 | 1.6 ± 0.1 |
| 2 | 1.88 ± 0.03 | 2.0 ± 0.2 |
| 6 | 1.90 ± 0.03 | 2.1 ± 0.2 |
| 10 | 1.88 ± 0.03 | 2.0 ± 0.2 |

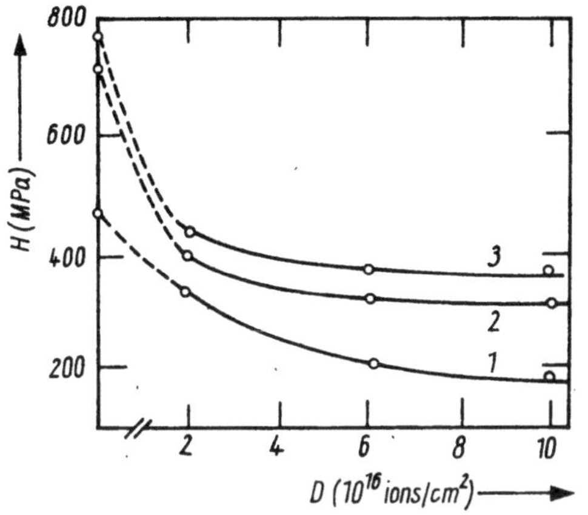


Fig. 2. Dose dependence of microhardness for various indenter penetration depths; (1) 5, (2) 7, (3) 9 μm

The loss of strength of doped samples manifests itself also in the decrease of their microhardness. Fig. 2 shows the dose dependence of the microhardness H for various penetration depths h of the indenter. The relative reduction of the microhardness at a depth h = 5 μm is 70%.

The distribution of impurities in a surface layer before and after ion implantation of the test alloy D16 was studied using the Rutherford backscattering of 1.4 MeV helium ions and X-ray spectroscopy analysis to elucidate the reason for the above effects. In Fig. 3 the Rutherford backscattering spectra are given for the initial samples and after implantation of $10^{17}$ $Sb^+/cm^2$. The spectra show that doping decreases the concentration of copper and of oxygen, which seems to be available in a bound state ($Al_2O_3$) in non-implanted samples. The layer antimony concentration for the above dose is $5.7x10^{16}$ atoms/$cm^2$ with a layer of 50 nm. The X-ray microspectral analysis of the initial and doped samples has shown that the initial samples contain copper and iron both in solutions and in compounds (possibly, $FeAl_3$, $CuMgAl_2$, $CuAl_2$). These impurities diffuse deeply into bombarded samples: the characteristic emissivity of $CuK_\alpha$ and $FeK_\alpha$ is about an order of magnitude decreased for solid solutions and decreased

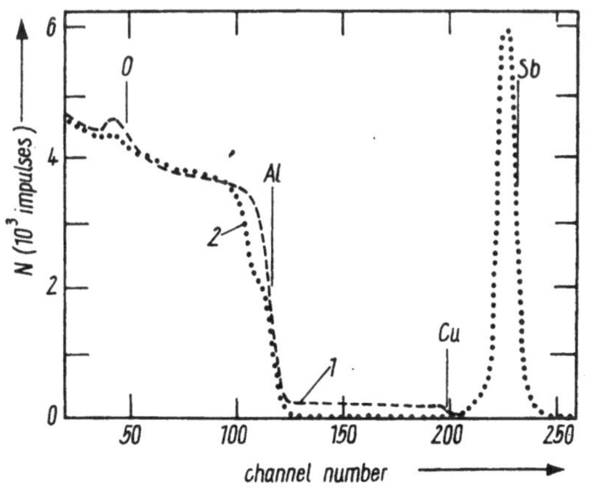


Fig. 3. RBS spectra of 1.4 MeV helium ions for the initial sample (1) and after implantation of $10^{17}$ $Sb^+/cm^2$ (2)

by a factor 5 for hardening compounds. Moreover, the hardening compound size decreased by a factor 2. The antimony distribution throughout the sample is seen to be uniform and to have a homogeneity level of 12%. These findings somewhat differ from the data reported in /3/ where antimony 200 keV ion bombardment of pure aluminium caused copper and iron diffusion correspondingly from and to the sample surface.

Thus, the loss of strength of the aluminium alloy D16 implanted with antimony ions may be related to partial destruction of hardening compounds and further diffusion of doping impurities from the surface into bulk of the crystal. Similar results were obtained in /4/ for different age-hardenable alloys.

## References

/1/ J. HIRVONEN, Ion Implantation, Izd. Metallurgiya, Moscow 1985 (in Russian).

/2/ A. BURENKOV, F. KOMAROV, M. KUMAKHOV, and M. TEMKIN, Tabulated Parameters of Spatial Distribution of Ion-Implanted Impurities, Izd. BSU, Minsk 1980 (in Russian).

/3/ P. MADAKSON, J. appl. Phys. 55, 3308 (1984).

/4/ P. P. GRINCHUK and E. M. LOBODA, Voprosy at. Nauki i Tekhniki, Ser. Topl. i Konstr. Mater. 1, 26 (1976).

(Received January 18, 1988; in revised form May 25, 1988)

phys. stat. sol. (a) 108, K5 (1988)

Subject classification: 64.75 and 66.30; 61.70; S1.2

Institute of Material Science and Applied Mechanics, Technical University of Wrocław[1)]

# Activation Energy for Precipitation in Undeformed and Deformed Cu-Cr Alloys

By

J. SZABLEWSKI and B. KUŻNICKA

Introduction The kinetics of the precipitation process in Cu-Cr alloys has been investigated mostly by resistivity measurements /1 to 4/. Two decidedly different values of the precipitation activation energy $E_p$ in undeformed alloys can be found in published data: 130 kJ/mol /1, 2/ or 200 kJ/mol /3, 4/. The purpose of this investigation was to verify the value of the precipitation activation energy and to evaluate the influence of prior cold work on it. As in the precipitation process the energy $E_p$ is closely connected with the activation energy for diffusion $E_D$, the $E_D$ values are also discussed in the light of the obtained results and literature data.

Experimental procedure A Cu-0.43 wt% Cr alloy was made from 99.95% Cu and Cr. Specimens were solution treated under an argon atmosphere for 1 h at 1000 °C, water quenched and cold rolled to a deformation of 0, 20, 50, 80% (these four different states will be referred to as conditions Q0, Q20, Q50, and Q80). In order to determine changes in electrical resistance as a function of the extent of precipitation (the resistivity of Cu-Cr alloys is a linear function of the amount of Cr in solid solution /5/), the measurements were made in situ during isothermal annealing under an argon atmosphere in the temperature range 350 to 500 °C. The error was not greater than 0.3%. At each temperature the transformed fraction y was computed as a function of time assuming

$$y = \frac{R_i - R_t}{R_t - R_f} , \qquad (1)$$

where $R_i$, $R_t$, and $R_f$ were the resistances found initially, at instant t, and after completion of transformation, respectively.

Results According to the formal theory of precipitation kinetics /6/ the time $t_Y$ required for a specified fraction y = Y to transform and the precipitation

1) Smoluchowskiego 25, PL-50-370 Wrocław, Poland.

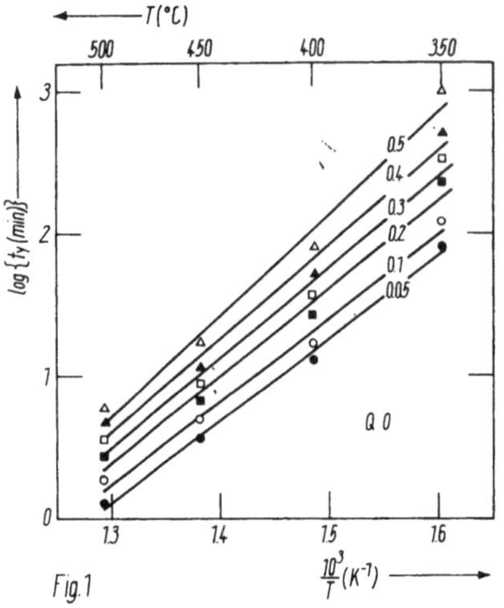


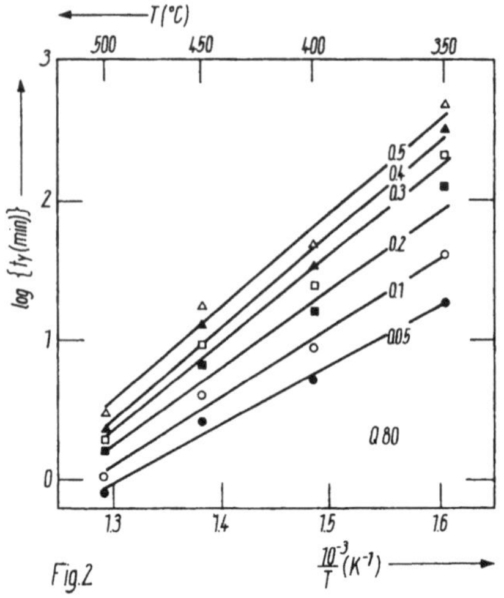


Fig. 1. log $t_Y$ versus 1/T plots for different degrees of precipitation in the undeformed solid solution; Y = ● 0.05, ○ 0.1, ■ 0.2, □ 0.3, ▲ 0.4, △ 0.5

Fig. 2. log $t_Y$ versus 1/T plots for different degrees of precipitation in the 80% cold deformed solid solution; Y = ● 0.05, ○ 0.1, ■ 0.2, □ 0.3, ▲ 0.4, △ 0.5

activation energy $E_P$ are connected by the relation

$$\log t_Y = \text{const} - \log A + E_P/2.3\,R\,T\,, \tag{2}$$

where A frequency factor in the Arrhenius equation, R gas constant, and T absolute temperature. Hence, if $E_P$ and A are independent of T the value of $E_P$ can be evaluated from the slope of a graph of log $t_Y$ against 1/T. Some examples of such $t_Y$ versus 1/T plots for conditions Q0 and Q80 for the fraction transformed Y ranging from 0.05 to 0.5 are drawn in the form of regression lines in Fig. 1 and 2. The evaluated $E_P$ values for the whole examined range of Q and Y are collected in Fig. 3. As shown in Fig. 3, the $E_P$ value of the undeformed solid solution (Q0) ranges from 110 to 136 kJ/mol with the precipitation process advancing. The statistically evaluated mean value of this energy is (120 ± 7) kJ/mol for the Y range of 0.05 to 0.5. For the high-deformation conditions (Q50, Q80), however, a large expansion of the $E_P$ range is observed, caused by considerable lowering of the original $E_P$ value (e.g. to 80 kJ/mol for Q80).

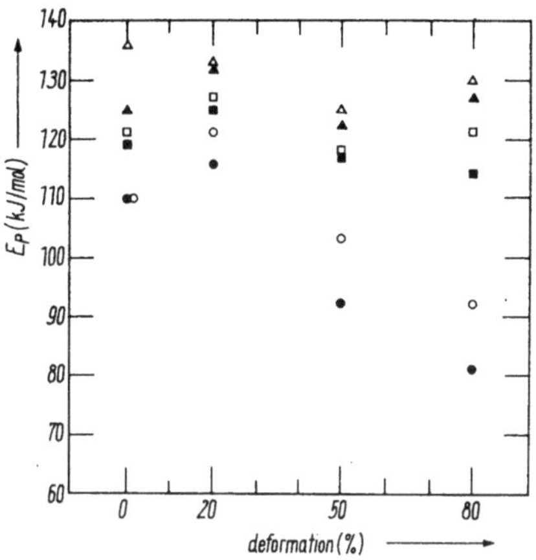


Fig. 3. Influence of cold working on the precipitation activation energy $E_P$ for the fraction transformed range 0.05 to 0.5; Y = ● 0.05, ○ 0.1, ■ 0.2, □ 0.3, ▲ 0.4, △ 0.5

Discussion The empirical activation energy for precipitation deduced from the formal theory of precipitation kinetics /6, 7/ is

$$E_P = E_D + E_N + \frac{1}{T}\frac{\partial E_N}{\partial(1/T)} , \qquad (3)$$

where $E_N$ is the nucleation activation energy. At low temperatures when the nucleation C-curve can be approximately taken as a straight line, the second and third terms are, respectively, positive and negative and $E_N \ll E_D$. It is generally accepted that the growth of Cr-rich particles is diffusion controlled with zero nucleation rate /2, 7, 8/. In this case the value of the third term depends on the form of the equilibrium diagram. In the Cu-Cr system the Cr solvus line varies sharply with temperature and the third term can be considered negligibly small /7, 9/. The calculation based on the classical theory of precipitate nucleation for Cr spherical particles has proved that the value of this term is not greater than 5 kJ/mol /10/. It follows that in Cu-Cr alloys $E_D \approx E_P$.

The precipitation activation energy for the undeformed state deduced from the presented experimental data is (120 ± 7) kJ/mol being close to the published value /1, 2/ and considerably smaller than the value obtained by the authors of /3, 4/. Considering the above reasoning, our results confirm the $E_D$ value (equal 105 kJ/mol) published in /8/ whereas $E_D$ = 207 kJ/mol noticed by Rezek /11/ is too high and disagrees with the formal theory of precipitation. The $E_P$ value for Q20 is (126 ± 7) kJ/mol, close to its value for Q0. The observed significant decrease in the original $E_P$ values by about 30 kJ/mol (compared with the $E_P$ for the undeformed condition) must be attributed to an increase in the structural defect density resulting from high deformation of the supersaturated solution that gave rise to lower $E_D$ values.

References

/1/ B. LYNCH, Ph.D. Thesis, Univ. of Aston, Birmingham 1968.
/2/ P.W. KNIGHTS and P. WILKES, Metallurg. Trans. 4, 10 (1973).
/3/ W. KÖSTER and W. KNORR, Z. Metallk. 45, 6 (1954).
/4/ B. SIMMEN, Neue Hütte 17, 5 (1972).
/5/ W.F. SCHIRIN, V.M. ROSENBERG, and N.P. BELOUSOV, Tsvetn. Metal. 45, 12 (1971).
/6/ J.W. CHRISTIAN, The Theory of Transformations in Metals and Alloys, Pergamon Press, London 1965.
/7/ J. BURKE, The Kinetics of Phase Transformations in Metals, Pergamon Press, London 1965.
/8/ G.M. FEDOROV, V.N. FEDOROV, and J.V. PIGUZOV, Izv. Akad. Nauk SSSR, Ser. Metal. 19, 5 (1978).
/9/ D.J. CHAKRABARTI and D.E. LAUGHLIN, Bull. Alloy Phase Diagrams 5, 1 (1984).
/10/ J. SZABLEWSKI, Doctoral Thesis, Technical Univ. of Wrocław, 1986.
/11/ J. REZEK, Canad. Metallurg. Quart. 17, 5 (1969).

(Received February 29, 1988; in revised form June 6, 1988)

phys. stat. sol. (a) 108, K9 (1988)
Subject classification: 62.20; S12

Departamento Física Aplicada (C-4) and Instituto de Ciencia de Materiales, sede B. Consejo Superior de Investigacoines Científicas, Universidad Autonoma de Madrid[1)]

# Brillouin Scattering Determination of the Elastic Constants of Tris-Sarcosine Calcium Chloride at Room Temperature

By

C. PRIETO, R. RAMIREZ, J.A. GONZALO, and W. WINDSCH[2)]

Tris-sarcosine calcium chloride, $(CH_3NHCH_2COOH)_3CaCl_2$ (TSCC), is a uniaxial ferroelectric with a Curie temperature of approximately 130 K /1/. It has orthorhombic symmetry in both phases characterized by the space group $D_{2h}^{16}$ in the paraelectric phase and $C_{2v}^{9}$ in the ferroelectric one /2/. TSCC is a very interesting ferroelectric material because of the quantum effect on the phase transition which has been observed in mixed crystals of calcium chloride and bromide compounds /3/.

The TSCC ferroelectric phase transition has been partially studied by ultrasound /4/ and Brillouin scattering /5/ but a complete determination of the nine elastic constants has not been made, even at room temperature.

The crystals were grown from water solution at the laboratories of the Karl-Marx-Universität, Leipzig, and they were of excellent optical quality. A single-domain sample was cut with faces perpendicular to the three crystallographic axes, and they were polished by using 1 μm alumina powder suspended in liquid paraffin.

Brillouin scattering experiments were taken with a He-Ne laser (Spectra Physics 125) at 631.8 nm (power 90 mW). The light was analyzed with a piezoelectrically scanned Fabry-Perot interferometer with automatical stabilisation (Burleigh RC-110 and DAS-10) operating in triplepass. Two free spectral ranges were selected: 60 and 37.5 GHz. The finesse of the measurements was 34 and a contrast of $\approx 10^5$. In this work the right angle scattering geometry was used to obtain the elastic constants $C_{44}$, $C_{55}$, $C_{66}$, $C_{12}$, $C_{13}$, and $C_{23}$.

---

1) E-28049 Madrid, Spain.

2) Permanent address: Sektion Physik der Karl-Marx-Universität, DDR-7010 Leipzig, GDR.

The elastic constants $C_{11}$, $C_{22}$, and $C_{33}$ could be easily obtained from the previous work by Hikita et al. /5/ (see below).

The differential equation for the acoustic waves can be written as

$$\rho \ddot{U}_i = C_{ijkl} (\partial^2 U_l/\partial x_j \partial x_k) , \tag{1}$$

where $C_{ijkl}$ is the elastic-constant tensor, $U_l$ is the volume element displacement, and $\rho$ is the density. Substitution in (1) of a sinusoidal wave gives

$$\gamma U_i = \Gamma_{il} U_l ; \tag{2}$$

$\gamma$ is related to the sound velocity (V) by $\gamma = \rho V^2$, and $\Gamma_{ij}$ is

$$\Gamma_{ij} = C_{ijkl} Q_l Q_k . \tag{3}$$

$Q_j$, $Q_l$ are the direction cosines of the unit vector normal to the wave plane ($\vec{Q}$ is the phonon propagation direction).

For a given $\vec{Q}$, the three values of $\gamma$ are the eigenvalues of (3), given by

$$\det \left[ \Gamma_{ij} - \gamma \delta_{ij} \right] = 0. \tag{4}$$

The Brillouin shifts ($\delta \nu_B$) are related with the sound velocity by

$$\delta \nu_B = (V/\lambda)(n_i^2 + n_s^2 - 2 n_i n_s \cos\Theta)^{1/2} , \tag{5}$$

where $\lambda$ is the wavelength of incident light, $n_i$ and $n_s$ are the refractive indices for the incident and scattered light, and $\Theta$ is the scattering angle.

By comparison of ultrasonic data /4/ the sound velocity along the b-axis ($C_{22} = \rho V^2$) with the Brillouin data /5/ at room temperature, it is possible to evaluate a mean value of the refractive index for TSCC. Taking into account (5) for right angle scattering ($\Theta = 90^0$),

$$\delta \nu_{ij} = (V_{ij}/\lambda)(n_i^2 + n_s^2)^{1/2} . \tag{6}$$

In the case of /5/ $n_i$ and $n_s$ correspond to the refractive indices along (1, 1, 0) and (-1, 1, 0). With $\rho = 1.533$ g cm$^{-3}$ the value obtained for the refractive index is $\langle n \rangle = (n_i^2 + n_s^2)^{1/2} = 1.59$. Using $\langle n \rangle$ as a mean value for

Table 1

| $\vec{Q}$ | $\vec{k}_i$ | $\vec{k}_s$ | $\vec{e}_i = e_s$ | mode | $\delta\nu_B$ (GHz) | $C_{ij}$ ($10^{10}$ N m$^{-2}$) | V ($10^3$ m s$^{-1}$) |
|---|---|---|---|---|---|---|---|
| (1, 0, 0, ) | (1, 1, 0) | (-1, 1, 0) | (0, 0, 1) | L $C_{11}$ | 16.0 | $C_{11}$ = 2.1 | 3.66 |
| (0, 1, 0) | (1, 1, 0) | (1, -1, 0) | (0. 0. 1) | L $C_{22}$ | 21.0 | $C_{22}$ = 3.5 | 4.81 |
| (0, 0, 1) | (0, 1, 1) | (0, 1, -1) | (1, 0, 0) | L $C_{33}$ | 21.0 | $C_{33}$ = 3.5 | 4.81 |
| $2^{-1/2}$(1, 1, 0) | (1, 0, 0) | (0, -1, 0) | (0, 0, 1) | L $\gamma^{10}$<br>qT $\gamma^{11}$ | 18.14<br>9.55 | $C_{66}$ = 1.3<br>$C_{12}$ = 1.5 | 2.68 |
| $2^{-1/2}$(0, 1, 1) | (0, 1, 0) | (0, 0, -1) | (1, 0, 0) | L $\gamma^{13}$<br>qT $\gamma^{14}$ | 17.925<br>9.26 | $C_{44}$ = 1.4<br>$C_{23}$ = 1.5 | 2.61 |
| $2^{-1/2}$(1, 0, 1) | (0, 0, 1) | (-1, 0, 0) | (0, 1, 0) | L $\gamma^{16}$<br>qT $\gamma^{17}$ | 17.44<br>9. 30 | $C_{55}$ = 1.2<br>$C_{13}$ = 1.3 | 2.61 |

$\vec{Q}$ phonon wave vector, $\vec{k}_i$ incident light wave vector, $\vec{k}_s$ scattered light vector, $\vec{e}_i$, $\vec{e}_s$ polarization of light beam, $\gamma^n = \gamma^n(C_{ij})$ see below, $\delta\nu_B$ Brillouin shift, $C_{ij}$ elastic constants, V sound velocity, L longitudinal, qT quasi-transverse

$$\gamma^{10(11)} = \frac{1}{4}\left(C_{11} + C_{22} + 2C_{66} +(-) \left((C_{11} - C_{22})^2 + 4(C_{12} + C_{66})^2\right)^{1/2}\right),$$

$$\gamma^{13(14)} = \frac{1}{4}\left(C_{22} + C_{33} + 2C_{44} +(-) \left((C_{22} - C_{33})^2 + 4(C_{23} + C_{44})^2\right)^{1/2}\right),$$

$$\gamma^{16(17)} = \frac{1}{4}\left(C_{11} + C_{33} + 2C_{55} +(-) \left((C_{11} - C_{33})^2 + 4(C_{13} + C_{55})^2\right)^{1/2}\right)$$

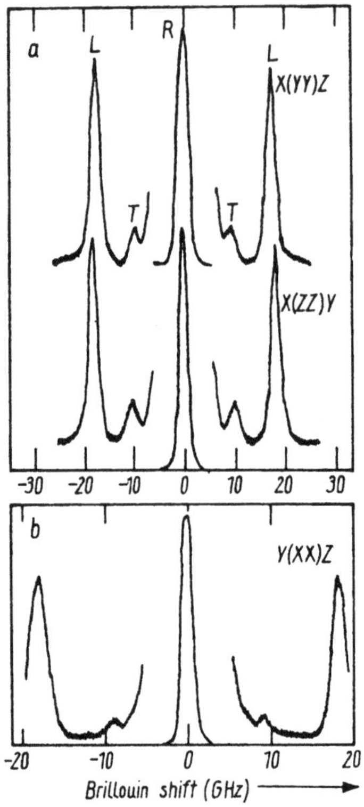


Fig. 1. Brillouin spectra of the TSCC at room temperature (geometries are indicated); R Rayleigh line, L longitudinal and T transverse modes. Note the different free spectral ranges of the a) and b) parts

other directions, it is possible to determine the elastic constants $C_{11}$ and $C_{33}$ from the data of /5/ (see Table 1).

To determine the other six elastic constants, it is necessary to use scattered phonons propagating along other directions (e.g. (1, 1, 0), (0, 1, 1), and (1, 0, 1). Fig. 1 shows the longitudinal and transverse Brillouin shifts of the spectra at the X(ZZ)Y, X(YY)Z, and Y(XX)Z right angle scattering geometry.

Table 1 shows the Brillouin for the three geometries performed, the velocity values for the longitudinal and transverse modes and the six elastic constants for this crystal. The notation of /6/ has been used.

At 0 K, the characteristic Debye temperature $\Theta_D$ is given by

$$\Theta_D(0) = (h/k_B)(9\,N/4\pi V_a\, I)^{1/3} \; ; \tag{7}$$

h and $k_B$ are the Planck and Boltzmann constants, N the number of vibrating units in the volume $V_a$, and I is

$$I = \int_0^{4\pi} \sum_{i=1}^{3} (V_i^{-3})(d\Omega/4\pi) \approx \langle V_L\rangle^{-3} + 2\langle V_T\rangle^{-3} \; .$$

$V_i$ are the three sound velocities. We have made the approximation considering that there is one longitudinal phonon ($V_L$) and two transverse ones ($V_T$) and $\langle\,\rangle$ is the average between the three axes. The Debye temperature obtained is 210 K.

On the other hand, assuming that the $T^3$ Debye law is followed by the specific heat in TSCC below 10 K: $C_p = (12\pi^4/5)(T/\Theta_D)^3 \, nk_B$, with n = 6 as can be inferred by extrapolation to $T \rightarrow 0$ of the experimental data for $C_p$ /1/ ($C_p^{saturation} \approx 3nk_B = 150$ J/Kmol). From the specific heat data /1/, (after substrating the linear high temperature anharmonic contribution), the obtained Debye temperature is $\Theta_D = 182$ K, which is of the same order as the previous value from the elastic constant data.

This work has been supported by CAICyT Grant No. 2017/83.

## References

/1/ Y. MAKITA, J. Phys. Soc. Japan 20, 2073 (1965).

/2/ T. ASHIDA, S. BANDO, and M. MAKUDO, Acta cryst. B28, 1560 (1972).

/3/ W. WINDSCH and H. BRAETER, Ann. Phys. 43, 369 (1986).

/4/ G. SORGE and U. STRAUBE, Ferroelectris 21, 533 (1986).

/5/ T. HIKIDA, P. SCHNACKENBERG, and V.H. SCHMIDT, Phys. Rev. B 31, 299 (1985).

/6/ R. VACHER and L. BOYER, Phys. Rev. B 6, 639 (1972).

(Received May 3, 1988)

phys. stat. sol. (a) 108, K15 (1988)
Subject classification: 64.10

Department of Physics, Anhui Normal University, Wuhu[1)]

# On the Equation of Solid State at High Pressure

By
LI-RONG CHEN and TZU-TUNG YEN

For some substances of solid state, the equation of state at high pressure is suggested by Couchman and Reynolds /1/ as

$$\frac{VP}{|\Delta V|} = A + DP \ , \tag{1}$$

where A and D are constants relating to the substance. Equation (1) is called the Couchman-Reynolds equation.

As wellknown, there are several empirical formulae for the equation of solid state at high pressure. The purpose of this note is to prove the equivalence of the Couchman-Reynolds equation with the Murnaghan equation and the Anderson equation.

From (1) we have

$$\frac{V_o}{V} = 1 + \frac{P}{A+DP} = 1 + \frac{P}{A}\ \frac{1}{1+D\frac{P}{A}} \ . \tag{2}$$

The Couchman-Reynolds equation is tested by the authors /2/ for 57 kinds of monatomic crystals. We obtained that the condition $D/A \ll 1$ can be satisfied for the majority of monatomic crystals. Therefore, (2) can be expanded in powers of P/A as follows:

$$\frac{V_o}{V} = 1 + \frac{P}{A} - D\left(\frac{P}{A}\right)^2 + D^2\left(\frac{P}{A}\right)^3 - \dots$$

or

$$\frac{V}{V_o} = 1 - \frac{P}{A} + D\left(\frac{P}{A}\right)^2 - D^2\left(\frac{P}{A}\right)^3 + \dots \ . \tag{3}$$

According to the definition of the isothermal bulk modulus $B = -V(\partial P/\partial V)_T$, using (1) we obtained the following relation between B and P:

$$B = A\left[1 + \frac{2D-1}{A}\ P + \frac{D(D-1)}{A^2}\ P^2\right] \ . \tag{4}$$

---

1) Wuhu, Peoples' Republic of China.

We can see that if we neglect the second and third term on the right of (4) when $D/A \ll 1$, then A is the bulk modulus at zero pressure $B_0$; the factor $2D - 1$ is the pressure derivative of the bulk modulus at zero pressure $B_0'$, i.e.

$$A = B_0 , \quad 2D - 1 = B_0' . \tag{5}$$

If we substitute (5) into (3), then we obtain the Couchman-Reynolds series in powers of $P/B_0$:

$$\frac{V}{V_0} = 1 - \frac{P}{B_0} + \frac{1}{2}(1 + B_0')\left(\frac{P}{B_0}\right)^2 - \frac{1}{4}\left(1 + B_0'\right)^2\left(\frac{P}{B_0}\right)^3 + \ldots . \tag{6}$$

Murnaghan /3/ assumed that the bulk modulus is linear with pressure, i.e.

$$B = B_0 + B_0' P . \tag{7}$$

He derived the following famous Murnaghan equation:

$$P = \frac{B_0}{B_0'}\left[\left(\frac{V_0}{V}\right)^{B_0'} - 1\right] \tag{8}$$

or

$$\frac{V_0}{V} = \left(1 + \frac{B_0'}{B_0} P\right)^{1/B_0'} . \tag{9}$$

From the experimental data /4/ it may be observed that the condition $V_0/V < 2$ is satisfied for nearly all monatomic crystals. Therefore, under the condition $(B_0'/B_0)P < 1$, we can expand the Murnaghan equation (9) in powers of $B_0'P/B_0$,

$$\frac{V_0}{V} = 1 + \frac{P}{B_0} + \frac{1}{2B_0'}\left(\frac{1}{B_0'} - 1\right)\left(\frac{B_0'}{B_0} P\right)^2 + \frac{1}{6B_0'}\left(\frac{1}{B_0'} - 1\right)\left(\frac{1}{B_0'} - 2\right)\left(\frac{B_0'}{B_0} P\right)^3 + \ldots$$

or

$$\frac{V}{V_0} = 1 - \frac{P}{B_0} + \frac{1}{2}(1 + B_0')\left(\frac{P}{B_0}\right)^2 - \frac{1}{6}(3 B_0' + 2 B_0'^2 - 5)\left(\frac{P}{B_0}\right)^3 + \ldots . \tag{10}$$

It is worth noticing that the equation of state in series of powers has been derived by Anderson /5/ as

$$\frac{V}{V_0} = 1 - \frac{P}{B_0} + m\left(\frac{P}{B_0}\right)^2 - n\left(\frac{P}{B_0}\right)^3 + \ldots , \tag{11}$$

Table 1

Compressibility parameters of metals

| | Anderson /5/ equation | | Murnaghan /6/ equation | | Couchman-Reynolds equation (this work) | | experimental /7/ values of $B_o$ (kbar) |
|---|---|---|---|---|---|---|---|
| | $B_o$(kbar) | $B'_o$ | $B_o$ (kbar) | $B'_o$ | $B_o$ (kbar) | $B'_o$ | |
| Ag | 1107.3 | 4.6810 | 1208.83 | 2.484 | 1099.6 | 5.78 | 1007 |
| Al:6061T6 | 814.27 | 10.6600 | 875.85 | 3.214 | 846.7 | 4.92 | |
| Al | 788.77 | 3.1647 | 778.97 | 4.26 | 781.9 | 3.99 | 722 |
| Ba | 94.62 | 1.7252 | 94.32 | 2.187 | 94.9 | 1.90 | 103 |
| Bi(I) | 309.73 | 4.0901 | 296.93 | 6.742 | 304.5 | 5.70 | 315 |
| Ca | 182.63 | 2.6157 | 186.82 | 2.524 | 184.2 | 2.49 | 152 |
| Cd | 473.04 | 4.0830 | 447.6 | 7.307 | 464.4 | 5.86 | 467 |
| Cu | 1510.3 | 4.0115 | 1625.35 | 0.426 | 1504.7 | 4.65 | 1370 |
| Fe | 1711.1 | 7.7886 | 1757.56 | 7.669 | 1702.1 | 9.14 | 1683 |
| In | 388.97 | 5.0951 | 390.77 | 5.239 | 390.4 | 4.93 | 411 |
| La | 246.17 | 2.8630 | 244.73 | 3.150 | 246.8 | 2.89 | 243 |
| Mo | 2531.0 | 13.2879 | 2660.37 | 11.963 | 2468.1 | 18.60 | 2725 |
| Ni | 1802.6 | 16.5761 | 1904.8 | 20.157 | 1776.8 | 21.87 | 1860 |
| Pb | 423.53 | 3.7435 | 399.82 | 6.76 | 414.6 | 5.51 | 430 |
| Sn | 550.81 | 2.7152 | 549.20 | 3.651 | 548.6 | 3.38 | 1110 |
| Ta | 2019.0 | 3.8421 | 2056.90 | 2.759 | 2002.3 | 5.01 | 2000 |
| Tl(I) | 375.29 | 3.1191 | 366.03 | 4.795 | 379.5 | 3.04 | 359 |
| Zn | 614.59 | 3.2639 | 597.91 | 4.880 | 610.3 | 4.18 | 598 |

where

$$m = \frac{1}{2}(1 + B'_o), \quad n = \frac{1}{6}\left[1 + 3B'_o + 2(B'_o)^2 - B_oB''_o\right],$$

etc. $B''_o$ is the second order pressure derivative of the bulk modulus at zero pressure.

From (6), (10), and (11) we discovered that if the third power terms in $P/B_o$ are neglected, the Couchman-Reynolds equation is equivalent to the Murnaghan equation and the Anderson equation.

Let us test their equivalence using the experimental data of compressibility of 18 metals up to 45 kbar by Vaidya and Kennedy /6/. $B_0$ and $B_0'$ evaluated from the polynomial regression (for the Anderson equation) and the Murnaghan equation fit of experimental data are given by Vaidya and Kennedy. We obtained $B_0$ and $B_0'$ evaluated from the linear regression of experimental data for the Couchman-Reynolds equation. These results together with the experimental values of $B_0$ are summarized in Table 1 which shows that this equivalence between the three equations of state is very good. The agreement of their values of $B_0$ with the experimental values is also good (with the exception of Sn).

References

/1/ P.R. COUCHMAN and C.L. REYNOLDS, Jr., J. appl. Phys. 47, 5201 (1976).

/2/ TZU-TUNG YEN, YI-JING DO, and LI-RONG CHEN, Acta Mech. Sinica 16, 299 (1984) (in Chinese).

/3/ F.D. MURNAGHAN, Proc. Nat. Acad. Sci. USA 30, 244 (1944).

/4/ G.C. KENNEDY and R.N. KEELER, in: American Institute Physics Handbook, 3rd ed., Ed. D.E. GRAY, McGraw-Hill, New York 1972 (p. 4 to 38).

/5/ O.I. ANDERSON, J. Phys. Chem. Solids 27, 547 (1966).

/6/ S.N. VAIDYA and G.C. KENNEDY, J. Phys. Chem. Solids 31, 2329 (1970).

/7/ C. KITTEL, Introduction to Solid State Physics, 5th ed., Wiley, New York 1976 (p. 85).

(Received April 21, 1988)

phys. stat. sol. (a) 108, K19 (1988)

Subject classification: 64.75 and 68.48; S4

Department of Metallurgical Engineering,
Indian Institute of Technology, Kharagpur[1]

# A Resistometric Study of Migrating Boundary Characteristics in Pb-Sn Alloys

By

I. MANNA and S.K. PABI

The structure and diffusivity of mobile boundaries (MBs) vis-a-vis static boundaries (Sbs) have evoked considerable controversy in recent times. While metallographic observations on few boundaries at the reaction temperatures (say, method M1) report diffusivity in the MBs 2 to 4 orders of magnitude higher than in the SBs /1, 2/, measurements of the boundary velocities at ambient temperature following periodic freezing of the reaction front by quenching (method M2) usually indicate identical diffusivities in both /3 to 6/. Investigations by method M1 are relatively poor in statistics and the boundaries under observation can be intrinsically faster moving or having higher diffusivity /7, 8/. On the other hand, periodic arrest of the boundary movement in M2 may change its (postulated) mobile configuration into a static one /9/.

In the present investigation, the effects of periodic arrest of the MBs (by pulse heating/cooling) on the isothermal discontinuous precipitation kinetics have been studied by a bulk-averaging technique, namely, electrical resistivity. This would be a test for the validity of method M2 and also it may point out whether MBs, on an average, have different structural characteristics than SBs.

Two Pb-Sn alloys containing 8.82 (alloy A) and 14.65 (alloy B) at% Sn were prepared from high purity Pb(99.99%) and Sn(99.999%). Cast billets were homogenized at 523 K for 50 h and extruded at 433 K to wires of 1 mm diameter. For resistivity measurements 1 m long wires were mounted on bakelite sheets, solution treated at 443 K for 12 h, quenched in water, and isothermal precipitation at $T_P \geqq 298$ K was carried out in oil and water baths controlled to $\pm 0.5$ K. After precipitation at $T_P$ for time t, the resistance ($R_t$)

---

1) Kharagpur 721302, India.

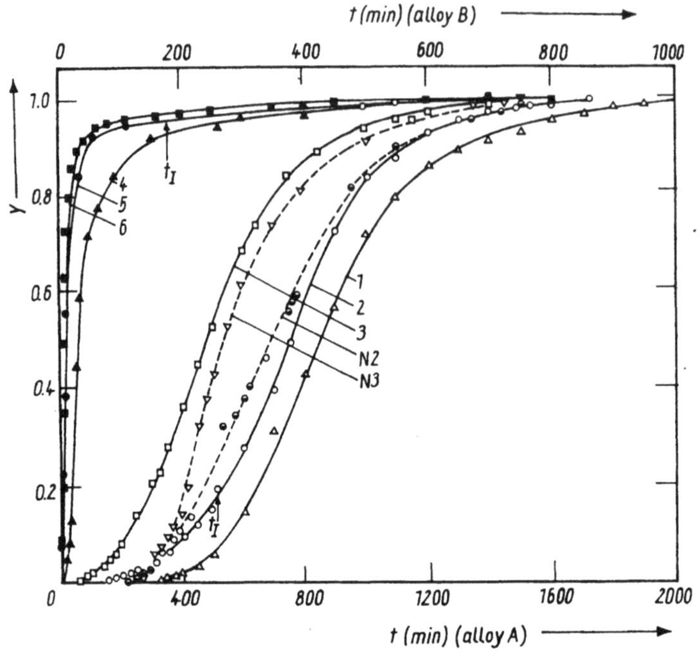


Fig. 1. Precipitation isotherms of alloy A at $T_P$ = (1) 318 K, (2) 308 K, (3) 298 K, and alloy B at $T_P$ = (4) 363 K, (5) 308 K, (6) 298 K. Curves N2 and N3 are precipitation isotherms of alloy A at $T_P$ = 308 K and 298 K, respectively, but measured at $T_M$ = 77 K. Arrows on (2) and (5) indicate $t_I$ for pulse treatments (see text)

measured at a temperature $T_M$ by a General Radio 1666 dc Kelvin bridge yields the fraction transformed Y from the relation /10/

$$Y = \frac{\Delta R_t}{\Delta R_{max}} \; ; \quad \Delta R_t = R_o - R_t \; ; \; \Delta R_{max} = R_o - R_f \; , \tag{1}$$

where $R_o = R_t$ at t = 0, and $R_f = R_t$ when $(dR_t/dt) < 0.04\,(dR_t/dt)_{max}$.

Precipitation isotherms (1) to (6) in Fig. 1 measured at $T_M = T_P$ support the two stage reaction mechanism in Pb-Sn alloys /10/. It is known that the rapid reaction rate in stage I ($Y \lessapprox 0.75$) is caused by the discontinuous mode of precipitation, while the residual supersaturation between the interlamellar spaces is relieved by volume diffusion in stage II /11/. Due to its higher degree of supersaturation, precipitation rates in alloy B are faster, the transition from stage I to stage II is more well defined, and the incubation period is nearly zero as compared to those in alloy A (Fig. 1). Precipitation rates in both alloys decrease with increasing $T_P$ ($\gtrsim$ 298 K). The temperature $T_{DP}$, above which discontinuous precipitation ceases, has been determined by extrapolating the $\Delta R_{max}$ versus $T_P$ plot to $\Delta R_{max} = 0$. $T_{DP}$ is 333 K for alloy A and 381 K for alloy B. It may be pointed out that the degrees of solute supersaturation $\Delta C_{DP}$ at $T_{DP}$ calculated from the equilibrium solubility data /12, 13/ are nearly identical for alloys A and B, which are 4.72 and 4.65 at% Sn, respectively. Perhaps, the driving force ($\sim\Delta C_{DP}$) required to keep the discontinuous reaction fronts in motion /14/ is not markedly dependent on the alloy composition.

The isothermal precipitation kinetics in alloy A measured with the same specimen at $T_M$ = 77 K show significant divergence from those recorded at $T_M = T_P$ (cf. isotherms N2 versus 2, N3 versus 3; Fig. 1), which casts some doubt over the validity of method M2. In order to simulate M2 type measurements in resistometric studies, the isothermal precipitation at ($T_P$=) 308 K has been interrupted in a number of experiments at a given instant $t = t_I$ by means of pulse heating/cooling for a short while ($t_D$) to another temperature $T_I$, as per following schedule: $T_I$ = 348 K ($t_D$ = 1, 2 min), 333 K (1, 2, 3 min), 318 K (1, 2, 5 min), 233 K (10 min), 77 K (15 min) for alloy A; and $T_I$ = 383 K ($t_D$ = 1, 2 min), 343 K (1, 2, 3 min), 318 K (1, 2 min), 274 K (1, 5 min), 77 K (10 min) for alloy B. For alloy A precipitation has been interrupted in stage I at $t_I \approx 510$ min, while alloy B has been pulse treated in stage II at $t_I \approx 180$ min (cf. curves 2 and 5, Fig. 1). Further, for $T_I < T_{DP}$, $t_D$ is short enough to preclude any precipitation at $T_I$, that requires fresh nucleation of lamellae /15/. These experiments essentially manifest that (i) the precipitation rates just before and after pulse treatment are identical and (ii) the resistance of the specimen increases by an amount $\delta R$ (i.e. $\Delta R_t$ in (1) diminished) due to such pulse treatments (e.g. in Fig. 2). Perhaps, the deviation of curves N2 from 2 or N3 from 3 in Fig. 1 can be attributed to $\delta R$, which affects Y via equation (1). The results also reveal that $\delta R$ is more pronounced in stage I (i.e. in alloy A) than in stage II (i.e. in alloy B) and its magnitude usually increases with $|T_I - T_P|$ (cf. Fig. 2). It may be postulated that $\delta R$ originates from the thermal stresses in pulse treatments, e.g., the stress can bring about decohesion at some precipitate-matrix interfaces.

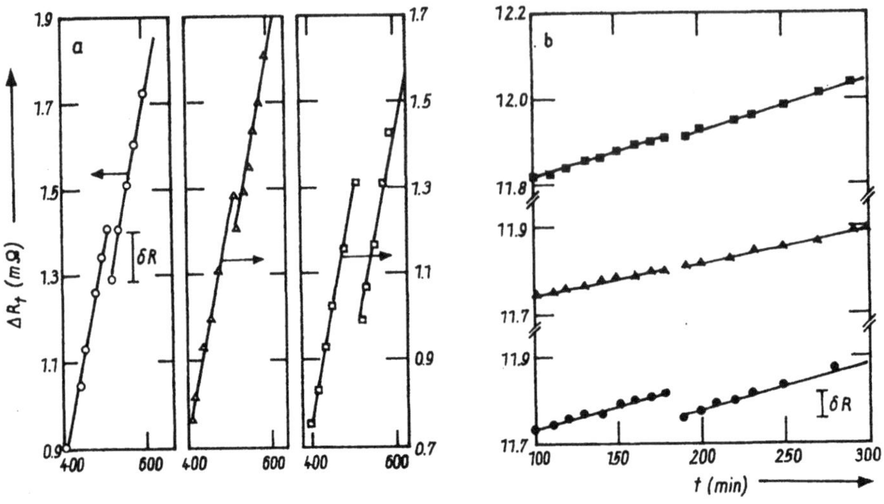


Fig. 2. Effect of pulse heating/cooling on precipitation isotherms;a) alloy A: $T_I$ = 348 K, $t_D$ = 1 min ( o ); $T_I$ = 318 K, $T_D$ = 5 min ( Δ ); $T_I$ = 77 K, $t_D$ = = 15 min ( □ ); b) alloy B: $T_I$ = 383 K, $t_D$ = 1 min (●); $T_I$ = 318 K, $t_D$ = = 1 min (▲); $T_I$ = 77 K, $t_D$ = 10 min (■)

Motion of the MBs in stage I (i.e. in alloy A) is arrested during the pulse heating interval $t_I$ /15/, when MBs can relax (except at $T_I$ = 77 K) /7/ to their perceptible static configurations. In this case, a decrease in the rate ($dR_t/dt$) following pulse treatments is expected /15/, if the incoherent reaction fronts have to revive their mobile configurations /9/. Since this is not borne out by the experiments, the present study does not evidence any diversity between the structure of MBs and SBs. Finally, these results also indicate that the metallographic measurements (which are insensitive to thermal stress effects) of the discontinuous reaction kinetics in a Pb-Sn alloy by method M2 may not be in error; although some structural changes (evident from $\delta R$) may result due to repeated quenching in M2.

The authors wish to thank Dr. H.S. Maity for providing the resistance bridge.

References

/1/ K. SMIDODA, W. GOTTSCHALK, and H. GLEITER, Acta metall. 26, 1833 (1978).

/2/ M. HILLERT and G.R. PURDY, Acta metall. 26, 333 (1978).

/3/ J. OBERSCHMIDT, K.K. KIM, and D. GUPTA, J. appl. Phys. 53, 5672 (1982).

/4/ C.R.M. GROVENOR, Scripta metall. 16, 317 (1982).

/5/ W. GUST, M.B. HINTZ, R. LUCIC, and B. PREDEL, in: Phase Transformation in Solids, Ed. T. TSAKALAKOS, Mater. Res. Soc. Symp. Proc. 21, 513 (1984), North Holland Publ., Co. Amsterdam 1984.

/6/ W. GUST, J. BEUERS, J. STEFFEN, S. STILTZ, and B. PREDEL, Acta metall. 34, 1671 (1986).

/7/ R.W. BALLUFFI, in: Diffusion in Crystalline Solids, Ed. G.E. MURCH and A.S. NOWICKI, Academic Press, New York 1984 (p. 319).

/8/ W. GUST, M.B. HINTZ, A. LODDING, H. ODELIUS, and B. PREDEL, Acta metall. 30, 75 (1982).

/9/ R. WIRTH and H. GLEITER, Acta metall. 29, 1825 (1981).

/10/ D. TURNBULL and H.N. TREAFTIS, Acta metall. 3, 43 (1955).

/11/ W. GUST, PhaseTransformations, Vol. 1, Inst. Metallurgists, London 1979 (p. II).

/12/ J.W. CAHN and H.N. TREAFTIS, Trans. AIME 218, 376 (1960).

/13/ Y.C. LIU and H.I. AARONSON, Acta metall. 16, 1343 (1968).

/14/ M. HILLERT, Mechanism of Phase Transformation in Crystalline Solids, Inst. of Metals, London 1969 (p. 231).

/15/ K.N. TU and D. TURNBULL, Acta metall. 15, 1317 (1967).

(Received March 4, 1988)

phys. stat. sol. (a) 108, K25 (1988)

Subject classification: 73.25; S5.11

Istituto di Fisica Superiore, Universita degli Studi, Torino[1] (a) and
Istituto Elettrotecnico Nazionale Galileo Ferraris, Torino[2] (b)

# Frenkel Effect in SIPOS Films

By

C. MANFREDOTTI (a), F. FIZZOTTI (a), and G. AMATO (b)

Introduction In the recent years, semi-insulating polycrystalline silicon (SIPOS) has been of great interest in power /1/ and IC /2,3/ device passivation.

In fact, such layers inhibit the insurgence of high electric fields at the silicon surface because the equipotential lines are relaxed with respect to the case of an $SiO_2$ passivation. A problem arises when such resistive films are applied to power devices: in fact conduction through the film becomes important. In many cases the leakage current through the film exceeds specifications /4/.

This current depends on the electric field and when it approaches the breakdown value, the behaviour is well described by the Frenkel model.

In this work we investigate this process in films of different oxygen content and propose a model which explains the voltage dependence of the current in the region near breakdown. The question relative to the contribution of this effect to the lowering of the dielectric strength is presented and discussed.

Experimental The samples are grown by the LPCVD method using $N_2O$ and $SiH_4$ at pressure of 150 mTorr in the reactor and deposition temperature of 650 $^{o}C$.

The oxygen content depends on the gas volumic flow ratio $\gamma = (N_2O)/(SiH_4)$.

The films are grown on silicon wafers, ⟨111⟩ oriented, n-type /5/, optically polished, and of resistivity ranging between 0.01 and 0.1 Ω cm.

The wafers are placed on a quartz boat containing also 100 dummy wafers in a zone of the silica tube in which the temperature profile is reasonably flat ($\pm 1$ $^{o}C$).

Different values of $\gamma$ are obtained by keeping constant the $SiH_4$ flow at 40 s $cm^3$ and by varying the $N_2O$ flow.

The pressure in the tube slightly depends on the $\gamma$ value.

An extreme care is employed in cleaning the substrates before deposition in order to avoid both surface contamination and formation of an $SiO_2$ layer between Si and SIPOS: in that case, the electric properties of the film will be influenced

1) Corso M. D'Azeglio 46, I-10125 Torino, Italy.

2) Strada delle Cacce 91, I-10135 Torino, Italy.

by interface effects. A buffered HF solution is employed followed by rinsing in deionized water.

After deposition, the wafers are masked on one side and etched by an $HNO_3$:$CH_3COOH$ solution on the other till to the complete etch of the SIPOS film.

Infrared measurements are made on samples of known thickness in order to determine by interference fringes the refraction index and the dielectric constant.

Several Au contacts, 300 nm thick and with a diameter of 2 mm are deposited by an electron beam at each wafer side in order to obtain an Au-SIPOS-Si-Au structure. Maximum care is employed in aligning opposite contacts.

The sample is contained in a copper box connected to ground, and a Keithley 616 multimeter with a Fluke power-supply are used to determine the J-U characteristics.

A typical characteristic is obtained with a ramp voltage of approximately 10 V/min, so that the current reaches a steady state at each voltage step.

Results and discussion Typical I-U characteristics are shown in Fig. 2 for different samples grown at different $\gamma$ ratios.

We have taken into account two regions of these curves, respectively, the region at low voltages (i.e. 0 to 1 V) and the region at high voltages just before the breakdown process.

In the first region, characteristics behave as ohmic: in fact the plots of log I versus log U show a slope near 1 (Fig. 1). This occurs either for a reverse bias or a forward bias. This fact can be explained by observing that at low voltages the conduction in the film has the same behaviour either for electron injection or hole injection. This "symmetric" behaviour is not conserved at high voltages: in fact we detect lower values of the dielectric strength when conduction is forward biased, as to say that we have a field-enhanced thermal excitation of trapped electrons.

In Table 1 are reported the values of resistivity of the samples in the ohmic region.

As we expect, there is an increase of resistivity of the film with the percentage of oxygen in the film.

At high voltages the model that explains the experimental result is the Frenkel effect /5 to 7/.

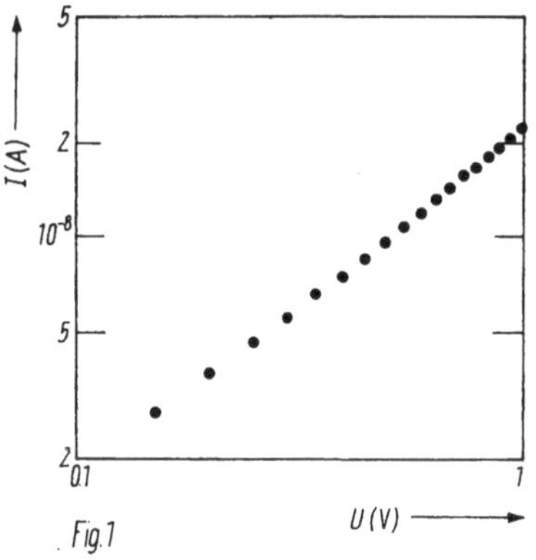


Fig.1

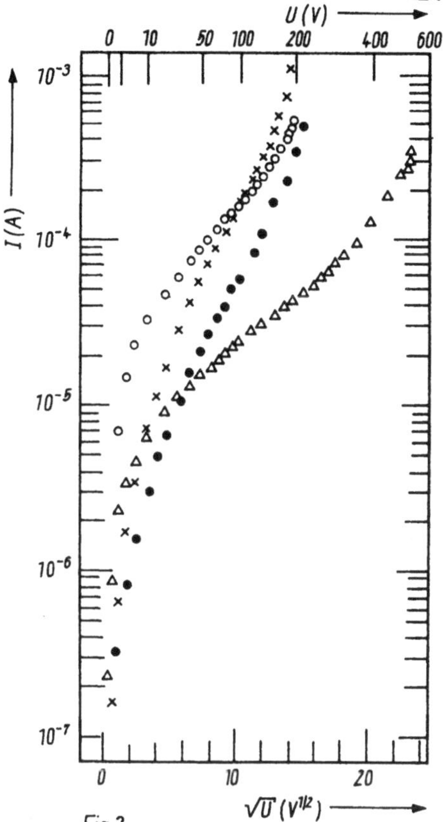


Fig.2

Fig. 1. A logarithmic plot of I-U characteristics of a sample ($\gamma$ = 0.2) for low bias. The slope near 1 shows that the conduction mechanism in SIPOS is ohmic at low voltages

Fig. 2. I-U characteristics at high voltage for four samples of different thicknesses and $\gamma$ ratios: $\gamma$ = 0.1 ( o ), 0.15 ( × ), 0.2 ( Δ ), 0.3 ( • ). The plot shows the linear dependence of log I versus $\sqrt{U}$, at high voltages according to the Frenkel model

In Fig. 2 are reported the values of I versus $\sqrt{U}$ for four samples of the same $\gamma$ values of those previously reported. In these diagrams the last points are fitted with the well-known Frenkel law /10, 11/:

$$J = C\, E \exp(-U_o/kT) \exp\left[(e^3 E/\pi\varepsilon_o \varepsilon_r)^{1/2}/kT\right], \qquad (1)$$

where E is the electric field, $U_o$ the barrier potential of the traps, kT is the thermal energy, C the conductivity which depends on trap concentration, e the electron charge, $\varepsilon_0$ permittivity of free space, $\varepsilon_r$ dielectric constant of SIPOS at high frequency.

Just after the last U value reported in Fig. 1 destructive breakdown occurs in the film and a very high current is detected. So the last U value normalized to film thickness is defined as the dielectric strength.

The dielectric strength values obtained are reported in Table 1.

Fits of the experimental data following (1) have very good correlation coefficients ranging between 0.95 ($\gamma$ = 0.3) and 0.99 ($\gamma$ = 0.15) and the values of $\varepsilon_r$ calculated by the slope of ln I/U versus $\sqrt{U}$ are reported in Table 1.

Comparison with the refractive indices calculated by interference fringes in IR is reported.

For the present case we consider traps defined by a Coulomb potential /7/, substituting at the electric field the dielectric strength values in order to evaluate the dimension of traps. Thus

$$r_0 = \left( \frac{e}{4\pi \varepsilon_0 \varepsilon_r E_{max}} \right)^{1/2} , \quad (2)$$

where $r_0$ is the radius of traps.

Using this approximation we find that the diameter of the traps is roughly 2 to 2.5 nm for all the samples. Assuming /7/ a carrier velocity of the order of $10^7$ cm/s, transit times for this distance should be about $10^{-14}$ s corresponding to a wavelength of 6 $\mu$m.

IR dielectric constant values are measured by interference fringes in the range between 4 and 16 $\mu$m, so the two measurements are perfectly consistent.

Table 1

Summary of main features of SIPOS films at different $\gamma$ ratios

| $\gamma$ | $\varrho$($\Omega$cm) | $E_{max}$(V/cm) | $\varepsilon_r^F$ | $n^F$ | $n_{IR}$ | $U_0$ (eV) |
|---|---|---|---|---|---|---|
| 0.10 | $6.9\times10^7$ | $1.3\times10^6$ | 11.98 | 3.46 | 3.4 | 0.54 |
| 0.15 | $5.4\times10^8$ | $8.0\times10^5$ | 12.59 | 3.55 | 3.3 | 0.56 |
| 0.20 | $4.2\times10^8$ | $7.4\times10^5$ | 11.51 | 3.39 | 3.3 | 0.56 |
| 0.30 | $5.3\times10^9$ | $9.0\times10^5$ | 27.48 | 5.24 | 2.7 | 0.50 |

It is important to notice that agreement is very good, taking into account that fitting by the Frenkel law one often obtains only an estimate.

The second important quantity in the Frenkel effect is the barrier height $U_o$ in (1).

Writing (1) in the following form:

$$\ln I/U = C_1 U^{1/2} + \ln C \frac{S}{d} - U_o/kT \quad , \tag{3}$$

where $C_1 = (e^3/\pi\varepsilon_o \varepsilon_r d)^{1/2}/kT$ (d is the thickness of the sample and S is the area of the contacts) from the intercept it is possible to obtain the value of $U_o$, if the C value is known.

If we suppose /7/ its value to be of the order of $10^{-1}$ A/mV we obtain the barrier height $U_o$ reported in Table 1.

The values are consistent with the hypothesis that oxygen in the film creates deep traps in the middle of the Si energy gap.

The current transport mechanism in SIPOS films is found to be controlled by deep oxygen traps.

We make the assumption that the trap concentration increases with the $\gamma$ value. Therefore the increase of the low-field resistivity with $\gamma$ (Table 1) is consistent with the hypothesis that the low-field current is limited by traps.

At high fields Frenkel emission occurs and experimental results are consistent with other measurements.

The Frenkel effect gives a dramatic contribution to the lowering of the dielectric strength of the film. In fact, we consider for the high-field conductivity the following relationship /6/:

$$\delta = Ne\mu \left\{ 1 + (N_T/N) \exp\left[ (e^3 E/\pi\varepsilon_o \varepsilon_r)^{1/2}/kT \right] \exp\left[ -U_o/kT \right] \right\}, \tag{4}$$

where $\delta = E/J$, $N_T$ is the trap concentration, Ne $\mu$ the charge concentration and mobility product.

There is a direct dependence between J and $N_T$, i.e., in the Frenkel region, breakdwoen is reached more easily in films with a higher trap concentration. This fact can explain the values of $E_{max}$ reported in Table 1 except for the sample with $\gamma = 0.3$.

This sample shows near breakdown a behaviour that is substantially different from the others. For this $\gamma$ value the simple model in which oxygen traps are embedded in a Si matrix cannot reasonably explain the electrical conduction.

In IR measurements this sample shows "$SiO_2$ character" /9/. The peak at K = 1050 $cm^{-1}$ corresponding to the Si-O bonding energy is much larger than in the other samples, i.e. the film shows an amorphous structure, and the refractive index tends to the $SiO_2$ value.

Conclusions We have shown, in good agreement with other authors /5, 6, 8/, that for $\gamma$ ranging between 0.1 to 0.2, SIPOS maintains at low voltages a constant conductivity. This conductivity assumes a dependence on the trap concentration at high field according to the Frenkel model.

These traps have a dimension of roughly 4 to 5 Si lattice constants and can be considered as insulator islands embedded in the Si matrix.

Our model is not confirmed in samples with $\gamma$ = 0.3 and this fact is due to a structural difference of these films which may assume an "$SiO_2$ character".

Thermal excitation of trapped electrons have a dramatic importance in lowering the dielectric strength of the film.

References

/1/ R. DENNING and J. WHITE, RCA Engng. 24, 66 (1979).

/2/ T. MATSUSHITA, T. AOKI, T. OTSU, H. YAMOTO, H. AYASHI, M. OKAYAMA, and W. KAWANA, Proc. 7th Conf. Solid State Devices, Tokyo 1975, in: Japan. J. appl. Phys., Suppl. 15, 35 (1976).

/3/ H. MOCHIZUKI, T. AOKI, H. YAMOTO, M. OKAYAMA, M. ABE, and T. ANDO, see /2/ (p. 41).

/4/ R.B. COMIZZOLI, H.T. WESTON, Y.H. WONG, and J.E. KOHL, J. Electrochem. Soc. 133, 8 (1986).

/5/ J.N. SANDOE and J.R. HUGHES, Proc. Internat. Conf. INFOS 85, Toulouse 1986 (p. 271).

/6/ R.B. COMIZZOLI and R.L. OPILA, J. appl. Phys. 61, 1 (1987).

/7/ S.M. SZE, J. appl. Phys. 38, 7 (1967).

/8/ D.W. TONG, J.L. BENJAMIN, and W.R. VAN DELL, IEEE Trans. Electron. Devices 33, 6 (1986).

/9/ C. MANFREDOTTI, G. AMATO, S. LAVAGNINO, G. BERTANA, and A. BELLONE, 4th French, Italian, Swiss Workshop, Zurich, October 1985.

/10/ C.A. MEAD, Phys. Rev. 128, 5 (1962).

/11/ J.C. SIMMONS, Phys. Rev. 155, 3 (1967).

(Received Februar 8, 1988; in revised form May 11, 1988)

phys. stat. sol. (a) 108, K31 (1988)

Subject classification: 61.80 and 79.20; S12

Department of Physics, Bharathiar University, Coimbatore[1)]

# Laser-Induced Damage Study of Vacuum Deposited Perylene Thin Films

By

A. SUBBARAYAN[2)], C. BALASUBRAMANIAN,
and SA.K. NARAYANDASS

Optically induced damage to the laser material and to components of systems generally determine the limit of useful performance of high power solid state lasers /1/. Nowadays increasing interest is seen in the fabrication of organic dye lasers /2/ and the study of laser induced damages in thin films /3/. Perylene is a promising active laser material due to the four-level fluorescing system of that molecule /4/. Damage can be defined as a permanent induced change in a bare or coated optical surface which degrades the merit of the surface with regard to the specifications such as scattering, reflection, or transmission. The weakest link in a well designed laser is the surface of the optical components. The survivability of thin film coatings to laser beam is of great importance in the reliable and efficient operation of high power laser systems. Due to surface imperfections and contamination, it is the surface of materials which generally fails first /5/. Therefore a knowledge of accurate and reliable measurements of damage thresholds of peryline thin films can be useful in the design of organic dye lasers in thin film form.

Perylene (99 +%, Aldrich Chemicals Company, USA) was evaporated from a molybdenum boat onto pre-cleaned glass substrates, under vacuum of $1.33\times 10^{-3}$ Pa in a conventional vacuum coating unit. The thicknesses of the films were measured by a multiple beam interferometer (Fizeau fringes) and the structure was identified as crystalline by X-ray diffraction.

Three stages are involved in the measurement of the laser damage threshold. The first stage is to irradiate the sample at several flux levels, some of which induce damage; the second is to measure the absolute characteristics of the pulse; and the third is to determine which shot in the sequence caused damage.

---

1) Coimbatore 641046, India.

2) Permanent address: Department of Physics, Kongunadu Arts and Science College, Coimbatore 641 029, India.

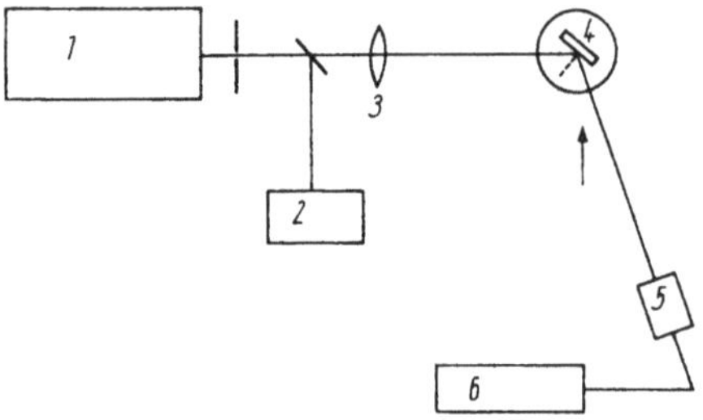


Fig. 1. Schematic diagram of laser-induced damage threshold measurement; 1 Q-switched Nd:glass laser, 2 calorimeter, 3 convex lens, 4 sample, 5 photo diode, 6 oscilloscope

It is essential that the spatial and temporal profile of the laser remains a constant during the entire operations since there is no way to determine whether an unexpected result obtained, was due to the fluctuation in the laser or an exceptional property of the damage sample.

Fig. 1 shows a schematic diagram of the experimental set-up to measure the damage threshold /6/. The laser used in this study was a Nd:glass laser Q-switched by Kodak 14015 dye in 1,2 dichloroethane with a pulse width of about 25 ns (FWHM). The beam structure was multimode and was approximately Gaussian in profile with a diameter of about 4 mm. The laser output was passed through a biconvex lens of 17 cm focal length, onto the thin film sample. The sample mounted on a rotating platform was oriented so that the angle of incidence of the laser beam was $56^{o}$, thereby avoiding the multiple beam interference in the glass substrate. The laser pulse width was monitored on a Tektronix Model 466 IM 44 storage oscilloscope with a Hewlett Packard $HP_2$ 4207 photo diode. The energy incident on the thin film sample was monitored by a 1 inch Scientech disc calorimeter Model 38-0101.

Throughout the study, the laser output energy was kept constant. With the present set-up absolute energy measurement was not possible for every damage attempt. However, an energy calibration was made before and after each set of damage tests. To begin with, the distance between the sample and the lens was adjusted such that the shot impinging on the film damages it. Then the film sample was pushed in the direction indicated by the arrow in Fig. 1, to enable the next shot to impinge on a site adjacent to the previous one. The film position is thus shifted, as it moves away from the focus of the lens the power density incident on the sample is decreased. Once the damage threshold is attained, with the subsequent shots there will be no damage at all. On completion of the test run, the damage sites were examined with an optical microscope and

Table 1

Threshold energy density of the samples

| No. | film thickness (nm) | threshold energy density ($J/cm^2$) |
|---|---|---|
| 1 | 147 | 19.880 |
| 2 | 204 | 7.170 |
| 3 | 295 | 4.971 |

the threshold damage location was made. Once the shot which caused the threshold damage is identified, one can calculate the threshold energy for the film material by knowing the area of the damaged site from the geometrical considerations.

The threshold energy densities estimated for different thicknesses are shown in Table 1.

It is observed from the table that the threshold energy density increases with decrease in thickness. For dielectric films /7, 8/, it has been reported that the threshold energy density increases with absorptance and decreases with film thickness /9/. As the most noticeable feature, unique to the impurity model is the prediction of an increase in damage threshold with a decrease in film thickness, the mechanism followed in this film material may be an impurity dominated one. In this model a spherical absorbing particle embedded in a host material is considered. This impurity absorbs the incident radiation and its temperature rises, which ultimately produces melting, vapourization, or stress fracture of the film material around the impurity. Hence it is quite reasonable that the maximum size of an impurity is limited by the film thickness, i.e. as the film thickness increases as does the impurity size. Hence the threshold energy density increases with decrease in film thickness.

The authors wish to thank Dr. M.G Krishna Pillai (Cochin University of Science and Technology, India) and Dr. P. Radhakrishnan (Cochin College, India) for their valuable help and suggestions throughout the investigation.

References

/1/ W.H LOWDERMILK, Coatings for Laser Fusion, SPIE Proc. 325, 2 (1982).

/2/ M. SADRAI and G.R. BIRD, Optics Commun. 51, 62 (1984).

/3/ K.M DYUMAEV, A.A MANENKOV, A.P. MASLYUKOV, et al., Soviet Phys. - J. Quantum Electronics 13, 503 (1983).

/4/ L.I. BOGUSLAVSKII and A.V. VANNIKOV, Organic Semiconductors and Biopolymers, Plenum Press, New York 1970 (p. 207).

/5/ P. RADHAKRISHNAN, K SATHIANANDAN, and C.P.G. VALLABHAN, Bull. Electrochem. 1, 319 (1985).

/6/ N. SUBHASH and K. SATHIANANDAN, J. appl. Phys. 54, 423 (1983).

/7/ W.H LOWDERMILK, D. MILAM, and F. RAINER, Proc. Symp. Laser-Induced Damage in Optical Materials, NBS Special Publ. No. 568, 1980 (p. 391).

/8/ T.W. WALKER, A.H. GUENTHER, C.G. FRY, and P. NIELSON, see /7/ (p. 405).

/9/ T.W. WALKER, Ph.D. Thesis, Part II, Air Force Inst. Technol., Ohio 1979 (p. 21).

(Received December 7, 1987; in revised form May 3, 1988)

phys. stat. sol. (a) 108, K35 (1988)

Subject classification: 61.70 and 68.48; 78.65; S15.11

Central Research Institute for Physics,
Hungarian Academy of Sciences, Budapest[1)]

# Ellipsometric and Ion Backscattering Measurements of the Properties of Silicon-On-Insulator Structure Formed by Thermally Activated Redistribution of High-Dose Ion Implanted Nitrogen

By

N.Q. KHANH, M. FRIED, G. BATTISTIG, Z. LACZIK, T. LOHNER, E. JÁROLI, V. SCHILLER, and J. GUYLÁI

Introduction In recent years the fabrication of buried insulating layers in silicon has attracted increasing attention. Silicon-on-insulator (SOI) structures are currently being evaluated for use in (VLSI) circuits /1, 3, 4/. Potential advantages of SOI for advanced MOS devices are numerous. These include higher speed, lower dynamic power consumption, greater packing density, increased radiation tolerance, a simpler fabrication sequence in comparison with the bulk approach, and for CMOS, a freedom from latch-up /1 to 4/. One of the promising techniques for SOI fabrication which has been investigated by several groups /3 to 6/, is ion implantation, where a high dose of reactive ions ($^{16}O^+$ or $^{14}N^+$) is implanted at high energies (150 to 500 keV) into silicon. During a successive anneal, these ions react with silicon to form the insulating layer in silicon. For this SOI technology to be viable, it is necessary to achieve self-annealing of the damage caused by implantation in order to retain the single crystal character of the top silicon layer. As some authors reported, the operation of the devices fabricated on SOI structures is a direct function of the top silicon thickness and/or the buried insulator thickness /8, 9/. These thicknesses typically are measured by SEM, TEM imaging, RBS, and lately may be measured by an electrical method /10/.

We applied an alternative method for investigation of the multilayer structure. In this note we report the formation of SOI structure consisting of a buried layer of silicon nitride and the results of measurement of the refractive indices and the thicknesses of SOI layers by multiple-angle-of-incidence (MAI) ellipsometry /14/, which is fast, contactless, non-destructive, and inexpensive. Moreover, the refractive index can be used as a quality control

1) P.O.Box 49, H-1525 Budapest, Hungary.

parameter of the silicon nitride layer /15/. Finally we compare the results of ellipsometry measurements with other methods (RBS).

Experimental Nitrogen ions ($^{14}N^{+}$) were implanted at 200 or 400 keV into ⟨100⟩, p-type (12 Ω cm) single crystal silicon using the 500 kV ion implanter at the Central Research Institute for Physics, Budapest. To ensure hot implantation conditions a special heater and sample holder was designed and manufactured. The substrate temperature was maintained at about 600 °C during implantation and was measured by a chromel-alumel thermocouple. The samples were given a dose of $7.5\times10^{17}$ $N^{+}/cm^{2}$ with a current density of 2 $\mu A/cm^{2}$ on a 2x2 $cm^{2}$ surface. After implantation, thermal annealing was performed at 1300 °C for 2 h in dry Ar atmosphere, which may eliminate a nitrogen related heavy doping of the silicon surface layer adjacent to the higher nitrogen content insulating region /11/. Both of the as-implanted and annealed samples were examined by Rutherford backscattering spectroscopy (RBS) with 2 MeV $He^{+}$ ions and by ellipsometry (LEM-2) with He-Ne laser source ( $\lambda$ = = 632.8 nm) at five different angles of incidence (50°, 55°, 60°, 65°, 70°).

Results and discussion Fig. 1a shows RBS spectra of 2 MeV $He^{+}$ ions from a sample (No. 1) implanted with dose of $7.5\times10^{17}$ $N^{+}$ $cm^{-2}$ at 200 keV energy. The random RBS spectrum for a sample prior to annealing is marked with (●). The nitrogen depth distribution may be inferred from the shape of the dip between channels 187 and 255, and from the shape of the peak between channels 70 and 135. This distribution is non-Gaussian and it has a higher concentration nitrogen tail toward the surface of silicon than toward the bulk of silicon. The channeled (aligned) spectrum(+) is from an as-implanted sample confirming the silicon overlayer to be highly defective (# 235 to 266). However, due to heating of the target during implantation, a partial self-annealing took place leaving a thin layer of silicon (# 266 to 270) in form of a good single crystal. After annealing at 1300 °C for 2 h, the redistribution of the implanted nitrogen has occurred. Both of the interfaces moved toward the nitrogen rich layer. This movement was more pronounced at the upper interface. Thus, in the layer corresponding to channels 204 and 235, the nitrogen concentration increased to form a uniform, well defined nitride layer with more abrupt interfaces. This is supported by the results of ellipsometric measurements, since the obtained refractive index is close to 2.02. The redistribution of nitrogen is also observable on the nitrogen peaks (#75 to 120). The annealed nitrogen profile is smoother and sharper than the as-implanted one. Comparing curves (+) and (×) a significant improvement of the upper silicon layer is observed. Most of the defects have disappeared. The minimum yield between channels 250 and 265 is 0.06 (×). This is a measure of the reasonably improved crystalline quality.

From RBS spectra the thickness of the upper silicon layer was calculated to be 350 nm /12/. The determination of the exact thickness of the nitride layer is a problem due to the lack of an adequate atomic density value in the literature ture. Generally, the thicknesses of SOI structures are measured by TEM or SEM imaging. In our study we have used the multiple-angle-of-incidence (MAI) ellipsometry /14/ to measure these thicknesses. For evaluation of ellipsometry data, a fitting program was used on the five-phase optical model with abrupt interfaces: air, native oxide, the top single crystal silicon, the buried nitride layer, and bulk silicon. For the top layer, the 354.5 nm thickness and the 3.89-j0.04 complex refractive index were obtained (in contrast to the 3.88-j0.02 complex refractive index of virgin silicon). This reveals a nearly perfect quality of this layer. For the buried nitride layer the thickness and the refractive index are 176.5 nm and 2.03-j0.0, respectively. The experimental data characteristic of the annealed samples are listed in Table 1. The evaluated ellipsometric results are good to such a degree as the applied five-phase optical model is good. For example the "native oxide" layer may be a several nm thick rough surface covered by silicon oxide. However, the reliability of ellipsometric results can be judged by the layer thicknesses measured with independent methods. For comparison, we insert the thickness values measured by Hemment et al. on a sample processed under almost identical conditions /13/. There is a reasonable agreement between the thickness data obtained by ellipsometry, RBS, and their TEM investigations even though the results were obtained in two laboratories (CRIP, Budapest and University, Surrey, Guilford) and with two implanters.

The atomic density can be determined from RBS and ellipsometric measurements. RBS can tell the amount of atoms, ellipsometry can tell the thickness of the layers. From these, one can calculate the atomic density of this nitride layer. The result is $7.2\times10^{22}$ atoms/cm$^3$. This means 2.4 g/cm$^3$ mass density compared to (2.4 to 3.1)g/cm$^3$ in the literature /17/.

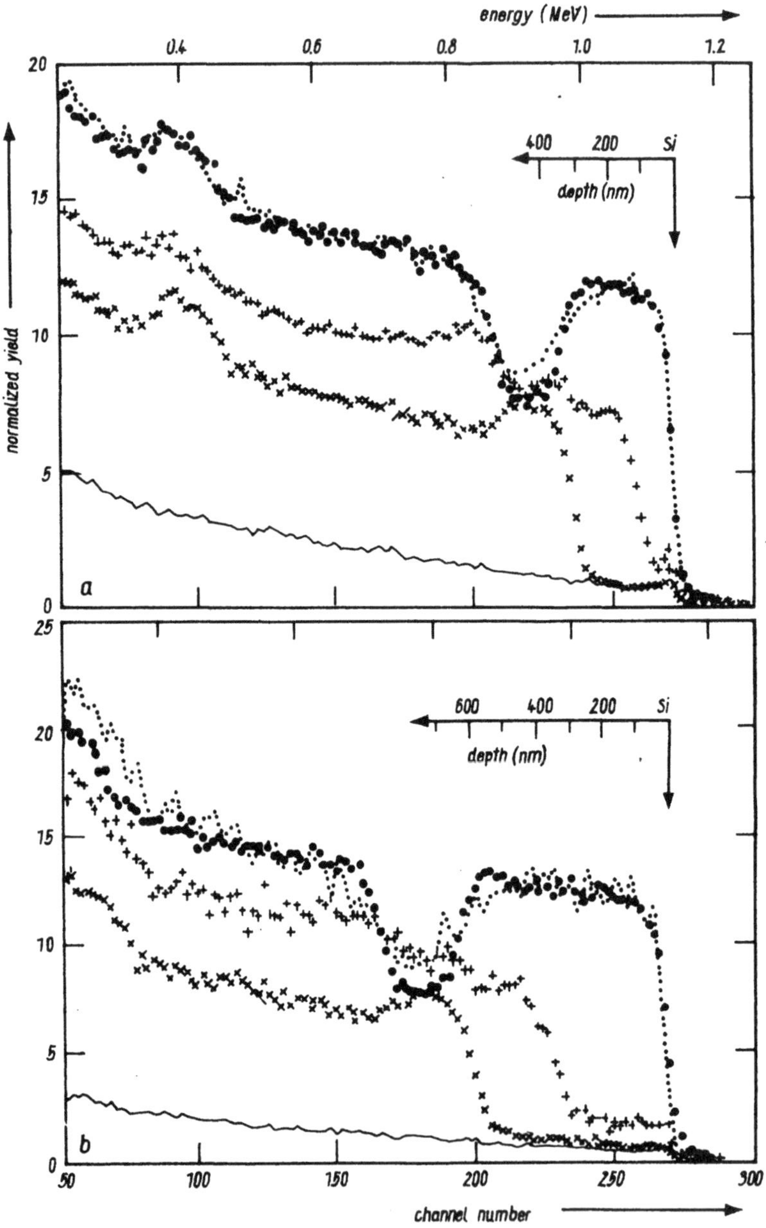


Fig. 1. RBS spectra of a) sample No. 1, b) sample No. 2. a) 200 keV, $0.75 \times 10^{18}$ $N^+ cm^{-2}$, $T_i$ = 600 °C, $T_a$ = 1300 °C in Ar for 2 h; b) 400 keV, $0.75 \times 10^{18}$ $N^+ cm^{-2}$, $T_i$ = 600 °C, $T_a$ = 1300 °C in Ar for 2 h; • as-implanted random, + as-implanted aligned, o annealed random, × annealed aligned, —— Si aligned

Table 1

Collected results by ellipsometry, RBS, and TEM on the sample No. 1 (upper data) and the sample No. 2 (lower data), rspectively

| | ellipsometry | RBS | TEM /13/ |
|---|---|---|---|
| thickness of top silicon (nm) | 354.5<br>603 | 350<br>600 | 320 |
| complex refractive index of top silicon | 3.89-j0.04<br>3.89-j0.02 | | |
| thickness of buried nitride (nm) | 176.5<br>250 | | 180 |
| complex refractive index of buried nitride | 2.03-j0.0<br>2.57-j0.0 | | |

Fig. 1b shows RBS spectra of a sample (No. 2) implanted at 400 keV with the same dose as sample No. 1. This sample is an example of an incomplete buried nitride layer. The as-implanted nitrogen depth profile is deeper and wider than in the case of sample No. 1. Because of this the top is thicker. The thickness of the nitride layer is also larger, therefore the N:Si ratio is lower. The ellipsometric measurement confirms this, the refractive index of this nitride layer being 2.57 (see Table 1). Presumably, it is a mixture of silicon and silicon nitride.

We have shown that multiple-angle-of-incidence ellipsometry with a simple five-phase optical model (air, surface oxide, c-silicon, silicon-nitride, bulk silicon) is suitable for a non-destructive, rapid checking of SOI structures formed by ion beam synthesis. To control the as-implanted samples it is necessary to measure at different wavelengths or to use spectroscopic ellispometry /16/, since the layers in this case are inhomogeneous and absorbing.

The authors are grateful for the precise sample preparation to Dr. J. Gyimesi, Mr. J. Waizinger, and Mr. F. Gazdácska. The authors have used the program RUMP by L. Doolittle to produce the graphs used in this note. The work was partially supported by the Hungarian Academy of Sciences OTKA grant No. 1794.

## References

/1/ H.W. LAM, R.F. PINIZOTTO, and A, F. TASCH, Jr., in: VLSI Handbook, Academic Press, 1985 (p. 503).

/2/ H.W. LAM, A.F. TASCH, Jr., and R.F. PINIZOTTO, in: VLSI Electronics: Microstructure Science, Vol. 4, Academic Press, 1982 (p. 1).

/3/ K. IZUMI, M. DOKEN, and H. ARIYOSHI, Electronics Letters 14, 593 (1978).

/4/ G. ZIMMER and H. VOGT, IEEE Trans. Electron. Devices 30, 1515 (1983).

/5/ O.W. HOLLAND, T.P. SJOREEN, D. FATHY, and J. NARAYAN, Appl. Phys. Letters 45, 108 (1984).

/6/ G.J. REESON, P.L.F. HEMMENT, R.F. PEART, C.D. MEEKISON, C. MARSH, G.R. BOOKER, R.J. CHATER, J.A. KILNER, and J. DAVIS, Radiat. Eff. 99, 71 (1986).

/7/ L. NESBIT, G. SLUSSER, R. FRENÈTTE, and R. HALBACH, J. Electrochem. Soc. 133, 1186 (1986).

/8/ H.K. LIM and J.G. FOSSUM, IEEE Trans. Electron. Devices 30, 1244 (1983).

/9/ J.P. COLINGE, Electronics Letters 22, 187 (1986).

/10/ J. WHITFIELD and S. THOMAS, IEEE Trans. Electron. Devices Letters 7, 47 (1986).

/11/ D.E. DAVIES, J.A. ADAMSKI, and E.F. KENNEDY, Appl. Phys. Letters 48, 347 (1986).

/12/ E. KOTAI, Program for Measure, Evaluation, and Simulation of RBS Spectra in PCs, 1985.

/13/ P.L.F. HEMMENT, R.F. PEART, M.F. YAO, K.G. STEPHENS, R.J. CHATER, J.A. KILNER, D. MEEKISON, G.R. BOOKER, and R.P. ARROWSMITH, Appl. Phys. Letters 46, 952 (1985).

/14/ R.M.A. AZZAM and N.M. BASHARA, Ellipsometry and Polarized Light, North-Holland, New York 1977.

/15/ H. WATANABE, K. KATOH, and S. IMAGI, Thin Solid Films 136, 77 (1986).

/16/ J. NARAYAN, S.Y. KIM, K. VEDAM, and R. MANUKONDA, Appl. Phys. Letters 51, 343 (1987).

/17/ A.C. ADAMS, in: VLSI Technology, Ed. S.M. SZE, McGraw-Hill, New-York 1983 (p. 93).

(Received March 29, 1988; in revised form June 17, 1988).

phys. stat. sol. (a) 108, K41 (1988)

Subject classification: 73.40; S7.11

Centre Lorrain d'Optique et d'Electronique des Solides, ESE, Metz[1] (a), Solid State Section, University of Athens[2] (b), and Laboratoire d'Electronique et de Physique des Interfaces, ESSTIN, Vandoeuvre[3] (c)

# Determination by Optical DLTS of the Distribution of States near the Valence Band of Plasma Oxidized n-Type InP

By

B. LEPLEY (a), A. BATH (a), C. CARABATOS-NEDELEC (a), G.J. PAPAIOANNOU (b), P.C. EUTHYMIOU (b), and S. RAVELET (c)

The indium phosphide devices are promising for high speed optoelectronic integrated circuits /1/. Interface stability is the key for the development of MISFET transistors /2/. It is of prime interest to know the interface state distribution between the semiconductor and the gate insulator. Such a determination has been extensively carried out in the upper half of the InP band gap, mainly with the Terman method, in order to analyze the pinning of the Fermi level and its consequence on the operation of the n-channel inversion MISFET. Many authors have measured the interface state distribution across the band gap, especially for deposited thick oxides on InP and only few authors have published experimental data for the lower half of the band gap and precisely for the near valence band states; among them, Hasegawa and Sawada have suggested the existence of a non-localized state region in a limited energy range, 0.05 to 0.1 eV near the valence band edge, leading to a band-gap reduction at the interface /3/. They have used anodic $Al_2O_3$ insulator on InP. All other results in the literature show an increase of the density of states as the energy approaches the valence band value, whatever the dielectric used /4 to 7/. Such a high density of states, also found in GaAs, has been related by Hasegawa to the formation of a highly disordered non-stoichiometric semiconductor region giving that particular interface state distribution.

Plasma bilayer native oxide has been shown to give an interesting approach and promising results for InP MISFET gate insulator /8/; hence the purpose of

1) 2 Rue E. Belin, F-57078 Metz, France.

2) 104 Solonos Str., G-10680 Athens, Greece.

3) Parc Robert Bentz, F-54500 Vandoeuvre, France.

this note is to study by low temperature optical DLTS the interface state distribution in the lower part of the energy gap on MIS structures elaborated with rf plasma oxidation. For such structures, the density of states near the conduction band has been studied by the conductance method and has pointed out the interest of the bilayer oxide /9/.

Commonly used on Si-MOS structures for minority carrier trap distribution /10/, optical DLTS has been used on these InP MIS devices at very low temperature to analyze states near the valence band and to compare the results to published data with different insulating layers. UPS measurements on InP surface have shown that the edge value of the U-shaped density of states was close to $10^{14}$ $cm^{-2}$ $eV^{-1}$ depending on the surface treatment /11/. These states play an important role for the luminescence efficiency because they have been shown to govern the recombination process. Such a high density would appear on the low temperature optical emission transient for holes.

Native oxide of 13 nm thickness has been grown on (100) n-type InP at 573 K by low power rf plasma oxidation in pure oxygen. Au-Ge is evaporated to make the back ohmic contact and evaporated gold is used for the semitransparent top electrode. The native oxide process involves two oxidation sequences with an intermediate annealing. The first oxide layer is phosphorus rich and the top one is aluminium rich /9/. The net donor density in the undoped InP substrate is $N_D = 1.0x10^{22}$ $m^{-3}$.

The gate electrode is illuminated with a pulse of light having a photon energy higher than the band-gap energy, obtained by a 3 mW solid state AlGaAs laser emitting at a central wavelength of 820 nm and concentrated on the sample. Only the surface depletion region is probed with such an energy, owing to the high value of the absorption coefficient in InP, $\alpha = 3x10^4$ $cm^{-1}$. Minority carriers photogenerated within a diffusion length of the surface may be captured by the interface states and the emission of holes from these states gives the transient capacitance signal which is analysed by the lock-in DLTS technique. The different sequences are shown in Fig. 1. The total gate off time is $2t_{go} = T_0/10$ and the optical pulse width is $T_0/20$, $T_0$ being the period of the laser pulse. Complete filling of the interface states was achieved by changing the illumination intensity with neutral filters and observing the saturation of the DLTS signal. ODLTS spectra have been obtained in the 20 to 330 K range for

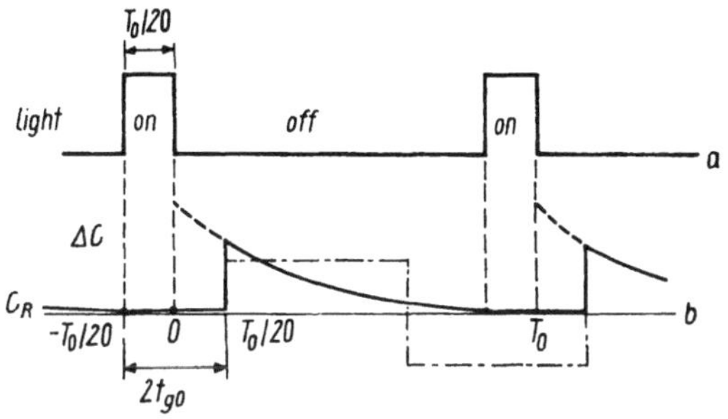


Fig. 1. Illustration of the ODLTS sequences: a) light intensity, b) capacitance transient and lock-in reference signal as the weighting function

different frequencies between 6 and 86 Hz. A typical spectrum is given in Fig. 2 and shows a continuous increase of the signal as the temperature is decreasing.

The density of states $N_{ss}(E)$ is calculated from the DLTS signal Y

$$Y = AS_1 f_o \int_{E_v}^{E_c} N_{ss}(E) J(E) dE \ , \tag{1}$$

where $S_1$ is the illuminated transparent area and

$$A = C_R^3/\varepsilon_s N_D C_{ox} S_T^2 \tag{2}$$

with $C_R$ the steady-state capacitance, $C_{ox}$ the oxide capacitance, $S_T$ the total

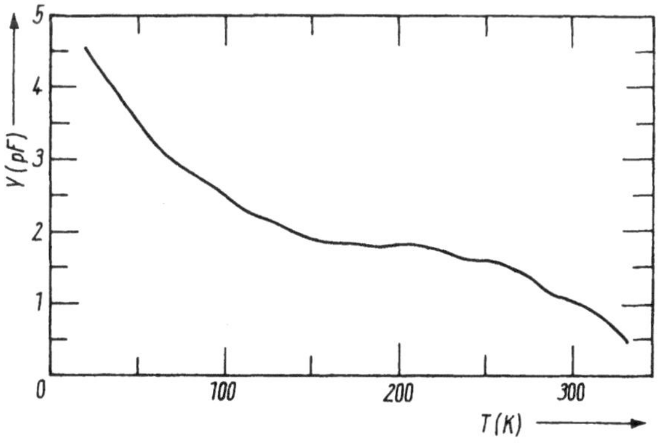


Fig. 2. ODLTS spectrum obtained on n-type InP plasma bilayer native oxide. Frequency 86 Hz

area of the gate, and $\varepsilon_s$ the permittivity of the semiconductor.

$$J(E) = (1/e_P)\exp(-e_P t_{go})(1 - \exp(-e_P T_o/2))^2 \tag{3}$$

is the probe function. $e_P$ is the hole emission rate given by

$$e_P = \sigma_P \nu_{th} N_v \exp(-(E - E_v)/k_B T)) = f_o/0.46 \tag{4}$$

with $\sigma_P$ the capture cross section, $\nu_{th}$ the thermal velocity of the carriers, $N_v$ the effective density of states in the valence band, and $k_B$ the Boltzmann constant. The interface state distribution is deduced from (1) assuming a slowly varying function, as we can expect from the curve in Fig. 2, and given by

$$N_{ss}(E) = Y \varepsilon_s N_D C_{ox} S_T^2/0.559 C_R^3 k_B T S_1 . \tag{5}$$

The energy scale is related to the temperature and the capture cross section by (4). Only few authors have studied the capture cross section properties for interface states on InP. For deep levels the values of $\sigma_P$ are found of the order of $10^{-24}$ to $10^{-19}$ $m^2$. Only the photoionization cross section has been measured by Hasegawa who has given an empirical relationship between its value and the photon energy. A constant value $\sigma_P = 10^{-20}$ $m^2$ is assumed for our calculations, therefore the energy scale is an approximate one. The maximum horizontal shift is around 0.07 eV near the midgap if $\sigma_P$ is ten times higher or smaller. Near the valence band the error in energy scale is reduced to 0.01 eV for the same ratio at $E - E_v = 0.03$ eV.

The distribution of interface states is plotted in Fig. 3 and compared with four curves found in the literature for various oxides on InP. The maximum value of $2\times10^{14}$ $cm^{-2}$ $eV^{-1}$ at $E - E_v = 0.025$ eV is in good agreement with the value reported by Moison et al. using photoemission on different samples for which the surface was oxygen treated or amorphized by $3\times10^{13}$ $cm^{-2}$ ion bombardment /11/. The shape of the curve shows a sharp increase close to the valence band giving the profile of a region of high density of states with values higher than $5\times10^{13}$ $cm^{-2}$ $eV^{-1}$ for $E - E_v$ smaller than 0.1 eV. This result clearly agrees with Hasegawa's model where a non-localised state region is proposed near the valence band. The origin of such a region is presumably related to the formation of a disordered non-stoichiometric InP surface due to rf plasma, although it has been used with low power /8/. These results

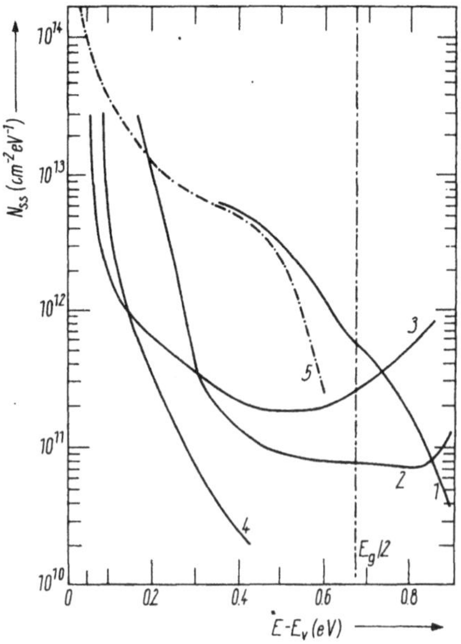


Fig. 3. Interface state distribution in the lower half of the band gap: (1) /4/ $InP_xO_y$ by LPCVD, (2) /5/ $SiO_2$/native oxide by PECVD, (3) /6/ $SiO_2$ by CVD, (4) /7/ BN by LTCVD, (5) present work

show small differences in the shape of the $N_{ss}$ distribution near the valence band when they are compared to other curves in Fig. 3. It is interesting to point out that all curves are joining near the valence band. The differences which appear significant are found in the middle of the lower half of the band gap, however in that region, the error due to the capture cross section uncertainty is maximum. The best agreement is found with the results of Chang et al. /4/ for which a native oxide is also concerned, $InP_xO_y$ and not a deposited one as in the other cases.

All these results seem to be consistent with the model of a disordered semiconductor surface region inducing a high increase of the density of states near the valence band to a value corresponding for InP to one state per surface atom.

In conclusion, optical DLTS at low temperature has shown that the interface state distribution for minority carriers in plasma bilayer native oxide on n-InP near the valence band edge is quite similar to the distribution obtained for deposited gate insulators. The main differences in magnitude appear in the middle of the lower half of the band-gap.

We wish to thank Mr. Bouchikhi (LEPI, Nancy) for providing the samples used in this work.

References

/1/ A. ANTREASYAN, P.A. GARBINSKI, V.D. MATTERA, Jr., and H. TEMKIN, Appl. Phys. Letters 49, 513 (1986).

/2/ S.K. KRAWCZYK, L'Onde Electrique 67, 41 (1987).

/3/ H. HASEGAWA and T. SAWADA, J. Vacuum Sci. Technol. 21, 457 (1982).

/4/ H.L. CHANG, L.G. MEINERS, and C.J. SA, Appl. Phys. Letters 48, 37 (1986).

/5/ K.P. PANDE and G. GUTTIERREZ, Appl. Phys. Letters 46, 416 (1985).

/6/ L.G. MEINERS, J. Vacuum Sci. Technol. 21, 655 (1982).

/7/ E. YAMAGUCHI and M. MINAKATA, J. appl. Phys. 55, 3098 (1984).

/8/ B. BOCHIKHI, C. MICHEL, S. RAVELET, and B. LEPLEY, phys. stat. sol. (a) 101, 173 (1987).

/9/ B. BOCHIKHI, C. MICHEL, O. AITSAADA, S. RAVELET, and B. LEPLEY, Proc. Conf. E-MRS, Strasbourg 1986, Vol. XII, Ed. Phys., Les Ulis 1986 (p. 361).

/10/ Y. TOKUDA and A. USAMI, Japan. J. appl. Phys. 21, L165 (1982).

/11/ J.M. MOISON, M. VAN ROMPAY, and M. BENSOUSSAN, Appl. Phys. Letters 48, 1362 (1986).

(Received April 6, 1988)

phys. stat. sol. (a) 108, K47 (1988)
Subject classification: 71.55; 78.60; S8.12

Institute of Physics, Technical University of Łódź[1)]

# Comments on the Interpretation of Thermally Stimulated Luminescence and Conductivity in Conductive ZnSe Crystals

By
A. OPANOWICZ

In 1981 Oczkowski reported the results of a study of thermoluminescence (TL peak at 180 K containing the red emission band) in conductive n-type ZnSe crystals /1/. In his interpretation of the TL phenomenon the unusually long lifetime of the trapped hole is attributed to the metastable state of donor-acceptor pairs (MDA) /2/. He assumed also that the thermal excitation of electrons from the donor changes the state of the pair and enables the radiative annihilation of the captured hole at the acceptor. In further papers /3 to 5/ Oczkowski developed a theoretical description of the TL curves in conductive semiconductors on the basis of the MDA model.

Another model for the TL in conductive semiconductors, which assumes that the TL is connected with the thermal release of the trapped hole from the deep hole trap (DHT) was proposed in /6/. Because it is not clear why the trapped hole can exist in the metastable state for a long time after excitation in an n-type semiconductor with high density of free electrons, in /7/ it was proposed that the TL occurs in the semi-insulating layer at the surface of the conductive semiconductor. In this model the TL is interpreted by the well-known mechanism of electron release from the deep electron trap (DET) and subsequent recombination in the recombination centre. Using the model with DHT or DET one can obtain as many emission bands of the TL as kinds of radiative recombination centres are active. In particular, these models can also describe the case of the single red emission band of the TL in conductive ZnSe /1/ when the short-wavelength emission bands are thermally quenched.[2)]

---

1) Ul. Wólczańska 219, PL-93-005, Lodz, Poland.

2) Comparing the TL emission spectra with the photoluminescence (PL) spectra measured at approximately equal temperature (see /1/) it should be taken into accound that the temperature range of thermal quenching of the PL shifts to lower temperatures with decreasing excitation intensity (see e.g. /8/). Besides, it is not known what relative intensities of the TL and PL were measured in /1/.

Recently Oczkowski presented new results of calculations of the shape and the position of the TL and the thermally stimulated conductivity (TSC) curves applying the modified kinetic model with MDA and shallow donor /9/. The nature of the centre which is responsible for the TL in conductive ZnSe is not determined in /9/.

It is well-known that the characteristic values of the capture cross-sections of the recombination centre for the free electron ($S_n$) and hole are helpful for the determination of the nature of this centre. In this note we present a simple estimation of the $S_n$ value for the centre (unidentified H-centre, possibly MDA, or DHT, or other centre) which keeps the captured hole for a long time after removing the optical excitation from the conductive n-type ZnSe crystal at 77 K.[3] For this estimation we take the data based on the experimental results of Oczkowski: the density of equilibrium electrons ($n_{oe}$) is taken as $10^{15}$ cm$^{-3}$, and the lifetime ($\tau_a$) of the hole trapped at the H-centre is assumed to be equal to $10^3$ s. Using these data one can calculate the recombination rate constant $\varphi_n$ for the recombination of the free electron with the hole captured by the H-centre at 77 K: $\varphi_n = (\tau_a n_{oe})^{-1} = 10^{-18}$ cm$^3$ s$^{-1}$. Now one can estimate the cross-section of the H-centre for a free electron: $S_n = \varphi_n v_n^{-1} = 10^{-25}$ cm$^2$, where $v_n = 10^7$ cm s$^{-1}$ is the thermal electron velocity at 77 K. The values of $\varphi_n$ and $S_n$ are therefore extremely small. These are about four orders of magnitude smaller than the typical values for the negatively charged recombination centre (see e.g. /11/). It is also clear that if the H-centre is localized in a semi-insulating material having an equilibrium electron density $n_{oe} < 10^{10}$ cm$^{-3}$, a reasonable value of $S_n > 10^{-20}$ cm$^2$ is obtained. Such semi-insulating regions in a conductive semiconductor are connected with energetic barriers (see e.g. /7/ and /11 to 13/). For example, in the region of a relatively low barrier with height $E_b = 0.08$ eV the density of free electrons is reduced by a factor $\exp(E_b/kT) \approx 2\times10^5$ at T = 77 K.

The interpretation of the TSC in conductive ZnSe described in /9/ also raises some doubts. Firstly, it is not explained why a thick conductive ZnSe crystal excited optically in a thin surface layer (see /7/) shows the TSC

---

3) According to measurements by Oczkowski /10/ the strong TL peak at 180 K is detected one day after excitation and keeping of the ZnSe crystal in darkness at 80 K.

phenomenon. Secondly, Oczkowski /9/ made some assumptions which are not compared with the experimental results /1/. He discusses the kinetic model for a conductive semiconductor containing MDA and the shallow donor having an ionization energy $E_d$ = 0.1 to 0.2 eV. These values of $E_d$ are arbitrarily assumed because the donors which most strongly influence the conductivity of ZnSe, have an ionization energy of about 0.04 eV /1/. Oczkowski assumes that the TSC is caused by thermal release of excess electrons from shallow donors which are filled with excess electrons by optical excitation at 77 K. But if these shallow donors are localized in conductive material like ZnSe with equilibrium density of electrons $n_{oe} = 10^{15}$ $cm^{-3}$ at 77 K /1/, they are neutral and cannot act as electron traps. It follows from the position of the equilibrium Fermi level $E_F = kT \ln(N_c/n_{oe}) = 0.034$ eV below the conduction band of semiconductor at T = 77 K ($N_c = 2.3 \times 10^{17}$ $cm^{-3}$ at 77 K /5/). On the other hand, when the equilibrium Fermi level is localized at the shallow donor levels, (i.e. $E_d = E_F$), these donors are partly empty and they can act as electron traps. In this case the density of equilibrium electrons is $n_{oe} = N_c \exp(-E_d/kT) = 3.5 \times 10^7$ $cm^{-3}$ at T = 77 K if $E_d$ = 0.15 eV. Then, this semiconductor is semi-insulating!

According to the Oczkowski model /9/ the neutrality condition for a semiconductor in equilibrium at low temperature has the form $n_{oe} + N_A = N^+_{doe}$, where $N_d$ and $N_A$ are densities of the shallow donors and deep acceptors, respectively. In this equation it is arbitrarily assumed that at low temperature (e.g. T = 77 K) the deep donors associated in MDA are neutral, the acceptors are negatively charged, and the shallow donors are almost totally ionized, i.e. $N^-_A = N_A$, $n_{doe} \ll N^+_{doe} \approx N_d$, where $n_{doe}$ is the density of neutral shallow donors. It should be noted that this neutrality equation is valid for a partly compensated semiconductor at sufficiently high temperatures /12/. For example, in conductive ZnSe crystal with shallow donors ($E_d$ = 0.04 eV) the condition $N^+_{doe} \approx N_d$ is fulfilled at temperature T > 200 K /1/. In addition, Oczkowski /9/ assumed that in steady state after excitation at low temperature the following neutrality equation is valid: $\Delta n_{do} = a_o = N_A$, where $a_o$ is the density of trapped holes by MDA, $\Delta n_{do}$ is the density of excess electrons captured by shallow donors. Because in conductive ZnSe the shallow donors ($E_d$ = 0.1 to 0.2 eV) are in equilibrium occupied by electrons, then the neutrality equations proposed by Oczkowski and discussed above cannot be fulfilled for this material. In conclusion, it seems that the kinetic model with MDA and shallow donor considered by Oczkowski in /9/ is not very effective for describing the TL and TSC in a conductive semiconductor like ZnSe /1/. On the other hand, a critical analysis of this model yields some indications that the TL and TSC in conductive ZnSe /1/ can be explained on the basis of the kinetic model with DET (or DHT or MDA) developed for a semi-insulating material. This explanation seems to be reasonable if the conductive ZnSe crystals studied by Oczkowski contain semi-insulating regions, i.e. if they have a heterogeneous spatial distribution of the equilibrium electron density. These semi-insulating regions may be localized at the surface /7/ and/or in the bulk of the crystal. For example, it is known that conductive ZnSe crystals contain a large number of twins /5/ which create planar defects. These planar defects can easily accumulate native and impurity point defects, and thereby they may produce a heterogeneous electronic structure of the crystal (see also Chapter 2 in /12/ and the review article /13/). It seems, therefore, that in order to explain the TL and TSC in conductive ZnSe crystals further experimental investigations of the electronic structure of these crystals and the TL and TSC phenomena in these crystals are necessary.

References

/1/ H.L. OCZKOWSKI, phys. stat. sol. (a) 68, 199 (1981).

/2/ C.J. GUNSUL and F.E. WILLIAMS, Bull. Amer. Phys. Soc. 11, 227 (1966).

/3/ H.L. OCZKOWSKI, phys. stat. sol. (a) 74, 65 (1982).

/4/ H.L. OCZKOWSKI, phys. stat. sol. (a) 82, 213 (1984).

/5/ H.L. OCZKOWSKI, Termoluminescencja kryształów ZnSe, Uniwersytet Mikłaja Kopernika, Rozprawy, Toruń 1984.

/6/ A. OPANOWICZ, phys. stat. sol. (a) 90, 675 (1985).

/7/ A. OPANOWICZ, phys. stat. sol. (a) 93, K65 (1986).

/8/ M.K. SHEINKMAN, B. ERMOLOVICH, and G.L. BELENKII, Fiz. tverd. Tela 10, 2628 (1968).

/9/ H.L. OCZKOWSKI, phys. stat. sol. (a) 103, 499 (1987).

/10/ H.L. OCZKOWSKI, private communication.

/11/ A.G. MILNES, Deep Impurities in Semiconductors, John Wiley and Sons, New York 1973.

/12/ J.S. BLAKEMORE, Semiconductor Statistics, Pergamon Press, Oxford 1962.

/13/ M.K. SHEINKMAN and A.YA. SHIK, Fiz. Tekh. Poluprov. 10, 209 (1976).

(Received March 21, 1988)

phys. stat. sol. (a) 108, K53 (1988)

Subject classification: 71.55 and 72.80; S8

Department of Physics, Kurume University[1] (a) and
Department of Electronic Engineering, Miyazaki University[2] (b)

# Electrical Properties of Cd-Doped p-InSe

By

S. SHIGETOMI (a), T. IKARI (b), Y. KOGA (a), and S. SHIGETOMI (a)

Much attention has been focused on the optical and electrical properties of the layered III-VI compound InSe, which can be applied to a photovoltaic solar energy converter /1/. Most of the investigations of the electrical properties have been made on not purposely doped n-type InSe and their properties were now known /2 to 5/. There are few studies available for the acceptor level in InSe. Although the two elements of Zn and Cd have been known as dopants for p-type InSe, only the acceptor level associated with Zn has been investigated. An activation energy of 0.28 eV for the Zn acceptor level was found by Hall measurements /6/. We have already reported the emission band due to the Zn acceptor-donor pair recombination in Zn-doped InSe by the photoluminescence measurements /7/.

In this note, we report the results on the temperature dependences of the hole concentration and the Hall mobility in Cd-doped p-InSe, and the acceptor level related to Cd atoms is characterized. Furthermore, the relationship between the mobility and the impurity concentration is discussed.

Single crystals of p-type InSe were grown by a conventional Bridgman technique. Cd of 99.999% purity was doped from 0.1 to 1 at% to the stoichiometric melt of InSe. The results of X-ray measurements showed that these p-type InSe belong to the $\gamma$-polytype as in the case for not purposely doped n-type InSe /8/. The samples were prepared by cleaving an ingot parallel to the layer which is the crystallographic c-plane, and typical sample dimensions were 12x3x0.3 $mm^3$.

Ohmic electrodes for the electrical measurements were made by In evaporation in a vacuum of $2\times10^{-3}$ Pa. The measurements of resistivity and Hall voltage were carried out in a standard configuration and by using a high impedance electrometer in the temperature range from 170 to 390 K. The

1) 1635 Mii-machi, Kurume, Fukuoka 830, Japan.

2) 7710 Kumano, Miyazaki 889-12, Japan.

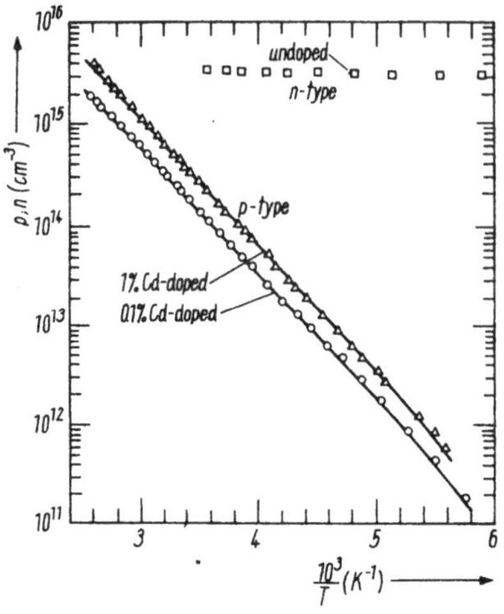


Fig. 1. Carrier concentration as a function of reciprocal temperature for undoped and Cd-doped InSe. The solid lines indicate the calculated hole concentrations from (1)

current flowed parallel to the c-plane and the magnetic field of 0.5 T was applied perpendicular to the layer. The ratio of length to width of the samples was about 4, and no correction was made for the shortening effect of the current electrodes.

The carrier concentrations of undoped and Cd-doped InSe are shown as a function of reciprocal of temperature in Fig. 1. The carrier concentrations of holes p and electrons n were calculated from the Hall coefficient by assuming a Hall factor of unity. For the undoped InSe, n decreased slightly with decreasing temperature. The conduction of undoped InSe is dominated by the shallow donor center at 0.023 eV below the conduction band /4/.

After doping of Cd to the stoichiometric InSe, p-type conduction was observed. The hole carrier concentration p drastically decreased with decreasing temperature. Since the undoped InSe behaves as an n-type semiconductor, the donor level seems to be still active in the p-type samples. Therefore, we analyze the experimental results by assuming that the Cd-doped samples behave as partially compensated p-type semiconductors. In this model, the donor and the acceptor levels are taken into consideration at the same time. Further assumption is made on the expression for the density of states of the valence band $N_V$. Despite the strong structural anisotropy of layer InSe, the anisotropy of the hole effective mass is not large /9/. Therefore, we use the usual three-dimensional expression for $N_V$. The dependence of p on temperature T is thus given by /10/

$$p = \frac{2(N_A - N_D)}{\left[1 + \frac{\beta N_D}{N_V}\exp\left(-\frac{E_A}{kT}\right)\right] + \left[\left\{1 + \frac{\beta N_D}{N_V}\exp\left(-\frac{E_A}{kT}\right)\right\}^2 + \frac{4\beta(N_A - N_D)}{N_V}\exp\left(-\frac{E_A}{kT}\right)\right]^{1/2}} \tag{1}$$

Table 1

Values of $N_A$, $N_D$, and $E_A$ obtained from (1)

| Cd concentration (%) | $N_A$ ($10^{17}$ $cm^{-3}$) | $N_D$ ($10^{11}$ $cm^{-3}$) | $E_A$ (eV) |
|---|---|---|---|
| 0.1 | 2.3 | 1.8 | 0.45 |
| 1 | 7.3 | 5.4 | 0.45 |

with

$$N_V = 4.82 \times 10^{15}\ T^{3/2}\ (m_h/m_o)^{3/2}\ (cm^{-3}),$$

where $E_A$ is the acceptor activation energy, k is Boltzmann's constant, $N_A$ and $N_D$ are the concentrations of the acceptor and the donor, respectively. The hole effective mass ratio of $m_h/m_o = 1.3$ and the degeneracy factor for the acceptor level of $\beta = 2$ were used in these calculations. The best fitted curves for p by using (1) are drawn by solid lines in Fig. 1. The values of $N_A$, $N_D$, and $E_A$ for the fit are shown in Table 1. The deep acceptor level for Cd-doped samples is introduced at 0.45 eV above the valence band.

The Hall mobilities $\mu_H$ in Cd-doped InSe are shown as a function of temperature in Fig. 2. For 0.1% Cd-doped samples, $\mu_H$ increased with decreasing temperature. The mobility $\mu_H$ of the 1% Cd-doped sample shows a maximum at 230 K. The values of $\mu_H$ decrease with the increase of the acceptor and donor concentrations which are shown in Table 1. These facts suggest that the experimental results of $\mu_H$ in Cd-doped samples can be analyzed by combining the optical phonon and the ionized impurity scatterings. The combined mobility $\mu_T$ for Mathiessen's rule can be expressed by

$$1/\mu_T = 1/\mu_S + 1/\mu_I \ , \qquad (2)$$

where $\mu_S$ and $\mu_I$ are the mobilities for the optical phonon and the ionized impurity scatterings, respectively. The phonon scattering is determined by using the model developed by Schmid /11/ who assumed a short-range interaction with homopolar optical phonons polarized normal to the layers. The values of phonon energy of 0.023 eV /12/ and the phonon coupling constant of 0.25 /5/ were used in the present calculation of $\mu_S$. The temperature variation of $\mu_S$ is drawn by the dashed line in the figure. The mobility for the ionized impurity

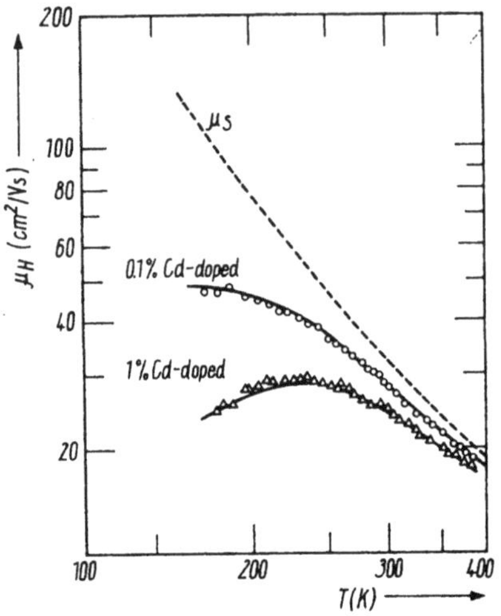


Fig. 2. Hall mobility as a function of temperature for Cd-doped samples. The dashed line indicates the calculated hole mobility by Schmid's model. The solid lines indicate the calculated mobility from (2)

scattering is given by the Brooks-Herring equation for nondegenerate semiconductors /13/. The temperature dependence of $\mu_I$ is calculated from

$$\mu_I = \frac{3.28\times10^{15}(m_0/m_h)^{1/2}\varepsilon^2 T^{3/2}}{(2N_D+p)\left[\ln(a+1) - (a/(a+1))\right]} \ (\mathrm{cm^2/Vs}) \ , \tag{3}$$

where

$$a = 1.29\times10^{14}(m_h/m_0)\varepsilon T^2/p^* .$$

The effective screening density $p^*$ is $p+\left[(p+N_D)(N_A-N_D-p)/N_A\right]$ and the static dielectric constant $\varepsilon$ is 10.6 /5/. The temperature variation of $\mu_I$ was evaluated by using the hole concentration p and $N_A$ and $N_D$ in Table 1, which were determined from (1). The combined mobility $\mu_T$ was calculated from (2) and the calculated curves are drawn by solid lines in Fig. 2. The experimental values of $\mu_H$ agree well with the calculated mobility of $\mu_T$.

In conclusion, the carrier transport of Cd-doped p-InSe is dominated by the acceptor level at 0.45 eV above the valence band. The temperature dependence of the hole mobility can be well fitted by the combined mobility of the optical phonon and the ionized impurity scatterings.

References

/1/ M. DI GIULIO, G. MICOCCI, A. RIZZO, and A. TEPORE, J. appl. Phys. 54, 5839 (1983).
/2/ S.M. ATAKISHIEV and G.A. AKHUNDOV, phys. stat. sol. 32, K33 (1969).
/3/ C. DE BLASI, G. MICOCCI, A. RIZZO, and A. TEPORE, Phys. Rev. B 27, 2429 (1983).

/4/ S. SHIGETOMI, T. IKARI, Y. KOGA, and S. SHIGETOMI, phys. stat. sol. (a) 86, K69 (1984).

/5/ A. SEGURA, F. POMER, A. CANTARERO, W. KRAUSE, and A. CHEVY, Phys. Rev. B 29, 5708 (1984).

/6/ A. SEGURA, K. WÜNSTEL, and A. CHEVY, Appl. Phys. A 31, 139 (1983).

/7/ T. IKARI, S. SHIGETOMI, Y. KOGA, and S. SHIGETOMI, phys. stat. sol. (b) 103, K81 (1981).

/8/ T. IKARI and Y. KOGA, J. Phys. Soc. Japan 47, 1017 (1979).

/9/ J.C. PORTAL, R.J. NICHOLAS, E. KRESS-ROGERS, A. CHEVY, J. BESSON, J. GALIBERT, and P. PERRIER, Proc. XV. Internat. Conf. Phys. Semicond., Kyoto 1980, in: J. Phys. Soc. Japan 49, Suppl. A, 874 (1980).

/10/ J.S. BLAKEMORE, Semiconductor Statistics, Pergamon, New York 1962 (p. 139).

/11/ PH. SCHMID, Nuovo Cimento B21, 258 (1974).

/12/ S. SHIGETOMI, T. IKARI, Y. KOGA, and S. SHIGETOMI, Japan. J. appl. Phys. 20, L343 (1981).

/13/ E.H. PUTLEY, The Hall Effect and Semiconductor Physics, Dover, New York 1060 (p. 146).

(Received April 6, 1988)

phys. stat. sol. (a) 108, K59 (1988)
Subject classification: 71.55 and 72.80; S7.11

Institute of Applied Physics, Academy of Sciences of the Moldavian SSR, Kishinev[1)]

# Electrical Properties of InP:Fe Single Crystals Implanted by Phosphorus Ions

By
S.I. RADAUTSAN, I.M. TIGINYANU, and N.B. PYSHNAYA

Iron doped semi-insulating indium phosphide has an increasing application in electronics. In this connection it seems to be important to study ion implantation in InP crystals for producing layers having predetermined parameters /1/. This note is devoted to the investigations of phosphorus ion implantation and to the influence of the post-implantation annealing process upon the electrical properties of InP:Fe crystals grown by the liquid encapsulated Czochralski method under argon pressure.

Iron doping has been performed during growth. The electron concentration in original samples was $n = (3 \text{ to } 4) \times 10^8\ cm^{-3}$, the mobility $\mu_n = 1550\ cm^2/Vs$ (T = 300 K), the dislocation density $N_D = 5 \times 10^4\ cm^{-2}$. 40 keV phosphorus ion implantation was realized at room temperature into (100) oriented 0.5 mm thick wafers. The implantation doses were in the range $(1 \text{ to } 5) \times 10^{14}\ cm^{-2}$. After implantation the face wafer side was coated by an $Al_2O_3$ film with a thickness of $\approx 0.2\ \mu m$. The post-implantation annealing was being carried out in nitrogen atmosphere during 15 min followed by protecting film etching. The electrical parameters of the samples have been determined by Hall effect measurements using the Van der Pauw method.

The dependence of the electron surface concentration $n_s$ and the mobility $\mu_s$ in InP:Fe crystals on the post-implantation annealing temperature is shown on Fig. 1. It can be seen that the sample annealing temperature in the range of (500 to 600) °C leads to a significant increase (up to 3 orders of magnitude) of the effective electron surface concentration, the value of the electron mobility being changed little. Note that the bulk value of the electron concentration in implanted layers reaches $10^{16}$ to $10^{17}\ cm^{-3}$ at annealing temperature of 600 °C (supposing that the thickness of the conducting layer equals 0.1 μm /2/). This

1) Grosula 5, SU-277028 Kishinev, USSR.

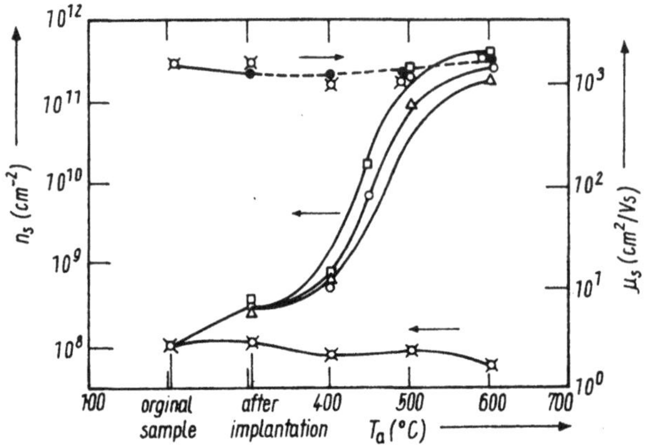


Fig. 1. Surface electron concentration and mobility in InP:Fe layers implanted by $P^+$ ions at different doses versus annealing temperature. Dose (cm$^{-2}$) ¤ 0; △ $4\times10^{13}$; ○ $1.2\times10^{14}$; □ and ● $2\times10^{14}$. T = 300 K

value is by 8 to 9 orders of magnitude higher than that for original crystals.

To estimate the depth of donor levels arising during the post-implantation annealing of samples the temperature dependence of the surface Hall coefficient $R_{Hs}$ has been investigated in the range 100 to 400 K. The thermal electron activation energy has been determined to be 0.09 eV. It should be underlined that donor levels with a depth of $E_c$ - 0.09 eV have been found earlier in specially non-doped indium phosphide crystals irradiated by 1 MeV electrons. Taking this into account, the connection of the above mentioned

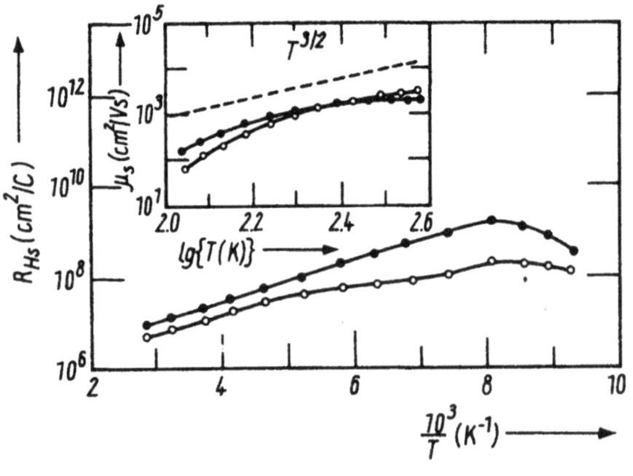


Fig. 2. Temperature dependence of the surface Hall coefficient and electron mobility (inset) for InP:Fe crystals after implantation phosphorus ions at a dose of ~$10^{14}$ cm$^{-2}$ and post-implantation annealing at 500 °C (●) and 600 °C (○)

donor levels with native lattice defects seems to be most probable

The dependence of $\mu_s = f(\lg T)$ for conducting InP:Fe layers enriched by phosphorus is shown in the inset of Fig. 2. As is seen from the plot, the experimental curves approximately follow the $\mu_s \sim T^{3/2}$ law indicating that the scattering on charged defects exerts the main influence on the electron mobility in the given temperature range.

Thus, phosphorus ion implantation into semi-insulating InP:Fe single crystals followed by annealing leads to the formation of n-type conducting layers, the thermal donor activation energy being 0.09 eV.

References

/1/ V.D. VERNER, A.A. PSHENICHNYI, and G.I. RADAUTSAN, Zarubezhnaya Elektronnaya Tekh. 9, 73 (1983).

/2/ M. LEVINSON, J.L. BENTON, H. TEMKIN, and L.C. KIMERLING, Appl. Phys. Letters 40, 990 (1982).

(Received March 14, 1988)

phys. stat. sol. (a) 108, K63 (1988)
Subject classification: 72.80; S8.16

Institute of Solid State and Semiconductor Physics, Academy of Sciences of the Belorussian SSR, Minsk[1)]

# Electrical Properties of Cadmium and Zinc Doped $CuAlS_2$

By
I.A. AKSENOV, L.A. MAKOVETSKAYA, and V.A. SAVCHUK

Of the large group of ternary $A^{I}B^{III}C_2^{VI}$ chalcopyrite semiconductors $CuAlS_2$ is of considerable interest because it has a direct energy gap of about 3.5 eV /1/, which corresponds to the ultraviolet spectral range. $CuAlS_2$ among many $A^{I}B^{III}C_2^{VI}$ chalcopyrite semiconductors is the only perspective candidate for blue to ultraviolet light emitting device application /2/.

$CuAlS_2$ single crystals have first been obtained by chemical vapour transport /3/ and some data on the electrical properties of this compound have been reported /3, 4/. The conductivity control in $CuAlS_2$ compound, just like in other $A^{I}B^{III}C_2^{VI}$ compounds, is an extremely difficult task which might stem from both strong compensation effects and non-stoichiometry. But a better knowledge of the properties of intrinsic defects (vacancies, interstitials, and antisite disorder) and extrinsic impurities is indispensable to device fabrication. Therefore, the search for suitable dopants and techniques of doping is must challenging.

The aim of this note is to study the electrical properties of cadmium and zinc doped $CuAlS_2$ single crystals.

Copper (99.999%), aluminium and sulphur (99.9999%) were used as starting materials for obtaining the $CuAlS_2$ ternary compound. The stoichiometric quantities of the starting elements were sealed under high vacuum (up to $10^{-3}$ Pa) in a NB crucible held in quartz ampoule. The synthesis was carried out by the direct melting method in a vertical furnace with vibration. To achieve equilibrium the compound obtained was homogenised by subsequent annealing. X-ray diffraction studies showed that the resulting material was of single phase.

This material was then used to grow single crystals by the chemical vapour transport method in a closed system with iodine as a transport agent. A temperature variation method of chemical vapour transport was used /5/.

---

1) Podlesnaya 17, SU-220726 Minsk, USSR.

The resulting crystals were green and blue. Both needle-like crystals up to 1.5 cm long with well-characterized $(11\bar{2})$ and (101) faces and plate-like crystals several mm wide with developed $(11\bar{2})$ surface were obtained, both having reflecting mirror-like faces. X-ray powder analysis showed that the crystals obtained had a chalcopyrite tetragonal structure with the lattice parameters $a = (0.5333 \pm 0.0002)$ nm and $c = (1.0435 \pm 0.0005)$nm.

All single crystals obtained by the chemical vapour transport method were p-type conducting as measured by a thermal probe. The ohmic contacts to the needle-like crystals were performed by electrolytic deposition of copper, the resistivities of the plate-like crystals were determined by the four-point probe technique. The resistivities of the as-grown crystals were in the range $\rho_0 = (1.4$ to $5) \times 10^4\ \Omega$cm.

The as-grown $CuAlS_2$ single crystals were polished with an alumina grit, then cleaned using standard reagents and placed in silica ampoules with 50 mg of cadmium, cadmium/aluminium, and zinc. The evacuated and sealed ampoules were heated to the annealing temperature in the range 100 to 800 $^{\circ}$C, held there for 20 h and then quenched in water bath. The conductivity type and resistivity of the annealed crystals were measured after polishing the samples in order to avoid errors due to possible surface effects.

Fig. 1 shows the temperature dependence of the variation of resistivity $(\rho/\rho_0')$ for $CuAlS_2$ single crystals, annealed in the presence of Cd, Cd/Al, and Zn. The filled dots in the figure denote the data for p-type conductivity samples,

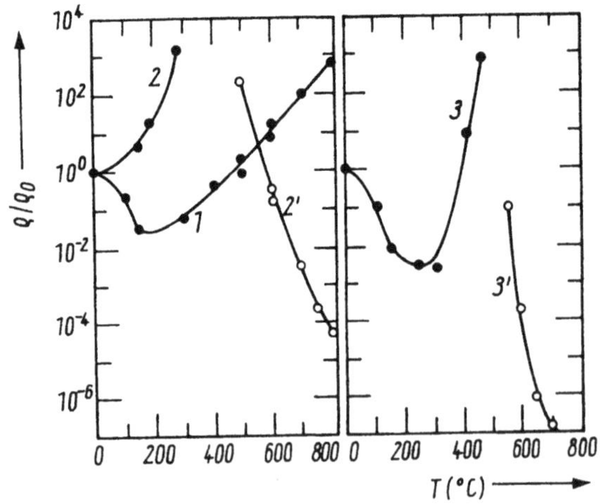


Fig. 1. The temperature dependence of the variation of resistivity $(\rho/\rho_0)$ for $CuAlS_2$ single crystals, annealed in the presence of cadmium (1), cadmium/ /aluminium (2, 2'), and zinc (3, 3 )

and the unfilled dots denote data for n-type conductivity samples. The decrease in the resistivity of the p-type $CuAlS_2$ single crystals after annealing with Cd at temperatures less than 150 °C may be due to the fact that the probability of formation of $Cd_{Al}$ defects (where Cd is an acceptor) is greater than of $Cd_{Cu}$ defects (where Cd is a donor). With the rise of the annealing temperature the probability of formation of $Cd_{Cu}$ defects increases, which leads to an increase of the resistivity of the samples by a factor of $\approx 10^3$ (see Fig. 1). The annealing of the $CuAlS_2$ single crystals in the presence of cadmium at 100 to 800 °C did not result in an inversion of the conductivity type, i.e. in the preparation of n-type conductivity samples, which contradicts the data /4/.

As seen from the figure, the $CuAlS_2$ single crystals with n-type conductivity can be obtained by annealing of the samples in the presence of cadmium/aluminium as well as zinc. The preparation of n-type conductivity samples by annealing in the presence of cadmium with aluminium is caused by the fact that Cd atoms in this case substitute preferably the sites in the Cu sublattice, since Al vapor prevents the formation of Al vacancies and, consequently, the formation of $Cd_{Al}$ defects.

The n-type conductivity single crystals with resistivity up to $3\times10^{-3}$ Ωcm can be obtained while annealing in the presence of zinc. With regard for the small values of carrier mobilities in the $CuAlS_2$ single crystals /3/, the preparation of low resistivity samples of this compound upon doping with zinc gives evidence of a high concentration of Zn atoms dissolved in the crystal lattice. The temperature dependence of the electrical conductivity of the $CuAlS_2$ single crystals, doped by zinc, has a metallic character. This fact also speaks for a high concentration of donors (Zn atoms) in the crystals and a low activation energy of the donor levels formed by zinc.

References

/1/ N. YAMAMOTO, H. HORINAKA, K. OKADA, and T. MIYACHI, Japan. J. appl. Phys. 16, 1817 (1977).

/2/ N. YAMAMOTO, Japan. J. appl. Phys. 19, Suppl. 3, 95 (1980).

/3/ W.N. HONEYMAN, J. Phys. Chem. Solids 30, 1935 (1969).

/4/ P.C. DONOHUE, J.D. BIERLEIN, J.E. HANLON, and H.S. JARETT, J. Electrochem. Soc. 121, 829 (1974).

/5/ C. PAORICI, L. ZANOTTI, and G. ZUCCALLI, J. Crystal Growth 43, 705 (1978).

(Received March 4, 1988)

phys. stat. sol. (a) 108, K67 (1988)

Subject classification: 76.80; 75.50; S1.1

Institute of Physics, Warsaw Technical University[1] (a) and SGGW, Academy of Agriculture, Warsaw[2] (b)

# The Influence of Transition Metal Substitution and $B_{3/2}$ Values for Some Amorphous Alloys

By

J. GWIAZDA (a), E. MARIAŃSKA (b), and W. ZYCH (a)

As we know already from the theoretical work of Gubanov /1/, for the existence of magnetism the periodic arrangement of atoms on a crystalline lattice is not necessary. For a material to be magnetic, all that is necessary is the existence of a magnetic moment on an atom and the requirement that when these atoms are assembled in a solid, they interact via the exchange forces /2/. It is the reason that we have ferromagnetism in amorphous solids. The amorphous alloys studied which exhibit ferromagnetism, have been obtained by liquid quenching (Institute of Material Science of the Warsaw Technical University) and are of the type $Fe_{80-x}TM_xB_{20}$, where TM is a transition metal like Ti, Mn, Cr, Co, Ni. We have used Mössbauer spectroscopy to study the magnetic properties of amorphous substances at temperatures from 4.2 K to about 800 K. The results were analysed by Window's method /3/. The influence of TM substitution on the hyperfine magnetic field and also the temperature dependence of the reduced hyperfine field is discussed. Mössbauer spectroscopy became already a conventional method for the determination of magnetic properties of amorphous substances /4/. One of the most important results obtained with this method is the determination of the hyperfine magnetic field and its distribution. Another important question concerns the temperature dependence and many publications are devoted to this question. We reported some preliminary results concerning those questions for amorphous alloys in /5/. The Mössbauer spectra and the corresponding hyperfine field distribution P(H) at selected temperatures for amorphous $Fe_{74}Cr_6B_{20}$ are shown in Fig. 1 and the corresponding H value in Table 1. Because of the prolongated heating at temperatures near $T_C$, Mössbauer measurements are not adequate for determining the Curie temperature $T_C$ for amorphous alloys. The amorphous

1) Ul. Koszykowa 75, PL-00-662 Warsaw, Poland.

2) Ul. Rakowiecka 8, Warsaw, Poland.

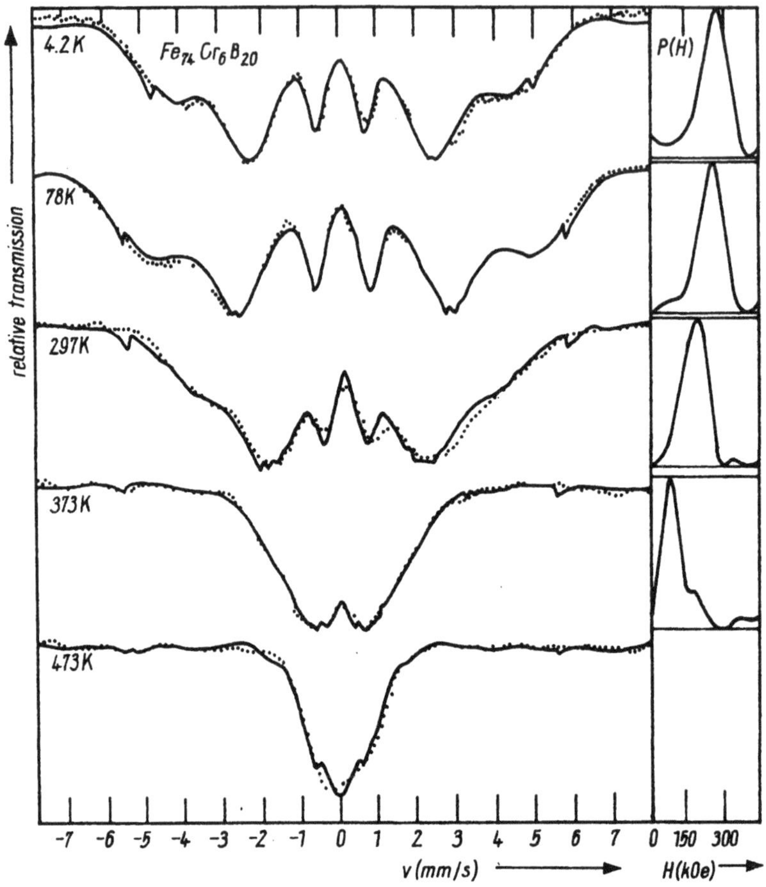


Fig. 1. Mössbauer spectra and hyperfine field distribution P(H) at selected temperatures for $Fe_{74}Cr_6B_{20}$. The last spectrum is for temperature above $T_C$

state has no rapid transformation into the crystalline state, but changes in some temperature interval depending on TM and composition /6/.

We have got $T_C$ /6/ from DTMG measurements /5/ and for $Fe_{74}Cr_6B_{20}$ the value $T_C$ = 450 K follows. As we can see from Fig. 1, the six-line structure

Table 1

Mean hyperfine magnetic field at several temperatures for $Fe_{74}Cr_6B_{20}$

| T(K) | 4.2 | 80 | 297 | 373 |
|---|---|---|---|---|
| ⟨H⟩(kOe) | 227 | 225 | 170 | 38 |

of the spectrum is already not well defined at T = 373 K, and above the magnetic ordering temperature $T_C$ consists of two overlapping lines suggesting quadrupole splitting. According to considerations of Chien and Hasegawa /7/ we assume that the calculated reduced hyperfine magnetic field has the same temperature dependence as that of the reduced magnetization $M(T)/M(0)$. As the authors write, the validity of this approximation has been established experimentally in many crystalline magnetic solids and for non-crystalline magnetic solids, there are experimental evidences of the validity of the approximation by independent measurements of $H_{eff}(T)$ and $M(T)$.

Fig. 2 gives the temperature dependence of the mean hyperfine magnetic field for samples containing Cr, and Fig. 3 gives the temperature dependence of the reduced hyperfine field as function of the reduced temperature. Therefore the hyperfine magnetic field H(T) at low temperatures decreases as

$$H(T)/H(0) = 1 - BT^{3/2} - CT^{5/2} + \dots \quad (1)$$

or

$$H(T)/H(0) = 1 - B_{3/2}(T/T_C)^{3/2} - C_{5/2}(T/T_C)^{5/2} + \dots , \quad (1a)$$

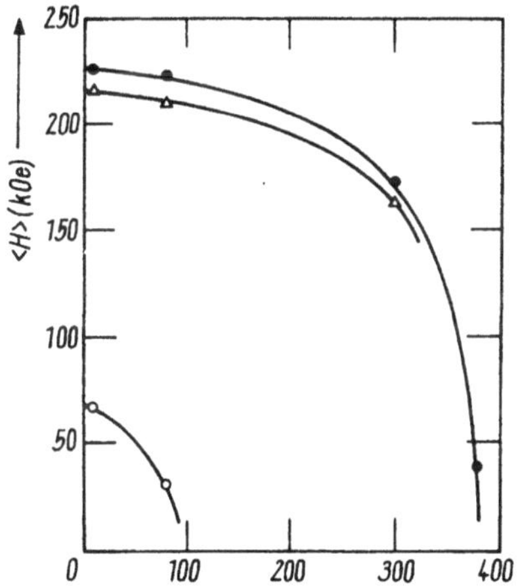


Fig.2

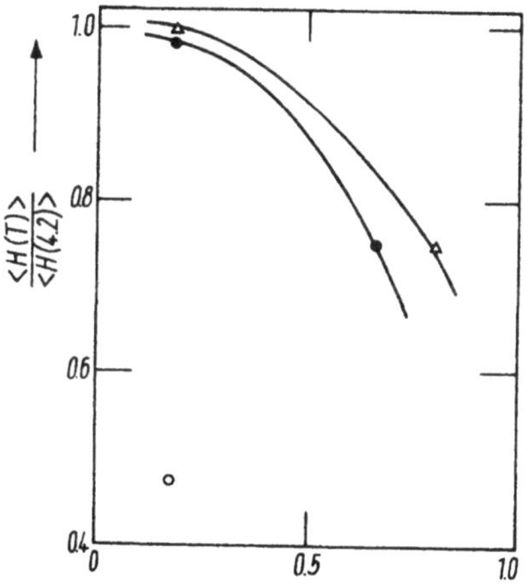


Fig.3

Fig. 2. Temperature dependence of the mean hyperfine magnetic field for samples containing Cr. ● $Fe_{74}Cr_6B_{20}$, △ $Fe_{70}Cr_{10}B_{20}$, O $Fe_{60}Cr_{20}Si_8B_{12}$

Fig. 3. Temperature dependence of the reduced hyperfine magnetic field as function of the reduced temperature. ● $Fe_{74}Cr_6B_{20}$, △ $Fe_{70}Cr_{10}B_{20}$, O $Fe_{60}Cr_{20}Si_8B_{12}$

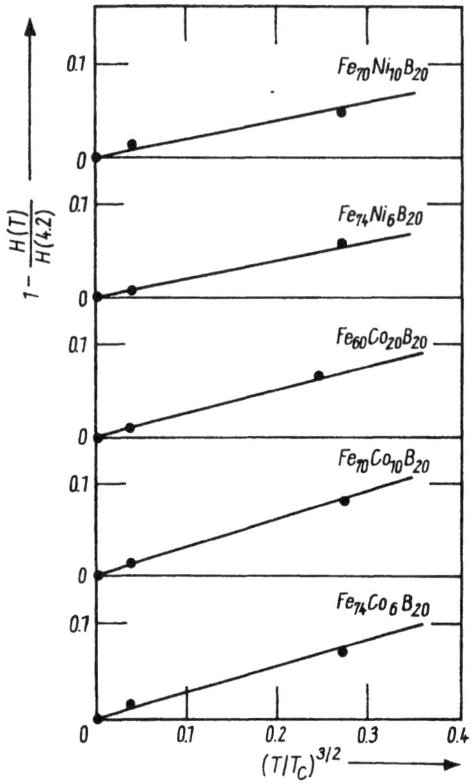


Fig. 4. A plot of 1-H(T)/H(0) versus $(T/T_C)^{3/2}$.

The leading term gives Bloch's law. We have assumed also that the leading $T^{3/2}$ term dominates in a very large temperature range and we calculated the value of $B_{3/2}$ taking the measurements at 4.2, 80, 300 K for alloys $Fe_{80-x}TM_xB_{20}$, where TM = Co, Ni and at 4.2 and 80 K for alloys with TM = Mn, Cr because the Curie temperature for these last alloys is 450 and 563 K, respectively. The transition metals such as Mn and Cr with less 3d electrons than Fe, greatly reduce the hyperfine field and the Curie temperature $T_C$.

If we neglect the $T^{5/2}$ term and write (1a) as $1-H(T)/H(0) = B_{3/2}(T/T_C)^{3/2}$, then a plot of 1-H(T)/H(0) versus $(1/T_C)^{3/2}$ gives a straight line with slope equal to $B_{3/2}$.

Fig. 4 gives the results for samples with Co and Ni what confirms our assumption about the leading role of the $T^{3/2}$ term at low temperatures. In our calculations we used the approximation H(T = 0 K) = H(T = 4.2 K).

Table 2 gives $B_{3/2}$ values for our samples. It is interesting to mention that the values of $B_{3/2}$ from these measurements are comparable with the data of /7/.

One can assume that H(T)/H(0) versus $T/T_C$ seems to have a universal behaviour, i.e. H(T)/H(0) decreases more rapidly with increasing temperature than for corresponding crystalline elements like iron and nickel. It can be concluded that this statement must somehow be related to the random distribution of atoms in amorphous systems resulting in varying exchange interactions between atoms. As we know, the short-range interaction between a magnetic ion and its surrounding can be described by the Heisenberg Hamiltonian $H = -\alpha J(r_{i,j})\vec{S}_i\vec{S}_j$ where the exchange constant $J(r_{i,j})$ depends upon the distance

Table 2

Curie temperatures $T_C$ and $B_{3/2}$ values

| sample | $Fe_{74}Co_6B_{20}$ | $Fe_{70}Co_{10}B_{20}$ | $Fe_{70}Co_{20}B_{20}$ |
|---|---|---|---|
| $T_C$ (K)<br>$B_{3/2}$ | 730<br>0.279 | 721<br>0.295 | 763<br>0.241 |
| sample | $Fe_{74}Ni_6B_{20}$ | $Fe_{70}Ni_{10}B_{20}$ | $Fe_{60}Ni_{20}B_{20}$ |
| $T_C$(K)<br>$B_{3/2}$ | 713<br>0.209 | 715<br>0.164 | 738<br>0.098 |
| sample | $Fe_{74}Cr_6B_{20}$ | $Fe_{70}Cr_{10}B_{20}$ | |
| $T_C$ (K)<br>$B_{3/2}$ | 450<br>0.295 | 365<br>0.091 | |
| sample | $Fe_{74}Mn_6Si_{12}B_8$ | $Fe_{70}Mn_{10}Si_{12}Be$ | |
| $T_C$ (K)<br>$B_{3/2}$ | 563<br>0.079 | 513<br>0.217 | |

$r_{i,j}$ between interacting i-th and j-th atoms. According to the authors of /8/, due to the disordered structure of an amorphous substance, $J(r_{i,j})$ is no more discrete and well defined as is the case in crystalline solids. This randomness introduces a distribution of values of the exchange constant and therefore the exchange constant fluctuates from site to site and is essentially responsible for the observed faster decrease of H(T) with increase in temperature.

We conclude that the hyperfine magnetic field measured at the Fe sites by Mössbauer spectroscopy in a number of glassy ferromagnets of the type $Fe_{80-x}TM_xB_{20}$ exhibits a temperature dependence according to Bloch's $T^{3/2}$ law at low temperatures. The leading $T^{3/2}$ term dominates for a relatively great temperature range for samples with Co and Ni, but in a smaller range for samples with Mn and Cr.

The spectra analysed by Window's method showed the existence of only one peak in the distribution of the magnetic hyperfine field.

Further work in this direction is in progress and will be reported in the future. The authors wish to thank Mr. M. Peryt for his help in Mössbauer measurements.

This work was sponsored by the Institute of Physics of the Polish Academy of Science in the frame of the Scientific Program CPBP 01.04.II.1.8.

References

/1/ A.I. GUBANOV, Fiz. tverd. Tela 2, 502 (1960).
/2/ K. MAORIANI and J.M.D. COY, Magnetic Glasses, Elsevier, 1984.
/3/ B. WINDOW, J. Phys. E 4, 401 (1971).
/4/ U. GONSER, Atomic Energy Rev., Suppl. No. 1 (1981).
/5/ J. GWIAZDA, M. MARIAŃSKA, M. PERYT, W. PERYT, and W. ZYCH, Acta Phys. Polon. (1987), in the press.
/6/ R. MOSKALEWICZ and W. ZYCH, phys. stat. sol. (a) 97, K43 (1986).
/7/ C.J. CHIEN and R. HASEGAWA, Phys. Rev. B 16, 2115 (1977).
/8/ A.K. BHATNAGAR et al., Solid State Commun. 44, 905 (1982).

(Received July 24, 1987)

phys. stat. sol. (a) 108, K73 (1988)

Subject classification: 77.80; S11.1

Department of Applied Material Science,
Institute of Technology, Muroran[1)]

# Temperature Dependence of the Dielectric Constant in Samarium Nitrate Crystals

By

R. KAWASHIMA and T. MATSUDA

Recently, investigations /1 to 3/ of rare earth compounds have received considerable attention to study the physical properties associated with rare earth spectroscopy /1/, valence instability /2/, high temperature superconductivity (rare earth oxides) /3/, and ferroelectricity and ferroelasticity ($Gd_2(MoO_4)_3$ family) /4/. In the material system, a rare earth ion plays an important role for the characteristic properties. The nitrates are among the few soluble and easily prepared inorganic compounds of the rare earths /5/. The crystal symmetry of samarium nitrate crystal, $Sm(NO_3)_3 \cdot 6H_2O$, is triclinic, the space group $P1^-$ with lattice parameters of a = 0.920 nm, b = 1.17 nm, c = 0.678 nm, $\alpha = 91^\circ$, $\beta = 112^\circ$, and $\gamma = 109^\circ$ /6/ at room temperature. In a previous note /7/, one of the authors has reported preliminary data on the electric and optical measurements on the samarium nitrate crystal. The existence of a polymorphic transition in the crystal has been given in the previous work. In the present note, to study the phase transition of the samarium nitrate crystal, the authors have improved the scheme of both the electric measurement and data proceeding, and reported refined results on the temperature variation of the dielectric constant.

We have measured automatically the temperature dependence of the ac impedance of the crystal at 1 kHz by using a digital two-phase lock-in amplifier (NF Electronic Instrum., 5610) and a $\mu$ computer (NEC PC9801E) at an applied voltage of 300 mV. The single crystals of samarium nitrate were grown from a saturated solution by evaporation at 26 $^\circ$C for 14 days. Specimens were prepared by cutting the crystal parallel to the (001) plane. A specimen was used for the present measurements with silver electrodes: the dimension of sample is (0.072 $\pm$ 0.005) cm in thickness and (0.206 $\pm$ 0.005) $cm^2$ in area.

---

1) Muroran 050, Japan.

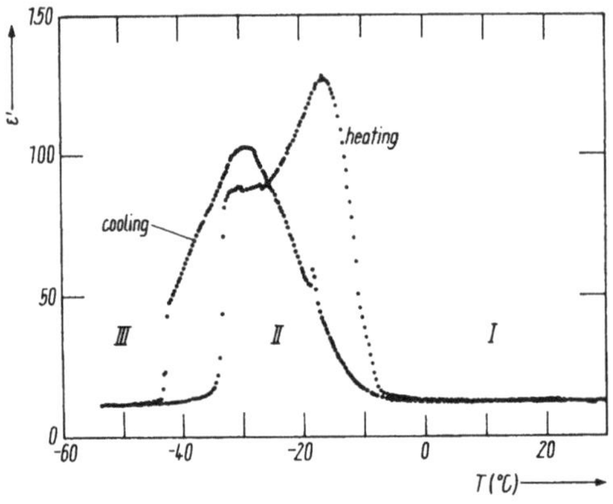


Fig. 1. Real part ε' of the complex dielectric constant ε* along the c-axis of samarium nitrate crystal at temperatures in the range between room temperature and -55.0 °C in both the cooling and warming processes; 1 kHz

The specimen was mounted in a massive copper block, of which the weight is 10 kg, and submersed in a silicon oil bath to avoid the decomposition of the sample and evaporation of the crystal water. Temperature measurement was carried out by a copper-constantan thermocouple mounted in the sample cell. The oil bath was controlled to ±0.1 K. The sample was annealed at 70 °C just below the melting point during 24 h, and repeated a few times of careful heating and cooling runs. The measurements were performed in the course of warming and cooling with the rate of temperature variation, 0.021 to 0.031 K/min. The experimental results were analyzed by using the micro-computer system (NEC PC9801E) and appropriate application software.

The temperature dependence of the real ε' and the imaginary part ε'' at 1 kHz are given in Fig. 1 and 2 of both heating and cooling run in the range from room temperature to -55 °C. From the figures, two singularities with a thermal hysteresis are found in the temperature variation of the dielectric constant. The hysteresis loop in the temperature variation of the dielectric constant and resistivity is found to be closed. The phenomena could not be observed in the previous experimental scheme. The fact would be attributed to the invariance in the quality of the specimen in the course of the present measurements.

The anomalous temperature dependence of the dielectric constant between ≈-45 °C and ≈0 °C would be related to polymorphic transitions with a change of the crystal structure in the samarium nitrate crystal. The results in this study may suggest the possibility of two of the three phases I, II, and III, given in Fig. 1 and 2, for the samarium nitrate crystal. The detailed structural analyses on the crystal and the temperature variation have not been reported by X-ray or neutron diffraction methods /6, 8/.

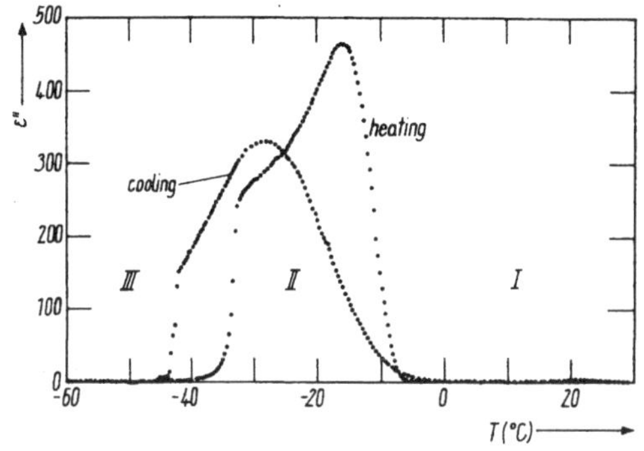


Fig. 2. Imaginary part ε'' of the complex dielectric constant ε* along the c-axis of samarium nitrate crystal at temperatures in the range between room temperature and -55.0 °C in both the cooling and warming processes; 1 kHz

The electric behaviors near the polymorphic transitions, given in the figures, however, are different from those found in the previous one /7/. The fact would be attributed to the characteristics of the phase transitions or to the instability of the intermediate phase II. The physical phenomena observed in our studies would be related to the variation in the complex structure by the polymorphic phase transitions or to the characteristic electronic processes in the rare earth ion, $Sm^{3+}$, in the samarium nitrate crystal.

Additional studies, however, will be necessary in order to fully understand the phase transition occurring in the samarium nitrate crystal.

References

/1/ JEZOWSKA-TRZEBIATAWSKA, J. LEGENDRIEWICZ, and W. STREK (Ed.), Proc. Internat. Symp. Rare Earths Spectroscopy, World Scientific Publ. Co. Pte. Ltd., 1985.

/2/ P. WACHTER and H. BOPPART (Ed.), Proc. Internat. Conf. Valence Instabilities, North-Holland Publ. Co., Amsterdam/New York/Oxford 1982.

/3/ J.G. BEDNORZ and K.A. MÜLLER, Z. Phys. B 64, 189 (1986).

/4/ T. MITSUI (Ed.), Landolt-Börnstein, Vol. 16, Ferroelectrics: Oxides, Springer-Verlag, Berlin/Heidelberg/New York 1982.

/5/ K.A. GSCHNEIDER, Jr. and L. EYRING (Ed.), Handbook on the Physics and Chemistry of Rare Earths, Vol. 8, North-Holland Publ. Co., Amsterdam 1986 (p. 302).

/6/ V.I. IVERONOVA, V.P. TARASOVA, and Z.K. ZOLINA, Zh. fiz. Khim. 29, 314 (1955).

/7/ R. KAWASHIMA, J. Phys. Soc. Japan 56, 415 (1987).

/8/ M. QUARTON and D. SVORONOS, J. Solid State Chem. 42, 325 (1982).

(Received April 25, 1988)

phys. stat. sol. (a) 108, K77 (1988)
Subject classification: 78.20; S8.15

Physics Department, Yonsei University, Seoul[1)]

# Determination of x in $Hg_{1-x}Zn_xTe$ Single Crystals by the Temperature Coefficient of the Energy Gap

By
H.W. JEON, K.M. KIM[2)], H.K. KIM, H.L. PARK[3)], and C.H. CHUNG

The $Hg_{1-x}Zn_xTe$ (MZT) is considered as a potential candidate material for infrared detectors and currently active researches are involved in this material /1 to 3/. The existing $Hg_{1-x}Cd_xTe$ (MCT) infrared detector has a drawback of poor chemical stability due to the weak Hg-Te bond which is destabilized by Cd incorporation in MCT /4/. It was theoretically predicted that Zn would stabilize the Hg-Te bond in $Hg_{1-x}Zn_xTe$ /5/.

The great difference in lattice parameters of HgTe and ZnTe induces strong interactions between cations, but this difference also helps to determine the composition, x, in $Hg_{1-x}Zn_xTe$ through Vegard's law using the conventional X-ray technique. In this short communication, we are going to present the determination of composition, x, in $Hg_{1-x}Zn_xTe$ single crystals using the temperature coefficient of the energy gap, i.e., $dE_g/dT$.

A tungsten-iodine lamp was employed as a light source for transmission measurements. A cryostat (Air Product Displex) was used for low temperature measurements and a Spex 3/4 m monochromator was also used to obtain the transmission spectra.

A typical transmission spectrum of $Hg_{1-x}Zn_xTe$ single crystal is shown in Fig. 1. The temperature coefficients of the energy gap of HgTe and ZnTe were reported as $+5\times10^{-4}$ /8/ and $-5\times10^{-4}$ eV/K /9/, respectively. Our approach is to use these coefficients and assuming the temperature coefficient of the energy gap as a linear function of composition, x, in $Hg_{1-x}Zn_xTe$. The rationale behind our assumption is that the band gap of $Hg_{1-x}Zn_xTe$ is a linear function of composition, x, in the first approximation. Our result is shown in Fig. 2. The uncertainty was shown in Fig. 2 as an error bar. This error bar indicates the maximum deviation of $\Delta x \approx 0.04$.

1) Seoul, Korea.

2) Present address, Department of Physics, Soowon University.

3) To whom all correspondence should be addressed.

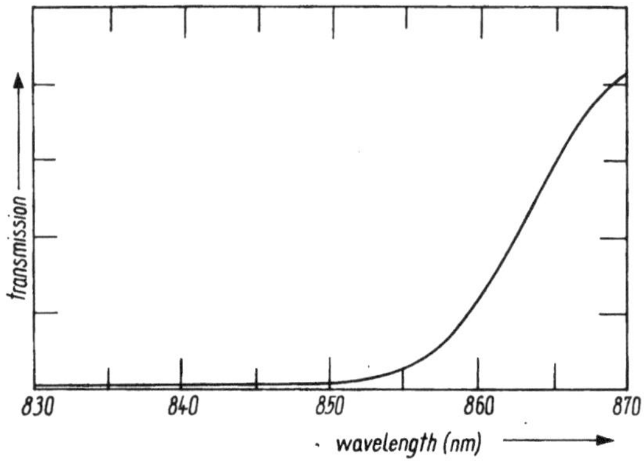


Fig. 1. Transmission spectrum of $Hg_{1-x}Zn_xTe$ (x = 0.705) at 20 K

Even though the best technique for determining the composition x in $Hg_{1-x}Zn_xTe$ is the EPMA technique, the $dE_g/dT$ approach is an alternative technique to get the approximate composition, x, in $Hg_{1-x}Zn_xTe$. The composition dependent $dE_g/dT$ of $Hg_{1-x}Zn_xTe$ is found to be $dE_g/dT = -1.03 \times 10^{-3} x + 5.00 \times 10^{-4}$ (eV/K) in this study.

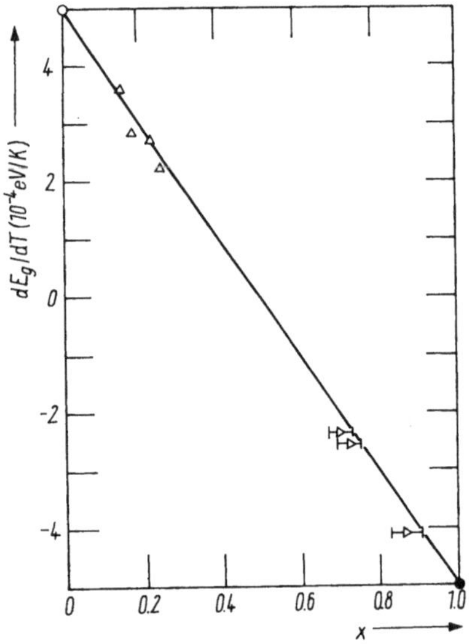


Fig. 2. Composition dependent temperature coefficient of $Hg_{1-x}Zn_xTe$ crystals; ○ HgTe /8/, ● ZnTe /9/, △ Sher et al. /5/, ▽ present work

References


/1/ A. SHER, D. EGER, A. ZEMEL, H. FELDSTEIN, and A. RAIZMAN, J. Vacuum Sci. Technol. A4, 2024 (1986).

/2/ R. TRIBOULET, A. LASBLEY, B. TOULOUSE, and R. GRANGER, J. Crystal Growth 79, 695 (1986).

/3/ J.P. FAURIE, J. Crystal Growth 81, 483 (1987).

/4/ W.E. SPICEP and J.A. WILSON, J. Vacuum Sci. Technol. A1, 1735 (1983).

/5/ A. SHER, A.B. CHEN, W.E. SPICER, and C.K. SHIN, J. Vacuum Sci. Technol. A3, 105 (1985).

/6/ H.U. CHUN, H.C. KO, S.R. HAHN, H.L. PARK, K.M. KIM, and C.H. CHUNG, Bull. Korean Phys. Soc. 5, 47 (1987).

/7/ T. BOECK and P. RUDOLPH, J. Crystal Growth 79, 105 (1986).

/8/ R.S. KIM and S. NARITA, phys. stat. sol. (b) 73, 741 (1976).

/9/ R.E. NAHORY and H.Y. FAN, Phys. Rev. Letters 17, 251 (1966).


(Received December 30, 1987)

phys. stat. sol. (a) 108, K81 (1988)
Subject classification: 61.70 and 78.60; S1.61

Departamento de Física de Materiales, Facultad de Físicas, Universidad Complutense, Madrid[1)]

## SEM-CL of Reaction Bonded SiC

By

J. PIQUERAS, F. DOMÍNGUEZ-ADAME, and B. MENDEZ

Reaction bonded silicon carbides have a complex polycrystalline structure formed by bonding together SiC powders with SiC produced by a reaction between silicon and carbon. The latter SiC has been often described to form an epitaxial coating on the original $\alpha$-SiC (hexagonal polytype) grains and to nucleate also independently as fine grained $\beta$-SiC (cubic polytype). The structure of the epitaxial deposit has been investigated by several authors /1 to 6/. It has been found /5, 6/ that the deposit grows as $\beta$-SiC but almost fully transforms to $\alpha$-SiC during the high temperature processing. The resulting grains have a composite nature with different levels of impurity in the original grains and the epitaxial coating. Some interstitial regions are occupied by residual silicon. In the above mentioned investigations, scanning electron microscopy (SEM) and transmission electron microscopy (TEM) have been frequently used. In particular the secondary electron SEM images clearly show /1, 5/ the different parts - original grains, coating, silicon regions of the grain structure. In the present work the application of the SEM cathodoluminescence (CL) mode to the microstructural characterization of reaction bonded SiC is investigated.

Samples with a thickness of about 1 mm were cut from a 5 mm diameter reaction bonded SiC rod (Goodfellow Metals Ltd.). The samples were mechanically polished and observed in a S4-10 SEM in the CL mode at temperatures between 300 and 100 K. The experimental method for CL measurements, in the range 350 to 850 nm, has been previously described /7/.

Fig. 1 (see on the photo pages before the Short Notes part) shows the CL images of the same specimen area at different temperatures. The CL emission is highly inhomogeneous through the sample. Some regions are highly luminescent as compared with other regions which either show a moderate emission or appear dark in the CL image. Decreasing the temperature from

1) E-28040 Madrid, Spain.

300 to 215 K causes a marked increase of the total CL intensity. Some regions with low emission at room temperature appear very intense at 215 K while dark regions remain unaltered. No changes are observed in the range 215 to 190 K but further cooling to about 140 K produces a luminescence decrease. Comparison of Fig. 1b and c indicates that this effect is not due to the luminescence decrease of all emitting regions but mainly to the decrease in some high emitting regions, as those marked A in Fig. 1b. In addition, a small number of regions, as B in Fig. 1c, show a luminescence increase. Decreasing the temperature to 100 K causes a further rearrangement of the luminescent areas.

The temperature dependence of the CL images is associated to spectral changes. Fig. 2a shows the CL spectrum at 273 K. By decreasing the temperature a broadening of the spectrum and a relative increase of the emission at wavelengths at both sides of the peak is observed. This is shown in the spectrum at 180 K of Fig. 2b. At lower temperatures a band at 410 nm emerges and the spectrum of Fig. 2c, recorded at 100 K, is obtained. These results indicate that the contrast in Fig. 1a (room temperature) is mainly due to the regions with emission in the broad band centered at 530 nm shown in Fig. 2a. At lower temperatures (about 200 K) regions with emission at other wavelengths contribute additionally to the CL causing the intensity increase observed in Fig. 1b. Below 130 K the luminescence is localized in regions emitting at 410 nm while some regions with high room temperature luminescence show a poor emission at low temperature.

Some sample regions appear dark in the CL image at all temperatures. The lack of emission and their spare distribution in the sample suggest that these regions correspond to the residual silicon observed /1, 5/ in backscattered and secondary electron SEM images. On the other hand, the SiC grains cannot be readily recognized in the CL image. This is probably due to the impurity distribution within the grains with a concentration which is higher in the epitaxial coating than in the grain core. The impurity distribution would cause an inhomogeneous luminescence emission in the SiC grains. Close to room temperature the CL spectrum shows a broad complex band with intensity maximum at 520 to 550 nm. Several shoulders and sub-bands, more visible at 180 K (Fig. 2b) indicate that a number of impurity, and possibly defect, levels contribute to this emission. It is known that impurities as boron, nitrogen, and aluminium /8 to

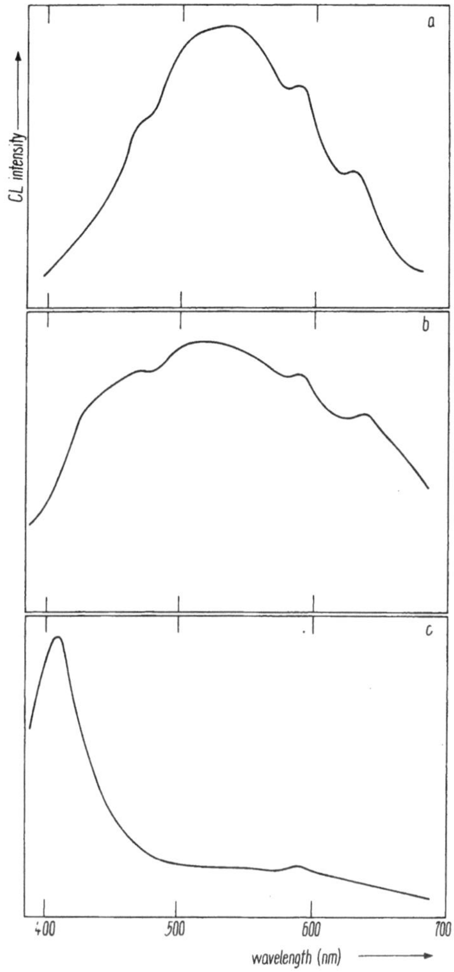


Fig. 2. CL spectrum at a) 273 K, b) 180 K, c) 100 K

10/ are involved in the green and yellow emission of β-SiC: In /10/ a broad CL band peaked at about 540 nm in Al implanted β-SiC is reported. Al has been found /5/ to be the main impurity in the epitaxial layer of the reaction bonded SiC while in the grain core is present in smaller amount. We suggest that in the upper range of temperatures used in this work, impurity related CL emission dominates and the CL images show mainly the impure (100 ppm /5/) regions. At temperatures below 130 K the CL is localized in regions emitting at 410 nm. This wavelength corresponds to the band gap energy of 6H-SiC. The regions imaged at these temperatures would be the purer parts - grain cores - of the SiC grains.

The present results show that SEM-CL provide information, additional to the information obtained by other electron microscopy techniques, on the structure of reaction bonded SiC.

This work was supported by Ministerio de Educación y Ciencia (Proyecto PB 86 - 0151).

References

/1/ G.R. SAWYER and T.F. PAGE, J. Mater. Sci. 13, 885 (1978).

/2/ A.H. HEUER, G.A. FRYBURG, L.U. OGBUJI, and T.E. MITCHELL, J. Amer. Ceram. Soc. 61, 406 (1978).

/3/ T.F. PAGE and G.R. SAWYER, J. Mater. Sci. 15, 1850 (1980).

/4/ L.U. OGBUJI, T.E. MITCHELL, and A.H. HEUER, J. Amer. Ceram. Soc. 64, 100 (1980).

/5/ J.N. NESS and T.F. PAGE, J. Mater. Sci. 21, 1377 (1986).

/6/ L.U. OGBUJI, T.E. MITCHELL, and A.H. HEUER, J. Mater. Sci. 14, 2267 (1979).

/7/ J. LLOPIS and J. PIQUERAS, J. appl. Phys. 54, 4570 (1983).

/8/ A.P. ANDREEV, E.E. VIOLIN, YU.M. TAIROV, and O.A. FAYANS, Soviet Phys. - Semicond. 7, 208 (1973).

/9/ S.H. HAGEN and A.W.C. VAN KENENODE, phys. stat. sol. (a) 33, 97 (1976).

/10/ A.P. ANDREEV, E.E. VIOLIN, K.D. DEMAKOV, YU.M. TAIROV, and V.E. UDALTSOV, Soviet Phys. - Semicond. 15, 119 (1981).

(Received March 24, 1988)

# *Pre-Printed Titles*

Papers to be published
in "physica status solidi (a) - applied research"
Vol. 108, No. 2

Original Papers

L. L. YUAN and S.W.S. MC KEEVER, Stillwater, Oklahoma
Impurity Clustering Effects in Magnesium-Doped Lithium Fluoride

K.N. MADHUSOODANAN and J. PHILIP, Kerala
Optical Energy Gap and Thermal Diffusivity of Ge-Se Semiconducting Glasses

S.K. ASADOV, E.A. ZAVADSKII, YU.D. ZAVOROTNEV, and B.M. TODRIS, Donetsk
Low-Temperature Magnetic Transformations in Ferrous Fluorsilicate Induced by Pressure

SH.M. EFENDIEV, V.E. BAGIEV, R.A. ALIEV, and E.R. MUSTAFAEV, Baku
Nonlinear I-U Characteristics and Photoelectret State in Sillenite-Type Crystals

HONG ZHANG, XIAO-GAN WANG, and YAO-XIAN FU, Shanghai
A Study of the Movement of Oxygen Vacancies in the Orthorhombic Phase of $YBa_2Cu_3O_{7-x}$

G. KOSCHEK, H. LAKNER, and E. KUBALEK, Duisburg
On the Improvement of the Semiinsulating GaAs Substrate Homogeneity Due to Annealing

W. SIEGEL, H. WITTE, G. KÜHNEL, H.A. SCHNEIDER, and T. FLADE, Freiberg
Thermal Activation Energies of Undoped Semi-Insulating LEC GaAs

M. KRANJČEC, B. ČELUSTKA, B. ETLINGER, and D. DESNICA, Zagreb
The Indirect Allowed Optical Transitions in $(Ga_{0.3}In_{0.7})_2Se_3$

D. LAPRAZ, P. IACCONI, Y. SAYADI, P. KELLER, and J. BARTHEL, Nice and Fontenay aux Roses
Some Thermoluminescence Properties of an $\alpha$-$Al_2O_3$ Sample

MING MAO, YUANSHENG WANG, and YIZHEN LI, Hangzhou and Hefei
Crystallization of Metallic Glass $Fe_{46}Ni_{31}VSi_8B_{14}$

D. DĂDÂRLAT, R.M. CÂNDEA, R. TURCU, L.P. BIRÓ, I.I. ZASAVITSKII, M.V. VALEIKO, and A.P. SHOTOV, Cluj-Napoca and Moscow
Size Effects in Polycrystalline PbSe Films Obtained by Chemical Deposition

JIN-SONG ZHU, R. SCHALLER, and W. BENOIT, Nanjing and Lausanne
Internal Friction Transitory Effects Associated with Martensite Transformation in Ni-Ti Alloys

H.R. HÖCHE and J. NIEBER, Halle (Saale)
Transformation of the Angular Dependent Form of X-Ray Diffraction into the Photon-Energy-Dependent Form for Grazing Incidence

J.M. NETO and T. ABRITTA, Rio de Janeiro
Luminescence and Site Occupation of $Cr^{3+}$ in Sodium- $\beta''$-Alumina

Y. KIKUCHI, K. WATANABE, S. NAITOH, K. HIRATSUKA, and H. YAMAGUCHI, Tokyo
HREM Image of a Vacancy in Copper

V.M. KOLESHKO and I.V. KIRYUSHIN, Minsk
Deformation Mechanism Maps and Gettering Diagrams for Single-Crystal Silicon

A.R. ALI and M.A. FAHIM, Cairo
On the Influence of Internal Stress and Magnetizing Frequency on Domain Wall Motion in High Purity Ni

V. PARIZEK and T.A. CHUONG, Prague
Influence of In on Transport Properties and Phase Transition in $Pb_{1-x}Sn_xTe$ Thin Films

A. ARMIGLIATO, A. BOURRET, S. FRABBONI, and A. PARISINI, Bologna and Grenoble
Contrast of Small SiX Particles in Silicon by Computed HREM Images

R. MANAILA, V. FLORESCU, A. JIANU, and A. BADESCU, Bucharest
Icosahedral Phases in the Al-Fe-Cr Alloy System

CH. SCHNITTLER and G. TEICHMANN, Ilmenau
A Kinetic Investigation of Cluster Formation in Semiconductors at Elevated Temperatures

YUHUAN XU, ZHONGRONG LI, CHAORUI LI, XIANDUAN LIN, WU LI, WAN-MING LAO, HONG WANG, and GUOWEI CHEN, Guangzhou
The Internal Strain Effect on $T_c$ in the Y-Ba-Al-Cu Oxide

R. AGRAWAL, R. GOPAL, R. DWIVEDI, and S.K. SRIVASTAVA, Varanasi
Analytical Modelling of a MOSFET with Non-Uniform Impurity Profile

A.V. ZARETSKII, V.F. PETRENKO, and V.A. CHESNAKOV, Chernogolovka
The Protonic Conductivity of Heavily KOH-Doped Ice

R. MATTHEIS and D. HESSE, Jena and Halle (Saale)
A Reinvestigation of $Ni_2Si$ Thin Film Growth on Si (111) by TEM and RBS

O.S. RAJORA and A.E. CURZON, Burnaby, British Columbia and Brighton
A Study of the Diffusion of In into $TaS_2$ and $TaSe_2$ in a Scanning Electron Microscope

I.A. BACHILO, F.F. KOMAROV, A.P. NOVIKOV, and S.A. PETROV, Minsk
Structural Analysis of Silicon Doped with High Doses of $C^+$ Ions

V.V. ARISTOV, U. WINTER, A.YU. NIKULIN, S.V. REDKIN, A.A. SNIGIREV, P. ZAUMSEIL, and V.A. YUNKIN, Chernogolovka and Franfurt (Oder)
Interference Thickness Oscillations of an X-Ray Wave on Periodically Profiled Silicon

P.J. RATCLIFFE and R.A. COLLINS, Lancaster
Thermal Oxidation of Ion Implanted Copper

JUN CHEN and DUAN FENG, Nanjing
TEM Study of Phases and Domains in $NaNbO_3$ at Room Temperature

G. GRUMMT, Leipzig
Calculation of the C-U Profiles of Graded Heterojunctions

S. FUJITA, K. MAEDA, and S. HYODO, Tokyo
Radiation Enhanced Dislocation Glide Effect on Indentation-Induced Fracture of GaAs Single Crystals

Papers to be published
in "physica status solidi (b) - basic research"
Vol. 148, No. 2

## Original Papers

ZIJING LIN and KELIN WANG, Hefei
A Simple Scheme for Studies of Surface Phonons of Covalent Semiconductors (II)

M. KASNER and W. WELLER, Leipzig
Localization in a System of N Coupled Chains (II)

A.G. MAKHANEK, V.S. KOROLKOV, and L.A. YUGURYAN, Minsk
Analysis of the Intensity and Polarization Dependence of Two-Photon Absorption in the Channel $E_{3/2}(4f)-E_{3/2}(5d)$ for $Ce^{3+}$ in $CaF_2$

V.P. ZHUKOV, N.I. MEDVEDEVA, D.L. NOVIKOV, and V.A. GUBANOV, Sverdlovsk
The Study of Energy Band Structure and Chemical Bonding in Yttrium Carbide, Yttrium, and Strontium Dicarbides by the LMTO and $X_\alpha$ DV Method

S.S. KOTSUR and I.V. STASYUK, Lvov
The Microscopic Approach to the Description of the Optical Activity in Dielectric Crystals

A. KUMAR and I. SINGH, Roorkee
Electronic Structure of the Ge-ZnSe, Ge-GaAs, and Ge-AlAs(110) Interfaces and Superlattices

H. LEON and F. COMAS, Havana
On the Scattering of Electrons by Confined LO Phonons in a Semiconductor Quantum Well

E. HEINER, Dubna
Nonequilibrium Thermalized State of Noble Metals

J. CHARITONIDES and D.S. KYRIAKOS, Thessaloniki
Magnetoresistance Behavior in Orthorhombic Crystals

S.S. APLESNIN, Krasnoyarsk
Monte-Carlo Study of Two-Dimensional Antiferromagnets with Competing Anisotropies

C. MAJUMDAR, A.B. MAITY, and A.N. CHAKRAVARTI, Calcutta
Quantized Photoemission from Quantum Wells in Ultrathin Films of Wide-Gap Semiconductors in a Tilted Magnetic Field

S.A. MOSKALENKO, A.H. ROTARU, YU.M. SHVERA, and V.A. ZALOI, Kishinev
Quantum Fluctuations and Time Structures of High Density Polaritons

M. INOUE, S. ŌHARA, S. HORISAKA, M. KOYANO, and H. NEGISHI, Hiroshima
Transport Properties of Quasi-Two-Dimensional $Mo_4O_{11}$ Crystals

W. KULLMANN, H. RAUH, R. GEICK, U. STEIGENBERGER, and W.F. FUHS, Würzburg
Thermal Parameters and Probability Density Functions of the Narrow Gap Semiconductor $Bi_2Te_3$ under Hydrostatic Pressure

YU.V. ZAKHAROV, YU.I. MANKOV, and L.S. TITOV, Krasnoyarsk
Electric Conductivity of a Metallic Ferromagnet and Magnetization Process

J. NANDI and D. NEOGY, Burdwan
Magnetic Behavior and Crystal Field of $Pr^{3+}$ in Praseodymium Selenate Octahydrate

M. HAAS, V. HIZHNYAKOV, E. REALO, and J. JŌGI, Tartu
Nuclear Polaritons in the Mössbauer Absorber

I.A. GOICHUK, V.V. KUKHTIN, and E.G. PETROV, Kiev
Orientational Vibrations of Two-Dimensional Dipole Lattice on Support

D.B. TRAN THOAI and E. ZEITLER, Berlin (West)
Dynamical Image Potential of a Point Charge Approaching a Metal Sphere

G. DAS, B.K. PANDA, D.P. MAHAPATRA, and H.C. PADHI, Bhubaneswar
Simple Numerical Method for the Estimation of Multiple Scattering in the Compton Effect

L. SAMAJ, Bratislava
A Theory of Higher Order Correlations (I)

N. SINGH, N.S. BANGER, and S.P. SINGH, Rohtak
Elastic Constants and Phonon Spectra of Alkali Metals

K.L. VODOPYANOV, L.A. KULEVSKII, CS. TÓTH, GY. FARKAS, and Z.GY. HORVÁTH, Moscow and Budapest
Laser-Induced Electron Emission from Gold Surface Irradiated by Picosecond Pulses at $\lambda = 2.8\ \mu m$

V.R. BELOSLUDOV, M.YU. LAVRENTIEV, and S.A. SYSKIN, Novosibirsk
Dynamical Properties of the Molecular Crystals with Electrostatic Interaction Taken into Account

J.-P. ZÖLLNER, G. PAASCH, G. GOBSCH, and H. ÜBENSEE, Dresden
Theory for n-Surface Charge Layers in $Hg_{1-x}Cd_xTe$ MIS Structures

S.S. KUSHWAHA and S.K. SRIVASTAVA, Varanasi
A Model for Lattice Dynamics of Lanthanides

H. NAKANISHI, M. ISOMAE, S. ENOD, and T. IRIE, Tokyo
Piezoresistance in $CdIn_2S_4$

B. SCHÖNFELD, L. REINHARD, G. KOSTORZ, and W. BÜHRER, Zürich and Würenlingen
Short-Range Order and Atomic Displacements in Ni-20 at% Cr Single Crystals

T.YU. BILGILDEEVA and T.A. POLYANSKAYA, Leningrad
The Quantum Corrections to the Conductivity in Strong Spin-Orbit Coupling Semiconductors

G. CIOBANU, I. CORCOTOI, and A.B. FAZAKAS, Bucharest
Theory of the Superconductivity in Two Energy Band Systems

Z. DREGER, J. KALINOWSKI, I. DAVOLI, S. STIZZA, and S. FELIZIANI, Gdańsk and Camerino
Thermoluminescence Study of Charge Traps in Polycrystalline Layers of Phenanthrene

N.A. SHILKOVA and V.P. SHIROKOVSKII, Izhevsk
Calculation of Dipole Transition Matrix Elements in Crystals by the Relativistic Green's Function Method

V.M. YARTSEV, Chelyabinsk
Optical Properties of Quasi-1D n-Merized Molecular Semiconductors

W. KRAAK, G. NACHTWEI, R. HERRMANN, and M. GLINSKI, Berlin and Wrocław
Properties of the Quantum Hall Effect ot the Two-Dimensional Electron Gas in the n-Inversion Layer of InSb Grain Boundaries under High Hydrostatic Pressure

J. OTTO, Münster
Stress-Induced Dichroism of the Vibrational Lines and Phonon Sidebands of $SH^-$ Centres in CsBr

J.R. FALEIRO FERREIRA, Oxford
A Variation Linear Approach for the Free Energy Applied to the Ising Model

V.N. BAIER, V.M. KATKOV, and V.M. STRAKHOVENKO, Novosibirsk
Incoherent Radiation and Pair Creation in Crystals

T. HALICIOGLU, O. PAMUK, and S. ERKOC, Stanford and Ankara
Interatomic Potentials with Multi-Body Interactions

I.A. VAKARCHUK and I.F. MARGOLYCH, Lvov
Magnon Spectrum of Two-Component Amorphous Ferromagnets

## Substance Classification*)

S1. Transition metals and their alloys and compounds (for oxides see S.10)
- S1.1. Fe
- S1.2. Other 3d metals
- S1.3. Noble metals
- S1.4. Lanthanides
- S1.5. Actinides
- S1.61. Carbides, silicides, germanides
- S1.62. Nitrides, phosphides, arsenides
- S1.63. Sulfides, selenides, tellurides

S2. Alkali metals and their alloys and compounds (for alkali halides see S9.11)

S3. Alkaline earth metals and their alloys and compounds (for alkaline earth halides, see S9.12)

S4. Other metals and their alloys

S5. Nonmetallic elements
- S5.11. Silicon
- S5.12. Germanium
- S5.2. Solidified Gases

S6. Group IV compounds (for transition metal compounds, see S1.61)

S7. Group V compounds (for transition metal compounds, see S1.62)
- S7.1. III–V compounds
- S7.11. Phosphides
- S7.12. Arsenides
- S7.13. Antimonides
- S7.15. Mixed crystals
- S7.16. Higher compounds

S8. Chalcogenides (for transition metal compounds, see S1.63)
- S8.1. II–VI compounds
- S8.11. Sulfides
- S8.12. Selenides
- S8.13. Tellurides
- S8.15. Mixed crystals
- S8.16. Higher compounds

S9. Halides
- S9.11. Alkali halides
- S9.12. Alkaline earth halides
- S9.15. Mixed crystals
- S9.16. Higher compounds

S10. Oxides
- S10.1. Simple oxides
- S10.15. Mixed crystals

S11. Other higher compounds
- S11.1. Salts of oxy-acids
- S11.2. Spinels, garnets, ferrites

S12. Organic compounds
- S12.1. Quasi-one-dimensional systems

## *Attention*

The *Classification Scheme* on cover two is reproduced and used with permission of International Council for Scientific and Technical Information (ICSTI).

Because of the lack of space only the main points of the Classification Scheme are printed here. The *detailed subdivided scheme* is to be found in the January issues of each annual set and in the Index Volumes from 1986 on.

The *Subject Classification* given at the head of each paper begins with the main subject matter of the paper, and this is followed, if necessary, by figures referring to subsidiary subject matter and the type of solids investigated.

The papers published in this issue are listed in the order of their subject classification in the *Systematic List* at the end of the table of contents.

*) Outside the ICSTI Classification for Physics.

## Manuscripts and letters for physica status solidi (b) — basic research and physica status solidi (a) — applied research

should be sent to

Leipziger Straße 3—4, P. B. 1233, DDR-1086 Berlin

or

Kunst und Wissen, Erich Bieber OHG, Postfach 330 104, D-1000 Berlin 33.

The object of "physica status solidi (b)" and "physica status solidi (a)" is to publish rapidly important and topical results concerning fundamental research in the field of solid state physics and aspects of application and experimental techniques of fundamental solid state research, respectively. The issues of "physica status solidi (b)" and "physica status solidi (a)" will be published on the 1st and the 16th day of each month, respectively. Either series will contain the following categories:

1. Review articles written by arrangement with the editors. They will be remunerated.
2. Original papers selected from those which have not been submitted for publication elsewhere, after being reviewed. (The latest date for acceptance of contributions is the 8th and the 23rd day of each month for "physica status solidi (b)" and "physica status solidi (a)", respectively, 50 days before the delivery of the journal.)
3. Short notes published in the off-set part of the journal. They are, if necessary after being reviewed, selected by editorial staff. (The latest date for acceptance of manuscripts for this part is the 12th and the 27th day of each month for "physica status solidi (b)" and "physica status solidi (a)", respectively, 16 days before the delivery of the journal.)
4. Pre-printed titles of papers to be published in the next issue of both series.

## Editorial Information

Manuscripts should be submitted preferentially in English. Besides, papers in German, French, and Russian are admitted. The final decision on the incorporation of the papers in the series (b) or (a) is taken by the editorial staff after consultation of the referee. Papers on liquids are only published in physica status solidi if the problems treated are also relevant for the solid state. Authors are asked to observe the following rules:

1. Manuscripts have to be written as consisely as possible. At most, 20 and 4 pages (28 lines per page; abstracts, tables, references, figures, and captions of figures included, where 3 figures are counted as about one page) are admitted for original papers and short notes, respectively. All manuscripts should be typewritten on one side only. Two copies are required. Badly arranged manuscripts are returned.
2. The author's name, initials, and institution should be given.
3. One abstract of about ten lines in English and one in German, French, or Russian are required for original papers.
4. Because of lack of space conclusions (if any) must not have the character of a summary (i. e. substantially a repetition of the contents of the abstract).
5. Formulae should not be written too small and not with pencil. Separate lines for formulae are desirable. Indistinct writing will lead to a return of the manuscript.
6. Drawings should be enclosed on separate sheets of paper, not exceeding the maximum and minimum sizes of $20 \times 30$ and $10 \times 15$ cm$^2$, respectively. Photographs should be submitted, on white paper, only for review articles and original papers. Duplicates are required.
7. Captions for the figures should be submitted on separate sheets of paper, and special instructions with regard to the size of illustrations should be given on these.
8. References should be numbered in order of appearance and should be given, on a separate sheet, in the form used in physica status solidi.
9. SI-units should be used.
10. Each paper must contain a proposal for the subject classification according to the classification scheme appearing in the January issues of each annual set.

The authors will be sent galley proofs of review articles and original papers. Proof reading of short notes and pre-printed titles will be done by the editorial staff.

For each original paper and short note a total of 75 reprints is available free of charge. The number of reprints free of charge per review article is 25. Requests for additional reprints (against payment) should be given together with the submitted manuscripts. **The deadline for orders of additional reprints is the 15th and the 30th of each month** for "physica status solidi (b)" and "physica status solidi (a)", respectively, two weeks before the delivery of the journal.

Bei Fragen zur Produktsicherheit wenden Sie sich bitte an:
If you have any questions regarding product safety,
please contact:

Walter de Gruyter GmbH
Genthiner Straße 13
10785 Berlin
productsafety@degruyterbrill.com